Data-Driven Pharmaceutical Processing
and Drug Development

Data-Driven Pharmaceutical Processing and Drug Development

Edited By

Md. Faiyazuddin, PhD

Abdul Wasy Zia, PhD

WILEY

This edition first published 2026

Registered Office(s)
John Wiley & Sons, Inc., 111 River Street, Hoboken, NJ 07030, USA
John Wiley & Sons Ltd, New Era House, 8 Oldlands Way, Bognor Regis, West Sussex, PO22 9NQ, UK

For details of our global editorial offices, customer services, and more information about Wiley products visit us at www.wiley.com.

The manufacturer's authorized representative according to the EU General Product Safety Regulation is Wiley-VCH GmbH, Boschstr. 12, 69469 Weinheim, Germany, e-mail: Product_Safety@wiley.com.

Wiley also publishes its books in a variety of electronic formats and by print-on-demand. Some content that appears in standard print versions of this book may not be available in other formats.

Library of Congress Cataloging-in-Publication Data has been applied for:

Print ISBN: 9781394344765
ePub ISBN: 9781394344772
ePdf ISBN: 9781394344789
obook ISBN: 9781394344796

Cover Design: Wiley
Cover Image: Corona Borealis/stock.adobe.com

Typesetting: Set in 9.5/12.5 STIX Two Text by Lumina Datamatics

Contents

List of Contributors

Fawad Abidi
Institute for Materials and Processes
School of Engineering
University of Edinburgh
Edinburgh
UK

Naveed Ahmad
Institute of Pharmaceutical Science,
Faculty of Life Science and Medicine
King's College London
London
UK

Sareer Ahmad
Department of Applied Physical and
Material Sciences
University of Swat
Swat
Pakistan

Md Akbar
School of Pharmacy, Al-Karim University
Katihar
India

Mandana Akhavan
Cell and Molecular Biology, Department of
Biology, Science and Research Branch
Islamic Azad University
Tehran
Iran

and

Division of Molecular Biology
Medical Mycology and Parasitology
Department
School of Public Health Research Center
Tehran University of Medical Sciences
Tehran
Iran

Javed Akhter
Faculty of Mechanical and Aeronautical
Engineering
University of Engineering and Technology,
Taxila
Rawalpindi
Pakistan

Muhammad Asim Sajid Al-Hanafi
Institute of Health and Social Care
School of Pharmacy, Optometry and
Medical Sciences
University of Bradford
West Yorkshire
UK

Hasan Ali
Department of Pharmacy
Meerut Institute of Technology
Meerut
India

Jawairia Ali
Department of Applied Physical and Material Sciences
University of Swat
Swat
Pakistan

Safia Ali
Department of Applied Physical and Material Sciences
University of Swat
Swat
Pakistan

Aisha Badruddin
Department of Business Management
Integral Business School
Integral University
Lucknow
India

Dr. Machindra Chavan
Department of Pharmacognosy
Amrutvahini College of Pharmacy
Sangamner
India

Wenai Chen
State Key Laboratory of Fine Chemicals
Frontiers Science Center for Smart Materials Oriented Chemical Engineering
Department of Pharmaceutical Sciences
Institute of Chemical Process Systems Engineering
School of Chemical Engineering
Dalian University of Technology
Dalian
China

Jian Du
State Key Laboratory of Fine Chemicals
Frontiers Science Center for Smart Materials Oriented Chemical Engineering
Department of Pharmaceutical Sciences
Institute of Chemical Process Systems Engineering
School of Chemical Engineering
Dalian University of Technology
Dalian
China

and

Ningbo Institute of Dalian University of Technology
Ningbo
China

F. Fareedhul Fahmitha
PG and Research Department of Microbiology
Sacred Heart College (Autonomous)
Tirupattur
India

Md. Faiyazuddin
Department of Pharmacy
ARAM Institute
Ranchi
India

Amol D. Gholap
Department of Pharmaceutics
St. John Institute of Pharmacy and Research
Palghar
India

and

Department of Pharmaceutics
Amrutvahini College of Pharmacy
Sangamner
India

Lixiang Guo
State Key Laboratory of Fine Chemicals
Frontiers Science Center for Smart Materials Oriented Chemical Engineering
Department of Pharmaceutical Sciences
Institute of Chemical Process Systems Engineering
School of Chemical Engineering
Dalian University of Technology
Dalian
China

Navnath Hatvate
Department of Pharmaceutical Sciences and Technology
Institute of Chemical Technology Mumbai
Marathwada Campus
Jalna
India

Talaya Hidayatullah
Department of Pharmacy
University of Peshawar
Peshawar
Pakistan

Zahid Hussain
Department of Pharmaceutics and Pharmaceutical Technology
College of Pharmacy
University of Sharjah
Sharjah
United Arab Emirates

and

Research Institute for Medical and Health Sciences
University of Sharjah
Sharjah
United Arab Emirates

Anam Ijaz
Institute of Mechanical Process and Energy Engineering (IMPEE)
School of Engineering and Physical Sciences
Heriot-Watt University
Edinburgh
UK

L. Charlie Jelura
PG and Research Department of Microbiology
Sacred Heart College (Autonomous)
Tirupattur
India

Bruno Arnaud Dzokwe Kene
Institute of Health and Social Care
School of Pharmacy, Optometry and Medical Sciences
University of Bradford
West Yorkshire
UK

Shahzeb Khan
Institute of Health and Social Care
School of Pharmacy, Optometry and Medical Sciences
University of Bradford
West Yorkshire
UK

Muzna Ali Khattak
Department of Pharmacy
Iqra University
Peshawar
Pakistan

Pankaj R. Khuspe
Department of Pharmaceutics
College of Pharmacy
Paniv
Solapur
India

Deepak Kumar
Department of Pharmaceutical Chemistry
School of Pharmaceutical Sciences
Shoolini University
Solan
India

Rakesh Kumar
Amity Institute of Pharmacy
Amity University
Gurugram
India

Jiaqi Lin
MOE Key Laboratory of Bio-Intelligent Manufacturing
School of Bioengineering
Dalian University of Technology
Dalian
China

Qilei Liu
State Key Laboratory of Fine Chemicals
Frontiers Science Center for Smart Materials Oriented Chemical Engineering
Department of Pharmaceutical Sciences
Institute of Chemical Process Systems Engineering
School of Chemical Engineering
Dalian University of Technology
Dalian
China

and

Ningbo Institute of Dalian University of Technology
Ningbo
China

Tahir Mahmood
School of Computing, Engineering and Physical Sciences
University of the West of Scotland
Paisley
UK

M. Manigandan
PG and Research Department of Microbiology
Sacred Heart College (Autonomous)
Tirupattur
India

Mubbashar Mehmood
Department of Mechanical Engineering
School of Engineering and Physical Sciences
Heriot-Watt University
Edinburgh
UK

Qingwei Meng
State Key Laboratory of Fine Chemicals
Frontiers Science Center for Smart Materials Oriented Chemical Engineering
Department of Pharmaceutical Sciences
Institute of Chemical Process Systems Engineering
School of Chemical Engineering
Dalian University of Technology
Dalian
China

and

Ningbo Institute of Dalian University of Technology
Ningbo
China

Raiha Mumtaz
National Centre for Bioinformatics
Quaid-i-Azam University
Islamabad
Pakistan

Vivek Panwar
Department of Pharmaceutical Chemistry
Chitkara School of Pharmacy
Chitkara University
Solan
India

Naeem Ramzan
School of Computing, Engineering and Physical Sciences
University of the West of Scotland
Paisley
UK

R. Rithika
PG and Research Department of Microbiology
Sacred Heart College (Autonomous)
Tirupattur
India

Rida Sagheer
Department of Chemistry
Integral University
Lucknow
India

P. Saranraj
PG and Research Department of Microbiology
Sacred Heart College (Autonomous)
Tirupattur
India

Eman M. Sarhan
Biochemistry Department
Faculty of Science
Alexandria University
Alexandria
Egypt

Sadikali F. Sayyad
Department of Pharmaceutics
Amrutvahini College of Pharmacy
Sangamner
India

Piyush Sewal
Amity School of Engineering and Technology
Amity University Punjab
Mohali
India

Ankur Sharma
Hyalo Technologies
Somerset
New Jersey
USA

Shrestha Sharma
Amity Institute of Pharmacy
Amity University
Gurugram
India

Samra Siddiqui
Department of Computer Science and Engineering
Integral University
Lucknow
India

S. Swetha
PG and Research Department of Microbiology
Sacred Heart College (Autonomous)
Tirupattur
India

M. S. Swetha
Department of Microbiology
Padmavani Arts and Science College for Women (Autonomous)
Salem
India

Nikita Thakur
Chitkara School of Pharmacy
Chitkara University
Solan
India

Sriti Thakur
Department of Mathematics
Jaypee University of Information Technology
Waknaghat
Solan
India

Salah Uddin
Department of Applied Physical and Material Sciences
University of Swat
Swat
Pakistan

Kefayat Ullah
Department of Applied Physical and Material Sciences
University of Swat
Swat
Pakistan

Nitin Verma
Chitkara School of Pharmacy
Chitkara University
Solan
India

R. Yuvaraj
PG and Research Department of Microbiology
Sacred Heart College (Autonomous)
Tirupattur
India

Lei Zhang
State Key Laboratory of Fine Chemicals
Frontiers Science Center for Smart Materials Oriented Chemical Engineering
Department of Pharmaceutical Sciences
Institute of Chemical Process Systems Engineering
School of Chemical Engineering
Dalian University of Technology
Dalian
China

and

Ningbo Institute of Dalian University of Technology
Ningbo
China

Yujing Zhao
State Key Laboratory of Fine Chemicals
Frontiers Science Center for Smart
Materials Oriented Chemical Engineering
Department of Pharmaceutical Sciences
Institute of Chemical Process Systems
Engineering
School of Chemical Engineering
Dalian University of Technology
Dalian
China

and

MOE Key Laboratory of Bio-Intelligent
Manufacturing
School of Bioengineering
Dalian University of Technology
Dalian
China

Abdul Wasy Zia
Institute of Mechanical Process and Energy
Engineering (IMPEE)
School of Engineering and Physical
Sciences
Heriot-Watt University
Edinburgh
UK

Preface

Pharmaceutical manufacturing is entering into a transformative era defined by the convergence of data, computation, and intelligence. Artificial intelligence (AI), supported by vast data resources and advanced computational architectures, is reshaping every facet of the drug development pipeline—from molecular discovery and formulation design to large-scale manufacturing, performance evaluation, and regulatory oversight. This book, *Data-driven Pharmaceutical Processing and Drug Development*, has been conceived as both a comprehensive reference and a forward-looking guide to this transformation, capturing how data-driven technologies are redefining the pharmaceutical enterprise from molecule to market.

The novelty of this book lies in its holistic treatment of AI applications, encompassing design, formulation, manufacturing, performance evaluation, ethics, and compliance. It moves beyond isolated examples to demonstrate how AI can orchestrate a seamless continuum of innovation—where data from laboratory research informs manufacturing optimization, where digital twins enable real-time process control, and where predictive analytics shape regulatory decision-making. Drawing on real-world case studies of small-molecule drug development, AI-assisted peptide design, and nanotechnology-driven repurposing strategies, the chapters highlight the tangible translation of AI from theoretical potential to practical impact. The book also brings an important dimension often overlooked in technical discussions: the evolving domain of data governance, ethical AI, and regulatory adaptation, which together form the foundation for trustworthy and sustainable digital transformation in the life sciences.

The first part of the book explores the application of artificial intelligence in drug design and formulation. It demonstrates how machine learning and data analytics are enabling precise molecular modeling, peptide design, and formulation optimization, allowing for faster, safer, and more personalized drug discovery. The second part turns to pharmaceutical manufacturing, addressing the emergence of intelligent manufacturing systems supported by data-centric engineering, digital twins, and real-time monitoring. It discusses how AI is redefining process efficiency, quality assurance, and nanotechnology-based therapeutic innovation through data-driven control and adaptive process design. The final part focuses on performance evaluation and regulatory compliance, where the integration of AI provides new methods for quality monitoring, advanced process control, and ethical data management. By examining regulatory frameworks, responsible AI adoption, and governance practices, this section emphasizes the importance of aligning technological innovation with safety, accountability, and societal trust.

The overarching impact of this work lies in its ability to frame AI not as an isolated tool but as a transformative force in the creation of intelligent pharmaceutical systems. The integration of AI into every stage of drug development shortens research cycles, enhances product quality, supports personalized medicine, and fosters a culture of transparency and efficiency. This convergence of science and intelligence signals a shift from traditional, siloed practices to a more dynamic, data-driven model of pharmaceutical innovation.

Ultimately, this book is a vision statement for the future of medicine. It invites the scientific and industrial communities to view AI not merely as a disruptive technology, but as a collaborative partner in the pursuit of safer, smarter, and more sustainable healthcare. In doing so, it hopes to inspire a new generation of pharmaceutical researchers and innovators to build a future where data and intelligence coalesce to accelerate discovery, ensure quality, and advance human well-being.

Editors
Md. Faiyazuddin, PhD
Abdul Wasy Zia, PhD

List of Abbreviations

ABC	Accelerated Blood Clearance
ADF	Allotrope Data Format
ADMET	Absorption, Distribution, Metabolism, Excretion, and Toxicity
ADRs	Adverse Drug Reactions
AFT	Acoustofluidic Transwell
AI	Artificial Intelligence
ALCOA	Attributable Legible Contemporaneous Original Accurate
AMR	Antimicrobial-resistant
ANN	Artificial Neural Network
APIs	Active Pharmaceutical Ingredients
AR	Augmented Reality
ATMPs	Advanced Therapeutic Medical Products
BBB	Blood–Brain Barrier
BIRC5	Baculoviral Inhibitor of Apoptosis Repeat-Containing 5
BO	Bayesian Optimization
CAR	Chimeric Antigen Receptor
CCEA	Complete, Consistent, Enduring, and Available
CDISC	Clinical Data Interchange Standards Consortium
CFD	Computational Fluid Dynamics
CFR	Code of Federal Regulation
cGMP	Current Good Manufacturing Practice
CIP	Ciprofloxacin
CM	Continuous Manufacturing
CN	Cinnamon Aldehyde
CNN	Convolutional Neural Network
CNS	Central Nervous System
CPPs	Critical Process Parameters
CPS	Cyber-physical System
CQAs	Critical Quality Attributes
CRISPR	Clustered Regularly Interspaced Short Palindromic Repeats
CuET NCs	CuET Nanocrystals
CuET	Copper Ethylthiocarbamate
DCE	Data-centric Engineering
DCS	Distributed Control Systems
DEM	Discrete Element Method

DL	Deep Learning
DNA	Deoxyribonucleic Acid
DNNs	Deep Neural Networks
DNS	Direct Numerical Simulation
DOX	Doxorubicin
DQA	Data Quality Assurance
DSF	Disulfiram
DTs	Digital Twins
ECM	Natural Extracellular Matrix
EDC	Electronic Data Capture
EE	Euler–Euler
EHRs	Electronic Health Records
EL	Euler–Lagrange
EMA	European Medicines Agency
EPR	Enhanced Permeability and Retention
ERP	Enterprise Resource Planning
EUG	Eugenol
EU	European Union
FAERS	Food and Drug Administration's Adverse Event Reporting System
FDA	Food and Drug Administration
FEA	Finite Element Analysis
FHIR	Fast Healthcare Interoperability Resources
FMEA	Failure Mode and Effects Analysis
FPGAs	Field-programmable Gate Arrays
FTIR	Fourier Transform Infrared Spectroscopy
GAMP	Good Automated Manufacture Practices
GANs	Generative Adversarial Networks
GBSA	Generalized Born Surface Area
GCP	Good Clinical Practice
GDP	Good Documentation Practice
GDPR	General Data Protection Regulation
GLP	Good Laboratory Practices
GMP	Good Manufacturing Practices
hBMEC	Human Brain Microvascular Endothelial Cells
HIPAA	Health Insurance Portability and Accountability Act
HL7	Health Level Seven
HPC	High-performance Computing
HPLC	High-performance Liquid Chromatography
HTE	High-throughput Experimentation
HTS	High-throughput Screening
HYDRA	Hybrid Resonant Acoustic Device
IBDs	Inflammatory Bowel Illnesses
ICH	International Council for Harmonization
IDS	Intrusion Detection Systems
IDT	Interdigital Transducer
IoMT	Internet of Medical Things
IoT	Internet of Things
ISO	International Organization for Standardization

IVM	Ivermectin
Kd	Dissociation Constant
KPIs	Key Performance Indicators
LBM	Lattice-Boltzmann Method
LDL	Low-density Lipoprotein
LES	Large Eddy Simulation
LIMS	Laboratory Information Management Systems
LNPs	Lipid Nanoparticles
LoRaWAN	Long-range Wide Area Network
MD	Molecular Dynamics
MDR	Multidrug Resistance
MDR	Multidrug-resistant
MedDRA	Medical Dictionary for Regulatory Activities
MEMS	Microelectromechanical Systems
MES	Manufacturing Execution System
MET	Metformin
MFA	Multi-factor Authentication
MHRA	Medicines and Healthcare Products Regulatory Agency
ML	Machine Learning
MM	Multiple Myeloma
MN	Matching Network
MOF	Metal–Organic Framework
MPS	Mononuclear Phagocyte System
MRI	Magnetic Resonance Imaging
MS	Mass Spectrometry
MSNs	Mesoporous Silica Nanoparticles
MSPC	Multivariate Statistical Process Control
MVA	Multivariate Analysis
NARX	Nonlinear Autoregressive
NF	Composite Nanofiber
NIR	Near-infrared
NLC	Nanostructured Lipid Carrier
NLP	Natural Language Processing
NPs	Nanoparticles
OOS	Out-of-specification
OPC UA	Open Platform Communications Unified Architecture
PaaS	Platform-as-a-service
PAT	Process Analytical Technology
PCA	Principal Component Analysis
PCB	Printed Circuit Board
PD	Pharmacodynamics
PDAC	Pancreatic Ductal Adenocarcinoma
PDMS	Polydimethylsiloxane
PEG	Polyethylene Glycol
PET	Positron Emission Tomography
PI	Pharmaceutical Industry
PII	Personally Identifiable Information
PK	Pharmacokinetics

PLCs	Programmable Logic Controllers
PLD	PEGylated Liposomal Doxorubicin
PLGA	Poly (lactic-co-glycolic acid)
PMDA	Pharmaceuticals and Medical Devices Agency
PMS	Post-market Surveillance
PNPs	Polymeric Nanoparticles
PTX	Paclitaxel
QA	Quality Assurance
QbD	Quality by Design
QC	Quality Control
QMSR	Quality Management System Regulation
QRM	Quality Risk Management
QSAR	Quantitative Structure–Activity Relationship
R&D	Research and Development
RANS	Reynolds-averaged Navier–Stokes
RCA	Root Cause Analysis
RCCP	Rough-cut Capacity Planning
RF	Random Forest
RL	Reinforcement Learning
RNA	Ribonucleic Acid
RNN	Recurrent Neural Network
RTD	Residence Time Distribution
RTRT	Real-time Release Testing
RUL	Remaining Useful Life
RWD	Real-world Data
RWE	Real-world Evidence
SaaS	Software-as-a-service
SAR	Structure–Activity Relationship
SAW	Surface Acoustic Waves
SBAT	Single Beam Acoustic Tweezer
SCADA	Supervisory Control And Data Acquisition
SDTM	Study Data Tabulation Model
SEM	Scanning Electron Microscopy
SiRNA	Small Interfering Ribonucleic Acid
SLNs	Solid Lipid Nanoparticles
SMEs	Small- and Medium-sized Businesses
SNP	Single Nucleotide Polymorphism
SoS	Science of Scale
SPC	Statistical Process Control
SPH	Smoothed Particle Hydrodynamics
SVM	Support Vector Machine
TEER	Transendothelial Electrical Resistance
THY	Thymoquinone
TPUs	Tensor Processing Units
U.S.	United States
US-FDA	United States Food and Drug Administration
V&V	Validation and Verification
XAI	Explainable AI

Part I

AI for Drug Design and Formulation

1

Integrating AI in Drug Formulation: Influential Factors and Practical Applications

*Vivek Panwar[1], Nikita Thakur[1], Piyush Sewal[2], Nitin Verma[1], and Deepak Kumar[3]**

[1]*Department of Pharmaceutical Chemistry, Chitkara School of Pharmacy, Chitkara University, Solan, India*
[2]*Amity School of Engineering and Technology, Amity University Punjab, Mohali, India*
[3]*Department of Pharmaceutical Chemistry, School of Pharmaceutical Sciences, Shoolini University, Solan, India*

*Corresponding author: guptadeepak002@gmail.com

CONTENTS

1.1 Introduction

The Discovery and preliminary development of artificial intelligence (AI) have evolved through a continuous series of intellectual, conceptual, and technological milestones overpassing centuries. The conceptual underpinnings of AI can be utilized from the findings of Aristotle, Descartes, and Leibniz, postulated about reasoning, logic, and mechanical aspects [1, 2].

In the 1950s, Alan Turing introduced the Turing Test to assess machine intelligence and published in "Computing Machinery and Intelligence" [3]. Later, in 1956, the term "Artificial Intelligence" was coined at the Dartmouth Conference, organized by pioneers including John McCarthy, Marvin Minsky, Claude Shannon, Allen Newell, and Herbert Simon. The critical advancements and milestones in the following decades include logical theorists (in 1950–1960s), expert systems (1980s), IBM's Deep Blue (in 1977), deep learning (DL) revolution (in 2012), DeepMind's AlphaGo (in 2016) and many more. With ongoing research and continuous advancements in fields including machine learning (ML), natural language processing (NLP), robotics, and the quest for artificial general intelligence (AGI), leading research institutes and corporations like OpenAI, Google DeepMind, Anthropic, Meta, and Microsoft are still developing AI today [4–6].

AI is transforming drug discovery and formulation by optimizing excipient selection and nanocarrier design, enhancing solubility and release kinetics, predicting the bioavailability and stability, ensuring improved therapeutic efficacy, targeted delivery systems, and patient-specific drug delivery utilizing AI-driven predictive modeling [7, 8]. ML and DL algorithms interpret and analyze vast datasets to predict physicochemical properties of drug molecules, forecast degradation pathways to predict the biotransformation, and refine formulation strategies, including preformulation parameters, thereby accelerating the drug development process and reducing the chances of pharmaceutical formulation failures [9–11].

Moreover, AI facilitates precision medicine by enabling personalized drug formulations tailored to individual genetic, environmental, and physiological factors as well as empowers traditional medicine concepts like Chinese traditional medicine [12, 13]. Through computational simulations and big data analytics, AI enhances the stability and efficacy of pharmaceutical compounds, streamlining regulatory approvals and manufacturing processes. As AI continues to integrate into drug formulation, it paves the way for more efficient, cost-effective, and innovative therapeutic solutions, ultimately improving patient outcomes and advancing pharmaceutical sciences [9, 14].

AI-enabled drug discovery and lead optimization include AI-based drug discovery (target identification, drug repurposing, and molecular design), preclinical development (predicting safety and toxicity, optimizing drug properties), clinical trials (trial monitoring and data analysis, predicting trial outcomes), precision medicine, biomarker identification, and accelerated drug approvals (regulatory affairs, and simulating clinical trials). AI can analyze large biological datasets to identify new drug targets (specific proteins or genes related to diseases). ML-based algorithms find applications in identifying patterns in genetic data and thus suggest suitable targets for specific diseases (target identification).

AI also finds applications in drug repurposing of the existing medications, and thus might accelerate the discovery and development of drugs for different ailments (drug repurposing). By analyzing the databases of the approved drugs, including their adverse effects, AI finds applications to predict the effectiveness of the same drug for other diseases as well as the novel compounds designed can also be used for the prediction of the activity against the suitable target using AI algorithms [15]. Such information can be useful in designing new moieties by identifying the pharmacophoric features and the active site residues (molecular design) [16–18].

Verifying the safety of a new medication is one of the most challenging and important tasks in drug development. By evaluating data from analogous drugs and applying models to forecast the pharmacokinetic aspects of the drugs, AI can be utilized to predict possible toxicity and safety prior to clinical trials. AI can improve the stability, bioavailability, and capacity of the drug to pass across biological barriers (such as the blood–brain barrier). Healthcare formulation can be modified by using AI models to model how the drug behaves in the body [10, 16].

Clinical trials often struggle to find the right patients. AI can analyze patient data and identify individuals who are most likely to benefit from participating in specific trials, ensuring more efficient recruitment (patient recruitment) and reducing trial time. AI is increasingly used to analyze clinical trial data in real time (trial monitoring and data analysis). This can help identify issues early, such as unexpected side effects or patterns in patient responses. AI can also identify which treatment regimens are most effective, speeding up the process of bringing drugs to market. AI models can predict the success of clinical trials by analyzing past trial data (predicting trial outcomes), helping researchers identify which trials are most likely to succeed before they even start [19, 20].

AI enables the applications of precision medicine, which tailor treatments based on individual genetic, environmental, and lifestyle factors. AI can help identify patients with treatments that are most likely to work for them based on their unique profiles, optimizing drug efficacy and minimizing adverse effects. AI enables the identification of biomarkers, which are essential for disease diagnosis and thus determine patient responsiveness to a particular treatment. AI does so by screening and analyzing the complex biological data sets and thus facilitates the selection of personalized therapeutic regimens. AI also facilitates the regulatory approval for the pharmaceutical companies by streamlining the filing process of new pharmaceuticals for the evaluation of clinical trial-related data and regarding previous approvals. AI, in addition, can also be utilized for mimicking clinical trials and thus allow the prediction of drug performance in lesser time and cost [21, 22]. This may result in faster drug approvals and their market release.

1.2 AI and Drug Formulation

For the approved medications, AI streamlines the production procedures. AI-based models help and ensure the issues related with quality control, address the potential quality issues, and thus enhance the efficiency in large-scale medicine manufacturing. Some of the key AI-based technologies having applications in drug development includes ML, NLP, generative adversarial networks (GANs), and DL. ML-enabled analysis of vast libraries of biological, chemical, and clinical data undergoes such analysis. ML models are able to spot trends and forecast the appropriate medication that could be most effective for a given illness. NLP analyzes the scientific literature, outcomes of clinical trials, and the drug-related information and thus facilitates the screening of best suited ideas for streamlining the clinical research [23]. AI imparts significant potential and is influencing the process of drug formulation by enhancing multiple factors of the medicinal manufacturing. Some of the key AI-related factors targeting pharmaceutical development include AI-based active pharmaceutical ingredient (API) optimization and manufacturing parameters, AI-enabled tools for predicting the pharmacokinetics, possible toxicity, and physicochemical compatibility, AI and personalized therapy, AI and process optimization, AI in quality assurance (QA) regarding the regulatory, and ethical considerations [23, 24]. The different factors impacting AI-based drug formulation (in Figure 1.1) are discussed in brief in the chapter.

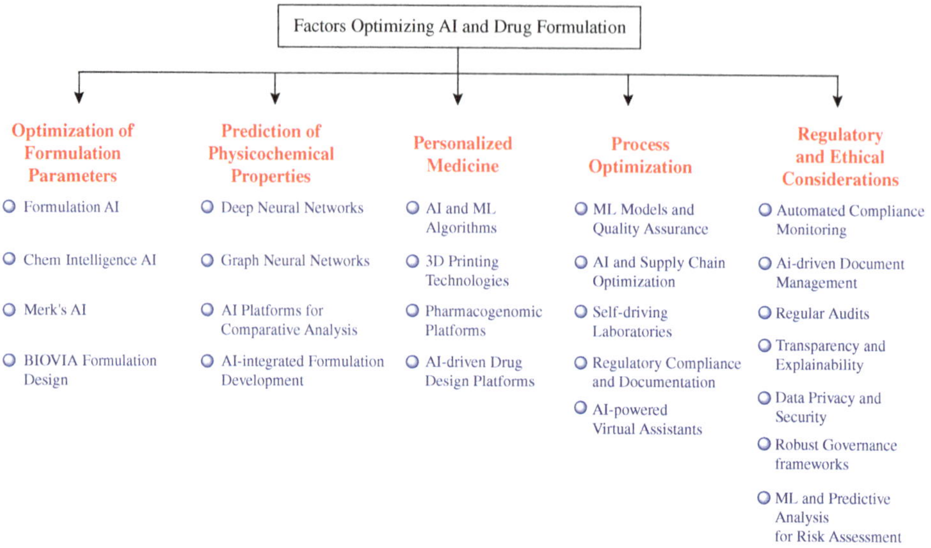

Figure 1.1 Factors optimizing artificial intelligence and drug formulation.

1.3 AI-based Drug Optimization and Pharmaceutical Formulation Parameters

Utilizing AI-driven techniques may help toward formulation development with better effectiveness and improved productivity. Formulation AI, Merck's AI tool, BIOVIA Formulation Design, and AI-Powered Batch Recipe Optimization are some of the AI-based tools for drug and formulation optimization. Polymer drug ratios and encapsulation methods are the most optimal formulation parameters in drug development. This intricate dataset of the formulation parameters is utilized by the AI systems for drug optimization and thus enhances the drug delivery and drug release, resulting in improved efficacy of the formulation [25, 26]. The different factors optimizing AI and drug formulation are represented in Figure 1.1.

Formulation AI uses the algorithms to predict and improve the solubility and bioavailability of the medicinal formulations and thus predicts the best formulation design possible. This feature-rich web-based platform streamlines the drug development process and minimizes the risk of trial-and-error-based failures. Chem Intelligence also utilizes AI inputs and provides solutions related to drug formulation using ML-based techniques like Bayesian optimization. Such AI-driven technologies help researchers to effectively capitalize on R&D more efficiently as compared to the conventional approaches like trial-and-error approaches or procedures based on design of experiments (DoE).

Merk has developed the AI-enabled tool for the determination and prediction of solubility-related parameters for poorly or moderately soluble drugs [27]. This technique thus accelerates the drug development process, thereby allowing fast medication supply to the individual. BIOVIA Formulation Design provides a reliable method and efficient approach toward formulation development, quality testing, and drug optimization.

AI-based algorithms are efficient for screening large biological datasets and thus determine the best formulation process parameters in lesser time. This AI-based approach helps with improved efficiency and confirms the continual product quality regarding the formulation and development of pharmaceutical products [26, 28].

Thus, such AI-driven tools facilitate the novel and particular methods to optimize and regulate the formulation and drug development parameters, which enables improved therapeutic outcomes, demonstrating the precise and accurate output of AI on pharmaceutical formulation.

1.4 AI-driven Models and Tools for Predicting the Physicochemical Properties of Drugs

AI-driven ML tools and techniques utilize the formulation components and the chemical structures of the molecules for predicting significant physicochemical aspects, including stability and solubility [29]. Such predictive specifications are helpful for drug designing and formulations. AI-driven tools and technologies impart significant potential toward screening and optimizing the characteristic features of the API and excipients. For forecasting the physicochemical characteristics of the API and excipients, AI has evolved into an important tool. The AI-enabled models find significant applications with improved efficiency and accuracy in drug development by providing the structural characteristics of the drugs and excipients at molecular level [30–33].

Bioavailability and solubility are the two most important factors toward determining the therapeutic potential of the medication. Poor drug absorption of the drug into the gastrointestinal tract (GIT) or poor aqueous solubility of the drug may result in poor therapeutic outcomes. The AI-based tools for optimization of formulation parameters and ML tools may provide better insights and solutions with improved therapeutic outcomes. AI uses large-scale datasets of the structural characteristics of the molecules and the solubility patterns, and thus the trained ML-based models can provide the missing loops that those conventional methods might miss. Such models can be used for the determination of suitable solubilization principles, like solid dispersions, canonization, complexations, along with the application of surfactants. AI, in addition, can be used to predict the pharmacokinetic and pharmacodynamic aspects of the drug molecules by utilizing the data from in vitro, in vivo, and computational research. Thus, using such information AI designs appropriate formulations with good bioavailability and solubility [30, 34, 35]. Different applications of AI [36] in drug formulation are given in Figure 1.2.

To predict the physicochemical aspects of the drugs and excipients, including solubility, stability, and permeability, the AI-enabled deep neural networks (DNNs) and related tools can also be used. Such networks are very well trained with the evaluation of structural characteristics of the small molecules as drugs with modified drug design and decision-making. Just like DNNs, the graph neural networks (GNNs) use and describe the chemical structures with atoms and bonds as nodes and edges. This representation enables the prediction of aqueous solubility and bioavailability of the drug candidates [37]. The model incorporates self-attention mechanisms and thus increasing the accuracy and interpretability. Many novel AI platforms have been developed for determining the changes in the

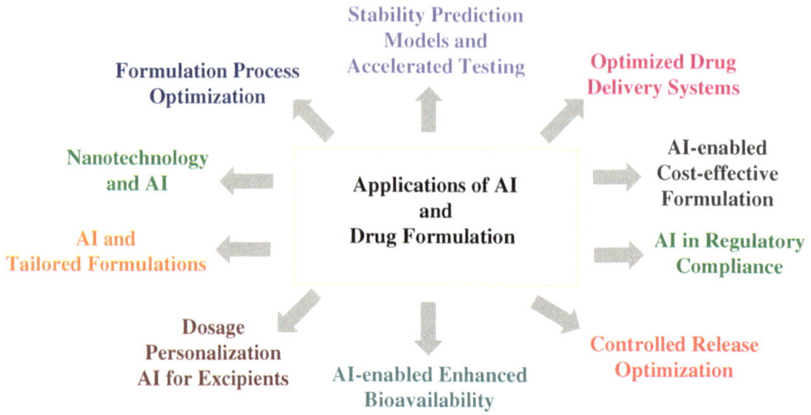

Figure 1.2 Applications of artificial intelligence and drug formulation.

properties on their own or from the variances. Such platforms provide researchers with more precise and highly effective tools toward developing treatments with desired outcomes [38, 39]. The developments confer safer and highly effective pharmaceutical manufacturing in addition to continuing the drug discovery and development.

AI-based tools help in creating and optimizing the appropriate nanoformulation design and related drug delivery by means of evaluating the complex datasets [40]. AI-based tools also predict the cellular absorptions and biodistribution-related factors by evaluating the key characteristics associated, like size, shape, surface charge, and composition of the nanomaterials. Al-based tools are designed to analyze the complex receptor patterns. AI also helps predict the appropriate ligand combinations for effective targeted treatment for cancer therapy [41, 42].

AI methods like artificial neural networks (ANNs) and neuro-fuzzy networks find applications toward improving the self-emulsifying drug delivery systems (SEDDS), which use drugs, oils, surfactants, or cosolvents in isotropic mixtures. For instance, AI model helps in optimizing the crucial parameters impacting the size of the droplet and predicts the biological behavior of the lipid-based medications, thereby improving the strength of the SEDDS [43, 44].

1.5 Implementing AI in Regulatory and Ethical Aspects

To improve the therapeutic outcomes and minimize the trial failure risks, pharmaceutical industries are rapidly using AI-based tools. While using the AI tools, one must consider the ethical considerations and legal aspects of the related drugs or formulations [45, 46]. For integrating the field of drug discovery and development with AI technology, some key factors must be considered. AI-powered platforms provide easier ways for submission, processing, and management of regulatory documents. Because of the automation, the process of documentation becomes more efficient and the chances of errors get reduced [47, 48].

ML tools and algorithms evaluate the available data and implement the mitigation solutions to anticipate possible issues. Although it is very crucial to ensure the confidentiality and the data security of the patient, and to preserve the integrity, the AI-based findings and outcomes must adhere strictly to the data protection laws. AI-based algorithms can produce biased results unintentionally, and thus to make sure the fair treatment of the individuals, regular audits and updates are mandatory [48, 49].

Strong governance frameworks, frequent audits, and interdisciplinary cooperation should be created by the organizations in order to successfully integrate the AI tools into the drug discovery and development. Regulations specifically targeting and supervising the use of AI in relation to following the moral principles and legal procedures must be obeyed strictly [50–53]. Also, the regular cooperation between the technologists, ethicists, and regulatory specialists must be maintained. The pharmaceutical organizations must maintain the patient safety and confidentiality along with utilizing the AI-based tools with special emphasis on maintaining the ethical integrity and regulatory compliance [53, 54].

1.6 Conclusion

The development of AI and the integration of AI-based tools with pharmaceutical development greatly impact the process of drug discovery and development. AI-based screening of vast library of data sets and algorithms facilitates the efficient and economic development of personalized treatment options with defined pharmaceutical parameters like bioavailability, drug distribution, and drug development. Present difficulties related to AI and drug development include biased or low-quality data that might lead to predictions with errors, unclear decision-making processes, in case of DL models, difficulty of incorporating the AI with ongoing workflow and regulations of the drug development process. The stages and processes of drug development are becoming more and more reliant on AI-based tools, by means of transforming the stepwise process of finding, creating, and introducing new and improved medications with more specificity and lesser side effects. Certain issues like data accessibility, privacy, bias, and regulatory concerns must need to be resolved.

References

1 Mak, KK, Wong YH, and Pichika MR. Artificial intelligence in drug discovery and development. In: *Drug Discovery and Evaluation: Safety and Pharmacokinetic Assays* (pp. 1461–1498). Cham: Springer; 2024.

2 Wang H, Fu T, Du Y, et al. Scientific discovery in the age of artificial intelligence. *Nature* 2023;620(7972):47–60.

3 Turing AM. *Computing Machinery and Intelligence*. Springer Netherlands; 2009.

4 Rasheed K, Zaland A, Saad S, Ammad S, and Rostami A. *History of AI InAI in Material Science* (pp. 15–46). CRC Press; 2024.

5 Kaul V, Enslin S, and Gross SA. History of artificial intelligence in medicine. *Gastrointestinal Endoscopy* 2020;92(4):807–812.

6 Mainzer K and Mainzer K. A short history of the AI. *Artificial Intelligence-when do Machines Take Over* 2020: 7–13.

7 Visan AI and Negut I. Integrating artificial intelligence for drug discovery in the context of revolutionizing drug delivery. *Life* 2024;14(2):233.

8 Khan IU, Ouaissa M, Ouaissa M, Fayaz M, and Ullah R, editors. *Artificial Intelligence for Intelligent Systems: Fundamentals, Challenges, and Applications.* CRC Press; 2024.

9 Noorain, Srivastava V, Parveen B, and Parveen R. Artificial intelligence in drug formulation and development: Applications and future prospects. *Current Drug Metabolism* 2023;24(9):622–634.

10 Ali KA, Mohin SK, Mondal P, Goswami S, Ghosh S, and Choudhuri S. Influence of artificial intelligence in modern pharmaceutical formulation and drug development. *Future Journal of Pharmaceutical Sciences.* 2024;10(1):53.

11 Serrano DR, Luciano FC, Anaya BJ, et al. Artificial intelligence (AI) applications in drug discovery and drug delivery: Revolutionizing personalized medicine. *Pharmaceutics* 2024;16(10):1328.

12 Mottaghi-Dastjerdi N and Soltany-Rezaee-Rad M. Advancements and applications of artificial intelligence in pharmaceutical sciences: A comprehensive review. *Iranian Journal of Pharmaceutical Research: IJPR* 2024;23(1):e150510.

13 Song Z, Chen G, and Chen CY. AI empowering traditional Chinese medicine. *Chemical Science* 2024;15(41):16844–16886.

14 Bannigan P, Aldeghi M, Bao Z, Häse F, Aspuru-Guzik A, and Allen C. Machine learning directed drug formulation development. *Advanced Drug Delivery Reviews* 2021;175:113806.

15 Da Silva-Júnior EF. You've got the Body I've got the Brains-Could the current AI-based tools replace the human ingenuity for designing new drug candidates. *Bioorganic & Medicinal Chemistry* 2023;94:117475.

16 Deng J, Yang Z, Ojima I, Samaras D, and Wang F. Artificial intelligence in drug discovery: applications and techniques. *Briefings in Bioinformatics* 2022;23(1):bbab430.

17 Walters WP and Barzilay R. Critical assessment of AI in drug discovery. *Expert Opinion on Drug Discovery* 2021;16(9):937–947.

18 Jiménez-Luna J, Grisoni F, Weskamp N, and Schneider G. Artificial intelligence in drug discovery: Recent advances and future perspectives. *Expert Opinion on Drug Discovery* 2021;16(9):949–959.

19 Harrer S, Shah P, Antony B, and Hu J. Artificial intelligence for clinical trial design. *Trends in Pharmacological Sciences* 2019;40(8):577–591.

20 Chen E, Prakash S, Janapa Reddi V, Kim D, and Rajpurkar P. A framework for integrating artificial intelligence for clinical care with continuous therapeutic monitoring. *Nature Biomedical Engineering* 2023;9(4):445–454.

21 Johnson KB, Wei WQ, Weeraratne D, et al. Precision medicine, AI, and the future of personalized health care. *Clinical and Translational Science* 2021;14(1):86–93.

22 Pantanowitz L, Hanna M, Pantanowitz J, et al. Regulatory aspects of AI-ML. *Modern Pathology* 2024;37:100609.

23 Fanni SC, Febi M, Aghakhanyan G, and Neri E. Natural language processing. In: *Introduction to Artificial Intelligence* (pp. 87–99). Cham: Springer International Publishing; 2023.

24 Mundhra S, Kadiri SK, and Tiwari P. Harnessing AI and machine learning in pharmaceutical quality assurance. *The Journal of Pharmaceutical Quality Assurance and Quality Control* 2024: 19–29.

25 Vora LK, Gholap AD, Jetha K, Thakur RR, Solanki HK, and Chavda VP. Artificial intelligence in pharmaceutical technology and drug delivery design. *Pharmaceutics* 2023;15(7):1916.

26 Suriyaamporn P, Pamornpathomkul B, Patrojanasophon P, Ngawhirunpat T, Rojanarata T, and Opanasopit P. The artificial intelligence-powered new era in pharmaceutical research and development: A review. *AAPS PharmSciTech* 2024;25(6):188.

27 Han Y and Tao J. Revolutionizing pharma: Unveiling the AI and LLM trends in the pharmaceutical industry. *arXiv preprint arXiv:2401.10273*; 2024.

28 Singh R, Arya P, and Dubey SH. Artificial intelligence in pharmaceutics: Revolutionizing drug formulation and optimization. *Journal of Drug Discovery and Health Sciences* 2024;1(03):138–145.

29 Dey H, Arya N, Mathur H, Chatterjee N, and Jadon R. Exploring the role of artificial intelligence and machine learning in pharmaceutical formulation design. *International Journal of Newgen Research in Pharmacy & Healthcare* 2024: 30–41.

30 Tummala SR and Gorrepati N. AI-driven predictive analytics for drug stability studies. *Journal of Pharma Insights and Research* 2024;2(2):188–198.

31 Yingngam B. AI models for biopharmaceutical property prediction. In: *Artificial Intelligence and Machine Learning in Drug Design and Development* (pp. 407–449). Wiley; 2024.

32 Chou WC, Chen Q, Yuan L, et al. An artificial intelligence-assisted physiologically-based pharmacokinetic model to predict nanoparticle delivery to tumors in mice. *Journal of Controlled Release* 2023;361:53–63.

33 Gholap AD, Uddin MJ, Faiyazuddin M, Omri A, Gowri S, and Khalid M. Advances in artificial intelligence in drug delivery and development: A comprehensive review. *Computers in Biology and Medicine* 2024;178:108702.

34 Chou WC and Lin Z. Machine learning and artificial intelligence in physiologically based pharmacokinetic modeling. *Toxicological Sciences* 2023;191(1):1–4.

35 Yang X, Wang Y, Byrne R, Schneider G, and Yang S. Concepts of artificial intelligence for computer-assisted drug discovery. *Chemical Reviews* 2019;119(18):10520–10594.

36 Bhat VN, Bharati S, Bothiraja C, Sangshetti J, and Gaikwad V. A review on intervention of ai in pharmaceutical sector: Revolutionizing drug discovery and manufacturing. *Intelligent Pharmacy* 2025;3:342–349.

37 Huuskonen J, Livingstone DJ, and Manallack DT. Prediction of drug solubility from molecular structure using a drug-like training set. *SAR and QSAR in Environmental Research* 2008;19(3-4):191–212.

38 Taskinen J and Yliruusi J. Prediction of physicochemical properties based on neural network modelling. *Advanced Drug Delivery Reviews*. 2003;55(9):1163–1183.

39 Lee S, Lee M, Gyak KW, Kim SD, Kim MJ, and Min K. Novel solubility prediction models: Molecular fingerprints and physicochemical features vs graph convolutional neural networks. *ACS Omega* 2022;7(14):12268–12277.

40 Shang Z, Chauhan V, Devi K, and Patil S. Artificial intelligence, the digital surgeon: Unravelling its emerging footprint in healthcare–The narrative review. *Journal of Multidisciplinary Healthcare* 2024;17:4011–4022.

41 Serov N and Vinogradov V. Artificial intelligence to bring nanomedicine to life. *Advanced Drug Delivery Reviews* 2022;184:114194.

42 Kasture K and Shende P. Amalgamation of artificial intelligence with nanoscience for biomedical applications. *Archives of Computational Methods in Engineering* 2023;30(8):4667–4685.

43 Singh I, Kaur J, Kaur S, Barik BR, and Pahwa R. Artificial neural networks and neuro-fuzzy models: Applications in pharmaceutical product development. *Brazilian Archives of Biology and Technology* 2023;66:e23210769.

44 Fatouros DG, Nielsen FS, Douroumis D, Hadjileontiadis LJ, and Mullertz A. In vitro–in vivo correlations of self-emulsifying drug delivery systems combining the dynamic lipolysis model and neuro-fuzzy networks. *European Journal of Pharmaceutics and Biopharmaceutics* 2008;69(3):887–898.

45 Walz A and Firth-Butterfield K. Implementing ethics into artificial intelligence: A contribution, from a legal perspective, to the development of an AI governance regime. *Duke Law & Technology Review* 2019;18:176.

46 Agbese M, Mohanani R, Khan A, and Abrahamsson P. Implementing AI ethics: Making sense of the ethical requirements. In: *Proceedings of the 27th International Conference on Evaluation and Assessment in Software Engineering*; 2023. pp. 62–71.

47 Matyuk YS. Ethical and legal aspects of development and implementation of artificial intelligence systems. In: *European Proceedings of Social and Behavioural Sciences* (pp. 689–696). European Publisher; 2022.

48 Alahmed Y, Abadla R, Ameen N, and Shteiwi A. Bridging the gap between ethical AI implementations. *International Journal of Membrane Science and Technology* 2023;10(3):3034–3046.

49 Mennella C, Maniscalco U, De Pietro G, and Esposito M. Ethical and regulatory challenges of AI technologies in healthcare: A narrative review. *Heliyon* 2024;10(4):e26297.

50 Seeram E and Kanade V. Ethical and regulatory considerations. In: *Artificial Intelligence in Medical Imaging Technology: An Introduction* (pp. 151–167). Cham: Springer Nature Switzerland; 2024.

51 Fosch-Villaronga E, Chokoshvili D, Vallevik VB, Ienca M, and Pierce RL. Implementing AI in healthcare: An ethical and legal analysis based on case studies. *Data Protection and Privacy* 2021;13:187.

52 Hasan HE, Jaber D, Khabour OF, and Alzoubi KH. Ethical considerations and concerns in the implementation of AI in pharmacy practice: A cross-sectional study. *BMC Medical Ethics* 2024;25(1):55.

53 Prem E. From ethical AI frameworks to tools: A review of approaches. *AI and Ethics* 2023;3(3):699–716.

54 Anishchenko MA, Gidenko I, Kaliman M, Polyvaniuk V, and Demianchuk Y. Artificial intelligence in medicine: Legal, ethical and social aspects. *Acta Bioethica* 2023;29(1):63–72.

2

Artificial Intelligence-driven Innovations in Drug Design, Delivery, and Personalization

Amol D. Gholap[1,2], Pankaj R. Khuspe[3], Sadikali F. Sayyad[2*], Navnath Hatvate[4], and Machindra Chavan[5]*

[1]*Department of Pharmaceutics, St. John Institute of Pharmacy and Research, Palghar, India*
[2]*Department of Pharmaceutics, Amrutvahini College of Pharmacy, Sangamner, India*
[3]*Department of Pharmaceutics, College of Pharmacy, Paniv, Solapur, India*
[4]*Department of Pharmaceutical Sciences and Technology, Institute of Chemical Technology Mumbai, Marathwada Campus, Jalna, India*
[5]*Department of Pharmacognosy, Amrutvahini College of Pharmacy, Sangamner, India*

*Corresponding authors: amolgholap16@gmail.com; sadiksayyad@gmail.com

CONTENTS

2.1 Introduction

Determining the optimal composition, stability, and delivery route of pharmaceutical substances requires extensive testing, which is a complex and resource-intensive procedure. Prolonged development schedules and increased costs are the results of conventional formulation procedures, which mostly rely on empirical methodologies, iterative optimization, and extensive in vitro and in vivo testing [1]. The advent of data-driven approaches brought about by artificial intelligence (AI) as a paradigm shift in pharmaceutical sciences has greatly improved the effectiveness, accuracy, and economy of drug manufacture. Rapid analysis of high-dimensional datasets, including physicochemical features, molecular interactions, pharmacokinetic profiles, and patient-specific factors, is made possible by AI-driven techniques such as machine learning (ML), deep neural networks (DNNs), and predictive analytics [2]. By making it easier to anticipate excipient-drug compatibility, identify ideal formulation conditions, and simulate drug release kinetics, these sophisticated computer models reduce the need for traditional trial-and-error techniques. Furthermore, by clarifying intricate drug-excipient interactions and refining targeted delivery methods, AI enhances the logical design of drug delivery systems [3]. The combination of digital twins (DTs) and in silico modeling has also been accelerated by recent developments in AI, enabling real-time simulation and optimization of pharmaceutical formulations prior to experimental validation. Notwithstanding these ground-breaking developments, issues like algorithmic interpretability, data heterogeneity, and regulatory restrictions still exist, calling for interdisciplinary cooperation between regulatory agencies, pharmaceutical scientists, and AI specialists. This paper examines the crucial role AI plays in contemporary drug formulation, offering a thorough examination of its recent developments, current applications, inherent difficulties, and potential to completely transform pharmaceutical creation in the future as presented in Figure 2.1 [4, 5].

2.2 AI in Predictive Modeling for Drug Formulation

One of the mainstays of AI applications in drug development is predictive modeling. Large-scale datasets are analyzed by ML algorithms to forecast the stability, bioavailability, and physicochemical characteristics of medication formulations. Among the major innovations are described below.

2.2.1 Quantitative Structure-activity Relationship Models

A key technique in computational drug creation, quantitative structure-activity relationship (QSAR) models allow for accurate prediction of the relationship between chemical structure and biological activity. These models help researchers optimize drug formulation and design by using statistical and mathematical methods to convert chemical descriptors into biologically useful insights. The introduction of AI has greatly improved the predictive power of QSAR models, which previously depended on statistical modeling and linear regression. Advanced ML techniques, including random forest (RF), support vector

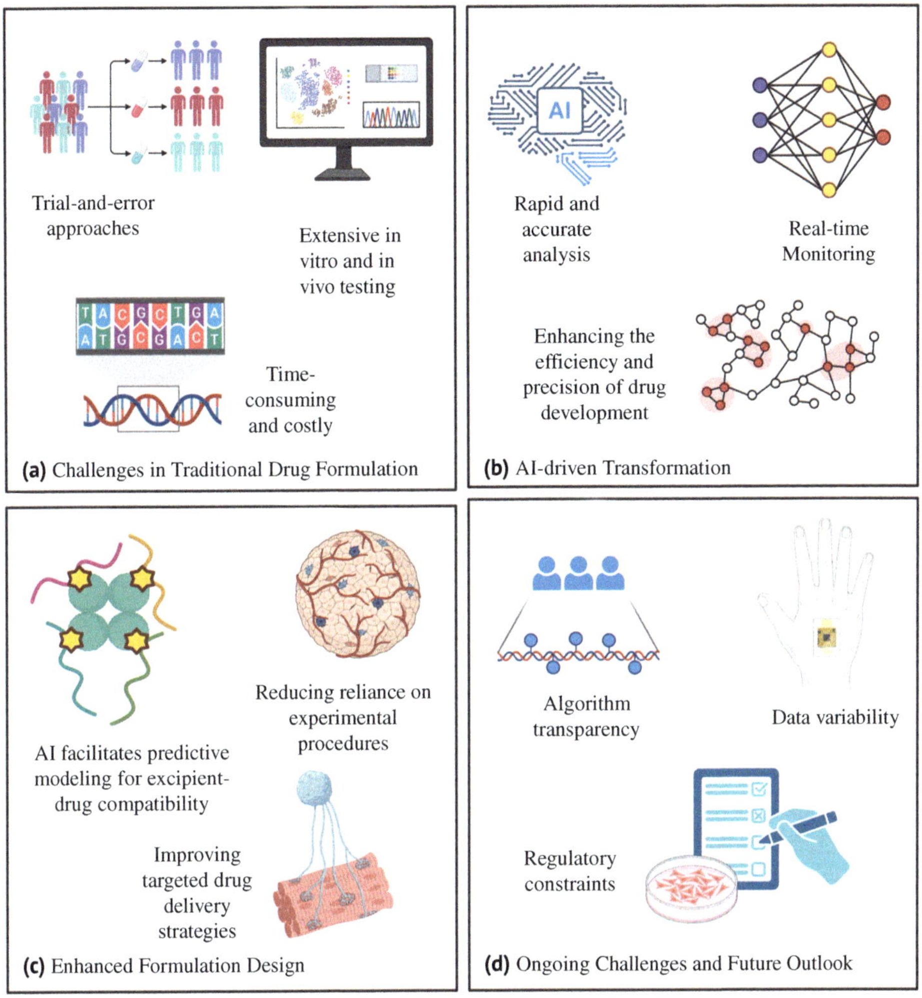

Figure 2.1 This figure delineates the evolution from conventional drug formulation to innovations driven by artificial intelligence (AI). (a) Traditional methodologies are characterized by trial-and-error techniques and extensive testing and are both time-intensive and costly. (b) The transformation facilitated by AI allows for rapid and precise analysis alongside real-time monitoring, thereby enhancing the precision and efficiency of drug development. (c) The improved formulation design utilizes AI for predictive modeling of excipient-drug compatibility, which reduces the necessity for experimental procedures and supports advanced drug delivery strategies. (d) Despite these advancements, challenges persist, including issues related to algorithm transparency, data variability, and regulatory constraints. Collectively, these panels underscore the revolutionary impact of AI integration in pharmaceutical formulation while also identifying areas that require further development and governance for broader implementation. Created in BioRender by Md. Faiyazuddin.

machines (SVM), and DNN, are integrated into AI-driven QSAR models to improve pattern detection and generalization from high-dimensional molecular datasets. Because they can precisely forecast crucial physicochemical characteristics, including solubility, permeability, stability, lipophilicity, and bioavailability, important factors affecting the choice of lead compounds, these models are very useful in preformulation research [6].

By using convolutional neural networks (CNNs) and recurrent neural networks (RNNs) to automatically extract molecular characteristics from extensive chemical libraries, deep learning architectures have improved QSAR modeling even more. For instance, CNN-based QSAR models use 3D structural representations and molecular graphs to identify complex atomic interactions that affect biological activity. RNN models, on the other hand, are very good at studying temporal drug-receptor interactions and metabolic stability because they are very good at processing sequential data. By making it easier to choose the best excipients, guaranteeing formulation stability, and improving drug delivery systems, these AI-enhanced QSAR approaches go beyond simply forecasting pharmacological activity. For example, water solubility and membrane permeability, two important variables influencing medication absorption, systemic circulation, and overall therapeutic efficacy, have been accurately predicted using AI-driven QSAR models as presented in Figure 2.2 [7].

QSAR models face obstacles that prevent their widespread use in drug development, despite their revolutionary promise. Since reliable AI models need high-quality experimental datasets to be trained, data scarcity continues to be a major obstacle. Furthermore, model interpretability is a problem, especially with deep learning-based QSAR techniques where the decision-making procedures are still mostly unknown. In order to improve model trustworthiness, explainable AI (XAI) methodologies must be developed, as regulatory agencies stress the necessity for transparency in AI-driven predictions. Furthermore, in order to guarantee the accuracy and suitability of QSAR predictions in actual pharmaceutical applications, thorough validation using experimental datasets is essential [8].

The combination of hybrid ML techniques, transfer learning strategies, and quantum mechanics-based molecular descriptors is probably going to propel future developments in AI-driven QSAR modeling. Descriptors of quantum mechanics improve prediction accuracy for complicated medication formulations by facilitating a deeper knowledge of atomic-level molecular interactions. By utilizing knowledge from current QSAR models, transfer learning, which enables pretrained models to be applied to new datasets with less retraining, shows promise in overcoming data limits. Furthermore, QSAR models will be further improved by AI-driven multi-objective optimization techniques that balance important formulation factors, including stability, toxicity, and efficacy [9].

By lowering the experimental burden, speeding up lead optimization, and simplifying pharmaceutical development, AI-driven QSAR models are set to play an increasingly important role in transforming drug formulation as they continue to advance. Combining these models with other AI-powered frameworks, such as reinforcement learning (RL) and generative adversarial networks (GANs), may open up new possibilities for predictive formulation design and hasten the development of safer and more efficient treatments [6, 7].

2.2.2 Virtual Screening

Drug formulation has been transformed by virtual screening, an AI-driven computational method that speeds up the identification and optimization of highly compatible excipients and active pharmaceutical ingredients (APIs). AI-powered virtual screening offers a high-throughput, effective substitute for conventional experimental screening by combining ML, DNN, and state-of-the-art computational methods like molecular docking, molecular dynamics (MD) simulations, and QSAR modeling. By mimicking important

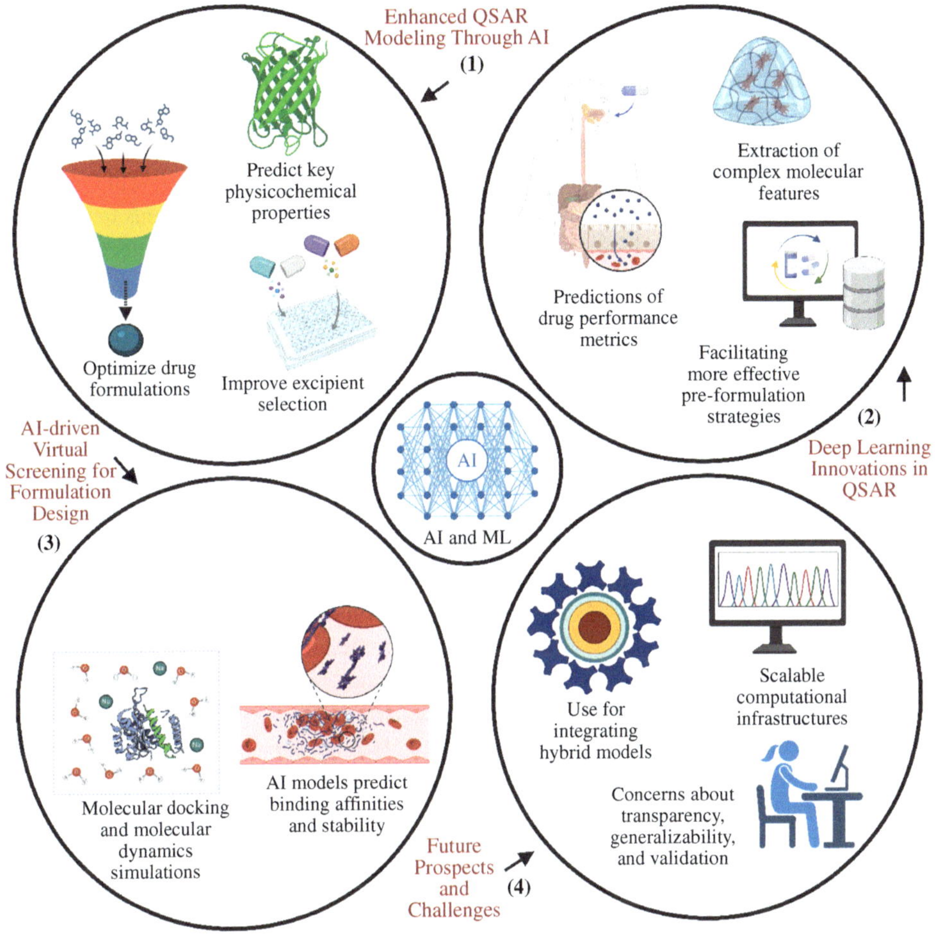

Figure 2.2 AI-driven advancements in QSAR modeling and drug formulation. The image illustrates four key areas: (1) enhanced QSAR modeling through AI, predicting physicochemical properties and improving excipient selection; (2) deep learning innovations in QSAR, extracting complex molecular features and predicting drug performance metrics; (3) AI-driven virtual screening for formulation design, utilizing molecular docking and dynamics simulations; (4) future prospects and challenges, including the integration of hybrid models and concerns about transparency and validation. Central to these advancements is the application of AI and machine learning techniques, represented by a neural network diagram, driving progress in drug discovery and development processes. Created in BioRender by Md. Faiyazuddin.

physicochemical interactions such as hydrogen bonding, hydrophobic forces, van der Waals contacts, and steric hindrance, molecular docking methods forecast the binding affinity, interaction energy, and stability of API-excipient complexes. By using docking studies to rate possible excipients according to how well they work with APIs, excipients that improve medication solubility, stability, and bioavailability can be chosen [10].

MD simulations are essential for offering dynamic insights into the structural behavior of drug-excipient complexes under physiological settings, going beyond static docking simulations. By anticipating how excipients affect the dissolution rate, polymorphic transitions, and aggregation tendencies of APIs in various biological contexts, MD simulations improve formulation techniques by accounting for temperature differences, solvent effects, and molecular flexibility. By utilizing extensive pharmaceutical databases that include physicochemical characteristics, solubility profiles, and pharmacokinetic factors, AI integration in virtual screening goes beyond molecular-level predictions to comprehensive formulation optimization. These datasets are used by AI-driven predictive models to improve formulation designs, reducing experimental effort and increasing excipient selection accuracy [11].

By making it possible to construct molecular structures with specific attributes from scratch, recent developments in GANs and RL have further revolutionized virtual screening techniques. While GANs aid in the development of new excipients by producing molecular structures with improved compatibility profiles, RL algorithms repeatedly improve API-excipient pairings by optimizing structural features that affect medication performance. The transition from drug discovery to commercialization is accelerated by these AI-driven methods, which also drastically cut down on research and development expenses and formulation development durations [12].

The complete integration of AI-powered virtual screening in pharmaceutical workflows is hampered by a number of issues, despite its revolutionary potential. Because black-box models might hide the reasoning behind particular predictions, algorithm transparency is still a major worry. To address this, XAI frameworks must be developed. Another issue is model generalizability, since AI models that have been trained on certain datasets do not necessarily extrapolate correctly to new drug candidates with unique physicochemical characteristics. Furthermore, experimental verification is still necessary to validate AI-generated predictions, which calls for the creation of hybrid computational-experimental frameworks that smoothly combine wet-lab research and in silico modeling [13].

The confluence of multimodal data sources, combining structural, genetic, and pharmacokinetic data to produce incredibly precise predictive models, is probably where AI-enhanced virtual screening will go in the future. Cloud and quantum computing integration is also anticipated to improve AI-driven screening's speed and accuracy, allowing for previously unheard-of real-time simulations of intricate molecular interactions. AI-powered virtual screening is set to become a key technique in medication formulation as it develops further, fostering creativity, effectiveness, and accuracy in pharmaceutical R&D [14].

2.3 Personalized Medicine and AI

By customizing medication formulas to each patient's unique profile, AI significantly advances personalized medicine. Examples of innovations are discussed below.

2.3.1 Patient-specific Formulation

Drug formulation has been transformed by the incorporation of AI into personalized medicine, which makes it possible to develop patient-specific therapeutic approaches that maximize effectiveness and reduce side effects. Conventional drug formulation takes a generic approach, frequently ignoring inter-individual differences in drug response, metabolic pathways, and genetic composition. The creation of personalized formulations is made easier by AI-driven approaches that use ML, DNN, and predictive analytics to analyze large and intricate patient datasets, such as pharmacogenomic profiles, metabolic biomarkers, and physiological parameters. Pharmaceutical experts can forecast each drug's metabolism, absorption, and excretion rates by using AI algorithms. This enables them to precisely modify dosages to meet the demands of each patient [15].

In order to create optimal treatment plans for patients with unique genetic polymorphisms impacting drug-metabolizing enzymes (such as cytochrome P450 variants), AI-powered platforms use pharmacogenomics, the study of genetic variations influencing medication response. For example, formulation scientists can adjust excipient composition, release kinetics, and bioavailability to match patient-specific pharmacokinetics (PK) and pharmacodynamics (PD) by using ML models to analyze single nucleotide polymorphism (SNP) data and forecast an individual's reaction to a particular drug. AI systems also take into account adherence trends and patient-reported preferences, which guarantees better treatment results and compliance. Advanced AI frameworks, such as RL models, enable adaptable and precision-driven therapeutic approaches by continuously improving medication compositions based on clinical data and real-time patient feedback [16].

Notwithstanding its revolutionary promise, issues like algorithm openness, data privacy, and regulatory compliance need to be resolved to guarantee the moral and secure application of AI in patient-specific medication development. Interdisciplinary partnerships between bioinformaticians, pharmacologists, and regulatory bodies will be crucial as AI-driven personalized medicine develops in order to advance the field, guarantee its smooth integration into clinical practice, and improve patient-centric therapeutic outcomes [17].

2.3.2 Digital Twins in Formulation Development

DTs, which are high-fidelity virtual representations of patients or biological systems that allow simulation-based testing of drug formulations, have been made possible by the introduction of AI in pharmaceutical research. These AI-powered computational models combine pharmacokinetic, genetic, and physiological data unique to each patient to produce a dynamic virtual world that simulates actual medication interactions. In the field of drug formulation, DTs are effective instruments for forecasting how pharmaceutical substances would behave in various biological settings, which lessens the need for traditional experimental methods. DTs use DNN and ML algorithms to constantly improve their predictions by integrating real-time clinical and experimental data. This iterative optimization process improves formulation precision [18].

Customizing drug delivery systems is one of the main uses of DTs in formulation development. Inter-individual variations in medication metabolism, absorption, and clearance are frequently overlooked by conventional formulation techniques. By mimicking patient-specific PK and PD profiles, DTs overcome this restriction and allow for customized modifications to drug composition, dosage forms, and release mechanisms. These virtual models also make it easier to optimize drug stability, bioavailability, and excipient selection, guaranteeing that formulations are both efficient and biocompatible. The number of physical trials needed is greatly decreased by using in silico modeling driven by AI-driven DTs to simulate excipient-drug interactions, dissolution kinetics, and tailored delivery methods [19].

Prior to clinical trials, DTs reduce formulation-related safety concerns by improving predictive toxicology and risk assessment. Only the most promising formulations move on to experimental validation thanks to DTs' integration of data from preclinical models, patient records, and molecular simulations, which offer insights into potential side effects. This method reduces the moral and monetary costs related to late-stage failures while simultaneously speeding up the formulation development process [20].

The full-scale incorporation of DTs into medicine formulation is hampered by a number of issues, despite their revolutionary promise. The durability of AI-driven predictive algorithms and the availability of high-quality, standardized datasets are prerequisites for these models' accuracy. Guidelines for model validation, data security, and ethical considerations must be established because regulatory frameworks controlling the application of AI in pharmaceutical sciences are still in their infancy. To fully utilize DTs in pharmaceutical research and development, several issues must be resolved [21].

By enabling real-time, patient-specific simulations, cutting down on drug development times, and improving treatment efficacy, DTs have the potential to completely transform formulation development as AI technologies progress. To enable the smooth integration of DTs in precision medicine and medication development, future research should concentrate on enhancing algorithmic interpretability, incorporating multi-omics data, and cultivating regulatory approval [20].

2.4 Automation and Optimization in Drug Design

Automation powered by AI improves drug formulation processes' efficiency. The major advancements in this area are discussed below.

2.4.1 Automated High-throughput Experimentation

By drastically lowering manual intervention and improving experimental precision, the combination of robotics and AI in high-throughput experimentation (HTE) has transformed medication formulation as presented in Figure 2.3. Conventional medication formulation uses labor-intensive techniques that are prone to human error and inconsistency in addition to being time-consuming [22]. These restrictions have been lessened by the development of AI-driven HTE, which makes it possible to screen pharmaceutical formulations quickly and thoroughly with previously unheard-of accuracy. Thousands of formulations may be synthesized, evaluated, and optimized in a fraction of the time needed by traditional methods thanks to automated workflows orchestrated by robotics and AI algorithms. These systems use ML-driven predictive modeling to dynamically build experimental protocols, modifying variables like solvent compositions, concentration gradients, and excipient interactions in response to real-time data analytics. One prominent use of AI-driven HTE is in the optimization of drug delivery systems based on nanoparticles (NPs), where AI platforms methodically investigate different formulation parameters to

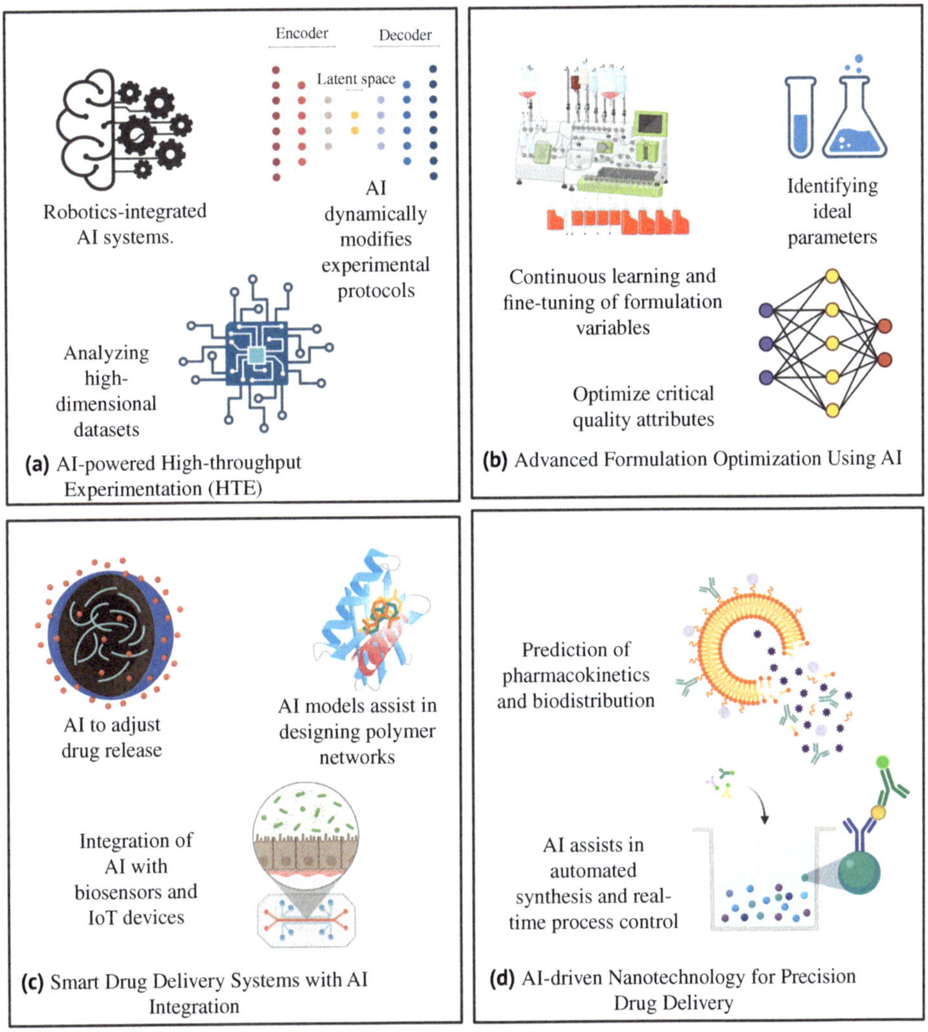

Figure 2.3 Four key applications of artificial intelligence (AI) in pharmaceutical research and development. (a) Depicts AI-powered high-throughput experimentation, featuring robotics-integrated AI systems for analyzing complex datasets and dynamically modifying protocols. (b) Showcases advanced formulation optimization using AI, highlighting continuous learning, fine-tuning of variables, and identification of ideal parameters to optimize critical quality attributes. (c) Demonstrates smart drug delivery systems with AI integration, showing AI's role in adjusting drug release, designing polymer networks, and integrating with biosensors and IoT devices. (d) Presents AI-driven nanotechnology for precision drug delivery, emphasizing AI's capabilities in predicting PK, biodistribution, and assisting in automated synthesis and real-time process control. Created in BioRender by Md. Faiyazuddin.

improve drug loading effectiveness, stability, and targeted delivery to sick tissues. Finding the best formulations for site-specific drug release is made easier by sophisticated AI models that model and forecast the physicochemical interactions between NPs and biological surroundings [23]. For example, lipid-based NPs for mRNA vaccine formulations have been developed using AI-guided robotic platforms, which has sped up the optimization process while guaranteeing batch-to-batch repeatability. Furthermore, real-time discovery of good formulation candidates is made possible by AI's capacity to assess high-dimensional experimental datasets, which lessens the need for laborious trial-and-error testing. To effectively utilize AI in HTE-driven medication design, however, issues including data consistency, integrating AI with current lab infrastructure, and obtaining regulatory approval for AI-generated formulations must be resolved [24].

2.4.2 Formulation Optimization

Nanotechnology has significantly reshaped the domain of human therapeutics by offering precise, targeted, and efficient methodologies for diagnosing, treating, and preventing a wide array of diseases [25–30]. By manipulating materials at the nanoscale, typically less than 100 nanometers, researchers and clinicians can harness unique physicochemical properties that are not present in bulk materials, resulting in notable advancements in drug delivery, cancer therapy, tissue engineering, and regenerative medicine. A particularly transformative application of nanotechnology is in drug delivery systems [31–35]. Traditional drug administration methods often face challenges such as poor solubility, rapid degradation, systemic toxicity, and nonspecific distribution. Nanocarriers, including liposomes, dendrimers, solid lipid NPs, and polymeric NPs, provide controlled and sustained drug release, thereby enhancing therapeutic efficacy while minimizing adverse effects. These nanocarriers can be engineered to target specific tissues or cells, such as cancerous tumors, through surface modifications or ligand conjugation, thereby improving selectivity and reducing harm to healthy tissues. In the field of oncology, nanotechnology is pivotal in both diagnosis and treatment [36–42]. NPs can be designed to preferentially accumulate in tumor tissues through the enhanced permeability and retention (EPR) effect. Additionally, multifunctional NPs are being developed that integrate therapeutic agents with diagnostic imaging modalities, leading to theranostics platforms that enable simultaneous disease diagnosis, monitoring, and treatment [43–49]. For example, gold NPs and magnetic NPs are being explored for their ability to deliver drugs, enhance imaging contrast, and facilitate photothermal or magnetic hyperthermia therapies. Nanotechnology is also advancing gene therapy, where nanoscale carriers protect genetic material such as DNA or RNA from degradation and facilitate their transport into target cells [50–54]. Lipid-based NPs have played a crucial role in delivering mRNA vaccines, as evidenced during the COVID-19 pandemic. These platforms lay the groundwork for future therapies aimed at correcting genetic disorders or modulating gene expression in diseases such as cancer, cardiovascular conditions, and neurodegenerative disorders [55–60].

Beyond drug and gene delivery, nanomaterials are extensively utilized in tissue engineering and regenerative medicine. Nanoscale scaffolds and hydrogels create biomimetic environments conducive to cell growth, differentiation, and tissue regeneration [61–66]. For example, nanofiber-based scaffolds replicate the natural extracellular matrix (ECM) of tissues, thereby enhancing cellular interactions and promoting healing in applications such as skin grafts, bone repair, and neural regeneration [67–72]. Furthermore, biosensors and nanodiagnostics are advancing disease detection and monitoring. Nanosensors are capable of detecting biomarkers at ultra-low concentrations, facilitating early diagnosis of diseases such as cancer, diabetes, and infectious diseases. These tools hold significant promise for point-of-care testing and personalized medicine, where treatments can be tailored based on an individual's unique biological profile [73–80]. Despite its vast potential, the clinical translation of nanotechnology continues to face challenges, including issues of toxicity, biocompatibility, scalability, and regulatory hurdles. Ongoing research endeavors to address these challenges to ensure that nanotechnology-based therapies are safe, effective, and accessible [81–85]. AI, ML, and nanotechnology are revolutionizing human therapeutics by enhancing diagnostics, drug delivery, and personalized medicine. AI and ML facilitate the analysis of large datasets to uncover patterns in disease progression, predict patient responses, and identify potential drug candidates. In clinical settings, ML algorithms enhance early diagnosis, optimize treatment plans, and monitor therapeutic outcomes in real-time. Conversely, nanotechnology enables precise targeting and controlled release of therapeutics at the cellular or molecular level. NPs can traverse biological barriers, delivering drugs directly to diseased tissues, thereby minimizing side effects and improving efficacy. In cancer therapy, for instance, nanocarriers such as liposomes or polymeric NPs are engineered to deliver chemotherapeutics specifically to tumor cells. The integration of AI/ML with nanotechnology further augments these capabilities. AI models assist in designing nanomaterials with desired properties and predicting their interactions with biological systems. This synergy accelerates the development of smart nanocarriers that respond to specific stimuli or disease markers, ensuring precision treatment. Overall, the convergence of AI, ML, and nanotechnology offers promising advancements in diagnostics, targeted therapies, and regenerative medicine, paving the way toward highly personalized and efficient healthcare solutions [51].

In order to achieve the best possible therapeutic efficacy and bioavailability, formulation optimization, which is a crucial stage in drug design, further involves adjusting important factors such as particle size, dissolving rate, excipient compatibility, and drug release kinetics. This procedure has historically required a lot of work and relied on statistical techniques and many experimental trials to determine appropriate formulation conditions. This field has changed with the introduction of AI, especially RL algorithms, which allow for dynamic, data-driven optimization techniques that continuously adjust formulation parameters in response to real-time feedback. By independently determining the ideal conditions that optimize drug stability and efficacy while minimizing undesirable attributes like aggregation or poor solubility, RL models, which learn through iterative interactions with the formulation environment, have shown remarkable promise in improving the precision of pharmaceutical formulations [2, 12, 16, 23].

Bayesian optimization (BO), a probabilistic model-based technique that effectively traverses high-dimensional formulation spaces to find ideal medication compositions with the fewest number of experimental trials, is one of the most promising uses of AI-driven formulation optimization. BO has been used extensively in the creation of sophisticated drug carriers, including as liposomal formulations, where it optimizes important factors such as particle size distribution, lipid content, and encapsulation efficiency to promote increased cellular absorption and controlled drug release. Through the use of surrogate modeling approaches, BO improves precision while using fewer resources by iteratively adjusting formulation parameters based on previous experimental data. This method has proven very helpful when designing delivery systems based on NPs, where even little changes to formulation parameters can have a big impact on PK and therapeutic results.

The full-scale integration of RL and BO in pharmaceutical formulation is still hampered by issues including data scarcity, model interpretability, and regulatory approval, notwithstanding recent developments. To overcome these obstacles and guarantee the dependability and clinical application of AI-optimized formulations, strong validation procedures, interdisciplinary cooperation, and flexible regulatory frameworks are needed. In order to create a completely automated and intelligent medication formulation pipeline, AI must be integrated synergistically with high-throughput screening, computer simulations, and real-time analytics. This is where formulation optimization is headed.

2.5 AI in Drug Delivery Systems

AI is transforming medicine delivery systems by enhancing their functionality and design, and some breakthroughs in this area are described below.

2.5.1 Smart Drug Delivery

Pharmaceutical sciences have undergone a revolution thanks to the incorporation of AI into smart drug delivery systems, which allow for accurate, regulated, and patient-specific drug administration. Therapeutic drugs can be released via smart drug delivery systems in response to environmental or physiological cues like pH, temperature, enzyme activity, or particular biomolecular interactions. The logical design of these intelligent systems is greatly aided by AI-driven predictive modeling and optimization algorithms, which analyze enormous datasets on target site-specific circumstances, degradation rates, and drug-polymer interactions. AI improves the precision and effectiveness of intelligent medication delivery systems by utilizing DNN, RL algorithms, and sophisticated ML models. This ensures maximum therapeutic efficacy while reducing side effects [86].

AI-optimized hydrogels for regulated and stimuli-responsive drug release are among the most promising AI-enabled smart drug delivery technologies. Because hydrogels are made of networks of hydrophilic polymers, they can change structurally in response to external stimuli, which makes them ideal for applications involving targeted and prolonged drug delivery. By mimicking drug diffusion dynamics and forecasting the hydrogel's response under physiological settings, AI algorithms make it easier to choose the best polymer compositions, crosslinking densities, and drug-loading efficiencies. AI-powered models, for example, can create temperature-sensitive hydrogels that adjust medication release in response to local inflammation levels or pH-sensitive hydrogels that release pharmaceuticals preferentially in acidic tumor microenvironments. These AI-optimized hydrogels decrease systemic toxicity, increase medication half-life, and greatly improve therapeutic agent bioavailability [87].

AI is essential to smart medicine delivery systems' adaptive control and real-time monitoring. By combining AI with biosensors and Internet of Things (IoT) technology, physiological factors may be continuously tracked, allowing for dynamic modifications to drug release profiles. In order to meet the unique needs of each patient and the course of their illness, this adaptive mechanism guarantees responsive and customized drug administration.

The widespread clinical translation of smart drug delivery systems is still hampered by issues including material biocompatibility, regulatory compliance, and the intricacy of AI-driven predictive models, notwithstanding recent developments. To overcome these obstacles and improve patient-centered treatment outcomes, interdisciplinary cooperation between AI experts, biological engineers, and pharmaceutical scientists is needed. As AI develops further, its combination with cutting-edge smart drug delivery systems has enormous promise to transform customized healthcare and precision medicine [88].

2.5.2 Nanotechnology Integration

The design and development of sophisticated drug delivery systems has been completely transformed by the combination of AI with nanotechnology, providing focused, effective, and precise therapeutic interventions. Through the optimization of their physicochemical characteristics, biocompatibility, and functionalization procedures, AI-driven computer models such as ML algorithms, DNN, and predictive analytics allow for the logical design of NPs. The identification of critical characteristics that affect the PK and biodistribution of NPs, such as particle size, surface charge, and hydrophobicity, is made easier by AI's capacity to analyze large and complicated datasets. Furthermore, drug–NP interactions can be predicted with the use of AI-driven molecular docking and modeling approaches, which guarantees improved therapeutic index, sustained release profiles, and encapsulation efficiency [89].

AI-guided production of polymeric nanoparticles (PNPs) for medication delivery is one of the most exciting uses of AI in nanotechnology. Conventional PNP manufacturing techniques frequently call for a great deal of optimization through experimental iterations, which is expensive and time-consuming. By using sophisticated prediction models to identify the ideal polymer composition, crosslinking density, and degradation kinetics for particular therapeutic applications, AI expedites this process. Additionally, quick selection of biocompatible polymers is made possible by AI-enhanced high-throughput screening, which maximizes drug loading capacity and controlled release while guaranteeing minimal cytotoxicity. Researchers can create PNPs with adjustable characteristics that enable targeted medication administration, especially for illnesses like cancer, neurological diseases, and infectious disorders, by combining AI within silico modeling and simulations based on quantum mechanics. Additionally, real-time modifications during NP production are made possible by AI-driven predictive modeling, which enhances scalability and repeatability [90].

Data heterogeneity, a lack of defined protocols, and regulatory complexity are some of the obstacles that AI-assisted NP creation must overcome despite its revolutionary promise. To guarantee safety, effectiveness, and regulatory compliance, experimental and clinical research are still necessary to validate AI-generated models. The development of next-generation drug delivery platforms, however, which offer individualized and precision-based treatment solutions with increased efficacy and less systemic toxicity, is extremely promising due to the convergence of AI and nanotechnology. Multimodal data integration, real-time adaptive learning, and AI-augmented robotic synthesis are anticipated to be used in future developments in AI-driven nanomedicine to further improve drug delivery systems based on NPs for better clinical results [91].

2.6 Challenges and Limitations

AI in medication development confronts a number of obstacles despite its revolutionary potential. Few of challenges and limitations are described below.

2.6.1 Data Quality and Availability

The quality, diversity, and comprehensiveness of the datasets utilized for training and validation are intrinsically linked to the predicted accuracy and efficacy of AI models in drug design. The lack of sufficient, inconsistent, and sometimes biased datasets, however, is one of the biggest obstacles to AI-driven pharmaceutical research and can seriously impair model performance and dependability. Complex physicochemical interactions, pharmacokinetic and pharmacodynamic profiles, and patient-specific reactions are all part of drug development, which makes large, well-curated datasets necessary to create reliable AI-driven predictive models. Sadly, a lot of the pharmaceutical datasets that are now available have flaws like inconsistent experimental circumstances, inadequate records, different measurement methods, and a lack of representation of various chemical structures and excipient interactions. The generalizability of AI models across various formulation settings may be diminished as a result of these data limitations, which may result in biased or incorrect predictions [92].

Data scarcity is made worse by the proprietary nature of pharmaceutical data and legal restrictions on clinical and preclinical datasets. Private pharmaceutical companies have a sizable amount of crucial formulation data, which restricts access for more extensive AI research and development. Furthermore, it is difficult to create universal AI frameworks that can be applied to a variety of formulation studies since publicly available datasets sometimes lack standards and have differences in data formats, terminologies, and reporting procedures. The underrepresentation of particular drug classes, treatment categories, or patient demographics in datasets might introduce bias, which can distort AI model predictions and produce less-than-ideal or unreproducible results in practical applications [93].

The creation of high-quality, standardized, and interoperable datasets that thoroughly capture drug-excipient compatibility, stability profiles, solubility parameters, and bioavailability characteristics must be given top priority by the pharmaceutical industry and AI researchers in order to meet these challenges. One potential way to get around proprietary data restrictions is to integrate federated learning techniques, which allow AI models to learn from several decentralized datasets without sacrificing data privacy. Furthermore, the use of synthetic data generation, data augmentation techniques, and optimization strategies based on RL can help reduce data scarcity and increase the accuracy of AI-driven predictions. To improve the quality and accessibility of datasets for AI applications in drug formulation, it is essential to create cooperative data-sharing consortia, support open-access repositories, and follow legal requirements for standardized data curation. In the end, addressing these issues will help to progress precision medicine and tailored therapy development by guaranteeing the dependability, openness, and clinical applicability of AI-driven formulation methodologies [92].

2.6.2 Regulatory Hurdles

A major obstacle to the broad use of AI in drug formulation is the lack of precise and uniform regulatory standards, despite the fact that it represents a paradigm shift in pharmaceutical research and development. Regulatory organizations, such as the United States. Comprehensive frameworks that specifically address the validation, approval, and ethical implications of AI-driven drug formulation procedures have not yet been established by the food and drug administration (FDA), the European medicines agency (EMA), and other international organizations. In contrast to AI-driven techniques, which incorporate complicated, algorithm-based decision-making that lacks transparency and interpretability, traditional regulatory pathways are made for traditional drug development methodologies that rely on empirical trial and well-specified validation criteria. DNN and other advanced ML models are "black-box" in nature, which raises questions about bias, accountability, and reproducibility in drug formulation predictions. While simultaneously adjusting to the dynamic and changing nature of AI technologies, regulatory bodies must make sure that AI-generated formulations comply with accepted pharmaceutical quality standards, such as good manufacturing practices (GMP) and International Council for Harmonization (ICH) guidelines. Furthermore, the lack of widely recognized standards for model performance, data quality, and forecast accuracy makes it more difficult to approve AI models, which makes validation a significant difficulty. The incorporation of AI into good laboratory practices (GLP) and good clinical practices (GCP) is further impacted by the absence of established regulatory methods, which raises questions regarding clinical translation and patient safety evaluations. Furthermore, disparities in international regulatory rules impede the smooth implementation of AI-driven medication formulation, calling for standardized international standards to expedite market approvals and guarantee cross-jurisdictional compliance. Addressing these regulatory hurdles requires a concerted effort among policymakers, AI researchers, pharmaceutical scientists, and regulatory authorities to establish transparent, adaptable, and risk-based guidelines that promote the safe and effective implementation of AI-driven innovations in drug formulation.

2.6.3 Integration with Existing Systems

The integration of AI with traditional drug development processes provides a substantial challenge due to the intrinsic contrasts between conventional pharmaceutical approaches and AI-driven computational frameworks. Drug formulation has typically depended on well-established experimental methods, regulatory-compliant analytical techniques, and empirical optimization tactics that are deeply buried within the industry's standard operating procedures. A paradigm change in the way data is gathered, processed, and interpreted in pharmaceutical R&D is required to integrate AI-driven methodologies like ML, DNN, and predictive analytics [94]. The compatibility of AI-based models with legacy systems used for formulation design, data administration, and quality control is one of the main integration problems. The computational capacity of the infrastructure used by many pharmaceutical businesses is insufficient to handle sophisticated AI algorithms, which results in inefficiencies in predictive analytics and real-time data processing. Furthermore, large, high-quality datasets are necessary for training and validating AI-driven models, yet conventional workflows frequently produce inconsistent and insufficient data that might not be immediately compatible with AI techniques. Since different AI models may interpret the same dataset differently, resulting in inconsistent predictions, the absence of established methods for combining AI with traditional formulation techniques further hampers seamless adoption [95]. The requirement for interdisciplinary competence is another significant barrier, since successful AI adoption requires cooperation between data scientists, computational biologists, and pharmaceutical scientists, all of whom work within different methodological frameworks. Significant obstacles are also presented by regulatory restrictions, since AI-generated formulation predictions need to be verified by stringent experimental procedures in order to guarantee adherence to GMP and regulatory standards established by organizations like the United States. The EMA and the FDA. Furthermore, the interpretability and transparency of AI algorithms continue to be significant issues in pharmaceutical research and development. This is because black-box models can produce formulation recommendations that lack a clear rationale, making it challenging for industry experts and regulatory bodies to evaluate their dependability. Standardized AI integration frameworks, strong data governance regulations, and hybrid modeling strategies that blend conventional experimental methods with AI-driven predictive analytics are urgently needed to meet these issues. Future developments in cloud-based infrastructure, AI-driven automation, and regulatory-compliant AI algorithms could help remove these integration obstacles and eventually make it easier for AI to be smoothly incorporated into drug formulation processes [96].

2.7 Future Directions

AI in medication formulation has a bright future ahead of it, with possible developments including the following.

2.7.1 AI-augmented Drug Discovery Platforms

The smooth integration of AI with quantum computing and multi-omics data is set to transform drug formulation in the future and facilitate a paradigm shift toward comprehensive and extremely effective formulation techniques. Advanced ML methods, DNN, and predictive analytics are used by AI-augmented drug discovery platforms to examine large and complex datasets from transcriptomics, proteomics, metabolomics, and genomes. Quantum computing greatly increases computational capacity when integrated into AI-driven platforms, allowing for the processing and optimization of intricate chemical interactions at a never-before-seen scale. The rational design of highly effective pharmaceutical formulations with improved bioavailability and therapeutic efficacy is made possible by quantum-enhanced AI models, which enable real-time simulations of drug-excipient compatibility, molecular docking, and predictive stability assessments [97].

The development of customized medicine is promoted by the identification of patient-specific formulation strategies made possible by the convergence of AI and omics data. AI can interpret complex biological processes, forecast therapeutic responses specific to a disease, and improve formulation parameters to guarantee precision-targeted drug delivery by combining multi-omics datasets. The time needed to find lead compounds with the best pharmacokinetic and pharmacodynamic profiles is significantly decreased by using AI-driven systems in conjunction with quantum computing, which also speeds up high-throughput screening. Furthermore, by dynamically modifying formulation parameters in response to real-time experimental feedback, AI-augmented platforms enable adaptive formulation design, therefore reducing the drawbacks of conventional static formulation techniques [98].

Notwithstanding its revolutionary promise, a number of obstacles need to be overcome before AI-augmented drug discovery platforms can be widely used. These include creating regulatory frameworks for AI-driven pharmaceutical advances, guaranteeing the interpretability of quantum-enhanced AI models, and standardizing the integration of multi-omics data. To get beyond these obstacles and realize AI's full potential in medication creation, cooperation between computer scientists, pharmaceutical experts, and regulatory bodies is essential. Intelligent, data-driven, and flexible approaches that improve drug development effectiveness, maximize therapeutic results, and open the door for next-generation pharmaceutical breakthroughs will define the future of AI-driven formulation science [99].

2.7.2 Collaborative AI Ecosystems

In order to provide standardized frameworks that guarantee dependability, reproducibility, and adherence to international pharmaceutical standards, the integration of AI in drug formulation calls for a cooperative approach including academics, industry, and regulatory agencies. By utilizing the combined knowledge of researchers, pharmaceutical companies, and regulatory bodies, a collaborative AI ecosystem promotes multidisciplinary innovation and optimizes AI-driven medication formulation procedures. In order to develop new ML algorithms, DNN, and predictive modeling strategies specifically suited for pharmaceutical applications, academia is essential to the advancement of basic AI methodology. Research organizations provide contributions by producing high-quality datasets, conducting experimental research to validate AI models, and clarifying mechanistic insights into stability profiles, pharmacokinetic characteristics, and drug-excipient interactions. By using AI-driven process optimization, automation, and quality control in formulation creation, industry, on the other hand, offers the translational link between theoretical AI models and actual pharmaceutical manufacture. Pharmaceutical businesses use AI-powered data analytics to improve the precision of drug delivery methods, forecast formulation stability under different conditions, and speed up drug discovery pipelines [86, 89].

To guarantee efficacy, safety, and transparency, regulatory harmonization is necessary before AI can be widely used in drug formulation. Regulatory agencies like the United States. To address issues with bias, reproducibility, and ethical compliance, the FDA and the EMA must create thorough criteria for AI model validation, data integrity, and algorithm interpretability. The development of standardized AI protocols can be facilitated by collaborative frameworks like regulatory sandboxes, industry-academia partnerships, and AI governance consortia. This builds confidence and speeds up the incorporation of AI into pharmaceutical R&D. Furthermore, by providing a variety of excellent training datasets and guaranteeing adherence to legal and ethical requirements, open-access programs and shared data repositories can improve model robustness. Stakeholders may promote innovation in drug formulation while maintaining strict regulatory standards by cultivating transparent AI ecosystems, which will eventually turn AI into a key component of next-generation pharmaceutical research [2–4, 13].

2.7.3 Real-time Monitoring and Feedback

Real-time monitoring and adaptive feedback systems have been made possible by the combination of AI with cutting-edge sensor technologies in drug formulation and production, greatly improving process accuracy and product quality. High-resolution data on crucial formulation parameters, including temperature, pH, viscosity, particle size distribution, and chemical composition, is continuously gathered by AI-enabled sensors integrated into manufacturing systems. Predictive analytics and automated decision-making are made possible by these smart sensors, which use DNN and ML algorithms to evaluate massive data streams in real time. These systems can identify deviations from ideal formulation circumstances by using sophisticated signal processing techniques and AI-driven pattern recognition. This allows for prompt remedial action to preserve batch-to-batch uniformity and compliance with strict regulatory criteria [6, 8, 9, 11].

In continuous manufacturing processes, where AI-driven feedback loops constantly modify formulation parameters to maximize medication stability, bioavailability, and therapeutic efficacy, real-time monitoring is very important [100, 101]. In order to provide continuous quality assurance (CQA), AI-powered control systems work in tandem with process analytical technology (PAT) frameworks to anticipate formulation problems and start remedial actions without interfering with production. Additionally, real-time AI-driven optimization speeds up the transition from lab to industrial production while lowering operating expenses and material waste. By simulating process variances and anticipating possible faults before they materialize, the use of DTs virtual copies of the production environment further improves real-time monitoring [15, 17, 18].

The smooth integration of AI-enabled real-time monitoring systems into pharmaceutical workflows requires addressing issues including sensor calibration, data integrity, and algorithmic interpretability, notwithstanding its revolutionary promise. These systems will be further refined by upcoming developments in edge computing and XAI, which will improve transparency, dependability, and regulatory compliance. AI-driven real-time monitoring has the potential to revolutionize medication formulation and manufacture as it develops, ushering in a period of intelligent, flexible, and precision pharmaceutical manufacturing [2, 11, 12].

2.8 Conclusion

Through the introduction of data-driven approaches that improve accuracy, efficiency, and scalability, AI has completely transformed the field of medication formulation. The drug development process has been greatly sped by the combination of AI, computational modeling, ML, and automation. This has reduced the need for traditional trial-and-error methods and allowed for the quick identification of ideal formulations. Predictive models powered by AI, including deep neural networks, RL, and QSAR modeling, have revolutionized preformulation research by precisely predicting important physicochemical and pharmacokinetic characteristics. Furthermore, by customizing medication formulations to each patient's unique genetic, metabolic, and illness profiles, AI promotes personalized medicine by guaranteeing increased therapeutic efficacy and reduced side effects. AI-powered automation, such as real-time adaptive manufacturing and robotic high-throughput screening, has further optimized formulation procedures, improving reproducibility and regulatory compliance. To properly utilize AI in medication creation, a number of obstacles need to be overcome despite these revolutionary developments. The broad use of AI-driven approaches is still hampered by problems including data scarcity, algorithm interpretability, and regulatory ambiguity. Improving the generalizability and dependability of AI models requires high-quality, standardized datasets, and the advancement of XAI methods will increase openness and promote confidence among regulatory bodies. Furthermore, to close the gap between theoretical AI advancements and practical pharmaceutical applications, it is essential to promote multidisciplinary collaboration among AI specialists, pharmaceutical scientists, and regulatory agencies. Future developments in federated learning, multi-objective optimization, and AI-driven generative models hold potential for significantly improving drug formulation techniques as AI technologies develop. The next generation of intelligent, data-centric medication formulations with enhanced efficacy, safety, and accessibility will be driven by AI, which has the potential to revolutionize pharmaceutical research and development by resolving present constraints and encouraging cross-disciplinary integration.

References

1 Allam H. Prescribing the future: The role of artificial intelligence in pharmacy. *Information [Internet]* 2025;16(2):131. Available from: https://www.mdpi.com/2078-2489/16/2/131.

2 Wong A, Wentz E, Palisano N, et al. Role of artificial intelligence in pharmacy practice: A narrative review. *JACCP Journal of the American College of Clinical Pharmacy [Internet]* 2023;6(11):1237–1250. Available from: https://accpjournals.onlinelibrary.wiley.com/doi/10.1002/jac5.1856.

3 Cain J, Malcom DR, and Aungst TD. The role of artificial intelligence in the future of pharmacy education. *American Journal of Pharmaceutical Education [Internet]* 2023;87(10):100135. Available from: https://linkinghub.elsevier.com/retrieve/pii/S0002945923001420.

4 Jarab AS, Abu Heshmeh SR, and Al Meslamani AZ. Artificial intelligence (AI) in pharmacy: An overview of innovations. *Journal of Medical Economics [Internet]* 2023;26(1):1261–1265. Available from: https://www.tandfonline.com/doi/full/10.1080/13696998.2023.2265245.

5 Khan O, Parvez M, Kumari P, Parvez S, and Ahmad S. The future of pharmacy: How AI is revolutionizing the industry. *Intel Pharmaceuticals [Internet]* 2023;1(1):32–40. Available from: https://linkinghub.elsevier.com/retrieve/pii/S2949866X23000084.

6 Mao J, Akhtar J, Zhang X, et al. Comprehensive strategies of machine-learning-based quantitative structure-activity relationship models. *iScience [Internet]* 2021;24(9):103052. Available from: https://linkinghub.elsevier.com/retrieve/pii/S2589004221010208.

7 Soares TA, Nunes-Alves A, Mazzolari A, Ruggiu F, Wei G-W, and Merz K. The (Re)-Evolution of quantitative structure–Activity relationship (QSAR) studies propelled by the surge of machine learning methods. *Journal of Chemical Information and Modeling [Internet]* 2022;62(22):5317–5320. Available from: https://pubs.acs.org/doi/10.1021/acs.jcim.2c01422.

8 Gautam V, Gupta R, Gupta D, et al. deepGraphh: AI-driven web service for graph-based quantitative structure–activity relationship analysis. *Brief Bioinform [Internet]* 2022;23(5). Available from: https://academic.oup.com/bib/article/doi/10.1093/bib/bbac288/6648791.

9 Staszak M, Staszak K, Wieszczycka K, Bajek A, Roszkowski K, and Tylkowski B. Machine learning in drug design: Use of artificial intelligence to explore the chemical structure–biological activity relationship. *WIREs Computational Molecular Science [Internet]* 2022;12(2). Available from: https://wires.onlinelibrary.wiley.com/doi/10.1002/wcms.1568.

10 Lin Y, Zhang Y, Wang D, Yang B, and Shen Y-Q. Computer especially AI-assisted drug virtual screening and design in traditional Chinese medicine. *Phytomedicine [Internet]* 2022;107:154481. Available from: https://linkinghub.elsevier.com/retrieve/pii/S0944711322005700.

11 Turon G, Hlozek J, Woodland JG, Kumar A, Chibale K, and Duran-Frigola M. First fully-automated AI/ML virtual screening cascade implemented at a drug discovery centre in Africa. *Nature Communications [Internet]* 2023;14(1):5736. Available from: https://www.nature.com/articles/s41467-023-41512-2.

12 Zhou G, Rusnac D-V, Park H, et al. An artificial intelligence accelerated virtual screening platform for drug discovery. *Nature Communications [Internet]* 2024;15(1):7761. Available from: https://www.nature.com/articles/s41467-024-52061-7.

13 Raza MA, Aziz S, Noreen M, et al. Artificial Intelligence (AI) in pharmacy: An overview of innovations. *Inova Pharmaceutical [Internet]* 2022;13(2):13. Available from: https://pubs.lib.umn.edu/index.php/innovations/article/view/4839.

14 Arul Murugan N, Ruba Priya G, Narahari Sastry G, and Markidis S. Artificial intelligence in virtual screening: Models versus experiments. *Drug Discovery Today [Internet]* 2022;27(7):1913–1923. Available from: https://linkinghub.elsevier.com/retrieve/pii/S1359644622002033.

15 Lin B and Wu S. Digital transformation in personalized medicine with artificial intelligence and the internet of medical things. *OMICS: A Journal of Integrative Biology [Internet]* 2022;26(2):77–81. Available from: https://www.liebertpub.com/doi/10.1089/omi.2021.0037.

16 Taherdoost H and Ghofrani A. AI's role in revolutionizing personalized medicine by reshaping pharmacogenomics and drug therapy. *Intel Pharmaceuticals [Internet]* 2024;2(5):643–650. Available from: https://linkinghub.elsevier.com/retrieve/pii/S2949866X2400087X.

17 Johnson KB, Wei W, Weeraratne D, et al. Precision medicine, AI, and the future of personalized health care. *Clinical and Translational Science [Internet]* 2021;14(1):86–93. Available from: https://ascpt.onlinelibrary.wiley.com/doi/10.1111/cts.12884.

18 Matić J, Stanković-Brandl M, Bauer H, et al. Pharmaceutical hot melt extrusion process development using QbD and digital twins. *International Journal of Pharmaceutics [Internet]* 2023;631:122469. Available from: https://linkinghub.elsevier.com/retrieve/pii/S0378517322010249.

19 Moreno-Benito M, Lee KT, Kaydanov D, Verrier HM, Blackwood DO, and Doshi P. Digital twin of a continuous direct compression line for drug product and process design using a hybrid flowsheet modelling approach. *International Journal of Pharmaceutics [Internet]* 2022;628:122336. Available from: https://linkinghub.elsevier.com/retrieve/pii/S0378517322008912.

20 Szilágyi B, Eren A, Quon JL, Papageorgiou CD, and Nagy ZK. Digital design of the crystallization of an active pharmaceutical ingredient using a population balance model with a novel size dependent growth rate expression. *From Development of a Digital Twin to In Silico Optimization and Experimental Validation. Cryst Growth Des [Internet]* 2022;22(1):497–512. Available from: https://pubs.acs.org/doi/10.1021/acs.cgd.1c01108.

21 Davidopoulou C and Ouranidis A. Pharma 4.0-artificially intelligent digital twins for solidified nanosuspensions. *Pharmaceutics [Internet]* 2022;14(10):2113. Available from: https://www.mdpi.com/1999-4923/14/10/2113.

22 Murthy T and Lim J. Life sciences discovery and technology highlights. *SLAS Technology [Internet]* 2024;29(4):100169. Available from: https://linkinghub.elsevier.com/retrieve/pii/S2472630324000517.

23 Lu J-M, Pan J-Z, Mo Y-M, and Fang Q. Automated intelligent platforms for high-throughput chemical synthesis. *Artificial Intelligence Chemistry [Internet]* 2024;2(1):100057. Available from: https://linkinghub.elsevier.com/retrieve/pii/S2949747724000150.

24 Benavides-Hernández J and Dumeignil F. From characterization to discovery: Artificial intelligence, machine learning and high-throughput experiments for heterogeneous catalyst design. *ACS Catalysis [Internet]* 2024;14(15):11749–11779. Available from: https://pubs.acs.org/doi/10.1021/acscatal.3c06293.

25 Magar Tushar V. MDS. Advances in the field of nanooncology: An overview. *Inventi Impact NDDS [ISSN 0976-3791] [Internet]* 2011;4(1):1–3. Available from: https://www.inventi.in/journal/article/2/1078/Inventi Impact: NDDS/Pharmaceutical.

26 Anand M. Smart crystals technology: A review. *Pharmacologyonline [Internet]* 2011;1(1):238–243. Available from: https://pharmacologyonline.silae.it/files/newsletter/2011/vol3/028.amol.pdf.

27 Borude S. Aquasomes: A potential drug delivery carrier. *Pharmacologyonline [Internet]* 2011;1(3):230–237. Available from: https://pharmacologyonline.silae.it/files/newsletter/2011/vol3/027.amol.pdf.

28 Borude SS and Mahajan AM. Recent updates on Q fever: Diagnosis and treatment. *Pharmacologyonline [Internet]* 2011;1(3):254–264. Available from: https://pharmacologyonline.silae.it/files/newsletter/2011/vol3/030.gholap.pdf.

29 Magar Tushar V. MDS. Elastic vesicles as drugs carriers through the skin. *Inventi Rapid NDDS [ISSN 0976-3791] [Internet]* 2011;1(4):1–2. Available from: https://www.inventi.in/journal/article/127/9611/Inventi Rapid: NDDS/Pharmaceutical#.

30 Borude Santosh S and Mahajan AM. POEMS syndrome: A review. *Pharmacologyonline [Internet]* 2011;1(3):265–274. Available from: https://pharmacologyonline.silae.it/files/newsletter/2011/vol3/031.gholap.pdf.

31 Rajurkar VG. Recent technologies in nanomedicine for treatment of cancer. *Inventi Rapid Pharm Tech [Internet]* 2012;4(1):1–3. Available from: https://www.inventi.in/journal/article/126/9966/Inventi Rapid: Pharm Tech/Pharmaceutical.
32 Rajurkar VG. Lipid nanocapsules: A novel promising tool as drug delivery system. *Inventi Rapid NDDS [ISSN 0976-3791] [Internet]* 2012;1(1):1–4. Available from: https://www.inventi.in/journal/article/127/9969/Inventi Rapid: NDDS/Pharmaceutical.
33 Gholap AD. Translational nanomedicine: Status assessment and opportunities. *International Journal Research Pharmaceutical biological Sciences* 2012;4(3):1454–1463.
34 Rajurkar VG. New threat of clinical toxicities evoked from nanocarrier systems. *Inventi Rapid NDDS [ISSN 0976-3791] [Internet]* 2012;1(4):1–4. Available from: https://www.inventi.in/journal/article/127/9982/Inventi Rapid: NDDS/Pharmaceutical.
35 Rajurkar VG. Potential applications of capillary electromigration methods for the improvement of separation efficiency. *Inventi Rapid: Pharm Analysis Quality Assurance [ISSN 0976-3813] [Internet]* 2012;1(4):1–3. Available from: https://www.inventi.in/journal/article/136/9968/Inventi Rapid: Pharm Analysis /Pharmaceutical.
36 Dattatray T. Potential applications of botanical extract. *Inventi Rapid Ethnopharmacol [ISSN 0976-3805] [Internet]* 2012;1(4):1–4. Available from: https://www.inventi.in/journal/article/135/9972/Inventi Rapid: Ethnopharmacolo/Pharmaceutical.
37 Dighe SB. Dandruff: Pathogenesis and treatment. *Inventi Rapid Cosmeceuticals [ISSN 0976-3864] [Internet]* 2012;1(4):1–4. Available from: https://www.inventi.in/journal/article/141/9975/Inventi Rapid: Cosmeceuticals/Pharmaceutical.
38 Tupe Dattatray. The threat of photoaging: Clinical conditions, evaluation and treatment. *Inventi Rapid Cosmeceuticals [ISSN 0976-3864] [Internet]* 2012;1(4):1–3. Available from: https://www.inventi.in/journal/article/141/9971/Inventi Rapid: Cosmeceuticals/Pharmaceutical.
39 Dighe SB. Role of sebum in human health & acne. *Inventi Rapid Cosmeceuticals [ISSN 0976-3864] [Internet]* 2012;1(4):1–3. Available from: https://www.inventi.in/journal/article/141/9980/Inventi Rapid: Cosmeceuticals/Pharmaceutical.
40 Dattatray T. Artificial viral envelopes: Emerging vehicle for gene delivery. *Pharmatutor-Art-1439* 2012;1(4):1–3.
41 Rajurkar VG. Potential applications of superparmagnetic nanoparticles technology. *Inventi Rapid NDDS [ISSN 0976-3791] [Internet]* 2012;1(4):1–3. Available from: https://www.inventi.in/journal/article/127/9983/Inventi Rapid: NDDS/Pharmaceutical.
42 Bhagat V. Gold nanorods: A promising drug delivery vehicle. *Inventi Rapid NDDS [ISSN 0976-3864] [Internet]* 2012;1(4):1–4. Available from: https://www.inventi.in/journal/article/127/9964/Inventi Rapid: NDDS/Pharmaceutical.
43 Rajurkar VG. DSB. Phonophoresis: Ultrasound enhanced transdermal drug delivery. *Inventi Rapid NDDS [ISSN 0976-3791] [Internet]* 2012;1(4):1–3. Available from: https://www.inventi.in/journal/article/127/9981/Inventi Rapid: NDDS/Pharmaceutical.
44 Khedkar Ajay TD. Cubosomes: Emerging drug delivery system. *Inventi Rapid NDDS [ISSN 0976-3791] [Internet]* 2012;1(4):1–3. Available from: https://www.inventi.in/journal/article/127/9987/Inventi Rapid: NDDS/Pharmaceutical.
45 Khedkar Ajay TD. Pharmacosomes: An emerging vesicular system. *Inventi Rapid NDDS [ISSN 0976-3791] [Internet]* 2012;1(4):1–3. Available from: https://www.inventi.in/journal/article/127/9986/Inventi Rapid: NDDS/Pharmaceutical.

46 Bothiraja C, Gholap KS, and Pawar AP. Investigation of ethyl cellulose microsponge gel for topical delivery of eberconazole nitrate for fungal therapy. *Therapeutic Delivery [Internet]* 2014;5(7):781–794. Available from: https://www.future-science.com/doi/10.4155/tde.14.43.

47 Singh T, Ravi Kumar V, Sanap G, Anjaiah B, and Roja K. In vitro antipyrial activity of psidium guajava leaf extract. *European Journal Pharmaceutical and Medical Research* 2016;3(1):312–315.

48 Puri AV. Comparative evaluation of papaya and banana starch as disintegrant in diclofenac sodium tablet formulation. *International Journal of Pharmacy and Biological Sciences* 2018;8(3):512–522.

49 Gholap AD, Sayyad SF, Hatvate NT, et al. Drug delivery strategies for avobenzone: A case study of photostabilization. *Pharmaceutics [Internet]* 2023;15(3):1008. Available from: https://www.mdpi.com/1999-4923/15/3/1008.

50 Hatvate NT, Khuspe PR, Mandhare TA, Kashid P, and Gaikwad VD. Proteomics in oncology: Retrospect and prospects. In: Kulkarni S, Haghi AK, and Manwatkar S, editors. *Novel Technologies in Biosystems, Biomedical & Drug Delivery [Internet]*, (pp. 243–269). Singapore: Springer Nature Singapore; 2023. Available from: https://link.springer.com/10.1007/978-981-99-5281-6_10.

51 Vora LK, Jetha K, Thakur RRS, Solanki HK, and Chavda VP. Artificial intelligence in pharmaceutical technology and drug delivery design. *Pharmaceutics [Internet]* 2023;15(7):1916. Available from: https://www.mdpi.com/1999-4923/15/7/1916.

52 Sawant SD, Sayyad SF, Hatvate NT, Chavan M, Rojekar S, and Faiyazuddin M. Food biopackaging for human benefits: Status and perspectives. In: Ahmed S, editor. *Biobased Packaging Materials [Internet]*, (pp. 1–35). Singapore: Springer Nature Singapore; 2023. Available from: https://link.springer.com/10.1007/978-981-99-6050-7_1.

53 Hatvate NT, Sawant SD, Khemka PN, and Vora LK. Unlocking the potential of nanobiohybrids to combat environmental pollution. In: Lens P, editor. *Nanobiohybrids for Advanced Wastewater Treatment and Energy Recovery [Internet]*, (pp. 169–90). UK: IWA Publishing; 2023. Available from: https://iwaponline.com/ebooks/book/886/chapter/3516052/Unlocking-the-potential-of-nanobiohybrids-to.

54 Pardeshi S, More M, Togre N, Rebello N, and Giram P. Dendrimers based antibacterial and antiviral materials. In: *Antibacterial and Antiviral Functional Materials [Internet]*, (pp. 139–69). USA: ACS Publications; 2023. Available from: https://pubs.acs.org/doi/abs/10.1021/bk-2023-1458.ch005.

55 Pardeshi S and Giram P. Preparation, antibacterial and antiviral activity measurements and detection methods. In: Deshmukh K, editor. *Antibacterial and Antiviral Functional Materials [Internet]*, (pp. 33–64). USA: ACS Publications; 2023. Available from: https://pubs.acs.org/doi/abs/10.1021/bk-2023-1458.ch002.

56 SJ, Kamandar PA, Banchhod GV, and Hatvate NT. Antibody-drug conjugates for cancer therapy: An up-to-date review on the chemistry and pharmacology. In: Verma SK and Das AK, editors. *Bioconjugated Materials Part 2 Applications in Drug Delivery, Vaccine Formulations and Important Conjugates for Cancer Therapy [Internet]*, (pp. 105–190) (Comprehensive Analytical Chemistry; vol. 103). Amsterdam, Netherlands: Elsevier; 2023. Available from: https://www.sciencedirect.com/science/article/pii/S0166526X23000259.

57 Bhowmik DD, Deshmukh AY, and Hatvate NT. Quintessential impact of dendrimer bioconjugates in targeted drug delivery. In: Verma SK and Das AK, editors. *Bioconjugated Materials Part 2 Applications in Drug Delivery, Vaccine Formulations and Important Conjugates for Cancer Therapy [Internet]*, (pp. 257–302) (Comprehensive Analytical Chemistry; vol. 103). Amsterdam, Netherlands: Elsevier; 2023. Available from: https://www.sciencedirect.com/science/article/pii/S0166526X23000272.

58 PR, Pawar RD, Ambore GS, and Hatvate NT. Importance of carbohydrate-drug conjugates in vaccine development: A detailed review. In: Verma SK and Das AK, editors. *Bioconjugated Materials Part 2 Applications in Drug Delivery, Vaccine Formulations and Important Conjugates for Cancer Therapy [Internet]*, (pp. 191–256) (Comprehensive Analytical Chemistry; vol. 103). Amsterdam, Netherlands: Elsevier; 2023. Available from: https://www.sciencedirect.com/science/article/pii/S0166526X23000296.

59 Rojekar S, Kapare HS, Vishwakarma N, et al. Chitosan scaffolds: Expanding horizons in biomedical applications. *Carbohydrate Polymers [Internet]* 2024;323:121394. Available from: https://linkinghub.elsevier.com/retrieve/pii/S0144861723008597.

60 Kapare HS, Pagar S, Kamandar P, et al. Exploring modified chitosan-based gene delivery technologies for therapeutic advancements. *International Journal of Biological Macromolecules [Internet]* 2024;260:129581. Available from: https://linkinghub.elsevier.com/retrieve/pii/S0141813024003842.

61 Pardeshi SR, Hatvate NT, Dhorkule N, Sayyad SF, Faiyazuddin M, and Khalid M. Environmental implications and nanotechnological advances in octocrylene-enriched sunscreen formulations: A comprehensive review. *Chemosphere [Internet]* 2024;358:142235. Available from: https://linkinghub.elsevier.com/retrieve/pii/S0045653524011287.

62 Gupta J, Kamandar P, Bhowmik DD, et al. Harnessing nanovaccines for effective immunization—A special concern on COVID-19: Facts, fidelity, and future prospective. *ACS Biomaterials Science Engineering [Internet]* 2024;10(1):271–297. Available from: https://pubs.acs.org/doi/10.1021/acsbiomaterials.3c01247.

63 Parit S, Manchare A, Mundhe P, Hatvate N, Rojekar S, and Patravale V. Antibody-drug conjugates: A promising breakthrough in cancer therapy. *International Journal of Pharmaceutics [Internet]* 2024;659:124211. Available from: https://linkinghub.elsevier.com/retrieve/pii/S0378517324004459.

64 Hatvate NT, Khuspe PR, Mandhare TA, Kashid P, and Vora LK. Decoding of nanoneutraceuticals and their delivery. In: Mohanta YK, editor. *Sustainable Green Nanotechnology [Internet]*, (pp. 284–309). Florida, New York: CRC Press; 2024. Available from: https://www.taylorfrancis.com/books/9781003389408/chapters/10.1201/9781003389408-16.

65 Uddin MJ, Faiyazuddin M, Omri A, Gowri S, and Khalid M. Advances in artificial intelligence for drug delivery and development: A comprehensive review. *Computers in Biology and Medicine [Internet]* 2024;178:108702. Available from: https://linkinghub.elsevier.com/retrieve/pii/S001048252400787X.

66 Khuspe PR, Pardeshi SR, Uddin MJ, et al. Achieving optimal health with host-directed therapies (HDTs) in infectious diseases—A new horizon. *Advances Therapy [Internet]* 2024; 2400169. Available from: https://onlinelibrary.wiley.com/doi/10.1002/adtp.202400169.

67 Rojekar S, Togre N, Bhoj P, et al. Current status of mannose receptor-targeted drug delivery for improved anti-HIV therapy. *Journal of Controlled Release [Internet]* 2024;372:494–521. Available from: https://linkinghub.elsevier.com/retrieve/pii/S0168365924003444.

68 Shaikh MS, Faiyazuddin M, Khan MS, et al. Chikungunya virus vaccine: A decade of progress solving epidemiological dilemma, emerging concepts, and immunological interventions. *Frontiers Microbiology [Internet]* 2024; 15. Available from: https://www.frontiersin.org/articles/10.3389/fmicb.2024.1413250/full.

69 Rojekar S, Parit S, Manchare A, et al. Revolutionizing eye care: Exploring the potential of microneedle drug delivery. *Pharmaceutics [Internet]* 2024;16(11):1398. Available from: https://www.mdpi.com/1999-4923/16/11/1398.

70 Nimmy S, Ibrahim SA, and Nasr MHA, editor. Manipulating the rhizosphere microbiome for plant health. In: *Microbiome Engineering*, 1st ed. (p. 19). Florida: CRC Press; 2024.

71 Mesut B, Al-Mohaya M, Yeşilkaya E, et al. Demystifying the potential of lipid-based nanocarriers in targeting brain malignancies. *Naunyn-Schmiedeberg's Archives of Pharmacology [Internet]* 2024;397(12):9243–9279. Available from: https://link.springer.com/10.1007/s00210-024-03212-6.

72 Pardeshi S, Pillai D, Giram P, Pardeshi C, and Naik J. Nanocarriers for ocular drug delivery. In: *Nanocarrier Drug Delivery Systems [Internet]*, (pp. 469–500). Germany: De Gruyter; 2024. Available from: https://www.degruyter.com/document/doi/10.1515/9783111320847-015/html.

73 Patil M, Dhumal D, Rojekar S, Jain H, Pingale P, and Kumar Prajapati M. Nanocarriers for antimicrobial therapy. In: *Nanocarrier Drug Delivery Systems [Internet]*, (pp. 561–602). Germany: De Gruyter; 2024. Available from: https://www.degruyter.com/document/doi/10.1515/9783111320847-018/html.

74 Sawant SD, Pardeshi SR, Khuspe PR, et al. The role of inulin in altering the animal microbiome: A comprehensive review. *Microbiome Engineering* 2024;1:79–102.

75 Sawant SD, Pardeshi SR, Khuspe PR, et al. Omics tools for determining microbial composition: A systematic review. *Microbiome Engineering* 2024;1:35–50.

76 Kumar D, Kumar S, Kumari R, et al. A Comprehensive review on the biomass-mediated synthesis of silver nanoparticles: Opportunities and challenges. *Journal of Environmental Chemical Engineering [Internet]* 2024; 115133. Available from: https://linkinghub.elsevier.com/retrieve/pii/S2213343724032652.

77 Faiyazuddin M, Maqsood S, Abbas S, Chaudhari YS, Sharmila S, and Webster TJ. Prospects and ethics of ChatGPT in biomedical research and medical practice: An emphasis on nanotechnology. *Journal Biomedical Nanotechnology [Internet]* 2024;20(11):1661–1678. Available from: https://www.ingentaconnect.com/content/10.1166/jbn.2024.3925.

78 Pardeshi S, Kapare H, More A, Rebello N, and Giram P. Coating approaches of functionalized magnetic nanoparticles for theranostic applications. In: *Functionalized Magnetic Nanoparticles for Theranostic Applications [Internet]*, (pp. 131–171). Wiley; 2024. Available from: https://onlinelibrary.wiley.com/doi/10.1002/9781394172917.ch5.

79 Pardeshi SR, Hatvate NT, Gharat KD, Naik JB, and Omri A. Advances in dorzolamide hydrochloride delivery: Harnessing nanotechnology for enhanced ocular drug delivery in glaucoma management. *Discover Nano [Internet]* 2024;19(1):199. Available from: https://link.springer.com/10.1186/s11671-024-04154-x.

80 Vardikar A, Das U, Mandal S, et al. Surface engineered multimodal magnetic nanoparticles for neurodegenerative diseases. In: *Targeted Therapy for the Central Nervous System [Internet]*, (pp. 121–53). USA; Elsevier; 2025. Available from: https://linkinghub.elsevier.com/retrieve/pii/B9780443238413000078.
81 Hatvate NT, Bhowmik D, Sawant SD, Londhe KM, Rojekar S, and Faiyazuddin M. Engineered artificial MicroRNA in plant: An overview. In: *MicroRNA Advances and Application in Plant Biology [Internet]*, (pp. 311–341). Elsevier; 2025. Available from: https://linkinghub.elsevier.com/retrieve/pii/B9780443216824000130.
82 Uddin MJ, Faiyazuddin M, Saeed SM, Sadiq SB, and Koly IAC. Lipid-based drug delivery for systemic effect through pulmonary route of administration. In: *Lipids in Pulmonary Drug Delivery [Internet]*, (pp. 255–299). Elsevier; 2025. Available from: https://linkinghub.elsevier.com/retrieve/pii/B9780443223747000128.
83 Khan AR, Grewal NS, Jun Z, Khalid M, and Zhang H-J. Advances in smart hybrid scaffolds: A strategic approach for regenerative clinical applications. *Engineered Regeneration [Internet]* 2025;6:85–110. Available from: https://linkinghub.elsevier.com/retrieve/pii/S2666138125000039.
84 Khuspe PR, Faiyazuddin M, Uddin MJ, et al. Current scenario and challenges of lipid-based nanomedicines for pulmonary drug delivery. In: *Lipids in Pulmonary Drug Delivery [Internet]*, (pp. 21–47). Elsevier; 2025. Available from: https://linkinghub.elsevier.com/retrieve/pii/B9780443223747000025.
85 Khuspe PR, Faiyazuddin M, Uddin MJ, et al. Lipid-based vaccines against viruses including COVID-19. In: *Lipids in Pulmonary Drug Delivery [Internet]*, (pp. 409–433). Elsevier; 2025. Available from: https://linkinghub.elsevier.com/retrieve/pii/B9780443223747000190.
86 Gormley AJ. Machine learning in drug delivery. *Journal of Controlled Release [Internet]* 2024;373:23–30. Available from: https://linkinghub.elsevier.com/retrieve/pii/S0168365924004024.
87 Egorov E, Pieters C, Korach-Rechtman H, Shklover J, and Schroeder A. Robotics, microfluidics, nanotechnology and AI in the synthesis and evaluation of liposomes and polymeric drug delivery systems. *Drug Delivery Translational Research [Internet]* 2021;11(2):345–352. Available from: https://link.springer.com/10.1007/s13346-021-00929-2.
88 Alshawwa SZ, Kassem AA, Farid RM, Mostafa SK, and Labib GS. Nanocarrier drug delivery systems: Characterization, limitations, future perspectives and implementation of artificial intelligence. *Pharmaceutics [Internet]* 2022;14(4):883. Available from: https://www.mdpi.com/1999-4923/14/4/883.
89 Heydari S, Masoumi N, Esmaeeli E, Ayyoubzadeh SM, Ghorbani-Bidkorpeh F, and Ahmadi M. Artificial intelligence in nanotechnology for treatment of diseases. *Journal of Drug Targeting [Internet]* 2024;32(10):1247–1266. Available from: https://www.tandfonline.com/doi/full/10.1080/1061186X.2024.2393417.
90 Zhang P, Guo Z, Ullah S, Melagraki G, Afantitis A, and Lynch I. Nanotechnology and artificial intelligence to enable sustainable and precision agriculture. *Nature Plants [Internet]* 2021;7(7):864–876. Available from: https://www.nature.com/articles/s41477-021-00946-6.

91 Das KP and J C. Nanoparticles and convergence of artificial intelligence for targeted drug delivery for cancer therapy: Current progress and challenges. *Frontiers in Medical Technology [Internet]* 2023; 4. Available from: https://www.frontiersin.org/articles/10.3389/fmedt.2022.1067144/full.

92 Whang SE, Roh Y, Song H, and Lee J-G. Data collection and quality challenges in deep learning: A data-centric AI perspective. *VLDB Journal [Internet]* 2023;32(4):791–813. Available from: https://link.springer.com/10.1007/s00778-022-00775-9.

93 Liang W, Tadesse GA, Ho D, et al. Advances, challenges and opportunities in creating data for trustworthy AI. *Nature Machine Intelligence [Internet]* 2022;4(8):669–677. Available from: https://www.nature.com/articles/s42256-022-00516-1.

94 Lee A, Orr W, Johnson WG, Harb JI, and Henne K. Barriers to regulating AI: Critical observations from a fractured field. In: *Handbook of Critical Studies of Artificial Intelligence [Internet]*, (pp. 413–423). Edward Elgar Publishing; 2023. Available from: https://www.elgaronline.com/view/book/9781803928562/book-part-9781803928562-44.xml.

95 Mennella C, Maniscalco U, De Pietro G, and Esposito M. Ethical and regulatory challenges of AI technologies in healthcare: A narrative review. *Heliyon [Internet]* 2024;10(4):e26297. Available from: https://linkinghub.elsevier.com/retrieve/pii/S2405844024023284.

96 Mezrich JL. Is Artificial Intelligence (AI) a pipe dream? Why legal issues present significant hurdles to AI autonomy. *American Journal of Roentgenology [Internet]* 2022;219(1):152–156. Available from: https://www.ajronline.org/doi/10.2214/AJR.21.27224

97 Zhang K, Yang X, Wang Y, et al. Artificial intelligence in drug development. *Nature Medicine [Internet]* 2025;31(1):45–59. Available from: https://www.nature.com/articles/s41591-024-03434-4.

98 Mirakhori F and Niazi SK. Harnessing the AI/ML in drug and biological products discovery and development: The regulatory perspective. *Pharmaceuticals [Internet]* 2025;18(1):47. Available from: https://www.mdpi.com/1424-8247/18/1/47.

99 Kokudeva M, Vichev M, Naseva E, Miteva DG, and Velikova T. Artificial intelligence as a tool in drug discovery and development. *World Journal of Experimental Medicine [Internet]* 2024;14(3). Available from: https://www.wjgnet.com/2220-315X/full/v14/i3/96042.htm.

100 Gholap AD and Omri A. Advances in artificial intelligence-envisioned technologies for protein and nucleic acid research. *Drug Discovery Today [Internet]* 2025; 104362. Available from: https://linkinghub.elsevier.com/retrieve/pii/S1359644625000753.

101 Gholap AD, Khuspe PR, Bharati DK, et al. Computational neuropharmacology in psychiatry. In: *Computational Neuropharmacology [Internet]*, (pp. 207–244). Wiley; 2025. Available from: https://onlinelibrary.wiley.com/doi/10.1002/9781394242474.ch7.

3

Data-driven Approaches in Pharmaceutical Formulation and Drug Development: Case Studies and Insights

Eman M. Sarhan[1] and Mandana Akhavan[2,3]**

[1]*Biochemistry Department, Faculty of Science, Alexandria University, Alexandria, Egypt*
[2]*Cell and Molecular Biology, Department of Biology, Science and Research Branch, Islamic Azad University, Tehran, Iran*
[3]*Division of Molecular Biology, Medical Mycology and Parasitology Department, School of Public Health Research Center, Tehran University of Medical Sciences, Tehran, Iran*

*Corresponding authors: ema.sarhan86@gmail.com; mandanaakhavan1990@gmail.com

CONTENTS

3.1 Introduction

The integration of data-driven methodologies in pharmaceutical formulation and drug development has revolutionized a transformative shift in the biomedical healthcare industry and clinical trial (CT) research. Recent advancements in this field have been extensively documented in peer-reviewed literature, showcasing the potential of artificial intelligence (AI), machine learning (ML), computational modeling, big data analytics, and other data-centric approaches to streamline the drug development process. The key impact of these data-driven approaches is transforming the pharmaceutical industry by enhancing drug formulation, product development, and operational efficiency through an accelerated drug discovery and development process. These approaches facilitate the rapid analysis of complex biological and chemical data, enabling the identification of potential drug candidates and optimization of formulations with unprecedented speed and accuracy. They analyze vast datasets, including genomic information, clinical records, and chemical properties, to identify patterns and/or correlations that traditional methods might overlook and inform decision-making throughout the drug development life cycle (Figure 3.1) [1].

Automation and AI/ML predictive modeling facilitate pharmaceutical formulation development by improving bioavailability, stability, and excipient selection [2]. Deep learning (DL) has shown better performance than conventional ML techniques for predicting pharmaceutical formulations. It excels in identifying complex correlations between formulation components and their in vitro characteristics, even with small or imbalanced datasets [3, 4]. They enable the development of more efficient and targeted drug formulations by analyzing vast amounts of data generated during the development process. For instance, ML algorithms can predict the pharmacokinetic properties and toxicity of drug candidates, allowing researchers to prioritize and optimize lead compounds more effectively [5]. This not only accelerates the drug development timeline but also reduces the reliance on costly and time-consuming experimental methods [6].

Beyond drug discovery and formulation, data-driven strategies play a crucial role in CT optimization. AI-powered patient recruitment, real-time monitoring, and adaptive trial designs have been shown to reduce costs and improve trial success rates [7, 8]. Additionally, cutting-edge digital technologies, including big data analytics, blockchain, ML, and the Internet of Things (IoT), are reshaping pharmaceutical manufacturing through predictive quality control (QC), supply chain optimization, and smart manufacturing techniques aligned with Industry 4.0 [9]. This revolutionized how drugs are formulated and developed to be more efficient, precise, and cost-effective, improve product quality, as well as ultimately, better patient outcomes [1, 5, 10]. Thus, the traditional methods, often reliant on inefficient,

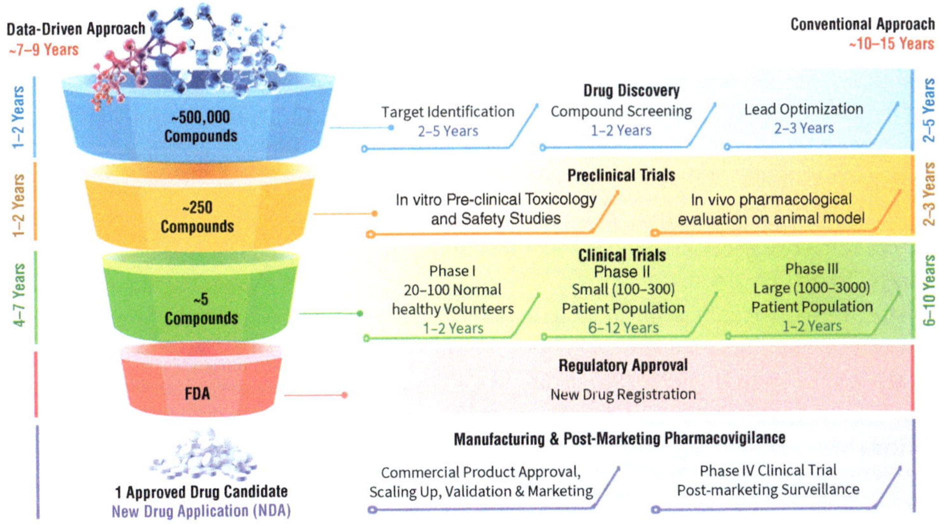

Figure 3.1 Drug development life cycle. Timeline of conventional vs. data-driven approaches in drug discovery and development process.

time-consuming trial-and-error experimentation, are being replaced by predictive modeling, optimization algorithms, and AI-driven insights, which accelerate the identification of drug candidates, enhance formulation design, and improve therapeutic outcomes.

Despite the advancements in data-driven drug formulation, some challenges need to be addressed. One significant issue is the limited guidance on modeling practices specific to drug product development, leading to suboptimal modeling that may not yield reliable predictions in practice [6]. In the pharmaceutical industry, regulatory compliance is paramount, especially as companies increasingly adopt data-driven approaches in formulation and drug development. However, these methodologies introduce several challenges that necessitate robust guidelines. One critical consideration is the transparency and validation of data-driven models. Traditional statistical methods have long provided a framework for ensuring the safety and efficiency of drugs through clear validation processes. In contrast, newer techniques, such as ML, often lack the interpretability required for regulatory approval. This gap highlights the urgent need for established guidelines that address these aspects, ensuring that data-driven methodologies can be trusted and effectively utilized in compliance with regulatory standards [11, 12].

This chapter explores the emerging trends for data-driven methods application in pharmaceutical manufacturing and drug development, highlighting real-world case studies, challenges, and insights into their impact.

3.2 Data-driven Algorithms in Pharma

The conventional drug discovery and development process (Figure 3.1) is lengthy, complex, costly, and multidisciplinary [13]. It takes over 10–15 years to get a new therapeutic drug (NTD) approved for clinical use with an overall average cost exceeding $US 2 billion [14–17]. However, 90% of drug research projects fail at the CT stage, mostly because of poor drug-likeness (10–15%), unmanageable toxicity (30%), lack of clinical efficacy (40–50%), lack of commercial demand, and insufficient strategic planning (10%) [15]. Data-driven-based technologies have been increasingly used at different stages of the drug development process to shorten the average timeline to 7–9 years instead of 10–15 years. They can also reduce the overall research cost and the failure rate of CTs. Thus, research and development (R&D) sectors of many pharmaceutical companies including, Pfizer, Novartis, GlaxoSmithKline (GSK), Merck, Genentech, Sanofi, and Takeda, rely on the automation and AI/ML-integrated cutting-edge tools to manage the massively generated data and provide cost-effective solutions. The rise in cross-industry partnerships and collaborations to manage drug discovery research costs is one of the main factors expanding the AI market growth in pharmaceutical development [18]. Moreover, pharmaceutical companies are increasingly relying on integrated AI/ML strategies to decrease not only financial costs but also failure risks. Experts expected that a 40% growth rate in the pharmaceutical and biomedical industries from 2017 to 2024 was because of the AI impact. Many pharmaceutical firms have invested in AI technologies, where AI market growth was estimated to reach $US 200 million in 2015, $US 700 million in 2018, and $US 5 billion in 2024 [19].

ML, a subcategory of AI, is concerned with building algorithms and statistical models learning from data to predict outcomes and help make decisions [1]. The process of integrating ML models in the pharmaceutical industry is comprised of six main steps

(Figure 3.2) starting with data collection and preprocessing, followed by algorithm selection, validation, and deployment, then ending with performance monitoring and model updating once needed.

Different types of ML algorithms (Figure 3.3) can be selected and implemented according to the type of data collected and the model's purpose. Supervised ML is used for training labeled data either by classification, where data is assigned to categories, or regression, where predicted outcomes are numerical values. Unsupervised ML is used with unlabeled data, where clustering groups these data points to detect anomalies or create new categories, and dimensional reduction handles to visualize complex datasets. Furthermore, another subset of ML called DL relies on multilayered neural networks to model complex patterns [1]. Many ML algorithms, including random forest (RF) and support vector machine (SVM), have been used in drug design to bridge the gap between rational and random drug discovery [17].

These ML algorithms are also being used throughout the whole drug and development stages to hasten the process [17] and revolutionize healthcare and pharmaceuticals, enhancing disease diagnosis and drug discovery [20]. AI/ML enables faster identification of drug targets, de novo drug design, and virtual screening of molecules. These methods analyze massive datasets from CTs, biological databases, and academic research to predict drug–target (DT) interactions and optimize drug design [10, 21]. This functionality allows for a more effectively focused strategy in drug discovery, thereby enhancing the chances of

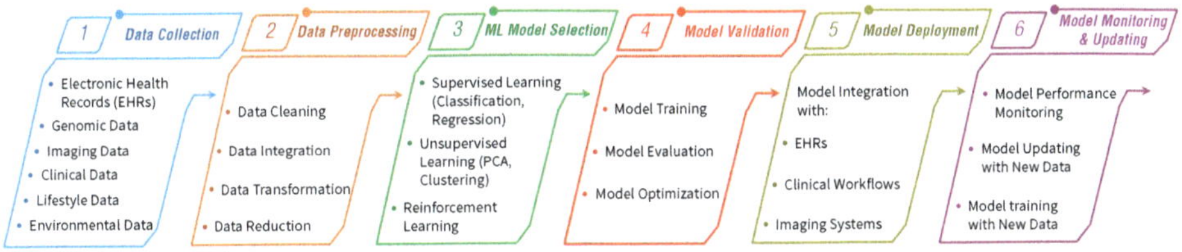

Figure 3.2 Data-driven AI/ML workflow in drug discovery and development process.

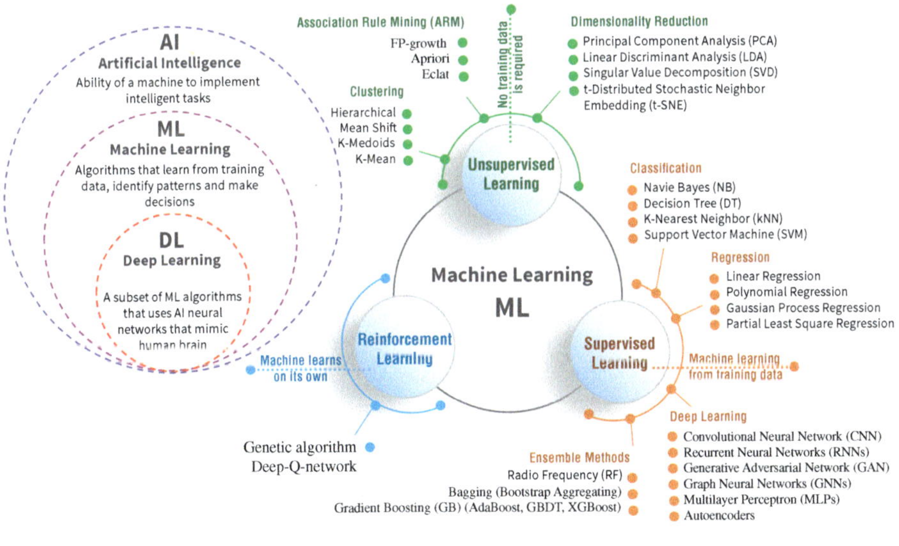

Figure 3.3 Types of AI/ML algorithms.

securing successful drug approvals [5]. They can also analyze vast datasets to identify patterns and correlations that inform formulation strategies, thereby expediting the development process. For instance, ML techniques can predict critical physicochemical properties of active pharmaceutical ingredients (APIs) and optimize the combination of drugs with excipients to improve bioavailability and therapeutic efficacy [20, 22]. This not only speeds up the new drug candidates discovery but also optimizes existing formulations by predicting their stability, solubility, and release kinetics based on empirical data rather than solely relying on trial-and-error methods [23].

3.3 Data-driven Application in Healthcare and Pharmaceutical Industry

Multipurpose data-driven approaches widely applied in different stages of drug development (Figure 3.4), include drug target identification and validation, biomarker development, new drug designing, drug repurposing, R&D efficiency enhancement, biomedical data gathering, as well as analyzing, and improving the decision-making process to select and recruit eligible patients for CTs [24–26]. These AI potentials would reduce bias and human intervention while addressing the inefficiencies and uncertainties encountered in traditional drug development approaches [27].

3.3.1 Data-driven Drug Discovery

Rational drug discovery relies on different data from multidisciplinary resources, including bioassays, CTs, and pharmaceutical and structural biology. These generated data compromise a diverse array of large datasets where AI can play a significant role in manipulating and resolving their complexity. Furthermore, this ongoing big data incrementation requires advanced computational resources and algorithms to analyze the resulting complex data. The accumulated big data can be categorized according to the data type into different databases or libraries (Table 3.1) [18]. Screening large chemical space libraries comprising vast

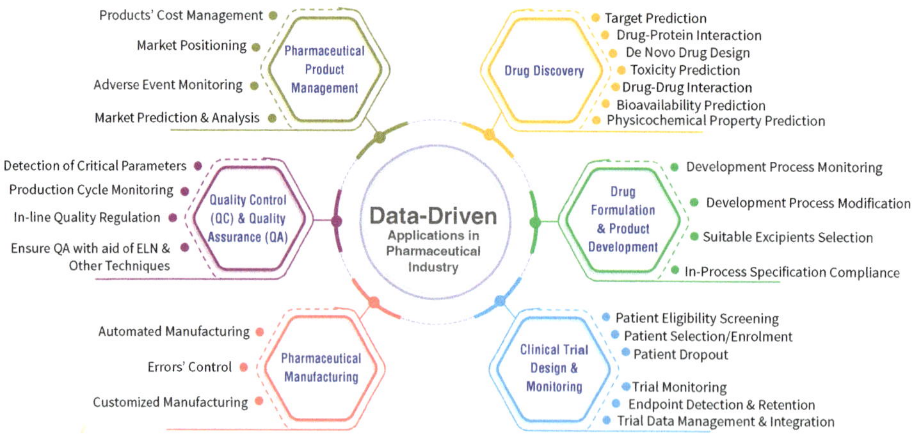

Figure 3.4 Applications of data-driven approaches in different stages of the pharmaceutical industry.

Table 3.1 List of databases commonly used in drug discovery.

	Database name	Description	Reference
1	ChEMBL	A curated database of bioactive molecules with drug-like properties, used for target validation and lead optimization.	[33]
2	PubChem	Open-access repository of chemical structures, physicochemical prosperities, bioactivity, and toxicity data for virtual screening.	[34]
3	DrugBank	AI-powered platform with curated data of drug–target interactions, pharmacokinetics, and metabolomics for drug repurposing.	[35]
4	ZINC	A free database of commercially available compounds for virtual screening and molecular docking.	[36]
5	BindingDB	A publicly accessible database of experimental data focuses on protein–ligand binding affinities for structure-based drug design (SBDD).	[37]
6	CMap	The world's largest perturbation-driven database for gene expression and proteomic datasets.	[38]
7	PharmaGKB	A publicly available platform of curated pharmacogenomics data to predict drug responses based on disease genetic variations.	[39]
8	TTD	Target-related database exploring therapeutic targets, drugs, and clinical trial data for target prioritization.	[40]
9	Tox21	High-throughput toxicity screening data for predictive toxicology and safety profiling.	[41]
10	STITCH	A text mining and network analysis repository mapping chemical-protein interactions.	[42]
11	PDB	Repository of experimentally determined 3D protein and nucleic acid structures. Critical for SBDD and molecular dynamics simulations.	[43]
12	KEGG	A manually curated database resource of large-scale molecular datasets generated by high-throughput experimental technologies like genome sequencing. It integrates enrichment analysis of biological pathways, diseases, and drug interactions supporting pharmacology studies.	[44]
13	LINCS	Contains gene expression profiles under drug perturbations. Used for drug repurposing and mechanism-of-action (MoA) studies.	[45]
14	AlphaFold DB	An AI-based system that predicts the 3D protein structure from its amino acid sequence with competitive accuracy to experiment.	[46, 47]

CMap: Connectivity Map; TTD: Therapeutic Target Database; PDB: Protein Data Bank; KEGG: Kyoto Encyclopedia of Genes and Genomes.

drug-like chemical molecules exceeding 10^{60} [28] offers significant potential for drug development [29]. However, limited technologies hinder the conventional drug discovery process making it costly and time-consuming [30, 31]. Thus, recent advancements in AI/ML-based technologies can provide rapid, effective, and innovative drug discovery solutions. Several AI, ML, and big data approaches are widely used for drug discovery applications, including de novo drug designing, high-throughput virtual screening (HTVS), pharmacophore modeling, drug repurposing, protein folding prediction, and protein–protein interaction. Quantitative structure-activity relationships (QSAR), molecular docking, and molecular dynamics simulation [32]. Some of these AI/ML/DL-based tools and platforms are summarized in Table 3.2.

3.3.1.1 Targets Identification

Data-driven frameworks have significantly transformed the predictive methodologies of pathway and/or target identification by incorporating informative genomics data, biochemical features, and target tractability [27, 58]. AI algorithms can analyze proteomics and genomics data to identify diagnostic biomarkers and potential therapeutic targets. It can also examine diseases' genetic mutations to identify molecules that may modulate disease-related protein activities [59].

3.3.1.2 De Novo Drug Design

De novo molecular design is a process in which novel drug-like compounds or chemical structures that most effectively match a particular molecular profile can be automatically proposed [60] without a starting template [61]. A virtual chemical library of computational testing is created and followed by experimental synthesis and characterization. The variation autoencoder and generative adversarial networks (GANs) are two significant techniques in the deep generative modeling of molecules. Also, recursive neural networks (RNNs) that are most commonly applied for modeling and creating sequencing have been successfully used for de novo design [17].

3.3.1.3 Hit Identification

AI is significantly advancing in hit identification, where large chemical libraries can be screened by predictive ML models to identify the most effective and highly safe virtual target-specific molecules [27, 29, 59]. It can reduce the number of synthesized compounds that have to be further in vitro or in vivo investigated, reducing the attrition rates and the R&D expenditure [27, 62–64]. This potential drastically reduces both the cost and time of the conventional high-throughput screening (HTS) methods, speeding up the promising novel drug candidate identification [59].

3.3.1.4 Physicochemical Properties

Various in silico profile selection approaches, including virtual ligands or structure-based designs, can be employed using the data available on small-molecule modulator probes and/or their structural biology. Also, Coulomb matrices and molecular fingerprinting techniques of drug design consider toxicological and physicochemical characteristics while selecting compounds [29, 62]. Furthermore, the drug design logarithm incorporates chemical descriptors including simplified molecular input line entry system (SMILES) strings,

Table 3.2 Data-driven pipelines/tools/platforms used in drug discovery and development.

	Tool/Platform	Description	Algorithms[a]	Company	Reference
1	AlphaFold	Predicts 3D biomolecular structures of protein, nucleic acids, ligands, ions, and chemical modifications.	DNN	Google DeepMind	[46]
2	Atomwise (AtomNet)	AI-enabled virtual screening platform to predict small molecules' bioactivity and identify hits' structure-based binding affinity.	DCNNs	Atomwise Inc.	[48]
3	Insilico Medicine (GENTRL)	Generative tensorial reinforcement learning (GENTRL) for de novo design of novel molecules with optimized properties.	GANs, RL	Insilico Medicine	[49]
4	Benevolent[AI]	Integrates biomedical clinical and preclinical data for target discovery and drug repurposing.	NLP, GNNs	BenevolentAI	[50]
5	Schrödinger	Physics-based ML platform for molecular simulation, free energy calculations, absorption, distribution, metabolism, excretion, and toxicity (ADMET), and descriptor prediction.	MD, Monte Carlo simulations, ML classifiers	Schrödinger Inc.	[51, 52]
6	AIDDISONTM	AI/ML-integrated ligand-based and structure-based platform for de novo retrosynthesis and predicting drug-likeness, stability, and toxicity in early-stage discovery.	RF, SVM	Merck KGa	[53]
7	Open Targets	Target/disease-centric workflow integrates genomic, proteomic, and chemical data to prioritize therapeutic targets.	NLP	Open Targets[b]	[38, 54]
8	OpenEye Orion®	Cloud-based platform for molecular design, virtual screening, and ADMET prediction, and combines ML with cheminformatics.	RF, GB, Molecular Fingerprinting	OpenEye Scientific (Cadence Molecular Sciences)	[55]

(Continued)

Table 3.2 (continue)

	Tool/Platform	Description	Algorithms[a]	Company	Reference
9	DeepChem	Open-source library for drug discovery tasks, including toxicity prediction and virtual screening.	GCNs, RL	DeepChem Community	[56]
10	OpenChem	Deep learning toolkit for computational chemistry	GCNs, RL	GitHub Repository	[57]

[a]DNN: Deep neural network; DCNNs: Deep convolutional neural networks; GANs: Generative adversarial networks; RL: Reinforcement learning; NLP: Natural language processing; GNNs: Graph neural networks; MD: Molecular dynamics; RF: Random forest; SVM: Support vector machines; GCNs: Graph convolutional networks; GB: Gradient boosting.
[b]A consortium partnership with EMBL-EBI, Genentech, GSK, MSD, Pfizer, Sanofi, and Wellcome Sanger Institute.

3D atomic coordinates, potential energy scales, and electron densities around molecules [29]. DL is very helpful when structural data are insufficient. Thus, deep neural network (DNN)-based algorithms utilizing biological, chemical, or disease phenotypic data can replace the conventional de novo drug design approach that suffers from the drawbacks of difficult bioactivity prediction and complicated synthesis pathways.

Once the potential drug candidates are identified, AI-powered structure-activity relationship (SAR) modeling plays an essential role in lead compound optimization. AI algorithms can recommend optimal molecular structural modifications for enhancing efficiency and reducing side effects by analyzing the relationship between chemical structure and biological activity [59]. Various DL-powered applications in activity or property predictions, including ADMET or physicochemical properties, have strengthened its use in QSAR or quantitative structure-property relationships (QSPRs) [65].

3.3.1.5 Lead Prioritization

Only the validated AI-powered techniques can accelerate the success rates in the drug development process. Synthesis of chosen molecules is considered to be the most crucial stage in the drug development process. In this context, AI approaches can prioritize molecules based on how easily they are synthesized or develop tools that are efficient for the optimal synthetic roadmap [27, 62, 66].

3.3.2 Data-driven Drug Formulation and Product Development

Drug formulation is crucial in developing new medicines, allowing scientists to optimize properties like bioavailability and targeted delivery. Traditionally, this process often involves costly, time-consuming, and extensively iterative trial-and-error experimentations [20, 22]. The integration of data-driven approaches such as ML, AI, and digital technologies has significantly impacted drug formulation and product development. Data-driven drug formulation refers to the integration of data-centric methodologies in the pharmaceutical development process, significantly improving the potency and effectiveness of drug formulation as well as addressing the limitations of its conventional approaches. Data-driven techniques can streamline this process, reducing its timeline and cost by enabling rapid data analysis and predictive modeling, therefore leading to more informed decisions regarding formulation design and optimization [5]. Ultimately, this paradigm shift not only raises the success rates of drug candidates but also contributes to improved patient outcomes by ensuring that medications are more effective and safer [23, 67].

Automated systems for data generation and formulation testing play a critical role in modern pharmaceutical development. These systems utilize advanced robotics and AI to streamline different stages of drug formulation, from initial compound synthesis to final product testing. For instance, automated liquid handling systems can precisely measure and mix ingredients, significantly reducing human error and enhancing reproducibility. This automation allows for HTS, where thousands of formulations can be tested simultaneously under controlled conditions, accelerating the optimal drug candidate identification. Additionally, automated analytical instrumentation integrates seamlessly with these systems to perform complex analyses rapidly, ensuring that the data generated is both accurate and reliable. Such automation not only increases productivity but also facilitates compliance with regulatory standards by maintaining meticulous records of all processes and results, thereby enhancing overall QC in drug formulation [68]. The benefits of HTS and data curation in drug development are profound. HTS platforms enable researchers to evaluate numerous compounds against biological targets within a fraction of the time required by traditional methods, significantly speeding up the drug discovery process. By automating tasks such as sample preparation, compound handling, and data acquisition, HTS generates extensive datasets that provide insights into compound interactions and pharmacological profiles. Furthermore, the integration of data curation practices ensures that this vast amount of information is systematically organized and analyzed, allowing for informed decision-making in subsequent stages of development. This combination not only makes it more likely to identify viable drug candidates but also optimizes resource allocation by focusing efforts on the most promising formulations. Moreover, digital transformation tools, in formulation development, are driving the shift toward data-driven methodologies in pharmaceutical technology. For example, AI-based platforms streamline formulation design by automating predictions of critical properties without requiring physical experiments [5, 22]. Continuous manufacturing lines integrate real-time data analytics for process control, ensuring consistent product quality [69]. Advanced frameworks for biopharmaceutical manufacturing incorporate AI to optimize both formulation and production processes [4]. As a result, the adoption of automated systems and HTS methodologies is transforming the pharmaceutical research landscape, leading to faster market entry for new therapies and improved patient outcomes [70, 71].

3.3.2.1 Formulation Design and Optimization

AI/ML algorithms have been increasingly utilized in drug formulation and preformulation optimization, demonstrating significant advancements in pharmaceutical development. They can also provide novel solutions in biomedical nanotechnology applications, where the lack of qualitative and quantitative understanding regarding nanomaterial characteristic features and biological responses precluded the nanomedicines' wide application [61]. Several computational approaches can resolve the challenging issues encountered in pharmaceutical formulation design regarding solubility, stability, porosity, tensile strength, dissolution, excipient selection, encapsulation efficiency [72], and in vivo performance, including drug permeability, absorption, and mechanistic release patterns. The excipients' type, nature, and quantity are assessed by decision-support tools using rule-based systems in accordance with the drug's physicochemical attributes [31, 73, 74]. Optimization algorithms including genetic algorithms (GAs) and simulated annealing are also used to define the optimal combination of formulation parameters. Furthermore, the experimental validation techniques that might be employed include the design of experiments (DOE) for confirming model predictions, and the iterative refinement loops between experimental data and AI/ML models for improving the prediction accuracy over time [75]. AI/ML integration with scalability and manufacturability considerations facilitates a seamless transition from lab trials to commercial production, reducing costs and time-to-market [2, 75].

3.3.2.2 Solubility

Enhancing the solubility of poorly water-soluble drugs remains a critical challenge in pharmaceutical formulation, especially for biopharmaceutical classification system (BCS) Class II and IV compounds (Figure 3.5). AI/ML algorithms can predict the compound's ability to dissolve and its potential efficacy as a drug based on the molecular descriptors and experimental datasets of the vast chemical space to explore [75]. For instance, RF, gradient boosting (GB), and extreme gradient boosting (XGBoost) have been employed to predict solubility in diverse solvent systems. Moreover, XGBoost coupled with Molecular Access System (MACCS) chemical fingerprints has demonstrated superior predictive accuracy for drug solubility in various solvents [76]. RF and NNNs can recognize patterns between molecular features and solubility, leveraging descriptors such as polarity, lipophilicity, size/shape, flexibility, rigidity, and hydrogen bonding capacity [77]. These models can accurately predict a compound's solubility based solely on its molecular structural information.

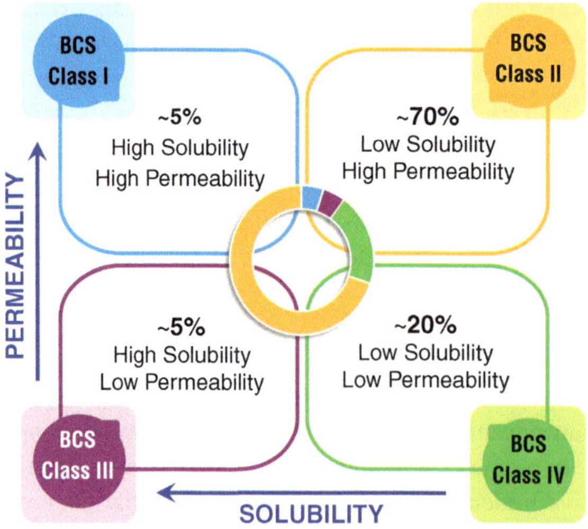

Figure 3.5 Biopharmaceutical classification system (BCS).

Integration of solubility-predictive AI/ML models into drug discovery workflow streamlines lead compound identification and accelerates therapeutics development [75].

3.3.2.3 Permeability Optimization

Drug permeability prediction and optimization are essential for assessing its ability to pass through biological barriers, influencing its pharmacokinetics absorption, distribution, metabolism, and excretion (ADME) properties [75]. Thus, absorption-modifying excipients (AMEs) are increasingly employed to enhance the absorption of low-permeability drug molecules of class III and IV (Figure 3.5) [78]. AI/ML models including artificial neural network (ANN) [79], combined ANN-genetic algorithm (ANN-GA) [80], RF [81, 82], SVM [83], and GB [84] have been widely adopted to predict drug permeability due to their ability to handle complex drugs' experimental and structural datasets and identify patterns. Data-driven approaches, such as QSPR models, rely on molecular descriptors, such as surface tension, polarity, and hydrogen bond acceptors, to estimate drug permeability. For instance, QSPR models were developed to predict the permeability of drug-like compounds through polydimethylsiloxane (PDMS) membranes, with SVM models showing the best performance [83]. Therefore, data-driven permeability detection can promote compound selection and optimization, provoking the likelihood of CT success [75].

3.3.2.4 Stability Prediction

Chemical and physical stability prediction can assess the compound's endurance under a diversity of environmental circumstances. AI/ML algorithms are increasingly being used for predicting drug stability, enabling early issue detection, and guiding the optimization of drug formulation [75]. These models analyze complex large datasets of different degradation parameters (such as humidity, temperature, and exposure to light) gathered from stability studies, historical data, and real-time stability testing. AI/ML models can identify characteristic patterns and trends within these datasets, forecasting stability profiles, degradation kinetics, and shelf life based on compound structure and environmental factors. The predicted stability profiles can inform decisions of formulation adjustments, recommending strategies to improve stability and prolong shelf life, thus maintaining pharmaceutical product safety and efficacy [2]. For instance, ML algorithms have been successfully applied to predict the chemical stability of amorphous solid dispersions (ASDs) with high accuracy, ensuring the robustness of formulations [85]. Similarly, they have also been used for predicting critical properties such as droplet size, polydispersity index, and drug loading, ensuring the stability and efficacy of the self-microemulsifying drug delivery systems (SMEDDS) formulation [86].

3.3.3 Data-driven Healthcare and CT Integration

CTs are the longest stage (~6–10 years) in the drug development and manufacturing process (Figure 3.1). It is composed of three phases (I, II, and III) according to the type and number of participating individuals. It is estimated that only ~10% of phase I tested drugs are ultimately approved with an overall R&D average cost exceeding $US 2.5 billion for every NTD approved [87]. AI can significantly speed up, streamline, and enhance CTs' efficiency by optimizing the trial design, facilitating patient selection and recruitment, and analyzing both trial outcomes and real-world data (RWD) [59]. It can also reduce the cost of clinical trial protocol (CTP) development while improving clinical outcomes. Moreover, vast datasets and data-driven analytics can streamline an extremely efficient personalized medicine development without the need for long-lasting and highly costly CTs. These transformative advancements have revolutionized the healthcare industry by allowing for more targeted and efficient personalized medicine and reducing the logistical and financial burden of conventional CT procedures while improving patient clinical outcomes.

3.3.3.1 Improving Trial Design, Patient Selection, and Real-time Response Monitoring

The CT design could be greatly enhanced by ML algorithms and data-driven analytics of previous research studies. This can easily identify the variables/aspects most relevant to the study outcomes and recommend suggestions for the CTP design, including the participants' numbers, the inclusion and/or exclusion criteria, the estimated study duration, the endpoint selection, and the optimal dosage regimen schedule. ML algorithms can also analyze the RWD to match patients to trials based on medical, socioeconomic, demographic, and behavioral traits. Finding patients who fulfill the inclusion criteria of a given study plan and who have a more likely adherence, and lower drop-out rates can improve the efficacy and efficiency of the selection and recruitment process among more diverse patient populations, faster their enrollment, and ultimately reduce the CTs' cost and time. Furthermore, AI/ML and data analysis of RWD and other relevant data sources can predict the patient features' indications of treatment response, including diagnostic biomarkers, genetic profiles, or clinical measurements, which allows for more targeted and effective CT designs. They can also identify the potential safety issues that may not have been previously detected early in the study design phase, enabling the adjustment of the study CTP design accordingly to reduce the risk of adverse events and ensure patient safety throughout the trial.

Personal healthcare data comprises everything ranging from demographics and medical records to data from genetic testing or wearable technology [88]. Advanced AI algorithms can leverage participants' clinical data via wearables, apps, and sensors to gather and follow up on real-time insights into treatment safety and efficacy. Multiple digital biomarkers can be integrated by AI to assess the patient's response to the treatment and detect dose adjustments if needed. FDA-approved AI-augmented devices for remote monitoring of patients' healthcare and CT settings allow patients to stay informed and supported in real time. Other methods were developed to enhance adherence, including incentives based on behavioral economics, virtual pillboxes, and intelligent pillboxes. DL and ML algorithms can analyze real-time patient data via wearables and video monitoring and predict a patient's likelihood of dropping out by detecting the onset of behaviors that have previously led to nonadherence.

3.3.3.2 Reducing Trial Errors with AI-driven Digital Data Flow

The conventional flow of the massive data volume associated with CTs is often hampered by the inefficient manual data entry of case forms and study reports into multiple systems leading to inconsistencies, errors, the need for rework, and a slowdown in trial execution [89]. AI-powered cognitive automation affords a promising administrative solution to enhance CT efficiency and streamline data gathering and management [7]. It can intelligently interpret data elements, feed the downstream infrastructure, auto-populate the required reports and analysis, improve decision-making, and speed up a more efficient and less prone to errors in CTs at a lower cost. These developments can transform the potency and efficiency of CT design and execution, improve patient outcomes, and accelerate the overall drug approval and marketing entry process.

3.3.3.3 Cost Optimizations

Cost-effectiveness analysis (CEA) has become a critical tool for evaluating healthcare services based on incremental cost-effectiveness ratios (ICERs). This approach helps allocate resources to more cost-effective treatments without compromising health outcomes. For instance, studies have indicated that certain cholesterol-lowering therapies and HPV (Human Papillomavirus) testing strategies can reduce costs while improving patient outcomes, influencing national guidelines [90].

3.3.4 Data-driven Pharmaceutical Manufacturing, QC, and Supply Chain

The pharmaceutical manufacturing process (Figure 3.6) comprises multiple stages, proceeding from raw materials preparations to APIs, formulation, and ultimately the desired drug product. Traditionally, drug development and pharmaceutical manufacturing processes have been costly and time-consuming [23]. Moreover, the drug manufacturing process complexity and the generated data volume increase pose considerable challenges to maintaining data integrity and ensuring robust quality management [9]. Nevertheless, the continuous advancements in AI/ML technologies, including chatbots, advanced

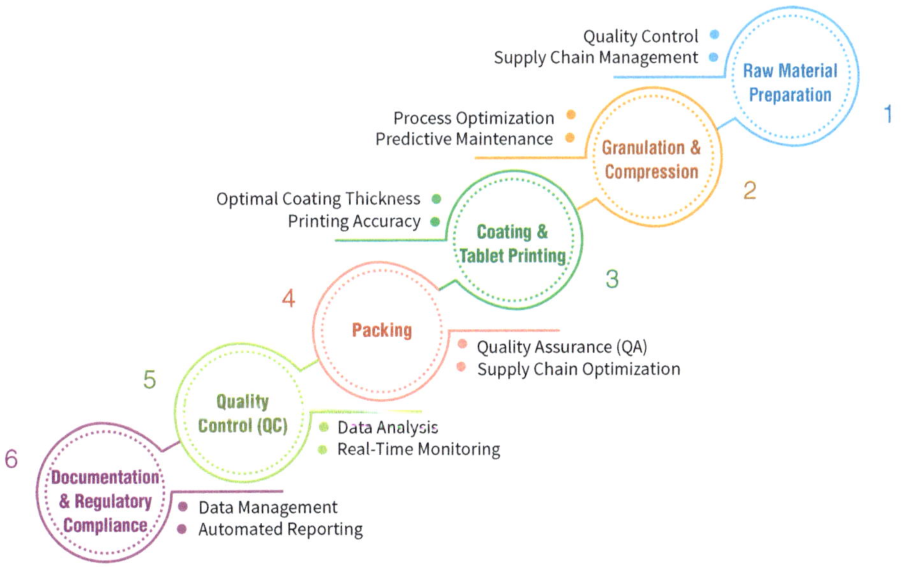

Figure 3.6 Pharmaceutical manufacturing process.

communication methods, and high-speed algorithms, have prompted the transition of the whole pharmaceutical manufacturing industry to a more streamlined and efficient development approach [23].

3.3.4.1 Optimizing Manufacturing Processes

AI/ML has transformed pharmaceutical manufacturing by optimizing production processes, reducing waste, maximizing throughput, and ensuring consistent drug quality [1]. Predictive modeling and simulation tools can analyze complex formulation development, optimizing drug delivery methods. Virtual models of physical processes called digital twins are increasingly being used for real-time simulation and adjustment, enhancing process control and product quality [91]. For instance, AI-driven digital twins, such as the conditional synthetic data engine (COSYNE), generate high-fidelity digital twins to predict batch failures and aid timely intervention, reducing time-to-value by 10% [92].

3.3.4.2 Quality Control and Predictive Maintenance

Despite the efficiency, conventional QC methods remain laborious and prone to human errors. AI techniques, especially ML and computer vision, are applied to automate QC processes, ensuring compliance with robust regulatory requirements [93]. AI-driven image analysis using machine vision systems streamlines QC procedures by detecting defects in drug formulations, packaging, and labeling, thereby maintaining compliance with regulatory standards [1]. Data-driven modeling methods such as DoE, multivariate analysis, Bayesian inference, and ML are increasingly used to optimize pharmaceutical processes. These tools enable dynamic reaction modeling, catalyst kinetics prediction, and identification of critical process variables. ML techniques like supervised learning and DL further enhance predictive capabilities, improving decision-making in process design and control [11]. As the pharmaceutical sector transitions to Industry 4.0, ensuring data integrity becomes critical [94]. The adoption of process analytical technology (PAT) and quality by design (QbD) frameworks allows for real-time process control and quality assurance. However, managing large datasets generated by advanced automation and ML tools presents challenges related to data integrity risks [95]. Mitigation strategies include validated computerized systems, cloud-based platforms for secure data storage, and adherence to regulatory guidelines like ALCOA+ principles [94].

Predictive maintenance attempts to address challenging issues by anticipating potential failures before their occurrence, enabling instant interventions that avoid extensive disruptions. Predictive maintenance, enabled by AI/ML models, detects potential equipment breakdowns, reducing downtime and ensuring continuous production. These advancements have significantly enhanced manufacturing efficiency and product quality [1, 93].

3.3.4.3 Supply Chain Management

The pharmaceutical supply chain is a complex framework involving everything from material sourcing and warehousing to manufacturing and final distribution logistics, including pharmacies and patients [93]. This framework may be long-lasting from months to years or longer, reducing the response to market demands (e.g., epidemics). Moreover, lengthy supply chains can complicate the CT development [95]. Thus, AI/ML supply chain optimization is an essential component of effective pharmaceutical operations, enhancing the prompt delivery of critical healthcare products while reducing costs and mitigating risks [1]. Generative AI and large language models (LLMs) have been used to optimize pharmaceutical supply chains by addressing challenges in drug inventory management, procurement, distribution, and dispensing. Unlike natural language processing (NLP) models, LLM uses self-attention mechanisms to recognize long-range dependencies, analyze contextual information, and automate processes, leading to better decision-making and overall optimization of pharmaceutical supply chain management [96].

3.3.5 Data-driven Pharmaceutical Management (Marketing)

Data-driven approaches have become integral to pharmaceutical marketing, allowing companies to tailor their strategies based on real-time data and analytics. These approaches can analyze customer data to understand preferences and behaviors, thereby enabling more targeted marketing campaigns and enhancing the relevance and effectiveness of marketing efforts.

3.3.5.1 Personalized Marketing Strategies

Pharmacy marketing strategies have been transformed with integrated data analytics, offering more efficient and personalized approaches to improve patient engagement and medication adherence [97]. Data analytics allows pharmaceutical firms to create tailored marketing campaigns based on patient demographics, behaviors, and clinical needs. For example, stratified persona modeling, such as the InsighTCROSS® framework, evaluates marketing effectiveness, identifies competitors, and develops targeted strategies. This approach has been effective in promoting specific drugs like anticoagulants by leveraging patient-specific data to enhance engagement and market penetration [98].

3.3.5.2 RWD Utilization

The integration of RWD from electronic health records, insurance claims, and patient registries provides actionable insights into patient preferences and treatment outcomes. RWD is increasingly being used to bridge gaps between research, development, and clinical applications. It helps assess intrinsic (e.g., genetic polymorphisms) and extrinsic factors (e.g., food–drug interactions) that affect drug clearance, enabling better dose optimization and personalized medicine strategies [99]. It not only informs product positioning but also assesses the marketing efforts' impact on real-world treatment patterns. This permits pharmaceutical companies to refine their strategies based on actual usage and outcome data [99].

3.3.5.3 Predictive Analytics for Market Trends

ML models analyze historical data to predict future market trends and consumer behaviors. These insights guide resource allocation, optimize promotional activities, and anticipate competitor moves. For instance, decision tree analysis has been used to identify the strengths and weaknesses of competing drugs, allowing companies to focus on markets where their products have a competitive edge [98].

3.3.5.4 Digital Marketing Innovations

Digital platforms are increasingly central to pharmaceutical marketing. Tools like AI-driven customer relationship management (CRM) systems analyze online interactions to personalize the communication between healthcare doctors and patients. These innovations improve product visibility and foster stronger engagement by delivering targeted content through preferred digital channels [98].

3.3.5.5 Active Pharmacovigilance for Trust Building

Data-driven active pharmacovigilance systems monitor adverse drug reactions in real time using big data analytics. This transparency strengthens the patients' and healthcare providers' trust while also serving as a marketing tool by demonstrating a commitment to safety and efficacy [100].

3.4 Tools and Case Studies

Several case studies illustrate the in silico-based platforms, pipelines, and tools' trials successfully applied in pharmaceutical development and their potential to enhance patient outcomes and assess the efficacy of new medications.

3.4.1 AI-powered Molecular Simulation Platforms

AI-driven molecular simulations have accelerated drug discovery by enabling rapid screening of millions of compounds. Atomwise uses DL (via its AtomNet platform) to predict molecular interactions, identifying candidates for diseases like Ebola and multiple sclerosis. BenevolentAI leverages AI knowledge graphs to uncover novel drug targets, while Insilico Medicine employs GANs to design molecules with reduced side effects. Pfizer has also adopted AI to predict patient enrollment rates and dropout risks in CTs, optimizing trial timelines [101].

3.4.2 Open Targets Platform

It is a computational prediction framework of gene-disease-associated data. It has been employed in one study to assess the viability of therapeutic target prediction, where it reported that the most predictive power was observed in animal models with a disease-relevant phenotype and a neural network classifier exceeding 71% accuracy [27, 102]. This evidence suggested that genes linked to disease data are sufficient for accurately predicting novel therapeutic targets, supporting its essential use to strengthen hypotheses in the target discovery process, accelerate the automated target prioritization, and decrease the cost and time of bringing new medications to patients [102].

3.4.3 IBM's Watson® for Drug Discovery

It is an AI platform that has identified five novel RNA-binding proteins (RBPs) associated with a neurodegenerative pathogenesis of amyotrophic lateral sclerosis (ALS) [27, 103]. These findings recommend the use of AI-based tools to accelerate complex disease discovery like ALS.

3.4.4 EPA's Estimation Program Interface (EPI) Suite™

It is a standalone windows®-based QSPR screening level tool. The program was developed by the Environmental Protection Agency (EPA) and Syracuse Research Corp. (SRC) to estimate the physicochemical and environmental fate properties [29, 104, 105].

3.4.5 DeepDTA

It is a DL-based model of convolutional neural networks (CNNs) that uses only 1D representations of both target protein sequence and drug SMILES to predict DT interaction binding affinities [106]. The model calculates the binding affinity based on the characteristic feature-based interactions between the targets and the drug's chemical moieties rather than the external features or 3D structures of the binding complexes [29, 106].

3.4.6 Reinforced Adversarial Neural Computer

It is a DNN architecture [29] based on the reinforcement learning (RL) and GAN paradigm for de novo design of novel small organic molecules. Reinforced adversarial neural computer (RANC) trained on the molecules' SMILES string representation excels its first DNN-based rival ORGANIC by many drugs discovery-relevant parameters, including the number of novel structures, passing medicinal chemistry filters (MCFs), high QED scores, and Muegge criteria. It can generate structures matching the SMILES string length in the training dataset and the main chemical descriptors' distribution including molecular weight (MW), logP, and TPSA. Thus, it is reasonable to consider RANC a promising start for the development of novel molecules with potential activities against various biological targets and/or pathways. Additionally, RANC not only saves scientists time but also covers a broad chemical population with novel and diverse compounds [107].

3.4.7 ML-based Pharmacokinetics Model

In a comparative study, six AI prediction models including SVM, k-nearest neighbor (kNN), probabilistic neural network (PNN), ANN, partial least square (PLS), and linear discriminant analysis (LDA) were tested to predict the chemical compounds' intestinal absorption. Among the overall predicted accuracies of these six tested classifiers, SVM-based model was the highest (91.54%), followed by kNN (88.33%), ANN (86.51%), PNN (84.30%), and LDA (80.08%). While PLS showed the lowest prediction accuracy of 79.07% for the considered dataset. Thus, the SVM-based prediction model can be effectively used in the preliminary stages of drug design and development, to predict whether a compound's likelihood of being highly or poorly absorbed [108].

3.4.8 AquaSol

It is an online web-based solubility predictor, available through the ChemDB chemoinformatics portal [109]. It can predict the aqueous solubility (logS) of small molecules using undirected graph recursive neural network (UG-RNN) ensembles [110].

3.4.9 DeepDDI Platform

It is a DNN-based computational framework that uses name and structural data (SMILES) to predict the types of drug–drug interaction (DDI) for given drug pairs or drug–food ingredient pairs. It can also predict the adverse drug events (ADEs) of DDI mechanisms, prescribe alternative medications with the same desired pharmacological impact without negative side effects, and predict the food constituents' bioactivities and their effects on interacting medications [111].

3.4.10 FormulationAI

It is the first freely available AI web-based platform for in silico drug formulation design. It allows researchers to quickly predict essential properties of pharmaceutical formulations, facilitating the design of effective delivery systems without extensive laboratory work. This innovative web-based tool was developed to aid in the design of drug formulations by leveraging extensive datasets collected over a decade. It utilizes these comprehensive datasets and AI algorithms to predict critical formulation parameters such as solubility, stability, and drug–excipient interactions. This eliminates the reliance on trial-and-error methods, reducing time and costs [4, 22]. The platform employs several AI algorithms and new approaches for drug-excipient pairwise representation to predict and evaluate critical properties of the tested formulations without the demand for complex theoretical calculations or extensive laboratory experiments. By simply inputting basic information about drugs and excipients, users can rapidly obtain predictions, facilitating faster and more precise formulation design [22]. With a vast repository of up-to-date data and advanced AI models, FormulationAI delivers precise predictions through an intuitive user interface. Thus, FormulationAI strives to lead the way in offering top-tier solutions for in silico drug formulation design.

3.4.11 ANN-based Nanoformulation Composition Optimization

A study by Li et al. was conducted to investigate the controlled release optimization of polymer–lipid hybrid nanoparticles (PLNs) loaded with an ionic water-soluble drug called verapamil hydrochloride (VRP). Li et al. have proposed that the combined application of factorial design, ANN, and continuous GAs is an effective and efficient approach for modeling and multi-objective optimization to facilitate PLN development with the desired properties [112].

3.4.12 AI/ML-based 3D Printing Nanomaterial Formulation Pipeline

It has been developed by Muñiz Castro et al. [113] and can predict the extrusion temperature, filament mechanical characteristics, and dissolution time of nanomaterials [61]. Muñiz Castro et al. [113] used AI/ML predictive models to investigate the literature-mined data of formulations' composition and other enhancing features of drug release, including surface area/volume, infill percentage, drug solubility, pH, weight, and volume of dissolution media, for predicting the drug release profile of the fused deposition modelling (FDM™) printed formulations. The best prediction result was obtained by the ANN algorithm with the ability to predict the formulation drug release dissolution times with a mean error of ±24.29 min, suggesting the efficiency of literature data mining in modeling 3D printing (3DP) workflow [113].

3.4.13 ANN-based Platform for Nanoparticle Cellular Internalization

It has been developed by Alafeef et al. The proposed model that implements an ML algorithm with the characteristic cellular uptake responses of individual cancer cell types can successfully classify different cancer types. It can also predict the nanoparticles' cellular internalization in different cancer stages [114]. As the nanomedicine efficacy depends on its cellular uptake, this cellular uptake prediction model is extremely valuable for predicting nanomedicine potential [61] and optimizing the nanomaterials formulation to obtain an optimal structure–internalization response for a given particle [114].

3.4.14 Oligonucleotide-drugs Conjugate Design

Oligonucleotide therapeutics of shorted strand DNA or RNA, including small interfering RNA (siRNA), antisense oligonucleotides (ASO), clustered regularly interspaced short palindromic repeats (CRISPR), and CRISPR-associated protein (Cas), have become a novel class of drugs that enable the precise treatment of diverse diseases. Because of their costly and extensive experimental design, AI/ML approaches have been employed to identify and design oligonucleotide-based medications [61]. eSkip-Finder [115], an ML-based application and database, was developed by Chiba et al. to rapidly analyze a given set of ASOs and intron/exon sequences to identify effective exon-skipping ASOs [116]. Another ML-based web server called SMEpred workbench has been developed by Dar et al. to predict the effectiveness of normal siRNAs and chemically modified siRNAs (cm-siRNAs) [117].

3.4.15 Quantitative Systems Pharmacology Model of Immuno-oncology Platform

It is a mechanistic framework developed to investigate the antitumor immune response in a virtual patient model. The Quantitative Systems Pharmacology Model of Immuno-oncology (QSP-IO) model has been tested on different tumor types, including triple-negative breast cancer (TNBC) [118], non-small cell lung cancer (NSCLC) [119, 120], and neoadjuvant hepatocellular carcinoma (HCC) through many studies funded by CytomX Therapeutics, Inc., South San Francisco, CA, and partially supported by NIH grants R01CA138264 and U01CA212007 (A.S.P.). It can simulate tumor growth and predict population and individual response to different therapeutics, including immuno-, chemo-, and combinatorial medications [118]. The model's algorithm can predict how individual patients would respond to different treatment regimens based on the input patients' data, such as genetic mutations and tumor features. This customized approach allowed oncologists to optimize treatment plans for maximizing efficacy, minimizing adverse effects, and eventually improving patient outcomes [59].

3.4.16 Cardiovascular Risk Evaluation Frameworks

Thanks to the US Food and Drug Administration (FDA) and the European Medicines Agency (EMA) initiatives stressing the need to adopt in silico computational modeling assessment as part of the medical regulatory process [121]. An in silico modeling project has been conducted to assess the potential safety impact of new cardiovascular drugs [122]. By combining clinical and computational simulation data, researchers were able to predict the likelihood of unfavorable cardiovascular events attributed to particular treatments. This innovative approach would speed up the evaluation process and promote regulatory decision-making upon the robust proof support of the new drugs' safety [59]. Another ongoing collaborative project called SimCardioTest (Grant agreement ID: 101016496) [123], comprising 10 organizations from the United States and other 6 European countries is funded by the European Commission (EU H2020). This simulation project for testing medications and medical devices is supposed to provide new insight into designing new predictive tools in cardiac pathologies [124].

3.4.17 Novartis-SIRIUS

It is the first in silico multi-arm trial program (NCT05974345) that aims to predict the potential inclusion efficacy on adverse cardiovascular events (ACE) in atherosclerotic cardiovascular disease (ASCVD) patients [125]. This in silico knowledge-based simulation model can efficiently predict the prospective outcomes of randomized CTs. However, it is unlikely to take the place of prospective trials unless being developed with reasonable assumptions offering helpful adjunctive information [126].

3.4.18 Novartis-Biovista Platform

It is a systematic drug repurposing (SDR) tool [127] for developing potent, innovative, and affordable medications [128]. Novartis is one of the pharmaceutical companies that embraced in silico trials in its drug developmental strategies. It utilized in silico modeling, in collaboration with the AI-powered Biovista platform, to explore the existing drug's potential for repurposing in new therapeutic contexts. The in silico platform also enhanced the analysis of multiple large datasets of drug interactions and biological pathways to identify the prospective candidates that may be effectively impactful in treating conditions exceeding their original indications. This approach allowed Novartis to accelerate the drug repurposing procedure and open new research avenues [59].

3.4.19 Human Universal Immune System Simulator (UISS)

It is an in silico trial platform that is used to predict the vaccination strategies' outcome against SARS-CoV-2 and speed up their potential discovery pipeline [129]. The COVID-19 pandemic has provoked an upsurge in the application of in silico trials where several computational models have been developed worldwide to simulate the impact of potential vaccines and therapeutics against the virus. In silico modeling can also prioritize various vaccine candidates for clinical testing based on their predicted immuno-response. This flexibility affirmed the significance of in silico trials in responding to urgent public health challenges, further in pharmaceutical development [59].

3.4.20 Heartline™ Wearables App

Integrating wearables in CTs enhances participant engagement, enables remote monitoring, and removes obstacles like time and travel constraints. A virtual research initiative called the Heartline Study has been launched by a partnership of Apple [130] and Johnson & Johnson's Janssen pharmaceutical companies [131]. The study intends to investigate how early detection of atrial fibrillation (AFib) via the Heartline Study app [132] on the iPhone and the heart health features on the Apple Watch can enhance health outcomes and lower the risk of stroke. This collaboration utilizes digital health tools like the Apple Watch's irregular rhythm notification feature and ECG app to promote proactive heart health behavior and enable AFib early detection. AFib diagnostics is often challenging because of the lack of noticeable symptoms unless a severe cardiovascular event such as a stroke [133, 134]. Researchers intend to address this issue and enhance timely detection and intervention for stroke prevention through the integrated Heartline Study app and Apple Watch features. Electrocardiograms (ECGs) can be analyzed by the Apple Watch's ECG app to reveal if they exhibit a normal sinus rhythm or irregular heart rhythms, indicating AFib [135]. This remote innovative app-based approach allows participants to engage conveniently in the study, using their iPhone or Apple Watch without having to travel to the trial site, potentially saving costs and time compared to conventional CTs [136, 137].

3.4.21 Pharmacology-AI Platform

Another case study project exploring the integration of ML automation in pharmaceutical development is the collaboration between REPROCELL, IBM, and the STFC Hartree Centre, which led to the creation of the Pharmacology-AI platform. This innovative platform is designed to streamline the analysis of large datasets generated during drug development studies, particularly for precision medicine. By utilizing AI algorithms, Pharmacology-AI [138] enables researchers to identify inter–individual differences that significantly influence drug responses and related clinical outcomes. The platform simplifies the extraction of actionable insights from complex datasets, making it accessible even to those without extensive bioinformatics expertise. This capability not only speeds up the discovery of relevant genomic or clinical features driving drug responses but also enhances the overall efficiency of drug development processes [139].

3.4.22 IBM's Watson for Oncology

It is an AI-clinical decision-support system (CDSS) that has been designed to assist oncologists in making better decisions for cancer treatment [30]. Other similar CDSSs for oncology are the American Society for Clinical Oncology's CancerLinq [140] and OncoDoc [141].

Watson for Oncology (WFO) analyzes the acquired clinical data from vast database networks and then provides evidence-based treatment options. In a blind investigation, WFO treatment recommendations were highly concordant for 638 breast cancer cases previously prescribed between 2014 and 2016 at the Manipal Comprehensive Cancer Center, Bengaluru, India. The study revealed the potential use of AI CDSSs WFO for decision-making in breast cancer treatment, especially at centers where expert resources are limited [142].

3.4.23 UCSF Robot Pharmacy

It is a robotic technology used by UCSF Medical Center for preparing and tracking medications. It has successfully prepared around 350,000 medication doses without any error. Its ability to prepare both oral and injectable medications, including toxic chemotherapies, allows pharmacists and nursing staff to pay more focus on direct patient care. Once this automated pharmacy medication system receives electronically the prescription from the UCSF physicians and/or pharmacists, robots pick, package, and dispense the individual pill doses. The machines assemble doses onto a bar-coded thin plastic ring containing all of the prescriptions that must be taken within 12 hours. Then, both chemotherapy and non-chemotherapy doses are prepared and filled within intravascular (IV) syringes or bags under sterile circumstances. The fully automated facilities also contain an inventory management system that keeps tracking all products along with both refrigerated and non-refrigerated pharmacy warehouses that provide the storage and retrieval of supplies and medications [30, 143].

3.4.24 AI–CT Design and Recruitment Systems

AI enhances CT efficiency through patient recruitment and trial design. Deep 6 AI uses natural language processing to analyze medical records, accelerating candidate identification. Antidote Technologies matches patients with trials globally using structured and unstructured data, improving diversity and engagement [101]. AstraZeneca has streamlined target identification and development processes via AI, though specific trial-focused implementations are less detailed in the provided sources [144, 145].

3.4.25 AI-powered Manufacturing and QC Systems

AI-powered computer vision systems, such as those developed by DevisionX, detect defects in drug production (e.g., shape and color anomalies) and reduce contamination risks by minimizing human intervention. Merck employs ML models to recognize defects during production, improving product quality. AstraZeneca reduced equipment downtime by forecasting failures with AI [144, 145].

3.4.26 AI-powered Regulatory Compliance Tools

AI tools enhance adherence to regulatory standards. GSK automated compliance systems to minimize documentation errors during audits, while Bayer used regulatory intelligence tools to streamline compliance processes [145].

3.5 Challenges, Limitations, and Ethical Considerations

Despite the rapid advancements of AI/ML algorithm data-driven technologies implemented and/or integrated into the drug development, pharmaceutical industry, and manufacturing process, several challenges and limitations can hinder progress. These challenges (Figure 3.7) span technical, organizational, and ethical domains, impacting the effectiveness and efficiency of data-driven methodologies. Among these challenging obstacles, data quality, data privacy, cybersecurity, systemic barriers within organizations, and the demand for skilled personnel remain critical. Addressing these issues is essential for fully realizing and effectively leveraging the potential of data-driven methodologies in pharmaceutical R&D.

3.5.1 Technical Limitations

3.5.1.1 Data Quality, Integrity, and Complexity

High-quality data is essential for building reliable and accurate data-driven models [11]. Poor, inadequate data quality may cause inaccurate, misleading predictive models and unreliable outcomes, impacting decision-making processes [146]. Insufficient data integrity due to the datasets' source diversity would provide inappropriate data formats in terms of structure and quality, making the model output of the ML algorithm inaccurate. Thus, collecting and collating datasets before the drug discovery process must be required for efficient integrative analysis [147]. Automated parallel high-throughput experimentation could generate invaluable high-quality consistent data for a robust pharmaceutical

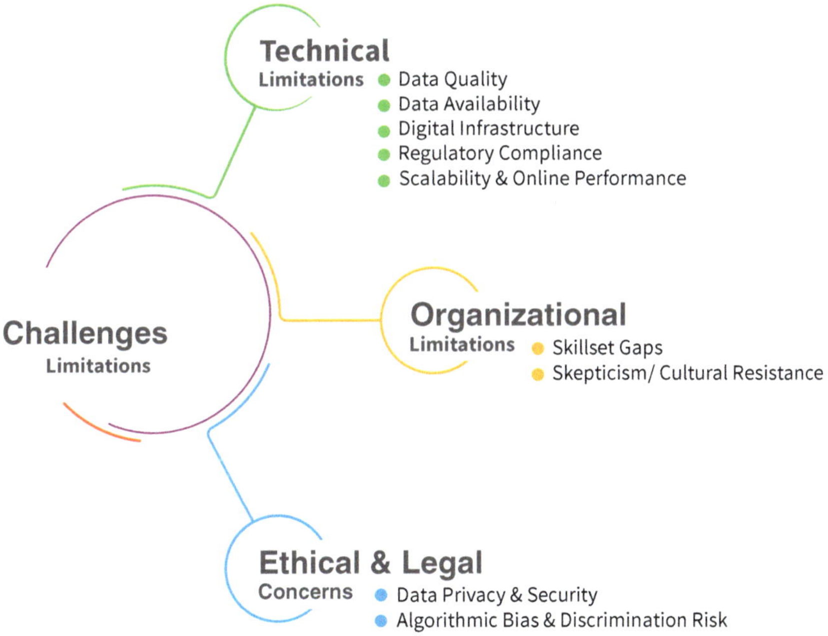

Figure 3.7 Challenges of data-driven approaches in the pharmaceutical industry.

development process [148]. More collaborative effort would be needed between the academic and industrial research communities to gather collective curated databases for adoption in different applications [11].

Data quality also plays a critical role in ensuring regulatory compliance. Challenges such as data cleansing, integration, and standardization are prevalent due to the diverse sources of data used in drug development. Effective data management practices are crucial for maintaining accuracy across datasets. For instance, the adoption of global data standards like those developed by the clinical data interchange standards consortium (CDISC) ensures consistency in reporting formats and facilitates seamless communication with regulatory authorities. These standards span various stages of clinical research, from data gathering to analysis and submission, enabling streamlined compliance processes [149]. data-driven approaches offer immense potential to improve regulatory compliance in pharmaceutical formulation by automating processes, ensuring model transparency, and maintaining high data quality. However, addressing challenges related to AI validation, ethical considerations, and data integration is crucial for fully realizing these benefits [11, 149, 150].

3.5.1.2 Digital Infrastructure

Insufficient IR-robust digital infrastructure can impede and struggle to collect and/or analyze the large datasets necessary for effective data-driven approaches, which is a common requirement in pharmaceutical development [146].

3.5.1.3 Scalability and Online Performance

Framework development of a scalable, accurate, data-driven model for handling large-scale pharmaceutical systems is still challenging. Modeling a complex bio/chemical reaction network could be sometimes difficult to solve sufficiently within the application time frame. Parallel computing strategies with highly efficient algorithms and distributed architectures are crucial for developing and widely adopting intelligent computational tools in pharmaceutical manufacturing [11].

3.5.1.4 Regulatory Compliance

Conventional methods such as statistical analysis and mechanistic models provide transparency and validation for drug safety and efficacy. While it is still difficult to address standards and guidelines aspects of model transparency, validation, and interpretability for data-driven techniques in the pharmaceutical industry [11]. Data-driven approaches are reshaping regulatory compliance in pharmaceuticals by enhancing efficiency and decision-making. The use of AI-ML streamlines workflows like dossier preparation and submission, expediting product approvals while maintaining compliance. Tools such as DocShifter and Veeva Vault reduce documentation errors and improve consistency. However, challenges like AI validation, data security, and ethical concerns persist, highlighting the demand for industry-regulator collaboration [11, 150]. Additionally, increasing model transparency by enhancing the interpretability and elucidating the predictive data-driven models is essential for trustful regulatory compliance [11] and demonstrating drug safety and efficacy [11, 150].

3.5.2 Organizational Challenges

3.5.2.1 Skepticism, Cultural Resistance, and Lack of Financing Investment

Organizations may struggle to embrace data-driven methodologies due to entrenched traditional practices and resistance to change [151]. The pharmaceutical industry's skepticism regarding ML and AI is owing to the "black box" phenomenon, where there is a lack of knowledge about algorithmic methods and mistrust in the outcomes produced [31, 152]. Skeptics might be reluctant to exploit the AI and ML-generated data, which would waste time and money, limit efficient operation, and hinder the industry's productivity. This mistrust could also result in a hesitancy to invest money in this technology, and ultimately, slower and less effective AI-related advancements in the pharmaceutical R&D processes in the pharmaceutical industry [147]. Thus, future studies should emphasize developing more interpretable models that provide transparent and trustworthy results, rather than relying solely on black-box methods. Additionally, exploring combined modeling approaches and leveraging data scraping techniques could further enhance the effectiveness of data-driven methodologies in drug formulation [6].

3.5.2.2 Occupational and Skillset Gaps

Many current employees in the pharmaceutical sector lack the necessary training and credentials required to operate AI systems [147]. There is often a shortage of skilled personnel capable of interpreting complex data and utilizing advanced analytical tools [153]. Experts with a skill set combination of proficiency in data science, molecular chemistry, and biology are considered to be skilled enough to apply AI in a pharmaceutical context [147].

3.5.3 Ethical and Legal Concerns

3.5.3.1 Data Privacy and Security

The use of personal health data raises significant ethical and legal issues, particularly regarding patient consent and data security [153]. AI and ML in drug discovery heavily rely on vast amounts of patient data, including genetic information, medical records, and CT data. While these datasets are critical for training AI models, they raise significant concerns about data privacy and security due to the lack of transparency in data usage and sharing. Patients may be unaware of how their data is being used, and breaches of sensitive information could have severe consequences [154–156]. Robust encryption and anonymization techniques are needed to avoid the risks of data breaches and misuse.

3.5.3.2 Algorithmic Bias and Discrimination Risks

Data-driven models can inadvertently perpetuate biases, leading to discriminatory practices in drug development and patient care [146]. AI algorithms are only as unbiased as the data they are trained on. If the training data is biased, the resulting models may perpetuate or even amplify existing inequalities in healthcare. For instance, algorithms trained on datasets that underrepresent certain populations may lead to less accurate predictions for those groups [157–159]. Regular auditing of AI models built on diverse and representative training datasets is needed to avoid disparities in healthcare outcomes due to biased algorithms.

3.6 Conclusion

In conclusion, the adoption of data-driven approaches in pharmaceutical formulation and drug development marks a significant shift, enhancing the industry's capacity for innovation and responsiveness to patient needs. These methodologies enable faster identification of drug targets and lead compounds' optimization, streamlining development timelines. By analyzing large biological datasets, researchers can pinpoint promising candidates and enhance predictions of drug stability and bioavailability, improving product quality and manufacturing efficiency. Additionally, data-driven strategies are crucial for advancing personalized medicine, allowing treatments to be tailored to individual genetic profiles for better efficacy and safety. Their role extends to post-market surveillance, where advanced analytics facilitate the early detection of safety signals, ensuring patient safety and regulatory compliance. Despite challenges such as data availability and regulatory hurdles, the potential benefits—from accelerated timelines to improved patient outcomes—underscore the importance of embracing these strategies for pharmaceutical companies aiming to thrive in a competitive landscape while delivering safe and effective therapies.

References

1 Kandhare P, Kurlekar M, Deshpande T, and Pawar A. A review on revolutionizing healthcare technologies with AI and ML applications in pharmaceutical sciences. *Drugs Drug Candidates* 2025;4(1):1–32.

2 Dangeti A, Bynagari DG, and Vydani K. Revolutionizing drug formulation: Harnessing artificial intelligence and machine learning for enhanced stability, formulation optimization, and accelerated development. *International Journal of Pharmaceutical Sciences and Medicine* 2023;8(8):18–29.

3 Yang Y, Ye Z, Su Y, Zhao Q, Li X, and Ouyang D. Deep learning for in vitro prediction of pharmaceutical formulations. *Acta Pharmaceutica Sinica B* 2019;9(1):177–185.

4 Chavda VP. Artificial intelligence and machine learning-based formulation and process development for drug products. In: *Bioinformatics Tools for Pharmaceutical Drug Product Development*, (pp. 183–195). Wiley Online Library; 2023.

5 Vora LK, Gholap AD, Jetha K, Thakur RRS, Solanki HK, and Chavda VP. Artificial intelligence in pharmaceutical technology and drug delivery design. *Pharmaceutics* 2023;15(7):1916.

6 Murray JD, Lange JJ, Bennett-Lenane H, et al. Advancing algorithmic drug product development: Recommendations for machine learning approaches in drug formulation. *European Journal of Pharmaceutical Sciences* 2023;191:106562.

7 Mohammed S. The impact of AI on clinical trial management. *International Journal of Advanced Research in Computer and Communication Engineering* 2024;13(6):57–60.

8 Zeb S, Fnu N, Abbasi N, and Fahad M. AI in healthcare: Revolutionizing diagnosis and therapy. *International Journal of Multidisciplinary Sciences and Arts* 2024;3(3):118–128.

9 Ullagaddi P. Digital transformation strategies to strengthen quality and data integrity in pharma. *International Journal of Business and Management* 2024;19(5):16–16.

10 Yadav S, Singh A, Singhal R, and Yadav JP. Revolutionizing drug discovery: The impact of artificial intelligence on advancements in pharmacology and the pharmaceutical industry. *Intelligent Pharmacy* 2024;2(3):367–380.

11 Dong Y, Yang T, Xing Y, Du J, and Meng Q. Data-driven modeling methods and techniques for pharmaceutical processes. *Processes* 2023;11(7):2096.

12 Castaldello C, Facco P, Bezzo F, Georgakis C, and Barolo M. Data-driven tools for the optimization of a pharmaceutical process through its knowledge-driven model. *AIChE Journal* 2023;69(4):e17925.

13 Modi T. Bioanalysis in drug development. RegulatoryFocus.org; 2023.

14 Hinkson IV, Madej B, and Stahlberg EA. Accelerating therapeutics for opportunities in medicine: A paradigm shift in drug discovery. *Frontiers in Pharmacology* 2020;11:770.

15 Sun D, Gao W, Hu H, and Zhou S. Why 90% of clinical drug development fails and how to improve it? *Acta Pharmaceutica Sinica B* 2022;12(7):3049–3062.

16 Abbas MKG, Rassam A, Karamshahi F, Abunora R, and Abouseada M. The role of AI in drug discovery. *ChemBioChem* 2024; 25(14):p. e202300816(1–21).

17 Dhudum R, Ganeshpurkar A, and Pawar A. Revolutionizing drug discovery: A comprehensive review of AI applications. *Drugs and Drug Candidates* 2024;3(1):148–171.

18 Tripathi MK, Nath A, Singh TP, Ethayathulla AS, and Kaur P. Evolving scenario of big data and Artificial Intelligence (AI) in drug discovery. *Molecular Diversity* 2021;25(3):1439–1460.

19 Ali KA, Mohin SK, Mondal P, Goswami S, Ghosh S, and Choudhuri S. Influence of artificial intelligence in modern pharmaceutical formulation and drug development. *Future Journal of Pharmaceutical Sciences* 2024;10(1):1–15.

20 Bannigan P, Aldeghi M, Bao Z, Häse F, Aspuru-Guzik A, and Allen C. Machine learning directed drug formulation development. *Advanced Drug Delivery Reviews* 2021; 175.

21 Finelli LA and Narasimhan V. Leading a digital transformation in the pharmaceutical industry: Reimagining the way we work in global drug development. *Clinical Pharmacology and Therapeutics* 2020;108(4):756–761.

22 Dong J, Wu Z, Xu H, and Ouyang D. FormulationAI: A novel web-based platform for drug. *Briefings in Bioinformatics* 2024;25(1):1–10.

23 Vidhya KS, Sultana A, M NK, and Rangareddy H. Artificial intelligence's impact on drug discovery and development from bench to bedside. *Cureus* 2023;15(10):1–10.

24 Mamoshina P, Vieira A, Putin E, and Zhavoronkov A. Applications of deep learning in biomedicine. *Molecular Pharmaceutics* 2016;13(5):1445–1454.

25 Huang Z, Juarez JM, and Li X. Data mining for biomedicine and healthcare. *Journal of Healthcare Engineering* 2017; 2017:7107629.

26 Zhang Y, Zhang G, and Shang Q. Computer-aided clinical trial recruitment based on domain-specific language translation: A case study of retinopathy of prematurity. *Journal of Healthcare Engineering* 2017; 2017:7862672.

27 Mak KK and Pichika MR. Artificial intelligence in drug development: Present status and future prospects. *Drug Discovery Today* 2019;24(3):773–780.

28 Segler MHS, Kogej T, Tyrchan C, and Waller MP. Generating focused molecule libraries for drug discovery with recurrent neural networks. *ACS Central Science* 2018;4(1):120–131.

29 Kate A, Seth E, Singh A, et al. Artificial Intelligence for computer-aided drug discovery. *Drug Research* 2023;73(7):369–377.

30 Vyas M, Thakur S, Riyaz B, et al. Artificial intelligence: The beginning of a new era in pharmacy profession. *Asian Journal of Pharmaceutics* 2018;12(2):72–76.

31 Paul D, Sanap G, Shenoy S, Kalyane D, Kalia K, and Tekade RK. Artificial intelligence in drug discovery and development. *Drug Discovery Today* 2021;26(1):80–93.

32 Vatansever S, Schlessinger A, Wacker D, et al. Artificial intelligence and machine learning-aided drug discovery in central nervous system diseases: State-of-the-arts and future directions. *Medicinal Research Reviews* 2021;41(3):1427–1473.

33 Gaulton A, Hersey A, Nowotka ML, et al. The ChEMBL database in 2017. *Nucleic Acids Research* 2017; 45(D1):D945–D954.

34 Kim S, Chen J, Cheng T, et al. PubChem 2019 update: Improved access to chemical data. *Nucleic Acids Research* 2019; 47(D1):D1102–D1109.

35 Wishart DS, Feunang YD, Guo AC, et al. DrugBank 5.0: A major update to the DrugBank database for 2018. *Nucleic Acids Research* 2018; 46(D1):D1074–D1082.

36 Sterling T and Irwin JJ. ZINC 15 - Ligand discovery for everyone. *Journal of Chemical Information and Modeling* 2015;55(11):2324–2337.

37 Gilson MK, Liu T, Baitaluk M, Nicola G, Hwang L, and Chong J. BindingDB in 2015: A public database for medicinal chemistry, computational chemistry and systems pharmacology. *Nucleic Acids Research* 2016; 44(D1):D1045–D1053.

38 Whirl-Carrillo M, McDonagh EM, Hebert JM, et al. Pharmacogenomics knowledge for personalized medicine. *Clinical Pharmacology and Therapeutics* 2012;92(4):414–417.

39 Subramanian A, Narayan R, Corsello SM, et al. A next generation Connectivity Map: L1000 platform and the first 1,000,000 profiles. *Cell* 2017;171(6):1437–1452.e17.

40 Zhou Y, Zhang Y, Lian X, et al. Therapeutic target database update 2022: Facilitating drug discovery with enriched comparative data of targeted agents. *Nucleic Acids Research* 2022;50(1):D1398-D1407.

41 Huang R, Xia M, Nguyen DT, et al. Tox21 challenge to build predictive models of nuclear receptor and stress response pathways as mediated by exposure to environmental chemicals and drugs. *Frontiers in Environmental Science* 2016;3:1–9.

42 Kuhn M, Szklarczyk D, Franceschini A, Von Mering C, Jensen LJ, and Bork P. STITCH 3: Zooming in on protein-chemical interactions. *Nucleic Acids Research* 2012; 40(D1):876–880.

43 Burley SK, Bhikadiya C, Bi C, et al. RCSB Protein Data Bank: Powerful new tools for exploring 3D structures of biological macromolecules for basic and applied research and education in fundamental biology, biomedicine, biotechnology, bioengineering and energy sciences. *Nucleic Acids Research* 2021;49(1):D437–D451.

44 Kanehisa M, Furumichi M, Sato Y, Kawashima M, and Ishiguro-Watanabe M. KEGG for taxonomy-based analysis of pathways and genomes. *Nucleic Acids Research* 2023; 51(D1):D587–D592.

45 Koleti A, Terryn R, Stathias V, et al. Data Portal for the Library of Integrated Network-based Cellular Signatures (LINCS) program: Integrated access to diverse large-scale cellular perturbation response data. *Nucleic Acids Research* 2018; 46(D1):D558–D566.

46 Jumper J, Evans R, Pritzel A, et al. Highly accurate protein structure prediction with AlphaFold. *Nature* 2021;596(7873):583–589.

47 Varadi M, Bertoni D, Magana P, et al. AlphaFold protein structure database in 2024: Providing structure coverage for over 214 million protein sequences. *Nucleic Acids Research* 2024;52:D368–D375.

48 Wallach I, Dzamba M, and Heifets A. AtomNet: A deep convolutional neural network for bioactivity prediction in structure-based drug discovery. *arXiv preprint*: arXiv:1510.02855.2015:1–11.

49 Zhavoronkov A, Ivanenkov YA, Aliper A, et al. Deep learning enables rapid identification of potent DDR1 kinase inhibitors. *Nature Biotechnology* 2019;37(9):1038–1040.

50 Richardson P, Griffin I, Tucker C, et al. Baricitinib as potential treatment for 2019-nCoV acute respiratory disease. *The Lancet* 2020;395(10223):e30–e31.

51 Shivakumar D, Williams J, Wu Y, Damm W, Shelley J, and Sherman W. Prediction of absolute solvation free energies using molecular dynamics free energy perturbation and the OPLS force field. *Journal of Chemical Theory and Computation* 2010;6(5):1509–1519.

52 Sastry GM, Adzhigirey M, Day T, Annabhimoju R, and Sherman W. Protein and ligand preparation: Parameters, protocols, and influence on virtual screening enrichments. *Journal of Computer-Aided Molecular Design* 2013;27(3):221–234.

53 Rusinko A, Rezaei M, Friedrich L, Buchstaller HP, Kuhn D, and Ghogare A. AIDDISON: Empowering drug discovery with AI/ML and CADD tools in a secure, web-based SaaS platform. *Journal of Chemical Information and Modeling* 2024;64(1):3–8.

54 Ochoa D, Hercules A, Carmona M, et al. The next-generation open targets platform: Reimagined, redesigned, rebuilt. *Nucleic Acids Research* 2023;51:1353–1359.

55 McGann M. FRED Pose prediction and virtual screening accuracy. *Journal of Chemical Information and Modeling* 2011;51(3):578–596.

56 Zhan K, Lu ZH, and Zhang YQ. Performance optimization for feature extraction section of DeepChem. In: *Algorithms and Architectures for Parallel Processing, 20th International Conference, ICA3PP 202012452*, (pp. 290–304). Cham: Springer International Publishing; 2020.

57 Korshunova M, Ginsburg B, Tropsha A, and Isayev O. OpenChem: A deep learning toolkit for computational chemistry and drug design. *Journal of Chemical Information and Modeling* 2021;61(1):7–13.

58 Wang Q, Feng YH, Huang JC, Wang TJ, and Cheng GQ. A novel framework for the identification of drug target proteins: Combining stacked auto-encoders with a biased support vector machine. *PLoS ONE* 2017;12(4):1–18.

59 Olumide EI, Precious AU, Luqman AA, Victor A, Ejike IN, and Akachukwu OM. AI and digital health innovation in pharmaceutical development. *Computer Science & IT Research Journal* 2024;5(10):2301–2340.

60 Meyers J, Fabian B, and Brown N. De novo molecular design and generative models. *Drug Discovery Today* 2021;26(11):2707–2715.

61 Chen W, Liu X, Zhang S, and Chen S. Artificial intelligence for drug discovery: Resources, methods, and applications. *Molecular Therapy: Nucleic Acid* 2023;31:691–702.

62 Okafo G. Adapting drug discovery to artificial intelligence. *Drug Target Review* 2018;5(2):50–52.

63 Sudan SS, Marhatta U, and Sudan GS. Artificial intelligence in drug discovery. *International Journal of Creative Research Thoughts (IJCRT)* 2020;8(8):727–730.

64 Habeeba S. Use of artificial intelligence in drug discovery and its application in drug development. *Asian Journal of Research in Chemistry* 2023;16:83–90.

65 Hessler G and Baringhaus KH. Artificial intelligence in drug design. *Molecules* 2018;23(10):2520.

66 Segler MHS, Preuss M, and Waller MP. Planning chemical syntheses with deep neural networks and symbolic AI. *Nature* 2018;555(7698):604–610.

67 Li YY, Lenaghan SC, and Zhang M. A data-driven predictive approach for drug delivery using machine learning techniques. *PLoS ONE* 2012;7(2):e31724.

68 Macarron R, Banks MN, Bojanic D, et al. Impact of high-throughput screening in biomedical research. *Nature Reviews Drug Discovery* 2011;10(3):188–195.

69 Cao H, Mushnoori S, Higgins B, et al. A systematic framework for data management and integration in a continuous pharmaceutical manufacturing processing line. *Processes* 2018;6(5):1–21.

70 Inglese J, Auld DS, Jadhav A, et al. Quantitative high-throughput screening: A titration-based approach that efficiently identifies biological activities in large chemical libraries. *Proceedings of the National Academy of Sciences of the United States of America* 2006;103(31):11473–11478.

71 Schneider P, Walters WP, Plowright AT, et al. Rethinking drug design in the artificial intelligence era. *Nature Reviews Drug Discovery* 2020;19(5):353–364.

72 Mehta CH, Narayan R, and Nayak UY. Computational modeling for formulation design. *Drug Discovery Today* 2019;24(3):781–788.

73 Zhao C, Jain A, Hailemariam L, et al. Toward intelligent decision support for pharmaceutical product development. *Journal of Pharmaceutical Innovation* 2006;1(1):23–35.

74 Rudrurkar MN, Dharashive VM, Shafi S, et al. The impact of artificial intelligence on drug discovery; A comprehensive review. *Asian Journal of Pharmaceutical Research and Development* 2024;12(2):171–178.

75 Dey H, Arya N, Mathur H, Chatterjee N, and Jadon R. Exploring the role of artificial intelligence and machine learning in pharmaceutical formulation design. *International Journal of Newgen Research in Pharmacy & Healthcare* 2024;(1):30–41.

76 Wang S and Ji Y. A novel strategy of coupling artificial intelligence with chemical fingerprinting to predict drug phase behaviors in complex systems. *AIChE Journal* 2024;70(4):e18359.

77 Tosca EM, Bartolucci R, and Magni P. Application of artificial neural networks to predict the intrinsic solubility of drug-like molecules. *Pharmaceutics* 2021;13(7):1101.

78 Dahlgren D, Sjöblom M, and Lennernäs H. Intestinal absorption-modifying excipients: A current update on preclinical in vivo evaluations. *European Journal of Pharmaceutics and Biopharmaceutics* 2019;142:411–420.

79 Lefnaoui S, Rebouh S, Bouhedda M, and Yahoum MM. Artificial neural network for modeling formulation and drug permeation of topical patches containing diclofenac sodium. *Drug Delivery and Translational Research* 2020;10(1):168–184.

80 Rebouh S, Lefnaoui S, Bouhedda M, and Yahoum MM, editors. Artificial neural network-genetic algorithm based prediction of Metformin diffusion from transdermal films based on biopolymers for diabetes disease treatment. In: *2019 International Conference on Applied Automation and Industrial Diagnostics (ICAAID)*; 2019.

81 Falcón-Cano G, Molina C, and Cabrera-Pérez MÁ. Reliable prediction of Caco-2 permeability by supervised recursive machine learning approaches. *Pharmaceutics* 2022;14(10):1998.

82 Tan X, Liu Q, Fang Y, et al. Predicting peptide permeability across diverse barriers: A systematic investigation. *Molecular Pharmaceutics* 2024;21(8):4116–4127.

83 Shaik B, Gupta R, Louis B, and Agrawal VK. Prediction of permeability of drug-like compounds across polydimethylsiloxane membranes by machine learning methods. *Journal of Pharmaceutical Investigation* 2015;45(5):461–473.

84 Abdallah RM, Hasan HE, and Hammad A. Predictive modeling of skin permeability for molecules: Investigating FDA-approved drug permeability with various AI algorithms. *PLOS Digital Health* 2024;3(4):1–20.

85 Jiang J, Lu A, Ma X, Ouyang D, and Williams RO. The applications of machine learning to predict the forming of chemically stable amorphous solid dispersions prepared by hot-melt extrusion. *International Journal of Pharmaceutics: X* 2023;5:100164.

86 Aumklad P, Suriyaamporn P, Panomsuk S, Pamornpathomkul B, and Opanasopit P. Artificial intelligence-aided rational design and prediction model for progesterone-loaded self-microemulsifying drug delivery system formulations. *Science, Engineering and Health Studies* 2024;18:1–7.

87 Plenge RM. Disciplined approach to drug discovery and early development. *Science Translational Medicine* 2016;8(349):1–6.

88 Dawoodbhoy FM, Delaney J, Cecula P, et al. AI in patient flow: Applications of artificial intelligence to improve patient flow in NHS acute mental health inpatient units. *Heliyon* 2021;7(5):e06993.

89 Quennelle S, Douillet M, Friedlander L, et al. The smart data extractor, a clinician friendly solution to accelerate and improve the data collection during clinical trials. *Studies in Health Technology and Informatics* 2023;302:247–251.

90 Yong PL, Saunders RS, and Olsen L. *The Healthcare Imperative: Lowering Costs and Improving Outcomes*. Washington (DC): National Academies Press (US); 2010.

91 Soni SJ and Patel AN. Digital transformation and industry 4.0 in pharma manufacturing: The role of IOT, AI, and big data. *Journal of Integral Sciences* 2024;7(4):1–8.

92 Chandra S, Duvinage M, Prakash PKS, Davis P, and Timmer S. Improving process yield through manufacturing Digital Twin using Conditional Synthetic Data Engine (COSYNE). *Frontiers in Artificial Intelligence and Applications* 2024;392:4673–4680.

93 Huanbutta K, Burapapadh K, Kraisit P, et al. Artificial intelligence-driven pharmaceutical industry: A paradigm shift in drug discovery, formulation development, manufacturing, quality control, and post-market surveillance. *European Journal of Pharmaceutical Sciences* 2024;203:106938.

94 Alosert H, Savery J, Rheaume J, et al. Data integrity within the biopharmaceutical sector in the era of Industry 4.0. *Biotechnology Journal* 2022;17(6):1–8.

95 Rantanen J and Khinast J. The future of pharmaceutical manufacturing sciences. *Journal of Pharmaceutical Sciences* 2015;104(11):3612–3638.

96 Aguero D and Nelson SD. The potential application of large language models in pharmaceutical supply chain management. *Journal of Pediatric Pharmacology and Therapeutics* 2024;29(2):200–205.

97 Kassem RG, Mbata AO, Usuemerai PA, and Ogbewele EG. Leveraging data analytics for optimizing pharmacy marketing strategies: Enhancing patient engagement and medication adherence. *World Journal of Advanced Research and Reviews* 2022;16(01):1184–1195.

98 Hashimoto S, Motozawa Y, and Mano T. Digital marketing innovation: New business models for pharmaceutical and medical device product marketing. *Digital Health* 2024;10:20552076241279186.

99 Zhao X, Iqbal S, Valdes IL, Dresser M, and Girish S. Integrating real-world data to accelerate and guide drug development: A clinical pharmacology perspective. *Clinical and Translational Science* 2022;15(10):2293–2302.

100 Shin H, Cha J, Lee C, et al. The 2011–2020 trends of data-driven approaches in medical informatics for active pharmacovigilance. *Applied Sciences (Switzerland)* 2021;11(5):1–13.

101 Digitaldefynd. Agentic AI in the Pharma Industry [5 Case Studies]; Digitaldefynd; [cited 2025 Feb 14]. Available from: https://digitaldefynd.com/IQ/agentic-ai-in-the-pharma-industry/.

102 Ferrero E, Dunham I, and Sanseau P. In silico prediction of novel therapeutic targets using gene-disease association data. *Journal of Translational Medicine* 2017;15(1):1–16.

103 Bakkar N, Kovalik T, Lorenzini I, et al. Artificial intelligence in neurodegenerative disease research: Use of IBM Watson to identify additional RNA-binding proteins altered in amyotrophic lateral sclerosis. *Acta Neuropathologica* 2018;135(2):227–247.

104 Zang Q, Mansouri K, Williams AJ, et al. In Silico prediction of physicochemical properties of environmental chemicals using molecular fingerprints and machine learning. *Journal of Chemical Information and Modeling* 2017;57(1):36–49.

105 United States Environmental Protection Agency (U.S EPA). EPI Suite™-Estimation Program Interface. *U.S EPA Washington (DC)*. 2024. [cited 2025 Feb 17] Available from: https://www.epa.gov/tsca-screening-tools/epi-suitetm-estimation-program-interface.

106 Öztürk H, Özgür A, and Ozkirimli E. DeepDTA: Deep drug-target binding affinity prediction. *Bioinformatics* 2018;34(17):i821–i829.

107 Putin E, Asadulaev A, Ivanenkov Y, et al. Reinforced adversarial neural computer for de novo molecular design. *Journal of Chemical Information and Modeling* 2018;58(6):1194–1204.

108 Kumar R, Sharma A, Siddiqui MH, and Tiwari RK. Prediction of human intestinal absorption of compounds using artificial intelligence techniques. *Current Drug Discovery Technologies* 2017;14(4):244–254.

109 ChemDB Chemiinformatics Portal; 2012. https://www.uci.edu/: https://www.igb.uci.edu/; https://cdb.ics.uci.edu/.

110 Lusci A, Pollastri G, and Baldi P. Deep architectures and deep learning in chemoinformatics: The prediction of aqueous solubility for drug-like molecules. *Journal of Chemical Information and Modeling* 2013;53(7):1563–1575.

111 Ryu JY, Kim HU, and Lee SY. Deep learning improves prediction of drug–drug and drug–food interactions. *Proceedings of the National Academy of Sciences of the United States of America* 2018;115(18):E4304–E4311.

112 Li Y, Abbaspour MR, Grootendorst PV, Rauth AM, and Wu XY. Optimization of controlled release nanoparticle formulation of verapamil hydrochloride using artificial neural networks with genetic algorithm and response surface methodology. *European Journal of Pharmaceutics and Biopharmaceutics* 2015;94(1):170–179.

113 Muñiz Castro B, Elbadawi M, Ong JJ, et al. Machine learning predicts 3D printing performance of over 900 drug delivery systems. *Journal of Controlled Release* 2021;337:530–545.

114 Alafeef M, Srivastava I, and Pan D. Machine learning for precision breast cancer diagnosis and prediction of the nanoparticle cellular internalization. *ACS Sensors* 2020;5(6):1689–1698.

115 Chiba S, Lim KRQ, Sheri N, et al. ESkip-Finder: A machine learning-based web application and database to identify the optimal sequences of antisense oligonucleotides for exon skipping. *Nucleic Acids Research: University of ALBERTA*; 2024. Available from: https://eskip-finder.org/cgi-bin/input.cgi.

116 Chiba S, Lim KRQ, Sheri N, et al. ESkip-Finder: A machine learning-based web application and database to identify the optimal sequences of antisense oligonucleotides for exon skipping. *Nucleic Acids Research* 2021; 49(W1):W193–W198.

117 Dar SA, Gupta AK, Thakur A, and Kumar M. SMEpred workbench: A web server for predicting efficacy of chemicallymodified siRNAs. *RNA Biology* 2016;13(11):1144–1151.

118 Ippolito A, Wang H, Zhang Y, Vakil V, Bazzazi H, and Popel AS. Eliciting the antitumor immune response with a conditionally activated PD-L1 targeting antibody analyzed with a quantitative systems pharmacology model. *CPT: Pharmacometrics and Systems Pharmacology* 2024;13(1):93–105.

119 Wang H, Arulraj T, Kimko H, and Popel AS. Generating immunogenomic data-guided virtual patients using a QSP model to predict response of advanced NSCLC to PD-L1 inhibition. *npj Precision Oncology* 2023;7(1):1–14.

120 Ippolito A, Wang H, Zhang Y, Vakil V, and Popel AS. Virtual clinical trials via a QSP immuno-oncology model to simulate the response to a conditionally activated PD-L1 targeting antibody in NSCLC. *Journal of Pharmacokinetics and Pharmacodynamics* 2024: 747–757.

121 Viceconti M, Pappalardo F, Rodriguez B, Horner M, Bischoff J, and Musuamba Tshinanu F. In silico trials: Verification, validation and uncertainty quantification of predictive models used in the regulatory evaluation of biomedical products. *Methods* 2021;185:120–127.

122 The European High Performance Computing Joint Undertaking (EuroHPC JU). A EuroHPC success story: Human in-silico clinical trials to assess drug-induced arrhythmic risk. *Luxembourg: EuroHPC* 2024. [cited 2025 Feb 24] Available from: https://www.eurohpc-ju.europa.eu/eurohpc-success-story-human-silico-clinical-trials-assess-drug-induced-arrhythmic-risk_en.

123 SimCardioTest. Available from: https://www.simcardiotest.eu/wordpress/

124 Cordis EU. Simulation of cardiac devices & drugs for in-silico testing and certification. *Brussels: CORDIS*. 2024.. [cited 2025 Feb 24] Available from: https://cordis.europa.eu/project/id/101016496.

125 Angoulvant D, Granjeon-Noriot S, Amarenco P, et al. In-silico trial emulation to predict the cardiovascular protection of new lipid-lowering drugs: An illustration through the design of the SIRIUS programme. *European Journal of Preventive Cardiology* 2024;31(15):1820–1830.

126 Bonaca MP, Szarek M, and Schwartz GG. Is it time to get SIRIUS about in silico modelling of cardiovascular outcomes trials? *European Journal of Preventive Cardiology* 2024;31(15):1831–1832.

127 Biovista. Read Biovista's two contributed chapters in the book Drug Repositioning or read Biovista President's and 50 of the world's top futurists' reflections on Alvin Toffler's classic book "Future Shock" and insights into the future, or read Biovista CEO's interview on drug repositioning. Available from: https://www.biovista.com/.

128 Andreas P. Why systematic drug repurposing is here to stay. *Biovista Inc.: Drug Positioning and Prioritization*. 2016. [cited 2025 Feb 24] Available from: https://www.biovista.com/systematic-drug-repurposing-why-it-is-here-to-stay/.

129 Russo G, Pennisi M, Fichera E, Motta S, Raciti G, Viceconti M, et al. In silico trial to test COVID-19 candidate vaccines: A case study with UISS platform. *BMC Bioinformatics* 2020;21(17):1–16.

130 Apple. *Apple Inc. United States.* [accessed 2025 Feb 24]. Available from: https://www.apple.com/.

131 Johnson & Jouhnson. Available from: https://www.jnj.com/.

132 Johnson & Jouhnson. Welcome to Heartline. Heartline *TM Study app* [cited 2025 Feb 24] Available from: https://heartlinestudy.com/.

133 Morillo CA, Banerjee A, Perel P, Wood D, and Jouven X. Atrial fibrillation: The current epidemic. *Journal of Geriatric Cardiology* 2017;14(3):195–203.

134 Reiffel JA, Verma A, Kowey PR, et al. Incidence of previously undiagnosed atrial fibrillation using insertable cardiac monitors in a high-risk population: The REVEAL AF study. *JAMA Cardiology* 2017;2(10):1120–1127.

135 Johnson, Johnson SI. Johnson & Johnson launches Heartline™, the first-of-its-kind, virtual study designed to explore if a new iPhone App and Apple Watch can help reduce the risk of stroke; *Johnson & Johnson Services, Inc.* 2020 [cited 2025 Feb 24]. Available from: https://www.jnj.com/media-center/press-releases/johnson-johnson-launches-heartline-the-first-of-its-kind-virtual-study-designed-to-explore-if-a-new-iphone-app-and-apple-watch-can-help-reduce-the-risk-of-stroke?utm_source=chatgpt.com.

136 Gibson CM. The Heartline study: Fun, convenient, and educational; Heartline™ *Study app* 2021[cited 2025 Feb 24] Available from: https://heartlinestudy.com/.

137 Gibson CM, Steinhubl S, Lakkireddy D, et al. Does early detection of atrial fibrillation reduce the risk of thromboembolic events? Rationale and design of the Heartline study. *American Heart Journal* 2023;259:30–41.

138 Pharmacology-Ai. *REPROCELL Inc.* [accessed 2025 Feb 24] Available from: https://www.reprocell.com/pharmacology-ai .

139 Sokolova D. ReproCell, IBM, and HNCDI Debut Pharmacology-AI to boost efficiency of drug discovery; *BioPharmaTrend (BPT) Analytics LTD.* 2023. [cited 2025 March 4] Available from: https://www.biopharmatrend.com/news/reprocell-ibm-hncdi-pharmacology-ai-drug-discovery-681/.

140 Mayo RM, Summey JF, Williams JE, Spence RA, Kim S, and Jagsi R. Qualitative study of oncologists' views on the CancerLinQ rapid learning system. *Journal of Oncology Practice* 2017;13(3):e176–e184.

141 Séroussi B and Bouaud J. Using OncoDoc as a computer-based eligibility screening system to improve accrual onto breast cancer clinical trials. *Artificial Intelligence in Medicine* 2003;29(1-2):153–167.

142 Somashekhar SP, Sepúlveda MJ, Puglielli S, et al. Watson for Oncology and breast cancer treatment recommendations: Agreement with an expert multidisciplinary tumor board. *Annals of Oncology* 2018;29(2):418–423.

143 University of California San Francisco (UCSF). New UCSF Robotic Pharmacy aims to improve patient safety; *UCSF Medical Center.* 2011 [cited 2025 Feb 16] Available from: https://www.ucsf.edu/news/2011/03/98776/new-ucsf-robotic-pharmacy-aims-improve-patient-safety.

144 Karwa S; 2023. Available from: https://www.linkedin.com/pulse/use-cases-ai-pharmaceutical-industry-seema-karwa/.

145 Antonyuk S; 2024. Available from: https://litslink.com/blog/use-cases-of-ai-in-pharma-how-to-leverage-it.

146 Trade UN, Development. Challenges of data for development, In: Data for Development. (pp. 35–65). United Nations; 2024. Available from: https://www.un-ilibrary.org/content/books/9789211065428c006.

147 Patel V and Shah M. Artificial intelligence and machine learning in drug discovery and development. *Intelligent Medicine* 2022;2(3):134–140.

148 Selekman JA, Qiu J, Tran K, et al. High-throughput automation in chemical process development. *Annual Review of Chemical and Biomolecular Engineering* 2017;8:525–547.

149 Oronsky B, Burbano E, Stirn M, et al. Data management 101 for drug developers: A peek behind the curtain. *Clinical and Translational Science* 2023;16(9):1497–1509.

150 Ajmal CS, Yerram S, Abishek V, et al. Innovative approaches in regulatory affairs: Leveraging artificial intelligence and machine learning for efficient compliance and decision-making. *AAPS Journal* 2025;27(1):1–27.

151 Rashedi J. Challenges and barriers of the data-driven organization. In: Rashedi J, editor. *The Data-driven Organization. Business Guides on the Go*, (pp. 33–37). Cham: Springer Nature Switzerland; 2023.

152 Lamberti MJ, Wilkinson M, Donzanti BA, et al. A study on the application and use of artificial intelligence to support drug development. *Clinical Therapeutics* 2019;41(8):1414–1426.

153 Freitas AT. Data-driven approaches in healthcare: Challenges and emerging trends. In: Sousa Antunes H, Freitas PM, Oliveira AL, Martins Pereira C, Vaz de Sequeira E, and Barreto Xavier L, editors. *Multidisciplinary Perspectives on Artificial Intelligence and the Law. Law, Governance and Technology Series*, (pp. 65–80). Cham: Springer International Publishing; 2024.

154 Dousa R. Toward the clinic: Understanding patient perspectives on AI and data-sharing for AI-driven oncology drug development. In: Cassidy J, Taylor B, editors. Rijeka: IntechOpen; 2020.

155 Pitel E, Leașu F, Nicolau A, and Rogozea L. Ethical aspects in the use of artificial intelligence in the process of drug development. *Jurnal Medical Brasovean* 2024;(2):53–61.

156 Livieri G, Anastasopoulou C, Briola K, Nikolopoulou P, and Efthymiou IP. Ethical issues arising from the use of AI in drug discovery. *Journal of Politics and Ethics in New Technologies and AI* 2024;3(1):e37093.

157 Warokar P and Lote S, editors. Ethical and regulatory consideration in AI-assisted drug development. In: *2024 2nd DMIHER International Conference on Artificial Intelligence in Healthcare, Education and Industry (IDICAIEI)*; 2024.

158 Agrawal K. The ethics of computing in pharmaceutical research. In: *Computer Aided Drug Development. ThinkPlus Pharma Publications*. 2024. pp. 269–274. Available from: https://publications.jopir.in/wp-content/uploads/2024/11/CADD-Chapter-10.pdf

159 Rahate KP and Mondal R. Applications of AI in drug discovery: Its challenges, opportunities, and strategies. In: Grover V, Balusamy B, M.K N, Anand V, and Milanova M, editors. *Approaches to Human-Centered AI in Healthcare*, (pp. 86–120). Hershey, PA, USA: IGI Global; 2024.

4

Case Studies on Data-driven Drug Formulation/ Development

Yujing Zhao[1,2], *Wenai Chen*[1], *Lixiang Guo*[1], *Lei Zhang*[1,3], *Jian Du*[1,3], *Jiaqi Lin*[2], *Qilei Liu*[1,3*], *and Qingwei Meng*[1,3*]

[1]*State Key Laboratory of Fine Chemicals, Frontiers Science Center for Smart Materials Oriented Chemical Engineering, Department of Pharmaceutical Sciences, Institute of Chemical Process Systems Engineering, School of Chemical Engineering, Dalian University of Technology, Dalian, China*
[2]*MOE Key Laboratory of Bio-Intelligent Manufacturing, School of Bioengineering, Dalian University of Technology, Dalian, China*
[3]*Ningbo Institute of Dalian University of Technology, Ningbo, China*

*Corresponding authors: Qilei Liu liuqilei@dlut.edu.cn, Qingwei Meng mengqw@dlut.edu.cn

CONTENTS

4.1 Introduction

The development of novel pharmaceuticals stands as a critical endeavor in safeguarding public health and driving socioeconomic progress, particularly evident during global health crises like the COVID-19 pandemic. Small-molecule drugs remain essential therapeutic solutions, yet their discovery faces formidable challenges characterized by exorbitant costs exceeding \$2.5 billion, protracted timelines spanning 10–15 years [1], and dismal clinical success rates. Traditional drug discovery paradigms increasingly integrate computational approaches to mitigate these challenges, with computer-aided techniques demonstrating particular promise in accelerating early-stage research through virtual screening (VS) and *de novo* design strategies. While VS methods efficiently rank potential candidates through molecular docking or similarity algorithms [2], their effectiveness diminishes when confronting the astronomical chemical space of 10^{30} to 10^{60} feasible compounds [3]. This limitation has spurred interest in optimization-based *de novo* design methodologies that systematically explore chemical space through generative models, property predictors, and optimization algorithms.

Contemporary approaches leverage both evolutionary algorithms and deep learning architectures to navigate this complexity. Genetic algorithms enable heuristic-guided molecular assembly and optimization but suffer from local optima entrapment [4]. Deep generative models like variational autoencoders (VAEs) [5] and generative adversarial networks (GANs) [6] offer data-driven alternatives, yet their propensity to generate structurally infeasible molecules, limited extrapolation capabilities [7], and "black box" nature constrain practical utility. Parallel developments in computer-aided molecular design (CAMD) employ mixed-integer nonlinear programming (MINLP) to formulate drug design as an optimization problem with explicit structural and property constraints. Although MINLP-based CAMD provides interpretable solutions through mathematical formulations, its direct application to drug design falters when handling complex ring systems and results in anomalous scaffold generation, compounded by computational intractability for large-scale problems.

Recent paradigm shifts further emphasize the crucial role of binding kinetics in drug efficacy, particularly dissociation rate constants (k_{off}) and protein–ligand residence time (τ, $\tau = 1/k_{off}$). For example, some inhibitors have been reported to exhibit moderate binding activity to their target at the micromolar level, in stark contrast to their high antiproliferative potency in cell culture and their potent antitumor potency in animal models at the nanomolar level [8]. The high potency of these inhibitors in both cell culture and in vivo can be attributed to their prolonged binding to the target alone [8], which also impacts their dosing frequency. The use of computational methods is a worthy choice to obtain the kinetic properties of protein–ligand interactions. While molecular dynamics (MD) simulations offer atomic-level insights into protein–ligand unbinding pathways [9–11], their prohibitive computational costs limit practical application given the timescale disparity between simulated trajectories (μs) and therapeutic residence times (minutes–hours) [9]. This limitation has motivated the development of enhanced MD sampling algorithms [12] such as τ-random acceleration molecular dynamics (τ-RAMD) [13]. Despite offering a promising method for predicting protein–ligand dissociation kinetics, the enhanced sampling algorithms still face challenges in providing high-throughput drug screening or design due to their computational cost. Besides the above mechanism models, surrogate models using molecular mechanics-derived descriptors and machine learning algorithms have emerged as alternatives [14–25], yet their prediction efficiency cannot support the high-throughput drug screening or design tasks due to the molecular mechanics-based descriptors. These limitations underscore the need for robust predictive models that balance accuracy and efficiency to enable kinetics-aware drug design.

To address these multifaceted challenges, this chapter establishes an integrated optimization framework that integrates deep learning architectures with enhanced CAMD methodologies. The developed system incorporates a hybrid deep learning model combining molecular representations (Mol2vec, higher-order amino acid features), convolutional neural networks (CNNs), and gate-augmented attention mechanisms to predict protein–ligand binding affinity. A kinetics-informed extension employs a mixture-of-experts (MoE) architecture trained on an expanded dissociation kinetic database for high-throughput k_{off} prediction. The *de novo* drug design part features a scaffold-based CAMD method developed to ensure structural rationality while dramatically reducing problem dimensionality. Case studies demonstrate the versatility of our framework in designing anticoagulant drug candidates and HSP90 inhibitor candidates with optimized binding affinity and/or dissociation kinetics. Subsequent sections detail these innovations: Section 4.2 presents the deep learning architectures for affinity and kinetics prediction; Section 4.3 elaborates the optimized CAMD framework with scaffold processing algorithms; Section 4.4 validates the approach through case studies of drug candidate designs.

4.2 Deep Learning Models for Predictions of Binding Affinity and Dissociation Kinetics

4.2.1 Development of the Deep Learning Model for Predicting Binding Affinity

Protein–ligand interaction is a central issue in small-molecule drug discovery. The strength of such interactions is commonly described by binding affinity and is often characterized using properties such as K_i, K_d, and IC_{50}. In early-stage drug development, an important objective is to identify ligands that can bind tightly to a given protein target [26]. Existing computational approaches for affinity prediction have achieved useful results, but many of them depend on three-dimensional structural information for protein–ligand complexes [27]. In practice, however, sequence- and string-based information is usually easier to obtain than high-quality complex structures. For this reason, this section introduces an end to end deep learning framework based on text-derived representations to rapidly distinguish protein–ligand complexes with relatively high affinity from those with lower affinity. A schematic summary of the workflow is provided in Figure 4.1, including ligand encoding, protein encoding, and the overall prediction architecture.

For a given protein–ligand pair, the model takes the ligand simplified molecular input line entry system (SMILES) string and the protein amino acid sequence as its starting inputs, as illustrated in Figure 4.1. These two forms of textual information are then converted into

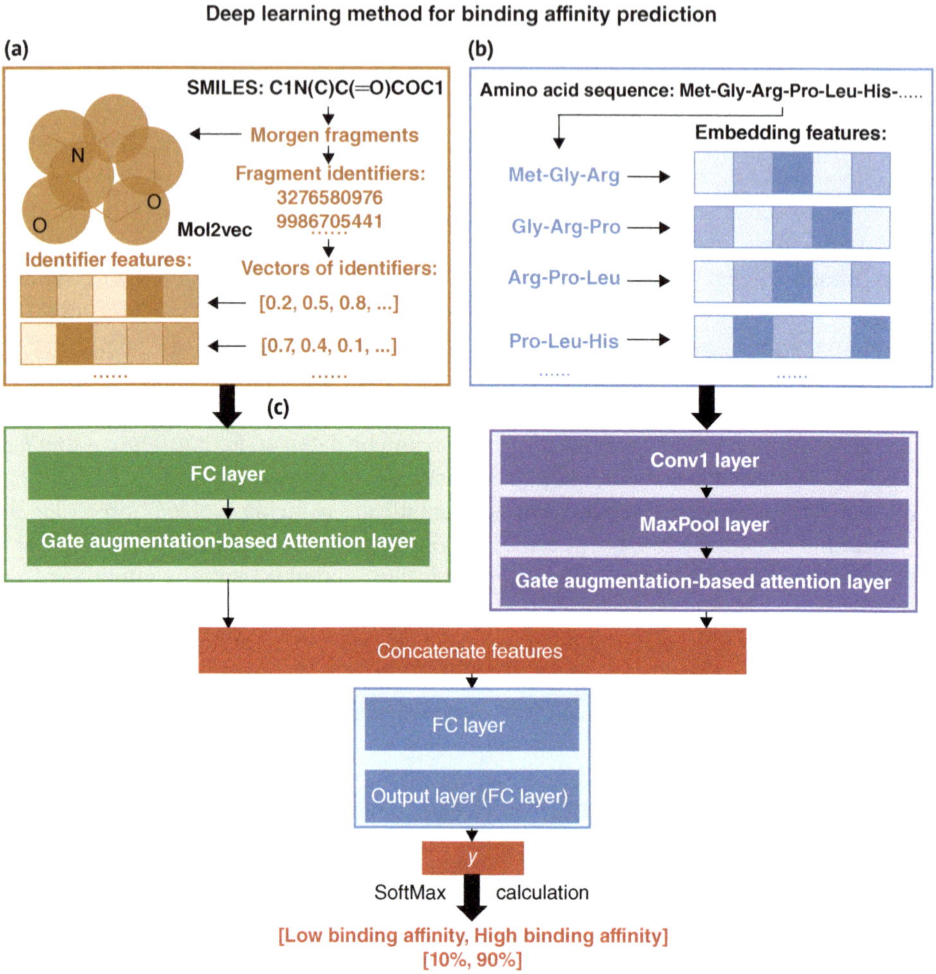

Figure 4.1 An overview of the deep learning model. (a) Ligand representation, (b) protein representation, and (c) architecture of the deep learning model.

matrix representations: the ligand is encoded as a Mol2vec-based fragment feature matrix, whereas the protein is represented by an embedded matrix constructed from higher-order amino acid units. After encoding, the ligand matrix passes through a fully connected (FC) layer followed by a gate-augmented attention module for feature extraction. In parallel, the protein matrix is first processed by a one-dimensional convolution layer and a max-pooling layer to compress the sequence representation before entering a similar attention module, thereby reducing the computational burden of subsequent operations. The resulting ligand and protein features are each aggregated across matrix rows and then concatenated for downstream FC layers, which output the predicted probability that the complex belongs to the high-affinity or low-affinity class. The following sections describe the dataset construction, input encoding strategy, and network architecture in detail.

4.2.1.1 Binding Affinity Database

In this work, the binding affinity dataset was constructed with reference to the 2019 release of the PDBbind database [28], which provides 17,679 protein–ligand complex entries together with basic annotations, experimental affinity measurements such as K_i, K_d, and IC_{50}, and corresponding structural files.

The 17,679 entries collected from the 2019 release of PDBbind originate from multiple research groups. To improve data consistency and make the dataset suitable for subsequent model development, these entries were further screened and preprocessed according to the criteria described below:

- Entries reporting IC_{50} values were excluded from the dataset because these measurements can vary substantially across experimental settings [29]. Since the underlying assay conditions are often inconsistent among different studies, directly combining IC_{50} records from multiple sources may introduce considerable noise into model training. In contrast, samples annotated with K_i or K_d were retained and used to construct the final binding-affinity dataset.
- Entries were also removed when the corresponding ligand structure could not be successfully parsed by RDKit [30], or when Mol2vec-based Morgan fragment extraction could not be completed. This step ensured that every retained sample could be converted into a valid ligand representation for downstream modeling.
- Entries were further filtered by sequence and fragment length. Specifically, samples were removed when the ligand contained more than 140 Morgan fragments or when the protein sequence exceeded 1,000 amino acids. Introducing these upper bounds helps control the size of the ligand and protein input matrices, which in turn improves training efficiency. The trade-off is that a small number of unusually large complexes are excluded from the final dataset.
- To formulate the affinity prediction task as a binary classification problem, complexes with K_i or K_d values below 1 μM were assigned to the high-affinity class and labeled as "1." In contrast, complexes with K_i or K_d values of 1 μM or above were assigned to the low-affinity class and labeled as "0."

After all preprocessing and filtering steps were completed, 8,393 unique protein–ligand pairs remained in the curated dataset. These retained samples constituted the binding-affinity database used for subsequent model training and evaluation.

4.2.1.2 Input Representation

Because reliable three-dimensional structures are often unavailable for many unknown protein–ligand complexes, the present model represents each complex using sequence- and string-based inputs rather than structure-derived features. Specifically, the input description contains two components: the ligand is described by its SMILES string, and the protein is represented by its amino acid sequence.

Ligand representation: In the ligand branch, each SMILES string is converted into a Mol2vec-based representation. Mol2vec is a text-derived molecular descriptor inspired by Word2vec [31] and is designed to encode molecular substructures as continuous vectors. In this procedure, the molecule is first decomposed into Morgan fragments with radius 1, and each fragment is then mapped to a pretrained embedding generated by the skip-gram training strategy in Mol2vec [32]. The default embedding dimension is 300. Because chemically related substructures tend to occupy nearby positions in the learned feature space, the descriptor provides a compact way to capture molecular similarity and fragment-level information. A molecular representation can then be assembled from the embeddings of its constituent fragments. Additional details of Mol2vec are available in Jaeger et al. [32]. Using the Mol2vec encoding, each ligand is organized as a two-dimensional matrix of size 140 × 300. In this matrix, rows correspond to fragment-level units and columns correspond to the associated embedding features. Since the number of fragments varies across ligands, zero padding is applied whenever a molecule contains fewer than 140 fragments so that all ligand inputs share a consistent matrix size.

Protein representation: Proteins are composed of a finite set of standard amino acids, each of which can be denoted by a three-letter abbreviation. For instance, "Gly" refers to glycine. Accordingly, a protein sequence may be written as an ordered series of amino acid tokens, which can then be converted into vector features through an embedding operation in a deep learning framework. A simple residue-by-residue embedding may be insufficient to capture local contextual relationships among neighboring amino acids. To incorporate such local sequence information, this work defines a higher-order amino acid unit by grouping every three consecutive residues in the protein sequence [32]. For example, the sequence "Met-Gly-Arg-Pro-Leu" can be decomposed into the overlapping triplets "Met-Gly-Arg," "Gly-Arg-Pro," and "Arg-Pro-Leu." Applying this strategy to the binding-affinity dataset produced 9,164 higher-order amino acid strings. Each triplet unit was then mapped to a 30-dimensional embedding vector for subsequent model input. Based on these higher-order sequence units, each protein is encoded as a 1,000 × 30 feature matrix. In this representation, rows correspond to higher-order amino acid units and columns correspond to their embedded features. Because protein lengths vary, zero padding is introduced when a sequence contains fewer than 1,000 higher-order units so that all protein inputs can be stored in a uniform matrix format.

4.2.1.3 Model Architecture

After defining the affinity labels and preparing the corresponding input representations, the next stage is to establish the deep learning architecture used for model training. A schematic illustration of the overall network is provided in Figure 4.1c, and the architectural details are described below. The ligand matrix $\mathbf{x}_L$ (140 × 300) is first passed through an FC layer $\mathbf{V}$ (300 × 30), which projects the ligand features from 300 dimensions to 30 dimensions. This dimensionality reduction helps lower model redundancy and reduce training cost. The resulting 140 × 30 ligand matrix $\mathbf{x}_L'$ is then processed by a gate augmentation-based attention layer. Attention mechanisms are now widely used for feature extraction in deep learning because they emphasize informative local patterns within complex inputs. Since their introduction in 2017 [33], such mechanisms have been adopted in many neural architectures, including recurrent neural networks (RNNs) [34], and so on.

In the attention layer, the ligand feature matrix $\mathbf{x}'_{\mathrm{L}}$ is first projected by a learnable matrix $\mathbf{W}$ (30 × 30) to obtain an updated representation $\mathbf{x}''_{\mathrm{L}}$. The pairwise attention score between the i-th and j-th fragments is then computed as:

$$e_{ij} = \mathbf{x}''_{\mathrm{L},i}\mathbf{x}''^{\mathrm{T}}_{\mathrm{L},j} + \mathbf{x}''^{\mathrm{T}}_{\mathrm{L},i}\mathbf{x}''_{\mathrm{L},j},$$

where e_{ij} measures the contribution of fragment j to fragment i. This symmetric form ensures $e_{ij} = e_{ji}$. The normalized attention weight is defined by:

$$a_{ij} = \frac{\exp(e_{ij})}{\sum_j \exp(e_{ij})}$$

Using these weights, each fragment feature is updated by aggregating information from neighboring fragments:

$$\mathbf{x}'''_{\mathrm{L},i} = \sum_j a_{ij}\mathbf{x}''_{\mathrm{L},j},$$

followed by a ReLU activation function. A gate augmentation step is subsequently introduced to combine the original feature $\mathbf{x}'_{\mathrm{L},i}$ with the updated feature $\mathbf{x}'''_{\mathrm{L},i}$, yielding the output of the attention layer:

$$\mathbf{x}^{\mathrm{out}}_{\mathrm{L},i} = z_i\mathbf{x}'_{\mathrm{L},i} + (1 - z_i)\mathbf{x}'''_{\mathrm{L},i},$$

where

$$z_i = \sigma\left(\left(\mathbf{x}'_{\mathrm{L},i} \| \mathbf{x}'''_{\mathrm{L},i}\right)\mathbf{U}\right).$$

Here, $\mathbf{U}$ is a learnable vector (60 × 1), σ denotes the sigmoid activation, and $(\cdot\|\cdot)$ represents vector concatenation. This gate augmentation strategy is conceptually related to residual learning [35] and helps improve the performance of deep learning models [36]. More details of the gate augmentation-based attention mechanism can be found in Lim et al. [37]. Finally, the ligand-level representation is obtained by summing all fragment-level outputs:

$$\mathbf{x}^{\mathrm{out}}_{\mathrm{L}} = \sum_{i \in \mathrm{fragments}} \mathbf{x}^{\mathrm{out}}_{\mathrm{L},i}.$$

The protein matrix $\mathbf{x}_{\mathrm{P}}$ (1,000 × 30) is processed by a one-dimensional convolution layer followed by a MaxPool layer, both configured with kernel size 3 and stride 3. Through parameter sharing, the convolution operation captures local sequence patterns across the protein representation [7]. This combination of convolution and pooling enables the model to identify informative higher-order amino acid features from the full sequence while reducing feature redundancy and lowering the computational burden for the subsequent attention module. The resulting protein representation is then passed to a gate augmentation-based attention layer, analogous to the ligand branch, to produce the final protein feature $(\mathbf{x}^{\mathrm{out}}_{\mathrm{P}} = \sum_{i \in \mathrm{higher-order\ amino\ acids}} \mathbf{x}^{\mathrm{out}}_{\mathrm{P},i})$.

The final ligand $(\mathbf{x}^{\mathrm{out}}_{\mathrm{L}})$ and protein features $(\mathbf{x}^{\mathrm{out}}_{\mathrm{P}})$ are combined through concatenation to form a unified representation of the protein–ligand complex. This joint feature is then fed into an FC layer (60 × 30) with a LogSoftMax activation, followed by an output layer (30 × 2) employing the same activation function. Empirical evaluation indicates that LogSoftMax contributes to improved classification performance. To mitigate overfitting, a dropout layer is inserted before the output stage [38]. Since LogSoftMax does not directly yield normalized probabilities, an additional SoftMax operation is applied after training to convert the model outputs into probabilities for high- and low-affinity classes. The hyperparameters of the deep learning model are determined through iterative trial-and-error procedures guided by empirical experience. The model is implemented based on the Pytorch library [39] and developed using the Python programming language [40].

4.2.1.4 Results and Discussion

The dataset of protein–ligand complexes is randomly split into training, validation, and test subsets with a proportion of 8:1:1. Specifically, the numbers of samples in these three subsets are 6,711, 841, and 841, respectively. Within the training set, 3,102 samples correspond to low binding affinity (label as "0"), while 3,609 samples correspond to high binding affinity (label as "1"), indicating a well-balanced dataset suitable for classification tasks. In this work, the cross-entropy loss (CEL) is adopted as the objective function, as it is widely applied in classification problems. Model parameters are optimized using the Adam algorithm [41], with the learning rate varying between 0.0005 and 0.00005. A decay strategy is applied, where the learning rate is reduced by a factor of 0.5 if the validation loss does not decrease during training. In binary classification tasks, a probability threshold of 50% is typically used to distinguish between positive and negative predictions. For instance, if a protein–ligand complex has a predicted probability of 70% for high binding affinity, it is classified as high affinity. The receiver operating characteristic (ROC) curve evaluates model performance across multiple threshold values by analyzing the corresponding true positive (TP) and false positive (FP) rates. The area under the ROC curve is widely used as a key metric to assess the classification performance of deep learning model. It represents the likelihood that a randomly selected positive sample will receive a higher prediction score than a randomly selected negative sample [42]. Thus, area under curve (AUC) can also serve as an indicator of the uncertainty in classification models. A value approaching 1 reflects improved accuracy and robustness in the model classification performance. Additional hyperparameters, such as the number of epochs, batch size, and dropout rate, are provided in Table A.1 in Appendix A of the supplementary material.

The performance of the developed deep learning model is summarized in Table 4.1 and Figure 4.2. As shown in Table 4.1, the CEL values for training, validation, and test set are 0.369, 0.442, and 0.401, while the corresponding AUC values are 0.914, 0.880, and 0.901, respectively. Figure 4.2a,b illustrate how the CEL and AUC metrics vary with the number of training epochs. These results suggest that the proposed deep learning model does not suffer from overfitting and demonstrates good generalization capability on unseen samples. It is worth noting that the incorporation of gate augmentation-based attention layers significantly improves training efficiency. The model achieves stable performance around the 40th training epoch, whereas more epochs are required to reach comparable performance when the attention mechanism is excluded.

Figure 4.2c presents a 2×2 confusion matrix for binary classification, which is used to assess the predictive performance of the developed model. In this matrix, the x-axis denotes the true labels ("0" for low affinity and "1" for high affinity), while the y-axis represents the corresponding predicted labels. The matrix consists of four categories: true negative

Table 4.1 The loss and metrics functions of the deep learning model.

Dataset	Loss function (CEL)	Metrics function (AUC)
Training	0.369	0.914
Validation	0.442	0.880
Test	0.401	0.901

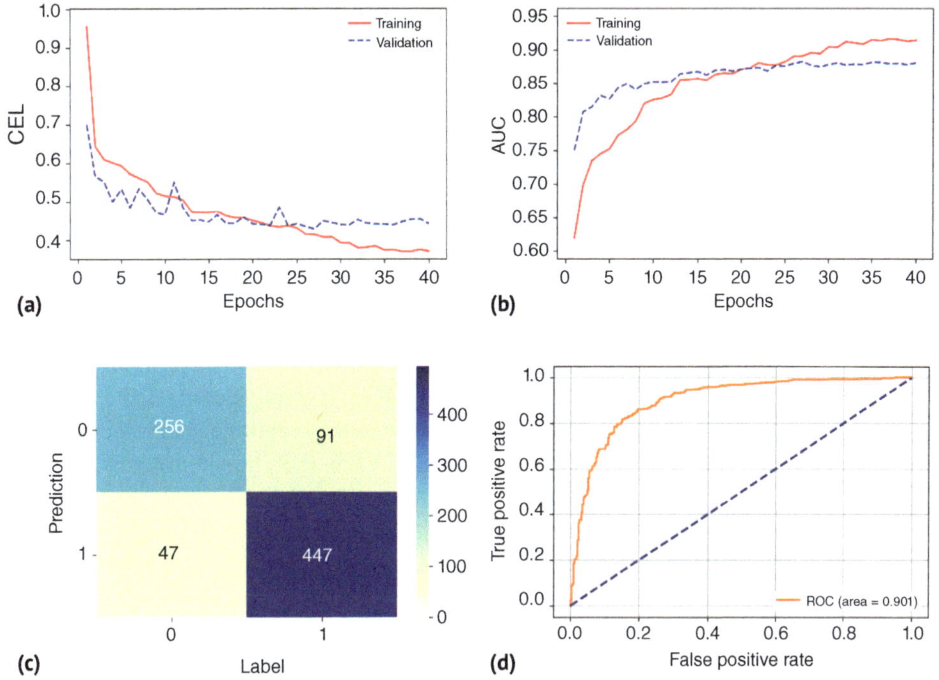

Figure 4.2 The training procedure, confusion matrix, and ROC curve of the developed deep learning model. (a) The CEL for the training and validation set. (b) The AUC for the training and validation set. (c) Confusion matrix for the test set. (d) ROC curve for the test set.

(TN), false negative (FN), TP, and FP. The test samples are distributed among these four categories based on their true and predicted labels. As illustrated in Figure 4.2c, the numbers of TN and TP cases are significantly higher than those of FN and FP, indicating that the proposed model performs effectively in distinguishing between high and low binding affinities. In addition, the ROC curve for the test set is shown in Figure 4.2d (orange solid line). The relatively large area under the curve (AUC = 0.901) further confirms the strong classification capability of the developed deep learning model.

It should be emphasized that the proposed deep learning model is relatively lightweight, containing only 15,634 trainable parameters, which is significantly fewer than those in many existing affinity-based models (e.g., approximately 1,000,000 parameters in K_{DEEP} [27]). This highlights its computational efficiency in both training and predictive generalization. A direct comparison with other deep learning models is not conducted in this work for two main reasons. First, many existing models incorporate IC_{50} data in their training sets, whereas such data are excluded here due to their inherent noise and strong dependence on experimental conditions. Second, the present model is formulated as a classification model, while most existing approaches are regression-based and aim at quantitative prediction of binding affinity, often with only moderate performance (e.g., Pearson correlation coefficients around 0.80).

4.2.2 Development of the Deep Learning Model for Predicting Dissociation Kinetics

4.2.2.1 Dissociation Kinetic Database

At present, publicly available datasets focusing specifically on protein–ligand kinetic properties remain limited in comparison to affinity-related data. To address this gap, a comprehensive dissociation kinetic database is constructed in this study by integrating data from four established sources, namely KIND [43], BindingDB [44], PDBbind-koff-2020 [24], and the work of Amangeldiuly et al. [23]. More details of these four databases and the preprocessed criteria of the dissociation kinetic data contributed by the above four datasets can be found in Appendix B of the supplementary material. After applying the aforementioned criteria, a curated dissociation kinetics dataset containing 2,773 unique protein–ligand pairs is obtained. In this dataset, ligands are encoded using SMILES strings (e.g., "CCC" represents propane), while proteins are described by FASTA sequences [45]. The pk_{off} values range from 0 to 5, covering the majority of experimentally reported values. Larger pk_{off} values correspond to longer residence times of inhibitors on their target proteins.

4.2.2.2 Input Representation

In the proposed dissociation kinetic model, protein–ligand complexes are represented using text-based formats, considering the significant computational cost associated with obtaining their three-dimensional structural information.

Consistent with the representation strategy described in Section 4.2.1.2, ligands are encoded using the Mol2vec descriptor. Accordingly, each ligand is transformed into a two-dimensional matrix of size 74 × 300, where rows correspond to Morgan fragments and columns denote their associated feature vectors. As the number of fragments varies across ligands, zero-padding is applied to ensure a fixed matrix size. Proteins are characterized using higher-order amino acid sequences combined with embedded feature vectors. Specifically, each higher-order unit is constructed from three consecutive amino acids derived from the FASTA sequence. In total, 6,791 such higher-order sequences are obtained from the dissociation kinetics dataset. These sequences are then mapped into feature vectors using the embedding method (nn.Embedding) implemented in PyTorch [39], with each vector having a dimension of 30. The overall protein representation is obtained by aggregating the feature vectors of all higher-order amino acid units. Accordingly, each protein is expressed as a two-dimensional matrix of size 500 × 30, where each row corresponds to a higher-order amino acid and each column denotes its feature representation. To handle proteins with fewer than 500 units, zero-padding is similarly applied to maintain a consistent input size.

4.2.2.3 Model Architecture

The MoE algorithm is used to develop the deep learning model for predicting dissociation kinetics. The overall framework of the dissociation kinetics model is illustrated in Figure 4.3.

The ligand matrix $\mathbf{x}_\mathrm{L}$ (74 × 300) is first reduced in dimensionality by passing through an FC layer $\mathbf{V}$ (300 × 30). This transformation converts the ligand representation into a 30-dimensional feature space, which helps reduce model redundancy and lowers computational cost during training. The resulting matrix $\mathbf{x}'_\mathrm{L}$ (74 × 30) is then fed into an attention layer, where it is further transformed via a learnable weight matrix $\mathbf{W}$ (30 × 30) to generate a refined ligand matrix $\mathbf{x}''_\mathrm{L}$ with enhanced feature representation. Subsequently, the attention coefficient e_{ij} is computed according to the equation $e_{ij} = \mathbf{x}''_{\mathrm{L},i}\mathbf{x}''^{\mathrm{T}}_{\mathrm{L},j} + \mathbf{x}''^{\mathrm{T}}_{\mathrm{L},i}\mathbf{x}''_{\mathrm{L},j}$, where e_{ij} quantifies the contribution of the jth fragment feature to the ith fragment feature. The symmetry of e_{ij} $\left(e_{ij} = e_{ji}\right)$ is ensured by combining the corresponding terms. The obtained e_{ij} values are then passed through a SoftMax function to produce normalized attention coefficients a_{ij}, defined as $a_{ij} = \exp\left(e_{ij}\right)/\sum_j \exp(e_{ij})$. Using a_{ij}, each fragment feature is updated as a weighted combination of its neighboring fragment features $\left(\mathbf{x}^{\mathrm{out}}_{\mathrm{L},i} = \sum_j a_{ij}\mathbf{x}''_{\mathrm{L},j}\right)$.

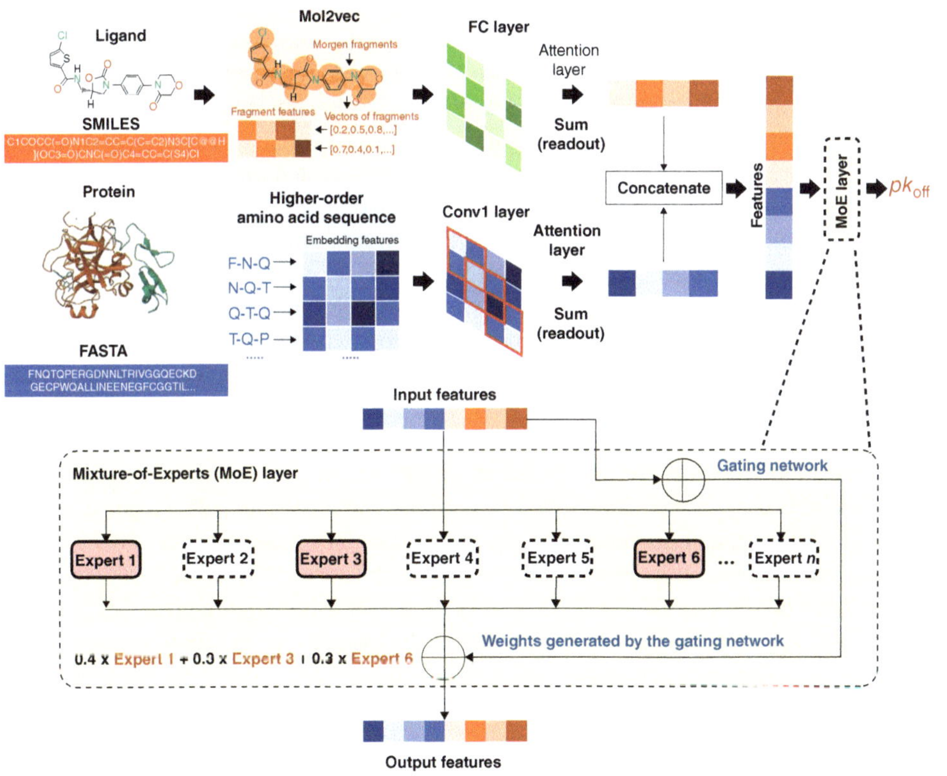

Figure 4.3 The architecture of the MoE-based generic dissociation kinetic model.

A Gelu activation function is subsequently applied to $\mathbf{x}_{L,i}^{out}$. Finally, all ligand fragment features are aggregated into a single vector to represent the overall ligand feature $\left(\mathbf{x}_{L}^{out} = \sum_{i \in \text{fragments}} \mathbf{x}_{L,i}^{out}\right)$.

The protein matrix $\mathbf{x}_P$ (500×30) is initially processed by a one-dimensional convolutional layer (Conv1) with 30 input channels, 30 output channels, a kernel size of 3, and a stride of 3, followed by a MaxPool layer with a kernel size of 3 and stride of 3. The convolutional kernels employ a parameter-sharing mechanism [7] to capture local structural patterns within the protein matrix and extract informative higher-order amino acid features, particularly those associated with the binding pocket of the protein–ligand complex. This step also contributes to reducing feature redundancy and decreasing the computational burden of the subsequent attention layer. The processed protein representation is then fed into an attention layer similar to that used in the ligand branch. Finally, the overall protein feature $\mathbf{x}_P^{out}$ is by aggregating all higher-order amino acid features $\mathbf{x}_P^{out} = \sum_{i \in \text{higher-order amino acids}} \mathbf{x}_{P,i}^{out}$.

Subsequently, the output ligand feature $\left(\mathbf{x}_L^{out}\right)$ and the output protein feature $\left(\mathbf{x}_P^{out}\right)$ are concatenated to form a unified representation of the protein–ligand complex, which is then fed into a MoE [46, 47] layer. The MoE layer is composed of two main components: an expert layer and a gating layer. The expert layer includes n individual experts, each consisting of an FC layer (60×30) followed by a Gelu activation function and an output layer (FC layer) (30×1) with an additional Gelu activation. A dropout layer is further applied before the output layer to alleviate overfitting during training [38]. The gating layer is constructed with two FC layers (60×30, 30×30) with two Gelu activations and a FC layer ($30 \times n$) with a SoftMax activation. This gating layer effectively allocates weights to assess the proficiency of each expert in handling the input features.

The hyperparameters of the deep learning model are selected through a trial-and-error approach guided by empirical knowledge. The dissociation kinetic model is implemented using the PyTorch library [39] within the Python programming environment [40]. For comparison purposes, a generic dissociation kinetic model without the MoE layer is also developed as a baseline. Apart from the absence of the MoE component, all other technical configurations remain consistent with those of the proposed MoE-based model. A detailed description of the baseline model is provided in Appendix C of the supplementary material.

4.2.2.4 Results and Discussion

The dissociation kinetic model adopts the mean square error (MSE) as the loss function, which is widely used in regression tasks. To optimize the model, an Adam algorithm [41] is applied with learning rates ranging from 0.0005 to 0.00005. A decay factor of 0.5 is introduced to adjust the learning rate when the validation loss does not decrease during training epochs. The Pearson correlation coefficient (r) and root mean square error (RMSE) are employed as key metrics to assess the predictive performance of the dissociation kinetic model for pk_{off}. Improved model performance is indicated by r values approaching 1 and RMSE values approaching 0. To reduce the impact of occasional training variability, the dissociation kinetics dataset is randomly partitioned 10 times into training (2,218 samples), validation (277 samples), and test sets (278 samples) by a ratio of 8 : 1 : 1. For each partition, the model is further trained through 10 independent runs. Additional hyperparameters, such as the number of epochs, batch size, and dropout rate, are provided in Table D.1 in

Appendix D of the supplementary material. It should be emphasized that the proposed dissociation kinetics model has a relatively low number of trainable parameters (46,774), which is significantly fewer than those of many existing deep learning models (e.g., approximately 1,000,000 in K_{DEEP} [27]). This highlights its efficiency in both training and prediction. The main reason why the MoE model has fewer parameters than the binding affinity model is due to the architecture advantages of our baseline model.

Based on 100 independent training runs, the variation of the loss function (MSE) and the evaluation metric (r) over training epochs for the MoE-based dissociation kinetic model are shown in Figure E.1 in Appendix E of the supplementary material. The training and validation results indicate that the developed MoE-based dissociation kinetic model has no overfitting issue. It is also observed that incorporating the attention mechanism accelerates the learning process of the model. During training, convergence is achieved at approximately the 40th epoch, whereas removing the attention layers requires about 70 epochs to reach comparable performance.

The average MSE and r values of our MoE-based dissociation kinetic model for training, validation, and test sets are 0.200 ± 0.012, 0.300 ± 0.045, 0.288 ± 0.028 and 0.890 ± 0.007, 0.821 ± 0.025, 0.835 ± 0.017, respectively. The predictive outcomes on the test set demonstrate that the established dissociation kinetic model possesses an acceptable predictive capacity for assessing the dissociation kinetics of protein–ligand complexes and is able to perform satisfactory predictive extrapolations on novel samples. Figure 4.4 is plotted to show the prediction results of 100 randomly trained MoE-based dissociation kinetics models across the overall dataset. The prediction errors are evaluated with determination

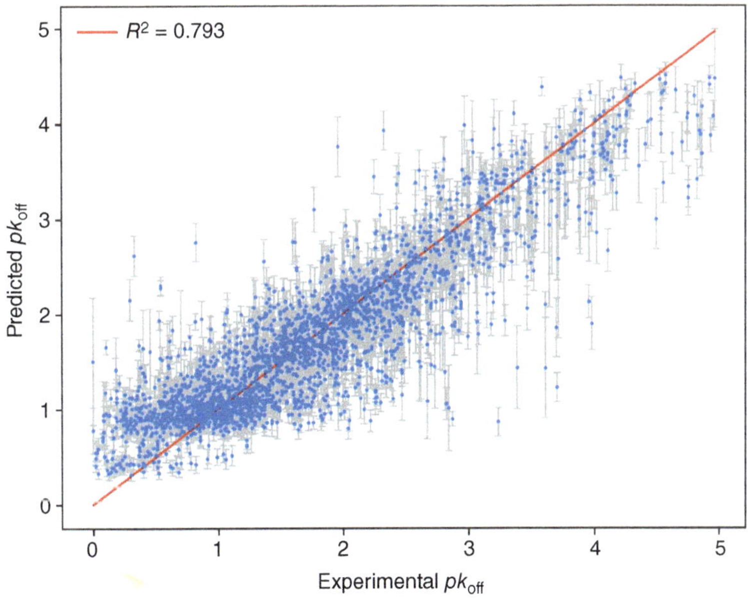

Figure 4.4 The prediction results of 100 randomly trained MoE-based dissociation kinetics models across the overall dataset.

coefficient (R^2), mean absolute error (MAE), and mean absolute percentage error (MAPE) by comparing the predicted pk_{off} (represented as mean ± standard deviation based on 100 predictions) and the experimental pk_{off}. The R^2, MAE, and MAPE for the entire dataset are 0.793, 0.324, and 38.4%, respectively. The absolute prediction error between the predicted pk_{off} (represented as mean ± standard deviation based on 100 predictions) and the measured pk_{off} ranges from 0 to 2.473. Approximately 96.5% of the samples exhibit an absolute prediction error of less than or equal to 1. Therefore, the prediction errors are not large when utilizing these 100 randomly trained MoE models. More analyses for the prediction errors of our MoE-based dissociation kinetics model are provided in Appendix E of the supplementary material.

An ablation study is carried out to highlight the superiority of incorporating the MoE architecture into our dissociation kinetic model. The model performances of the dissociation kinetic models with/without MoE architecture are compared and presented in Table 4.2 and Figure 4.5. As shown in Table 4.2, four evaluation criteria (MSE, r, RMSE, R^2)

Table 4.2 Performance comparisons between the dissociation kinetic models with/without MoE architecture.

Model	Dataset	MSE	r	RMSE	R^2
Model with MoE	Training	0.200 ± 0.012	0.890 ± 0.007	0.447 ± 0.014	0.790 ± 0.013
	Validation	0.300 ± 0.045	0.821 ± 0.025	0.546 ± 0.040	0.667 ± 0.045
	Test	**0.288 ± 0.028**	**0.835 ± 0.017**	**0.536 ± 0.026**	**0.693 ± 0.031**
Model without MoE (Baseline)	Training	0.245 ± 0.018	0.864 ± 0.010	0.495 ± 0.019	0.743 ± 0.020
	Validation	0.320 ± 0.047	0.808 ± 0.028	0.564 ± 0.041	0.646 ± 0.047
	Test	0.307 ± 0.034	0.825 ± 0.022	0.553 ± 0.030	0.673 ± 0.036

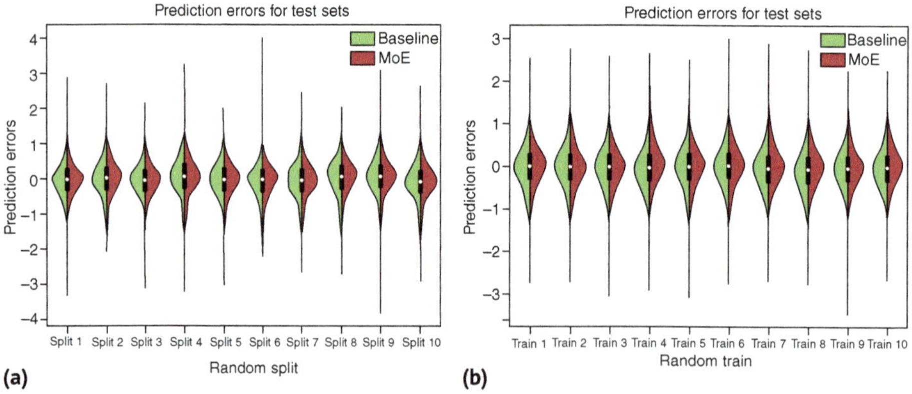

Figure 4.5 The prediction errors (predicted pk_{off} − experimental pk_{off}) for test sets using violin plots. (a) The prediction errors accessed using 10 random splits, with each split evaluated by 10 randomly trained MoE and baseline models. (b) The prediction errors for 10 randomly trained MoE and baseline models in the first random data split.

of the dissociation kinetic model with MoE architecture for the training, validation, and test sets all outperform than those of the dissociation kinetic model without MoE architecture. Figure 4.5 depicts the violin plots of the prediction errors (predicted pk_{off} minus experimental pk_{off}) for test sets based on 10 random data splits (Figure 4.5a) and 10 random model trainings in the first random data split (Figure 4.5b). The remaining nine random data splits are given in Figures F.1–F.9 in Appendix F of the supplementary material. From Figure 4.5 it is found that most of the prediction errors of both dissociation kinetic models with/without MoE architecture are located within ±1 (log unit). The prediction errors of the dissociation kinetic model with MoE architecture are more distributed near 0 than those of the dissociation kinetic model without MoE architecture. The above results indicate that the use of the MoE architecture is beneficial to the accurate predictions of k_{off}. Additionally, if the higher-order amino acid strings are replaced with ordinary amino acid strings and all other technical details are kept the same to the baseline model, the average r value of the ordinary amino acid string-based dissociation kinetic model for test sets is 0.821 ± 0.022, which is slightly lower than that of the baseline model, where $r = 0.825 \pm 0.022$. This result indicates that the use of higher-order amino acid strings yields similar prediction results for the dataset studied in this work compared to ordinary amino acid strings.

To evaluate the capability of the feature extraction layer (the layer before the MoE layer) in extracting protein–ligand features, as well as the role of the MoE layer to improving the predictive accuracy of the dissociation kinetics model, a chemical space is constructed based on the extracted features using the t-distributed stochastic neighbor embedding (t-SNE) cluster technique, as shown in Figure 4.6. Different colors in Figure 4.6a represent the top 10 proteins, sorted in descending order by sample size. Figure 4.6a reveals that nearly all samples are well-clustered according to their protein types, indicating that the feature extraction layer successfully differentiates protein–ligand features through a completely data-driven manner. In Figure 4.6b, the same chemical space is relabeled with expert categories, determined by the maximum weights of the gating network in the MoE layer for each sample. On the one hand, the prediction outcomes for certain proteins

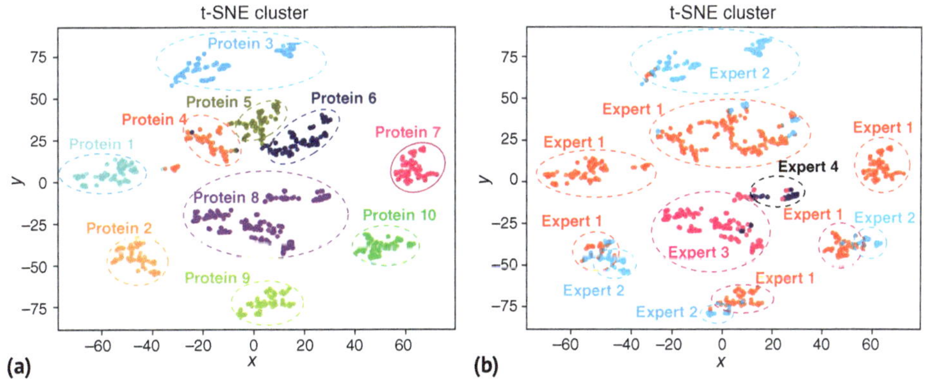

Figure 4.6 The chemical space created by the protein–ligand features and the t-SNE technique and labeled by (a) protein and (b) expert types.

(e.g., proteins 1 and 7) are predominantly influenced by a specific expert (e.g., expert 1). Additionally, proteins with similar protein–ligand features (e.g., proteins 4–6) are associated with the same expert (e.g., expert 1). These observations confirm that some experts are particularly adept at treating certain proteins, especially those samples with similar protein–ligand features (e.g., proteins 4–6). On the other hand, the prediction results for some proteins (e.g., protein 8) involve multiple experts (e.g., experts 3 and 4), highlighting the superiority of the collaborative mechanism of the MoE system in addressing complicated samples of a specific protein that exhibit distinct protein–ligand features. The above findings demonstrate the effectiveness of the gating network in allocating suitable experts to predict k_{off} based on the specific protein–ligand features they are most adept at treating. This capacity may help explain why the MoE-based dissociation kinetic model outperforms the model without the MoE layer in predicting k_{off}.

Several machine learning-based generic dissociation kinetic models have been reported in the literature [23–25]. For example, Amangeldiuly et al. [23] developed a generic random forest (RF) model for predicting pk_{off} values with $r = 0.78$ and RMSE = 0.82 on the validation set, and $r = 0.75$ and RMSE = 1.10 on the test set. Liu et al. [24] developed an RF-based model for pk_{off} estimation, where the optimal model yielded r and RMSE values of 0.968 and 0.474 for the training set, 0.706 and 0.986 for the validation set, and 0.501 and 0.891 for the test set. Here, 100 randomly trained MoE models are used to predict pk_{off} (represented as mean ± standard deviation based on these 100 predictions) for the samples from Amangeldiuly et al. [23] and Liu et al. [24]. Note that these two datasets are cleaned as some samples fail to be predicted by our MoE model due to some reasons. Details on the cleaning processes for these two datasets are presented in Appendix G of the supplementary material. Finally, the r values for the prediction results on samples from Amangeldiuly et al. [23] and Liu et al. [24] are 0.870 and 0.728, respectively, indicating the robustness and high prediction accuracy of our MoE model in estimating k_{off}. This result likely stems from the well-cleaned and extensive dataset compiled in this work. The overlap between our dataset and those of Amangeldiuly et al. [23] and Liu et al. [24] is shown in Figure G.1 in Appendix G of the supplementary material. Moreover, our model demonstrates increased computational efficiency due to its utilization of text descriptors, providing it with an advantage over the models that rely on three-dimensional structures of protein–ligand complexes [23–25]. The deep learning-based generic dissociation kinetic model developed in this work can be regarded as a benchmark for model comparisons with other future studies.

4.3 Integration of the MINLP-based Drug Design Model with Deep Learning Models

In this section, an optimization-based framework for *de novo* drug design is constructed by integrating the deep learning models developed in Section 4.2 with the MINLP-based drug design model. The overall procedure for constructing this framework is illustrated in Figure 4.7 and is summarized as follows.

Referring to Figure 4.7 (a) A drug database is constructed to enable the generations of drug scaffolds. (b) Drug scaffolds are produced using the Bemis–Murcko algorithm, serving as inputs for the subsequent MINLP-based drug design model. (c) The *de novo* drug

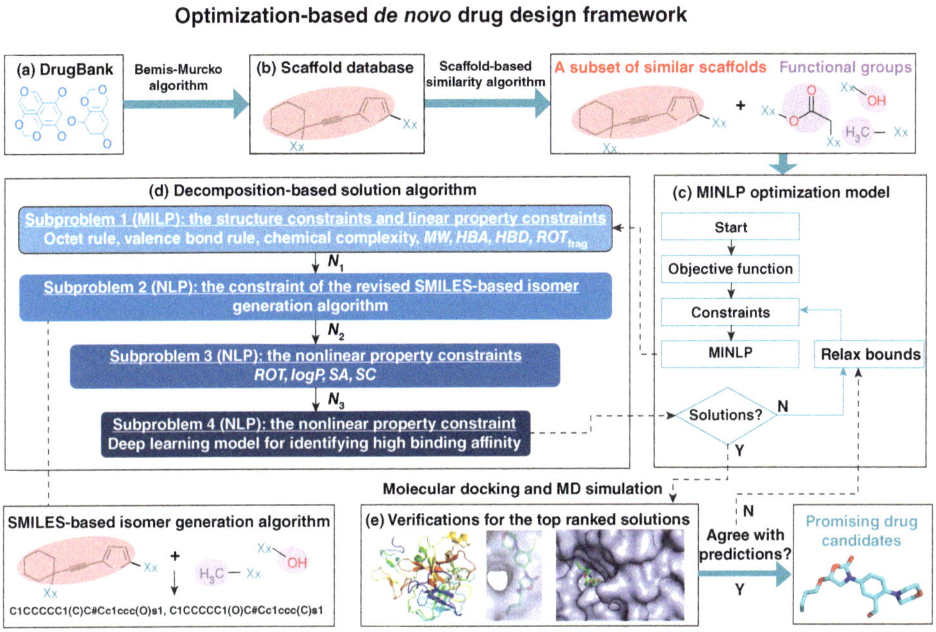

Figure 4.7 The procedure of developing the optimization-based *de novo* drug design framework.

design problem is formulated as an MINLP model that includes an objective function (e.g., maximizing pk_{off} or the probability of strong protein–ligand binding), along with drug structure constraints, property constraints, and other constraints. To limit the search space and reduce computational cost in candidate design, a scaffold-based similarity algorithm is applied beforehand to select a subset of scaffolds similar to those of known drugs prior to constructing the MINLP model. (d) As the MINLP model involves a large number of nonlinear constraints, a decomposition-based solution algorithm [48] is adopted to solve it. If no optimal drug candidate satisfies all the constraints, the procedure returns to step (c) to relax the constraint bounds. (e) A verification step is conducted to assess the effectiveness of designed drug candidates with respect to the binding affinities and binding modes of protein–ligand complexes, using physics-based approaches, namely molecular docking and MD simulations, respectively. Moreover, the dissociation kinetic property can be rigorously verified by the τ-RAMD method.

4.3.1 Drug Database

A total of 7,746 small-molecule drugs along with their Chemical Abstracts Service (CAS) numbers are initially collected from the DrugBank database V5.0 [49] (https://go.drugbank.com/). Subsequently, the Compound Identifier (CID) numbers (unique identifiers for chemicals in the PubChem database, https://pubchem.ncbi.nlm.nih.gov/search/) corresponding to these drugs are retrieved from PubChem using a web crawler based on their CAS numbers. After removing entries without CID numbers, 7,474 drugs remain for further screening. The Lipinski's "Rule of Five" (molecular weight (MW) $\leq$ 500, hydrogen bond donors (HBD) ≤ 5, hydrogen bond acceptors (HBA) ≤ 10, octanol-water partition coefficient (log*P*) ≤ 5, and number of rotatable bonds (ROT) ≤ 10) is then applied to identify drugs with favorable pharmacokinetic properties [50, 51]. The "Rule of Five" descriptors and isomeric SMILES of all 7,474 drugs are subsequently obtained via the PubChem API according to their CID numbers. Finally, a drug database is constructed, comprising 4,781 small-molecule drugs with their CAS numbers, CID numbers, isomeric SMILES, and corresponding "Rule of Five" properties.

4.3.2 Generation of Scaffolds

In medicinal chemistry, drug scaffolds have long played a crucial role. Although multiple definitions of scaffolds exist, the most commonly accepted one was proposed by Bemis and Murcko [52]. According to their approach, all side-chain atoms are removed from a drug molecule to obtain its scaffold, which consists of ring systems and the linkers connecting them. For instance, Figure 4.8 illustrates a drug molecule along with its scaffold (ring systems and linkers) and side chains. It has been reported that many scaffolds exhibit privileged features in their ability to modulate the activities of target proteins [53]. From this perspective, drug scaffolds serve as an effective starting point for drug candidate design. In addition, employing scaffolds to replace traditional ring functional groups in the CAMD method enables the development of more structurally rational drug candidates. Consequently, this section constructs a database comprising 2,898 Bemis–Murcko scaffolds for subsequent drug design. These scaffolds are derived from 4,781 drugs in our database using the Bemis–Murcko algorithm implemented in RDKit.

4.3.3 MINLP Optimization Model

In this section, an optimization-driven mathematical programming method is adopted for the intelligent and high-throughput design of promising drug candidates. The *de novo* drug design task is represented as an MINLP model comprising an objective function together

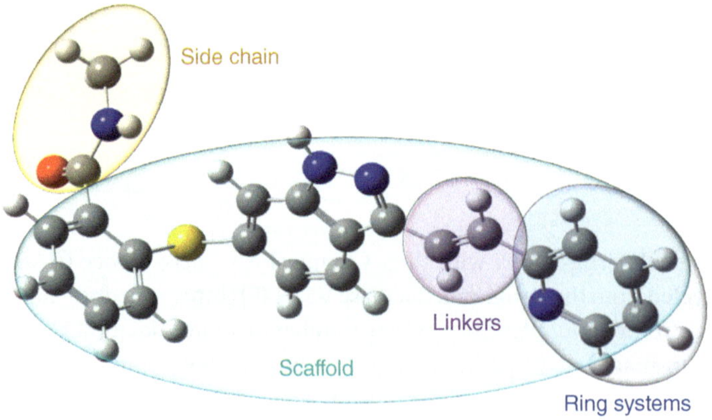

Figure 4.8 The diagrammatic sketch of a drug structure consisting of its scaffold (ring systems and linkers) and side chain.

with constraints on drug structures and physicochemical properties. Solving this MINLP formulation enables the selection of an optimal drug candidate that maximizes the probability of strong binding affinity from a feasible solution space, where candidates are constructed through combinations of scaffolds and functional groups subject to the imposed constraints. It is worth noting that each MINLP model is specifically formulated for an individual protein to generate its corresponding promising drug candidates.

However, incorporating all 2,898 scaffolds into a single MINLP-based drug design model would lead to an excessively large model size, thereby reducing computational efficiency. Based on the assumption that drug candidates sharing similar scaffolds may exert comparable effects on target protein activities [54], a scaffold-based similarity algorithm is introduced to extract a subset of similar scaffolds (G_1) from the full set of 2,898, with reference to the scaffold of a given drug identified in the case study. This strategy allows computational efforts to be concentrated on more promising scaffolds, effectively narrowing the search space and accelerating the drug design process. Within this algorithm, a total of 24 approaches are constructed by combining six similarity metrics (Tanimoto, Dice, Cosine, Sokal, Russel, and Kulczynski) with four types of molecular representations (topological fingerprints, MACCS keys, Extended-Connectivity Fingerprint (ECFP), and Functional-Class Fingerprint (FCFP)) [30], each of which is used to identify the three most similar scaffolds. After removing duplicate entries, a refined set of candidate scaffolds is obtained according to the specific problem, with the final number for each case study reported in Section 4.4. In addition, a set of commonly used functional groups (G_2) is selected for incorporation into the MINLP model. Consequently, the MINLP model can be formulated as follows:

$$\text{Objective function}: \max_{n_i(i \in G_1, G_2)} F_{\text{obj}}(n_i) = \text{Prob}_{\text{bind}} \text{ or } pk_{\text{off}}$$

Subject to:

- Deep learning constraints: The general formulation in Eq. (4.1) characterizes the deep learning model used to evaluate protein–ligand complexes with respect to high binding affinity or slow dissociation kinetics.

$$\text{Prob}_{\text{bind}} \text{ or } pk_{\text{off}} = f_{\text{deep}}(s) \tag{4.1}$$

- Drug structure constraints: The formulation in Eq. (4.2) describes structural restrictions, including the octet rule (m_2), and chemical complexity (m_3) [55, 56], which ensure the generation of structurally valid molecules through combinations of scaffolds and functional groups.

$$M_m^{\text{L}} \le f_m(n_i) \le M_m^{\text{U}}, n_i \in \mathbb{N}^+, m \in m_1, m_2, m_3 \tag{4.2}$$

- Drug property constraints: The formulations in Eqs. (4.3, 4.4) define property-related constraints, including the "Rule of Five" descriptors (MW, HBA, HBD, log$P$, ROT), as well as the synthetic accessibility (SA) score [57] and/or the synthetic complexity (SC) score [58].

$$P_k^{\text{L}} \le p_k(n_i) \le P_k^{\text{U}}, k \in \text{MW}, \text{HBA}, \text{HBD}, \text{ROT}_{\text{frag}} \tag{4.3}$$

$$P_k^{\text{L}} \le p_k(s) \le P_k^{\text{U}}, k \in \text{ROT}, \log P, \text{SA}, \text{SC} \tag{4.4}$$

- Other constraint: The formulation in Eq. (4.5) describes a revised SMILES-based isomer generation algorithm, which enables the automatic transformation of fragment sets (including scaffolds and functional groups) of drug candidates into their corresponding SMILES representations.

$$f_{\text{SMILES}}(n_i, s) = 0 \tag{4.5}$$

In the aforementioned general equations, F_{obj} denotes the objective function, $\text{Prob}_{\text{bind}}$ represents the probability of high binding affinity for the protein–ligand complex. The term pk_{off} corresponds to the logarithmic value of k_{off}. The index i refers to a fragment (including scaffolds and functional groups) involved in a drug candidate, and n_i indicates the number of fragments. The variable s denotes the SMILES representation of the molecule. The parameter m specifies the type of the structure constraint, with M_m^{U} and M_m^{L} representing its upper and lower bounds, respectively. Furthermore, p denotes a drug property, k indicates the property type, P_k^{U} and P_k^{L} define the corresponding upper and lower limits of p_k.

Detailed formulations of the structural constraints are available in previous studies [55, 56]. Compared with conventional MINLP-based solvent design models [56, 59], the primary distinction of our MINLP-based drug design model lies in the scaffold constraint $\left(1 \leq \sum_{i \in G_1} n_i \leq 1\right)$, which ensures that each drug candidate selects only a single scaffold from the predefined scaffold set.

Additional distinctions include the incorporation of deep learning models, drug property constraints, and an improved SMILES-based isomer generation algorithm. The deep learning models are described in detail in Section 4.2. The "Rule of Five" parameters (MW, HBA, HBD, log$P$, ROT) play a crucial role in ensuring that the generated molecules exhibit favorable drug-like characteristics and pharmacokinetic properties [51]. These parameters are evaluated using the quantitative estimate of drug-likeness (QED) [60] implemented in RDKit. Furthermore, the SA (ranging from 1 to 10) and SC (ranging from 1 to 5) are also considered to ensure that the designed drug candidates are amenable to synthesis. Lower values of SA and SC indicate greater ease of synthesis. More comprehensive explanations of these properties can be found in the works of Ertl et al. [55] and Coley et al. [58] respectively.

It should be noted that SMILES strings of drug candidates are required for predicting properties such as $\text{Prob}_{\text{bind}}$, ROT, log$P$, SA, and SC. In contrast, structural constraints rely on fragment-level information, including scaffolds and functional groups. Therefore, it is essential to establish a "bridge" linking SMILES representations with fragment descriptions; otherwise, the MINLP model cannot be effectively solved. To address this issue, a revised SMILES-based isomer generation algorithm is proposed to automatically transform fragment sets of drug candidates into their corresponding SMILES strings. This approach also enables the identification of isomers among candidates sharing identical fragment sets. Further details of the revised SMILES-based isomer generation algorithm can be found in Appendix H of the supplementary material.

4.3.4 Decomposition-based Solution Algorithm

In general, an MINLP model can be directly solved using the BARON solver [61] within the GAMS platform (https://www.gams.com/) when the nonlinear equations involved are not excessively complex. However, in our case, the SMILES-based property prediction models (e.g., deep learning models) introduce highly complex nonlinear relationships into the MINLP formulation, making it extremely challenging to obtain the optimal drug candidate through a direct solution approach. To overcome this difficulty, a decomposition-based solution algorithm [48] is adopted to handle the proposed MINLP optimization model. Specifically, the original MINLP problem is decomposed into a sequence consisting of one mixed-integer linear programming (MILP) subproblem and three nonlinear programming (NLP) subproblems. Subproblem 1 (MILP): Structural constraints, including the octet rule, valence bond rule, and chemical complexity, together with linear property constraints on MW, HBA, HBD, and $ROT_{frag,}$ are first imposed to generate a set of feasible solutions. Specifically, a certain number (N_1) of drug candidates, represented by fragment combinations, are obtained using the BARON solver within GAMS. It should be noted that the parameter N_1^{max} must be specified in GAMS to limit the maximum number of feasible solutions for the MILP problem.

Subproblem 2 (NLP): The revised SMILES-based isomer generation algorithm is then applied as a constraint to transform the N_1 fragment sets into corresponding SMILES strings, resulting in N_2 $(N_2 \geq N_1)$ candidate molecules. Subproblem 3 (NLP): Nonlinear property constraints, including ROT, log*P*, SA, and SC, are subsequently evaluated using nonlinear prediction models. These models compute the relevant properties for the N_2 SMILES strings, and candidates that do not meet the required criteria are filtered out. The remaining N_3 $(N_3 \leq N_2)$ SMILES strings are retained for further assessment. Subproblem 4 (NLP): Deep learning constraints are incorporated by employing nonlinear deep learning models to compute the $Prob_{bind}$ or pk_{off} (objective function) for N_3 protein–ligand complexes. Based on these computations, the SMILES strings of drug candidates are ranked according to their objective function values. A subset of the top-ranked candidates is subsequently validated using additional physics-based approaches, such as molecular docking and MD simulations.

Although the decomposition-based algorithm proceeds in a stepwise manner, its solutions are equivalent to those obtained through simultaneous optimization, provided that the parameter N_1^{max} is sufficiently large in GAMS and that the decomposed property constraints are independent. Notably, the mathematical programming approach used in Subproblem 1 is more efficient than a full enumeration strategy. Case study 1 in Section 4.4 demonstrates that the size of Subproblem 1 (MILP model) remains manageable when applying the scaffold-based similarity algorithm (see Appendix I of the supplementary material). The solution times for the MILP model are reported in the subsequent case studies. Furthermore, the SMILES-based property prediction models (log*P*, ROT, SA, SC, $Prob_{bind}$) can compute properties in milliseconds, making the computational cost of predicting these nonlinear properties reasonable even for large numbers of candidate SMILES strings. Since the NLP models are essentially mathematical functions, they can predict properties directly from MILP outputs, eliminating the need to solve the NLP models in GAMS. Overall, the decomposition-based solution algorithm provides an effective and efficient approach to handle the computationally intensive *de novo* drug design problem.

4.3.5 Verification of the Design Results

The final stage of our drug design framework involves validating the optimal or top-ranked candidates obtained from the MINLP model. In this step, two physics-based approaches—molecular docking and MD simulations—are employed to assess the effectiveness of the designed molecules in terms of their binding affinities and binding modes, respectively. Molecular docking utilizes the three-dimensional structural information of small ligands and protein macromolecules to identify binding sites and evaluate the binding affinities of stable protein–ligand conformations. This evaluation is based on principles such as energy matching, geometric alignment, and chemical environment complementarity. In this study, Auto dock 4.2.6 [62] is employed for molecular docking analysis. Detailed procedures for molecular docking can be found in Appendix J of the supplementary material.

Following docking, MD simulations are performed to extensively sample protein–ligand conformations and analyze binding modes. These simulations are carried out using GROMACS [63] with the 54a7 force field. More detailed descriptions of the MD simulation workflow are provided in Appendix K of the supplementary material. Regarding the dissociation kinetic property (pk_{off}), τ-RAMD is an enhanced sampling technique used to facilitate ligand unbinding in MD simulations of bound complexes [13]. This method involves performing a series of standard MD and RAMD simulations to compute τ. During the simulation, an additional force of constant magnitude and randomized orientation is applied to the ligand. At predefined time intervals, the displacement of the ligand mass center from its initial position is measured. If the displacement is below a threshold, a new random force direction is assigned for the next interval; otherwise, the same direction is maintained. This iterative process continues until the ligand moves beyond a predefined distance from its starting position, indicating dissociation from the macromolecule [64]. The simulation time can be modulated by adjusting the strength of the applied force, allowing ligand unbinding to be observed within relatively short simulations, usually lasting a few nanoseconds [65]. Additionally, using a consistent force magnitude for all ligands ensures that the perturbation effect is not influenced by ligand mass. As a result, the τ values for different compounds can be directly compared, offering valuable insights into their dissociation kinetics [64].

The uncertainty in the computed τ depends on the accuracy of sampling both the transition and bound states. To address this, multiple RAMD dissociation trajectories are employed to probe the transition state. Specifically, six distinct bound states (starting replicas) for a ligand are generated via MD simulations using independent trajectories initiated with different velocities. Each starting replica is then used to produce 20 RAMD dissociation trajectories, resulting in a total of 120 trajectories for calculating six dissociation times, which are subsequently used to determine relative τ values for the six replicas [64]. In the τ-RAMD, a force of 15.5 kcal/mol·Å is applied to accelerate dissociation. Importantly, varying the force magnitude does not significantly affect outliers or simulation accuracy. The reliability of the computed τ primarily depends on the quality of the MD simulations rather than the random force magnitude. Additional details on the MD and RAMD simulations are provided in Appendix L of the supplementary material.

4.4 Case Studies

In this section, we present two case studies—one on anticoagulant drug candidates and the other on HSP90 inhibitor candidates—to illustrate the application of the optimization-based *de novo* drug design framework.

4.4.1 Anticoagulant Drug Candidate Design

Factor Xa, an activated serine protease, plays a key role in the blood coagulation cascade by catalyzing the conversion of prothrombin to thrombin via the prothrombinase complex [66]. Thrombin performs multiple thrombotic functions, including converting fibrinogen to fibrin, activating platelets, and providing feedback activation of other coagulation factors, which amplifies its own production. Inhibiting Factor Xa reduces thrombin amplification, producing antithrombotic effects by decreasing thrombin-mediated activation of both coagulation and platelets while maintaining baseline thrombin levels sufficient for primary hemostasis, providing a favorable safety margin [67]. Consequently, Factor Xa represents a highly promising target for anticoagulant therapy and has garnered significant attention in recent years [68]. Small-molecule direct Factor Xa inhibitors offer the advantages of oral administration without requiring routine coagulation monitoring. Rivaroxaban, a potent and selective inhibitor with strong in vivo activity and good oral bioavailability, has been widely applied for preventing and treating thromboembolic disorders [68]. This case study aims to design potential alternatives to rivaroxaban as effective Factor Xa inhibitors, optimizing their binding affinity, compliance with the "Rule of Five," and SA using the proposed framework. The framework uses the scaffold database and the scaffold-based similarity algorithm to identify a subset of scaffolds (G_1) similar to rivaroxaban's scaffold for the MINLP model. Each combination of similarity method and molecular representation (24 in total) selects the top 3 matches, producing 72 scaffolds in total, with 14 duplicates removed. Additionally, a subset of functional groups (G_2) is chosen for the MINLP model based on prior work [69]. The G_1 and G_2 subsets are listed in Table M.1 of Appendix M of the supplementary material. This case study aims to maximize the probability of high binding affinity, with structural and property constraint bounds provided in Table 4.3.

Next, the MINLP model is solved using the decomposition-based strategy. In Subproblem 1, the mathematical programming approach generates $N_1 = 17{,}659$ feasible solutions (represented as fragment sets) under the structure and linear property constraints, taking 116 seconds on a desktop computer (Intel(R) Core (TM) i7-10700F CPU @ 2.90 GHz, 24.0 GB RAM). In Subproblems 2 and 3, the revised SMILES-based isomer generation algorithm and nonlinear property constraints are applied to produce $N_2 = 159{,}170$ SMILES strings from the N_1 fragment sets, filtering out unsatisfactory candidates (2,161 seconds). The remaining $N_3 = 42{,}932$ SMILES strings are selected for further analyses. These 42,932 designed candidates are first queried in the PubChem database via their SMILES strings, revealing that 2,261 (5.25%) already exist, demonstrating that the MINLP-based framework can identify known candidates while also generating novel molecules. Next, the chemical space is visualized (Figure 4.9) to assess the structural diversity of the 42,932 candidates

Table 4.3 List of structure and property constraints with upper and/or lower bounds.

Constraint	Upper and/or lower bound
Number of total scaffolds	$1 \le \sum_{i \in G_1} n_i \le 1$
Number of total fragments (scaffold and group)	$2 \le \sum_{i \in G_1, G_2} n_i \le 5$
Number of same fragments (scaffold and group)	$n_i \le 3, i \in G_1, G_2$
Number of groups (G_3 = $-CH_3$, $-CH_2$, $-CH$, $CH_2{=}CH-$, $-CH{=}CH-$, $CH_2{=}C<$, $-CH{=}C<$)	$1 \le \sum_{i \in G_3} n_i \le 4$
Number of groups (except for $-CH_3$, $-CH_2$, $-CH$, $CH_2{=}CH-$, $-CH{=}CH-$, $CH_2{=}C<$, $-CH{=}C<$)	$0 \le \sum_{i \in (G_2 - G_3)} n_i \le 2$
Molecular weight	$MW \le 500$
Number of hydrogen bond donors	$HBD \le 5$
Number of hydrogen bond acceptors	$HBA \le 10$
Octanol-water partition coefficient	$\log P \le 5$
Number of rotatable angles	$ROT(ROT_{frag}) \le 10$
Synthetic accessibility score	$SA \le 6$
Synthetic complexity score	$SC \le 3.4$
Objective function	$Prob_{bind}$

using ECFP fingerprints and principal component analysis (PCA), with the x- and y-axes representing the first two principal components, integers (0–7) in the color bar indicating the eight scaffolds, and point "4" denoting rivaroxaban. Figure 4.9 highlights a wide distribution of designed candidates, reflecting the capability of the MINLP-based framework to produce diverse molecules structurally similar to rivaroxaban.

In Subproblem 4, the deep learning model is employed to estimate the probabilities of high binding affinity for all 42,932 designed drug candidates and rank them in descending order. The ranking results indicate that four candidates surpass rivaroxaban, achieving a 98.76% probability of high binding affinity, as summarized in Table 4.4.

The highest ranked drug candidate, with a 98.78% probability of high binding affinity, was selected for molecular docking (414s on a desktop computer, Intel Core i5-10500 CPU @ 3.10 GHz) and MD simulation (about 8 hours on the Advanced Super Computing Center, AMD EPYC 7502 CPU @ 2.5 GHz, 64 cores) to validate the accuracy of the deep learning model and the effectiveness of the MINLP-based design framework.

Molecular docking was performed for the top-ranked drug candidate with Factor Xa (PDB ID: 2w26 [70]) to evaluate its binding affinity. The binding energy $\Delta G_{bind} = -42.39$ kJ/mol, corresponds to a $K_{i/d} = 0.037$ μmol/L (< 1 μmol/L indicates high binding affinity) using the equation of $\Delta G_{bind} = RT\ln K_{i/d}$, confirming strong binding between the designed candidate and Factor Xa. Conversely, a poorly performing candidate (C=CCOC(=O)c1ccc(C(=O)O)s1) with a 25.82% probability of high binding affinity was also analyzed. Its binding energy,

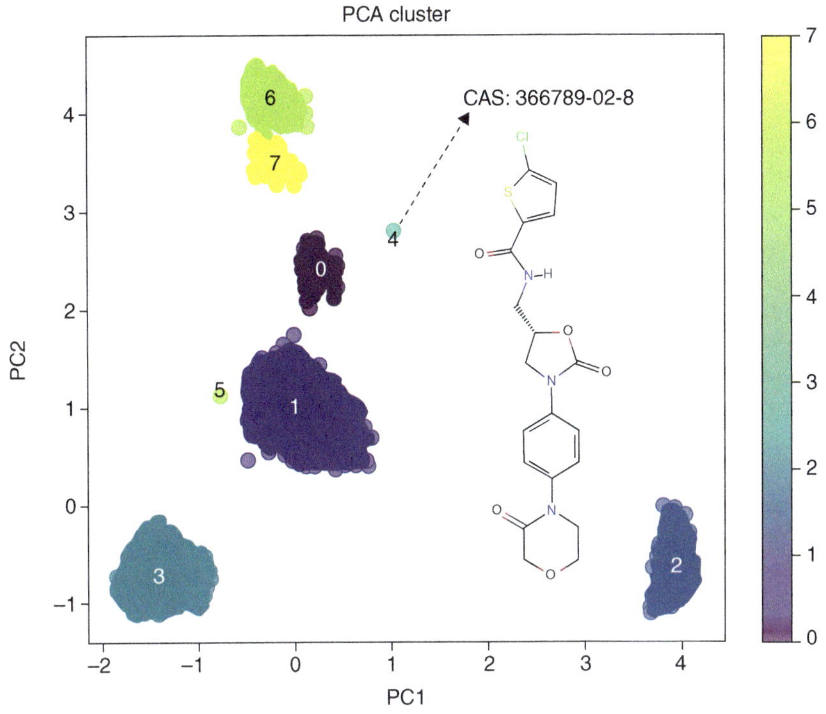

Figure 4.9 The chemical space of the designed drug candidates using the ECFP fingerprints and the PCA method (integers (0–7) in color bar represent 8 scaffolds, and the point "4" represents the rivaroxaban).

Table 4.4 The top four designed drug candidates for Factor Xa.

No.	SMILES	Probability of high binding affinity
1	C=CCCOC1CN(c2ccc(N3CCOCC3)c(C=O)c2)C(=O)O1	98.78%
2	CC(=CO)C1CN(c2ccc(N3CCOCC3)c([N+](=O)[O–])c2)C(=O)O1	98.78%
3	Cc1cc(N2CC(C=CCBr)OC2=O)ccc1N1CCOCC1	98.77%
4	O=C1OC(C=CCBr)CN1c1ccc(N2CCOCC2)c(O)c1	98.77%

$\Delta G_{bind} = -29.97$ kJ/mol, corresponds to $K_{i/d} = 5.61$ μmol/L (≥ 1 μmol/L indicates low affinity), demonstrating that the deep learning model reliably distinguishes between high- and low-affinity protein–ligand complexes.

MD simulations were conducted to investigate the binding mode between Factor Xa and the top-ranked designed drug candidate. The docking conformation of the complex served as the initial structure for a 20,000 ps MD simulation. To assess the stability of the initial complex, the root mean square deviation (RMSD) of the skeletal atoms was calculated, yielding values between 2.5 and 2.8 Å relative to the starting structure (Figure 4.10a). After the first 5,000 ps of equilibration, the RMSD stabilized around 2.7 Å and remained consistent throughout the simulation, indicating that the ligand binds stably and maintains a balanced complex conformation. In Figure 4.10b, the colored cartoon depicts Factor Xa, the rod-shaped structure represents the designed ligand, and the red circle marks the binding site. Figure 4.10c shows the entrance of the binding cavity, while Figure 4.10d illustrates an "L"-shaped conformation of the protein–ligand complex, with two sub-pockets (S1 and S2) forming the cavity. Rings A and B occupy sub-pocket S1, and ring C resides in sub-pocket S2.

Figure 4.11 illustrates the binding mode and high affinity of the optimal designed drug candidate. Key residues forming the binding cavity include hydrophobic amino acids **Ile227**, **Ala183**, **Ala190**, **Val213**, **Trp215** and hydrophilic amino acid residues **Ser195**, **Ser214**, **Asp194**, **Gln192**, **Gly216**, **His57**, **Tyr99**, **Cys191**, **Glu217**, **Thr185**, **Tyr228**, **Ser214**, **Gly184**, and **Gly226**. The S1 binding pocket of Factor Xa is hydrophobic, defined by **Trp215**, **Ala190**, **Val213**, **Ala183**, and **Ile227**. The nonpolar aryl ring of the optimal candidate interacts with **Trp215**. Notably, residues identified in the MD simulation are consistent with those observed in the X-ray crystal structure of the rivaroxaban–Factor Xa complex [68], including **Trp215**, **Ser195**, **His57**, **Tyr99**, and **Tyr228**. Experimental data also indicate the presence of a hydrophobic channel formed by the aromatic rings of **Trp215** and **Phe174**.

The optimal drug candidate forms multiple hydrogen bonds with Factor Xa. The carbonyl oxygen of the oxazolidinone core (ring C) forms a strong hydrogen bond with **Ser195** (—O· ·—N, 2.73 Å) and another with **Cys191** (—O· ·—O, 3.20 Å). The ether oxygen also forms a hydrogen bond with **Gly216** (—O· ·—O, 3.42 Å). These three hydrogen bonds facilitate an "L"-shaped binding conformation, serving as a template that positions substituents into the S1 and S2 sub-pockets, characteristic of synthetic direct Factor Xa inhibitors. The experimental X-ray structure of the rivaroxaban–Factor Xa complex confirms that the carbonyl oxygen of ring C forms a strong hydrogen bond with **Gly219** (—O· ·—N, 2.97 Å) [68].

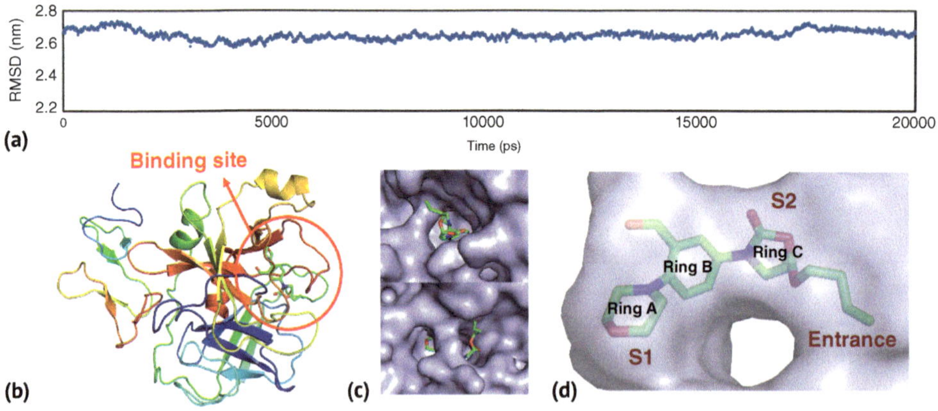

Figure 4.10 The MD simulation results of the complex between the Factor Xa protein and the optimal designed drug candidate. (a) Variation of RMSD values with simulation time. (b) The binding site of the protein–ligand complex. (c) The entrance of the binding cavity. (d) The "L"-shaped conformation of the protein–ligand complex.

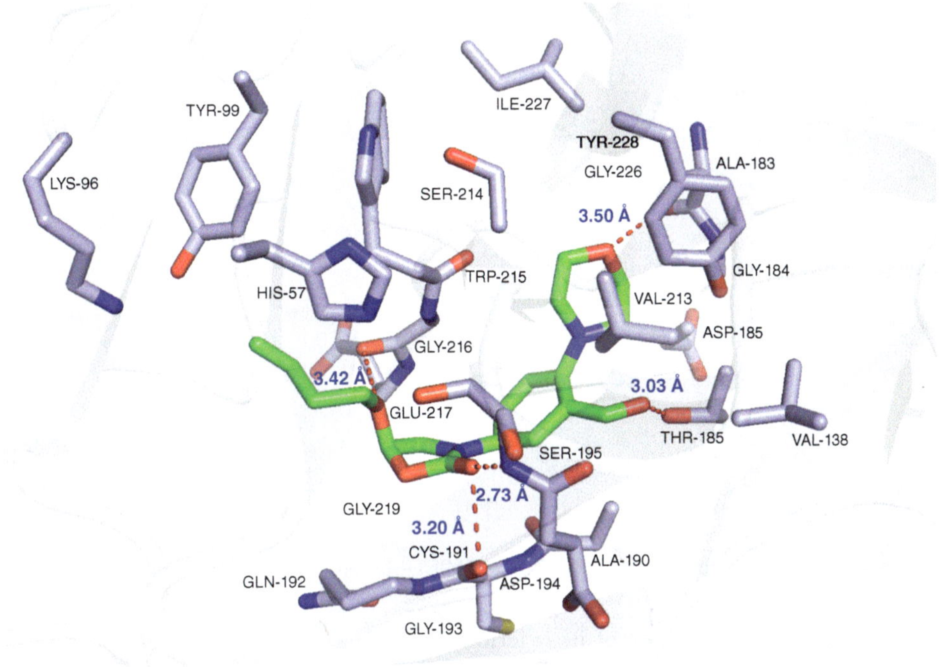

Figure 4.11 The binding mode of the complex between the optimal designed drug candidate and the Factor Xa protein based on the MD simulation (the amino acid residues in the binding site are represented as sticks; the molecular carbon and the amino acid residue carbon are represented by green and cyan, respectively; the N and O atoms are represented by dark blue and red, respectively; the hydrogen bonds are shown as red dotted lines).

Additional hydrogen bonds, including the O atom of morpholine (ring A) with **Ala183** (—O···—O, 3.50 Å) and the double-bond oxygen on ring B with **Thr185** (—O···—OH, 3.03 Å), are critical for stabilizing the bioactive conformation of the designed ligand. Overall, the candidate maintains a stable "L"-shaped conformation at the Factor Xa binding site, supported by hydrophobic interactions and hydrogen bonding.

4.4.2 HSP90 Inhibitor Candidate Design

In this section, the developed MoE-based dissociation kinetic model is incorporated into an MINLP-based *de novo* drug design framework to generate potential HSP90 inhibitor candidates with slow dissociation kinetics and favorable binding affinity in a high-throughput and intelligent manner. HSP90 has been extensively studied in terms of ligand dissociation [13, 71–75]. It is a small soluble protein that acts as a highly conserved molecular chaperone, playing a crucial role in the maturation of newly synthesized peptides and the turnover of misfolded or denatured proteins. In the context of cancer, the client proteins of HSP90 are involved in oncogenesis, and disrupting the complexes between HSP90 and its client proteins promotes proteasomal degradation of the clients. Consequently, HSP90 and its co-chaperones have emerged as promising targets for cancer therapeutics [76]. One strategy to inhibit the chaperone activity of HSP90 is by using small-molecule inhibitors

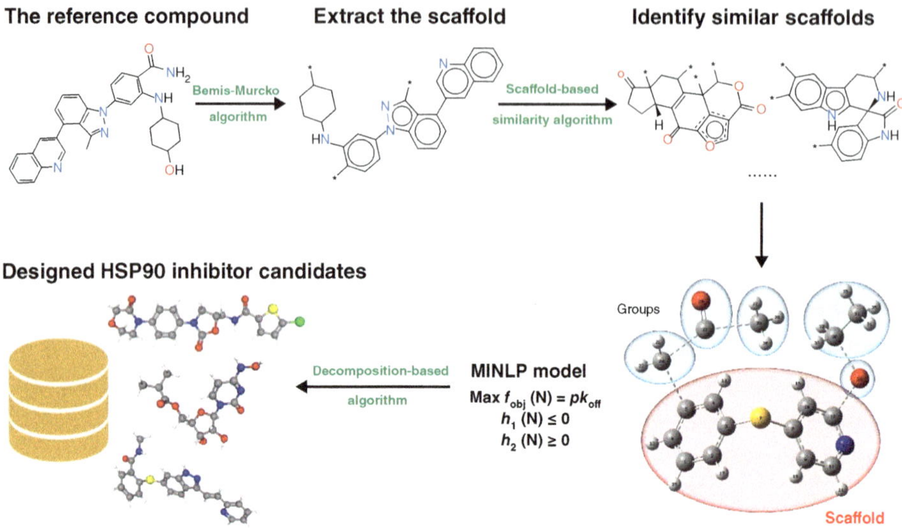

Figure 4.12 The workflow of the kinetics-guided MINLP-based *de novo* drug design framework.

to target the ATPase function of its N-terminal domain, resulting in the degradation of client proteins and a reduction in tumor growth [77]. The workflow of the kinetics-guided MINLP-based *de novo* drug design framework is shown in Figure 4.12.

First, an HSP90 inhibitor with slow dissociation kinetics, along with its experimental k_{off}, is collected from Ganotra et al. [13] and used as the reference compound. Assuming that candidates with similar scaffolds exhibit comparable protein–ligand interactions [54], the Bemis–Murcko algorithm and scaffold-based similarity algorithm were applied to extract the scaffold of the reference compound and identify a subset of similar scaffolds from our scaffold database [78]. This approach focuses computational resources on the most promising scaffolds, reducing the search space and accelerating the design of HSP90 inhibitor candidates for the subsequent MINLP model. As a result, 120 scaffolds were obtained, of which 37 duplicates were removed.

With 38 scaffolds (including that of the reference compound) and 25 commonly used functional groups, the HSP90 inhibitor design task was formulated as an MINLP model, including the objective function (pk_{off}), drug structure constraints, and property constraints (Lipinski's "Rule of Five," SA, and binding affinity predicted by a deep learning model using the model as described in the literature [78]). Lower and upper bounds for the structure and property constraints, along with the selected scaffolds and groups, are provided in Tables N.1, N.2 in Appendix N of the supplementary material. A decomposition-based algorithm [48] was applied to solve this highly nonlinear MINLP, generating a virtual inhibitor library by assembling scaffolds and groups within the model constraints. A total of 261,736 SMILES strings were successfully generated, representing compounds that satisfy the "Rule of Five," SA, and predicted binding affinity. Among these, 70,360 SMILES strings exhibited pk_{off} values higher than that of the reference compound ($pk_{off} = 2.598 \pm 0.281$), demonstrating the effectiveness of the kinetics-guided MINLP framework in designing candidates with slow dissociation kinetics and high binding affinity. The complete design procedure required approximately 5.4 hours on a desktop computer (Intel(R) Core (TM) i7-10700F CPU @ 2.90 GHz, 24.0 GB RAM). Data mining revealed that only 96 of the 261,736 designed compounds (0.037%) are present in the PubChem database, indicating that the MINLP-based framework can identify existing drugs while also generating novel molecules that have not been previously studied.

Furthermore, the chemical space was constructed to evaluate the structural diversity of 261,736 designed HSP90 inhibitor candidates using ECFP descriptors and the t-SNE method, with different colors representing different scaffolds (Figure 4.13a). Many designed candidates cluster near the reference compound (marked by the five-pointed star in Figure 4.13a), demonstrating the framework's ability to generate diverse molecules similar to the reference. At the same time, numerous candidates are located far from the reference, reflecting the framework capacity to design structurally diverse HSP90 inhibitors. This broad coverage is attributed to the scaffold-based similarity algorithm, which applies 24 distinct evaluation criteria to expand the search space for potential drug candidates.

To assess the reliability of our MoE model in predicting HSP90 family ligands, the HSP90 dataset from Liu et al. [24] is cleaned and evaluated using our MoE model, yielding r, R^2, MAE, and RMSE values of 0.895, 0.773, 0.359, and 0.458, respectively. In comparison, the model developed by Akhunzada et al. [79] reported R^2, MAE, and RMSE values of 0.80, 0.29, and 0.38, respectively, on the same dataset from Liu et al. [24]. While our MoE model demonstrates lower prediction accuracy compared to that of Akhunzada et al. [79], it should be noted that their method necessitates computationally intensive MD simulations, whereas our method uses text-based descriptors, resulting in significantly reduced

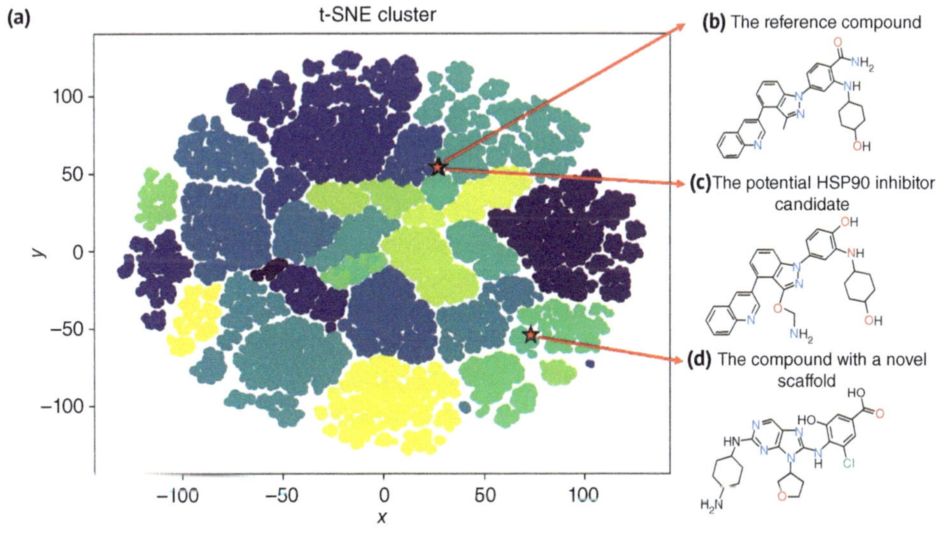

Figure 4.13 The designed results of HSP90 inhibitor candidates. (a) The chemical space of the designed HSP90 inhibitor candidates using the ECFP descriptors and the t-SNE method. (b) The two-dimensional molecular structure of the reference compound. (c) The two-dimensional molecular structure of the potential HSP90 inhibitor. (d) The two-dimensional molecular structure of the compound with a novel scaffold.

computational time. In addition, our MoE model is used to predict the pk_{off} values for the dataset from Kokh et al. [13]. After cleaning the dataset from Kokh et al. [13], the r and R^2 values of our MoE model are 0.876 and 0.767, respectively. In comparison, the prediction accuracy of the τ-RAMD method for τ is $R^2 = 0.86$ on the same dataset from Kokh et al. [13]. Although our MoE model has not outperformed the τ-RAMD method, it demonstrates high prediction accuracy and significantly greater computational efficiency than the τ-RAMD method. These results indicate that our MoE model is effective in estimating the pk_{off} values of HSP90 ligands. Furthermore, it is suitable for high-throughput evaluations of drug candidates, facilitating the identification of potential candidates for further τ-RAMD validation. Details on the cleaning processes for the HSP90 datasets from Liu et al. [24] and Kokh et al. [13] can be found in Appendix O of the supplementary material.

Furthermore, the analysis of ranking results reveals that the pk_{off} values of the compounds available in the PubChem database are universally lower than that of the reference compound. This finding illustrates that the compounds currently known are less likely to exhibit superior performance in terms of dissociation kinetics, highlighting the urgent need for designing novel drug candidates. Regarding other designed HSP90 inhibitor candidates that are not yet commercially available and for which limited organic synthesis routes are known, a rigorous mechanism method, τ-RAMD, is carried out in the next subsection to make comparisons among the reference compound (predicted $pk_{\text{off}} = 2.598 \pm 0.281$, as shown in Figure 4.13b), the potential HSP90 inhibitor candidate (predicted $pk_{\text{off}} = 2.608 \pm 0.284$, as shown in Figure 4.13c), and the compound with a novel scaffold (predicted $pk_{\text{off}} = 2.480 \pm 0.363$, as shown in Figure 4.13d).

In this section, the potential HSP90 inhibitor candidate is selected for further analysis using the τ-RAMD method. This method is able to calculate the relative τ of drug-like compounds for certain protein by conducting numerous RAMD simulations. These simulations not only offer insights into the pathways through which ligands exit from deep-seated binding sites within proteins but also provide the information on dissociation mechanisms [75]. The protein (PDB ID: 5LQ9) is selected in this work. The RAMD simulation results of the potential HSP90 inhibitor candidate, the reference compound, and the compound with a novel scaffold are presented in Figures P.1–P.3 in Appendix P of the supplementary material, respectively. The computed τ, averaged over six replicas, is 1.02 ± 0.20 ns for the potential HSP90 inhibitor candidate. In comparison, the computed τ for the reference compound is 0.70 ± 0.14 ns, averaged over six replicas. This result illustrates that the potential HSP90 inhibitor candidate has a longer τ and a lower dissociation rate compared to the reference compound. This demonstrates the powerful ability of our kinetics-guided MINLP-based *de novo* drug design framework in designing drug candidates with slow dissociation kinetics. Additionally, several HSP90 inhibitor candidates with novel scaffolds have also been identified to exhibit slow dissociation kinetics. For example, the computed τ averaged over six replicas of the compound with a novel scaffold (predicted $pk_{\text{off}} = 2.480 \pm 0.363$, as shown in Figure 4.13d) is 0.50 ± 0.16 ns. Although it has not outperformed the reference compound in terms of τ, its scaffold is entirely different from those of the potential HSP90 inhibitor candidate and the reference compound, indicating the potential for designing novel drug candidates with slow dissociation kinetics through our kinetics-guided MINLP-based *de novo* drug design framework. According to the τ-RAMD literature [13], the accuracy for predictions of τ using the τ-RAMD method is $R^2 = 0.86$ and AAPE = 36%. To check if the higher performance of the potential HSP90 inhibitor candidate stems from the prediction error of the τ-RAMD method or not, the relative errors for the values of 1.02 ns and 0.70 ns, as well as for 0.70 ns and 0.50 ns, are calculated, yielding 45.7% and 40.0%, respectively. Both relative errors exceed the prediction error of the τ-RAMD method (AAPE = 36%). This result indicates the effectiveness of the τ-RAMD method in verifying the ranking of the potential HSP90 inhibitor candidate, reference compound, and the compound with a novel scaffold. The verified results of the τ-RAMD method also demonstrate the reliability of our deep learning model in evaluating the pk_{off} values of compounds.

HSP90 is a challenging protein in predicting dissociation kinetics due to its flexible adenosine triphosphate (ATP) binding site, which is comprised of an unstable α-helix3 [64]. This α-helix3 segment (residues 101–122) can adopt either a "helix" conformation (Figure 4.14a) or a "loop" conformation (Figure 4.14b) depending on the bound ligand [64]. Note that the reference compound [13], the potential HSP90 inhibitor candidate, and the compound with a novel scaffold all belong to the "helix" category. The conformations of the latter two compounds are confirmed through docking methods. More calculation details can be found in Appendix P of the supplementary material. The "helix" conformation features an extra transient sub-pocket located next to the ATP binding site. This sub-pocket provides space for substitutions on the "helix-binders." The ligand segments highlighted by red and black circles in Figure 4.15a,b represent the differences between the potential HSP90 inhibitor candidate and the reference compound. Figure 4.15c,d show that the α-helix3 segments are stably combined with both the potential HSP90 inhibitor

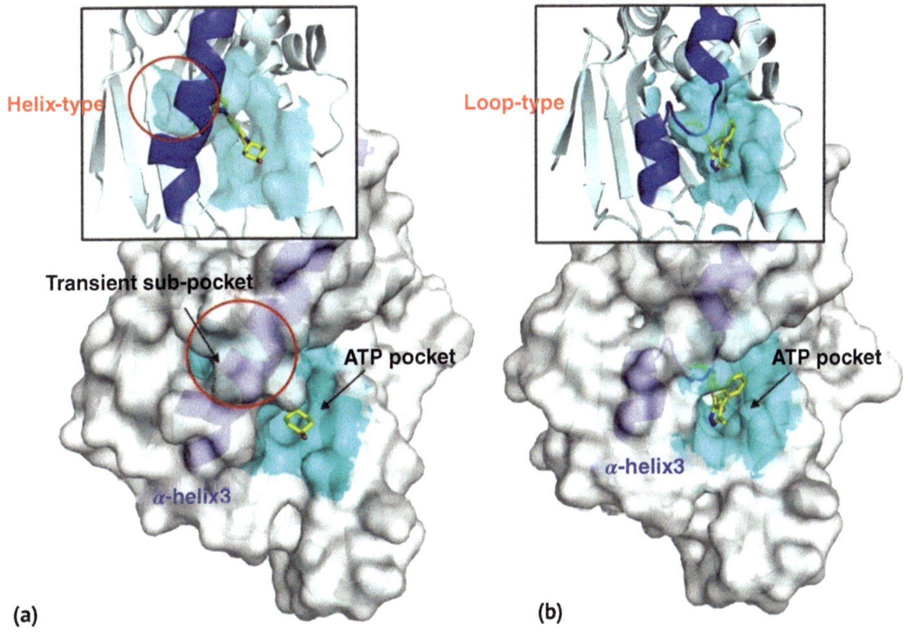

Figure 4.14 Two main conformations of the HSP90 binding site. (a) Helix-type and (b) loop-type; the α-helix3 region is shown in blue; the surface of the binding pocket is colored in cyan; the transient sub-pocket under α-helix3 is circled in red.

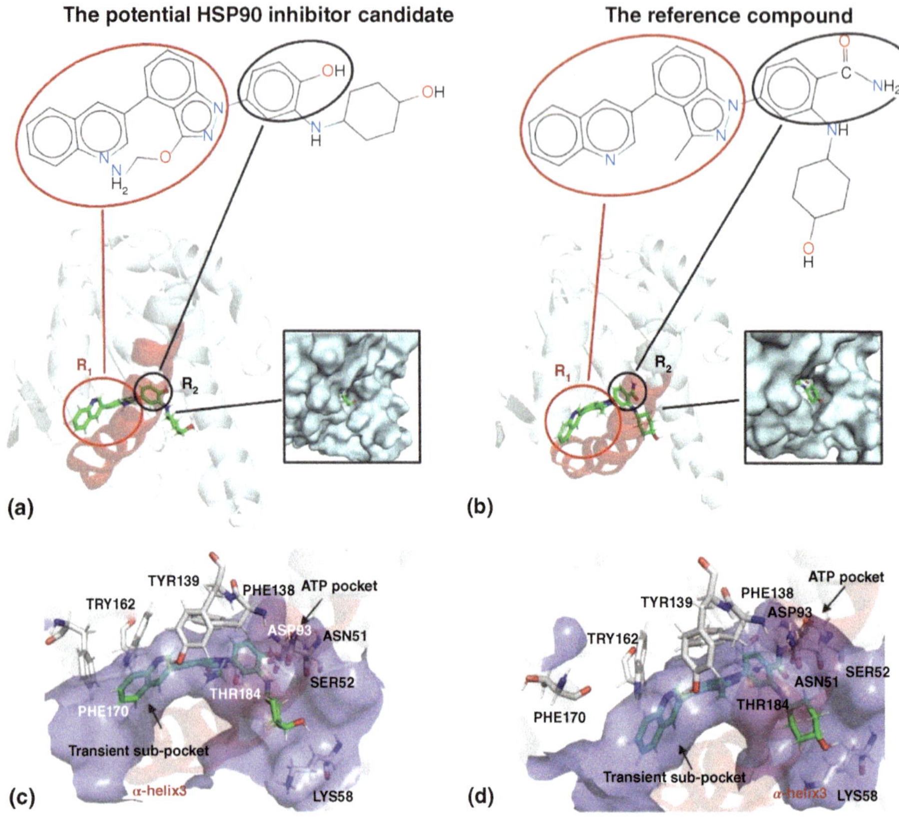

Figure 4.15 The MD simulation results for the bound states of the potential HSP90 inhibitor candidate and the reference compound. (a, b) Illustrations of the binding sites and binding poses; the entrance of the binding cavity is also presented. (c, d) Positions of two compounds in the binding pocket showing the occupation of the sub-pocket under α-helix3 by two compounds; the protein–ligand contacts for two compounds; protein structures are shown in a half-transparent cartoon representation with the ligands and key residues in stick representation, and the binding pocket is colored in blue; ligands are colored by atom type.

candidate and the reference compound, exhibiting "helix" conformations. The transient sub-pocket in the "helix" conformation offers room for substitutions on the "helix-binders" (represented by the red circle marked as R_1 in Figure 4.15a,b). The ligand occupies the transient sub-pocket adjacent to α-helix3 as illustrated in Figure 4.15c,d. The potential HSP90 inhibitor candidate differs from the reference compound by introducing a bulkier fragment (R_1 in Figure 4.15a) that occupies the sub-pocket. Recent studies have demonstrated a correlation between the transient sub-pocket in the bound complex and the dissociation kinetics of HSP90 inhibitors. It has been found that the introduction of a large R_1 substitution is able to effectively stabilize α-helix3, resulting in reduced binding and unbinding rates [74]. Particularly, ligands with larger R_1 substitutions generally exhibit slower binding and unbinding kinetics [75]. It is evident that the R_1 substitution of the potential HSP90 inhibitor candidate is larger than that of the reference compound, consequently leading to a longer τ. The compound with a novel scaffold also stably binds to the α-helix3 fragment in a "helix" conformation, and its binding mode is shown in Figure P.4 in Appendix P of the supplementary material.

4.5 Conclusion

This chapter establishes an optimization-based *de novo* drug design framework, in which two deep learning models are first developed to predict binding affinity and dissociation kinetics for protein–ligand complexes. For binding affinity, the models achieved AUC values of 0.914, 0.880, and 0.901 on the training, validation, and test sets, respectively, demonstrating their reliable ability to distinguish high- and low-affinity complexes. In the dissociation kinetics study, a MoE-based generic dissociation kinetic model is trained 100 times for fast predictions of k_{off}. The average MSE and r values of dissociation kinetic models for the training, validation, test sets are 0.200 ± 0.012, 0.300 ± 0.045, 0.288 ± 0.028, and 0.890 ± 0.007, 0.821 ± 0.025, 0.835 ± 0.017, respectively. These two deep learning models are then incorporated into the MINLP-based drug design framework, which includes the objective function, structure constraints, and property constraints. The Bemis–Murcko and scaffold-based similarity algorithms generate scaffolds and select essential subsets for the MINLP model, ensuring rational candidate design while significantly reducing the problem size. Nonconvex issues arising from complex nonlinear equations are addressed using a decomposition-based solution algorithm. By optimizing the chemical space in a high-throughput manner, anticoagulant candidates with high binding affinity and HSP90 inhibitors with slow dissociation kinetics are designed. The candidates are further validated through molecular docking, MD simulations, and τ-RAMD, confirming the broad applicability and effectiveness of this optimization-based framework for drug design.

Acknowledgments

The authors acknowledge financial support from the Dalian High-level Talents Innovation Support Program [2023RQ059], the National Natural Science Foundation of China [22208042], and the Liaoning Provincial Natural Science Foundation Doctoral Research Start-up Fund Project [2024-BSBA-08].

References

1 DiMasi JA, Grabowski HG, and Hansen RW. Innovation in the pharmaceutical industry: New estimates of R&D costs. *Journal of Health Economics* 2016;47:20–33.

2 Li Y, Pei J, and Lai L. Structure-based de novo drug design using 3D deep generative models. *Chemical Science* 2021;12(41):13664–13675.

3 Polishchuk PG, Madzhidov TI, and Varnek A. Estimation of the size of drug-like chemical space based on GDB-17 data. *Journal of Computer-Aided Molecular Design* 2013;27(8):675–679.

4 Winter R, Montanari F, Steffen A, Briem H, Noé F, and Clevert D-A. Efficient multi-objective molecular optimization in a continuous latent space. *Chemical Science* 2019;10(34):8016–8024.
5 Gómez-Bombarelli R, Wei JN, Duvenaud D, et al. Automatic chemical design using a data-driven continuous representation of molecules. *ACS Central Science* 2018;4(2): 268–276.
6 Méndez-Lucio O, Baillif B, Clevert DA, Rouquié D, and Wichard J. De novo generation of hit-like molecules from gene expression signatures using artificial intelligence. *Nature Communications* 2020;11(1):10.
7 Chen H, Engkvist O, Wang Y, Olivecrona M, and Blaschke T. The rise of deep learning in drug discovery. *Drug Discovery Today* 2018;23(6):1241–1250.
8 Gooljarsingh LT, Fernandes C, Yan K, et al. A biochemical rationale for the anticancer effects of Hsp90 inhibitors: Slow, tight binding inhibition by geldanamycin and its analogues. *Proceedings of the National Academy of Sciences* 2006;103(20):7625–7630.
9 Bruce NJ, Ganotra GK, Kokh DB, Sadiq SK, and Wade RC. New approaches for computing ligand–receptor binding kinetics. *Current Opinion in Structural Biology* 2018;49:1–10.
10 Plattner N and Noé F. Protein conformational plasticity and complex ligand-binding kinetics explored by atomistic simulations and Markov models. *Nature Communications* 2015;6(1):7653.
11 Miao Y, Bhattarai A, and Wang J. Ligand gaussian accelerated molecular dynamics (ligamd): Characterization of ligand binding thermodynamics and kinetics. *Journal of Chemical Theory and Computation* 2020;**16**(9):5526–5547.
12 Tiwary P, Limongelli V, Salvalaglio M, and Parrinello M. Kinetics of protein–ligand unbinding: Predicting pathways, rates, and rate-limiting steps. *Proceedings of the National Academy of Sciences* 2015;112(5): E386-E391.
13 Kokh DB, Amaral M, Bomke J, et al. Estimation of drug-target residence times by τ-random acceleration molecular dynamics simulations. *Journal of Chemical Theory and Computation* 2018;14(7):3859–3869.
14 Chiu SH and Xie L. Toward high-throughput predictive modeling of protein binding/unbinding kinetics. *Journal of Chemical Information and Modeling* 2016;56(6):1164–1174.
15 Zhang D, Huang S, Mei H, Kevin M, Shi T, and Chen L. Protein–ligand interaction fingerprints for accurate prediction of dissociation rates of p38 MAPK Type II inhibitors. *Integrative Biology* 2019;11(2):53–60.
16 Qu S, Huang S, Pan X, Yang L, and Mei H. Constructing interconsistent, reasonable, and predictive models for both the kinetic and thermodynamic properties of HIV-1 protease inhibitors. *Journal of Chemical Information and Modeling* 2016;56(10):2061–2068.
17 Huang S, Chen L, and Mei H, et al. In silico prediction of the dissociation rate constants of small chemical ligands by 3d-grid-based volsurf method. *International Journal of Molecular Sciences* 2020; 21. https://doi.org/10.3390/ijms21072456.
18 Ortiz AR, Pisabarro MT, Gago F, and Wade RC, Prediction of drug binding affinities by comparative binding energy analysis. *Journal of Medicinal Chemistry* 1995;38(14):2681–2691.
19 Pérez C, Pastor M, Ortiz AR, and Gago F. Comparative binding energy analysis of HIV-1 protease inhibitors: Incorporation of solvent effects and validation as a powerful tool in receptor-based drug design. *Journal of Medicinal Chemistry* 1998;41(6):836–852.

20 Ganotra GK and Wade RC. Prediction of drug–target binding kinetics by comparative binding energy analysis. *ACS Medicinal Chemistry Letters* 2018;9(11):1134–1139.

21 Nunes-Alves A, Ormersbach F, and Wade R.C. Prediction of the drug–target binding kinetics for flexible proteins by comparative binding energy analysis. *Journal of Chemical Information and Modeling* 2021;61(7):3708–3721.

22 Zhao Y, Liu Q, Du J, Meng Q, and Zhang L. Machine learning methods for developments of binding kinetic models in predicting protein-ligand dissociation rate constants. *Smart Molecules* 2023;1(3):e20230012.

23 Amangeldiuly N, Karlov D, and Fedorov MV, Baseline model for predicting protein–ligand unbinding kinetics through machine learning. *Journal of Chemical Information and Modeling* 2020;60(12):5946–5956.

24 Liu H, Su M, Lin HX, Wang R, and Li Y. Public data set of protein–ligand dissociation kinetic constants for quantitative structure–kinetics relationship studies. *ACS Omega* 2022;7(22):18985–18996.

25 Su M, Liu H, Lin H, and Wang R. Wang, Machine-learning model for predicting the rate constant of protein ligand dissociation. *Acta Physico-Chimica Sinica* 2020;36(1):1907006-0.

26 Wang K, Zhou R, Li Y, and Li M. DeepDTAF: A deep learning method to predict protein–ligand binding affinity. *Briefings in Bioinformatics*, 2021;22(5):bbab072.

27 Jiménez J, Škalič M, Martínez-Rosell G, and De Fabritiis G. KDEEP: protein–ligand absolute binding affinity prediction via 3D-convolutional neural networks. *Journal of Chemical Information and Modeling* 2018;**58**(2):287–296.

28 Liu Z, Su M, Han L, et al. Forging the basis for developing protein–ligand interaction scoring functions. *Accounts of Chemical Research* 2017;**50**(2):302–309.

29 Rezaei MA, Li Y, Wu D, Li X, and Li C. Deep learning in drug design: Protein-ligand binding affinity prediction. *IEEE/ACM Transactions on Computational Biology and Bioinformatics* 2020;19(1):407–417.

30 Landrum G. RDKit: Open-source Cheminformatics. 2006.

31 Mikolov T, Chen K, Corrado G, and Dean J. Efficient estimation of word representations in vector space. arXiv:1301.3781,2013.

32 Jaeger S, Fulle S, and Turk S. Mol2vec: Unsupervised machine learning approach with chemical intuition. *Journal of Chemical Information and Modeling*, 2018;58(1):27–35.

33 Vaswani A, Shazeer N, Parmar N, Uszkoreit J, Jones L, Gomez AN, Kaiser L, Polosukhin I. Attention is all you need. arXiv:1706.03762, 2017.

34 Liu Y, Gong C, Yang L, and Chen Y. DSTP-RNN: A dual-stage two-phase attention-based recurrent neural network for long-term and multivariate time series prediction. *Expert Systems with Applications* 2020;143:113082.

35 Tai Y, Yang J, and Liu X. Image super-resolution via deep recursive residual network. In: *Proceedings of the IEEE Conference on Computer Vision and Pattern Recognition*. 2017.

36 Kim Y, Kim J, Kim WJ, Lee EK, and Choi IS. Effects of pooling operations on prediction of ligand rotation-dependent protein–ligand binding in 3D graph convolutional network. *Bulletin of the Korean Chemical Society* 2021;42(5):744–747.

37 Lim J, Ryu S, Park K, Choe YJ, Ham J, and Kim WY. Predicting drug–target interaction using a novel graph neural network with 3D structure-embedded graph representation. *Journal of Chemical Information and Modeling* 2019;**59**(9):3981–3988.

38 Hinton GE, Srivastava N, Krizhevsky A, Sutskever I, and Salakhutdinov RR. Improving neural networks by preventing co-adaptation of feature detectors. arXiv:1207.0580, 2012.
39 Paszke A, Gross S, Massa F, et al. Pytorch: An imperative style, high-performance deep learning library. arXiv:1912.01703, 2019.
40 Oliphant TE. Python for scientific computing. *Computing in Science & Engineering* 2007;9(3):10–20.
41 Kingma DP and Ba J. Adam: A method for stochastic optimization. arXiv:1412.6980, 2014.
42 Fawcett T. An introduction to ROC analysis. *Pattern Recognition Letters* 2006;27(8):861–874.
43 Schuetz DA, Richter L, Martini R, and Ecker GF. A structure–kinetic relationship study using matched molecular pair analysis. *RSC Medicinal Chemistry* 2020;11(11):1285–1294.
44 Gilson MK, Liu T, Baitaluk M, Nicola G, Hwang L, and Chong J. BindingDB in 2015: A public database for medicinal chemistry, computational chemistry and systems pharmacology. *Nucleic Acids Research* 2016; 44(D1):D1045–D1053.
45 Pearson WR. Using the FASTA program to search protein and DNA sequence databases. In: Griffin AM and Griffin HG, editors *Computer Analysis of Sequence Data: Part I*, pp. (307–331). Totowa, NJ: Humana Press; 1994
46 Ma J, Zhao Z, Yi X, Chen J, Hong L, and Chi EH Modeling task relationships in multi-task learning with multi-gate mixture-of-experts, In: *Proceedings of the 24th ACM SIGKDD International Conference on Knowledge Discovery & Data Mining*, pp. 1930–1939. London, United Kingdom: Association for Computing Machinery; 2018.
47 Chen Z, Deng Y, Wu Y, Gu Q, and Li Y. Towards understanding the mixture-of-experts layer in deep learning. arXiv:2208.02813, 2022.
48 Karunanithi AT, Achenie LE, and Gani R. A new decomposition-based computer-aided molecular/mixture design methodology for the design of optimal solvents and solvent mixtures. *Industrial & Engineering Chemistry Research* 2005;44(13):4785–4797.
49 Wishart DS, Feunang YD, Guo AC, et al. DrugBank 5.0: A major update to the DrugBank database for 2018. *Nucleic Acids Research* 2018; 46(D1):D1074–D1082.
50 Lipinski CA, Lombardo F, Dominy BW, and Feeney PJ. Experimental and computational approaches to estimate solubility and permeability in drug discovery and development settings. *Advanced Drug Delivery Reviews* 1997;23(1):3–25.
51 Veber DF, Johnson SR, Cheng HY, Smith BR, Ward KW, and Kopple KD. Molecular properties that influence the oral bioavailability of drug candidates. *Journal of Medicinal Chemistry* 2002;45(12):2615–2623.
52 Bemis GW and Murcko MA. The properties of known drugs. 1. Molecular frameworks. *Journal of Medicinal Chemistry* 1996;39(15):2887–2893.
53 Reis J, Gaspar A, Milhazes N, and Borges F. Chromone as a privileged scaffold in drug discovery: Recent advances. *Journal of Medicinal Chemistry* 2017;60(19):7941–7957.
54 Eckert H and Bajorath J. Molecular similarity analysis in virtual screening: foundations, limitations and novel approaches. *Drug Discovery Today* 2007;12(5):225–233.
55 Churi N and Achenie LEK. Novel mathematical programming model for computer aided molecular design. *Industrial & Engineering Chemistry Research* 1996;35(10):3788–3794.
56 Liu Q, Zhang L, Liu L, et al. OptCAMD: An optimization-based framework and tool for molecular and mixture product design. *Computers & Chemical Engineering* 2019;124:285–301.

57 Ertl P and Schuffenhauer A. Estimation of synthetic accessibility score of drug-like molecules based on molecular complexity and fragment contributions. *Journal of Cheminformatics* 2009;1(1):8.

58 Coley CW, Rogers L, Green WH, and Jensen KF. SCScore: Synthetic complexity learned from a reaction corpus. *Journal of Chemical Information and Modeling* 2018;58(2):252–261.

59 Liu Q, Zhang L, Tang K, et al. Machine learning-based atom contribution method for the prediction of surface charge density profiles and solvent design. *AIChE Journal* 2021;67(2):e17110.

60 Bickerton GR, Paolini GV, Besnard J, Muresan S, and Hopkins AL, Quantifying the chemical beauty of drugs. *Nature Chemistry* 2012;4(2):90–98.

61 Tawarmalani M and Sahinidis NV. A polyhedral branch-and-cut approach to global optimization. *Mathematical Programming* 2005;103(2):225–249.

62 Morris GM, Huey R, Lindstrom W, et al. AutoDock4 and AutoDockTools4: Automated docking with selective receptor flexibility. *Journal of Computational Chemistry*, 2009;30(16):2785–2791.

63 Berendsen HJ, van der Spoel D, and van Drunen R. GROMACS: A message-passing parallel molecular dynamics implementation. *Computer Physics Communications* 1995;91(1):43–56.

64 Kokh DB, Doser B, Richter S, Ormersbach F, Cheng X, and Wade RC. A workflow for exploring ligand dissociation from a macromolecule: Efficient random acceleration molecular dynamics simulation and interaction fingerprint analysis of ligand trajectories. *The Journal of Chemical Physics* 2020;153(12):125102.

65 Berger BT, Amaral M, Kokh DB, et al. Structure-kinetic relationship reveals the mechanism of selectivity of FAK inhibitors over PYK2. *Cell Chemical Biology* 2021;28(5):686–698.e7.

66 Mann KG, Jenny RJ, and Krishnaswamy S. Cofactor proteins in the assembly and expression of blood clotting enzyme complexes. *Annual Review of Biochemistry* 1988;57(1):915–956.

67 Leadley RJ Jr. Coagulation factor Xa inhibition: Biological background and rationale. *Current Topics in Medicinal Chemistry* 2001;1(2):151–159.

68 Roehrig S, Straub A, Pohlmann J, et al. Discovery of the novel antithrombotic agent 5-Chloro-N-({(5S)-2-oxo-3- [4-(3-oxomorpholin-4-yl)phenyl]-1,3-oxazolidin-5-yl}methyl) thiophene- 2-carboxamide (BAY 59-7939): An oral, direct factor Xa inhibitor. *Journal of Medicinal Chemistry* 2005;48(19):5900–5908.

69 Liu Q, Jiang Y, Zhang L, and Du J. A computational toolbox for molecular property prediction based on quantum mechanics and quantitative structure-property relationship. *Frontiers of Chemical Science and Engineering* 2022;16(2):152–167.

70 Burley SK, Bhikadiya C, Bi C, et al. RCSB Protein Data Bank: powerful new tools for exploring 3D structures of biological macromolecules for basic and applied research and education in fundamental biology, biomedicine, biotechnology, bioengineering and energy sciences. *Nucleic Acids Research* 2021; 49(D1):D437–D451.

71 Wolf S, Amaral M, Lowinski M, et al. Estimation of protein–ligand unbinding kinetics using non-equilibrium targeted molecular dynamics simulations. *Journal of Chemical Information and Modeling* 2019;59(12):5135–5147.

72 Schuetz DA, Richter L, Amaral M, et al. Ligand desolvation steers on-rate and impacts drug residence time of heat shock protein 90 (hsp90) inhibitors. *Journal of Medicinal Chemistry* 2018;61(10):4397–4411.
73 Schuetz DA, Bernetti M, Bertazzo M, et al. Predicting residence time and drug unbinding pathway through scaled molecular dynamics. *Journal of Chemical Information and Modeling* 2019;59(1):535–549.
74 Amaral M, Kokh DB, Bomke J, et al. Protein conformational flexibility modulates kinetics and thermodynamics of drug binding. *Nature Communications* 2017;8(1):2276.
75 Kokh DB, Kaufmann T, Kister B, and Wade RC. Machine Learning Analysis of τRAMD Trajectories to Decipher Molecular Determinants of Drug-Target Residence Times. *Frontiers in Molecular Biosciences* 2019;6:36.
76 Zhang J, Li H, Liu Y, et al. Targeting hsp90 as a novel therapy for cancer: Mechanistic insights and translational relevance. *Cells* 2022; 11. https://doi.org/10.3390/cells11182778.
77 Tillotson B, Slocum K, Coco J, et al. Hsp90 (heat shock protein 90) inhibitor occupancy is a direct determinant of client protein degradation and tumor growth arrest in vivo. *Journal of Biological Chemistry* 2010;285(51):39835–39843.
78 Zhao Y, Liu Q, Wu X, Zhang L, Du J, and Meng Q. De novo drug design framework based on mathematical programming method and deep learning model. *AIChE Journal* 2022;68(9):e17748.
79 Akhunzada MJ, Yoon HJ, Braka A, Deb I, and Wu S. Using dynamic interaction fingerprints to derive baseline machine learning model for the prediction of protein-ligand dissociation rate constant. *Chemistry Research Archive* 2024. https://doi.org/10.26434/chemrxiv-2024-lx8dn.

5

Data-driven Pharmaceutical Processing and Drug Development

Fawad Abidi

Institute for Materials and Processes, School of Engineering, University of Edinburgh, Edinburgh, UK

Corresponding author: Fawad.A@ed.ac.uk

CONTENTS

5.1 Introduction

Optimizing bioprocesses can significantly enhance productivity, improve product quality, reduce energy consumption, and increase revenues. For real-time optimization, a nonlinear model is typically employed to address any nonlinearity within the process as discussed by Pantelides and Renfro [1]. This nonlinear model can be developed using one of three approaches: first principles, grey-box, or black-box.

In the first principles approach, the model is constructed based on a thorough understanding of the process, including reaction kinetics and transfer operations. Grey-box modeling involves using some process knowledge to initially parameterize a dynamic model. Subsequently, a parameter estimation problem is solved using data collected from the process to fine-tune the model parameters [2–4]. In contrast, the black-box approach does not assume any prior process knowledge; instead, the model is entirely estimated from process data using nonlinear function approximators such as neural networks (NNs) [5–7]. NNs have been widely used to model various biological processes, particularly when the biotic phase is considered as an unknown or unexplained component [8]. These data-driven models require minimal human intervention for parameter estimates and can be readily adapted for different applications. Although using NNs means that users have limited control over parameters such as weights and biases—since initial values are often predefined by the utilized library—this limitation can be mitigated by adjusting the learning rate. Kinetic models are also effective in describing bioprocess mechanisms and solving optimization issues but require a deep understanding of both the computational tools and specific processes like glycosylation [9, 10]. The parameterization and training of mechanistic kinetic models are often time-consuming and technically demanding [11].

On the other hand, employing machine learning techniques to describe processes such as monoclonal antibody (mAb) manufacturing necessitates only a basic understanding of the biological context and eliminates the need to construct detailed reaction networks. This approach simplifies the modeling process, though it sacrifices some degree of control over the model parameters. Referring to *1 N-linked glycosylation,* protein glycosylation is a crucial posttranslational modification of proteins in many biological processes, impacting protein folding, stability, solubility, and interaction with other proteins, such as in ligand–receptor signaling [12–14]. It is also a vital quality characteristic of glycoprotein medications. The quality of biopharmaceuticals is known to be influenced by correct glycosylation, as different glycoforms of a recombinant protein can have a major effect on the drug's half-life, potency, inflammatory and immunological responses, and safety [15]. The diverse mechanisms by which glycans might impact a protein's interactions with its surroundings render glycosylation a crucial attribute to regulate during the manufacturing of recombinant proteins. The glycans found at a glycosylation site can range from a single sugar molecule to a polymer consisting of over 200 sugars. These glycans have the potential to be modified with phosphate, sulfate, acetate, or phosphorylcholine. Significantly, glycans frequently exhibit branching as illustrated in Figure 5.1.

Research has demonstrated that N-linked glycosylation directly impacts the therapeutic effectiveness of mAbs. This includes factors like stability, immunogenicity, circulatory lifespan, and therapeutic efficacy, all of which are influenced by the structure and function of the antibodies [17]. Controlling glycosylation in recombinant protein production has proven challenging due to the intricate nature of glycans and the multiple elements involved in their synthesis. The available techniques for regulating the glycosylation of recombinant proteins have been thoroughly examined and encompass host cell manipulation, process parameters, and medium constituents. Examples encompass metabolite concentrations like glucose and lactose, the build-up of waste products such as ammonia, the reduction of intracellular nucleotide sugar donors, and potentially alterations in gene expression.

In this work, the goal is to optimize mAb production using data-driven approaches facilitated by NN models. The study's framework is established upon the utilization of a dataset pertaining to the synthesis of mAbs, obtained from a published source, Karst et al. [18]. Among the variables that are included in this dataset are metabolite concentrations, cell culture densities, and associated mAb yields. This dataset contains experimental data on

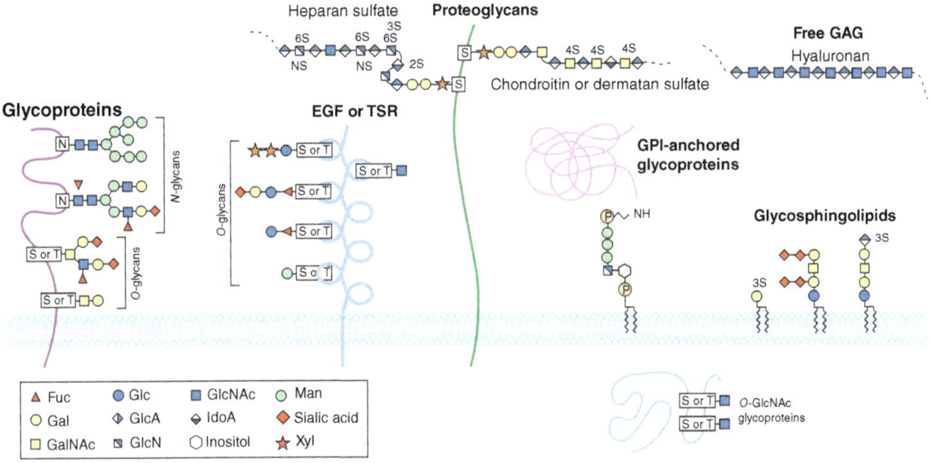

Figure 5.1 Attachment of various glycans to glycoproteins. *Source*: [16]/with permission of Springer Nature.

the manufacturing of mAbs. This kind of thorough data makes it possible to establish correlations between the variables that are input (metabolite concentrations and cell culture densities) and the variables that are output (mAb yield).

5.2 Methodology

5.2.1 Data Collection and Preprocessing

We collect process observations from a source having operational data for the manufacturing of mAbs. Nutrient concentrations and cell growth rates are two of the main characteristics that are included in these observations. Various strategies for data preprocessing are utilized in order to clean and standardize the dataset. The experimental (and simulation) data from the experiments conducted by [17], which are depicted in Figure 5.2, include experimental results from cell culture at 33.5 °C which span to 25 days (600 h), and simulation results from an integrated mathematical model to describe the dynamics of mAb glycosylation throughout cell culture. This data is then made available for the NN model development by digitizing these plots, and four datasets of comma-separated file format are prepared for each metabolite. Following the removal of outliers and the adjustment of units, the data that was produced is normalized.

5.2.2 Model Development

It is possible to develop multiple NN models by making use of the data obtained from [17]. These models have the following typical components in their architecture (Figure 5.3).

Input layer: this layer is responsible for receiving input features, which are the process variables that are known to have an effect on the production of mAb. These variables include different metabolite concentrations and cell culture density.

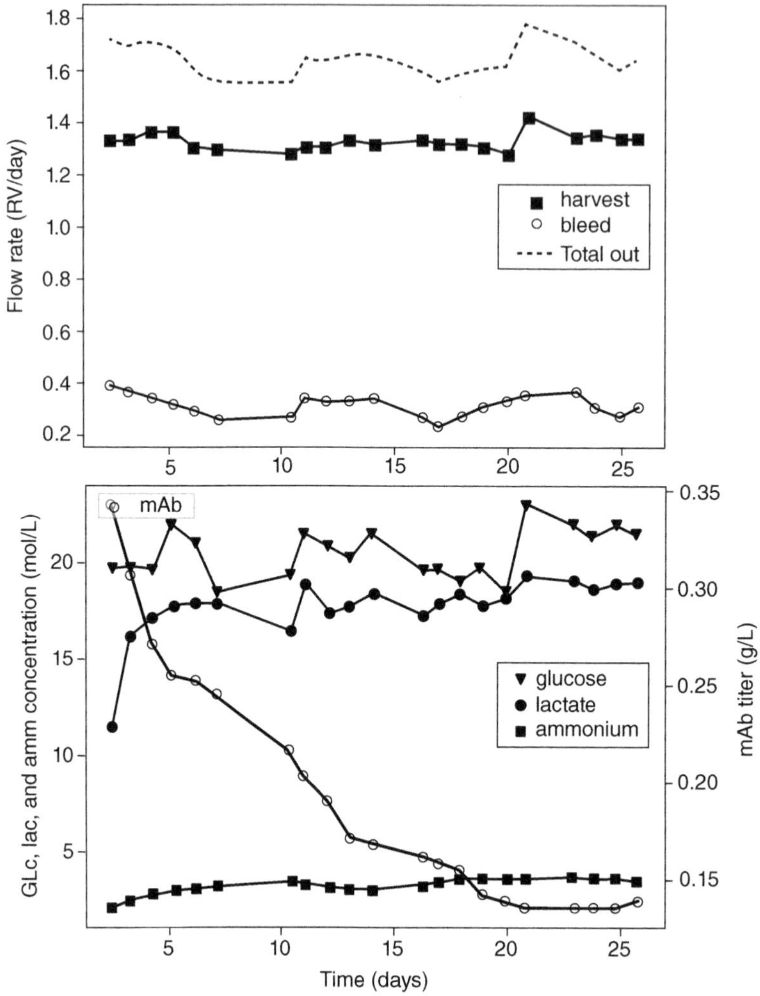

Figure 5.2 Cell culture dynamics. Flow rate (○ bleed, ▪ harvest), extracellular glucose, lactate, and ammonia concentrations, and mAb titer. *Source*: Adapted from [17].

Hidden layer(s): the model is able to learn complex patterns in the data through the use of hidden layers, which can consist of one or more layers. In order to process the information that is passed from the input layer and uncover intricate dependencies, these layers make use of nonlinear activation functions. Within the models, we make use of the sigmoid function (shown in Eq. 5.1).

$$f(x) = \frac{1}{1 + e^{-x}} \tag{5.1}$$

Output layer: finally produces the model's prediction for the mAb yield. This layer is known as the output layer.

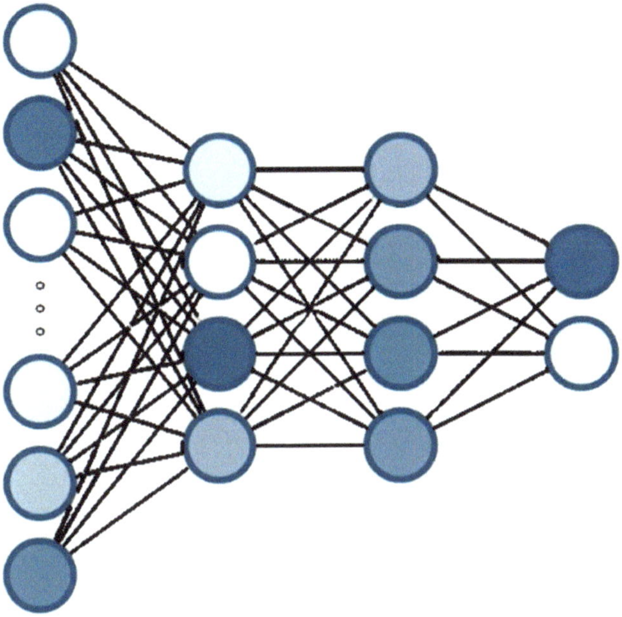

Figure 5.3 A visual depiction of passing image data through an NN, getting a classification.

The complexity of the data and the particular characteristics of the mAb production process are typically taken into consideration when making the decision regarding the network architecture, the number of layers, and the number of nodes that would be included in each layer. *MATLAB R2023a* is used here for the construction, training, and validation of the NN models used. The NN configurations used in model development, training, and validation included three, four, and five hidden layers with varied number of neurons.

For the initial scenario, three hidden layers are selected, each containing 15 neurons. This configuration aims to improve accuracy. In the second scenario, a comparable number of neurons is utilized as in the first scenario, but in a varied distribution—specifically, there are 4 hidden layers, each consisting of 10 neurons. The incorporation of more hidden layers poses the risk of overfitting and has been observed to substantially prolong the parameter estimation time, without contributing to enhanced model accuracy in this specific application. Additionally, a comparable number of neurons is utilized in the final scenario, just as it was in the first and second scenarios; however, the distribution of these neurons is again different. More specifically, there are five hidden layers, each of which is composed of ten neurons.

These configurations are used to make predictions, and the results of those predictions are measured and analyzed in order to identify and determine which configuration is the most appropriate. Plots of the metabolites glucose and lactate, as well as cell culture density and product titer, are displayed in Figures 5.4–5.7.

The data for glucose has the highest number of fluctuations, which is depicted here by relatively poor prediction results in comparison to the results for other metabolites (Figure 5.4). The NNs are found to be incapable of accurately capturing the original

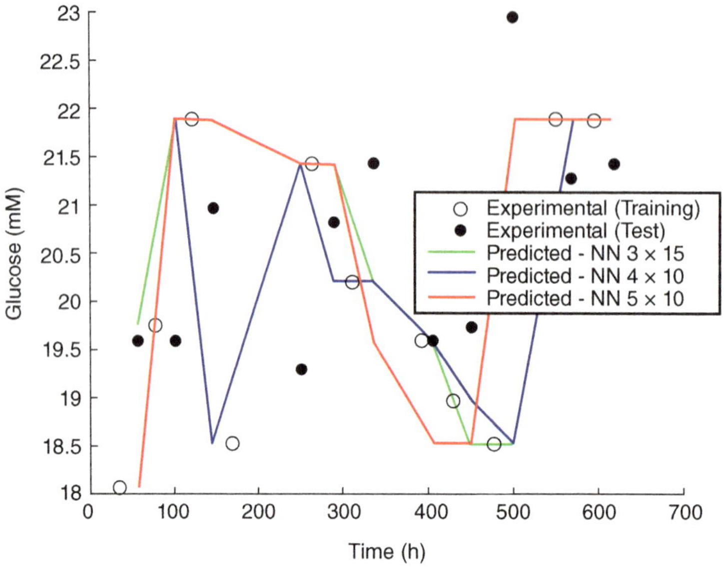

Figure 5.4 Comparison of NN model prediction vs. the experimental data for glucose. *Source*: Adapted from [17].

curve. It is not possible to determine the optimal configuration for prediction by using the results of glucose prediction measurements. In contrast, for the lactate experiments in Figure 5.5, all the models exhibit close predictive performance, nearly capturing every significant aspect. The 3 × 15 model covers a slightly wider range of points here than the others.

The data set for cell density in Figure 5.6 depicts all the models performing well throughout the span of data, almost perfectly catching the curve trend.

Plots in this section comprise predictions for mAb production rate. Figure 5.7 indicates that the model's predictive accuracy decreases for a subset of data points situated in the mid-range of the dataset, while the other two models are equally predicting quite closely. Based on these prediction results, we are now at the stage of choosing the appropriate configuration to continue with. With the backpropagation (BP) algorithm serving as the foundation, we are concentrating on formulating the optimization problem and then solving it using our NN model. After that, the NN model is retrained with the process observations that were collected.

5.2.3 BP Algorithm

The network architecture of the BP NN model (algorithm) is multilayered and is a gradient descent local optimization technique, which is related to the backward error correction of network weights (Figure 5.8). The multilayer structure makes the output of the model more accurate but may lead to the inability to find the global optimal solution.

Using the gradient descent principle, let the connection weights be corrected according to the specified learning rate. Gradient descent is a method of implementing backward

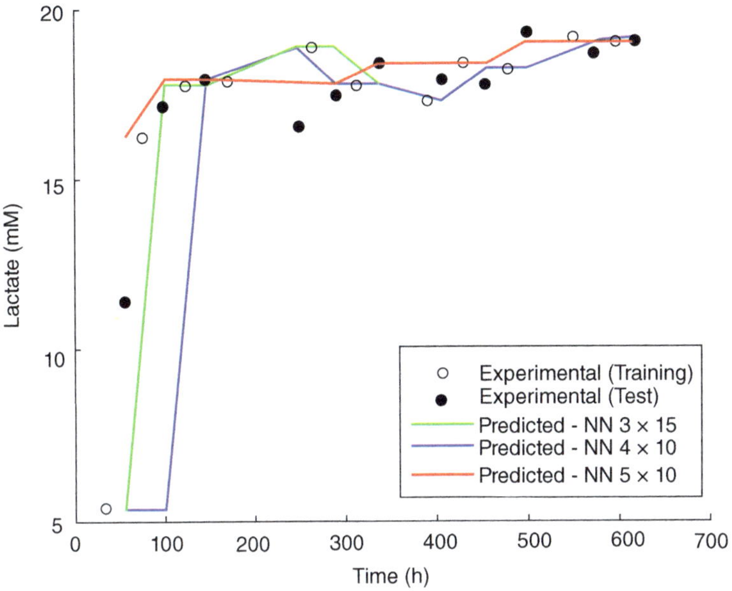

Figure 5.5 Comparison of NN model prediction vs. the experimental data for lactate. *Source*: Adapted from [17].

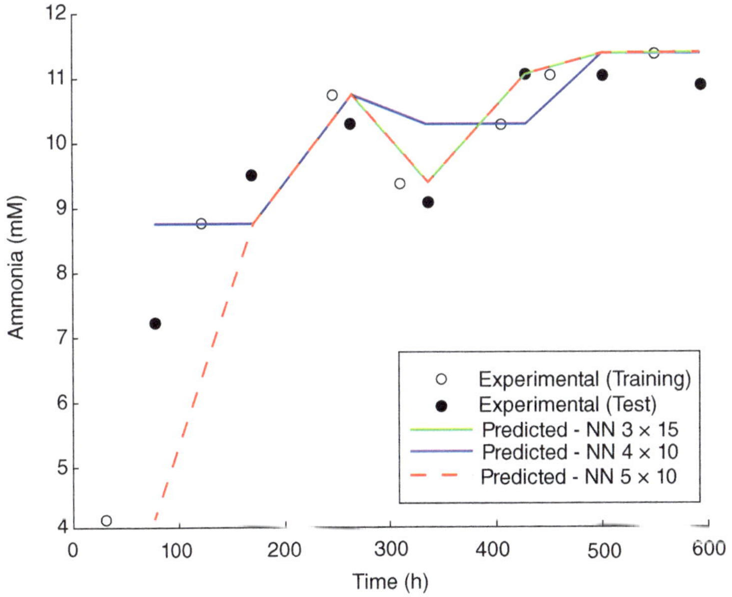

Figure 5.6 Comparison of NN model prediction vs. the experimental data for ammonia. *Source*: Adapted from [17].

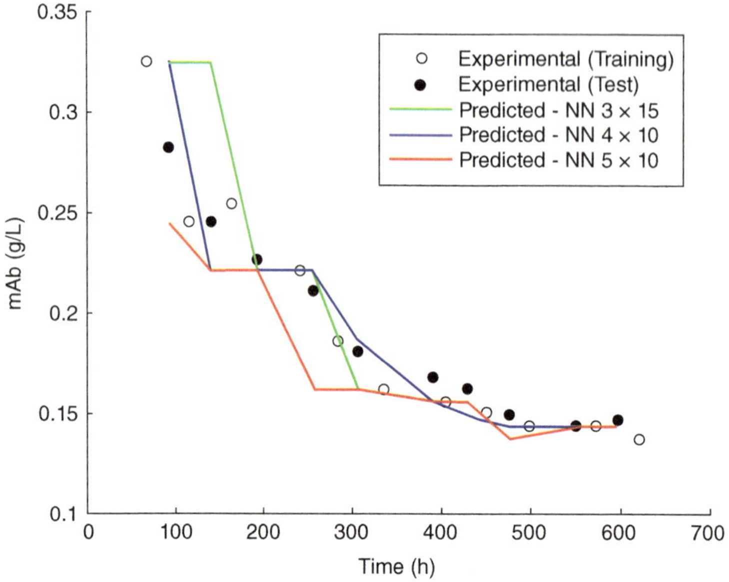

Figure 5.7 Comparison of NN model prediction vs. the experimental data for mAb. *Source*: Adapted from [17].

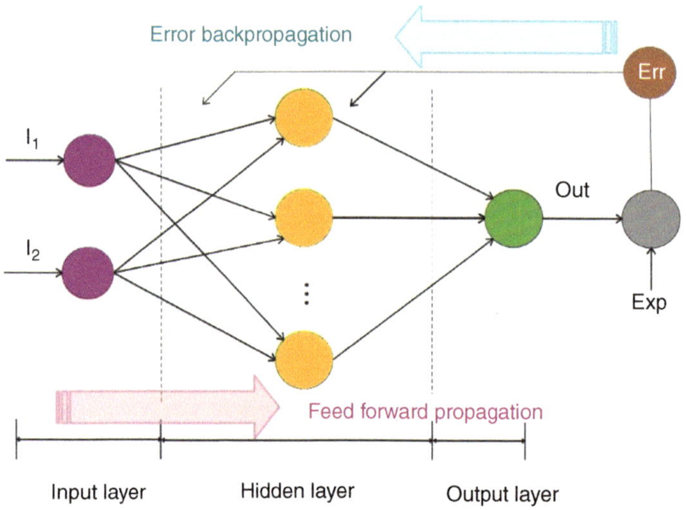

Figure 5.8 A typical BP NN structure. *Source*: [19]/with permission of Elsevier.

propagation of error that causes the weight of each training sample to change along a negative gradient until the error is reduced to its minimum. Genetic operators are utilized to complete the crossover and mutation processes, which ultimately results in the production of new decompositions. The selection procedure is validated based on the fitness level of each individual. In accordance with the concept of survival of the fittest (natural evolution mechanism), it is possible for the approximate solutions to be generated generation by generation after the main population has emerged via the use of genetic selection and processes. In the genetic algorithm, it is specified that the fitness function must be positive and that it must constantly evolve in the direction of the fitness function's maximum value.

5.2.4 Optimization Procedure

By modifying the parameters of the process, we intend to find a way to maximize the production of mAb. The NN model serves as a surrogate for the complex production process. We aim to maximize mAb yield while maintaining product quality, which is controlled by setting various constraints to our optimization problem.

The optimization procedure here is to find feasible process inputs for achieving optimal values of the objective function. This requires determination of the NN input activation values for maximizing the specified objective function. In this case, enhancement in mAb production is desired. So, the final concentration of mAb is the objective function here.

$$\text{Obj} = \max C_{\text{mAb}} \tag{5.2}$$

Assuming that the reactor operational volume does not change, the following balance equation can be used for a perfusion reactor:

$$F_{\text{in}} = F_{\text{out}} = F_{\text{Harvest}} + F_{\text{Bleed}} \tag{5.3}$$

The process constraints are selected based on the available data from other metabolites, including inlet flow rate, lactate, and ammonia concentration. We take their values as follows:

$$1.4 \le C_{\text{FR}} \le 2.0 \tag{5.4}$$

$$3.5 \le C_{\text{lac}} \le 14 \tag{5.5}$$

$$0.5 \le C_{\text{amm}} \le 16 \tag{5.6}$$

Here, C_{mAb} is mAb concentration, and C_{FR}, C_{lac}, and C_{amm} are values of flow rate and concentrations of lactate and ammonia, respectively.

In this work, three input variables (flow rate, and concentrations of lactate and ammonia), based on their potential impact on the output, are selected to maximize that one output (mAb titer). The study requires developing a mathematical model that accurately describes the relationship between mAb and the input variables. Given the complexity of the process, this relationship is often nonlinear, necessitating the use of nonlinear regression techniques. So, from the given data, one nonlinear mathematical function was approximated using nonlinear regression.

To model a nonlinear relationship between the dependent variable (C_{mAb}) and the independent variables (C_{FR} and C_{lac}, and C_{FR} and C_{amm}), we start with a polynomial model and then fit the model using nonlinear least squares. We take two sets of input pairs (C_{FR} and C_{lac}, and C_{FR} and C_{amm}) to avoid making the problem overcomplicated. A general polynomial with two independent variables may take the form (Eqs. 5.7, 5.8):

$$f(C_{\text{mAb}}) = a + b\,C_{\text{FR}} + c\,C_{\text{lac}} + d\,C_{\text{FR}}^2 + e\,C_{\text{lac}}^2 + f\,C_{FR}\,.\,C_{\text{lac}} + g\,C_{\text{FR}}\,.\,C_{\text{lac}}^2 \tag{5.7}$$

$$f(C_{mAb}) = a + b\, C_{FR} + c\, C_{amm} + d\, C_{FR}^2 + e\, C_{amm}^2 + f\, C_{FR} \cdot C_{amm} + g\, C_{FR} \cdot C_{amm}^2 \quad (5.8)$$

where *a*, *b*, *c*, *d*, *e*, *f*, and g are parameters to be estimated. These nonlinear model parameters are estimated using a process that minimizes the discrepancy between the predicted and observed values of *C*. Here, Python script for least squares fitting is utilized to determine the optimal parameters for the best fit (Figures 5.9 and 5.10).

Once the parameters have been estimated, the model is validated to assess its accuracy and predictive capability.

Figures 5.9 and 5.10 illustrate a comparison between the original curve and the regression models, which are quite closer to actual values, and hence, the models (and the parameters) are accepted. The first BP NN model comprises 45 hidden nodes, evenly distributed in 3 layers as can be seen in Figure 5.11.

In the second model, we set a total of 4 hidden layers, with each layer comprising 10 neurons. Figure 5.12 shows this schematic.

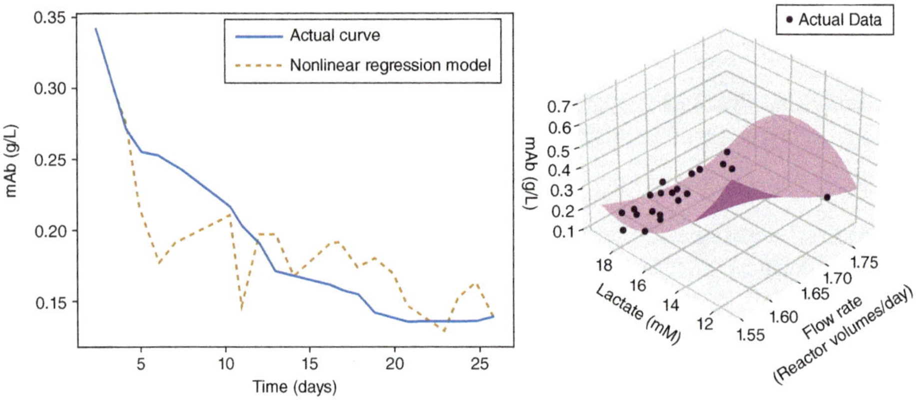

Figure 5.9 Comparison of the regression model with actual data for the first set of inputs C_{FR} and C_{lac}.

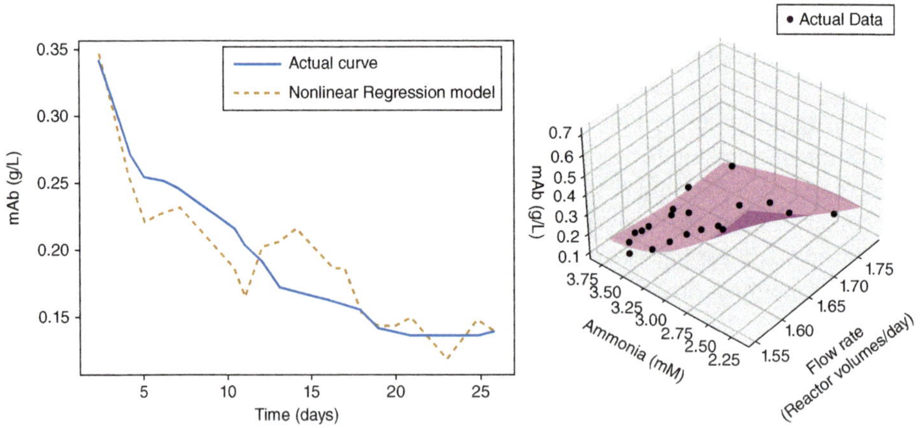

Figure 5.10 Comparison of the regression model with actual data for the second set of inputs C_{FR} and C_{amm}.

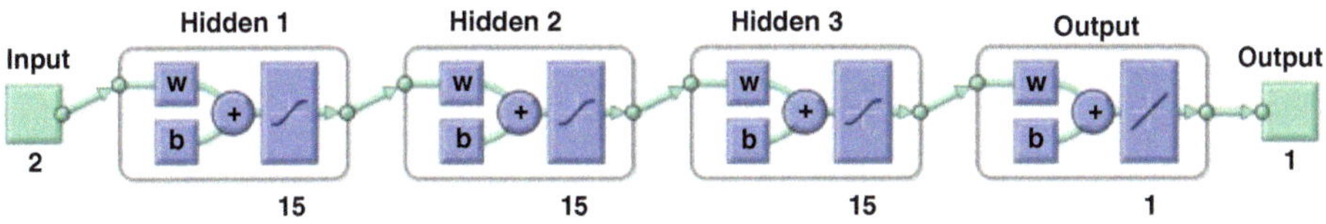

Figure 5.11 NN actual view in terms of layers/neurons BP NN Model 1.

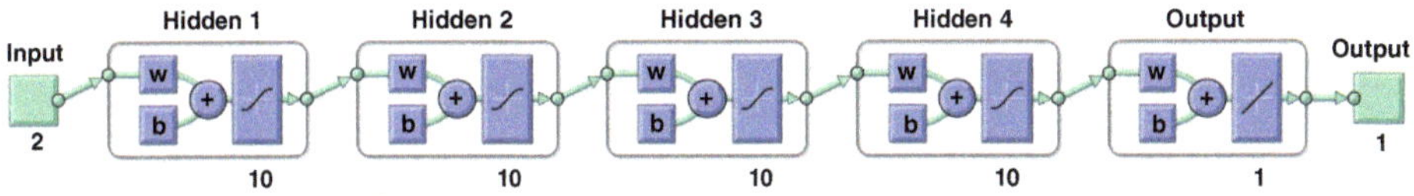

Figure 5.12 NN actual view in terms of layers/neurons BP NN Model 2.

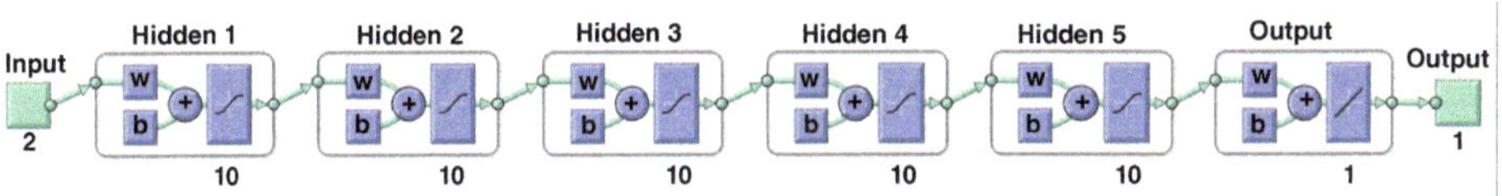

Figure 5.13 NN actual view in terms of layers/neurons BP NN Model 3.

The third BP NN model comprises 50 hidden nodes, evenly distributed in 5 layers as can be seen in Figure 5.13.

The process inputs are these concentration data from Figure 5.2 used as input layer nodes and the mAb concentration as 1 output layer node. In this way, the input vector includes a total of 2 nodes (Figures 5.11–5.13).

5.2.5 Training Strategy, Validation, and Testing

The feedforward (FF) network with 3 hidden layers, each having 15 neurons (BP NN model 1, Figure 5.11), the FF network with 4 hidden layers having 10 neurons in each layer (BP NN model 2, Figure 5.12), and the network with 5 hidden layers also having 10 neurons in each layer (BP NN model 3, Figure 5.13) using BP algorithm are developed in *MATLAB R2023a*. Following the methods outlined in Section 5.2, the datasets are partitioned into training, validation, and testing sets for the BP NN model. The model is trained using random 70% of the data, while the rest is reserved for validation and testing purposes (15% each). In order to ensure that the datasets do not contain any redundant information, the division proportion is an essential component that contributes to the overall efficiency of the model. After every training iteration, the performance of the model is meticulously evaluated on the test sets by computing the loss, which is a function of the predominant weight configuration. This is done in order to ensure that the model is performing as expected. The training performance is analyzed for training,

Table 5.1 Optimal hyperparameters for the BP NN models.

Hyperparameters	Model 1 values	Model 2 values	Model 3 values
Inputs	2	2	2
Number of hidden layers	3	4	5
Number of hidden layer neurons	45	40	50
Hidden layer activation functions	Sigmoid	Sigmoid	Sigmoid
Input/output layer transfer functions	Linear	Linear	Linear
Epochs	1,000	1,000	1,000
Learning rate	0.1	0.1	0.1
Outputs	1	1	1

validation, and testing data. The *MATLAB R2023a* libraries fitnet and feedforwardnet were used for training and testing the NN model. The nonlinear activation function used in the model was sigmoid function.

Finally, the optimization algorithm gradient descent (as function trained in *MATLAB R2023a*) is applied to find the optimum input parameters, which resulted in maximizing the output. Mesh grids are formed for the input data of both the inputs, and the optimum value input parameters are searched/calculated. The hyperparameters for training all three NN models are listed in Table 5.1.

5.3 Results

5.3.1 Optimization Results of BP NN Model 1

The gradient descent optimizer function is used for the BP NN training procedure, with a maximum iteration limit of 1,000. The evaluation of the model's performance was largely measured using accuracy and loss metrics.

Figures 5.14 and 5.15 illustrate the results of the optimization process that were produced for both sets of input pairs that were utilized. These numbers make it very clear that high lactate quantities are the conditions under which the highest possible yield is achieved. It is also recommended that flow rate be at a moderately higher level. Additionally, the concentration of ammonia is an important factor in this. Unlike the case above, higher amounts of ammonia at a moderately low flow rate are boosting the mAb output.

5.3.2 Optimization Results of BP NN Model 2

Regarding the second model, a higher number of hidden layers but less neurons overall are utilized throughout. In comparison to the earlier model, the outcomes of the optimization process are distinct.

One more time, a higher mAb output is observed when lactate levels are high with high inflow (Figure 5.16).

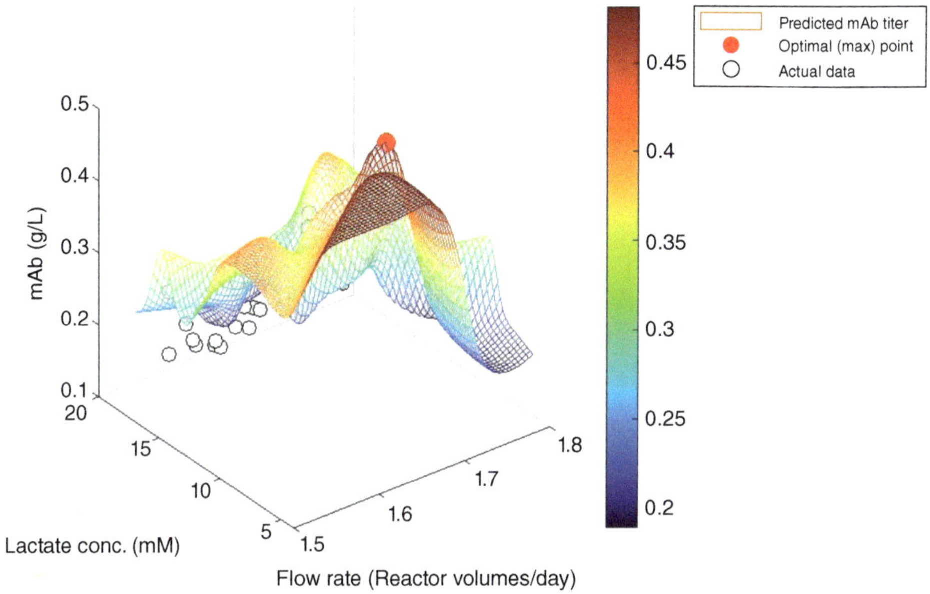

Figure 5.14 3D plot: mAb titer against flow rate and lactate concentration–BP NN Model 1. The red pointer specifies the maximum mAb concentration obtained.

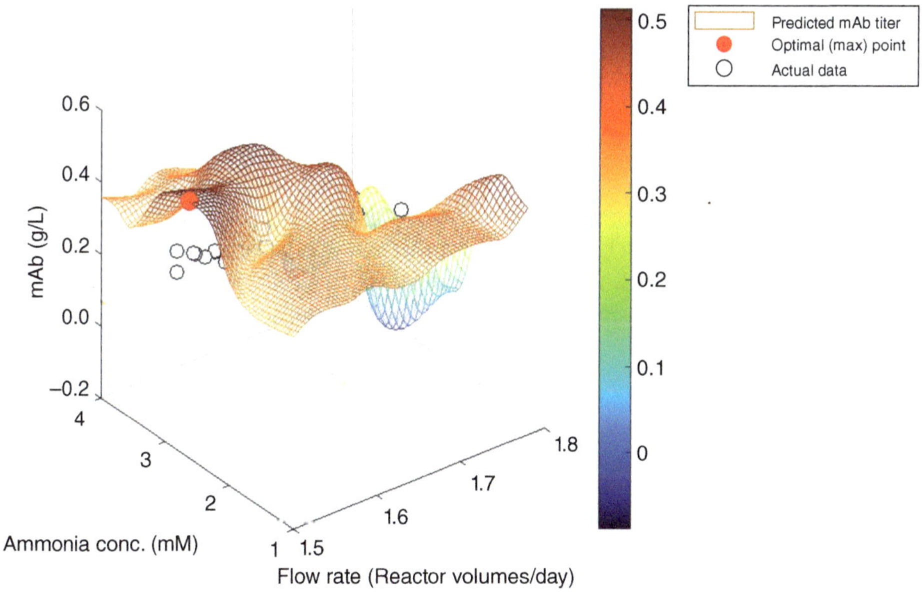

Figure 5.15 3D plot: mAb titer against flow rate and ammonia concentration–BP NN Model 1. The red pointer specifies the maximum mAb concentration obtained.

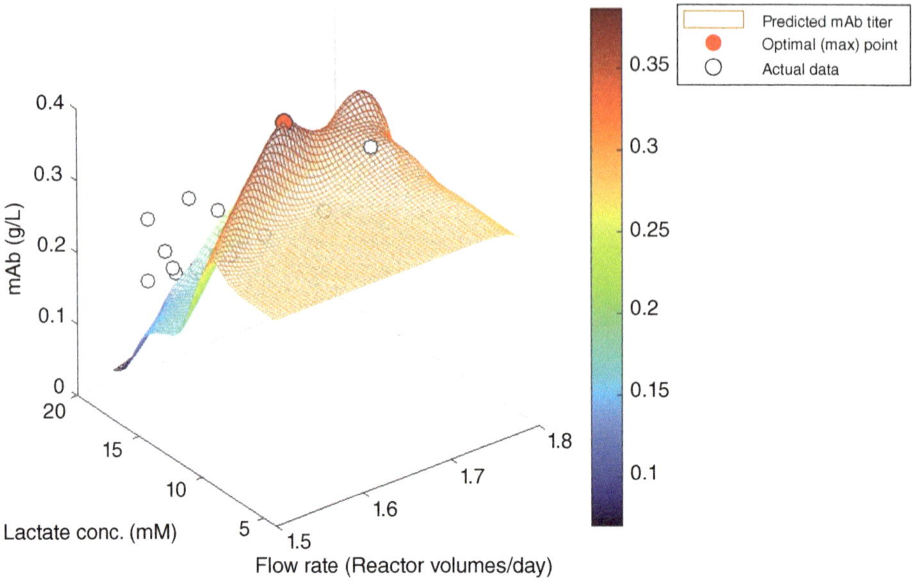

Figure 5.16 3D plot: mAb titer against flow rate and lactate concentration—BP NN Model 2. The red pointer specifies the maximum mAb concentration obtained.

However, in the second case, higher concentrations of ammonia at moderate flow are producing better yields (Figure 5.17).

5.3.3 Optimization Results of BP NN Model 3

In this third model, an even higher number of hidden layers (5) and higher number of neurons (50) than the previous models are utilized throughout. In comparison to the earlier models, the outcomes of the optimization process are quite similar to the first model.

Again, at high flow and high lactate, a higher mAb output is observed (Figure 5.18), same as with BP NN models 1 and 2. The case with ammonia, however, resembles the outcomes of BP NN model 1, with better mAb yield seen at low flow rate, keeping ammonia at higher levels (Figure 5.19).

5.3.4 Performance Comparison of Models

A comparison of the performance of BP NN models 1, 2, and 3, both of which were designed with optimal settings, was evaluated with the use of test datasets, as shown in Table 5.2. Following a total of one thousand iterations, the accuracy with which model 1 performed on the test sets is 0.45 g/L for the first set of inputs, while for the second set it is 0.42 g/L, accordingly. On the other hand, model 2 is able to reach a yield of 0.34 g/L only on the first set, while simultaneously increasing the production to 0.39 orders of magnitude. BP NN model 3 has even lower yields, with the first set producing a maximum of 0.33 g/L mAb, and the second input set yielding 0.32 g/L. More consistency was seen in the optimal values that were approximated by BP NN model 1. Improved performance can be due a lower

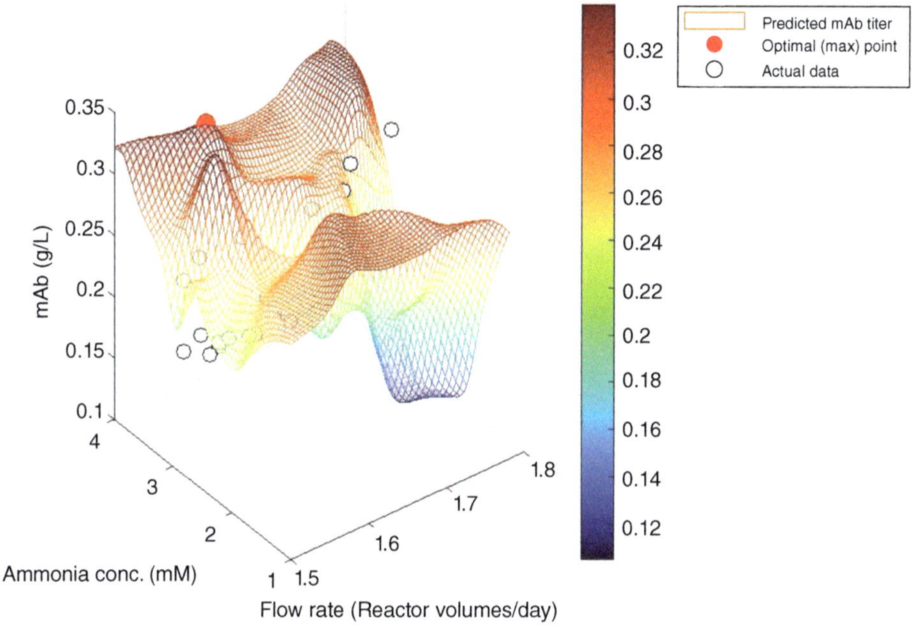

Figure 5.17 3D plot: mAb titer against flow rate and ammonia concentration—BP NN Model 2. The red pointer specifies the maximum mAb concentration obtained.

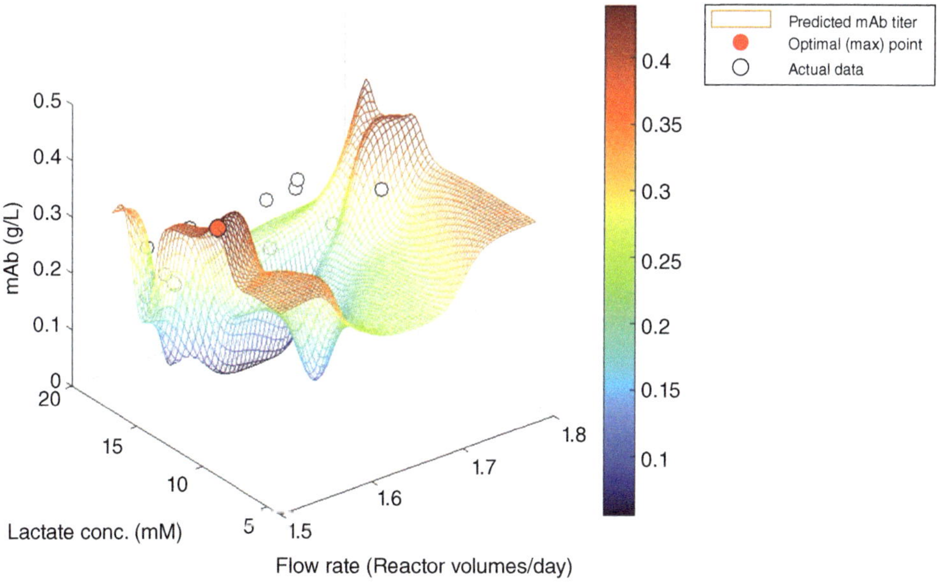

Figure 5.18 3D plot: mAb titer against flow rate and lactate concentration—BP NN Model 3. The red pointer specifies the maximum mAb concentration obtained.

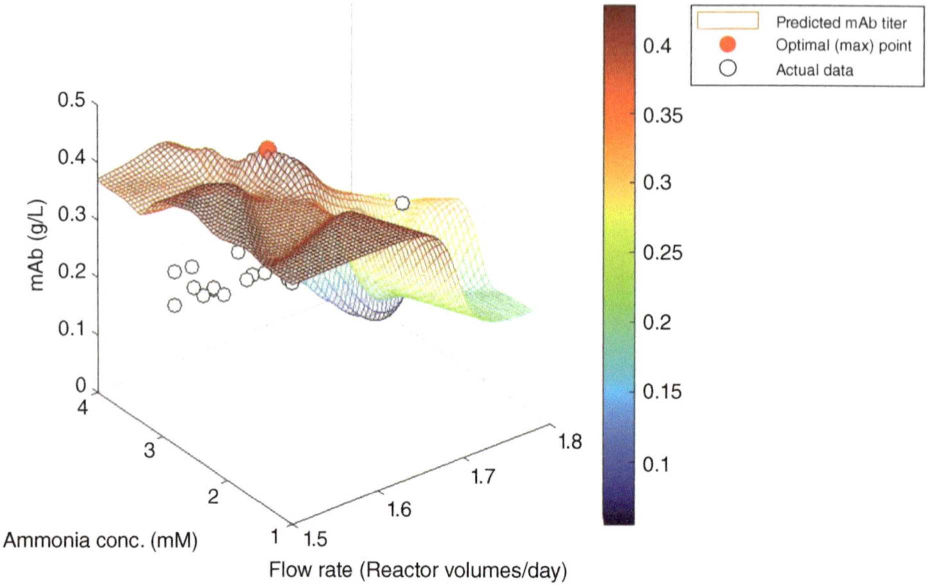

Figure 5.19 3D plot: mAb titer against flow rate and ammonia concentration—BP NN Model 3. The red pointer specifies the maximum mAb concentration obtained.

Table 5.2 Comparison of the approximated results for the test cases in models 1, 2, and 3.

	Approximations		
Model features	**Model 1**	**Model 2**	**Model 3**
Optimal flow rate	1.69 RV/day	1.65 RV/day	1.50 RV/day
Corresponding optimal lactate	9.73 mM	13.47 mM	8.85 mM
Maximum predicted mAb yield	0.48 g/L	0.39 g/L	0.44 g/L
Optimal flow rate	1.54 RV/day	1.57 RV/day	1.62 RV/day
Corresponding optimal ammonia	3.14 mM	3.44 mM	2.98 mM
Maximum predicted mAb yield	0.42 g/L	0.34 g/L	0.42 g/L

number of hidden layers incorporated or because of moderate number of total neurons. This ensures that the risk of early convergence into local minima is reduced, which in turn leads to improved performance.

5.4 Conclusion

The primary objective of this study was to create and evaluate two BP NN models in order to determine how well they approximate the optimal generation of mAbs. It has been discovered that the number of neurons present in the hidden layer has a substantial impact on the capability of the model to precisely match the data. In light of this, an analysis was carried out to investigate the impact that changing the number of neurons in the hidden layer, which may range from one to three, had on the performance of the model. In order to make reliable forecasts, the algorithms made use of essential traits while also taking into account nutrient levels and cell density.

The following is a list of the most important findings from the study:

1. The inlet flow rate, lactate concentration, and ammonia concentration were the three factors that were changed in order to build a thorough dataset. This dataset was created from four different experimental sets. Both of the BP NN models were trained on this dataset in order to make predictions about the results of the experiments.
2. There was a maximum yield of 0.48 g/L produced by BP NN Model 1, whereas there was a minimum loss of 2.97×10^{-2} percent.
3. In contrast, BP NN Models 2 and 3 produced yields that were substantially lower, coming in at 0.39 g/L and 0.44 g/L, respectively.
4. Model 3, on the other hand, shows the least mean squared error of 1.3×10^{-2} compared to the two.
5. When compared to Models 2 and 3, Model 1 displayed a higher level of approximation accuracy, which makes it a potentially useful choice for optimizing mAb production in this case.

According to the findings of this study, a greater number of hidden layers resulted in overfitting and did not offer any advantages in terms of performance. Model 1 produced the best results, despite having the fewest number of hidden layers (three), yet each of those layers included 15 neurons.

When the total number of neurons in the models was taken into consideration, Model 1 contained a total of 45 hidden neurons (15 in each of the 3 layers), Model 2 contained 40 neurons in hidden layers, and Model 3 contained the largest number of hidden neurons, which was 50. Additionally, the mAb yield that was produced by Model 2 was the lowest, due to the fact that it had the fewest hidden neurons. Model 3, which had the highest quantity, produced the same amount of mAb in the second set of inputs as Model 1 (0.42 g/L) but overall produced a significantly higher quantity of mAb (over 21%) compared to Model 2. On the other hand, Model 1 had a number of neurons that was somewhere in the middle of the other two models, but it produced the largest amount of mAb in both of the input sets. Model 1 produced 27% and 21% more mAb in the first and second input sets, respectively, than Model 2.

It is possible that the models that were produced as a result of this study have the ability to significantly improve the effectiveness of investigations and to lower the production costs that are linked with optimization strategies in the biopharmaceutical industry.

References

1 Pantelides CC and Renfro JG. The online use of first-principles models in process operations: Review, current status and future needs. *Computers & Chemical Engineering* 2013;51:136–148.

2 Zavala VM, Laird CD, and Biegler LT. A fast moving horizon estimation algorithm based on nonlinear programming sensitivity. *Journal of Process Control* 2008;8:876–884.

3 Kravaris C, Hahn J, and Chu Y. Advances and selected recent developments in state and parameter estimation. *Computers & Chemical Engineering* 2013;15:111–123

4 Kelly B. Industrialization of mAb production technology: The bioprocessing industry at a crossroads. *Taylor & Fransis* 2009;1:443–452.

5 Sjöberg J, Zhang Q, Ljung L, et al. Nonlinear black-box modeling in system identification: A unified overview. *Automatica* 1995;31(12):1691–1724.

6 MacMurray JC and Himmelblau D. Modeling and control of a packed distillation column using artificial neural networks. *Computers & Chemical Engineering* 1995;19(10):1077–1088.

7 Wu Z, Rincon D, Luo J, and Christofides PD. Machine learning modeling and predictive control of nonlinear processes using noisy data. *AIChE Journal* 2021;67(4):e17164.

8 Kumar P and Rawlings J. Structured nonlinear process modeling using neural networks and application to economic optimization. *Computers & Chemical Engineering* 2023;177:108314.

9 del Val IJ, Kontoravdi C, and Nagy JM. Towards the implementation of quality by design to the production of therapeutic monoclonal antibodies with desired glycosylation patterns. *Biotechnology Progress* 2010;26:1505–1527.

10 Kremkow BG, Lee K, and Glyco-Mapper. A Chinese hamster ovary (CHO) genome-specific glycosylation prediction tool. *Metabolic Engineering* 2018;47:134–142.

11 Medlock GL and Papin JA. Guiding the refinement of biochemical knowledgebases with ensembles of metabolic networks and machine learning. Cell Systems 2020;10(1):109–119.

12 Eon-Duval A, Broly H, and Gleixner R. Quality attributes of recombinant therapeutic proteins: An assessment of impact on safety and efficacy as part of a quality by design development approach. *Biotechnology Progress* 2012;28:608–622.

13 Hossler P. Protein glycosylation control in mammalian cell culture: Past precedents and contemporary prospects. *Advances in Biochemical Engineering Biotechnology* 2012;127:187–219.

14 del Val I, Fan Y, and Weilguny D. Dynamics of immature mAb glycoform secretion during CHO cell culture: An integrated modelling framework. *Biotechnology Journal* 2016;11:610–623.

15 Stanley P. Golgi glycosylation. *Cold Spring Harbor Perspectives in Biology* 2011;3(4):a005199.

16 Reily C, Stewart TJ, Renfrow MB, and Novak J. Glycosylation in health and disease. *Nature Reviews Nephrology* 2019;15(6):346–366.

17 Tharmalingam T, Wu CH, Callahan S, and Goudar CT. A frame- work for real-time glycosylation monitoring (RT-GM) in mammalian cell culture. *Biotechnology and Bioengineering* 2014;112:1146–1154.

18 Karst D, Scibona E, Serra E, et al. Modulation and modeling of monoclonal antibody N-linked glycosylation in mammalian cell perfusion reactors. *Biotechnology and Bioengineering* 2017;114(9):1978–1990.

19 Hai N, Wang S, Liu D, Fernandez C, and Guerrero J. A novel genetic weight-directed feed forward backpropagation neural network for state of charge estimation of lithium-ion batteries. *Journal of Energy Storage* 2024;88:111549.

Part II

AI for Pharmaceutical Manufacturing

6

Artificial Intelligence and Pharmaceutical Manufacturing

*R. Yuvaraj, R. Rithika, L. Charlie Jelura, and M. Manigandan**

PG and Research Department of Microbiology, Sacred Heart College (Autonomous), Tirupattur, India

*Corresponding author: senthilmanigandan@gmail.com

CONTENTS

6.1 Introduction

The pharmaceutical industry is expanding at a phenomenal pace owing to several reasons such as the need for cost-effective production, the prevalence of drug-resistant diseases, enhanced regulatory requirements, and enhanced demands for quality drugs. To counter such challenges and improve resources, compliance, and industrial efficiency, new technology and data-driven approaches are necessary. Instant access to a broad range of data in drug industry operations and improving laboratory efficiency have dramatically enhanced over the past few years owing to improvements in automation technology and increased processing capacities [1, 2]. To uncover routes for the synthesis of novel drugs, a range of high-throughput automation platforms have been developed for application in different stages of drug industry operations, including palladium-catalyzed cross-coupling reactions in room-temperature dimethyl sulfoxide that are amenable to automation [3]. They performed nanomole-scale modeling for different continuous variables and conducted more than 1,500 chemical experiments in a day using robots in combination with mass spectrometry-based high-throughput analytical methods.

An automated flow-based synthesis platform capable of performing both micromole-scale synthesis and nanomole-scale reaction screening was developed [4]. Such automated processes accelerate the design of medicinal compounds and the construction of solid synthetic routes. This new method accelerates drug discovery and encourages innovation in pharmaceutical design by enabling the easy creation of complex therapeutic formulations that were previously nearly impossible to produce [5]. The transition to individualized treatment is one of the largest challenges to the healthcare industry. Technology for three-dimensional (3D) printing is essential to solving this problem, since the "one-size-fits-all" method does not take into account each patient's particular demands. Traditional mass manufacture of medications usually fails to meet expectations. Clinical pharmacy are transformed when 3D printing is included into healthcare settings because it makes drug manufacture more patient-centered and accessible [6]. To make sure its successful acceptance, a number of obstacles must be overcome before 3D printing may be widely used in the pharmaceutical sector. The involvement of healthcare professionals, quality assurance, and regulatory compliance are major challenges that need immediate attention. For patient treatment to be stable and reliable, unified standards in the manufacture, safety, and efficacy of 3D-printed drugs need to be defined. A significant development that highlighted additive manufacturing's promise for the pharmaceutical sector was the food and drug administration (FDA) approval of the first 3D-printed medicine. This achievement, however, also highlights the ongoing need for rigorous regulatory guidelines capable of keeping pace with the rapid technological developments. Furthermore, these guidelines need to ensure that innovation in drug design and production is fostered while patient safety, dose consistency, and product stability are ensured. Establishing best practices and the expanded deployment of 3D-printed drugs in clinical and commercial settings will also rely on collaboration among regulatory bodies, drug companies, and medical practitioners. To emphasize the function of 3D printing in future pharmaceutical development, our review will focus on novel dosage forms enabled by this technology. Industry potential and challenges will be determined by evaluating the current needs, specifications, and manufacturing processes for pharmaceutical dosage forms tailored with 3D printing [7].

In addition, from enhancing real-time monitoring of important parameters to automation of complex manufacturing processes, AI is increasingly playing a significant role in the pharmaceutical manufacturing sector. Machine learning (ML) algorithms, particularly deep learning (DL) and reinforcement learning, have demonstrated remarkable promise in predicting production outcomes, optimizing resource allocation, and minimizing downtime. Pharmaceutical production is on the verge of a revolution due to AI-optimized process optimization, which will offer a more scalable, cost-effective, and efficient way of manufacturing drugs. This chapter discusses how pharmaceutical manufacturing has evolved over the years, reviewing design space ideas, simple tools for increasing productivity, process optimization methods, and emerging technology such as AI and 3D printing. It also discusses the challenges the industry is experiencing and the future that will shape the next generation of pharmaceutical manufacturing [8].

6.2 Design Space in Pharmaceutical Industry

International Council for Harmonisation (ICH) Q8(R2) guidelines have defined the design space as the multidimensional combination of important critical process parameters (CPPs) and critical material properties (CMPs) that ensure product quality. One of the fundamental elements of Quality by Design (QbD) is the establishment of a well-defined design space, which allows manufacturers to develop pharmaceutical processes that are robust, reproducible, and efficient. To establish and refine the design space, advanced statistical and mathematical modeling techniques such as kriging interpolation, dynamic kriging, and dynamic neural networks (DNNs) are now necessary. To demonstrate these methods practical application in pharmaceutical production, this paper covers them and presents a case study of a roller compactor. The importance of monitoring and control in real time in pharmaceutical production is also underscored by regulatory systems such as the FDA's process analytical technology (PAT) guidance [9].

6.2.1 Enhanced Modeling Techniques for Design Space Development

With advanced modeling techniques, the accuracy and efficiency of design space definition are enhanced. Dynamic kriging is a sophisticated interpolation method employed to predict missing values in a high-dimensional space based on available data points [10]. Dynamic kriging is an effective method for pharmaceutical process optimization since, unlike traditional kriging, it adapts immediately to varying conditions. On the other hand, ML algorithms known as DNNs have the ability to find complex, nonlinear relationships between inputs and outputs [11]. These networks are particularly beneficial in deciding in real time by predicting how changes in processes would impact final output quality. Additionally, Montgomery [11] discusses how experimental design methodologies are crucial for structuring data acquisition for process optimization. Manufacturers are able to systematically navigate design space and deliver effective and reliable process control through the integration of these statistical methods with pharmaceutical modeling techniques. Enhanced process control, reduced variability, and improved regulatory compliance are facilitated by using these models in pharmaceutical manufacturing [12].

6.2.2 Process Optimization of Roller Compactors

The roller compactor example is a significant demonstration of the use of advanced modeling techniques to implement design space. Roller compaction in the pharmaceutical industry is one of the typical dry granulations whose quality attributes of a tablet directly rely on as density, porosity, and disintegration/dissolution rate. Process-critical parameters, for instance, feed rate, gap setting, and roll speed, were kriging interpolation-modeled and optimized within this case study. By determining the optimal operating conditions, the model generated a predictive design space that minimized material loss and maximized product uniformity. In addition, the process model was revised in real time using DNNs and kriging interpolation. Pharmaceutical manufacturing processes become more dependable and efficient as a result of these techniques ability to precisely optimize operations, provide real-time flexibility, and improve regulatory compliance [13].

6.3 Fundamental Tools for Increased Process

In order to guarantee patient safety and product quality, quality risk management, or QRM, is an integrated strategy to identifying, evaluating, and controlling risks throughout a pharmaceutical product's lifespan [14]. Initiation, risk assessment (which determines the likelihood and seriousness of harm) [15], control measure execution, periodic review [16], and stakeholder communication are the five main steps of the QRM process. Failure mode and effects analysis (FMEA) for quantitative risk evaluation, hazard analysis and critical control points (HACCP) for identifying important process points, and Ishikawa diagrams for root cause analysis are examples of common risk assessment techniques. These techniques are used in a variety of pharmaceutical procedures, such as cleaning validation, process validation, and equipment qualification. Although QRM is crucial for preserving quality systems, its present drawbacks include its dependence on qualitative approaches and its disjointed knowledge management procedures [17–19].

6.3.1 Pharmaceutical Mechanistic Workflow Model

In pharmaceutical production, mechanistic process modeling has become a potent tool that allows for a deeper comprehension and optimization of crucial unit activities using computational methods based on physics. From improving basic process knowledge through qualitative “learning models” to facilitating reliable scale-up and technology transfer with precise digital representations, these models fulfill a number of crucial roles. Sensitivity studies are used extensively in QRM to help determine CPPs and guide failure mode evaluations. Current applications focus on four primary domains: fluidic systems modeling using computational fluid dynamics (CFD) techniques like RANS, LES, and DNS for bioreactors and crystallizers [20–22]; particle-based process simulation through discrete element methods (DEM) for powder handling operations [23].

6.3.2 Multiphase Flows and Fluidic Systems

Pharmaceutical unit operations involving fluidic and multiphase systems, such as stirred tanks, crystallizers, and bioreactors, are frequently simulated using CFD. Reynolds-Averaged Navier-Stokes (RANS) solvers are commonly used in CFD approaches for turbulence modeling, as demonstrated in research on the effects of agitation on protein stability [23] and biotechnology applications [24]. Micro-mixing models improve accuracy in rapid reactions. Advanced techniques include direct numerical simulations (DNS) and large eddy simulations (LES), which solve large-scale turbulence. Volume of fluid, Lagrangian, or front-tracking techniques are used to evaluate multiphase flows. Phase interactions and scattered phase tracking are handled by the Euler–Euler (EE) and Euler–Lagrange (EL) models, respectively. Alternative solutions are provided by particle-based techniques such as the Lattice-Boltzmann Method (LBM) [25] and smoothed particle hydrodynamics (SPH) [26].

6.3.3 Particle-based Systems

Particulate procedures like blending, granulation, milling, and tablet coating are frequently used in the pharmaceutical production industry. These procedures work with intricate granular fluxes made up of different sized, shaped, and mechanically characterized particles. Historically, continuum-based methods were employed; however, new developments enable mechanistic simulations through the application of the DEM [27].

DEM uses either the soft-sphere (enduring contacts) or hard-sphere (instantaneous collisions) technique to calculate collision forces and trajectories, modeling particles as separate pieces. The soft-sphere model is frequently applied to dense powder flows in pharmaceutical manufacture. Newton's second law is solved for every particle in DEM simulations, taking into account friction coefficients and material characteristics like Young's modulus [28].

6.4 AI Technologies in Pharmaceutical Manufacturing

In the process replicating human intellect in computers, AI makes it possible for them to carry out intricate tasks like pattern recognition, decision-making, and process optimization. AI is essential to pharmaceutical manufacturing because it automates processes, boosts productivity, and ensures adherence to strict regulatory requirements (such as those set forth by the FDA, EMA, and ICH). There is growing demand on the pharmaceutical sec tor to increase manufacturing efficiency while upholding the highest standards of quality. Manual supervision and sporadic quality checks are common in traditional manufacturing processes, which can result in irregularities and inefficiencies. AI solves these issues by offering ongoing, data-driven insights that improve output and product dependability [29].

6.5 AI Tools in Pharmaceuticals

AI includes a number of cutting-edge technologies, each with special advantages for use in pharmaceutical applications. Machine learning (ML): Predictive and process-optimizing algorithms that learn from past data. Unsupervised learning finds hidden patterns without labels, whereas supervised learning models categorize data and anticipate results [30]. Deep learning (DL): A branch of ML that processes complicated data, including images and sensor signals, using multi-layered neural networks. When it comes to visual inspection systems for identifying tablet flaws, DL works very well [31]. AI systems that learn the best course of action through trial and error while getting input from their surroundings are known as reinforcement learning (RL) systems. In real-time manufacturing, RL is useful for dynamic process modifications. By lowering human error and improving operational precision, the integration of these technologies facilitates more intelligent decision-making [32].

6.5.1 AI Systems and Algorithmic Procedures in Drug Manufacturing

Pharmaceutical manufacturing is changing as a result of AI's many algorithmic approaches that are suited to certain production problems. Yield prediction and equipment failure forecasting are made possible by regression analysis; batch output projections using linear regression models have an accuracy rate of 95% [33]. Quality control is automated by classification algorithms such as support vector machines (SVMs) and decision trees, which have shown 98% reliability in determining whether final goods pass or fail [34].

Convolutional neural networks (CNNs) evaluate tablet coatings at previously unheard-of speeds for visual inspection, achieving 99.3% accuracy in identifying minute flaws [35]. In order to avoid unscheduled downtime, recurrent neural networks (RNNs) analyze sequential process data and correctly forecast maintenance requirements with 90% accuracy [36]. Through dynamic compression modifications, Q-learning systems reduce tablet weight variance by 30%, demonstrating how reinforcement learning optimizes real-time parameters. In smart factories, these AI models are progressively integrating with IoT networks; together, they have reduced quality incidents by 40% and increased production efficiency by 25% [37].

6.5.2 Applications of AI in Pharmaceutical Production

By improving quality control, facilitating predictive maintenance, and streamlining production processes, AI is revolutionizing the pharmaceutical manufacturing industry. AI-driven algorithms in production optimization use real-time and historical data to automate scheduling, which improves resource allocation and cuts down on idle time by up to 40% [38]. For example, in the production of biologics, ML models dynamically modify fermentation parameters (such as oxygen levels and nutrient feed rates), resulting in a 25% increase in cell growth yields and a 30% decrease in batch failures. AI helps quality control by using automated visual inspection methods that use DL to identify tablet flaws (such cracks or discoloration) with 99.8% accuracy, which is a substantial improvement over manual examinations [39]. By processing thousands of units per minute, these systems guarantee adherence to strict regulatory requirements. Furthermore, AI-powered predictive maintenance reduces unscheduled downtime by 45% by analyzing sensor data (temperature, vibration) to detect equipment faults 72 hours in advance [40]. From optimizing lyophilization cycles (which reduce freeze-drying durations by 20%) to adaptive clean-in-place (CIP) systems that reduce water and solvent use by 35%, case examples highlight AI's adaptability. Pharmaceutical producers can achieve consistent product quality, reduced costs, and increased efficiency by combining AI with Internet of Things (IoT)-enabled smart factories.

6.5.3 AI in Pharmaceutical Production: Advantages, Difficulties, and Solutions

Increased operational efficiency through optimized workflows that cut batch processing times by 30% [41], significant cost savings from predictive maintenance systems that reduce equipment repair expenses by 25–30% [42], and improved regulatory compliance through automated data logging that guarantees complete GMP audit trails [43] are just a few of the many advantages that come with integrating AI in pharmaceutical manufacturing. Data silos from incompatible legacy systems, stringent model validation criteria like FDA's AI/ML framework, and the need for workforce adaption for AI system operation are some of the obstacles to deployment, though. Explainable AI (XAI) methodologies for transparent decision-making, standardized data architectures to facilitate cross-platform interoperability, and staggered AI rollouts alongside parallel manual systems are all effective mitigating strategies [44]. Together, these strategies maximize AI's transformational potential in pharmaceutical manufacture while addressing implementation difficulties [45, 46].

6.6 Predictive Maintenance in Pharmaceutical Manufacturing

Predictive maintenance in pharmaceutical manufacturing is a data-driven approach that uses advanced analytics, ML, and loT technologies to monitor equipment performance and predict potential failures before they occur. It helps manufacturers reduce downtime, optimize maintenance schedules, and ensure compliance with strict industry regulations [47].

6.6.1 The Role of Predictive Maintenance in Pharmaceutical Manufacturing

Predictive maintenance has emerged as a critical strategy in pharmaceutical manufacturing, enabling proactive equipment monitoring to prevent failures before they disrupt production. Unlike traditional scheduled maintenance, this approach uses real-time data from sensors measuring vibration, temperature, and pressure to assess equipment health. For instance, atypical vibration patterns in a tablet press can signal bearing wear up to 200 operating hours before failure, allowing planned interventions. The European Medicines Agency's Annex 15 guidelines now explicitly recommend such condition-based monitoring for sterile manufacturing processes, citing 30–40% reductions in unplanned downtime. A 2023 industrial study of 57 facilities demonstrated that predictive maintenance programs improve overall equipment effectiveness (OEE) by 18–25% while extending asset lifespans by 2–3 years [48, 49].

6.6.2 Predictive Analytics for Equipment Monitoring Enhanced by AI

By using ML algorithms to evaluate intricate equipment data, AI has revolutionized predictive maintenance. While RNNs use vibration time-series data to forecast failures more than 72 hours in advance, CNNs can now detect motor bearing faults using thermal pictures with 94% accuracy. AI lowered false alarms by 60% when compared to conventional threshold-based systems, according to a case study conducted at Roche's biologics plant. The requirement for validated AI models is emphasized by regulatory frameworks such as ICH Q10, which mandates documentation of all training datasets and decision logic [49].

6.6.3 Application and Results of AI-powered Predictive Maintenance in the Pharmaceutical Industry

Pharmaceutical manufacturing may successfully implement AI-based predictive maintenance by using a methodical, staged strategy that yields quantifiable operational advantages.

Phase 1: Sensor integration is the first stage of deployment, during which 20–30 IoT sensors per important machine gather data in real time at a frequency of 1 kHz to track variables including pressure, temperature, and vibration [50].

Phase 2: To ensure dependable inputs for AI models, data preprocessing uses sophisticated methods like Kalman filtering to remove noise and improve data quality. To comply with FDA compliance requirements (21 CFR Part 11), ML algorithms are validated using at least 12 months of historical data [51].

Phase 3: Model training. This framework was applied across 14 manufacturing sites, resulting in $2.8 million in annual maintenance cost savings and improved compliance documentation [52]. Industry case studies further demonstrate the observable advantages of such implementations: Novartis used AI vision systems to detect defects in solid dosage coatings with a 98.7% accuracy rate, exceeding the 91% accuracy of manual inspections and Pfizer reported a 42% reduction in bioreactor downtime through AI-powered vibration analysis [53]. Given that McKinsey's examination of 37 pharmaceutical sites showed an average ROI of 4.2:1 for predictive maintenance projects, these results are in line with general industry. Collectively, these illustrations highlight how methodically used AI solutions can maximize equipment dependability, lower expenses, and improve quality control in the manufacturing of pharmaceuticals [54].

6.6.4 Scientific and Administrative Aspects of AI Integration in Pharmaceutical Manufacturing Systems

To guarantee successful implementation, the integration of AI into pharmaceutical manufacturing systems requires a thorough assessment of both technological and operational aspects. Technically speaking, smooth integration necessitates that Al algorithms work with the current manufacturing infrastructure, which frequently calls for changes to legacy systems in order to achieve the best data interoperability [55]. AI systems must be able to fulfill demanding computing needs, which may include cloud-based solutions or hardware updates, while processing real-time data streams from industrial equipment [56]. Manufacturers must operationally evaluate how AI will affect their current workflows, making any necessary modifications to standard operating procedures and implementing thorough training programs to give staff members the ability to decipher AI-generated insights [57].

Pharmaceutical businesses are using phased implementation strategies that allow for progressive system adaption and middleware solutions to overcome integration difficulties [58]. Standardized cleaning and validation procedures are used by strong data governance frameworks to guarantee the consistency and quality of inputs for AI systems. To provide comprehensive integration solutions, cross-functional cooperation between manufacturing teams, compliance officials, and AI professionals has proven crucial [59].

A 98.7% defect detection rate attained by AI vision systems in solid dosage manufacturing and a 42% decrease in bioreactor downtime through predictive maintenance systems [60] are two examples of successful case studies that highlight AI's transformative potential. According to McKinsey [61], these initiatives improve compliance paperwork and product quality while generally yielding a 4.2:1 ROI. Future developments are anticipated to further transform pharmaceutical production paradigms by integrating edge computing for real-time process optimization and generative AI for drug creation [62].

6.7 3D Printing Techniques for Pharmaceutical

Its massive potential to create a wide range of medications has been highlighted by the ground-breaking research on the creative use of 3D printing technology in pharmaceuticals. According to these studies, 3D printing can be utilized to create flexible multi-drug combinations, controlled release preparations, gastroretentive tablets, suppositories, minitablets, medical devices, and formulations that dissolve quickly [63]. This demonstrates the important role that 3D printing will play in pharmaceutical development going forward. Fused deposition modeling (FDM), direct powder extrusion (DPE), semi-solid extrusion (SSE), binder jetting (BJ), selective laser sintering (SLS), and stereolithography (SLA) are examples of additive techniques. This cutting-edge industry has also made room for other methods like direct ink writing (DIW), inkjet printing (IP), and digital light processing (DLP) [5]. These technologies show the wide range of uses for 3D printing by helping to produce pharmaceutical items. FDM creates customized medication release profiles by layer-by-layer extrusion of hot melt polymers. DPE uses binders and medication powders to fuse and extrude them into solid dosage forms without the need of solvents [64]. To make medication formulations that can include hydrogels for controlled release, SSE extrudes semi-solid materials through a nozzle. By sintering powder materials with lasers, SLS produces porous structures that are perfect for specialized drug delivery systems. BJ-3DP forms intricate multi-drug complexes with varying release kinetics by depositing a liquid binder onto a powder bed. SLA produces very precise photopolymer resin layers for intricate drug delivery systems by using UV light to harden them. Using a digital light projector, DLP quickly cures photopolymer resins to create intricate medication delivery systems [65]. The creation of intricate, multifunctional pharmaceutical applications is made possible by ink, which directly deposits different materials onto a substrate and precisely deposits droplets of drug solutions or binders to build complicated multi-layered pharmaceutical products [66].

6.7.1 Fused Deposition Modeling

In the pharmaceutical industry, FDM is one of the most extensively studied and applied 3D printing technologies. A variety of materials, a reasonably simple and inexpensive setup, and the capacity to develop intricate dose forms that enhance patient compliance are the reasons for this significance [67]. With this technique, a drug-loaded polymer filament—first made by hot melt extrusion—is heated and forced through a nozzle tip. Then, following precise routes determined by a computer-controlled design, the heated extrusion nozzle moves. After being extruded layer by layer onto a build platform, the material solidifies upon coming into touch with the build plate, taking on the desired shape. The possible drug degradation carried on by the high temperatures of the extrusion and printing processes is a significant worry with this printing technique; however, low-temperature printing is possible with the correct excipients. A wide range of thermoplastic materials, each chosen for its own mechanical characteristics and stability to meet different purposes, can be processed using FDM [68]. The temperature setting is an important consideration in this operation. Setting the printing temperature too low may cause the printer's nozzle to become blocked, disrupting the material extrusion process and creating breaks along the printed paths. Setting the temperature too high, on other hand, may result in the plastic becoming overly soft or, worse, in the active pharmaceutical ingredient (API) degrading, which would reduce the end product's quality and effectiveness. Processes like polyvinyl alcohol (PVA), polylactic acid (PLA), hydroxypropyl cellulose (HPC), and various forms of Eudragit are best processed at temperatures between 120 and 250 °C, where FDM normally functions. Research on immediate-release 3D-printed formulations employing PVP and Eudragit EPO produce results at comparatively lower temperatures, between 110 and 135 °C [69].

These substances were selected due to their efficient extrusion at these temperatures. The majority of these polymers, with the exception of HPC and Eudragit E, are not ideal for making immediate-release dosage forms. FDM has attraction due to its many significant advantages. Both novice and expert users can utilize this technology because it is incredibly easy to use and requires little setup or upkeep. The cost of FDM materials and equipment is lower than that of other 3D printing technologies, making it more accessible to small enterprises and educational institutions [70]. As previously stated, a wide range of materials are compatible with this technique, each of which has special qualities that can be altered to satisfy certain needs. FDM is not without its limitations; however, one of the primary limitations is that, unlike other 3D printing technologies such as SLA, it has a coarser surface finish and lower resolution. Uses that need high precision or high-quality finishes might find this to be an issue [71].

6.7.2 Direct Powder Extrusion

The 3D printing method known as "direct powder extrusion" (DPE) allows powder-formulation-based drug substances to be printed directly into individualized dosage forms without the need for solvent or with minimal post-process. DPE converts dry powder mixtures into semi-molten state by applying heat, pressure, or mechanical stress. This makes it possible to control extrusion through the nozzle to create structured medicinal products. One of the major features of DPE is the ability to create high drug-load formulations, which makes it ideal for applications that demand potent APIs in defined dosages [72]. The process also avoids drying operations, simplifies the manufacturing, and reduces the risk of solvent-related stability problems because it doesn't use solvents. This allows DPE to achieve multi-layer drug delivery structures and complex geometries to achieve customized drug release profiles such as formulations with pulsatile, fast, or sustained release. By avoiding intermediate steps upon which other 3D printing technologies are founded, DPE creates drug-loaded prints directly from powder. It also works at low temperatures, which makes it ideal for heat-sensitive materials. However, this typically leads to less uniform dispersion of the API compared to Hot-Melt Extrusion (HME), which could affect the quality of the final product. DPE was effectively used by researchers in the development of paracetamol tablets and proved that the process can generate immediate-release tablets with the required mechanical properties and uniform drug distribution. DPE was used in one study for the development of sustained release theophylline tablets. The study proved how the method can be used for formulating properties to match medication release profiles to a desired therapeutic requirements [73].

6.7.3 Semi-solid Extrusion

SSE is a 3D printing method that constructs objects by extruding material in a series of layers from a syringe-like device in a semi-molten or semi-solid state. For the manufacture of drugs, the technology is highly versatile because it can be used to manufacture complex drug formulations that cannot be manufactured using conventional extrusion methods. The original materials, which are in the form of pastes or gels, are manufactured by mixing different combinations of different materials with solvents to achieve the desired viscosity for printing [74]. SSE has been utilized to manufacture a number of dosage forms, including solid lipid tablets, controlled release, infant chewable gummies, orodispersible pills, and immediate-release tablets. The benefits of SSE over solid extrusion include its capacity to handle a wider variety of materials, especially those that are sensitive to temperature. The reason for this is that SSE works at lower temperatures, protecting the potency and integrity of active chemicals that would be weakened by the heat required to melt solids. Compared to traditional extrusion methods, SSE is also more effective and produces less waste. By extruding materials in a semi-solid state, it reduces material waste, lowers energy consumption, and eliminates the need for substantial heating and post-processing. SSE does, however, come with certain difficulties. To enable efficient printing, the excipients must be carefully selected for compatibility and have the necessary viscosity and rheological characteristics. While too high a viscosity may completely obstruct the flow, problems like too low a viscosity might result in unregulated material flow. Also, post-processing procedures like drying or freezing are sometimes required, and the physical condition of the original material might affect these procedures. If the material is not hard enough, the structure may shrink, deform, or even collapse. Unlike other printing techniques that use finer extrusion orifices, SSE requires the printer's die heads to be large enough to withstand the high viscosity of the materials, which typically results in lower-resolution prints [75].

6.7.4 Additional 3D Printing Techniques

3D printing technologies advance quickly, more opportunities for their use in the pharmaceutical sector become available. There are more important techniques in addition to the ones that have already been covered, including direct writing (DW), IP, and DLP. In terms of resolution, material compatibility, and pharmacological innovation potential, each has special advantages.

6.7.4.1 Digital Light Processing

Digital micro mirror devices (DMD) and a light source are used in DLP to cure photoreactive polymers. Similar to SLA, this method cures complete layers at once, usually providing higher rates [76]. DLP can create extremely smooth and accurate medication administration systems, including transdermal drug delivery micro needle arrays. Because of its rapid creation of complex structures, it can be used to fabricate dosage forms in large quantities that are customized to meet the needs of specific patients. The nature of the layer curing process in DLP generally leads to nicer finishes and comparable or faster production times than SLA. It is perfect for iterative development processes in drug formulation because it provides a comparable resolution and material range while succeeding in a faster prototype turnover. Researchers manufactured silk fibroin-based microneedles employing a one-step DLP 3D printing process using low-concentration aqueous protein solutions. Using rhodamine B fluorescent dye, the resultant microneedles showed flexibility and effective skin delivery capabilities [76].

6.7.4.2 Direct Writing

DW includes a range of process technologies that are adaptable, multi-scale, and multidimensional. Another name for this method is digital writing or direct printing. DW creates complex structures by extruding drug-loaded materials or bioinks through a nozzle. In addition to handling a broad variety of viscosities, it frequently uses cells and biocompatible materials to mimic bio printing. This method is a general-purpose technology in high-order production, namely for scaffold or implant design with controlled release functionality and in regenerative medicine. DW is capable of processing "living" inks and provides material versatility unmatched compared to the standard 3D printing process. This capability is essential in biomaterials fabrication by writing directly on cells and is therefore central to future progress in medicine treatment and therapy. Although its resolution may not be as high as that of DLP's or SLA's, it has more potential to produce complex geometries and physiologically active structures with added growth agents or living cells [77].

6.8 Pharmaceutical Process Optimization

The development of predictive models for solids-based processes has led to an increase in the use of optimization techniques in pharmaceutical manufacture. These models, which include hybrid/reduced-order models, semi-empirical formulations, and population balance equations, are frequently quite intricate and computationally demanding. Thousands of differential, algebraic, or integral equations can be found in the flowsheets that are produced. Explicit mathematical expressions for process restrictions could not be available when hybrid or data-driven models are used. The size and black-box nature of these models make traditional deterministic optimization methods difficult to use, hence derivative-free optimization strategies that can deal with implicit restrictions are required [78].

6.8.1 Predictive Maintenance of Manufacturing Equipment

Complex models without explicit functions for objectives or constraints can be handled thanks to recent advancements in derivative-free optimization. Surrogate-based optimization has shown promise, substituting response surfaces for computationally costly models. Complex systems can be approximated with fewer evaluations using methods like kriging, artificial neural networks (ANNs), and high-dimensional model representation. A significant obstacle in surrogate models is striking a balance between computational expense and accuracy. With the aid of anticipated improvement (EI) functions, adaptive sampling facilitates effective model refinement. This method improves accuracy by focusing sampling on underrepresented areas or areas close to feasible borders. Uncertainty estimates are provided via kriging-based techniques, which enhance sampling tactics even further. These optimization techniques are frequently used in the design of pharmaceutical processes, such as feasibility testing and design space evaluation [79].

To guarantee robustness, they aid in defining viable operational zones, refining raw material specifications, and directing tests. These techniques also help with process design under uncertainty, which influences the choice of equipment and determines the best production routes. Product quality must be maintained when pharmaceutical procedures are optimized with financial goals in mind. This need is met by constraint-handling optimization approaches. Using surrogate models to first identify viable boundaries, suggested a derivative-free global optimization method that then uses a trust-region technique to find optimal operating locations. Global optimization is guaranteed by a multi-start method, which also guarantees cost-effectiveness without sacrificing product quality [80].

Pharmaceutical industry has adopted process systems engineering tools more slowly than other chemical sectors. This is because of the intricacy of solids-based process prediction models and regulatory requirements for excellent product quality in a short amount of time. But with advancements in hybrid/reduced-order techniques and particle process modeling, integrated flowsheet models are now possible, enabling uses such as process design, optimization, and dynamic simulation. Because of their computational complexity and black-box limitations, optimizing these models is still difficult; however, derivative-free optimization techniques offer useful answers. Specifically, surrogate-based techniques aid in defining design spaces and streamlining procedures while maintaining quality requirements. Process engineering in pharmaceutical manufacturing will be significantly improved by ongoing research into more precise models and sophisticated optimization strategies [81].

6.8.2 Real-time Quality Assurance and Defect Detection

Developments in derivative-free optimization have enabled the solution of complex process model issues where closed-from equations for the objective function or constraints are absent. A response surface is created in order to approximate the complex process model in surrogate-based optimization techniques. This less complicated approximation is then optimized, reducing the computing cost that comes from assessing the intricate process model at every optimization algorithm iteration. Kriging, ANNs, or high-dimensional model representation are examples of response surface methods that can be used to fit the outcomes of a small number of model examinations after the complex process model has been systematically sampled. The accuracy of the surrogate approximation and the quantity of intricate model evaluations needed are inherently traded off in surrogate-based methods. Finding a balance between the two can be helped by adaptive sampling. An EI function is used by adaptive sampling techniques to guide the search for novel places at which to assess the intricate process model. The EI function can be adjusted to focus sampling on areas of interest based on the type of problem. For example, it is possible to sample more points from regions that are underrepresented in the existing model. Kriging offers an assessment of prediction uncertainty at each model evaluation point, which makes it advantageous in this situation for response surface manufacturing. In pharmaceutical companies, surrogate-based optimization techniques have been used to solve design space evaluation and process optimization issues. For a variety of process models, including those with convex and nonconvex feasible regions, surrogate-based approaches have been shown to successfully tackle the feasibility test problem. These methods can be used to assess the design space for production processes in pharmaceutical models. Feasibility study results can help define raw material specifications, show the shape of the feasible zone, and guide experimental efforts to prove feasibility close to the design space boundaries. It can also be applied to process design issues where there's uncertainty. The outcomes of this kind of analysis can be used to choose the right equipment and even to decide whether manufacturing method is best for a specific API (e.g., direct compaction vs. wet or dry granulation). Meeting product quality requirements is crucial for optimizing pharmaceutical processes in order to achieve a financial goal. Pharmaceutical processes therefore need solutions for limited optimization. One such strategy is a derivative-free approach to global optimization that uses a feasibility stage to include restrictions. In this work, surrogate-based feasibility analysis is used to first identify the borders of the viable zone. Then, inside the defined feasible zone, optimal operational points are found using a trust-region technique. To make sure that global optima are found, a multi-start technique is followed. This approach can be used in pharmaceutical processes to make sure that the process's design space contains the best operating points in terms of cost or profit, maintaining that the requirements for product quality are maintained [82].

6.9 Challenges and Future Perspectives

Pharmaceutical procedures heavily rely on data-driven modeling, and still there are a number of obstacles to overcome. High-quality data, which is either infrequent or hard to get by, is necessary for accurate models. High-throughput technologies can improve data collecting, but in order to create collective databases, industry and academia must work together [83].

Model Interpretability and Generalization: Although validation within datasets has been thoroughly studied, complicated nonlinear interactions make it difficult to generalize to new conditions. Improving the interpretability of the model may increase its practicality.

Scalability and Online Performance: Effective computational techniques are necessary for real-time applications in large-scale pharmaceutical procedures. For broader usage, developments in distributed architectures and parallel computing are crucial [84].

Regulatory Compliance: The use of ML in the pharmaceutical industry presents questions about validation and transparency. Creating model explain ability guidelines is essential for regulatory approval. To address these challenges, two research directions have gained attention. Increasing accuracy and robustness by combining data-driven and mechanistic models. Data-driven models manage nonlinearities and uncertainties, whereas mechanistic models offer domain knowledge. Applications including real-time process control, hybrid model creation, and parameter estimation are supported by this integration [85].

To promote data-driven pharmaceutical modeling and guarantee its successful application, cooperation between academic institutions, business, and regulatory agencies is crucial. There are still certain challenges to overcome, despite the amazing advancements made in integrating AI and Industry 4.0 technology into pharmaceutical manufacturing. In order to fully realize the potential of smart manufacturing systems, several obstacles need to be overcome. The high implementation costs are among the main obstacles. It can occasionally cost a lot of money to adopt Industry 4.0 technologies like AI, IoT, robotics, and others. It is very costly for small pharmaceutical companies, for instance, to put up smart factories with sophisticated sensors, robotic systems, and data analytics resources. It may be necessary to pay for new software and hardware as well as personnel training to operate and maintain these nonbasic systems. Pharmaceutical companies have to account for the expenses of incorporating digital twin technologies, ML models, and AI-driven prediction systems into their existing operations. A skilled workforce is required. We require a highly skilled staff to comprehend and administer extremely complex digital systems in order to deploy industry 4.0 technology. Businesses must engage in internal staff training programs and collaborations with academic institutions to develop competence in AI, ML, data science, and robotics, as these fields will play an increasingly important role in pharmaceutical manufacturing. Pharmaceutical firms lack the qualified experts in data analytics, AI, and cyber physical systems needed to fully utilize Industry 4.0. Considering the future, Industry 5.0 will bring more human-centric approaches to pharmaceutical manufacture. To enhance decision-making and creativity, AI and automation will work together with human workers in this next phase. As AI-driven systems develop procedures, they will keep improving them as long as people are in charge of jobs requiring complex problem-solving, creativity, and empathy. In the future of pharmaceutical manufacturing, collaborative robots will play a crucial role by working alongside human operators to support their data-driven, quicker decision-making while also retaining human innovation. Additionally, the development of AI and ML models is projected to continue the rapid innovation in drug discovery and manufacturing. Researchers will be able to create more individualized treatments by using AI-powered simulations and predictive models to analyze enormous quantities of genomic, clinical, and real-world data. Finding the most effective treatment with the fewest negative effects will be made possible by the ability to generate customized medications based on each patient's unique genetic profile. The future of pharmaceutical manufacturing will also be influenced by quantum computing. On the other hand, quantum computers will be able to process exponentially more data and do tenfold more intricate molecular interaction simulations than classical computers are currently. Drug discovery could be significantly accelerated by that, making it possible to forecast the efficacy of drug candidates faster and more accurately than in the past. However, as the technology underlying quantum computing advances, the abilities offered by this research will be much more powerful to maximize manufacturing and promote customized healthcare development. Multi-scale modeling: Procedures range from molecular to full-scale production. Predicting drug interactions, streamlining unit operations, and enhancing large-scale production planning are all possible using data-driven approaches. For integration to be successful, model compatibility and dynamic adaptability are essential [86].

6.10 Conclusion

The pharmaceutical industry is experiencing a remarkable shift as AI reshapes how medicines are made. These smart technologies aren't just improving operations—they're helping companies make better medicines more efficiently while reducing waste. Imagine factories where machines can predict their own maintenance needs, cutting unexpected shutdowns by nearly half. Picture inspection systems that spot tiny flaws with near-perfect accuracy, far better than the human eye could manage. Companies using these tools are seeing dramatic improvements—medicines with more consistent quality, production lines running smoother, and significant cost savings. Some are completing batches in half the time while using fewer materials. Others have cut their energy use by a quarter through intelligent monitoring systems. But this transformation isn't without its challenges. The industry needs skilled workers who understand both pharmaceuticals and these new technologies. Regulators are working to keep pace with these advances while ensuring patient safety isn't compromised. Perhaps most importantly, there's a fundamental shift in philosophy underway—from testing quality into products to building quality in from the start. For patients worldwide, these changes could mean more reliable access to affordable medicines. For manufacturers, it represents both an opportunity and a challenge to rethink traditional approaches. The road ahead will require careful planning, investment in people, and collaboration across the industry. But the potential rewards—better medicines, more sustainable production, and improved global health—make this journey worthwhile.

References

1 Selekman JA, Qiu J, Tran K, et al. High-throughput automation in chemical process development. *Annual Review of Chemical and Biomolecular Engineering* 2017;8:525–547.

2 Coley CW, Eyke NS, and Jensen KF. Autonomous discovery in the chemical sciences part I: Progress. *Angewandte Chemie International Edition in English* 2020;59(51):22858–22893.

3 Buitrago Santanilla A, Regalado EL, Pereira T, et al. Nanomole-scale high-throughput chemistry for the synthesis of complex molecules. *Science* 2015;347(6217):49–53.

4 Perera D, Tucker JW, Brahmbhatt S, et al. A platform for automated nanomole-scale reaction screening and micromole-scale synthesis in flow. *Science* 2018;359(6374):429–434.

5 Milliken RL, Quinten T, Andersen SK, and Lamprou DA. Application of 3D printing in early phase development of pharmaceutical solid dosage forms. *International Journal of Pharmaceutics* 2024;653:123902.

6 Pravin S and Sudhir A. Integration of 3D printing with dosage forms: A new perspective for modern healthcare. *Biomedicine & Pharmacotherapy* 2018;107:146–154.

7 Javaid M, Haleem A, Singh RP, and Suman R. 3D printing applications for healthcare research and development. *Global Health Journal* 2022;6(4):217–226.

8 International Council for Harmonisation. ICH Q8(R2) – Pharmaceutical development [Internet]. 2009 [cited 2024 May 15]. Available from: https://www.ich.org/page/quality-guidelines

9 U.S. Food and Drug Administration. Guidance for industry: Process analytical technology [Internet]. 2004 [cited 2024 May 15]. Available from: https://www.fda.gov

10 European Medicines Agency. Guideline on continuous manufacturing [Internet]. 2020 [cited 2024 May 15]. Available from: https://www.ema.europa.eu

11 Montgomery DC. *Design and Analysis of Experiments*. 9th ed. Hoboken: Wiley; 2017.

12 Santosh K, Patel R, Sharma M, and Gupta A. Application of Kriging interpolation in pharmaceutical process optimization. *Journal of Pharmaceutical Sciences* 2021;110(3):1035–1047.

13 Smith J and Brown R. Neural networks for drug manufacturing process control. *AI Pharma Manufacturing* 2020;15(2):45–62.

14 International Organization for Standardization. ISO 31010: 2019 Risk management–Risk assessment techniques. Geneva: ISO; 2019.

15 Kaplan S and Garrick BJ. On the quantitative definition of risk. *Risk Anal* 1981;1(1):11–27.

16 International Council for Harmonisation. ICH Q9: Quality risk management. Geneva: ICH; 2005.

17 Naelapää K, Veski P, Kristensen HG, Rantanen J, and Bertelsen P. Building quality into a coating process. *Pharmaceutical Development and Technology* 2010;15(1):35–45.

18 Adam S, Suzzi D, Radeke C, and Khinast JG. An integrated Quality by Design (QbD) approach towards design space definition of a blending unit operation by Discrete Element Method (DEM) simulation. *European Journal of Pharmaceutical Sciences* 2011;42(1-2):106–115.

19 Baldinger A, Clerdent L, Rantanen J, Yang M, and Grohganz H. Quality by design approach in the optimization of the spray-drying process. *Pharmaceutical Development and Technology* 2012;17(4):389–397.

20 Hanjalić K, Popovac M, and Hadžiabdić M. A robust near-wall elliptic-relaxation eddy-viscosity turbulence model for CFD. *International Journal of Heat and Fluid Flow* 2004;25(6):1047–1051.

21 Durbin PA. Near-wall turbulence closure modeling without "damping functions." *Theoretical and Computational Fluid Dynamics* 1991;3(1):1–13.

22 Bai G, Bee JS, Biddlecombe JG, Chen Q, and Leach WT. Computational fluid dynamics (CFD) insights into agitation stress methods in biopharmaceutical development. *International Journal of Pharmaceutics* 2012;423(2):264–280.

23 Sharma C, Malhotra D, and Rathore AS. Review of computational fluid dynamics applications in biotechnology processes. *Biotechnology Progress* 2011;27(6):1497–1510.

24 Fox RO and Stiles HL. *Computational Models for Turbulent Reacting Flows.* Cambridge: Cambridge University Press; 2003.

25 Woo XY, Tan RB, and Braatz RD. Modeling and computational fluid dynamics-population balance equation-micromixing simulation of impinging jet crystallizers. *Crystal Growth & Design* 2008;9(1):156–164.

26 Bai G, Bee JS, Biddlecombe JG, Chen Q, and Leach WT. Computational fluid dynamics (CFD) insights into agitation stress methods in biopharmaceutical development. *International Journal of Pharmaceutics* 2012;423(2):264–280.

27 Sharma C, Malhotra D, and Rathore AS. Review of computational fluid dynamics applications in biotechnology processes. *Biotechnology Progress* 2011;27(6):1497–1510.

28 Eitzlmayr A and Khinast J. Co-rotating twin-screw extruders: Detailed analysis of conveying elements based on smoothed particle hydrodynamics. Part 1: Hydrodynamics. *Chemical Engineering Science* 2015;134:861–879.

29 Eitzlmayr A and Khinast J. Co-rotating twin-screw extruders: Detailed analysis based on smoothed particle hydrodynamics. Part 2: Mixing. *Chemical Engineering Science* 2015;134:880–886.

30 Zhang X, Li Y, and Wang H. Predictive modeling of batch yields in pharmaceutical manufacturing using machine learning. *Journal of Pharmaceutical Innovation* 2021;16(3):145–158.

31 Alshehri AS, Gani R, and You F. Deep learning and knowledge-based methods for computer-aided molecular design—Toward a unified approach: State-of-the-art and future directions. *Computers & Chemical Engineering* 2020;141:107005.

32 Boukouvala F and Ierapetritou MG. Surrogate-based optimization of expensive flowsheet modeling for continuous pharmaceutical manufacturing. *Journal of Pharmaceutical Innovation* 2013;8:131–145.

33 Lee J, Park S, and Kim M. High-speed visual inspection of pharmaceutical tablets using deep convolutional neural networks. *International Journal of Pharmaceutics* 2022;615:121492.

34 Colvin M and Maravelias CT. Modeling methods and a branch and cut algorithm for pharmaceutical clinical trial planning using stochastic programming. *European Journal of Operational Research* 2010;203(1):205–215.

35 Liang G, Fan W, Luo H, and Zhu X. The emerging roles of artificial intelligence in cancer drug development and precision therapy. *Biomedicine and Pharmacotherapy* 2020;128:110255.

36 Rios LM and Sahinidis NV. Derivative-free optimization: A review of algorithms and comparison of software implementations. *Journal of Global Optimization* 2012;56(3):1247–1293.

37 Boukouvala F and Ierapetritou MG. Derivative-free optimization for expensive constrained problems using a novel expected improvement objective function. *AIChE Journal* 2014;60(4):1332–1349.

38 Banerjee I, Pal S, and Maiti S. Computationally efficient black-box modeling for feasibility analysis. *Computers & Chemical Engineering* 2010;34(9):1515–1521.

39 Selekman JA, Qiu J, Tran K, et al. High-throughput automation in chemical process development. *Annual Review of Chemical and Biomolecular Engineering* 2017;8:525–547.

40 Vamathevan J, Clark D, Czodrowski P, et al. Applications of machine learning in drug discovery and development. *Nature Reviews Drug Discovery* 2019;18(6):463–477.

41 Yin J, Li J, Karimi IA, and Wang X. Generalized reactor neural ODE for dynamic reaction process modeling with physical interpretability. *Chemical Engineering Journal* 2023;452:139487.

42 Feinstein WP and Brylinski M. Structure-based drug discovery accelerated by many-core devices. *Curr Drug Targets* 2016;17(13):1595–1609.

43 Zhou T, Gani R, and Sundmacher K. Hybrid data-driven and mechanistic modeling approaches for multiscale material and process design. *Engineering* 2021;7(9):1231–1238.

44 Bradley W, Kim J, Kilwein Z, et al. Perspectives on the integration between first-principles and data-driven modeling. *Computers & Chemical Engineering* 2022;166:107898.

45 Marcato A, Marchisio D, and Boccardo G. Reconciling deep learning and first-principle modelling for the investigation of transport phenomena in chemical engineering. *The Canadian Journal of Chemical Engineering* 2023;101(6):3013–3018.

46 Wang W, Ye Z, Gao H, and Ouyang D. Computational pharmaceutics—A new paradigm of drug delivery. *Journal of Controlled Release* 2021;338:119–136.

47 Chen T, Wilson JA, and Roberts D. Recurrent neural networks for predictive maintenance in pharmaceutical manufacturing equipment. *Computers & Chemical Engineering* 2023;170:108103.

48 Patel R and Williams T. Predictive maintenance in Pharma 4.0: AI-driven equipment monitoring. *Computers & Chemical Engineering* 2023;172:108–120.

49 Zhang X and Patel R. Deep learning for industrial equipment monitoring. *IEEE Transactions on Industrial Informatics* 2023;19(5):4128–4140.

50 Shokrani A, Dogan H, Burian D, et al. Sensors for in-process and on-machine monitoring of machining operations. *CIRP Journal of Manufacturing Science and Technology* 2024;51:263–292.

51 Sharma SP, Bansal AK, and Chen L. Machine learning-based automated quality control systems for solid dosage forms. *IEEE Access* 2020;8:142311–142325.

52 Poozesh S and Bilgili E. Scale-up of pharmaceutical spray drying using scale-up rules: A review. *International Journal of Pharmaceutics* 2019;562:271–292.

53 MLPDS–Machine Learning for Pharmaceutical Discovery and Synthesis Consortium [Internet]. [cited 2022 Sep 18]. Available from: https://mlpds.mit.edu/

54 Saccomandi P, Schena E, Oddo CM, Zollo L, Silvestri S, and Guglielmelli E. Microfabricated tactile sensors for biomedical applications: A review. *Biosensors (Basel)* 2014;4:422–448.

55 Artificial Intelligence in Medical Imaging Market Report 2019: CAGR of 36.89%, By Offering, Technology, Deployment Type, Application, Leading Players | BenevolentAI, OrCam, Babylon, Freenome Inc., Ada Health GmbH. GlobeNewswire: Data Bridge Market Research; 2020.

56 Chow K, Tong HH, Lum S, and Chow AH. Engineering of pharmaceutical materials: An industrial perspective. *Journal of Pharmaceutical Sciences* 2008;97:2855–2877.

57 Lee J and Zhang X. AI-powered dynamic scheduling in pharmaceutical manufacturing. *International Journal of Production Research* 2023;61(8):2456–2472.

58 Schmidt E, Müller R, Wagner T, and Braun P. Deep learning for automated visual inspection of solid dosage forms. *Journal of Pharmaceutical Sciences* 2022;111(5):1328–1342.

59 Johnson M and Chen L. Data architecture solutions for AI integration in pharma manufacturing. *Journal of Pharmaceutical Innovation* 2023;18(2):145–160.

60 Williams R and Patel A. Predictive maintenance in Pharma 4.0. *Computers & Chemical Engineering* 2022;170:108103.

61 McKinsey & Company. Digital transformation in pharma operations. Industry Report. 2023.

62 Pesapane F, Codari M, and Sardanelli F. Artificial intelligence in medical imaging: threat or opportunity? Radiologists again at the forefront of innovation in medicine. *European Radiology Experimental* 2018;2(1):35.

63 Zhang X, Liu Y, Wang Q, and Chen Z. Regulatory compliance in AI-driven pharmaceutical systems. *IEEE Transactions on Automation Science and Engineering* 2023;20(1):412–425.

64 Liu Z, Li L, Zhang S, et al. Correlative image-based release prediction and 3D microstructure characterization for a long acting parenteral implant. *Pharmaceutical Research* 2021;38:1915–1929.

65 Funk NL, Leão JA, Oliveira TV, and Beck RC. Semi-Solid Extrusion (SSE) in Pharmaceuticals. In: *Advanced Manufacturing in Pharmaceutical Engineering*, (pp. 171–200). Springer; 2023.

66 Horst DJ. 3D printing of pharmaceutical drug delivery systems. *Advanced Organic and Inorganic Chemistry Science* 2018;1(2):1–5.

67 Nyavanandi D, Narala S, and Repka MA. Hot-melt extrusion paired fused deposition modeling 3D printing: Development of pharmaceutical medications. In: *3D Printing in Drug Delivery*, (pp. 169–194). Springer; 2023.

68 Katsiotis CS, Åhlén M, Strømme M, and Welch K. 3D-printed mesoporous carrier system for delivery of poorly soluble drugs. *Pharmaceutics* 2021;13(7):1096.

69 Okwuosa TC, Stefaniak D, Arafat B, Isreb A, Wan KW, and Alhnan MA. A lower temperature FDM 3D printing for the manufacture of patient-specific immediate release tablets. *Pharmaceutical Research* 2016;33(11):2704–2712.

70 Kurniawan A and Surini S. Utilization of PVP-PLA combination in fabricating theophylline-loaded filament for 3D printing with immediate release behavior. *Pharmaceutical Sciences and Research* 2023;10(2): 78–88.

71 Ilyés K. Three-dimensional printing by fused deposition modeling (3DP-FDM) in pharmaceutics. *Farmacia* 2020;68(4):586–596.

72 Sánchez-Guirales SA, Jurado N, Kara A, Lalatsa A, and Serrano DR. Understanding direct powder extrusion for fabrication of 3D printed personalised medicines: A case study for nifedipine minitablets. *Pharmaceutics* 2021;13(10):1583.

73 Aguilar-de-Leyva Á, Linares V, Domínguez-Robles J, Casas M, and Caraballo I. Extrusion-based technologies for 3D printing: A comparative study of the processability of thermoplastic polyurethane-based formulations. *Pharmaceutical Development and Technology* 2023; 1–32.

74 Funk NL, Leão JA, Oliveira TV, and Beck RC. Semi-Solid Extrusion (SSE) in Pharmaceuticals. In: *Advanced Manufacturing in Pharmaceutical Engineering*, (pp. 171–200). Springer; 2023.

75 Smith A and Johnson B. Advanced maintenance strategies for pharma equipment. *Journal Pharmaceutical Engineering* 2023;12(3):45–60.

76 Hadipoespito GW, Yang Y, Choi H, Ning G, and Li X. *Digital Micromirror Device based Microstereolithography for Micro Structures of Transparent Photopolymer and Nanocomposites*. Texas ScholarWorks; 2003.

77 Balani S, Ghaffar SH, Chougan M, Pei E, and Şahin E. Processes and materials used for direct writing technologies: A review. *Results in Engineering* 2021;11:100257.

78 European Medicines Agency. Annex 15: Qualification and validation. EMA Guidelines. 2022.

79 Chen L, Wang H, Zhang Y, and Li X. Vibration analysis for tablet press maintenance. *International Journal of Pharmaceutics* 2023;647:123456.

80 U.S. Food and Drug Administration. Guidance on AI/ML in manufacturing. CFR Title 21 Part 211. 2023.

81 Kumar D, Singh R, Sharma P, and Gupta A. Data preprocessing for industrial AI. *Computers & Chemical Engineering* 2022;158:107–118.

82 International Society for Pharmaceutical Engineering. GAMP 5 guide: Advanced manufacturing technologies. ISPE; 2023.

83 BioPharma International. Case study: Predictive maintenance in vaccine production. *BioPharm International* 2023;36(4):28–35.

84 Smith A and Johnson B. AI integration in pharma manufacturing. *Journal Pharmaceutical Engineering* 2023;15(2):112–125.

85 Chen L, Wang Y, Zhang H, and Liu J. Data interoperability challenges in industrial AI. *IEEE Transactions on Industrial Informatics* 2023;19(3):2015–2028.

86 European Medicines Agency. *Annex 11: Computerized systems. EMA Guidelines*; 2023.

7

Data-centric Engineering for Real-time Monitoring and Control in Pharmaceutical Manufacturing

Amol D. Gholap[1,2], Pankaj R. Khuspe[3], Sadikali F. Sayyad[2], Machindra Chavan[4], and Md. Faiyazuddin[5]*

[1]*Department of Pharmaceutics, St. John Institute of Pharmacy and Research, Palghar, India*
[2]*Department of Pharmaceutics, Amrutvahini College of Pharmacy, Sangamner, India*
[3]*Department of Pharmaceutics, College of Pharmacy, Paniv, Solapur, India*
[4]*Department of Pharmacognosy, Amrutvahini College of Pharmacy, Sangamner, India*
[5]*Department of Pharmacy, ARAM Institute, Ranchi, India*

*Corresponding author: amolgholap16@gmail.com

CONTENTS

7.1 Introduction

Manufacturing pharmaceuticals requires an unmatched degree of accuracy, quality assurance, and compliance with strict regulatory frameworks like the Quality by Design (QbD) paradigm and good manufacturing practices (GMP) [1]. Maintaining process efficiency, scalability, and regulatory compliance are intrinsic constraints of traditional drug formulation systems, which are frequently characterized by empirical methodologies and batch-based production. By facilitating data-driven decision-making, real-time process optimization, and predictive modeling for improved product consistency and therapeutic efficacy, AI has completely changed the drug formulation industry. Deep learning (DL), machine learning (ML), and neural networks are examples of AI-powered algorithms that make it easier to screen pharmaceutical excipients in large quantities, model molecular interactions computationally, and optimize formulation parameters to attain the required pharmacokinetics and pharmacodynamics [2]. Additionally, early detection of formulation failures is made possible by AI-driven predictive analytics, which speeds up the medication development lifecycle and minimizes trial-and-error testing. By building virtual models of formulation processes that are updated in real-time with sensor data, the combination of DTs and DCE enhances AI applications by simulating and improving crucial industrial factors. In order to provide robust formulation design with low variability, adaptive process control (APC) is currently utilizing advanced AI techniques like reinforcement learning (RL) and Bayesian optimization. Nevertheless, issues including data heterogeneity, algorithmic openness, and regulatory acceptability of AI-driven approaches still exist in spite of these developments. It is anticipated that future studies will concentrate on AI-enhanced formulation systems that integrate blockchain-based traceability, federated learning, and quantum computing to guarantee safe and extremely effective pharmaceutical production. The synergistic combination of AI, DTs, and DCE will be crucial in redefining drug formulation techniques for increased efficacy, safety, and sustainability as the industry moves toward Pharma 4.0 as presented in Figure 7.1 [3, 4].

7.1.1 Digital Twins

Digital Twins (DTs), which offer an advanced virtual depiction of physical systems, processes, or products, have become a paradigm shift in pharmaceutical manufacturing. In order to duplicate and forecast the behavior of a corresponding physical thing, a DT is an intelligent, data-driven model that is updated continually using real-time sensor data, ML algorithms, and sophisticated computer analytics. By bridging the gap between theoretical modeling and empirical experimentation, DTs enable a comprehensive approach to process design, optimization, and quality control (QC) in pharmaceutical formulation and manufacture. DTs enable real-time monitoring and adaptive control of crucial manufacturing parameters like drug solubility, dissolution rates, excipient interactions, and stability profiles by combining multi-physics simulations, AI-powered analytics, and IoT-based data acquisition. By lowering variability and enhancing batch-to-batch consistency, these capabilities guarantee increased precision in formulation development [5].

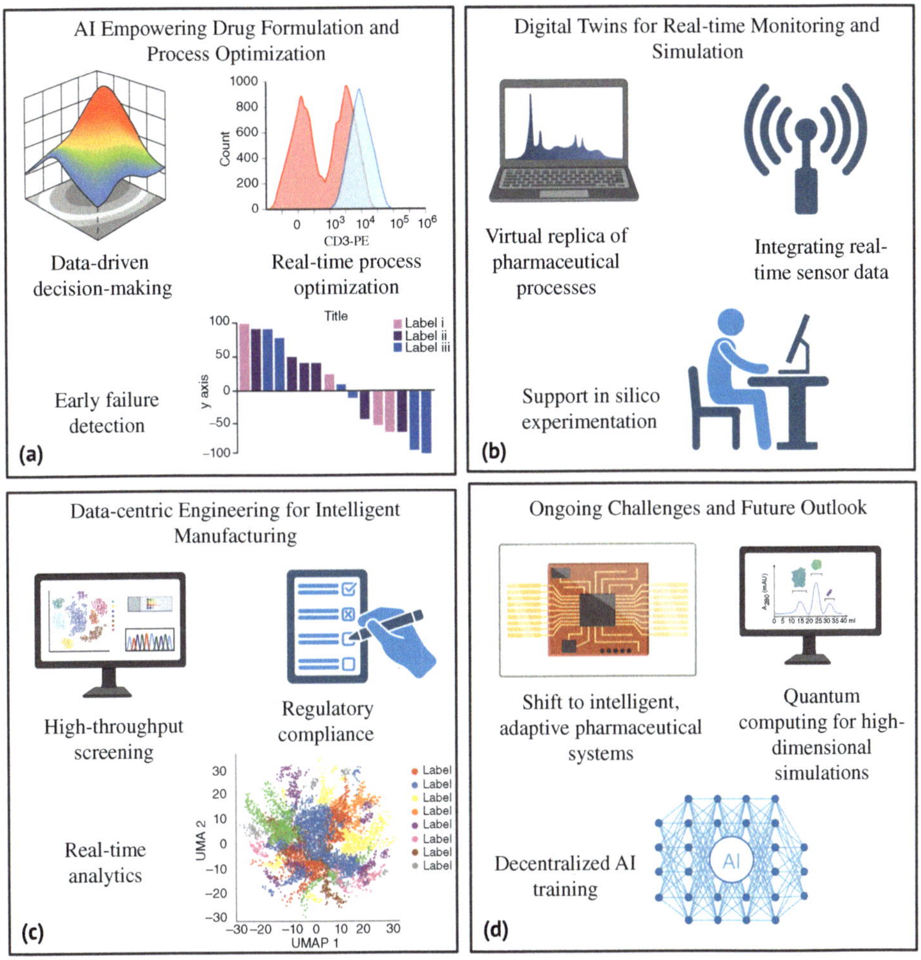

Figure 7.1 This figure illustrates the integration of artificial intelligence and digital technologies within pharmaceutical development. (a) Artificial intelligence facilitates drug formulation and process optimization through data-driven decision-making, early failure detection, and real-time analytics. (b) Digital twins simulate pharmaceutical processes by incorporating real-time sensor data and supporting in silico experimentation. (c) Data-centric engineering advances intelligent manufacturing through high-throughput screening, regulatory compliance, and real-time monitoring. (d) Future challenges encompass the adoption of intelligent adaptive systems, the utilization of quantum computing for complex simulations, and the implementation of decentralized AI training. Collectively, these innovations foster a transformative shift toward more intelligent, efficient, and personalized pharmaceutical development and manufacturing systems. Created in BioRender by Md. Faiyazuddin.

Beyond process modeling, DTs are used in risk assessment, predictive maintenance, and regulatory compliance to reduce deviations and improve adherence to GMPs. DTs facilitate in silico studies by utilizing QbD principles. This enables researchers to assess various formulation strategies prior to physical trials, hence minimizing time-to-market and resource consumption [6]. Additionally, AI-enhanced DTs are able to optimize formulation conditions on their own by examining past datasets, finding hidden trends, and suggesting changes to increase therapeutic efficacy and bioavailability [7]. Notwithstanding these developments, issues including the difficulty of data integration, cybersecurity flaws, and the demands on computational power call for more innovation in DT frameworks. In order to strengthen DTs as a key technology in AI-driven pharmaceutical formulation, future advancements are probably going to concentrate on integrating blockchain for safe data transactions, federated learning for decentralized model training, and cloud-based architectures for scalable real-time simulations [8].

7.1.2 Data-centric Engineering

A paradigm shift known as DCE places data at the center of drug formulation innovation, optimization, and decision-making. This method makes use of both structured and unstructured data from a variety of sources, including as sensor-integrated production systems, real-time spectroscopic analysis, computational modeling, and high-throughput screening (HTS). Through the use of sophisticated data acquisition frameworks, DCE guarantees the systematic collection, curation, and analysis of enormous amounts of heterogeneous data, ranging from pharmacokinetic modeling and dissolution profiles to the physicochemical characteristics of excipients. Real-time monitoring, predictive modeling, and APC are made possible by the integration of ML and AI algorithms into DCE [9]. This improves formulation accuracy and consistency from batch to batch. Cloud-based platforms and IoT connectivity enable real-time analytics, which minimizes deviations in critical quality attributes (CQAs) by enabling quick anomaly identification and proactive process modifications. Additionally, DCE encourages adherence to regulatory frameworks like the US continuous data-driven validation of formulation parameters is made possible by the Food and Drug Administration's (FDA) process analytical technology (PAT) effort and QbD principles. Notwithstanding its benefits, issues like cybersecurity threats, standardization, and data interoperability call for strong data governance frameworks and safe computing environments. To further improve drug formulation efficiency and regulatory compliance, future developments in DCE are anticipated to include blockchain-enabled traceability, federated learning for cooperative pharmaceutical research, and quantum computing for high-dimensional data analysis. Adopting DCE will be crucial to attaining intelligent, adaptable, and precision-driven drug development as pharmaceutical manufacturing moves toward Industry 4.0 [10].

7.2 Principles of DTs in Pharmaceutical Manufacturing

7.2.1 Definition and Framework

By providing a high-fidelity virtual model of physical systems, DTs have become a game-changing tool in the pharmaceutical manufacturing industry, facilitating real-time monitoring, predictive analytics, and process optimization. A DT functions as a cyber-physical system (CPS) that combines connection, data, and computational models to produce a digitally synchronized version of pharmaceutical production processes [11]. DTs enhance production efficiency, support data-driven decision-making, and guarantee adherence to regulatory frameworks like PAT and GMP by consistently simulating the state, behavior, and performance of the physical system. The physical entity, the virtual model, and data connectivity are the three interrelated parts that make up the basic framework of DTs. Each of these elements is essential to guaranteeing smooth communication between the digital and physical realms [12].

The tangible object is a representation of the actual pharmaceutical manufacturing facility, machinery, or procedure, including the synthesis of active pharmaceutical ingredients (APIs), formulation, and granulation, compression of tablets, coating, and packing. To guarantee constant drug quality and efficacy, these procedures necessitate strict control over CQAs and critical process parameters (CPPs), including temperature, pressure, pH, mixing speed, and material qualities. Historically, offline QC testing has been used to track these parameters, which has caused delays in process optimization. But with the introduction of DTs, proactive modifications and improved quality assurance are now possible thanks to real-time insights into process dynamics [13].

Using data-driven AI algorithms, mechanistic modeling, and simulations based on physics, the virtual model is an advanced digital representation of the physical system. Based on first-principles methods, mechanistic models use mathematical equations to explain reaction kinetics, fluid dynamics, and thermodynamics in pharmaceutical operations. AI-driven models, on the other hand, especially those that make use of ML and DL, examine enormous volumes of data from the past and present in order to identify trends, anticipate abnormalities, and improve production conditions. To improve the precision and predictive power of DTs, hybrid modeling approaches—which blend mechanical concepts with AI-based methodologies—are being used more and more. By simulating different process situations, identifying possible faults before they happen, and virtually prototyping new formulations, these virtual models greatly cut down on development time and expenses [14].

7.2.2 Functional Capabilities

The advanced functional capabilities of DTs in pharmaceutical manufacturing allow for predictive decision-making, improved QC, and real-time process optimization. To close the gap between virtual and physical production systems, these capabilities make use of sophisticated computational models, sensor integration, and data analytics [15].

Real-time monitoring, which guarantees ongoing surveillance of crucial process parameters including temperature, pressure, pH, and component concentration, is one of the main features of DTs. Pharmaceutical producers can maintain strict QC measures in accordance with regulatory requirements like GMP by utilizing DTs, which enable high-resolution data collecting through the integration of sensor networks and IoT technology. Real time data collection and analysis provide a thorough understanding of process dynamics, enabling the prompt identification of aberrations and guaranteeing product consistency [16].

Predictive analytics, which makes use of statistical modeling and ML algorithms to foresee possible anomalies, optimize process factors, and increase yield efficiency, is another essential skill. DTs minimize production failures and batch-to-batch variability by enabling proactive decision-making through the use of historical and real-time data. Predictive modeling also helps with process scale-up, which speeds up the creation of reliable and scalable medicine formulations by extrapolating small-scale experimental data to full-scale manufacture [17].

Feedback loops allow self-regulating processes based on model-driven predictions, they are essential to the autonomous operation of DTs. By enabling bidirectional data transmission between the digital and physical systems, these loops enable automatic process parameter modifications that reduce risks and boost productivity. Feedback loops, for example, can dynamically control bioreactor conditions in the production of biopharmaceuticals, guaranteeing ideal cell growth and metabolite production without the need for human involvement [18].

The virtual model, a high-fidelity digital counterpart that accurately replicates the real production environment, is the foundation of DT functionality. To forecast process behavior under various circumstances, this simulation combines mechanistic modeling, finite element analysis (FEA), and computational fluid dynamics (CFD). Pharmaceutical businesses can test new formulations, streamline process workflows, and verify equipment performance using the virtual model, which serves as a testbed for in silico experimentation, all without having to pay for physical materials [19].

The foundation of DT-enabled pharmaceutical manufacturing is data communication, which guarantees smooth transitions between the virtual and physical systems. In order to improve traceability and regulatory compliance, DTs create a constant exchange of structured and unstructured data through cloud computing, edge computing, and blockchain-secured data management. Multi-source data harmonization is made possible by advanced data fusion tools, which provide thorough insights into process irregularities and their underlying causes.

Notwithstanding these developments, obstacles including computing complexity, worries about data security, and the requirement for industry-wide standardization prevent broad adoption. The next generation of pharmaceutical DTs is anticipated to be driven by advancements in AI-enhanced DTs, federated learning for decentralized data processing and quantum computing for large-scale simulations. The functional capabilities of DTs will continue to develop as the industry moves toward Pharma 4.0, transforming drug manufacturing through accuracy, flexibility, and intelligent automation [20].

7.3 Applications of DCE

7.3.1 Process Optimization

Pharmaceutical production has changed as a result of DCE, which makes it possible to precisely monitor, analyze, and regulate intricate drug formulation processes. Manufacturers can optimize batch and continuous manufacturing (CM) processes by combining ML algorithms, real-time analytics, and structured data frameworks. DCE guarantees regulatory compliance while improving productivity, lowering variability, and cutting down on resource waste by utilizing high-dimensional datasets and predictive modeling as presented in Figure 7.2 [21].

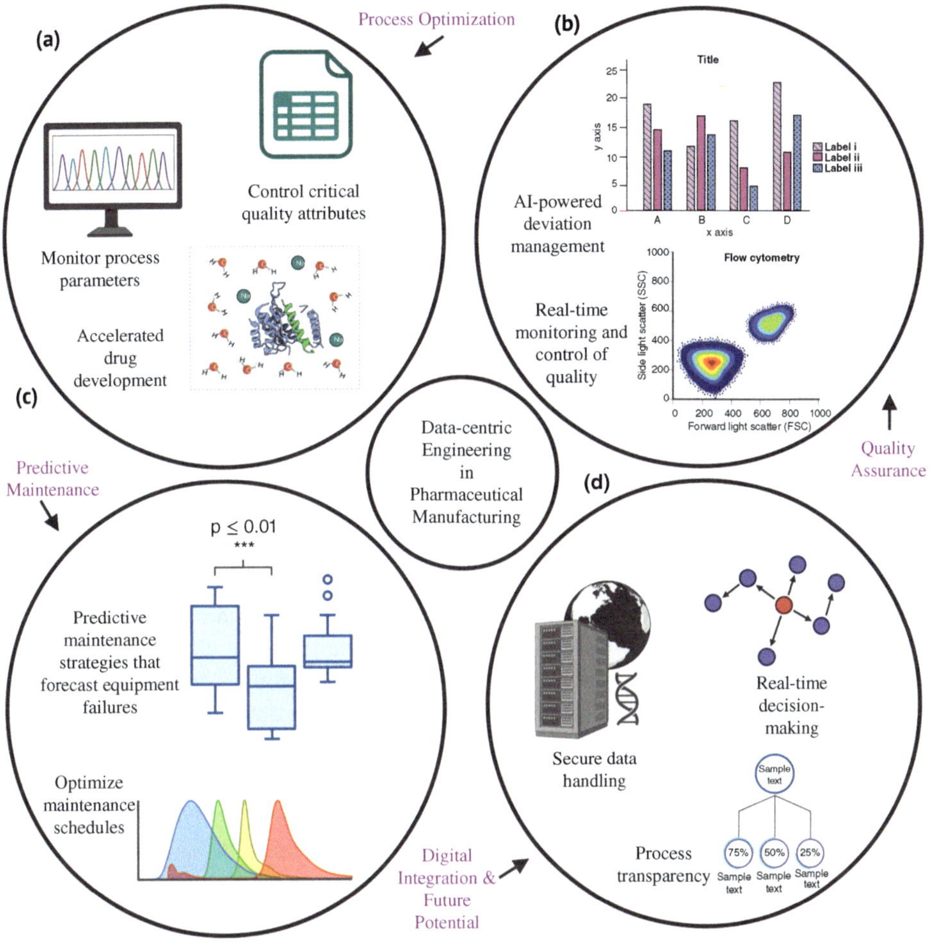

Figure 7.2 This figure showed the significance of data-centric engineering in the context of pharmaceutical manufacturing. (a) Process optimization is facilitated through the monitoring of process parameters and the regulation of critical quality attributes, thereby expediting drug development. (b) The implementation of AI-driven deviation management and real-time quality control augments quality assurance. (c) Predictive maintenance strategies employ statistical models to anticipate equipment failures and optimize maintenance schedules. (d) The prospective integration of secure data handling, real-time decision-making, and transparent digital processes is poised to enable seamless digital transformation. Collectively, these interconnected strategies underpin efficient, adaptive, and intelligent pharmaceutical manufacturing, promoting enhanced product quality, minimized downtime, and robust regulatory compliance. Created in BioRender by Md. Faiyazuddin.

7.3.1.1 Batch Manufacturing

To guarantee product consistency and adherence to regulatory frameworks like GMP and QbD, data-centric approaches are essential for monitoring and controlling CQAs and CPPs in batch manufacturing. Together with CPPs like temperature, pressure, mixing speed, and solvent composition, advanced data analytics enable real-time tracking of CQAs including particle size distribution, polymorphic shape, and dissolution rate. Predictive monitoring is made possible by the use of ML-driven models, which can spot deviations that could result in batch failures or quality issues. Process oversight is further improved by integrating IoT sensors and edge computing, which guarantees that deviations are identified early on to allow for prompt remedial action. AI-powered DTs also make it possible to simulate batch operations virtually, improving formulations prior to manufacturing, which minimizes trial-and-error testing and speeds up the time it takes for new treatments to reach the market [22].

7.3.1.2 Continuous Manufacturing

Continuous manufacturing (CM), in contrast to batch processing, offers real-time input modifications that guarantee the best possible product quality, yield, and cost-effectiveness. Advanced PAT frameworks are essential for the use of DCE in CM because they allow for the real-time monitoring and modification of crucial parameters such reaction kinetics, shear forces, and material flow rate. APC, in which ML models forecast process fluctuations and dynamically modify input parameters to preserve product consistency, is made possible by AI-driven control systems. Near-infrared (NIR) and Raman spectroscopy are two examples of inline spectroscopic techniques that offer constant input, enabling accurate compositional modifications. Furthermore, real-time simulation of process modifications is made possible by cloud-based DTs in CM, reducing the risks related to formulation transfers and scale-up. Even though CM has several benefits for lowering batch-to-batch variability and increasing operational effectiveness, issues like data harmonization, cybersecurity, and regulatory complexity are still important factors to take into account. The dependability and scalability of CM frameworks in pharmaceutical production will be improved by future developments in edge AI, blockchain for data security, and self-learning process control models [23].

The pharmaceutical sector can achieve increased process robustness, decreased waste, enhanced compliance, and accelerated drug development cycles by incorporating DCE-driven innovations into batch and continuous manufacturing. This will signal a paradigm shift toward a more data-driven, intelligent, and efficient production ecosystem.

7.3.2 Quality Assurance

In order to improve product reliability and regulatory compliance, DCE is rethinking pharmaceutical quality assurance by combining real-time monitoring technologies, AI, and advanced analytics. Quality assurance in pharmaceutical manufacturing demands rigorous validation of medication formulations and production processes to assure consistent efficacy, safety, and stability [24]. Historically, post-production laboratory analyses were used for batch-based quality testing, which was labor-intensive, resource-intensive, and prone to inefficiencies. However, with the advent of DCE, modern pharmaceutical manufacturing is transitioning toward real-time release testing (RTRT) and deviation management, both of which employ AI-driven predictive analytics, real-time data acquisition, and automated decision-making to mitigate risks and enhance production efficiency [25].

7.3.2.1 Real-time Release Testing

CQAs can be evaluated in real time with RTRT, a state-of-the-art method that eliminates the need for lengthy post-production laboratory testing. In order to assess product quality throughout the manufacturing process, it combines AI-driven prediction models with PAT instruments like mass spectrometry (MS), high-performance liquid chromatography (HPLC), Raman spectroscopy, and NIR spectroscopy. End-product testing is the foundation of traditional QC techniques, which results in major delays and inefficiencies. To ensure that every unit satisfies predetermined quality standards before to leaving the manufacturing line, RTRT, on the other hand, uses real-time data streams to provide continuous, nondestructive analysis of key formulation parameters [24].

By spotting intricate patterns in manufacturing data, ML and DL algorithms are essential to RTRT and allow for proactive QC. Real-time correlation of several variables is made possible by principal component analysis (PCA) and multivariate analysis (MVA), which makes it easier to identify departures from ideal conditions early on. Furthermore, DTs, which are virtual copies of manufacturing processes, can be integrated with AI-powered RTRT systems to replicate real-time circumstances and anticipate possible deviations before they have an impact on product quality. RTRT eliminates material waste, maximizes resource utilization, and increases production throughput by reducing the need for post-production testing, which is in conformity with US regulatory norms and GMP. The International Council for Harmonisation of Technical Requirements for Pharmaceuticals for Human Use (ICH), the European Medicines Agency (EMA), and the Food and Drug Administration (FDA) have strongly encouraged the adoption of advanced process analytical technologies and real-time quality assurance frameworks to improve pharmaceutical manufacturing efficiency, product consistency, and regulatory compliance [26].

Notwithstanding its many benefits, RTRT has drawbacks, such as the requirement for reliable data validation, sensor calibration, and regulatory approval of AI-powered quality assurance models. Additionally, a large investment in cybersecurity and data interoperability frameworks is necessary to ensure the smooth integration of RTRT systems with current industrial infrastructures and safeguard sensitive pharmaceutical data. Future developments are anticipated to concentrate on federated learning strategies that improve RTRT scalability and regulatory transparency, cloud-based AI frameworks, and blockchain-enabled data integrity solutions [27].

7.3.2.2 Deviation Management

Since process deviations can be brought on by inconsistent raw materials, broken equipment, operator error, or environmental changes, deviation management is another crucial component of quality assurance in the pharmaceutical manufacturing industry. Uncontrolled deviations can result in costly batch rejections, regulatory noncompliance, and even patient safety hazards by compromising product quality. Manual inspections and retrospective root cause analysis (RCA) were the mainstays of traditional deviation management, which frequently led to postponed remedial action. However, by providing real-time anomaly detection, automated root cause identification, and proactive corrective action implementation, AI-driven predictive analytics and statistical process control (SPC) approaches are currently revolutionizing deviation detection and resolution [28].

Modern deviation management systems may continually monitor manufacturing parameters and identify possible deviations before they become serious failures by utilizing anomaly detection techniques, Bayesian networks, and ML models. Manufacturers can put preventive measures in place and lessen the chance of reoccurring problems by using predictive analytics to identify patterns linked to past deviations. Furthermore, automated deviation reporting guarantees that nonconformities are recorded instantly, promoting prompt decision-making and adherence to legal requirements [29].

By building a digital model of the manufacturing process, the combination of DTs and IoT sensors improves deviation management even more. This allows manufacturers to model deviations under various circumstances and evaluates the effects of corrective measures prior to putting them into practice in the actual production setting. AI-driven RCA speeds up the resolution process and reduces production downtime by automating the discovery of underlying deviation sources [30].

Data standardization, model interpretability, and regulatory acceptance of AI-generated insights are some of the issues that AI-driven deviation management faces, despite its revolutionary promise. Industry stakeholders continue to place a high premium on ensuring data integrity, cybersecurity, and real-time interoperability among decentralized manufacturing units. To further improve deviation management techniques, future studies and technical developments are anticipated to concentrate on block chain-enabled audit trails, AI-enhanced regulatory reporting, and cloud-based predictive analytics platforms [31].

7.3.3 Predictive Maintenance

By using real-time data analytics, AI, and ML to predict equipment failures before they happen, predictive maintenance, an advanced application of DCE, significantly lowers maintenance costs, downtime, and operational disruptions in the pharmaceutical manufacturing industry. Predictive maintenance, as opposed to conventional reactive or preventive maintenance techniques, uses sensor-based monitoring, historical data analysis, and predictive modeling to evaluate the condition of the equipment, find early failure indications, and suggest prompt repairs. By guaranteeing that vital equipment, including tablet press machines, lyophilizers, bioreactors, and high-precision mixing systems, run within ideal parameters and preventing unplanned malfunctions that can jeopardize product quality and regulatory compliance, this strategy improves manufacturing reliability [32].

IoT-enabled sensors that continuously gather data on key performance indicators (KPIs) like vibration levels, temperature variations, pressure anomalies, and lubrication efficiency are essential to the use of predictive maintenance in pharmaceutical manufacturing. These enormous datasets are processed by sophisticated AI-driven analytics, which use DL models, anomaly detection algorithms, and predictive analytics to find minute departures from typical operating circumstances. The remaining useful life (RUL) of machinery components is predicted using methods including supervised and unsupervised ML, DT simulations, and probabilistic failure models. This enables dynamic optimization of maintenance schedules. This data-driven strategy improves overall equipment effectiveness (OEE), lowers the cost of spare parts inventory, and gets rid of needless preventive maintenance [33].

Predictive maintenance's ability to interface with enterprise resource planning (ERP) platforms and manufacturing execution systems (MES) is one of its key benefits; it guarantees smooth coordination between repair tasks and production schedules. By lowering the possibility of mechanical failures that could result in product variances, this integration promotes adherence to GMP, minimizes production delays, and avoids inconsistent drug formulation. Additionally, because maintenance staff may concentrate on high-priority interventions rather than regular inspections, real-time monitoring and automated diagnostics increase worker efficiency [34].

The use of DT-based predictive modeling, federated learning for decentralized data analysis, and blockchain technology for safe and unchangeable maintenance records are examples of recent developments in predictive maintenance. Widespread use is still significantly hampered by issues including data interoperability, cybersecurity risks, and the requirement for substantial processing power. To further optimize maintenance workflows, future advancements in AI-driven predictive maintenance are probably going to concentrate on boosting real-time edge computing capabilities, increasing sensor accuracy, and incorporating autonomous AI-driven decision-making [35].

The convergence of DCE, AI, and predictive maintenance will be essential to attaining operational excellence as pharmaceutical manufacturing moves toward Industry 4.0 [36, 37]. Predictive maintenance will remain a vital component of contemporary pharmaceutical manufacturing, bolstering efficiency, compliance, and sustainability in a sector that is becoming more and more data-driven by encouraging a proactive maintenance culture, minimizing unscheduled downtime, and guaranteeing consistent drug formulation procedures [38].

7.4 Real-time Monitoring and Control: Enabling Technologies

7.4.1 IoT and Sensor Integration

Real-time monitoring and control in pharmaceutical manufacturing, especially in drug formulation processes, has been transformed by the combination of IoT and sophisticated sensor technology. In order to ensure strict process control and compliance with regulatory standards, sensors integrated into production systems collect real-time data on crucial factors like temperature, pressure, flow rates, pH, viscosity, and humidity. By bridging the gap between digital analytics and physical equipment, these CPS allow for accurate process optimization through ongoing feedback mechanisms. The foundation of this infrastructure is made up of IoT platforms, which enable smooth data collection, transfer, and analysis in edge or cloud computing settings as presented in Figure 7.3 [39].

To improve measurement accuracy and responsiveness, contemporary IoT-enabled sensors make use of piezoelectric transducers, optical sensors, and microelectromechanical systems (MEMS). When placed within reactors, mixers, and bioreactors, these high-precision sensors offer real-time information on physicochemical changes that take place during drug development. For example, fiber-optic sensors evaluate medication homogeneity in CM systems, while noncontact infrared sensors track exothermic reactions in granulation and lyophilization procedures. To ensure low latency and high data

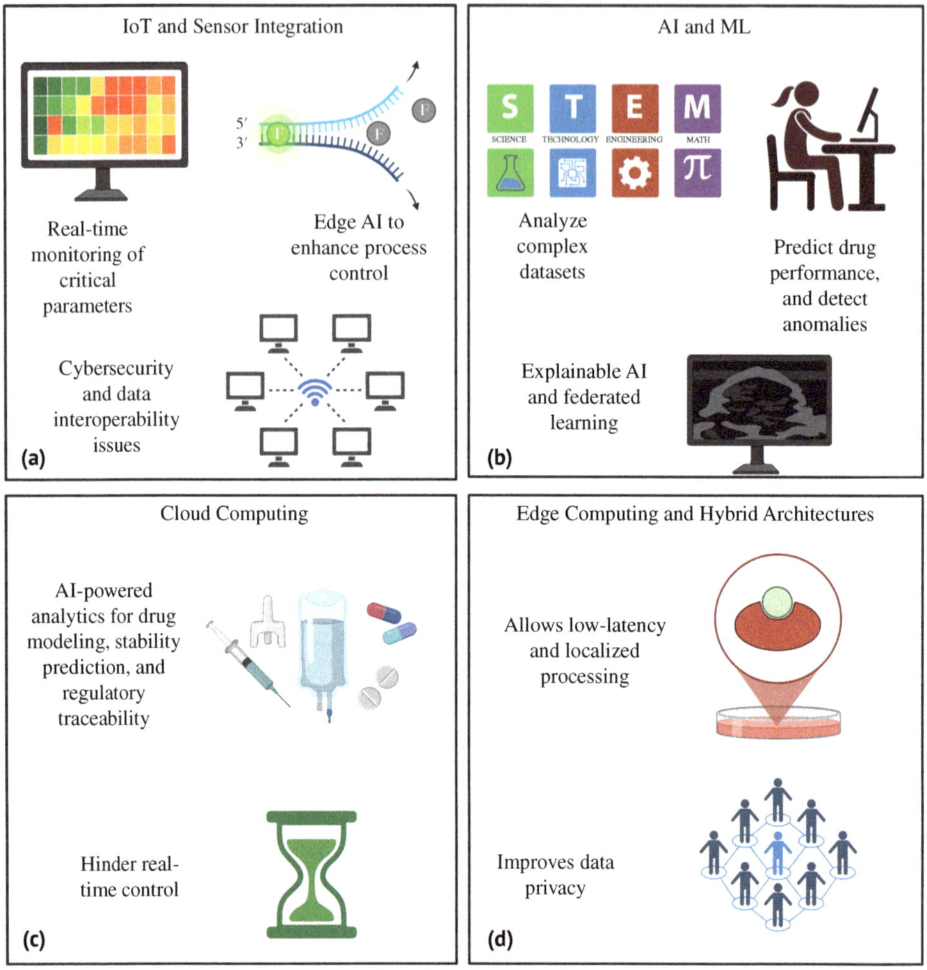

Figure 7.3 This figure elucidates the enablers of digital transformation within pharmaceutical manufacturing. (a) The integration of IoT and sensors facilitates real-time monitoring and process control, although it faces challenges related to cybersecurity and data interoperability. (b) Artificial intelligence and machine learning provide the capability to analyze complex datasets, predict drug performance, and detect anomalies, with support from explainable AI and federated learning. (c) Cloud computing underpins AI-driven analytics for drug modeling, stability, and traceability, though it may impede real-time control. (d) Edge computing and hybrid architectures enable low-latency, localized processing, thereby enhancing data privacy and responsiveness. Collectively, these technologies advance intelligent, connected, and secure pharmaceutical operations for next-generation manufacturing systems. Created in BioRender by Md. Faiyazuddin.

integrity, the collected sensor data is sent over wireless communication protocols such as Zigbee, Bluetooth Low Energy, LoRaWAN (Long Range Wide Area Network), and 5G networks [40].

AI and ML algorithms are used by IoT-enabled monitoring frameworks to identify abnormalities, forecast deviations, and activate adaptive control mechanisms. Multivariate statistical process control (MSPC) and DL models power predictive analytics, which helps anticipate and detect hazards including contamination, crystallization flaws, or changes in the quantities of APIs. Pharmaceutical companies can lower computational latency and bottlenecks and enable quicker decision-making without over-relying on cloud computing by including edge AI for decentralized processing.

Notwithstanding these developments, issues with data interoperability, cybersecurity, and sensor standardization still exist. Significant challenges include the requirement for 21 CFR Part 11 compliance, the heterogeneous nature of sensor outputs, and different data formats. Protecting IoT networks from cyberattacks requires strong encryption protocols, blockchain-based data integrity, and compliance with good automated manufacturing practice (GAMP) 5). Furthermore, IoT and DTs are expected to converge more in the pharmaceutical production industry in the future, allowing for real-time virtual simulations of formulation processes for improved predictive control. A paradigm shift in intelligent pharmaceutical manufacturing will be ushered in by the introduction of self-calibrating smart sensors, the incorporation of quantum computing for intricate drug modeling, and improved edge-to-cloud hybrid architectures, all of which will improve real-time monitoring capabilities [41].

7.4.2 ML and AI

AI and ML have become key technologies in the revolution of real-time medication formulation monitoring and control. Large-scale datasets, sophisticated algorithms, and predictive analytics are all used in these computational approaches to improve process effectiveness, optimize formulation parameters, and guarantee product consistency. Multidimensional data streams from production sensors are analyzed by AI-driven models, which include supervised and unsupervised learning approaches. These models uncover intricate correlations between formulation factors and CQAs. In order to extract complex patterns from high-dimensional datasets and make accurate real-time predictions about drug stability, dissolution rates, and bioavailability profiles, DL, a subset of ML, uses artificial neural networks (ANNs) [42].

By dynamically modifying formulation parameters in response to real-time data, the combination of AI and PAT has greatly enhanced adaptive process control. In order to improve medicine efficacy and stability, RL-based AI systems optimize production settings, excipient interactions, and constituent concentrations. Predictive models are further improved by Bayesian optimization approaches, which also speed up formulation creation and minimize trial iterations. Additionally, early anomaly detection is made easier by AI-enabled MVA, which reduces the chance of batch failures and guarantees regulatory compliance with the QbD and GMP frameworks [43].

Notwithstanding these developments, obstacles still exist in the use of AI in pharmaceutical applications, such as model interpretability, data heterogeneity, and strict legal requirements for AI-driven decision-making. In order to improve model transparency and regulatory acceptance, future advances are anticipated to incorporate explainable AI (XAI) and federated learning. Real-time process control will be further improved by the combination of AI, DTs, and cloud-based architectures, which will promote a paradigm shift toward completely autonomous drug formulation systems [44].

7.4.3 Cloud Computing and Edge Computing

AI-driven medication formulation has been transformed by the combination of cloud and edge computing in real-time monitoring and control, offering scalable, effective, and intelligent decision-making capabilities. These computational paradigms provide sophisticated data collecting, processing, and analytics that optimize formulation development, improve product consistency, and guarantee regulatory compliance as pharmaceutical production adopts Industry 4.0. Real-time responsiveness in crucial drug formulation workflows is made possible by the convergence of cloud and edge computing, which guarantees exact control over process variables, excipient interactions, and physicochemical parameters. Through the use of AI-powered algorithms that continuously adjust formulation parameters in response to real-time feedback from DTs, ML models, and IoT-enabled sensors, these technologies enable adaptive manufacturing techniques. This guarantees better scalability, increased process resilience, and predictive drug property optimization [45].

7.4.3.1 Cloud Computing: Centralized Data Storage and Scalable Analytics

Cloud computing helps pharmaceutical firms analyze massive volumes of heterogeneous data by providing a centralized platform for data aggregation, high-performance computing (HPC), and AI-driven analytics. Pharmaceutical businesses can use cloud-based systems for predictive modeling of drug-excipient interactions, virtual screening of APIs, and real-time simulations thanks to this paradigm. By examining molecular interactions, physicochemical stability, and bioavailability profiles, cloud infrastructure combined with AI-driven DL models greatly improves the accuracy of drug formulation predictions. Furthermore, cloud-based architectures ensure end-to-end traceability of formulation processes by offering smooth interaction with MES and laboratory information management systems (LIMS) that adhere to GMP [46].

Researchers may precisely forecast stability profiles and fine-tune CQAs by using cloud computing to enable multi-scale modeling of drug delivery systems using CFD simulations. Pharmaceutical companies can use cloud-based AI solutions for automated QC, anomaly detection, and regulatory compliance without having to make significant investments in on-premises equipment thanks to platform-as-a-service (PaaS) and software-as-a-service (SaaS) models. Notwithstanding these benefits, obstacles to broad adoption include issues including data latency, cybersecurity threats, and adherence to legal requirements (such as FDA 21 CFR Part 11 and EMA Annex 11). Furthermore, latency problems brought about by cloud-based data transfer might affect real-time process control, making edge computing a necessary supplementary option [47].

7.4.3.2 Edge Computing: Localized Data Processing for Real-time Responsiveness

By permitting localized data processing at or close to the data source, edge computing overcomes latency limitations and guarantees real-time responsiveness in AI-driven medicine development. By processing crucial formulation data closer to production lines, edge computing lessens reliance on high-bandwidth internet, in contrast to cloud computing, which depends on centralized data centers. This is especially helpful in CM settings where batch failures can be avoided by instantly correcting variations in process variables like temperature, pressure, and mixing efficiency. Pharmaceutical manufacturing facilities become more autonomous with the use of edge-AI models integrated into IoT-enabled smart sensors, programmable logic controllers (PLCs), and industrial automation systems. This enables real-time corrective measures without the need for cloud-based intervention [48].

The capacity of edge computing to facilitate decentralized AI decision-making is one of its core benefits; this enables adaptive control of medication formulation procedures, predictive maintenance, and real-time anomaly detection. For instance, by evaluating real-time spectroscopic and chromatographic data from inline sensors, edge-enabled AI models can dynamically optimize the choice of polymeric carriers in nanomedicine formulation. By doing this, cloud-based analytics delays are avoided and drug delivery systems are guaranteed to retain optimal release kinetics, particle size distribution, and stability. Additionally, edge computing improves cybersecurity and data privacy by processing patient data and sensitive material locally, reducing the possibility of data transmission vulnerabilities, cloud storage breaches, and regulatory noncompliance [49].

Notwithstanding its benefits, edge computing has drawbacks like low processing power, higher hardware expenses, and complicated maintenance requirements. In contrast to cloud computing, which makes use of infinitely scalable resources, edge devices need specialized hardware accelerators (such as neuromorphic processors, field-programmable gate arrays [FPGAs], and tensor processing units [TPUs]) and energy-efficient AI models in order to maintain real-time analytics. Furthermore, integration is difficult due to the absence of standardized edge-to-cloud interoperability frameworks, especially in large-scale pharmaceutical manufacturing settings that call for hybrid computing solutions.

A hybrid cloud-edge architecture, which combines the scalability of cloud computing with the real-time capabilities of edge computing, is emerging as the best way to fully realize the potential of AI-driven pharmaceutical manufacturing. Future developments in federated learning, blockchain-integrated data security, and 5G-enabled edge networks will make it possible to seamlessly combine cloud and edge infrastructures, guaranteeing fast data processing while maintaining strong regulatory compliance. Drug formulation will be further revolutionized by the deployment of intelligent edge devices that have self-learning AI models and autonomous decision-making capabilities. These devices will enable real-time process adjustments based on multi-dimensional data streams from monitoring systems that use biosensors, spectroscopic analysis, and chromatography [50].

Reducing development delays and guaranteeing repeatability in drug production, the use of DTs in hybrid cloud-edge environments will improve predictive modeling of formulation stability, drug-excipient compatibility, and process scalability. The pharmaceutical sector will transition to completely autonomous and intelligent manufacturing ecosystems by integrating AI-driven self-optimizing controllers in edge devices and simulations aided by quantum computing into cloud platforms. The confluence of AI, cloud computing, and edge computing will transform the field of drug formulation as regulatory bodies adjust to new technological developments, promoting more precision, sustainability, and creativity in pharmaceutical manufacturing [51].

7.5 Case Studies

The convergence of nanotechnology and AI is fundamentally transforming the field of human therapeutics [52–56]. These two innovative domains, in which one focused on the manipulation of matter at the nanoscale, and the other on ML and intelligent decision-making, are merging to provide highly personalized, precise, and effective medical interventions. Collectively, they hold the potential to revolutionize disease diagnostics, drug delivery, treatment monitoring, and regenerative medicine [57–63]. Nanotechnology offers tools to engineer materials and devices at the molecular and atomic levels. In therapeutics, nanoparticles ranging from liposomes and polymeric micelles to quantum dots and metallic nanostructures can be customized to transport drugs, genes, or imaging agents [64–70]. Their diminutive size enables them to traverse biological barriers, such as the blood-brain barrier, and target diseased cells with minimal harm to surrounding tissues. For instance, nanocarriers functionalized with specific ligands can bind to overexpressed receptors on cancer cells, facilitating targeted chemotherapy with reduced systemic toxicity [71–79]. Moreover, stimuli-responsive nanoparticles can release their payload in response to internal cues such as pH, temperature, or enzymes, or external triggers like light and magnetic fields. This "smart" behavior enhances therapeutic precision, reduces dosage frequency, and minimizes adverse effects. AI is revolutionizing the interpretation of complex biomedical data. ML algorithms can analyze vast datasets like genomics, proteomics, medical imaging, electronic health records to identify patterns that elude human cognition [80–87]. In therapeutics, AI is employed to predict disease progression, identify drug targets, and even generate novel drug candidates through DL models. A notable application is in precision oncology, where AI algorithms integrate genomic data with histopathological images to recommend tailored treatment regimens [88–96]. In clinical trials, AI aids in patient stratification, trial design optimization, and real-time monitoring of adverse effects, thereby enhancing success rates and reducing costs. The integration of AI with nanotechnology amplifies their individual potentials. AI can optimize nanoparticle design by predicting how variations in size, shape, surface charge, and composition affect biodistribution and cellular uptake [97–104]. It can also forecast interactions between nanomaterials and biological systems, expediting the development of safe and effective nanotherapeutics [105–112].

A versatile DT architecture was created by Shoshi et al. for the biomanufacturing of advanced therapeutic medical products (ATMPs), with an emphasis on the generation of chimeric antigen receptor (CAR) T cells. The framework's introduction of a hierarchical architecture that methodically arranges production into abstraction levels, such as processes, tasks and skills, and microservices, was intended to address the intrinsic complexity and dynamic nature of CAR T cell manufacturing. Both automatic and human processes may be easily monitored and controlled inside a single system thanks to this arrangement. Reusable components, such sensor data retrieval and notification services, could be connected across several production phases without redundancy because of the framework's incorporation of modular microservices. Furthermore, a task-to-skill matching mechanism made it possible to adjust automated and manual skills in a flexible way, guaranteeing the system's long-term scalability as manufacturing technologies advanced. Additionally, the DT framework offered a basis for smart industrial applications, such as data interpretation, intelligent operation planning, and control, which could be improved by including massive language models. Data anonymization techniques were used to address privacy and security concerns while maintaining regulatory compliance. In addition to improving the efficiency of CAR T cell production, this study showed that the suggested DT framework functioned as a flexible and durable digital infrastructure that could change to accommodate future developments in ATMP production [113].

By combining first-principles models, residence time distribution (RTD) simulations from the discrete element method (DEM), science of scale (SoS) tools, and data-driven models in a hybrid flowsheet approach, Moreno-Benito et al. created a DT model for a continuous direct compression line. Important qualities like tablet composition, weight, thickness, and hardness were estimated by this computerized framework, which also properly forecasted important process parameters across feeders, blenders, and tablet presses. The model reduced the need for extensive experimental trials and minimized the use of API by enabling early-stage de-risking of therapeutic product development through the simulation of steady-state operations, dynamic process responses, and parameter interactions. The DT also made it easier to optimize processes, which improved pharmaceutical manufacturing's robustness, efficiency, and quality. Its potential for wider use in CM and process digitization was highlighted by its capacity to facilitate real-time decision-making, address regulatory enquiries, and optimize production workflows, positioning it as a game-changing instrument in the pharmaceutical sector [114].

A DT for a continuous pharmaceutical powder blending process was created by Beke et al. using the principles of QbD and PAT. Continuous blending studies were used to examine a low-dosage system that included dextrose as the excipient and caffeine as the API. Process analytics based on NIR spectroscopy were applied, and quantitative spectral evaluation was made possible by multivariate data processing. The blending system was simulated using recurrent ANNs and mechanistic RTD models, which were trained and validated using experimental results. Because RTD models need fewer resources and are not affected by chemometric calibration errors at low concentrations, a comparative study of the two modeling methodologies showed that they were more effective for industrial applications. Both RTD and nonlinear autoregressive (NARX) neural networks had great predictive capability, with simulation-based validation confirming that the latter's high accuracy was reliant on the volume of training data. Further optimization research showed that ANN models needed a lot of processing power, but RTD models offered a more useful method for process monitoring. By accurately forecasting output concentration from input mass flows, the DT made it easier to create model-based control systems and laid the groundwork for incorporating continuous blending technology into pharmaceutical production lines from start to finish [115].

In order to forecast long-term capacity requirements across several operating domains, Santos et al. created a DT of a pharmaceutical company's internal supply chain by incorporating a simulation-based rough-cut capacity planning (RCCP) tool. The case study, which was carried out inside a pharmaceutical company, showed how well the DT worked at producing precise predictions for the distribution of resources over a 24-month period. With the help of the system's graphical user interface, managers were able to visualize supply chain dynamics in real time, evaluate planned tasks with ease, spot seasonal patterns, and allocate resources as efficiently as possible. Because simulation results were superior to those of traditional RCCP approaches, decision-makers were able to proactively negotiate changes to the master schedule, which reduced short-term capacity shortages and cost overruns. The approach demonstrated flexibility for wider industrial applications with only minor adjustments, mostly necessitating supply chain mapping specific to the industry. The program, however, was limited in its ability to handle qualitative disturbances that were not quantifiable and did not account for variations in the time frames for planned and actual job execution. Notwithstanding these obstacles, the DT showed great promise for improving pharmaceutical supply chain management's operational effectiveness and strategic planning [116].

In order to fill in the information gaps, Miozza et al. carried out a thorough literature analysis to investigate the effects of digital transformation in the pharmaceutical industry (PI) from the standpoint of management research. After 404 publications were first screened, 35 pertinent papers were found, which served as the foundation for a planned future study program. Operations management (I), strategic management (II), organization theory (III), and stakeholder theory (IV) were the four main research directions identified by the study. The analysis emphasized how important it is to comprehend the complex effects of digital transformation on PI, especially in the post-COVID-19 period when digital technologies have sped up industrial transformations. The results highlighted that although PI-digital transformation's practical applications have been thoroughly investigated, management research in this area has not yet reached its full potential since it is frequently overly practitioner-focused and ignores new developments in technology. The study laid the groundwork for future academic investigation into improving competitive tactics, streamlining operational procedures, restructuring organizational structures, and evaluating digital transformations impact on different stakeholders by presenting a thorough research methodology. In addition to highlighting the necessity of interdisciplinary cooperation to fully realize the potential of digital transformation in the PI, these insights advanced our understanding of PI-digital transformation management [117].

7.6 Challenges and Future Directions

7.6.1 Data Management and Security

AI in medication formulation is heavily dependent on large datasets that include toxicity, pharmacokinetics, molecular structures, and clinical trial results. Particularly under strict regulation like the FDA's 21 CFR Part 11, which requires electronic record-keeping, audit trails, and validation of data processing systems, it is critical to ensure data integrity and regulatory compliance. In AI-driven pharmaceutical research, data traceability, precision, and consistency are essential since even little discrepancies can result in inaccurate predictions that compromise the safety and effectiveness of drugs. To ensure compliance and improve the dependability of AI-generated insights, sophisticated data governance frameworks that include automated validation algorithms, real-time monitoring tools, and metadata standardization are crucial [73].

The cybersecurity flaws in networked AI systems used in drug formulation and pharmaceutical manufacturing provide another urgent problem. Due to the widespread use of cloud computing, IoT sensors, and block chain-based data sharing, private data is vulnerable to ransomware attacks, data breaches, and hostile AI manipulations. These risks can be reduced while maintaining data accessibility and confidentiality by putting end-to-end encryption, zero-trust security models, and federated learning techniques into practice. Furthermore, important medication formulation processes can be protected by proactive cyber threat identification and neutralization using real-time anomaly detection enabled by AI-driven intrusion detection systems (IDS) [118].

Strong regulatory frameworks, AI-optimized cybersecurity protocols, and decentralized data ecosystems must all come together for AI in medication development to continue on its current trajectory. Addressing new risks will require integrating blockchain-based traceability for compliance assurance, quantum encryption for improved data protection, and AI-powered risk assessment tools. A standardized approach to data management and security will be essential as AI continues to transform pharmaceutical innovation in order to maintain confidence, guarantee regulatory compliance, and speed up improvements in drug development and formulation [119].

7.6.2 Integration Complexity

Because of the complexity of legacy systems, data silos, and strict regulatory constraints, it is difficult to integrate DTs with current pharmaceutical production infrastructure and IT ecosystems. As dynamic cyber-physical replicas, DTs must constantly be in sync with actual manufacturing process, which calls for strong interoperability across a variety of hardware, software, and data management systems. However, traditional pharmaceutical manufacturing systems frequently depend on different automation technologies and proprietary architectures, which hinder real-time connectivity and unified data interchange. The requirement for standardized communication protocols to facilitate smooth data transfer between distributed control systems (DCS), ERP platforms, supervisory control and data acquisition (SCADA) systems, and IoT-enabled equipment is one of the main challenges in DT integration. The installation of DT necessitates considerable retrofitting or middleware solutions because many pharmaceutical facilities use older PLCs and batch processing systems that were not built to accommodate high-frequency data collecting and predictive analytics. Furthermore, in order to handle large-scale, high-dimensional information and guarantee low-latency replies for real-time decision-making, DT integration with PAT frameworks requires a significant amount of processing capacity [120].

Data governance and cybersecurity make DT integration much more difficult because pharmaceutical manufacture takes place in highly regulated settings that must adhere to good automated manufacture practices (GAMP), the US Guidelines on computerized systems from the EMA and the FDA 21 CFR Part 11. To safeguard sensitive manufacturing data while preserving accessibility for AI-driven analytics and optimization algorithms, safe and legal data exchange procedures must be put in place. Furthermore, the implementation of cloud-based DT systems raises issues with latency and data sovereignty, calling for a hybrid strategy that strikes a balance between cloud-based analytics for long-term predictive modeling and edge computing for real-time process control [15, 16].

A multifaceted strategy including open-source interoperability frameworks, modular system architectures, and automation powered by AI is needed to address these integration issues. More seamless connectivity between DT platforms and manufacturing infrastructure can be facilitated by the use of industry standards like Open Platform Communications Unified Architecture (OPC UA) and MES integration techniques. Additionally, it is anticipated that developments in digital thread technology, which offer an end-to-end data lineage throughout the product lifecycle, will be essential in expediting DT deployment in pharmaceutical environments. AI-driven self-adaptive integration frameworks that allow for autonomous DT interface configuration and optimization should be the main focus of future research and development activities. This will minimize manual intervention and hasten the shift to fully digitalized pharmaceutical manufacturing ecosystems [121].

7.6.3 Regulatory Acceptance

Since current frameworks were not initially intended to support real-time, data-driven decision-making and predictive modeling, the integration of DTs and DCE in pharmaceutical manufacturing poses major regulatory problems. Conventional regulatory pathways, such as those established by U.S. Food and Drug Administration, prioritize clearly defined and empirically validated processes, whereas digital twins (DTs) depend on continuous data acquisition, advanced simulations, and machine learning–driven analytics. Therefore, a paradigm change toward data-driven validation, in silico modeling approvals, and dynamic process management strategies that meet regulatory requirements is required in order to modify regulatory frameworks to embrace DT-enabled methodologies [122].

Since DT models' prediction powers rely on historical datasets, real-time sensor data, and AI-driven learning algorithms, their validation and verification (V&V) is one of the main obstacles to regulatory adoption. To make sure that forecasts match experimental results in accordance with GMP, regulators need standardized methods to evaluate the precision, repeatability, and dependability of DT models. Furthermore, in order to qualify DTs as a PAT tool under QbD principles, regulatory bodies must set precise guidelines. This will enable pharmaceutical companies to demonstrate process robustness and real-time quality assurance without the need for lengthy physical trials.

Data integrity, traceability, and cybersecurity issues related to DTs must be addressed by the regulatory environment. Data security and auditability depend on adherence to 21 CFR Part 11 and GAMP rules since DTs use cloud-based infrastructures and collect data from IoT devices. Emerging solutions that could promote regulatory trust in DT applications include the use of blockchain technology for immutable data logging and AI-driven anomaly detection techniques for real-time compliance monitoring [123].

7.6.4 Skill Development

A paradigm shift in workforce expertise is required for the integration of AI in drug formulation, necessitating extensive skill development programs to give professionals the technical know-how they need. As AI-driven approaches like data-driven predictive modeling, ML, and DL become more prevalent in drug development, pharmaceutical scientists, engineers, and regulatory specialists need to become proficient in data analytics, algorithmic optimization, and computational techniques [124]. Interdisciplinary programs that integrate pharmaceutical sciences with AI, bioinformatics, and computational chemistry must replace traditional pharmaceutical training, which mostly emphasizes chemical, biological, and process-oriented elements. Since pharmaceutical experts might not have received formal training in programming, statistical modeling, or AI-driven data interpretation, bridging the gap between domain expertise and AI competency is a crucial skill development challenge. To solve this, upskilling staff in AI systems specifically designed for drug development requires specialized training modules, certification programs, and cooperative academic-industry initiatives. Furthermore, the smooth implementation of AI-driven formulation strategies will be facilitated by practical training in AI platforms, cloud-based modeling tools, and automation technologies. To ensure that AI-integrated formulation processes adhere to changing regulatory frameworks like GAMP and AI-driven validation methods, it is equally critical to upskill regulatory personnel. A workforce skilled in utilizing AI for precision-driven drug design, formulation optimization, and real-time quality assurance will be fostered by future developments in AI-specific pharmaceutical training, which are anticipated to include immersive learning environments, augmented reality (AR)-based simulations, and AI-assisted decision support systems. In order to ensure industry-wide competency and innovation-driven advancement, closing the skill gap through organized educational reforms and ongoing professional development would be essential to speeding up the use of AI in pharmaceutical formulation [125].

7.7 Conclusion

Pharmaceutical manufacturing is being redefined by the convergence of AI, DTs, and DCE, which offers previously unheard-of levels of precision, efficiency, and adaptability in medication formulation and production processes. Pharmaceutical companies may improve process optimization, shorten development time frames, and produce higher-quality products by combining automation, real-time data analytics, and AI-driven predictive modeling. As high-fidelity virtual models of real-world systems, DTs facilitate predictive control, in silico experimentation, and real-time monitoring. They also allow for dynamic formulation parameter modifications, which improves batch-to-batch consistency. When combined with DCE, which stresses the use of structured data for process optimization, these technologies promote a change to intelligent manufacturing paradigms that place an emphasis on scalability and adaptability. To guarantee smooth adoption, issues including data interoperability, algorithmic transparency, regulatory compliance, and personnel upskilling must be resolved despite these developments. But ongoing advancements in AI-driven automation, advanced data analytics, and IoT connection are gradually removing these obstacles and easing the shift to more resilient and flexible pharmaceutical manufacturing ecosystems. Personalized medicine may benefit from the combination of AI with DTs and DCE since it will allow for the development of precision formulation techniques catered to the requirements of individual patients. To fully realize the potential of these game-changing technologies, the pharmaceutical business must adopt interdisciplinary partnerships, regulatory developments, and AI-specific skill development initiatives. In the era of Industry 4.0 and beyond, the industry is well-positioned to reach new heights in drug formulation as AI and data-driven approaches continue to advance, spurring innovation while guaranteeing sustainability, regulatory compliance, and operational excellence.

References

1 Marmolejo-Saucedo JA. Design and development of digital twins: A case study in supply chains. *Mobile Networks and Applications [Internet]* 2020;25(6):2141–2160. Available from: https://link.springer.com/10.1007/s11036-020-01557-9

2 Zobel-Roos S, Schmidt A, Uhlenbrock L, Ditz R, Köster D, and Strube J. Digital twins in biomanufacturing. *Advances in Biochemical Engineering Biotechnology* 2020;176:181–262. Available from: https://link.springer.com/10.1007/10_2020_146

3 Chen Y, Yang O, Sampat C, Bhalode P, Ramachandran R, and Ierapetritou M. Digital twins in pharmaceutical and biopharmaceutical manufacturing: A literature review. *Processes [Internet]* 2020;8(9):1088. Available from: https://www.mdpi.com/2227-9717/8/9/1088

4 Matić J, Stanković-Brandl M, Bauer H, et al. Pharmaceutical hot melt extrusion process development using QbD and digital twins. *International Journal of Pharmaceutics [Internet]* 2023;631:122469. Available from: https://linkinghub.elsevier.com/retrieve/pii/S0378517322010249

5 van Beek A, Nevile Karkaria V, and Chen W. Digital twins for the designs of systems: A perspective. *Structural and Multidisciplinary Optimization [Internet]* 2023;66(3):49. Available from: https://link.springer.com/10.1007/s00158-023-03488-x

6 Chiachío M, Megía M, Chiachío J, Fernandez J, and Jalón ML. Structural digital twin framework: Formulation and technology integration. *Automation in Construction [Internet]* 2022;140:104333. Available from: https://linkinghub.elsevier.com/retrieve/pii/S0926580522002060

7 Kuchemüller KB, Pörtner R, and Möller J. Digital twins and their role in model-assisted design of experiments. *Advances in Biochemical Engineering Biotechnology* 2020: 29–61. Available from: https://link.springer.com/10.1007/10_2020_136

8 Worden K, Cross EJ, Barthorpe RJ, Wagg DJ, and Gardner P. On digital twins, mirrors, and virtualizations: Frameworks for model verification and validation. *ASCE-ASME Journal of Risk and Uncertainty in Engineering Systems, Part B: Mechanical Engineering [Internet]* 2020;6(3). Available from: https://asmedigitalcollection.asme.org/risk/article/6/3/030902/1082000/On-Digital-Twins-Mirrors-and-Virtualizations

9 Pan I, Mason LR, and Matar OK. Data-centric Engineering: integrating simulation, machine learning and statistics. Challenges and opportunities. *Chemical Engineering Science [Internet]* 2022;249:117271. Available from: https://linkinghub.elsevier.com/retrieve/pii/S0009250921008368

10 Hamid OH. Data-centric and model-centric AI: Twin drivers of compact and robust industry 4.0 solutions. *Applied Sciences [Internet]* 2023;13(5):2753. Available from: https://www.mdpi.com/2076-3417/13/5/2753

11 Gerogiorgis DI and Castro-Rodriguez D. A digital twin for process optimisation in pharmaceutical manufacturing. *Computer Aided Chemical Engineering* 2021;50:253–258. Available from: https://linkinghub.elsevier.com/retrieve/pii/B9780323885065500413

12 Ritu TH, Nag A, Bairagi AK, and Rahman A. Digital twins for healthcare in industry 5.0: Fundamentals, pharmaceutical manufacturing applications, diagnostic precision, digital patient innovations, and drug discovery impact. In: *Soft Computing in Industry 50 for Sustainability [Internet]*, (pp. 357–376). Cham: Springer Nature Switzerland; 2024. Available from: https://link.springer.com/10.1007/978-3-031-69336-6_16

13 Fischer R-P, Volpert A, Antonino P, and Ahrens TD. Digital patient twins for personalized therapeutics and pharmaceutical manufacturing. *Frontiers in Digital Health [Internet]* 2024: 5. Available from: https://www.frontiersin.org/articles/10.3389/fdgth.2023.1302338/full

14 Erol T, Mendi AF, and Dogan D. The digital twin revolution in healthcare. In: *2020 4th International Symposium on Multidisciplinary Studies and Innovative Technologies (ISMSIT) [Internet]*, (pp. 1–7). IEEE; 2020. Available from: https://ieeexplore.ieee.org/document/9255249/

15 Katsoulakis E, Wang Q, Wu H, et al. Digital twins for health: A scoping review. *npj Digital Medicine [Internet]* 2024;7(1):77. Available from: https://www.nature.com/articles/s41746-024-01073-0

16 De Benedictis A, Mazzocca N, Somma A, and Strigaro C. Digital twins in healthcare: An architectural proposal and its application in a social distancing case study. *Journal of Biomedical and Health Informatics [Internet]* 2023;27(10):5143–5154. Available from: https://ieeexplore.ieee.org/document/9882337/

17 Malheiro V, Duarte J, Veiga F, and Mascarenhas-Melo F. Exploiting pharma 4.0 Technologies in the non-biological complex drugs manufacturing: Innovations and implications. *Pharmaceutics [Internet]* 2023;15(11):2545. Available from: https://www.mdpi.com/1999-4923/15/11/2545

18 Leng J, Wang D, Shen W, Li X, Liu Q, and Chen X. Digital twins-based smart manufacturing system design in Industry 4.0: A review. *Journal of Manufacturing Systems [Internet]* 2021;60:119–137. Available from: https://linkinghub.elsevier.com/retrieve/pii/S0278612521001151

19 Jiang Y, Yin S, Li K, Luo H, and Kaynak O. Industrial applications of digital twins. *Philosophical Transactions of the Royal Society A: Mathematical, Physical and Engineering Sciences [Internet]* 2021;379(2207):20200360. Available from: https://royalsocietypublishing.org/doi/10.1098/rsta.2020.0360

20 Portela RMC, Varsakelis C, Richelle A, et al. When is an in silico representation a digital twin? A biopharmaceutical industry approach to the digital twin concept. *Advances in Biochemical Engineering Biotechnology* 2020: 35–55. Available from: https://link.springer.com/10.1007/10_2020_138

21 Girolami M. Introducing data-centric engineering: An open access journal dedicated to the transformation of engineering design and practice. *Data Centric Engineering [Internet]* 2020;1:e1. Available from: https://www.cambridge.org/core/product/identifier/S2632673620000052/type/journal_article

22 Zha D, Bhat ZP, Lai K-H, et al. Data-centric Artificial Intelligence: A survey. *ACM Computing Surveys [Internet]* 2025;57(5):1–42. Available from: https://dl.acm.org/doi/10.1145/3711118

23 Kumar S, Datta S, Singh V, Singh SK, and Sharma R. Opportunities and challenges in data-centric AI. *IEEE Access [Internet]* 2024;12:33173–33189. Available from: https://ieeexplore.ieee.org/document/10444552/

24 Reis MS and Kenett R. Assessing the value of information of data-centric activities in the chemical processing industry 4.0. *AIChE Journal [Internet]* 2018;64(11):3868–3881. Available from: https://aiche.onlinelibrary.wiley.com/doi/10.1002/aic.16203

25 Di Francesco D. Risk management in the era of data-centric engineering. *Data-Centric Engineering [Internet]* 2025;6:e15. Available from: https://www.cambridge.org/core/product/identifier/S2632673625000103/type/journal_article

26 Kakhi M, Li J, and Dorantes A. Regulatory experience with continuous manufacturing and real time release testing for dissolution in new drug applications. *Journal of Pharmaceutical Sciences [Internet]* 2023;112(10):2604–2614. Available from: https://linkinghub.elsevier.com/retrieve/pii/S0022354923003234

27 Webster GK, Mankani B, Mozharov S, and Marquardt B. Opportunities and challenges: Process Raman for the real-time release testing (RTRT) of extended-release formulations. *Applied Spectroscopy Practica [Internet]* 2024;2(2). Available from: https://journals.sagepub.com/doi/10.1177/27551857241248477

28 Faßauer G. Linking deviation with innovation: Behavioral effects of management control through the lens of a theory of deviance. *Journal of Management Control [Internet]* 2018;29 (3–4):275–293. Available from: http://link.springer.com/10.1007/s00187-018-00271-8

29 Devineni SK, Kathiriya S, and Shende A. Machine learning-powered anomaly detection: Enhancing data security and integrity. *Journal of Artificial Intelligence and Cloud Computing [Internet]* 2023; 1–9. Available from: https://www.onlinescientificresearch.com/articles/machine-learningpowered-anomaly-detection-enhancing-data-security-and-integrity.pdf

30 Mattera G, Mattera R, Vespoli S, and Salatiello E. Anomaly detection in manufacturing systems with temporal networks and unsupervised machine learning. *Computers & Industrial Engineering [Internet]* 2025;203:111023. Available from: https://linkinghub.elsevier.com/retrieve/pii/S036083522500169X

31 Surucu O, Gadsden SA, and Yawney J. Condition monitoring using machine learning: A review of theory, applications, and recent advances. *Expert Systems Applications [Internet]* 2023;221:119738. Available from: https://linkinghub.elsevier.com/retrieve/pii/S0957417423002397

32 Ucar A, Karakose M, and Kırımça N. Artificial intelligence for predictive maintenance applications: Key components, trustworthiness, and future trends. *Applied Sciences [Internet]* 2024;14(2):898. Available from: https://www.mdpi.com/2076-3417/14/2/898

33 Molęda M, Małysiak-Mrozek B, Ding W, Sunderam V, and Mrozek D. From corrective to predictive maintenance—A review of maintenance approaches for the power industry. *Sensors [Internet]* 2023;23(13):5970. Available from: https://www.mdpi.com/1424-8220/23/13/5970

34 Murtaza AA, Saher A, Zafar MH, Moosavi SKR, Aftab MF, and Sanfilippo F. Paradigm shift for predictive maintenance and condition monitoring from Industry 4.0 to Industry 5.0: A systematic review, challenges and case study. *Results in Engineering [Internet]* 2024;24:102935. Available from: https://linkinghub.elsevier.com/retrieve/pii/S2590123024011903

35 Homaei M, Mogollón-Gutiérrez Ó, Sancho JC, Ávila M, and Caro A. A review of digital twins and their application in cybersecurity based on artificial intelligence. *Artificial Intelligence Review [Internet]* 2024;57(8):201. Available from: https://link.springer.com/10.1007/s10462-024-10805-3

36 Gholap AD and Omri A. Advances in artificial intelligence-envisioned technologies for protein and nucleic acid research. *Drug Discovery Today [Internet]* 2025: 104362. Available from: https://linkinghub.elsevier.com/retrieve/pii/S1359644625000753

37 Gholap AD, Khuspe PR, Bharati DK, et al. Computational neuropharmacology in psychiatry. In: *Computational Neuropharmacology [Internet]*, (pp. 207–244). Wiley; 2025. Available from: https://onlinelibrary.wiley.com/doi/10.1002/9781394242474.ch7

38 Kabashkin I. Digital twin framework for aircraft lifecycle management based on data-driven models. *Mathematics [Internet]* 2024;12(19):2979. Available from: https://www.mdpi.com/2227-7390/12/19/2979

39 Krishnamurthi R, Kumar A, Gopinathan D, Nayyar A, and Qureshi B. An overview of IoT sensor data processing, fusion, and analysis techniques. *Sensors [Internet]* 2020;20(21):6076. Available from: https://www.mdpi.com/1424-8220/20/21/6076

40 Balakrishna S, Thirumaran M, and Solanki VK. IoT sensor data integration in healthcare using semantics and machine learning approaches. In: *A Handbook of Internet of Things In Biomedical and Cyber Physical System*, (pp. 275–300). Springer; 2020. Available from: http://link.springer.com/10.1007/978-3-030-23983-1_11

41 Sejdiu B, Ismaili F, and Ahmedi L. Integration of semantics into sensor data for the IoT. *International Journal on Semantic Web and Information Systems [Internet]* 2020;16(4): 1–25. Available from: http://services.igi-global.com/resolvedoi/resolve.aspx?doi=10.4018/IJSWIS.2020100101

42 Helm JM, Swiergosz AM, Haeberle HS, et al. Machine learning and artificial intelligence: Definitions, applications, and future directions. *Current Reviews in Musculoskeletal Medicine [Internet]* 2020;13(1):69–76. Available from: https://link.springer.com/10.1007/s12178-020-09600-8

43 Badrulhisham F, Pogatzki-Zahn E, Segelcke D, Spisak T, and Vollert J. Machine learning and artificial intelligence in neuroscience: A primer for researchers. *Brain, Behavior, and Immunity [Internet]* 2024;115:470–479. Available from: https://linkinghub.elsevier.com/retrieve/pii/S0889159123003380

44 Nozari H, Ghahremani-Nahr J, and Szmelter-Jarosz A. AI and machine learning for real-world problems. *Advances in Computers* 2024;134:1 12. Available from: https://linkinghub.elsevier.com/retrieve/pii/S0065245823000153

45 Asghari A and Sohrabi MK. Server placement in mobile cloud computing: A comprehensive survey for edge computing, fog computing and cloudlet. *Computer Science Review [Internet]* 2024;51:100616. Available from: https://linkinghub.elsevier.com/retrieve/pii/S1574013723000837

46 Abughazalah M, Alsaggaf W, Saifuddin S, and Sarhan S. Centralized vs. Decentralized cloud computing in healthcare. *Applied Sciences [Internet]* 2024;14(17):7765. Available from: https://www.mdpi.com/2076-3417/14/17/7765

47 Demirbaga Ü, Aujla GS, Jindal A, and Kalyon O. Cloud computing for big data analytics. In: *Big Data Analytics [Internet]*, (pp. 43–77). Cham: Springer Nature Switzerland; 2024. Available from: https://link.springer.com/10.1007/978-3-031-55639-5_4

48 Sharma M, Tomar A, and Hazra A. Edge computing for industry 5.0: Fundamental, applications, and research challenges. *IEEE Internet Things Journal [Internet]* 2024;11(11):19070–19093. Available from: https://ieeexplore.ieee.org/document/10462492/

49 Nguyen T, Nguyen H, and Gia TN. Exploring the integration of edge computing and blockchain IoT: Principles, architectures, security, and applications. *Journal of Network and Computer Applications [Internet]* 2024;226:103884. Available from: https://linkinghub.elsevier.com/retrieve/pii/S1084804524000614

50 Al-Dulaimy A, Jansen M, Johansson B, et al. The computing continuum: From IoT to the cloud. *Internet of Things [Internet]* 2024;27:101272. Available from: https://linkinghub.elsevier.com/retrieve/pii/S2542660524002130

51 Andriulo FC, Fiore M, Mongiello M, Traversa E, and Zizzo V. Edge computing and cloud computing for internet of things: A review. *Informatics [Internet]* 2024;11(4):71. Available from: https://www.mdpi.com/2227-9709/11/4/71

52 Rajurkar VG. Recent technologies in nanomedicine for treatment of cancer. *Inven Rapid Pharm Tech [Internet]* 2012;4(1):1–3. Available from: https://www.inventi.in/journal/article/126/9966/Inventi Rapid: Pharm Tech/Pharmaceutical

53 Gholap AD and Musmade DS. Advances in the field of nanooncology: An overview. *Inven Impact NDDS [ISSN 0976-3791] [Internet]* 2011;4(1):1–3. Available from: https://www.inventi.in/journal/article/2/1078/Inventi Impact: NDDS/Pharmaceutical

54 Rajurkar VG. Lipid nanocapsules: A novel promising tool as drug delivery system. *Inven Rapid NDDS [ISSN 0976-3791] [Internet]* 2012;1(1):1–4. Available from: https://www.inventi.in/journal/article/127/9969/Inventi Rapid: NDDS/Pharmaceutical

55 Gholap AD. Translational nanomedicine: Status assessment and opportunities. *International Journal Research Pharmaceutical Bio-Medical Science* 2012;4(3):1454–63.

56 Rajurkar VG. New threat of clinical toxicities evoked from nanocarrier systems. *Inven Rapid NDDS [ISSN 0976-3791] [Internet]* 2012;1(4):1–4. Available from: https://www.inventi.in/journal/article/127/9982/Inventi Rapid: NDDS/Pharmaceutical

57 Anand M. Smart crystals technology: A review. *Pharmacologyonline [Internet]* 2011;1(1):238–243. Available from: https://pharmacologyonline.silae.it/files/newsletter/2011/vol3/028.amol.pdf

58 Borude S. Aquasomes: A potential drug delivery carrier. *Pharmacologyonline [Internet]* 2011;1(3):230–237. Available from: https://pharmacologyonline.silae.it/files/newsletter/2011/vol3/027.amol.pdf

59 Rajurkar VG. Potential applications of capillary electromigration methods for the improvement of separation efficiency. *Inventi Rapid: Pharm Analysis & Quality Assurance [ISSN 0976-3813] [Internet]* 2012;1(4):1–3. Available from: https://www.inventi.in/journal/article/136/9968/Inventi Rapid: Pharm Analysis /Pharmaceutical

60 Gholap AD and Tupe D. Potential applications of botanical extract. *Inven Rapid Ethnopharmacol [ISSN 0976-3805] [Internet]* 2012;1(4):1–4. Available from: https://www.inventi.in/journal/article/135/9972/Inventi Rapid: Ethnopharmacolo/Pharmaceutical

61 Dighe SB. Dandruff: Pathogenesis and treatment. *Inven Rapid Cosmeceuticals [ISSN 0976-3864] [Internet]* 2012;1(4):1–4. Available from: https://www.inventi.in/journal/article/141/9975/Inventi Rapid: Cosmeceuticals/Pharmaceutical

62 Gholap AD and Tupe D. The threat of photoaging: Clinical conditions, evaluation and treatment. *Inven Rapid Cosmeceuticals [ISSN 0976-3864] [Internet]* 2012;1(4):1–3. Available from: https://www.inventi.in/journal/article/141/9971/Inventi Rapid: Cosmeceuticals/Pharmaceutical

63 Dighe SB. Role of sebum in human health & acne. *Inven Rapid Cosmeceuticals [ISSN 0976-3864] [Internet]* 2012;1(4):1–3. Available from: https://www.inventi.in/journal/article/141/9980/Inventi Rapid: Cosmeceuticals/Pharmaceutical

64 Gholap AD and Tupe D. Artificial viral envelopes: Emerging vehicle for gene delivery. *PHARMATUTOR-ART-1439* 2012;1(4):1–3.

65 Rajurkar VG. Potential applications of superparmagnetic nanoparticles technology. *Inven Rapid NDDS [ISSN 0976-3791] [Internet]* 2012;1(4):1–3. Available from: https://www.inventi.in/journal/article/127/9983/Inventi Rapid: NDDS/Pharmaceutical

66 Puri AV. Comparative evaluation of papaya and banana starch as disintegrant in diclofenac sodium tablet formulation. *International Journal of Pharmacy and Biological Sciences* 2018;8(3):512–522.

67 Singh T, Vakkalagadda RK, Gholap AD, Sanap GS, Anjaiah B, and Roja K. In Vitro antipyrial activity of psidium guajava leaf extract. *European Journal Pharmaceutical and Medical Research* 2016;3(1):312–315.

68 Bothiraja C, Gholap KS, and Pawar AP. Investigation of ethyl cellulose microsponge gel for topical delivery of eberconazole nitrate for fungal therapy. *Therapeutic Delivery [Internet]* 2014;5(7):781–794. Available from: https://www.future-science.com/doi/10.4155/tde.14.43

69 Borude SS and Mahajan AM. Recent updates on Q fever: Diagnosis and treatment. *Pharmacologyonline [Internet]* 2011;1(3):254–264. Available from: https://pharmacologyonline.silae.it/files/newsletter/2011/vol3/030.gholap.pdf

70 Bhagat V. Gold nanorods: A promising drug delivery vehicle. *Inven Rapid NDDS [ISSN 0976-3864] [Internet]* 2012;1(4):1–4. Available from: https://www.inventi.in/journal/article/127/9964/Inventi Rapid: NDDS/Pharmaceutical

71 Gholap AD, Sayyad SF, Hatvate NT, et al. Drug delivery strategies for avobenzone: A case study of photostabilization. *Pharmaceutics [Internet]* 2023;15(3):1008. Available from: https://www.mdpi.com/1999-4923/15/3/1008

72 Hatvate NT, Khuspe PR, Mandhare TA, Kashid P, and Gaikwad VD. Proteomics in oncology: Retrospect and prospects. In: S. K, editor. *Novel Technologies in Biosystems, Biomedical & Drug Delivery [Internet]*, (pp. 243–269). Singapore: Springer Nature Singapore; 2023. Available from: https://link.springer.com/10.1007/978-981-99-5281-6_10

73 Vora LK, Gholap AD, Jetha K, Thakur RRS, Solanki HK, and Chavda VP. Artificial intelligence in pharmaceutical technology and drug delivery design. *Pharmaceutics [Internet]* 2023;15(7):1916. Available from: https://www.mdpi.com/1999-4923/15/7/1916

74 Magar TV, Gholap AD, and Musmade DS. Elastic vesicles as drugs carriers through the skin. *Inven Rapid NDDS [ISSN 0976-3791] [Internet]* 2011;1(4):1–2. Available from: https://www.inventi.in/journal/article/127/9611/Inventi Rapid: NDDS/Pharmaceutical

75 Gholap AD, Rajurkar VG, and Dighe SB. Phonophoresis: Ultrasound enhanced transdermal drug delivery. *Inven Rapid NDDS [ISSN 0976-3791] [Internet]* 2012;1(4):1–3. Available from: https://www.inventi.in/journal/article/127/9981/Inventi Rapid: NDDS/Pharmaceutical

76 Khedkar A, Tupe D, and Gholap AD. Cubosomes: Emerging drug delivery system. *Inven Rapid NDDS [ISSN 0976-3791] [Internet]* 2012;1(4):1–3. Available from: https://www.inventi.in/journal/article/127/9987/Inventi Rapid: NDDS/Pharmaceutical

77 Khedkar A, Tupe D, and Gholap AD. Pharmacosomes: An emerging vesicular system. *Inven Rapid NDDS [ISSN 0976-3791] [Internet]* 2012;1(4):1–3. Available from: https://www.inventi.in/journal/article/127/9986/Inventi Rapid: NDDS/Pharmaceutical

78 Gholap AD, Borude SS and Mahajan AM. POEMS syndrome: A review. *Pharmacologyonline [Internet]* 2011;1(3):265–274. Available from: https://pharmacologyonline.silae.it/files/newsletter/2011/vol3/031.gholap.pdf

79 Rojekar S, Kapare HS, Vishwakarma N, et al. Chitosan scaffolds: Expanding horizons in biomedical applications. *Carbohydrate Polymers [Internet]* 2024;323:121394. Available from: https://linkinghub.elsevier.com/retrieve/pii/S0144861723008597

80 Kapare HS, Pagar S, Kamandar P, et al. Exploring modified chitosan-based gene delivery technologies for therapeutic advancements. *International Journal of Biological Macromolecules [Internet]* 2024;260:129581. Available from: https://linkinghub.elsevier.com/retrieve/pii/S0141813024003842

81 Pardeshi SR, Hatvate NT, Dhorkule N, Sayyad SF, Faiyazuddin M, and Khalid M. Environmental implications and nanotechnological advances in octocrylene-enriched sunscreen formulations: A comprehensive review. *Chemosphere [Internet]* 2024;358:142235. Available from: https://linkinghub.elsevier.com/retrieve/pii/S0045653524011287

82 Gupta J, Kamandar P, Bhowmik DD, et al. Harnessing nanovaccines for effective immunization–A special concern on COVID-19: Facts, fidelity, and future prospective. *ACS Biomater Science & Engineering [Internet]* 2024;10(1):271–297. Available from: https://pubs.acs.org/doi/10.1021/acsbiomaterials.3c01247

83 Parit S, Manchare A, Mundhe P, Hatvate N, Rojekar S, and Patravale V. Antibody-drug conjugates: A promising breakthrough in cancer therapy. *International Journal of Pharmaceutics [Internet]* 2024;659:124211. Available from: https://linkinghub.elsevier.com/retrieve/pii/S0378517324004459

84 Sawant SD, Sayyad SF, Hatvate NT, Chavan M, Rojekar S, and Faiyazuddin M. Food biopackaging for human benefits: Status and perspectives. In: Ahmed S, editor. *Biobased Packaging Materials [Internet]*, (pp. 1–35). Singapore: Springer Nature Singapore; 2023. Available from: https://link.springer.com/10.1007/978-981-99-6050-7_1

85 Hatvate NT, Sawant SD, Khemka PN, and Vora LK. Unlocking the potential of nanobiohybrids to combat environmental pollution. In: Lens P, editor. *Nanobiohybrids for Advanced Wastewater Treatment and Energy Recovery [Internet]*, (pp. 169–190). UK: IWA Publishing; 2023. Available from: https://iwaponline.com/ebooks/book/886/chapter/3516052/Unlocking-the-potential-of-nanobiohybrids-to

86 Pardeshi S, More M, Togre N, Rebello N, and Giram P. Dendrimers based antibacterial and antiviral materials. In*: Antibacterial and Antiviral Functional Materials [Internet]*, (pp. 139–169). USA: ACS Publications; 2023. Available from: https://pubs.acs.org/doi/abs/10.1021/bk-2023-1458.ch005

87 Pardeshi S and Giram P. Preparation, antibacterial and antiviral activity measurements and detection methods. In: Deshmukh K, editor. *Antibacterial and Antiviral Functional Materials [Internet]*, (pp. 33–64). USA: ACS Publications; 2023. Available from: https://pubs.acs.org/doi/abs/10.1021/bk-2023-1458.ch002

88 Hatvate NT, Khuspe PR, Mandhare TA, Kashid P, and Vora LK. Decoding of nanoneutraceuticals and their delivery. In: Mohanta YK, editor. *Sustainable Green Nanotechnology [Internet]*, (pp. 284–309). Florida, New York: CRC Press; 2024. Available from: https://www.taylorfrancis.com/books/9781003389408/chapters/10.1201/9781003389408-16

89 Uddin MJ, Faiyazuddin M, Omri A, Gowri S, and Khalid M. Advances in artificial intelligence for drug delivery and development: A comprehensive review. *Computers in Biology and Medicine [Internet]* 2024;178:108702. Available from: https://linkinghub.elsevier.com/retrieve/pii/S001048252400787X

90 Gholap AD, Gupta JS, Kamandar PA, Banchhod GV, and Hatvate NT. Antibody-drug conjugates for cancer therapy: An up-to-date review on the chemistry and pharmacology. In: Verma SK and Das AK, editors. *Bioconjugated Materials Part 2 Applications in Drug Delivery, Vaccine Formulations and Important Conjugates for Cancer Therapy [Internet]*, (Comprehensive Analytical Chemistry; vol. 103 pp. 105–190). Amsterdam, Netherlands: Elsevier; 2023. Available from: https://www.sciencedirect.com/science/article/pii/S0166526X23000259

91 Bhowmik DD, Deshmukh AY, and Hatvate NT. Quintessential impact of dendrimer bioconjugates in targeted drug delivery. In: Verma SK and Das AK, editors. *Bioconjugated Materials Part 2 Applications in Drug Delivery, Vaccine Formulations and Important Conjugates for Cancer Therapy [Internet]*, (Comprehensive Analytical Chemistry; vol. 103 pp. 257–302). Amsterdam, Netherlands: Elsevier; 2023. Available from: https://www.sciencedirect.com/science/article/pii/S0166526X23000272

92 Gholap AD, Said PR, Pawar RD, Ambore GS, and Hatvate NT. Importance of carbohydrate-drug conjugates in vaccine development: A detailed review. In: Verma SK and Das AK, editors. *Bioconjugated Materials Part 2 Applications in Drug Delivery, Vaccine Formulations and Important Conjugates for Cancer Therapy [Internet]*, (Comprehensive Analytical Chemistry; vol. 103 pp. 191–256). Amsterdam, Netherlands: Elsevier; 2023. Available from: https://www.sciencedirect.com/science/article/pii/S0166526X23000296

93 Khuspe PR, Pardeshi SR, Uddin MJ, et al. Achieving optimal health with host-directed therapies (HDTs) in infectious diseases—A new horizon. *Advances in Therapy [Internet]* 2024; 2400169. Available from: https://onlinelibrary.wiley.com/doi/10.1002/adtp.202400169

94 Rojekar S, Togre N, Bhoj P, et al. Current status of mannose receptor-targeted drug delivery for improved anti-HIV therapy. *Journal of Controlled Release [Internet]* 2024;372:494–521. Available from: https://linkinghub.elsevier.com/retrieve/pii/S0168365924003444

95 Shaikh MS, Faiyazuddin M, Khan MS, et al. Chikungunya virus vaccine: a decade of progress solving epidemiological dilemma, emerging concepts, and immunological interventions. *Frontiers in Microbiology [Internet]* 2024: 15. Available from: https://www.frontiersin.org/articles/10.3389/fmicb.2024.1413250/full

96 Rojekar S, Parit S, Manchare A, et al. Revolutionizing eye care: Exploring the potential of microneedle drug delivery. *Pharmaceutics [Internet]* 2024;16(11):1398. Available from: https://www.mdpi.com/1999-4923/16/11/1398

97 Vardikar A, Das U, Mandal S, et al. Surface engineered multimodal magnetic nanoparticles for neurodegenerative diseases. In: *Targeted Therapy for the Central Nervous System [Internet]*, (pp. 121–153). USA: Elsevier; 2025. Available from: https://linkinghub.elsevier.com/retrieve/pii/B9780443238413000078

98 Srivastava N, Ibrahim SA, and Nasr MHA, editor. Manipulating the rhizosphere microbiome for plant health. In: *Microbiome Engineering*, 1st ed. (p. 19). Florida: CRC Press, Florida; 2024.

99 Mesut B, Al-Mohaya M, Yeşilkaya E, et al. Demystifying the potential of lipid-based nanocarriers in targeting brain malignancies. *Naunyn-Schmiedeberg's Archives of Pharmacology [Internet]* 2024;397(12):9243–9279. Available from: https://link.springer.com/10.1007/s00210-024-03212-6

100 Pardeshi S, Pillai D, Giram P, Pardeshi C, and Naik J. Nanocarriers for ocular drug delivery. In: *Nanocarrier Drug Delivery Systems [Internet]*, (pp. 469–500). Germany: De Gruyter; 2024. Available from: https://www.degruyter.com/document/doi/10.1515/9783111320847-015/html

101 Patil M, Dhumal D, Rojekar S, Jain H, Pingale P, and Kumar Prajapati M. Nanocarriers for antimicrobial therapy. In: *Nanocarrier Drug Delivery Systems [Internet]*, (pp. 561–602). Germany: De Gruyter; 2024. Available from: https://www.degruyter.com/document/doi/10.1515/9783111320847-018/html

102 Sawant SD, Pardeshi SR, Khuspe PR, et al. The role of inulin in altering the animal microbiome: A comprehensive review. *Microbiome Engineering* 2024;1:79–102.

103 Sawant SD, Pardeshi SR, Khuspe PR, et al. Omics tools for determining microbial composition: A systematic review. *Microbiome Engineering* 2024;1:35–50.

104 Kumar D, Kumar S, Kumari R, et al. A comprehensive review on the biomass-mediated synthesis of silver nanoparticles: Opportunities and challenges. *Journal of Environmental Chemical Engineering [Internet]* 2024; 115133. Available from: https://linkinghub.elsevier.com/retrieve/pii/S2213343724032652

105 Faiyazuddin M, Maqsood S, Abbas S, Chaudhari YS, Sharmila S, and Webster TJ. Prospects and ethics of ChatGPT in biomedical research and medical practice: An emphasis on nanotechnology. *Journal of Biomedical Nanotechnology [Internet]* 2024;20(11):1661–1678. Available from: https://www.ingentaconnect.com/content/10.1166/jbn.2024.3925

106 Pardeshi S, Kapare H, More A, Rebello N, and Giram P. Coating approaches of functionalized magnetic nanoparticles for theranostic applications. In: *Functionalized Magnetic Nanoparticles for Theranostic Applications [Internet]*, (pp. 131–171). Wiley; 2024. Available from: https://onlinelibrary.wiley.com/doi/10.1002/9781394172917.ch5

107 Pardeshi SR, Hatvate NT, Gharat KD, Naik JB, and Omri A. Advances in dorzolamide hydrochloride delivery: Harnessing nanotechnology for enhanced ocular drug delivery in glaucoma management. *Discover Nano [Internet]* 2024;19(1):199. Available from: https://link.springer.com/10.1186/s11671-024-04154-x

108 Hatvate NT, Bhowmik D, Sawant SD, Londhe KM, Rojekar S, and Faiyazuddin M. Engineered artificial MicroRNA in plant: an overview. In: *MicroRNA Advances and Application in Plant Biology [Internet]*, (pp. 311–341). Elsevier; 2025. Available from: https://linkinghub.elsevier.com/retrieve/pii/B9780443216824000130

109 Uddin MJ, Faiyazuddin M, Saeed SM, Sadiq SB, and Koly IAC. Lipid-based drug delivery for systemic effect through pulmonary route of administration. In: *Lipids in Pulmonary Drug Delivery [Internet]*, (pp. 255–299). Elsevier; 2025. Available from: https://linkinghub.elsevier.com/retrieve/pii/B9780443223747000128

110 Khan AR, Grewal NS, Jun Z, Khalid M, and Zhang H-J. Advances in smart hybrid scaffolds: A strategic approach for regenerative clinical applications. *Engineered Regeneration [Internet]* 2025;6:85–110. Available from: https://linkinghub.elsevier.com/retrieve/pii/S2666138125000039

111 Khuspe PR, Faiyazuddin M, Uddin MJ, et al. Current scenario and challenges of lipid-based nanomedicines for pulmonary drug delivery. In: *Lipids in Pulmonary Drug Delivery [Internet]*, (pp. 21–47). Elsevier; 2025. Available from: https://linkinghub.elsevier.com/retrieve/pii/B9780443223747000025

112 Khuspe PR, Faiyazuddin M, Uddin MJ, et al. Lipid-based vaccines against viruses including COVID-19. In: *Lipids in Pulmonary Drug Delivery [Internet]*, (pp. 409–433). Elsevier; 2025. Available from: https://linkinghub.elsevier.com/retrieve/pii/B9780443223747000190

113 Shoshi A, Xia Y, Fieschi A, et al. A flexible digital twin framework for ATMP production – Towards an efficient CAR T cell manufacturing. *Procedia CIRP [Internet]* 2024;125:124–129. Available from: https://linkinghub.elsevier.com/retrieve/pii/S2212827124003767

114 Moreno-Benito M, Lee KT, Kaydanov D, Verrier HM, Blackwood DO, and Doshi P. Digital twin of a continuous direct compression line for drug product and process design using a hybrid flowsheet modelling approach. *International Journal of Pharmaceutics [Internet]* 2022;628:122336. Available from: https://linkinghub.elsevier.com/retrieve/pii/S0378517322008912

115 Beke ÁK, Gyürkés M, Nagy ZK, Marosi G, and Farkas A. Digital twin of low dosage continuous powder blending – Artificial neural networks and residence time distribution models. *European Journal of Pharmaceutics and Biopharmaceutics [Internet]* 2021;169:64–77. Available from: https://linkinghub.elsevier.com/retrieve/pii/S0939641121002381

116 Santos JAM, Lopes MR, Viegas JL, Vieira SM, and Sousa JMC. Internal supply chain digital twin of a pharmaceutical company. *IFAC-PapersOnLine [Internet]* 2020;53(2):10797–10802. Available from: https://linkinghub.elsevier.com/retrieve/pii/S2405896320336302

117 Miozza M, Brunetta F, and Appio FP. Digital transformation of the Pharmaceutical Industry: A future research agenda for management studies. *Technological Forecasting and Social Change [Internet]* 2024;207:123580. Available from: https://linkinghub.elsevier.com/retrieve/pii/S0040162524003767

118 Kodumuru R, Sarkar S, Parepally V, and Chandarana J. Artificial intelligence and internet of things integration in pharmaceutical manufacturing: A smart synergy. *Pharmaceutics [Internet]* 2025;17(3):290. Available from: https://www.mdpi.com/1999-4923/17/3/290

119 Kapustina O, Burmakina P, Gubina N, Serov N, and Vinogradov V. User-friendly and industry-integrated AI for medicinal chemists and pharmaceuticals. *Artificial Intelligence Chemistry [Internet]* 2024;2(2):100072. Available from: https://linkinghub.elsevier.com/retrieve/pii/S2949747724000307

120 Roumeliotis C, Dasygenis M, Lazaridis V, and Dossis M. Blockchain and digital twins in smart industry 4.0: The use case of supply chain–A review of integration techniques and applications. *Designs [Internet]* 2024;8(6):105. Available from: https://www.mdpi.com/2411-9660/8/6/105

121 Kamini G, Swati W, Pooja S, and Nishita H. Review on digital twin in pharmaceutical and biopharmaceutical manufacturing. *International Journal of Scientific Research in Science Technology [Internet]* 2024;11(6):717–726. Available from: https://ijsrst.com/index.php/home/article/view/IJSRST241161127

122 Sen S and Kilic KD. Digital twins in medicine, AI-driven personalized healthcare, and predictive analytics. *Advances in Computational Intelligence and Robotics* 2025: 359–396. Available from: https://services.igi-global.com/resolvedoi/resolve.aspx?doi=10.4018/979-8-3373-0538-7.ch011

123 Niazi S. The coming of age of AI/ML in drug discovery, development, clinical testing, and manufacturing: The FDA perspectives. *Drug Design, Development and Therapy [Internet]* 2023;17:2691–2725. Available from: https://www.dovepress.com/the-coming-of-age-of-aiml-in-drug-discovery-development-clinical-testi-peer-reviewed-fulltext-article-DDDT

124 Serrano DR, Luciano FC, Anaya BJ, et al. Artificial Intelligence (AI) applications in drug discovery and drug delivery: Revolutionizing personalized medicine. *Pharmaceutics [Internet]* 2024;16(10):1328. Available from: https://www.mdpi.com/1999-4923/16/10/1328

125 Aritra S and Indu S. Harnessing the power of artificial intelligence in pharmaceuticals: Current trends and future prospects. *Intelligent Pharmacy [Internet]* 2025;3:181–192. Available from: https://linkinghub.elsevier.com/retrieve/pii/S2949866X24001217

8

Digital Twin in Pharmaceutical Manufacturing and Quality Control

F. Fareedhul Fahmitha[1], M.S. Swetha[2], S. Swetha[1], and P. Saranraj[1]*

[1]*PG and Research Department of Microbiology, Sacred Heart College (Autonomous), Tirupattur, India*
[2]*Department of Microbiology, Padmavani Arts and Science College for Women (Autonomous), Salem, India*

*Corresponding author: microsaranraj@gmail.com

CONTENTS

8.1 Introduction

New chemical engineering concepts and creative engineering solutions were not historically pioneered by the pharmaceutical and biopharmaceutical sectors. For many years, a regulatory framework that protected the final products quality and tested batch-based operations, raw material and end-product characteristics, fixed process conditions and, in-process material governed the manufacturing of pharmaceutical products [1]. It is generally accepted that testing thinking has limitations with regard to this quality for both small molecule and biopharmaceutical products. The development and application of manufacturing processes present the pharmaceutical industry with a number of significant challenges at the moment. ICH guidance Q8 introduced the Quality by Design (QbD) paradigm for process development, which regulatory bodies around the world are implementing. QbD promotes a science-based strategy for process understanding that ensures product quality. In this framework, enough process data must be gathered in order to demonstrate the connection between process performance and product quality. The viability and resilience of developed manufacturing processes must be proven [2]. In the meantime, businesses are under financial strain due to heightened worldwide competition, especially from generics producers. Process development needs to be accelerated to ensure profitability because of the limited period of time that drugs are free from competition due to patent expiration. For regulatory and economic reasons pharmaceutical manufacturing processes must be reliable, effective, and well-understood in order to guarantee product quality. In the development of pharmaceutical processes, modeling and optimization tools are crucial. The use of process modeling tools to enhance experimental studies in assessing process operability and sensitivity is advantageous for assessing process design space, identifying critical process parameters, and performing quality risk assessments. For a given product, a business case can be made for choosing one manufacturing route over another by using modeling tools to assess the relative costs of batch and continuous manufacturing (CM) processes [3]. In order to foster innovation, boost productivity, and boost profitability, modern competitive markets require the adoption of new digital technologies. A surge in the use of digital technologies in manufacturing sectors has resulted from the increased interest in these technologies and their promotion in various facets of economic activity. The development of digital technologies has brought about a number of changes in the manufacturing sector over the years, including but not limited to the substitution of computers for paper processing, and the development and promotion of the Internet and digital communication. From post-manufacturing product tracing and tracking to upstream supply chain management and shop floor control and management, the waves of digitalization have made a wide range of applications possible. Among the latest technological developments, governments agencies, academic institutions, and businesses have taken notice of the growth of artificial intelligence (AI) Internet of Things (IoT) devices and digital twins (DTs) [4]. The novel new processes are frequently continuous small-scale operations in contrast to the large-scale batch pharmaceutical manufacturing processes of the past, which had minimal automation and control. Continuous processes improve process reliability and product quality while reducing waste and processing rates. The advantages of small-scale processes are comparable, and they also enable rapid product switching and the possibility of producing dosage forms with different dosage amounts [5]. This chapter discusses DTs, data-centric engineering, real-time monitoring, and control in pharmaceutical manufacturing as well as quality and regulatory considerations.

8.2 DTs in Pharmaceutical Manufacturing

NASA first used the idea of making a twin of a process or product in the late 1960s when it combined two identical spacecraft for the Apollo project [6]. As a twin, one of the two was used to replicate every component and circumstance of the one that was launched into space. Here, the twin was used to mimic the counterparts behavior in real time. In 2002, during a presentation about product lifecycle management (PLM) at the University of Michigan Michael Grieves provided the first definition of a DT. According to Grieves, the DT is a digital informational construct of a physical system that is connected to the physical system and developed as an independent entity [7]. It is widely acknowledged that a full DT comprises a virtual component, a physical component, and automated data messages between the virtual and physical components. The potential of employing DTs to support smart manufacturing in pharmaceutical manufacturing is evident at various stages of process development and manufacturing. Because a DT can represent physical parts with different models, it can greatly speed up the process of choosing a manufacturing route and its unit operations during the process design stage. DT simulations can be used to understand process variations and predict productivity, quality, and process attributes, saving time and money compared to physical experiments [8]. Throughout the operation phase, the DT can continuously analyze the system to offer insights into process control and optimization. Due to its ability to simulate real-time scenarios and provide on-the-job feedback, DT can also be utilized as a training platform for engineers and operators. Among the pre- and post-manufacturing tasks that the DT platform can help with are material tracking serialization and quality assurance among others. Continuous data collection from equipment, intermediate and final products, real-time system monitoring and control via process analytical technology (PAT), and a continuous global modeling and data analysis platform are some essential prerequisites for accomplishing smart manufacturing with DT [9]. Flow-sheet modeling, CM, QbD, and PAT implementations are some of the methods the pharmaceutical industry has used to achieve this. The integration and development of DTs as a whole are still in their infancy although some of the tools have been thoroughly studied [10].

8.3 Data-centric Engineering in Pharmaceutical Manufacturing

Data-centric engineering is a comprehensive strategy that optimizes pharmaceutical manufacturing by combining data management analytics and engineering principles. The main difference between the paradigms of modern and traditional manufacturing is that the automation of conventional manufacturing can be thought of as model-based manufacturing. Through physical observations, such as visual inspection and noise recognition from manufacturing systems, experts gain experience. To better understand the mechanism, these experiences combined with human intelligence will create physical models through theoretical experimental and numerical methods. Despite significant advancements and applications in a variety of fields, including simulation and performance evaluation, these model-based approaches are still subpar due to their narrow accuracy and effective range [11]. One such advantage that can only be obtained through their data is the pharmaceutical industry, which is primarily a price inelastic industry. Pharmacies can optimize their inventory levels and schedule on-time medication delivery by anticipating demand fluctuations and seasonal data with proper demand forecasting and inventory management. This can guarantee that waste and stock-outs are minimized. Businesses like Merck have begun using analytics to improve their supply chain (95 percent OTIF) and prepare for potential disruptions. Gaining end-to-end visibility in supply chain operations such as improved supplier connectivity, real-time traceability across manufacturing warehouse and distribution centers, and identifying future vulnerabilities is possible with advanced analytics in pharmaceutical data. Due to their ability to utilize the large volume velocity and variety of data produced by contemporary systems, also known as big data AI digitization and data analytics, are the most appealing areas of study in both academia and business. In fact, (big) data analytics leverages computer science and mathematical theories methods, and instruments to glean useful insights and relevant information from massive data sets [12]. Information from connected devices or online publications is also included in data as is information from businesses themselves. Using machine learning algorithms to assist with medical diagnosis and prescriptions is a significant example of application. Information from medical studies and research can also shed light on the creation of new medications and vaccines. Apart from these applications in medicine, data analytics holds great promise for improving healthcare system monitoring. Data-enabled solutions, in particular, ought to significantly enhance pharmaceutical supply chains. When it comes to the production, storage, distribution, and dispensing of pharmaceuticals and medications, a pharmaceutical supply chain (PSC) is a socio-technical system made up of layers, specifically supplier, manufacturer, wholesaler, and pharmacy or hospital [13]. Figure 8.1 gives an overview of data-centric framework for material design.

Addressing a few major issues is necessary for a comprehensive design approach for the bi-directional traversal of progressive supranuclear palsy (PSP) relations. These issues include dimensionality reduction, microstructure representation and reconstruction, cost-effective processing techniques, and tractable optimization techniques [14]. Our capacity to address some of these issues with a data-centric approach for materials design is being accelerated by the rise of open-source materials databases and recent developments in machine learning techniques [15]. The core elements of this strategy as seen through the lens of design research can be divided into three categories: design representation, design evaluation, and design synthesis. Every one of these elements is directed by the insights gleaned from the PSP data kept in databases.

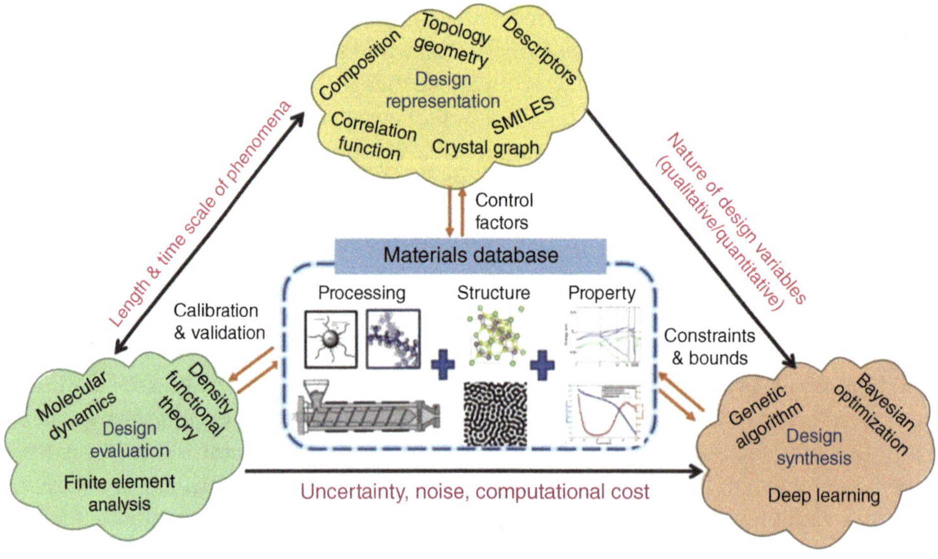

Figure 8.1 Data-centric framework for material design.

8.4 Real-time Monitoring and Control in Pharmaceutical Manufacturing

The production of pharmaceuticals entails a number of intricate procedures such as the procurement of raw materials, synthesis formulation, and packaging. Pharmaceutical product safety and quality must always come first, and regulatory bodies like the US FDA and EMA have put in place stringent rules to guarantee adherence. A crucial component of pharmaceutical manufacturing is real-time monitoring and control (RTM and C), which allows producers to monitor and modify procedures in real time to guarantee the highest possible level of product quality and safety. For measuring bioprocess variables and quality/performance indicators, a variety of online, at-line, and in-line probes and analyzers are available. The temperature level, pH, pressure, agitator speed, aeration rate, dissolved oxygen, and data acquisition systems are all present in a typical bioreactor. Among the available and preferred technologies are additional measurement systems like fluorometric sensors, mass spectroscopy in the off-gas, dissolved carbon dioxide, cell density, and online high-performance liquid chromatography (HPLC). Real-time predictions of performance endpoints can be produced mathematically through empirical and first principles modeling, according to research [16]. Using a generic framework, a soft-sensor is developed to produce an online quality estimate using process variables that are more commonly measured and easily accessible. A real-time prediction of the products' final titer in a cell culture production bioreactor case and a first principles model for real-time performance monitoring of a chromatography column are provided as industrial examples. Since the target protein is expressed in the production bioreactor, this stage is crucial to biopharmaceutical processes. Keeping a close eye on this unit's operation in real time is crucial. In this step, any departures from standard procedure could result in low productivity and possibly affect the quality of the final product. The performance parameter that is forecasted at the batch level while the process is being monitored in real time is the final titer. Many PAT opportunities, including off-line assay elimination (titer is usually determined via an HPLC method off-line) titer optimization, schedule optimization, and real-time control, are made possible by accurately predicting the final titer several days prior to harvesting. Final titer prediction in real time for a batch based solely on continuously measured inputs and outputs such as pH, oxygen flow, and bioreactor volume [17]. The aforementioned easily accessible process inputs and the final titer, as the response variable, are used to build a batch level partial least square (PLS) model. Several real-time monitoring technologies are used in pharmaceutical manufacturing, including [18]:

- Utilizing analytical tools to track process variables like pH pressure and temperature is known as process analytical technology or PAT.
- A nondestructive analytical method for tracking moisture content, particle size, and other process variables is near-infrared spectroscopy or NIR.
- The nondestructive analytical method known as Raman spectroscopy is employed to track the composition and structure of molecules.
- Machine vision helps to track parameters like tablet size and shape. It tracks by using a camera and image processing algorithms.

8.5 Integration of DTs, Data-centric Engineering, and Real-time Monitoring and Control

The introduction of DTs, Data-centric engineering, and Real-time monitoring and control is shifting how industries design, operate, and maintain refined systems. In principle, the system consists of a digitally enabled approach in which a twin, a virtual image of a real system which is needed to be developed, is employed to conduct the system's real-time behavior analysis and simulation. Technologies such as sensors, IoT devices, and other associated technologies that power with DT provide the stream of information regarding the system performance. Data-centric engineering analyzes this information and then summarizes system performance based on behavior utilizing analytics, machine learning, structures design, and operational decisions. These patterns are used to predict and diminish the need for maintenance, improved opportunities, and system performance maximization. The real-time monitoring and control automatically keeps track of the system's performance through usage of sensors, IoT devices, and other technologies and make modification simultaneously [19]. The information is used to improve the DT and assist in guiding data-centric engineering decisions. Integrated approaches require the highest standards of accuracy, speed, and reliability, which are very critical in complex systems and high-level decision-making for industries like manufacturing of oil, gas, and healthcare products. DTs help industries by simulating a physical system, which allows them to test and analyze the behavior of a system in real time without the need for physical prototypes or testing. This advancement makes the design and testing processes more efficient, minimizes costly errors, and speeds up the time frame to organize new products and services into the market. Conversely, data-centric engineering assists industries to apply their data to the fullest potential through pattern, trend, and insight recognition with the help of machine learning and advanced analytics. To sum up, the comprehensive system enables monitoring and control of operations in real time. As a result, companies are able to make changes as they occur, ensuring optimal performance, eliminating maintenance needs, and guaranteeing system reliability [20].

One of the greatest benefits of this approach is its ability to enable predictive maintenance, where maintenance is proactively planned based on analytic forecasts. Industries are able to analyze sensor data and identify trends that indicate maintenance will soon be necessary, which allow the businesses to obviate breakdowns. These approaches make it possible to reduce downtime, which saves money while improving system reliability and performance. Real-time system performance optimization is yet another benefit that can result from monitoring systems integrated with advanced analytics. This improvement enables organizations to reduce costs while increasing system efficiency. The twin phenomena of the digital revolution with data engineering and real-time control will greatly shape the next industrial reality. As industries continue changing and expanding, the need for advanced technologies that simulate, analyze, and optimize system behavior will only escalate. Industries can meet this challenge by developing new frameworks that are more reliable and sustainable within the context of a world undergoing rapid change. DT, data-centric engineering, as well as real-time monitoring and control systems, serve as powerful aids. Moreover, in subjects where an integrated perspective is possible, new business models such as "product-as-a-service," where instead of purchasing a product, customers pay to access it, can be created. It is also possible to achieve this fundamental understanding of product behavior through real-time monitoring and control, data-centric engineering, and DTs [21]. However, the combination of data-centric engineering, DTs, and real-time Monitoring and control presents a number of challenges. One of the biggest barriers is the lack of adequate data, which is essential for creating DTs and instructing data-centric engineering decisions. To be able to capture and transmit data, industries have to invest in sensors, IoT devices, and other technologies. They must also be strategically developed with appropriate tactics on how to implement analytics and machine learning into their everyday functions. Various sectors are undergoing a transformation in the design, operation, and maintenance of their sophisticated systems due to the merging of DTs, data-centric Engineering, as well as automatic monitoring and control. DTs enable industries to evaluate and forecast system performance and behavior in real time without the need for working models nor testing using physical analogues. Data-centric engineering, in contrast, enables industries to harness the full power of their data with the help of machine learning and advanced analytics that drive design and operational decisions. Finally, automatic monitoring and control helps industries adapt to changing conditions of the system in a responsive manner by changing parameters of processes in real time to limit the need for maintenance, enhance performance, and ensure reliability of the system [22].

8.6 Regulatory and Quality Considerations

Anyone regularly involved in the analysis of various industrial problems or challenges would therefore find a framework for evaluating the quality of information produced by empirical studies to be a vital tool. To do this, a plan for gathering data must be established along with the necessary analytical instruments and reporting techniques. Each project can have a preliminary risk assessment using the InfoQ framework to find gaps opportunities for improvement and priorities (expected cost-benefit analysis) before beginning [23]. It also makes it possible to diagnose past or current activities after the fact. The majority of data analysts have at least one story to share about spending time analyzing a problem or dataset only to come to the conclusion with a tone of resigned frustration that nothing could be done or accomplished with it while other pressing issues remained unresolved. As more experience is gained, issues like these and the limitations of big data initiatives become evident to take part in the discussions. Situations like these can be identified early on with greater likelihood when a systematic framework is in place [24]. Additionally, the weaknesses identified in the preliminary analysis will lead to effective actions to complete the project with missing components and enhance the quality of information it can potentially deliver.

8.7 Challenges

In recent years, the difficulties are brought mostly by big data. Even though big data holds enormous promise for technological advancement, its characteristics also pose a number of difficulties. The majority of studies derive data problems from the big data's inherent features. One of the studies determine the problems with big data including measurement errors that occurred by spurious correlation of noise accumulation with high dimensionality complex heterogeneity and incidental endogeneity. Heterogeneity and incompleteness, timeliness, and scale were all taken into consideration in another study as data-related concerns that needed to be addressed [25]. The authors of another study provide a thorough classification of data problems into three primary groups. Data problems pertaining to big data attributes like volume, variety, velocity, veracity, volatility, and variability fall under the first category. The second category addresses the challenges of data processing from collection to machine learning application. The third category includes issues with data management, involving privacy, security, and ethical standards [26].

8.8 Conclusion

Integrating DTs, data-centric engineering, and real-time monitoring and control is transforming the pharmaceutical manufacturing industry. Manufacturers may simulate and evaluate system behavior without using actual prototypes by employing DTs, which give virtual representations of physical systems. Thus, it renders its potential to develop operational perspectives in real time. Data-centric engineering optimizes the use of data that is accessible by enhancing design and operational choices by the use of machine learning and advanced analysis. Rapid reactions to modifying conditions are made feasible by real-time monitoring and control, which enhances the system's efficiency, reduces maintenance requirements, and assures overall durability. The combined approach improves system efficiency, eliminates waste, and promotes production, making it necessary for the development of the sector. Innovative business strategies such as product-as-a-service, in which clients pay for use rather than ownership, are made possible by it. As the pharmaceutical sector grows into complex form, the demand for technologies that model, assess, and optimize system behavior will rise. Successfully implementing this integrated approach requires a cultural shift toward data-driven decision-making, collaboration, and innovation, enabling manufacturers to fully capitalize on these technologies and fostering a more efficient and sustainable industry landscape.

References

1 Rathore AS and Winkle H. Quality by design for biopharmaceuticals. *Nature Biotechnology* 2009;27(1): 26–34.

2 Guideline, ICH Harmonised Tripartite. Pharmaceutical development. Q8 (2R), As revised in August 23, (2009).

3 Rogers A, Inamdar C, and Ierapetritou MG. An integrated approach to simulation of pharmaceutical processes for solid drug manufacture. *Indian Engineering and Chemical Research* 2014;53:28–47.

4 Bao J, Guo D, Li J, and Zhang J. The modelling and operations for the digital twin in the context of manufacturing. *Enterprise Information Systems* 2018;13:534–556.

5 Icten E, Giridhar A, Taylor LS, Nagy ZK, and Reklaitis GV, Dropwise additive manufacturing of pharmaceutical products for melt-based dosage forms. *Journal of Pharmaceutical Sciences* 2015;104(5):1641–1649.
6 Mayani MG, Svendsen M, and Oedegaard SI. Drilling digital twin success stories the last 10 years. In: *SPE Norway Subsurface Conference*, (p. D011S007R001); 2018.
7 Glaessgen E and Stargel D. The digital twin paradigm for future NASA and US Air Force vehicles. In: *53rd AIAA/ASME/ASCE/AHS/ASC Structures, Structural Dynamics and Materials Conference 20th AIAA/ASME/AHS Adaptive Structures Conference 14th AIAA*, (p. 1818); 2012.
8 Barrasso D. Developing and applying digital twins for continuous drug product manufacturing. In: *Proceedings of the PSE Advanced Process Modeling Forum*, (pp. 10–12). Tarrytown, NY, USA; 2019.
9 Barenji RV, Akdag Y, Yet B, and Oner L. Cyber-physical-based PAT (CPbPAT) framework for Pharma 4.0. *International Journal of Pharmacology* 2019;567:118445.
10 Kamble R, Sharma S, Varghese V, and Mahadik K. Process analytical technology (PAT) in pharmaceutical development and its application. *International Journal of Pharmaceutical Science Reviews and Research* 2013;23:212–223.
11 Mourtzis D, Doukas M, and Bernidaki D, Simulation in manufacturing: Review and Challenges. *Procedia CIRP* 2014;25:213–229.
12 Helleputte T, De Lannoy G, and Smyth P, Machine learning in the biopharma industry. *ESANN Proceedings* 2020: 533–540.
13 Silva RB and Mattos CA. Critical success factors of a drug traceability system for creating value in a Pharmaceutical Supply Chain (PSC). *International Journal of Environmental Research and Public Health* 2019;16(11):19–72.
14 Chen W, Iyer A, and Bostanabad R. Data centric design: A new approach to design of microstructural material systems. *Engineering* 2022;10:89–98.
15 Pan I, Mason LR, and Matar OK. Data-centric engineering: Integrating simulation, machine learning and statistics. Challenges and opportunities. *Chemical Engineering Science*, 2022;249:117–271.
16 Undey C, Ertunç S, Mistretta T, and Looze B. Applied advanced process analytics in Biopharmaceutical Manufacturing: Challenges and prospects in real-time monitoring and control. *Journal of Process Control* 2010;20(9):1009–1018.
17 Dumarey M, Hermanto M, Airiau C, et al. Advances in continuous Active Pharmaceutical Ingredient (API) manufacturing: Real-time monitoring using multivariate tools. *Journal of Pharmaceutical Innovation* 2019;14:359–372.
18 Bhaskar A and Singh R. Residence Time Distribution (RTD)-based control system for continuous pharmaceutical manufacturing process. *Journal of Pharmaceutical Innovation* 2019;14:316–331.
19 Bhambri P, Rani S, Kumar S, and Sinha VK. Big data analytics with digital twin for industrial applications. In: *AI-Driven Digital Twin and Industry 4.0*, (pp. 105–126); 2024.
20 Sacks R, Brilakis I, Pikas E, Xie HS, and Girolami M Construction with digital twin information systems. *Data-centric Engineering*, 2020;1:e14.
21 Freitas E, Oliveira Filho AT, do Carmo PR, et al. Data-centric industrial digital twin synchronization: Survey and architecture. Available at SSRN 4981777.

22 Liu L, Zeng N, Liu Y, Han D, and König M. Multi-domain data integration and management for enhancing service-oriented digital twin for infrastructure operation and maintenance. *Developments in the Built Environment* 2024;18:100–475.
23 Ermer J and Nethercote PW. *Method Validation in Pharmaceutical Analysis: A Guide to Best Practice*. John Wiley & Sons; 2025.
24 Norman J, Madurawe RD, Moore CM, Khan MA, and Khairuzzaman A. A new chapter in pharmaceutical manufacturing: 3D-printed drug products. *Advanced Drug Delivery Reviews* 2017:108:39–50.
25 Sarkis M, Bernardi A, Shah N, and Papathanasiou MM. Emerging challenges and opportunities in pharmaceutical manufacturing and distribution. *Processes* 2021;9(3):457.
26 Chen W, Park SJ, Cheng TN, et al. Pharmaceutical industry: Challenges and opportunities for establishing pharma 4.0. *Industry 4.0 Vision for Energy and Materials: Enabling Technologies and Case Studies* 2022;31:3–37.

9

Data-driven Acoustic Approaches to Pharmaceutical Efficacy

Mubbashar Mehmood[1] and Javed Akhter[2]*

[1]*Department of Mechanical Engineering, School of Engineering and Physical Sciences, Heriot-Watt University, Edinburgh, UK*
[2]*Faculty of Mechanical and Aeronautical Engineering, University of Engineering and Technology, Taxila, Rawalpindi, Pakistan*

*Corresponding author: m.mehmood@hw.ac.uk

CONTENTS

9.1 Introduction

Surface acoustic waves (SAWs) are normally generated using an interdigital transducer (IDT). When an alternating current (AC) is applied to an IDT, mechanical waves are generated due to the piezoelectric effect. Since these waves travel on the surface of the material, hence they are called SAWs [1].

Figure 9.1 illustrates the construction of an IDT and the generation of SAWs when they are patterned on the piezoelectric substrate [2]. The following section summarizes the types of SAWs.

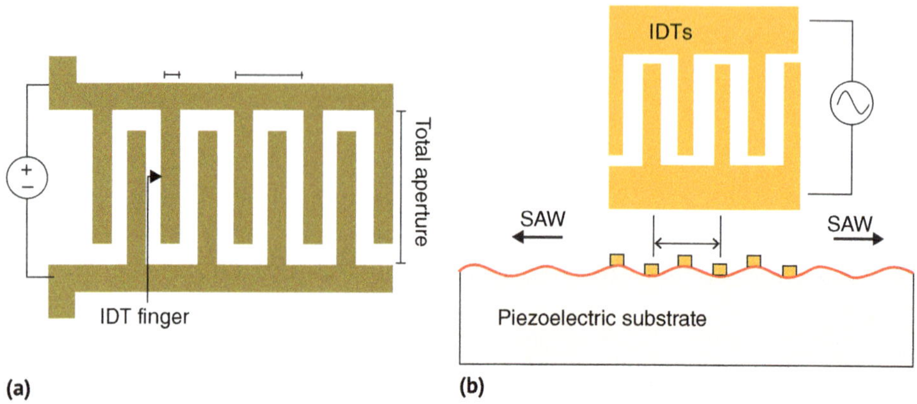

Figure 9.1 (a) Interdigital transducer (IDT) (b) A standard SAW device where IDTs are fabricated on a piezoelectric substrate. *Source*: Adapted from [2]/MDPI/CC BY 4.0.

9.1.1 Types of SAWs

Acoustic waves are primarily categorized into surface-generated acoustic waves (SGAWs) and bulk acoustic waves (BAWs) [3–5]. SGAWs encompass SAWs, pseudo surface acoustic waves (PSAWs), and leaky surface acoustic waves (LSAWs). On the other hand, BAWs include shear bulk acoustic waves (S-BAWs), longitudinal bulk acoustic waves (L-BAWs), film bulk acoustic resonators (FBARs), and Lamb waves. SAWs can be further divided into four modes: Rayleigh waves, Sezawa waves, shear-horizontal SAWs, and Love waves [6]. The use of SAWs to drive liquid or droplets for streaming, pumping, jetting, or nebulization [7] as well as to manipulate, pattern, and separate solid particles [8] or biological cells [9] has been the subject of much recent research. Waves propagating laterally and longitudinally along the surface are produced when SAWs produced by amplified radio frequency (RF) signals are applied to the IDTs on a piezoelectric substrate [10]. Traditional SAW devices are inflexible because they are frequently constructed on dielectric substrates like silicon and glass coated with piezoelectric films or on piezoelectric substrates like lithium niobate [11]. Their inability to adhere effectively to surfaces of any shape make them unsuitable for incorporation into wearable technology. Flexible acoustic wave devices (FSAW) are promising for wearable applications because of their good ductility and bendability. They are developed on polymer and metallic foil substrates, such as aluminum or polymer foils [12]. Piezoelectric thin film layers of FSAWs for wearable biosensors and lab-on-a-chip have been successfully demonstrated for zinc oxide (ZnO) and aluminum nitride (AlN) [13, 14]. A hybrid wave mode that combines Rayleigh and Lamb waves is frequently produced and used for various purposes by an FSAW device [8, 15].

9.1.2 Types of IDTs

IDT's designs can vary in shape and configuration to suit specific applications. Figure 9.2 illustrates various types of IDTs along with their most common uses. When IDTs are arranged to focus acoustic waves toward a central point, they are referred to as focused

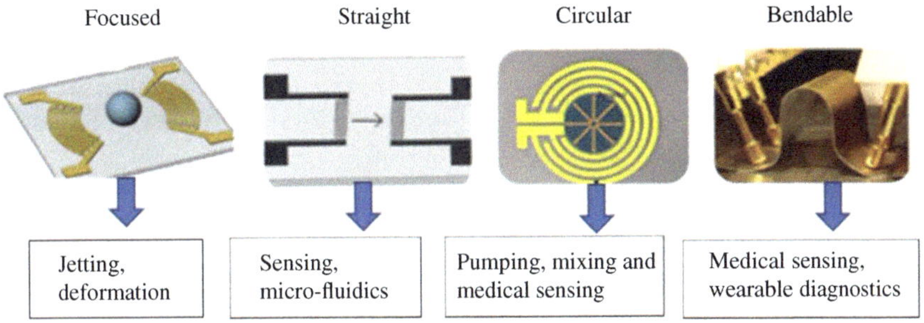

Figure 9.2 Different types of IDTs and their applications. *Source*: Adapted from [21]/SAGE Publications/CC BY 4.0.

IDTs (FIDTs) [16]. A variation of this configuration is the circular IDT, which concentrates acoustic pressure and energy at the center, enhancing droplet pumping and mixing in microfluidic systems, resulting in high droplet velocities and localized concentration effects [17].

Another design, known as the slanted or tapered IDT, features electrodes that gradually taper from one side to the other. This design allows for directional control of droplet movement by adjusting the operating frequency [17]. Additionally, IDTs can be fabricated on flexible piezoelectric substrates, resulting in flexible IDTs. These are particularly suited for biomedical applications such as drug delivery, wearable diagnostic devices, and minimally invasive surgical procedures [18–20].

9.1.3 Acoustic Streaming

Acoustic streaming has a wide range of applications in mixing, transporting, or manipulating particles/cells suspended in the liquid and in biosensing. The behavior of a Newtonian fluid under these conditions is governed by the continuity, momentum, and energy equations. There are two types of acoustic streaming: One is droplet-based, and the other is chamber-based, which mainly uses polydimethylsiloxane (PDMS). Figure 9.3 shows an example of droplet-based acoustic streaming where electrodes of the IDT are on the left side, a piezoelectric material where electrical signals are being converted to mechanical waves, and a droplet where waves enter and cause internal streaming.

The acoustic wave interacts with the fluid and undergoes refraction due to the difference in sound velocities between the piezoelectric substrate (C_s) and the fluid (C_f). The angle of refraction is determined by the relationship: $\theta = \sin^{-1} C_f/C_s$. As the acoustic wave propagates into the fluid, it causes radiation pressure, which causes particles to flow inside and hence streaming [22–24]. As shown in Figure 9.3, acoustic energy creates vortex-like internal streaming patterns inside the droplet [21].

Figure 9.4 shows chamber-based acoustic streaming consisting of an IDT device in a polydimethylsiloxane (PDMS) chamber. IDTs are fabricated on both sides of the piezoelectric substrate (128° $YXLiNbO_3$) and two sets of SAWs travelling towards each other. Due to the superposition effect, these waves are called standing surface acoustic waves (SSAWs).

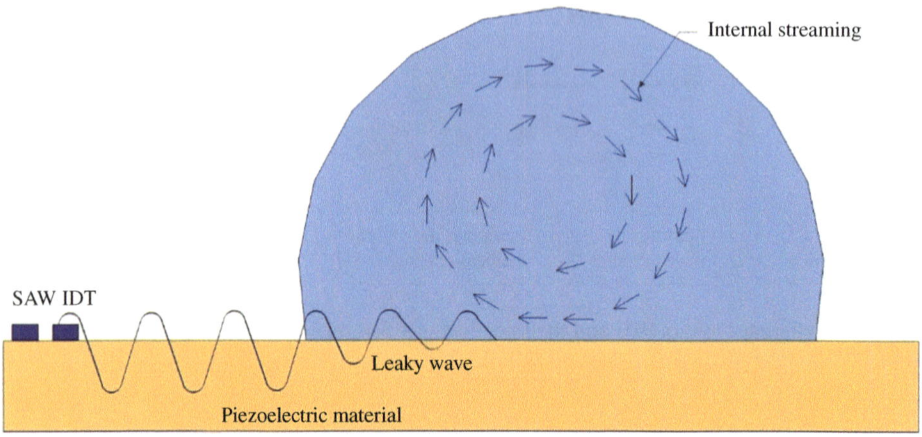

Figure 9.3 Internal streaming inside the droplet caused by surface acoustic waves (SAW). *Source*: Adapted from [21]/Sage Publications/CC BY 4.0.

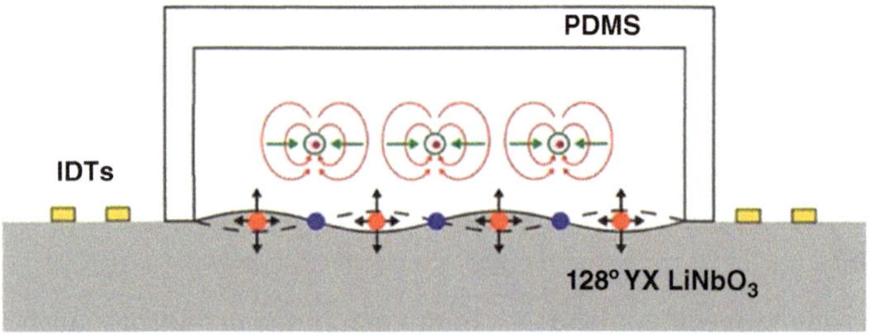

Figure 9.4 Visualization of standing surface acoustic waves' penetration in the polydimethylsiloxane (PDMS) microfluidic chamber. Both longitudinal and transverse waves are generated by the standing surface acoustic waves using a 128° YX lithium niobate substrate ($LiNbO_3$). *Source*: Adapted from [2]/MDPI/CC BY 4.0.

SSAWs have both longitudinal and transverse vibrations. After having interaction with the liquid in the chamber, they produce an acoustic field and hence acoustic streaming. This acoustic streaming inside the microfluidic chamber causes pushing and levitation of the suspended objects as shown in Figure 9.4 [2].

9.2 Applications of SAWs

9.2.1 Drug Delivery Using SAW

SAWs have emerged as a promising tool in drug delivery, offering precise, noninvasive, and efficient control over drug transport and release. SAWs create nanoscale vibrations on a substrate, facilitating fluid manipulation and enhancing drug dispersion [25].

9.2.1.1 Nebulization

Delivering therapeutic small interfering ribonucleic acid (siRNA) directly to the lungs offers promising potential for improving outcomes in patients with severe lung diseases. While recent advances in materials science have focused on overcoming barriers such as biodistribution, bioavailability, and cellular uptake, an equally critical challenge is creating an inhalation device capable of delivering siRNA effectively without compromising its integrity [26]. A nebulizer device for this purpose was patterned using the top surface of aluminum of 1.5 μm and chromium of 5 nm thickness on a 128° lithium niobate ($LiNbO_3$) piezoelectric substrate using a standard photolithography process. A resonant frequency of 29.78 MHz was used with an input power of <2 W. Aerosol mist was generated at an input power of 1.6 W.

As shown in Figure 9.5, successful nebulization of siRNA was performed using a compact, acoustomicrofluidic nebulization device. The system generates an aerosol mist with adjustable mean aerodynamic diameters around 3 μm, ideal for deep lung delivery via inhalation, and does so without causing significant degradation of the siRNA [26].

9.2.1.2 Transdermal Drug Delivery

Drug delivery using the skin is called transdermal drug delivery (TDD). TDD has been observed as a proven technology that has considerable clinical advantages over other conventional methods of drug delivery. It maintains a steady-state blood concentration and uses controlled drugs. It is a preferred method of drug delivery due to its clear advantages, such as convenient and pain-free self-administration, as well as bypassing hepatic first-pass metabolism and the gastrointestinal tract—making it especially beneficial for drugs with poor oral bioavailability [27].

TDD has emerged as a major noninvasive alternative drug administration method, especially for relatively smaller hydrophilic molecules such as nicotine and hormones. It offers many advantages over its alternative methods, such as oral and injectable drug delivery. However, at the same time, many challenges still need to be considered, specifically delivery that involves relatively larger molecules and water-soluble drugs through the

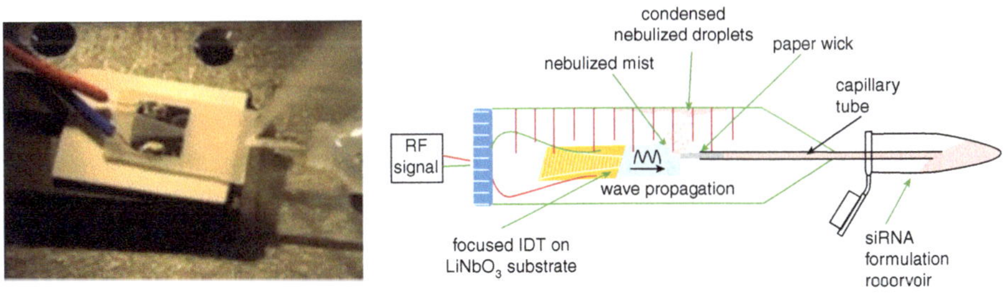

(a) (b)

Figure 9.5 (a) Nebulization of a liquid siRNA formulation using a microfabricated SAW device consisting of a focused IDT electrode photolithographically patterned onto a single-crystal piezoelectric ($LiNbO_3$ substrate). Electrical power was applied to the SAW device. After a threshold power, SAW causes the droplet to destabilize and form an aerosol mist that is suitable for inhalation. (b) Schematic diagram of experimental mist collection using a 50 ml centrifugal tube *Source*: [26]/with permission of Elsevier.

outermost skin layer called stratum corneum (SC) that consists of a 15–20-μm-thick layer of keratin-filled dead cells [28]. Several chemical and physical methods have been proposed to enhance permeability through the skin and diffuse small molecules of drugs into the blood. However, most of the chemical methods often cause unavoidable irritation or negatively affect the active cells under the SC. On the other hand, physical methods such as microneedles, electroporation, phonophoresis, ionophoresis, and thermal ablations have their own limitations for the penetration of larger or hydrophilic drug molecules; e.g., microneedles can cause localized erythema or edema and skin infections, while electrical methods could cause irritation, skin burns, or tissue scarring. Similarly, ultrasonic methods sometimes cause harm to adjacent healthy tissues or even cause cell deformation or DNA damage.

Like ultrasonic techniques, SAW excitation can penetrate the skin's layers for TDD. SAW technology has been explored as a minimally invasive method for drug administration, notably without inducing cavitation effects due to the use of high-frequency devices [28]. In contrast to conventional chemical or ultrasonic transdermal approaches, SAWs primarily interact with the epidermal layer, thereby minimizing harm to viable cells in the deeper skin layers and reducing the risk of irreversible tissue damage during drug delivery. The acoustic waves produced by the SAW device can propagate perpendicularly to the skin surface, facilitating both lateral diffusion and vertical penetration of drug molecules into the intercellular lipids, thereby overcoming the initial barrier of the skin, as shown in Figure 9.6. Upon application to the skin, the nanoscale, high-frequency mechanical vibrations generated by SAWs enhance the surface diffusion of drug molecules. Additionally, these vibrations can disrupt and deform the lipid structures within the SC by creating localized pressure and thermal effects, thereby effectively aiding the drug delivery process [29]. SAWs can be efficiently used to enhance TDD as an alternative to oral or injection methods. Recently, the wireless SAW technology has shown the potential of this technology as a compelling alternative for improving TDD, providing a safer, more effective, and patient-friendly therapeutic approach [30]. Another example has been observed by penetrating the skin layers using SAW excitation, which is comparable to ultrasonic techniques. Prior research has shown that SAW technology has the potential to be a minimally invasive drug delivery method without producing cavitation effects because of the high-frequency devices used [31]. SAWs primarily impact the epidermal layer, preventing harm to living cells found in the deeper layers of the skin and minimizing irreversible damage, in contrast to conventional chemical or ultrasonic TDD.

9.2.2 Surface Acoustic Waves for Blood–Brain Barriers

The blood–brain barrier (BBB) is a protective barrier that separates the blood in your circulatory system from the fluid that surrounds the brain (called cerebrospinal fluid). It's like a highly selective security checkpoint that protects your brain. It blocks harmful bacteria and viruses from entering the brain, and it allows essential nutrients (oxygen, glucose) to enter through [32]. Moreover, it regulates the chemical balance in the brain, which is very important for proper nerve function.

The BBB is made up of tightly packed cells lining the blood vessels in the brain. These cells are called endothelial cells, and they're much more tightly joined than in other parts

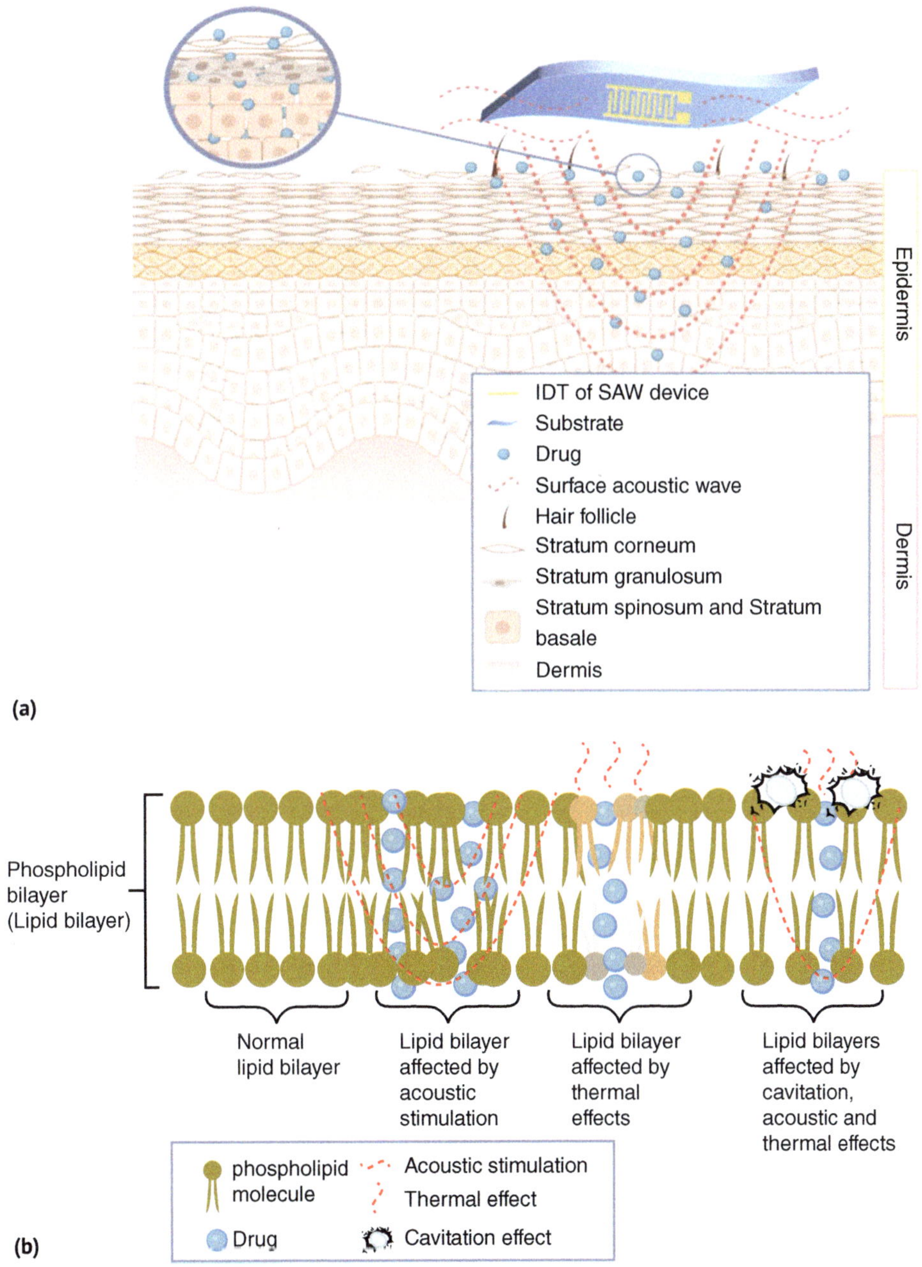

Figure 9.6 (a) Mechanisms of drug delivery and penetration in skin layers using the SAW technique. (b) Routes of drug delivery in the lipid bilayer using surface acoustic waves. *Source*: Adapted from [29]/Springer Nature/CC BY 4.0.

of the body. This tight seal only allows certain molecules to pass through—usually small, fat-soluble ones or those with special transport systems [33].

Figure 9.7 shows details of IDT electrodes patterned on the printed circuit board (PCB) with a reflection coefficient of the generated waves and a relation between input power and streaming velocity. Finger electrodes and their spacing are the key elements in generating mechanical vibrations. In this case, IDT width and spacing are 31.0 ± 1.1 μm and 67.6 ± 0.9 μm, respectively, as shown in Figure 9.7a. This results in the SAW wavelength of 197.1 ± 2.4 μm. Figure 9.7b shows the resonant frequencies of waves at S_{11} and S_{22} [34]. These are the frequencies at which the device will work. It is clear from Figure 9.7b that the reflection of the IDT was reduced from −2.5 to −53.5 dB and from −2.4 to −49.4 dB, respectively. It is also understandable from Figure 9.7b that the working frequency is 19.68 MHz. However, Figure 9.7c shows the relation between input power (dBm) and streaming velocity (μm/s). It is understood that between an input power of 25–30 dBm, streaming velocity

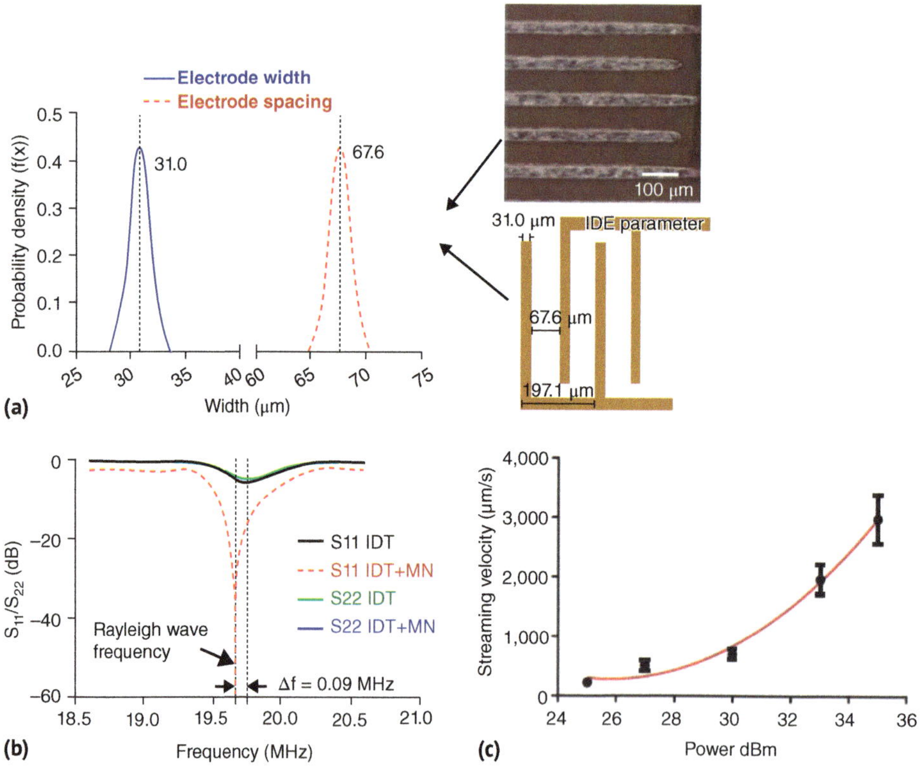

Figure 9.7 Characterization of acoustofluidic transwell (AFT). (a) The interdigital electrodes patterned on the printed circuit board (PCB) featured a width of 31.0 μm and a spacing of 67.6 μm, corresponding to a wavelength of 197 μm. (b) The reflection coefficient of the interdigital transducer was measured both with and without the use of a matching network (MN). (c) The relationship between input power and streaming velocity within the AFT was analyzed. *Source*: Adapted from [34]/American Association for the Advancement of Science/CC BY 4.0.

doesn't change much, but from 30 dBm onwards, there is a direct relationship between input power and the streaming velocity (μm/s) [34].

The transwell insert was coated with type IV collagen protein to simulate the basement membrane of the BBB. Ten human brain microvascular endothelial cells (hBMECs) were seeded onto the transwell insert to form a monolayer (Figure 9.8a). Typically, the in vitro BBB forms after 24 hours of cell culture, characterized by high Transendothelial electrical resistance (TEER), usually above 30 $\Omega \cdot cm^2$ in a 24-well transwell plate. After 24 hours of cell confluence, TEER continued to increase due to cell proliferation, and a tight BBB was established around 48 hours. The in vitro BBB was then treated with various SAW powers. BBB disruption was immediately observed following SAW stimulation. TEER of the hBMECs decreased with SAW agitation at powers of 20, 25, or 30 dBm, starting at the

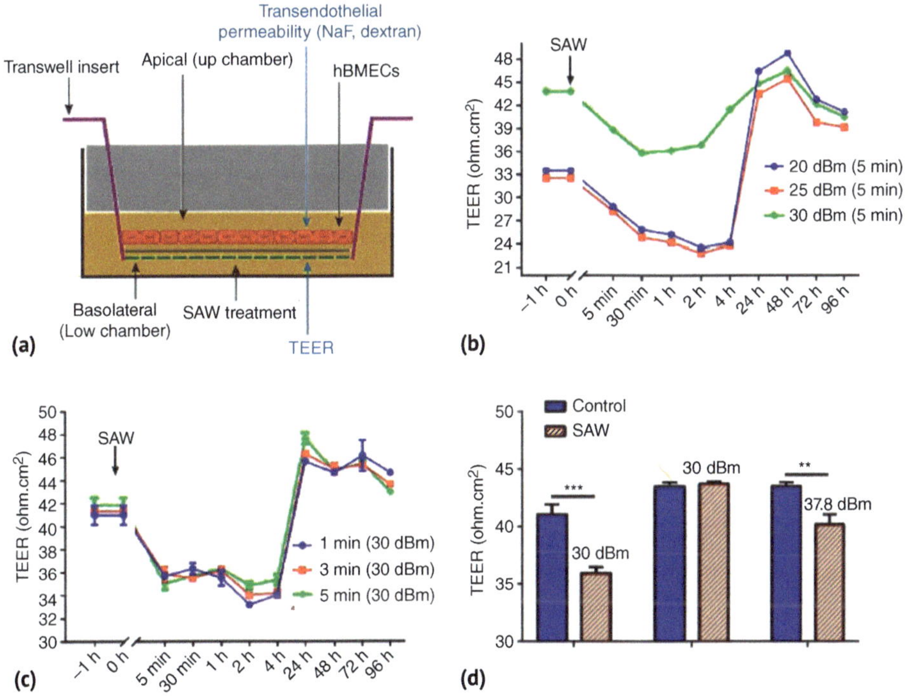

Figure 9.8 Blood–brain barrier (BBB) permeability under SAW treatment. (a) A transwell model was used to simulate the BBB. The setup included an insert suspended within a well plate, with a microporous membrane at the bottom of the transwell separating it into two compartments. The upper chamber (apical) represented the bloodstream, while the lower chamber (basolateral) represented the brain. (b) Human brain microvascular endothelial cells (hBMECs) were cultured in the transwell for 24 hours to establish a preliminary BBB model, then exposed to SAW at a frequency of 19.68 MHz for 5 minutes at power levels of 20, 25, and 30 dBm, respectively. (c) The BBB model was treated with a 19.68 MHz SAW at 30 dBm for varying durations of 1, 3, and 5 minutes, and transendothelial electrical resistance (TEER) was measured at specific time points. (d) hBMECs were seeded into the transwell, and the BBB model was subsequently exposed to SAW at different power levels. *Source*: Adapted from [34]/American Association for the Advancement of Science/CC BY 4.0.

end of a 5-minute SAW stimulation and continuing up to 4 hours (Figure 9.8b) [34]. After overnight incubation, TEER values of all groups recovered significantly at 24 hours. TEER showed a slow reduction after 72 hours, likely due to highly crowded cells consuming the medium nutrients. To explore the relationship between SAW duration and BBB disruption, a dose of 30 dBm was selected for further study. SAW duration was set for 1, 3, or 5 minutes, respectively. Similar results were observed, with TEER decreasing after 5 minutes of SAW agitation and sustaining up to 4 hours (Figure 9.8c). After 24 hours, TEER of hBMECs fully resumed, reflecting the recovery of barrier function. To further investigate BBB susceptibility to SAW agitation, different BBB statuses were exposed to SAWs. Figure 9.8d shows the BBB was sensitive to SAW at a low TEER; however, for a high TEER BBB, a higher SAW power (37 dBm) was required for BBB opening. These results clearly demonstrated that SAW treatment altered BBB TEER, but a tight BBB required higher SAW power to change permeability, and the BBB recovered more rapidly [34].

9.2.3 Acoustic Tweezers

Acoustic tweezers have emerged as an innovative and promising platform for accurately manipulating and controlling bioparticles of various sizes [35, 36]. These cutting-edge instruments use sound waves' interactions with solids, liquids, and gases to manipulate matter spatially and temporally.

The linear translation seen in frog eggs and trapped latex spheres inside an auditory environment was first explained by the acoustic tweezers [37]. Since then, a wide variety of acoustic tweezer configurations have been proposed to meet the needs of many scientific and engineering applications [37–39]. The optical tweezers that came before them served as the model for many of these acoustic tweezers.

Since their invention in 1986, optical tweezers have quickly established themselves as a vital tool in the domains of physics, chemistry, and biology [40, 41]. They have been widely used to trap a variety of objects, such as bacteria, viruses, and cells, allowing for force spectroscopy and biomolecular manipulation research. Over the years, many alternative techniques of optical tweezers have been developed to overcome the drawbacks of conventional optical tweezers and improve the accessibility and adaptability of particle manipulation technology without physical contact.

Acoustic tweezers offer a flexible and ground-breaking approach that addresses the common challenges faced by traditional manipulation techniques. By utilizing acoustic waves across a frequency range from kilohertz to megahertz, these tweezers allow for accurate and direct control over a broad length scale that extends beyond five orders of magnitude [42, 43]. Importantly, the levels of acoustic power and frequencies used in acoustic tweezers are similar to those found in ultrasonic imaging, a trusted and safe diagnostic technique, which further bolsters their reliability and credibility. Comprehensive studies on the biocompatibility of acoustic tweezers have produced encouraging outcomes, indicating the potential to fine-tune operational settings and reduce any harmful effects on cells and small-animal models. This natural biocompatibility, combined with the inherent flexibility and adaptability of acoustic tweezers, places them in a favorable position.

9.2.3.1 Operating Principles of SBATs

Figure 9.9 shows a mechanism of single-beam acoustic tweezer (SBAT). When a Gaussian ultrasound beam strikes a single cell, the rays bend as they go into the cell. The rays closer to the center of the beam (p1) have a greater intensity compared to those at the edges (p2). Every instance of refraction results in a shift in the direction of the ultrasound rays while maintaining momentum. This effect creates a gradient force that influences the cell [44].

9.2.3.2 Biomedical Applications of Acoustic Tweezers

Numerous biocompatibility tests on acoustic tweezers have produced encouraging findings, showing that they can optimize operating parameters and reduce the risk of damage to cells and small-animal models [46, 47]. Acoustic tweezers are a powerful tool for tackling current issues in the fields of biology and biomedicine because of their inherent biocompatibility as well as their inherent adaptability and versatility. Notably, acoustic tweezers have a great deal of promise for making it easier to identify and isolate circulating biomarkers for cancer diagnosis, which could spur improvements in precision medicine and early detection.

The acoustic tweezers serve as a powerful instrument for tackling modern issues in biology and biomedicine. In particular, they possess significant potential for the separation and identification of circulating biomarkers in cancer diagnostics, thus promoting progress in both early detection and precision medicine. By effectively merging with current biomedical methodologies, acoustic tweezers present an encouraging path for advancing scientific knowledge and enhancing patient care. In recent decades, there has been significant progress in the domain of SBATs, which marks a revolutionary leap in the ability to manipulate cells and organisms with precision, providing clear benefits compared to traditional methods such as standing waves [48, 49]. Acoustic tweezers present a promising way to advance science and enhance patient outcomes by blending in seamlessly with

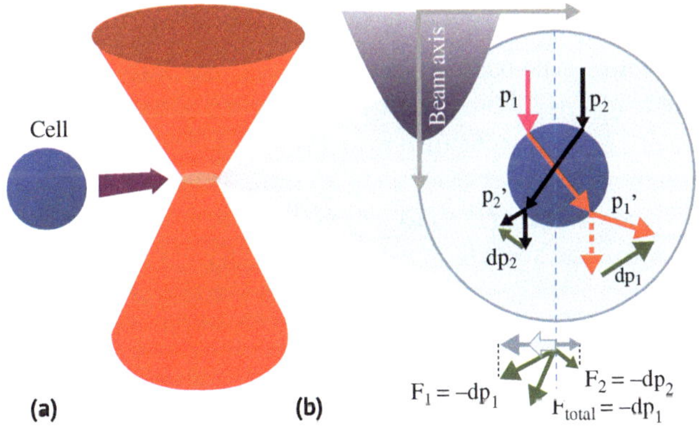

Figure 9.9 Mechanism of single-beam acoustic tweezers (SBATs): This involves the movement of cells through the propagation of two rays, characterized by distinct intensities (represented by red and black arrows). As a result, net force, represented as F_{total}, influences the direction of cell movement along the lateral axis. This directional movement is indicated by the white arrow, demonstrating the control achieved through SBATs. *Source*: Adapted from [45]/Springer Nature/CC BY 4.0.

current biomedical procedures. Significant progress has been made in the last few decades in the area of SBATs, which offers clear advantages over traditional methods like standing waves and represents a revolutionary development in the field of precisely manipulating cells and organisms [48, 49]. SBAT uses a single, easily controllable acoustic beam, which simplifies experiments and makes it accessible in a variety of research fields, in contrast to the intricate setups needed by standing wave techniques. Furthermore, SBAT makes it possible to precisely trap individual cells, guaranteeing that delicate samples that might be damaged by the higher energy of standing waves are handled gently. The exceptional adaptability demonstrated by SBAT makes them a valuable option to tackle existing issues in biology and medicine. Their functions encompass a broad spectrum of uses, including the separation of particles and cells as well as the regulated deformation of individual cells. Consequently, SBATs are becoming a substitute for traditional instruments used for particle and cell manipulation in various fields, from diagnostic applications to the complexities of single-cell research [44].

9.3 Conclusion

SAWs have tremendous applications in biological sciences and in medicines. Their applications are dependent on the types of waves, the types of IDT, the material of the IDTs, and the frequency of the waves. The efficiency of the device is dependent on the frequency data. Resonant frequency is where losses are minimum and energy produced is maximum. The frequency data varies from the first resonant frequency to higher resonant frequencies. In this chapter, we talked about wave interaction with the liquid, either in the form of a droplet or liquid in a chamber. Acoustic streaming can be generated in both the droplet and the chamber, like polydimethylsiloxane (PDMS), and is a key to applications like jetting, pumping, and nebulization. Amongst various biomedical applications, authors have focused on four applications: (i) nebulization, (ii) transdermal drug delivery (TDD), (iii) Blood-brain barrier (BBB), and (iv) acoustic tweezers.

Drug delivery using nebulization and TDD has emerged as the best alternative to oral and injection methods, where the controlled medicines are taken by the patient without the use of any needle. SAWs can also be used for the BBB with a device working frequency of 19.68 MHz, and the power input range varies from 20 to 30 dBm. However, acoustic tweezers have the capability to overcome challenges faced by conventional manipulation methods. The data for the frequency of the acoustic waves ranges from KHz to MHz. An example of this is a SBAT, which involves the movement of cells through the propagation of rays.

References

1 Galipeau DW, Story PR, Vetelino KA, and Mileham RD. Surface acoustic wave microsensors and applications. *Smart Materials and Structures* 1997;6(6):658.

2 Amorim D, Sousa PC, Abreu C, and Catarino SO. A review of SAW-based micro-and nanoparticle manipulation in microfluidics. *Sensors* 2025;25(5):1577.

3 Campanella H, Esteve J, Montserrat J, et al. Localized and distributed mass detectors with high sensitivity based on thin-film bulk acoustic resonators. *Applied Physics Letters* 2006;89(3): 33507.

4 Lakin KM. Thin film resonator technology. In: *Proceedings of the 2003 IEEE International Frequency Control Symposium and PDA Exhibition Jointly with the 17th European Frequency and Time Forum*; 2003 (pp. 765–778).

5 Du J, Harding GL, Collings AF, and Dencher PR. An experimental study of Love-wave acoustic sensors operating in liquids. *Sensors Actuators A Physical* 1997;60(1–3):54–61.

6 Rocha-Gaso M-I, March-Iborra C, Montoya-Baides Á, and Arnau-Vives A. Surface generated acoustic wave biosensors for the detection of pathogens: A review. *Sensors* 2009;9(7):5740–5769.

7 Tao X, Nguyen TD, Jin H, et al. 3D patterning/manipulating microparticles and yeast cells using ZnO/Si thin film surface acoustic waves. *Sensors Actuators B Chemical* 2019;299:126991.

8 Wang Y, Zhang Q, Tao R, et al. Flexible/bendable acoustofluidics based on thin-film surface acoustic waves on thin aluminum sheets. *ACS Applied Materials & Interfaces* 2021;13(14):16978–16986.

9 Dai Nguyen T, Fu YQ, Tran V-T, Gautam A, Pudasaini S, and Du H. Acoustofluidic closed-loop control of microparticles and cells using standing surface acoustic waves. *Sensors Actuators B Chemical* 2020;318:128143.

10 Fu YQ, Luo JK, Nguyen NT, et al. Advances in piezoelectric thin films for acoustic biosensors, acoustofluidics and lab-on-chip applications. *Progress in Materials Science* 2017;89:31–91.

11 Stringer M, Zeng Z, Zhang X, et al. Methodologies, technologies, and strategies for acoustic streaming-based acoustofluidics. *Applied Physics Reviews* 2023;10(1):011315.

12 Liu Y, Li Y, el-Hady AM, et al. Flexible and bendable acoustofluidics based on ZnO film coated aluminium foil. *Sensors Actuators B Chemical* 2015;221:230–235.

13 Zhou J, Guo Y, Wang Y, et al. Flexible and wearable acoustic wave technologies. A*pplied Physics Reviews* 2023;10(2):021311.

14 Arya SK, Saha S, Ramirez-Vick JE, Gupta V, Bhansali S, and Singh SP. Recent advances in ZnO nanostructures and thin films for biosensor applications. *Analytica Chimica Actavol* 2012;737:1–21.

15 Fu YQ, Luo JK, Du XY, et al. Recent developments on ZnO films for acoustic wave based bio-sensing and microfluidic applications: A review. *Sensors Actuators B Chemical* 2010;143(2):606–619.

16 Darmawan M and Byun D. Focused surface acoustic wave induced jet formation on superhydrophobic surfaces. *Microfluidics and Nanofluidics* 2015;18(5–6):1107–1114.

17 de Lima Jr MM, Alsina F, Seidel W, and Santos PV. Focusing of surface-acoustic-wave fields on (100) GaAs surfaces. *Journal of Applied Physics* 2003;94(12):7848–7855.

18 Kim DL, Jeong WH, Kim GH, and Kim HJ. Solution-processed indium–zinc oxide with carrier-suppressing additives. *Journal of Information Display* 2012;13(3):113–118.

19 Wu T-T and Chang I-H. Actuating and detecting of microdroplet using slanted finger interdigital transducers. *Journal of Applied Physics* 2005;98(2):24903.

20 Sun Y and Rogers JA. Inorganic semiconductors for flexible electronics. *Advanced Materials* 2007;19(15):1897–1916.

21 Mehmood M, Khan UF, Maka A, et al. A review of thermal impact of surface acoustic waves on microlitre droplets in medical applications. *Advances in Mechanical Engineering* 2022;14(8):16878132221116480.

22 Bradley CE. Acoustic streaming field structure: The influence of the radiator. *Journal of the Acoustical Society of America* 1996;100(3):1399–1408.

23 Nyborg WL. Acoustic streaming due to attenuated plane waves. *Journal of the Acoustical Society of America* 1953;25(1):68–75.

24 Das PK, Snider AD, and Bhethanabotla VR. Acoustic streaming in second-order fluids. *Physics of Fluids* 2020;32(12):123103.

25 Yeo LY and Friend JR. Surface acoustic wave microfluidics. *Annual Review of Fluid Mechanics* 2014;46:379–406.

26 Cortez-Jugo C, Masoumi S, Chan PPY, Friend J, and Yeo L. Nebulization of siRNA for inhalation therapy based on a microfluidic surface acoustic wave platform. *Ultrasonics Sonochemistry* 2022;88:106088.

27 Sachan R and Bajpai M. Transdermal drug delivery system: A review. *International Journal of Research and Development in Pharmacy and Life Sciences* 2013;3:748–765.

28 Seah BC-Q and Teo BM. Recent advances in ultrasound-based transdermal drug delivery. *International Journal of Nanomedicine* 2018;13:7749–7763.

29 Zhang J, Bahar D, Ong HL, et al. Flexible surface acoustic wave technology for enhancing transdermal drug delivery. *Drug Delivery and Translational Research* 2025;15(4):1363–1375.

30 Zhang J, Yang F, Wu H, et al. Wearable transdermal drug delivery system controlled by wirelessly powered acoustic waves. *Journal of Controlled Release* 2025;381:113619.

31 Ramesan S, Rezk AR, and Yeo LY. High frequency acoustic permeabilisation of drugs through tissue for localised mucosal delivery. *Lab on a Chip* 2018;18(21):3272–3284.

32 Daneman R and Prat A. The blood–brain barrier. *Cold Spring Harbor Perspectives in Biology* 2015;7(1):a020412.

33 Michiels C. Endothelial cell functions. *Journal of Cellular Physiology* 2003;196(3):430–443.

34 Wang K, Sun C, Dumčius P, et al. Open source board based acoustofluidic transwells for reversible disruption of the blood–brain barrier for therapeutic delivery. *Biomaterials Research* 2023;27(1):69.

35 Ozcelik A, Rufo J, Guo F, et al. Acoustic tweezers for the life sciences. *Nature Methods* 2018;15(12):1021–1028.

36 Zhou Z, Hou Z, and Pei Y. Reconfigurable particle swarm robotics powered by acoustic vibration tweezer. *Soft Robots* 2021;8(6):735–743.

37 Wu J. Acoustical tweezers. *Journal of the Acoustical Society of America* 199;89(5):2140–2143.

38 Silva GT and Baggio AL. Designing single-beam multitrapping acoustical tweezers. *Ultrasonics* 2015;56:449–455.

39 Guo F, Mao Z, Chen Y, et al. Three-dimensional manipulation of single cells using surface acoustic waves. *Proceedings of the National Academy of Sciences* 2016;113(6):1522–1527.

40 Ashkin A. Trapping of atoms by resonance radiation pressure. *Physical Review Letters* 1978;40(12):729.

41 Ashkin A, Dziedzic JM, Bjorkholm JE, and Chu S. Observation of a single-beam gradient force optical trap for dielectric particles. *Optics Letters* 1986;11(5):288–290.

42 Lim HG, Li Y, Lin M-Y, et al. Calibration of trapping force on cell-size objects from ultrahigh-frequency single-beam acoustic tweezer. *IEEE Transactions Ultrasonics Ferroelectrics Frequency Control* 2016;63(11):1988–1995.
43 Lim HG, Kim HH, and Yoon C. Evaluation method for acoustic trapping performance by tracking motion of trapped microparticle. *Japanese Journal of Applied Physics* 2018;57(5):57202.
44 Hae GL. Single beam acoustic tweezers for biomedical applications. *Journal of the Acoustical Society of Korea* 2023;42(5):452–459.
45 Lee O-J, Lim HG, Shung KK, Kim J-T, and Kim HH. Automated estimation of cancer cell deformability with machine learning and acoustic trapping. *Scientific Reports* 2022;12(1):6891.
46 Gesellchen F, Bernassau AL, Dejardin T, Cumming DRS, and Riehle MO. Cell patterning with a heptagon acoustic tweezer–application in neurite guidance. *Lab on a Chip* 2014;14(13):2266–2275.
47 Lim HG, Kim HH, Yoon C, and Shung KK. A one-sided acoustic trap for cell immobilization using 30-MHz array transducer. *IEEE Transactions Ultrasonics Ferroelectrics Frequency Control* 2019;67(1):167–172.
48 Lam KH, Li Y, Li Y, Lim HG, Zhou Q, and Shung KK. Multifunctional single beam acoustic tweezer for non-invasive cell/organism manipulation and tissue imaging. *Scientific Reports* 2016;6(1):37554.
49 Kim MG, Park J, Lim H.G, et al. Label-free analysis of the characteristics of a single cell trapped by acoustic tweezers. *Scientific Reports* 2017;7(1):14092.

10

Data-driven Practices for Pharmaceutical Coatings

*Anam Ijaz and Abdul Wasy Zia**

Institute of Mechanical Process and Energy Engineering (IMPEE), School of Engineering and Physical Sciences, Heriot-Watt University, Edinburgh, UK

*Corresponding author: a.zia@hw.ac.uk

CONTENTS

10.1 Introduction

Coatings play an important role in providing advanced performance characteristics to a product, like protection from degradation under adverse conditions like exposure to extreme temperatures, oxidation, effects of ultraviolet rays, moisture, etc. The environmental conditions have a minimum effect as long as the quality of coatings is being made better gradually. Coatings are used in various industries, including pharmaceuticals, electronics, and food, to protect the products and improve their functionality. In the pharmaceutical industry, the coatings are applied on solid dosage forms like tablets and capsules. The coating process was first introduced in the ninth century. Shaikh et al. [1] have discussed in their work that in early days some materials like talc, gelatin, and sugar were used for pill coating, with sugar coating being the most useful form. This coating process includes various steps where, first of all, a seal coat is applied on the tablet to protect it from water penetration. In the next step, subcoating is performed to round the tablet edges. Generally, the subcoat mixture consists of sucrose and a binder, which is applied over the tablet surface. The next step is that of smoothing the subcoated rough surfaces and increasing the tablet bulk to the desired size. The following step involves coloring, which is a significant step in a sugar-coating process. It gives the tablet improved appearance and stability. As a last step, the process of polishing is followed, where the appearance of the tablet is improved by giving it a shine and characteristic surface elegance.

Sugar coating may have numerous drawbacks, like an increase in tablet weight making it difficult for the patients to swallow. Another disadvantage is that it can reduce drug efficacy, as it may delay the release of active ingredients. In order to avoid these problems, the modern coatings often have polymer or polysaccharide films that provide more flexibility and additional functionalities than the former. Film coatings usually involve a refined process. O'Connor et al. [2] have presented their research where tablets are sprayed with a pigment-containing polymer solution while being rotated in a dry air stream. The main purpose of this drying air is to remove the moisture from the tablet surface. Multiple tablets are loaded on one end of a long cylinder where they are sprayed with multiple guns and then go down from the other end of the cylinder. This provides a uniform film coating on the surface of the drugs and improves its stability.

Modern-day film coatings provide better protection of the active pharmaceutical ingredients (APIs) present in drugs. These coatings protect them against the environmental factors like moisture, light, and oxygen. If these ingredients are not well preserved, they can degrade faster with the passage of time, making the use of the medicine limited and affecting its potency. Coatings are helpful in controlling the drug release in the body. There are some sustained-release coatings that gradually release the drug in the body to maintain a consistent level of medication throughout the blood stream. On the other hand, the enteric coatings prevent the drug from dissolving in the stomach and move it toward the more neutral environments like the intestine. Another advantage of the tablet coatings is the taste-masking effect. Most of the drugs have an unpleasant and bitter taste that makes their consumption difficult. Children are reluctant to these medicines, and it becomes difficult to provide the dosage to the child. The taste-masking effect reduces the bitterness of the medicine, often giving it a sweet taste that results in the easy utilization of the drug by the patient. Coatings also provide gastrointestinal protection where the irritation effect is prevented by making sure that the drugs like aspirin and other Non-steroidal Anti-inflammatory Drugs (NSAIDs) remain intact until they reach the intestine and then dissolve in the intestine and not the stomach.

With the world trends inclining toward data analysis and data-driven approaches, many industries, including the pharmaceutical industries, have shifted their paradigms toward the data-driven approach. In the data-driven approach, strategic decisions are made based on data analysis and interpretation. This approach enables the industries to organize their data with the goal of better serving their customers and consumers. The pharmaceutical

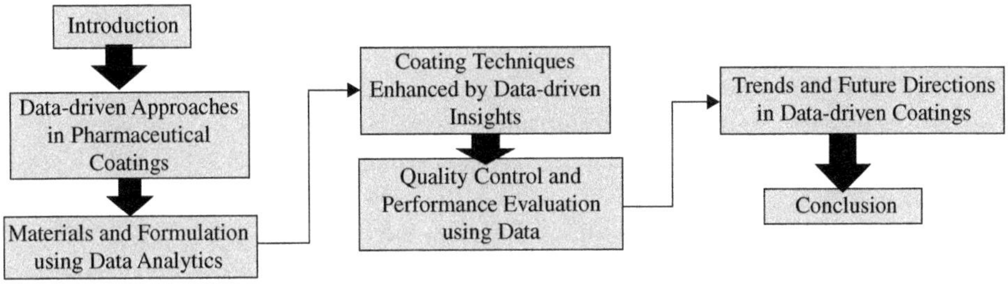

Figure 10.1 Overview of different sections of the chapter.

companies can use data-driven approaches to accelerate drug discovery and development, improve drug efficacy, enhance safety and risk management, and gain insight into target patient populations. This chapter gives a detailed overview and analysis of data-driven approaches in pharmaceutical coatings, materials, and formulation design using data analysis, coating techniques enhanced by data-driven methods, quality control and performance evaluation using data, and future directions in data-driven coatings. Figure 10.1 gives a block diagram indicating the organization of the chapter.

10.2 Data-driven Approaches in Pharmaceutical Industries

There are two types of data collection techniques in the coating process. The first technique is based on the data collected through sensors and real-time monitoring. In this area, Sacher et al. [3] have focused on a method called optical coherence tomography (OCT), which is being increasingly used for the fast analysis of high-resolution cross-sectional images to examine the thickness of coating and its quality. This technique gives real-time evaluation of the coating being deposited on the tablets. May et al. [4] used a terahertz sensor for monitoring the real time production of the tablets. This sensor gives an analysis of the coating thickness several times per minute instead of taking the average thickness value of a bulk of tablets being produced. Zeitler et al. [5], in their work, emphasized the importance of three-dimensional terahertz pulsed imaging, which is used to determine the spatial and statistical distribution of coating thickness in the production of pharmaceutical products like the tablets. It is a fully automated process.

The second approach for the data collection process includes analytical techniques like infrared spectroscopy and Raman imaging. Radtke et al. [6] showed in their work that Raman spectroscopy gives inline analysis during the production process. It is fast and does not affect the sample during its preparation. Mantanus et al. [7] used a near-infrared model to quantify the APIs in the drug reservoir. Critical quality attributes of the silicon-based drugs were effectively and efficiently evaluated using this model combined with Raman spectroscopy.

10.2.1 Modeling and Simulation Techniques

Different techniques are used for the modeling and simulation of conventional coating methods to improve the efficiency and effectiveness of the coating process. One such technique is computational fluid dynamics (CFD) for the spray coatings in the pharmaceutical industry. It is a computer-based simulation tool used to model and analyze the flow of the fluid. It involves three basic steps. In the first step, a system is designed based on the fluid flow and the physics involved in this process. Some equations are framed considering the laws of mass, momentum, and energy, and this phase is called the step of computational modeling. The second step involves the preprocessing where numerical equations are transfigured into digital systems. In the last step, the numerical and graphical results are collected and analyzed. By modeling the behavior of air and liquid sprays, CFD helps optimize coating uniformity, efficiency, and quality. Figure 10.2 represents a CFD-DEM simulation of the droplets being sprayed in a coating process.

Hemamanjushree et al. [9], in their work, have used the CFD approach to simulate the unit operations of the spray coating process. They suggest the impact of factors like flow pattern, temperature, and velocity in a coating process and how they can be optimized to improve the coating of the tablets. In another study, Böhling et al. [10] have modeled the Wurster coater that is used for the coating of particles and tablets. They emphasize optimizing the coating process by modifying the process parameters for controlling the inlet air flow rate and the particle-coating uniformity. Boehling et al. [11] presented a new coating process using the CFD technique. They proposed the design-space strategies to enhance process robustness and product quality.

Another useful simulation technique in practice is the finite element analysis (FEA) technique that divides the overall system into smaller systems and then gives analysis for each individual part. It combines the analysis of all subparts into the bigger system to form a mesh and give predictive analysis for the behavior of the system. It helps in understanding and optimizing factors like coating stress, adhesion, and mechanical integrity. Figure 10.3 gives a schematic representation of how this simulation is used in combination with the practical implementation.

Partheniadis et al. [12] have given a review about the feasibility of the FEA process and its potential to be implemented in pharmaceutical industries to gain important information about various factors like tablet formation, friction effect, density distributions, and failure of the tablet under stress. Yang et al. [13], in their study, developed the three-dimensional

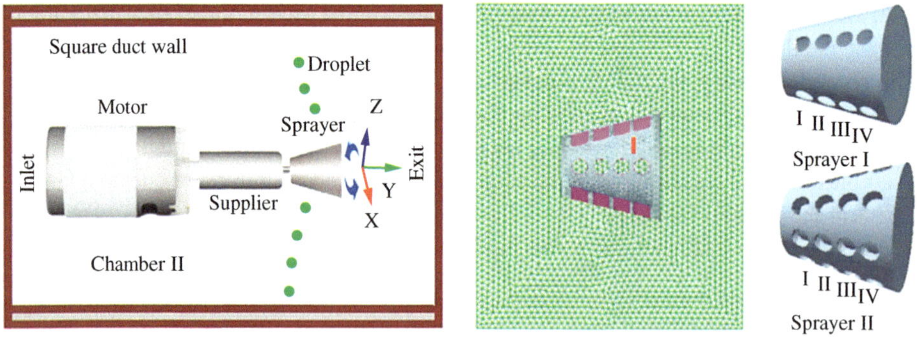

Figure 10.2 A CFD-DEM simulation of droplets in spray coating process. *Source*: Adapted from [8]/MDPI/CC BY 4.0.

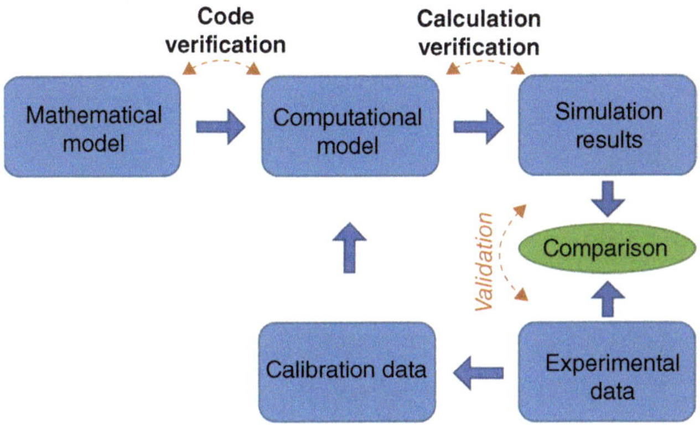

Figure 10.3 Flowchart of the finite element analysis simulation process. *Source*: Adapted from [12]/MDPI/CC BY 4.0.

FEA models to simulate and assess the burst behavior of the coated tablets. They identified the internal swelling pressure and inside corner radius as the key factors influencing stress distribution and burst behavior. In other research, Toschkoff et al. [14] have optimized the coating process to achieve desired film properties by incorporating a numerical simulation of the film formation process during tablet coating using FEA.

These studies highlight the importance of data-driven approaches in pharmaceutical companies, indicating the optimal simulation results achieved in less time. These techniques are helpful in predicting the outcomes of an experiment and how different factors can be manipulated in order to enhance the quality of the tablet coatings.

10.2.2 Machine Learning and Artificial Intelligence in Tablet Coatings

Pharmaceutical coatings serve essential functions like active ingredient protection, controlling drug release and avoiding the direct contact of the drug with the environmental bacteria. To check the effectiveness of the coating manually is a tiresome and time-consuming process. It is a hit-and-trial check by first developing the coating and then checking whether it is effective or not. After that, continuous improvements are done to improve the coating parameters. Another way to efficiently perform this task is to involve machine learning, artificial intelligence, and automated techniques to reduce the time and optimize the coating parameters quickly. In the solid dosage forms, the surface roughness directly affects the performance and quality of the final product influencing the drug dissolution, mechanical stability, and appearance. The machine learning models work by gathering information about coating parameters like air pressure, humidity and temperature, etc. from various datasets. The historical information contained in the dataset is helpful in learning about the previous trends and forecasting the behaviors of the future processes. Some important machine learning techniques include the regression models that use statistical processes for estimating the relationships between various properties to quantify the process parameters and surface roughness. Another method is the artificial neural networks that work by focusing on the nonlinear relationships among the various variables that the simpler models can miss.

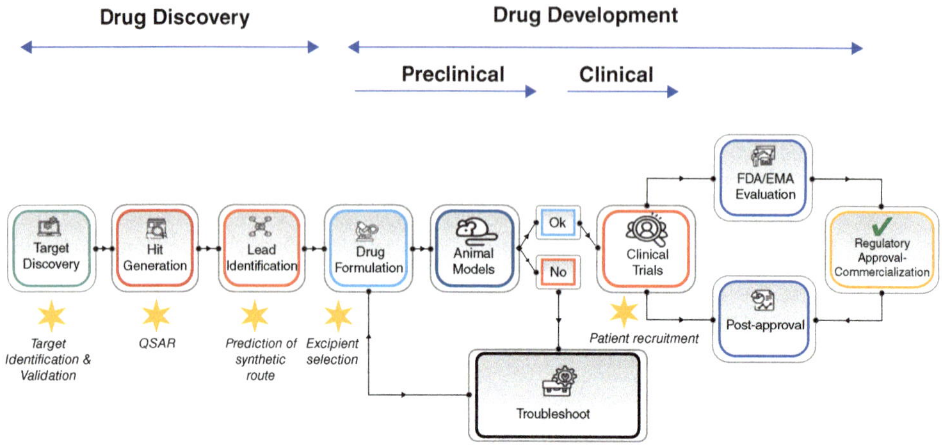

Figure 10.4 Machine learning and artificial intelligence in tablet coatings. *Source:* Adapted from [15]/MDPI/CC BY 4.0.

Support vector machines can also be used when dealing with complicated multidimensional data. They distinguish between tablets with acceptable and unacceptable roughness based on real-time sensor data. Figure 10.4 gives a depiction of artificial intelligence and machine learning techniques mimicking the human brain working.

Diaz et al. [16], in their work, have focused on the study of powder attributes like particle size, shape, and surface area. The datasets related to these properties are employed to train traditional machine learning models to predict the flow properties via flow function coefficient. This automated approach gives desirable results in bulk as compared to many other manual techniques. Gaudelet et al. [17] have focused on the use of machine learning graphs to review the methodologies within the drug discovery and development pipeline from an industrial perspective for the method developers in drug development applications.

The advancement in technology allows the optimization of coating process through different algorithms. The genetic algorithms, random forest, and decision trees trained on historical data could suggest spray gun angle, pan speed, and drying time to produce smooth coatings. In parallel, Bayesian optimization could fine-tune these parameters in real time, adapting the process for unexpected environmental shifts like humidity changes. In addition to these predictive algorithms, real-time sensors are also used for collecting data about the coating techniques and the properties of the complete process to ensure consistent roughness and quality. In a closed-loop feedback system the results of the process are continuously monitored and improved by changing the parameter compositions to improve the results collected from the coating process. This technique helps in self-optimization of the process. Artificial intelligence and machine learning are revolutionizing pharmaceutical tablet coatings by enabling predictive modeling, process optimization, and real-time control. In the context of surface roughness, these techniques enhance product quality, minimize the variability among the results, and reduce manufacturing and production costs. The collection of coating process results and the optimization work are done efficiently and effectively.

10.3 Material and Formulation Design Using Data Analytics

In pharmaceutical coatings the selection of suitable materials and optimizing formulations are of crucial importance. Data analytics helps achieve these tasks by providing efficient processes and informed decisions. The data-driven screening of polymers and plasticizers plays a crucial role in the selection of suitable materials required for tablet coatings. The datasets about physicochemical properties such as molecular weight, glass transition temperature, and solubility parameters can be collected, giving insight about the interaction results of different parameters, how they will react with each other, and how their combinations can be optimized in a better way. The desired process results can be achieved by optimizing the combinations of polymers and plasticizers used for different dosage coatings. Galí et al. [18], in their research, used 36 commercial batches to identify critical process parameters to enhance coating process robustness. They concluded in their work that analyzing existing commercial batch data can effectively optimize process robustness with minimal experimental investments.

In addition to the material selection, formulation optimization also plays a vital role in achieving desired product results. The appropriate coating thickness, adhesion, and surface roughness are crucial for the success of a coating process. In this regard, a technique called design of experiments (DoE) is used that systematically analyzes the factors affecting the coating process. This technique is further extended by response surface methodology (RSM) that employs mathematical models to describe the effects of independent variables on responses, facilitating the identification of optimal conditions. Cahyadi et al. [19], in their work, focused on optimizing the process parameters of a newly developed supercell quasi-continuous coater. They studied the varying effects of batch size, inlet temperature, atomizing pressure, and spray rate on the response variables like drug loading efficiency, coating thickness, and extent of tablet damage. They concluded that by using RSM the optimal conditions can be effectively identified. Integrating data analytics into the selection of coating materials and formulation optimization enhances the efficiency and effectiveness of developing pharmaceutical tablet coatings.

10.4 Coating Techniques Enhanced by Data-driven Insights

10.4.1 Spray Coating and Fluidized Bed Coating

During the process of spray coating, a solution containing pigments, polymers, and plasticizers is sprayed onto the rotating tablet bed, and the solvent, which is in the form of extra water, is evaporated with the help of air pressure. In the fluidized bed coating process, a particle suspension is created by entering air/gas in a distributor. This suspension is mixed with the wet coating mixture and then dried with the air/gas stream. Both these processes are commonly used as they have precise control over droplet size, spray patterns, and coating uniformity to ensure optimal film formation and drug stability. These techniques work well for producing quality coatings, but to accomplish the process with the most suitable parameters is a challenging and time-consuming task. Data-driven approaches and the use of advanced technologies can speed up this process and guarantee the improvement in results. Some machine learning models that can be used for this purpose include CFD. It is used to simulate spray trajectories, droplet impingement, and air distribution in coating systems. It can suggest suitable nozzle positions for effective spray and airflow patterns for uniform coating. Another important technique is the use of artificial neural networks that are trained on the existing process parameter datasets and then predict the optimal combination of these properties for achieving the best results in the final coating quality of the product. DoE and RSM are also used to develop empirical models for coating thickness and roughness. They are helpful in identifying critical process parameters and affect the uniformity of coating technique. The digital twin technology is another highly in-demand automated technique that creates a virtual model of the coating process to test parameter changes before implementation. Figure 10.5 gives an overview of the proposed modeling process of spray coating and fluidized bed coating.

Muñoz López et al. [21], in their study, used a Tucker3 model to capture and factorize the variability of the process, and subsequently a partial least-squares regression model was used to determine the impact of changing input variables on the particle size of the spray-dried drug products. The data-driven approach used in this research worked very well for predicting the median size of the particles. Liu et al. [22] used the approach of the discrete element method-discrete droplet method (DEM-DDM) for modeling the coating process of the tablets. The effect of different particle shapes and sizes on the inter-coating and intra-coating uniformity is discussed and analyzed. Boehling et al. [23], in their work, simulated the tablet coating process on the industrial scale using the eXtended particle system (XPS). They concluded that increasing the number of nozzles and decreasing the spray rate had a huge impact on the inter-tablet coating variation. The main factors like spray rate, number of nozzles, rotation rate, and drum load were examined. Another study by Schmitt et al. [24] made use of trained regression models for estimating initial values for process parameters that best achieve a target particle size for the particle formation process. An ensemble machine learning model was used to predict the dried particle size that impacts the bioavailability and manufacturability of the spray drying process and final dosage form.

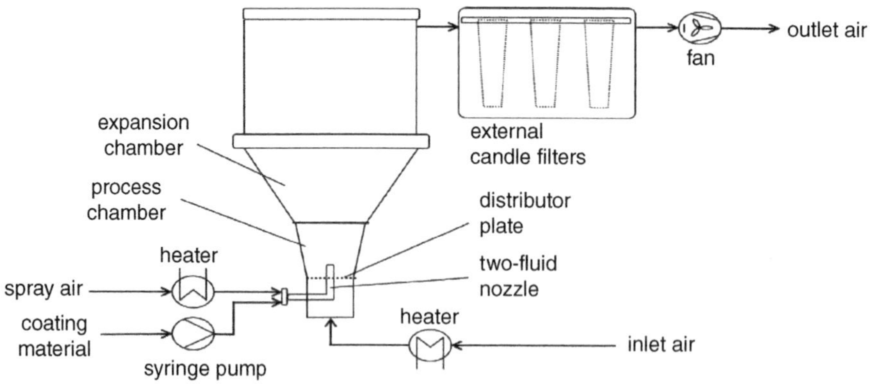

Figure 10.5 Proposed modeling strategy. *Source*: Adapted from [20]/MDPI/CC BY 4.0.

10.4.2 Pan Coating and Advanced Techniques

In a conventional pan coating process, the tablets are entered in a rotating drum. The tablets are sprayed by an atomized suspension while rotating through a spray nozzle at the center of the drum. In this way, they are coated with different coating materials. This traditional coating method can be improved to achieve better performance by merging them with machine learning and advanced data-driven techniques. Advancements in imaging systems, in-process data analytics, and automation have significantly enhanced the efficiency and quality of this process. The modern imaging technology like OCT is a noninvasive imaging technique. It uses light waves to take cross-sectional pictures of the area under observation without affecting the quality and structure of the underlying material. It is an effective technique for giving real-time assessment of coating thickness and uniformity, facilitating immediate adjustment to the parameters where required. In addition to this, another useful technology is the near-infrared spectroscopy that uses the near-infrared region of the electromagnetic spectrum. This technique, when combined with the process analytical technology (PAT) tools, provides continuous monitoring of critical quality attributes during coating process.

Another important approach is the automation driven by data for quality and consistency. Some automation systems being used are the TPR OPT coaters. This technology allows for the automatic adjustment control of the flow. It uses advanced sensors and control mechanisms to monitor critical parameters like humidity, temperature, and spray rate of the real-time coating process. The integration of machine learning, artificial intelligence, and deep learning-based methods with the traditional process of pan coating has revolutionized the performance of coating processes. These data-driven techniques are trained based on the existing data and then used to model and predict the behavior of real-time experiments. They are very useful in determining the defects on the surface of the tablets. They use images to recognize the irregularities on the surface of tablets, thus improving the reliability of quality inspections. Another approach is the application of quality by design principles that make use of analytical and statistical methods to improve the quality of the coating process. The simulation process for the pan coating process is described in Figure 10.6.

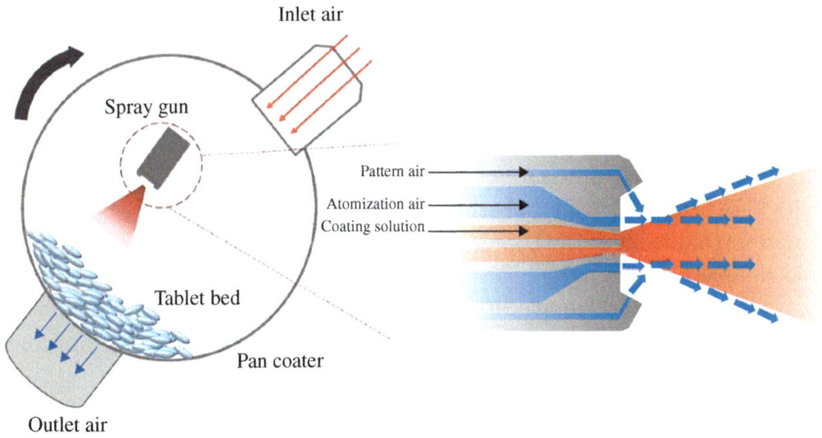

Figure 10.6 Tablet coating machine working principle. *Source*: Adapted from [25]/MDPI/CC BY 4.0

Pérez et al. [26], in their work, used the real-time measurements that were made through diffuse reflectance near-infrared probe placed inside the pan during the coating operation. Real-time spectra of replicate batches were used for modeling the film growth. When the desired coating thickness is achieved, the in-line measurements of the coating process allow the coating process to be stopped. Another study done by Pathak et al. [27] focused on the use of convolutional neural networks for detecting the defects on the surface of the tablets. They made a comprehensive dataset to train their system for detecting different forms of irregularities and nonuniformities on the surface of the tablets and then utilized this trained dataset for analyzing the new tablets through their model and suggesting the defect type for them. Dubey et al. [28], in their work, used the analytical model of laser-induced breakdown spectroscopy for detecting the variation in coating of the tablets. They developed a data-driven model based on the designed experiments to predict the best combination of coating parameters to ensure uniform thickness of the tablet coating. Another study by Niblett et al. [29] revealed that the process of tablet pan coating can be optimized and improved by using a dimensionless regime map. Their work was directed toward improving the appearance quality of the tablets by reducing defects on the surface.

10.4.3 Electrostatic and 3D Printing of Pharmaceutical Coatings Guided by Data Models

The electrostatic coating technique is basically adopted from the metal industry. The basic mechanism followed is that a powdered drug substance is charged with an electric charge, and then it is directed toward the oppositely charged tablet surface. This causes the deposition of a coating layer on the tablet surface due to the oppositely charged particles. This is a widely used technique, but it has a certain limitation of coating uniformity. The main challenge faced during this process is to maintain the distribution density of the coating powder, making sure that it is equally distributed on all parts of the tablet. Many researchers have developed the DEM to simulate the coating process. These models can predict the inter-tablet coating uniformity by considering factors such as particle charge, tablet movement, and electrostatic forces.

In addition to this method, another approach to 3D printing is also widely adopted in the pharmaceutical industry, where the models and shapes of different drugs and tablets are prepared through computer-aided design, and then their size and other dimensions are thoroughly worked upon before their final construction. This has increased on-demand manufacturing and reduced the time for pharmaceutical developments. Several 3D printing technologies have been explored for pharmaceutical applications, including binder jetting, fused deposition modeling (FDM), semi-solid extrusion (SSE), and stereolithography (SLA). Each technique presents unique advantages and challenges concerning material compatibility, resolution, and scalability. In the following Figure 10.7a, the simulation of electrostatic gun is highlighted, and Figure 10.7b indicates the modelling of 3D printing process.

Yang et al. [32], in their study, created a novel DEM, which is a mathematical model that details the factors controlling the inter-tablet coating parameters. It improves the uniformity of the coating and suggests better combinations of the process properties.

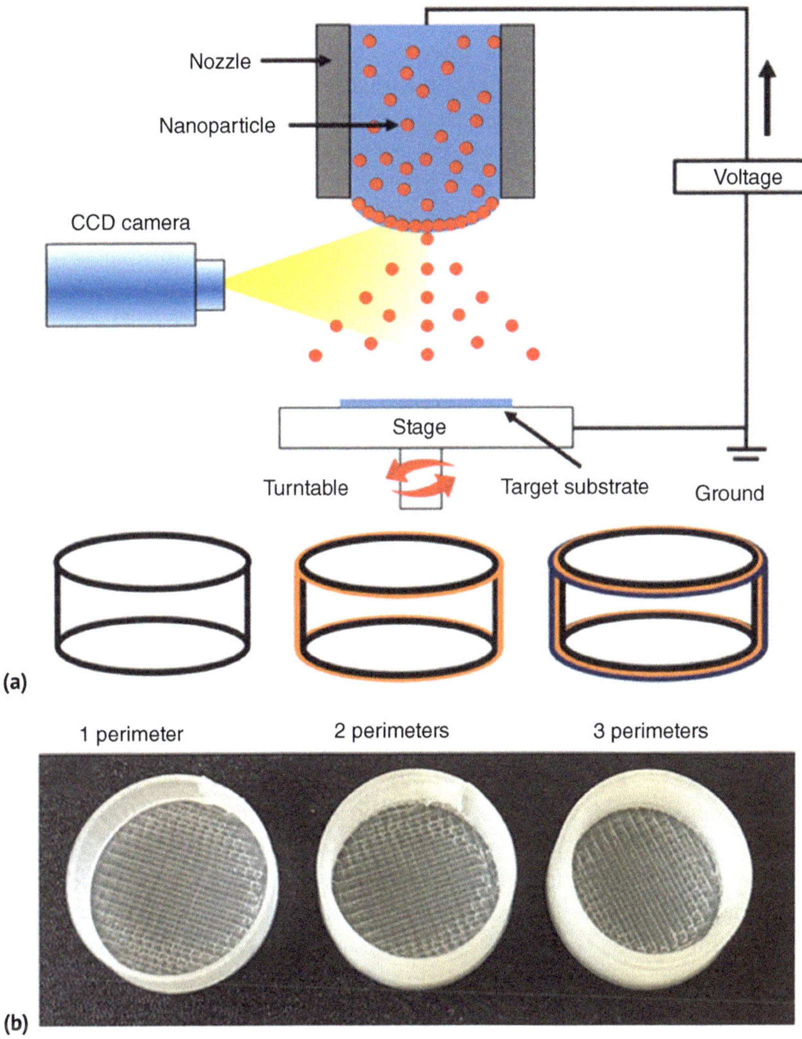

Figure 10.7 (a) Simulation of electrostatic tablet coating process. (b) 3D printing for tablet formation. *Source*: Adapted from [30, 31]/MDPI/CC BY 4.0.

Algahtani et al. [33], in their work, used a 3D printing technique to print a coating system in the form of shells with various designs to control and modify the drug release in immediate release tablets. They focused on improving the patient compliance with the produced drug. Another research by Qiao et al. [34] used electrostatic powder coating in a pan coater where a liquid plasticizer was sprayed onto the tablets' surface to increase their conductivity, and then the oppositely charged coating particles were sprayed on them. This resulted in an increased uniformity and smooth surface of the coated tablets. Adimulapu et al. [35], in their work, have given a detailed analysis of the various research areas in 3D printing and reviewed the benefits and prospects of utilizing this technology in the pharmaceutical industry.

The integration of data-driven technologies with the traditional coating methods has resulted in significant advancements in the pharmaceutical industry. These approaches enable precise control over the coating process parameters, facilitating the development of personalized and efficient drug delivery systems. Ongoing research and technological advancements are expected to further improve the quality of these processes by overcoming the current challenges and shortcomings.

10.5 Quality Control and Performance Evaluation Using Data

The check and balance for the quality and performance evaluation of the pharmaceutical coatings is critical for the safety of the patients and the reliability of the drug being produced. In the traditional quality check methods, tablets and other solid dosages are checked in batches in bulk, which is a highly inefficient method and involves a risk of missing the defect detection. The use of data-driven technologies ensures both efficiency and improved accuracy of the quality check process. Some of the data-driven technologies discussed in the following section. Data-driven techniques like PAT use advanced data analytics, sensors, and real-time process controls to ensure consistent pharmaceutical coatings. It integrates various monitoring techniques like near-infrared spectroscopy, Raman spectroscopy, and laser-induced breakdown spectroscopy to track the important coating parameters. In addition to this technique, imaging and laser-based methods can be used for monitoring the coating thickness, roughness, and uniformity in real time. The drug dissolution rate is very important as it influences a drug's bioavailability. The coating layer on a tablet controls the dissolution rate. The machine learning techniques like artificial neural networks and support vector machine algorithms are currently being used for predicting the dissolution behavior of the drugs based on their training and data collected from the real-time sensors. By monitoring the spray rate, drying temperature, and polymer distribution, these data-driven models analyze dissolution kinetics based on coating thickness, polymer composition, and tablet porosity.

Another important factor in determining the performance and quality of the drugs is the correct stability and shelf-life prediction for a drug. APIs are sensitive to environmental factors like moisture, oxidation, and mechanical stress. The artificial intelligence models can successfully predict the color changes in films due to oxidation. In addition to that, they can estimate moisture absorption in hydrophilic polymer coatings. They can also successfully identify the microcracks in tablet coating, which could lead to API exposure and potency loss. The improved and accurate predictions from these models help in improving the quality and shelf life of the solid dosage forms. The integration of data-driven models and advanced techniques with the traditional quality check methods has resulted in reliable dosage development, increasing the lifetime of the drugs being developed. The consumption rates of the drugs are improved as their patient compliance has increased. The machine learning models forecasting degradation and performance over time have proven to be cost-effective, efficient, and productive, contributing directly to the success of the pharmaceutical industry.

10.6 Trends and Future Directions in Data-driven Coating Technologies

10.6.1 Increased Use of AI and Big Data

The collection of extremely large and diverse structured, unstructured, and semi-structured data sets is called big data. This has revolutionized all industries, including the pharmaceutical industry. The diverse patient data, medication, and effects of the drugs are helpful in creating enormous data sets that can be analyzed by applying intelligent machine learning algorithms like AI. This data is used for better diagnosis, potential side effects of the drug, and its efficacy. The increased use of AI and big data will result in the creation of reliable predictive models that will simulate how the new drugs will interact with the human body.

10.6.2 Digital Twins and Smart Manufacturing

The virtual models of coating processes and coating conditions give a detailed picture about the challenges and shortcomings of the technique being followed and how the parameters can be manipulated in a good way to get the desired and optimized results. A digital twin of any experiment gives a clear picture of the correct proportions and combinations of ingredients for improving the current state-of-the-art drug development process. It can help in improving the performance of the process and the quality of the drug being developed.

10.6.3 Personalized Medicines and Coating

The integration of patient data into the medication process can result in the development of personalized coatings. Coatings can be customized to modify the release rate of APIs based on a patient's metabolism. The coating thickness and composition of the tablets can be optimized for specific populations, for example, considering the elderly patients with slower metabolism and pediatric patients requiring milder release. Other drug conditions can also be adjusted, like gastric pH variability or enzyme deficiencies that affect drug solubility and absorption. An AI model can predict the ideal polymer ratio in an extended-release coating for a patient with renal impairment, ensuring a steady plasma drug concentration without toxicity.

10.7 Conclusion

The integration of data-driven techniques into the pharmaceutical industry has proved to enhance its performance and improve the reliability of the drugs being manufactured. The machine learning algorithms and advanced artificial intelligence techniques like neural networks, support vector machines, and deep learning have revolutionized the process of coatings in pharmacy. The processes are made efficient and trustworthy in terms of different aspects like the proportion, composition, and combination of different ingredients in the coating process to achieve the desired results. With the inclusion of big data and personalized medicines, the new milestones can be achieved, giving a boost to the pharmaceutical industry and the development of tablet coatings.

References

1 Shaikh R, O'Brien DP, Croker DM, et al. Chapter 2 – The development of a pharmaceutical oral solid dosage forms. *Computer Aided Chemical Engineering* 2018;41:27–65. https://www.sciencedirect.com/science/article/pii/B9780444639639000026.

2 O'Connor T, and Lee S. Chapter 37 – Emerging technology for modernizing pharmaceutical production: Continuous manufacturing. *Developing Solid Oral Dosage Forms (Second Edition)*, 2017: 1031–1046. Academic Press. https://www.sciencedirect.com/science/article/pii/B9780128024478000376.

3 Sacher S, Wahl P, Weißensteiner M, et al. Shedding light on coatings: Real-time monitoring of coating quality at industrial scale. *International Journal of Pharmaceutics* 2019;566:57–66. https://doi.org/10.1016/j.ijpharm.2019.05.048.

4 May RK, Evans MJ, Zhong S, et al. Terahertz in-line sensor for direct coating thickness measurement of individual tablets during film coating in real-time. *Journal of Pharmaceutical Sciences* 2011;100(4):1535–1544. https://doi.org/10.1002/jps.22359.

5 Zeitler JA, Shen Y, Baker C, Taday PF, Pepper M, and Rades T. Analysis of coating structures and interfaces in solid oral dosage forms by three dimensional terahertz pulsed imaging. *Journal of Pharmaceutical Sciences* 2007;96(2):330–340. https://doi.org/10.1002/jps.20789.

6 Radtke J, Rehbaum H, and Kleinebudde P. Raman spectroscopy as a PAT-tool for film-coating processes: In-line predictions using one PLS model for different cores. *Pharmaceutics* 2020;12(9). https://doi.org/10.3390/pharmaceutics12090796.

7 Mantanus J, Rozet E, Van Butsele K, et al. Near infrared and Raman spectroscopy as process analytical technology tools for the manufacturing of silicone-based drug reservoirs. *Analytica Chimica Acta* 2011;699(1):96–106. https://doi.org/10.1016/j.aca.2011.05.006.

8 Li X, Chen X, Hong N, et al. A CFD-DEM simulation of droplets in an airless spray coating process of a square duct. *Coatings* 2024;14(3):282.

9 Hemamanjushree S and Tippavajhala VK. Simulation of unit operations in formulation development of tablets using computational fluid dynamics. *AAPS PharmSciTech* 2020;21(3):103. https://doi.org/10.1208/s12249-020-1635-1.

10 Böhling P, Khinast JG, Jajcevic D, et al. Computational fluid dynamics-discrete element method modeling of an industrial-scale Wurster coater. *Journal of Pharmaceutical Sciences* 2019;108(1):538–550. https://doi.org/10.1016/j.xphs.2018.10.016.

11 Boehling P, Jajcevic D, Davies C, et al. In Silico design-space analysis of a novel tablet coating process using advanced modeling. *Computational Particle Mechanics* 2024. https://doi.org/10.1007/s40571-024-00804-0.

12 Partheniadis I, Terzi V, and Nikolakakis I. Finite element analysis and modeling in pharmaceutical tableting. *Pharmaceutics* 2022;14(3):673.

13 Yang Y, Li J, Miao KS, Shan WG, Tang L, and Yu HN. Simulation and evaluation of tablet-coating burst based on finite element method. *Drug Development and Industrial Pharmacy* 2016;42(9):1384–1392. https://doi.org/10.3109/03639045.2015.1137304.

14 Toschkoffa G, Suzzia D, Reiterb F, Tritthartb W, Schlingmannb M, and Khinasta JG. Numerical simulation of film formation in tablet coating. *Scientia Pharmaceutica* 2010;78(3):663–663.

15 Serrano DR, Luciano FC, Anaya BJ, et al. Artificial intelligence (AI) applications in drug discovery and drug delivery: Revolutionizing personalized medicine. *Pharmaceutics* 2024;16(10):1328.

16 Diaz, LP. Machine learning models for the prediction of pharmaceutical powder properties. Stax, 2022. Available from: https://purl.org/coar/resource_type/c_db06.

17 Gaudelet T, Day B, Jamasb AR, et al. Utilizing graph machine learning within drug discovery and development. *Briefings in Bioinformatics* 2021;22(6):bbab159. https://doi.org/10.1093/bib/bbab159.

18 Galí A, García-Montoya E, Ascaso M, et al. Improving tablet coating robustness by selecting critical process parameters from retrospective data. *Pharmaceutical Development and Technology* 2016;21(6):688–697. https://doi.org/10.3109/10837450.2015.1045619.

19 Cahyadi C, Heng PW, and Chan LW. Optimization of process parameters for a quasi-continuous tablet coating system using design of experiments. *AAPS PharmSciTech* 2011;12(1):119–131. https://doi.org/10.1208/s12249-010-9567-9.

20 Orth M, Rotter S, Safdar W, et al. Fluidized bed spray coating for improved mechanical properties of particles. *Processes* 2023;11(2):314.

21 Muñoz López CA, Peeters K, and Van Impe J. Data-driven modeling of the spray drying process. Process monitoring and prediction of the particle size in pharmaceutical production. *ACS Omega* 2024;9(24):25678–25693. https://doi.org/10.1021/acsomega.3c08032.

22 Liu Z, Ma H, Zhou L, et al. DEM-DDM investigation of the tablet coating process using different particle shape models. *Industrial & Engineering Chemistry Research* 2023;62(1):829–840. https://doi.org/10.1021/acs.iecr.2c04030.

23 Boehling P, Toschkoff G, Knop K, et al. Analysis of large-scale tablet coating: Modeling, simulation and experiments. *European Journal of Pharmaceutical Sciences* 2016;90:14–24. https://doi.org/10.1016/j.ejps.2015.12.022.

24 Schmitt JM, Baumann JM, and Morgen MM. Predicting spray dried dispersion particle size via machine learning regression methods. *Pharmaceutical Research* 2022;39(12):3223–3239. https://doi.org/10.1007/s11095-022-03370-3.

25 Seo KS, Bajracharya R, Lee SH, and Han HK. Pharmaceutical application of tablet film coating. *Pharmaceutics* 2020;12(9):853.

26 Pérez-Ramos JD, Findlay WP, Peck G, and Morris KR. Quantitative analysis of film coating in a pan coater based on in-line sensor measurements. *AAPS PharmSciTech* 2005;6(1):E127–E136. https://doi.org/10.1208/pt060120.

27 Pathak KA, Kafle P, and Vikram A. Deep learning-based defect detection in film-coated tablets using a convolutional neural network. *International Journal of Pharmaceutics* 2025;671:125220. https://doi.org/10.1016/j.ijpharm.2025.125220.

28 Dubey A, Boukouvala F, Keyvan G, et al. Improvement of tablet coating uniformity using a quality by design approach. *AAPS PharmSciTech* 2012;13(1):231–246. https://doi.org/10.1208/s12249-011-9723-x.

29 Niblett D, Porter S, Reynolds G, et al. Development and evaluation of a dimensionless mechanistic pan coating model for the prediction of coated tablet appearance. *International Journal of Pharmaceutics* 2017:528(1):180–201. https://doi.org/10.1016/j.ijpharm.2017.05.060.

30 Lee G, Sim JH, Oh G, Won M, Mantry SP, and Kim DS. Electrostatic spray deposition of Al-doped ZnO thin films for acetone gas detection. *Processes* 2023;11(12):3390.

31 Skalická B, Matzick K, Komersová A, Svoboda R, Bartoš M, and Hromádko L. 3D-printed coating of extended-release matrix tablets: Effective tool for prevention of alcohol-induced dose dumping effect. *Pharmaceutics* 2021;13(12):2123.

32 Yang Q, Chen J, Zhou X, Zhou H, Yang G, and Zhu J. Modeling of inter-tablet coating uniformity of electrostatic dry powder coating by discrete element method. *Powder Technology* 2022;411:117929. https://doi.org/10.1016/j.powtec.2022.117929.

33 Algahtani MS, Mohammed AA, Ahmad J, and Saleh E. Development of a 3D printed coating shell to control the drug release of encapsulated immediate-release tablets. *Polymers (Basel)* 2020;12(6). https://doi.org/10.3390/polym12061395.

34 Qiao M, Zhang L, Ma Y, Zhu J, and Chow K. A novel electrostatic dry powder coating process for pharmaceutical dosage forms: immediate release coatings for tablets. *European Journal of Pharmaceutics and Biopharmaceutics: Official Journal of Arbeitsgemeinschaft für Pharmazeutische Verfahrenstechnik e.V* 2010;76:304–310. https://doi.org/10.1016/j.ejpb.2010.06.009.

35 Adimulapu AK, Jesudasan RE, Arya RK, et al. 3D printing coatings for pharmaceutical applications. *Functional Coatings for Biomedical, Energy, and Environmental Applications* 2024: 123–143. https://onlinelibrary.wiley.com/doi/abs/10.1002/9781394263172.ch6.

11

Nanotechnology and AI-driven Drug Repurposing and Combination Therapy

Naveed Ahmad[1], Muzna Ali Khattak[2], Talaya Hidayatullah[3], Muhammad Asim Sajid Al-Hanafi[4], Bruno Arnaud Dzokwe Kene[4], Zahid Hussain[5,6] and Shahzeb Khan[4]*

[1] *Institute of Pharmaceutical Science, Faculty of Life Science and Medicine, King's College London, London, UK*
[2] *Department of Pharmacy, Iqra University, Peshawar, Pakistan*
[3] *Department of Pharmacy, University of Peshawar, Peshawar, Pakistan*
[4] *Institute of Health and Social Care, School of Pharmacy, Optometry and Medical Sciences, University of Bradford, West Yorkshire, UK*
[5] *Department of Pharmaceutics and Pharmaceutical Technology, College of Pharmacy, University of Sharjah, Sharjah, United Arab Emirates*
[6] *Research Institute for Medical and Health Sciences, University of Sharjah, Sharjah, United Arab Emirates*

*Corresponding author: s.khan131@bradford.ac.uk

CONTENTS

11.1 Introduction

Nanomedicine is often defined as the science with the aim of nanotechnology application in medicine. It focuses on applying the nanoscale materials and devices technology in the healthcare unit. A couple of related research studies have been recently associated with nanotechnology for the enhancement of drug delivery, imaging, therapeutic techniques, or technology [1]. It is the use of substances that are up to one hundred nanometers in the health-related sector. This change to an atomic and molecular level enhances the biological reactions, performance of the drugs, and canceled side effects as compared to the normal large-scale modes [2]. Also, this is significant to consider that the materials involved in nanomedicine at the nanoscale are bound to exhibit some other physical, chemical, or even biological features that are different from those experienced in the usual larger scale. For instance, nanoparticles (NPs) can penetrate cell membranes given the fact that they are not very large in size. NPs can be also potentially introduced in specific cells or tissues in the body in an effort to increase the potency of the treatment. Apart from drug delivery, other important application of medicine at the nanoscale level is nanoscale medical diagnosis, which is considered more highly sensitive than normal standard diagnosis.

11.1.1 Mechanistic Insights into Nanotechnology in Medicine

It is used mainly to target and treat cells and tissues at the molecular level or cellular level and may be widely used in the medical science or physiology field for having greater functional selectivity with biological activities, and it can also be incorporated in vivo with structures that were not possible earlier. According to what has been mentioned, it must be noted that nanotechnology is not a separate emerging science, but is clustered from different sciences that could be described by chemists, physicists, material scientists, and biologists, which are required to advance the novel technologies [3]. What nanotechnology seems to promise is also complex and consists in providing better versions of existing methodologies and, in addition, new possibilities. At nanometer scale, the drugs and other material can be controlled in such a way that the materials' properties as well as activity associated with the material can also be controlled. Such tools may assist in managing one or several features of drugs or agents, including (a) alteration in solubility and (b) controlled discharge over short or long durations, i.e., inhibition of breakdown or slow or selective release in the environment or designed to release at a particular site [4].

11.1.2 Applications of Nanomaterials in Medicines

These are fluorescence biological labels, carriers of drugs and genes, biosensors for pathogens, identification of proteins, structural probing on DNA, identification of tumors, separation of the biological molecules and cells, contrast agents for MRI, and phagokinetic assessments. The most prospective of all the goals of nanomedicine is to stake clearly such quantitative characteristics as nanomachinery. It is therefore apparent that these targeted and strategic invasions of nanomachinery into the living cells can facilitate the identification and comprehension of the cell and hence aid the diagnosis as well as the therapy of the diseases. The importance of this field lies in advancing platform technologies for improving nanoscale imaging systems and enhancing molecular detection in living cells [5]. Molecular imaging is defined as the visualization of an illness in order to obtain proof of the problem's molecular functionality of an illness at times prior to the manifestation of the symptoms. Therefore, the realization of nanotechnology within the molecular imaging field can generate a new form of nanoprobes to biomarkers that is efficient in receptiveness, special to the disease sites, and might show secretion signals of human diseases [6].

Counting on some attributes of NPs for the improvement of the imaging techniques, there exist certain characteristics that favor the system due to high sensitivity, high-resolution and additional information at molecular and cellular levels. At a small scale, the magnetic particles can be used as an imaging contrast in MRI. These magnetic NPs can be then conjugate with other functional groups such as fluorescence markers, radioactive and other biomolecules, which are used in imaging and gene transfection plus cellular transport mechanisms. MRI can employ probes of magnetic NPs and adenoviruses that are attached to target cells and genes and also apply the gene expression and the delivery of genes, which can be seen by MRI with the help of green fluorescent proteins. Other methods, for instance, the positron-emission tomography (PET), can in fact be enhanced by sensitivities of the detection, which are a lot higher, and can afford to use the NPs in such low concentrations that are not possible in MRI. Additionally, in the PET/CT integration, if the PET is coupled with CT scans, which produce high anatomical detail, there is more visibility for further signaling and categorization depending on the vascular territories of the atherosclerosis [7, 8]. It is inevitable that the contrast agent goes to the particular site of interest, and as such, this can be done to a larger extent depending on the guidance of the NPs containing the contrast agent. This proves that target molecules have to be summarized whenever they are located outside the tissue barriers, and the only way through which this is possible is by the use of targeting groups. When the sensitivity of a specified imaging modality is low, then effects of summation are noted when dealing with multiple contrast groups. Some of these related NPs may be used at the same time to accomplish delivery of both the contrast medium and the drug so as to measure the biodistribution of the therapy and the efficiency of the therapy commonly known as theranostics [8].

11.1.3 Vitality of Drug Repurposing in the Contemporary Pharmacotherapy

In some cases, it can discover new compounds based on the enhancement in the phenotypic traits that may be associated with the drug's mechanism of action without reference to ME. It can be verified, especially in preclinical animal model studies, and such results are better for clinical purposes and research than using the preclinical diagnostic procedures only. It can go through the second phase of a clinical trial, skipping the first phase, and this has been supported by literature. This means that there is minimal likelihood of failure if a drug compound is used; thus, focusing on the drug has been in drug discovery, where it has been used as an example of an ideal repurposing drug. Since the drug would present the safety and toxicity data from the previous clinical trials, as well as have the approval of the authorities, the particular drug can avoid clinical trials if there is sufficient data to support the action, of course. In addition, the mode of action of each of the chosen drugs should be identified [9]. Thus, it is most ideal to opt for those compounds that have gone through several phases of clinical trials and have not performed as they were expected, maybe for other reasons apart from issues to do with safety issues of patients that make them unfit to be used as candidates for repositioning.

11.1.4 Synergistic Potential of Combination Therapies

Synergism is used here in the sense of working cooperatively in combination or in a group, in the literal sense. Synergism, therefore, refers to those cases in which compounds operate jointly, not independently, where the overall operation of the compounds intensifies the ability of the other beyond the actual sum of the abilities of the components [10]. The combined potency of two or more agents also reduces the feature such as toxicity and additional side effects linked with the high dosages of individual drugs since the body compensates for one drug by increasing the utilization of its related substance as well as other multi-target may only be achieved using high dosages of the individual drugs [11]. A facility to define the multiple targets with possible therapeutic potential can be rather peculiar if it is appraised with the help of both the experimental and theoretical [12] strategies. As suggested before, dealing with biological complexity is one of the potential advantages of using combinations as approaches to treat multifactorial disorders. Considering new indications of known compounds has become an attractive concept in the recent era because of the ability to save much time in the process of drug discovery and prominent costs. It has great potential when used alongside other treatment modalities; it may be useful in overcoming some of the barriers like drug resistance.

In this compound use, the old drugs affect many disease processes; hence, it increases the probability of recovery. For example, a work done on a viability test of Rapta-C, erlotinib, metformin, and parthenolide; it was found out that the compound possessed antitumor activity due to an ability to decrease energy in cancer cells and was not toxic for the normal cells [13]. Regarding the ability of repurposed drugs to overcome the resistance, authors found that: A synergistic mixture of the multiple antimicrobial agents can help to overcome pathogens resistance, including methicillin-resistant *Staphylococcus aureus* (MRSA) [14]. For example, when a number of drugs have been approved for other uses in small amounts, their strength is as mighty as the high-dosage single remedies; besides which, there is no question of side effects. Combination therapies and repositioning also enhance the possibility of availability of new treatments since these drugs have already been through the safety process [15].

11.1.5 Recent Advances in Combinations Therapy

In the recent advancement, CUSP9 is one where there is the drug combination approach for the treatment of glioblastoma (GBM), which has been widely researched. This later on combines nine noncytotoxic agents working alongside temozolomide to target each of the growth factor proteins that aids in the development of GBM. These works include the following: there was an option of using drug repurposing as one of the strategies for the quick identification of effective therapeutics to combat infectious diseases [16]. Machine learning (ML) models have been worked out based on synergistic drug combinations in order to identify novel, reused therapeutic agents [17].

11.2 Role of Nanotechnology in Addressing Challenges Associated with Repurposed Drugs and Combination Therapies

The directed evolution of drugs and concomitant therapy are the practical approaches for enhancing the efficiency of treatment and overcoming drug resistance. However, many of them have limitations, which include issues to do with bioavailability, side effects, and toxicity that rises with dose. However, these problems are surprisingly solved by nanotechnology by providing targeted delivery, improving solubility, and minimizing systemic toxicity.

11.2.1 Enhancing Bioavailability

Most repositioned drugs have poor water solubility and permeability, which consequently results in poor bioavailability. Liposomal, polymeric NPs and solid-lipid NPs are some of the nanotechnology-based delivery systems, which can increase solubility and bioavailability of drugs. For instance, the nanoformulations of the hydrophobic drugs are capable of increasing dissolution rates as well as improving the gastrointestinal (GI) permeability. For the similar reasons stated above, nanocarriers also improve controlled drug release, maintaining plasma drug levels and thus minimizing frequent administrations of the same [18].

11.2.2 Targeted Drug Delivery

One limitation of attempting to use or combine drugs in a repurposed manner primarily involves off-target effects of the drug. Ligand-receptor-mediated targeting in nanotechnology involves enhanced effects with decreased side effects and toxicity. The functionalized NPs have active surface molecules like antibodies, peptides, or aptamers to improve the drug concentration to disease. For example, targeted liposomes encapsulating repurposed anticancer drugs have shown increased tumor penetration with reduced systemic toxicity [19].

11.2.3 Overcoming Drug Resistance

Combination therapies often face resistance due to efflux pumps and metabolic adaptations in diseased cells. Nanotechnology can co-encapsulate multiple drugs within the single carrier, ensuring synchronized delivery and preventing premature drug degradation. Multifunctional nanoparticle systems have been engineered to simultaneously transport anticancer drugs and gene-silencing agents, aiming to bypass or mitigate therapeutic resistance mechanisms [20]. Such approaches improve synergistic interactions between drugs and enhance treatment efficacy.

11.2.4 Reducing Toxicity

Toxicity remains a significant concern in drug repurposing and combination therapy due to unpredictable pharmacokinetics and drug–drug interactions. Nanocarriers provide controlled drug release, reducing peak plasma concentrations and thereby minimizing toxic side effects. Moreover, surface modifications like polyethylene glycol PEGylation improve biocompatibility as well as prevent rapid clearance by the immune system [21].

11.3 Key Concepts in Nanotechnology-driven Drug Delivery

11.3.1 Definition and Principles of Nanotechnology

Nanotechnology acts as a link between the physical and biological sciences through the application of nanoscale structures and phases across diverse fields [22]. Such technology is particularly of interest in nanomedicine and nanobased systems for delivery of drugs [23]. Materials with dimensions ranging from 1 to 100 nm are defined as nanomaterials and are widely employed in tissue engineering, biosensors, microfluidics, and various drug delivery applications [24]. Nanomedicines are developed by using curative agents at nanoscale level [25]. In these systems, nanostructures function as carriers by either encapsulating drugs or attaching them to their surface, enabling controlled release and precise, site-specific delivery [26]. Figure 11.1 gives an overview of NP classifications.

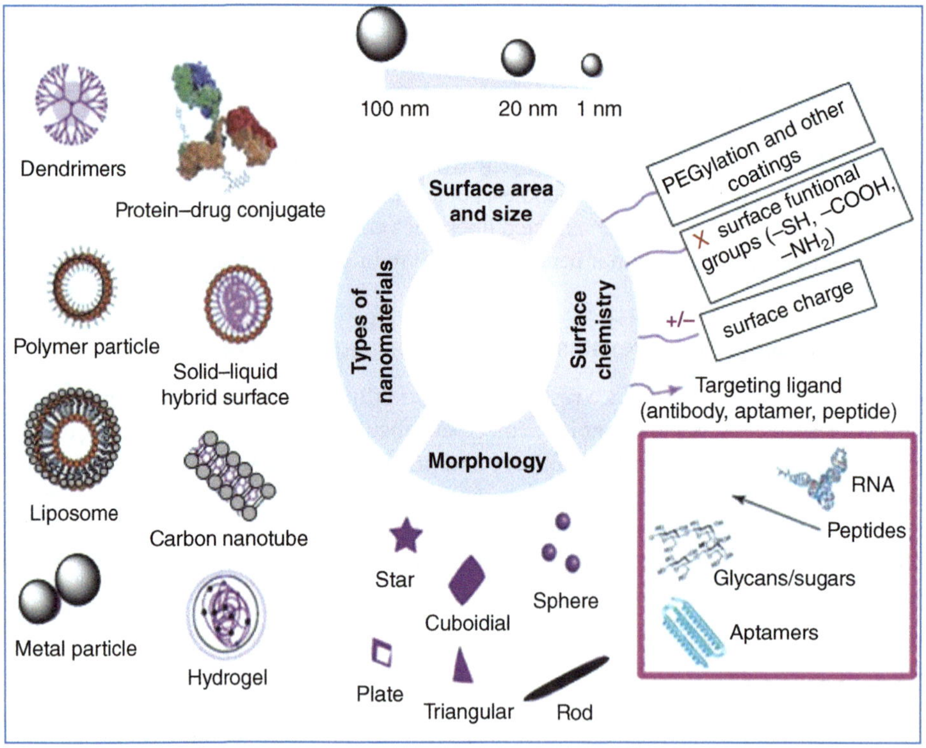

Figure 11.1 Classification of nanoparticles.

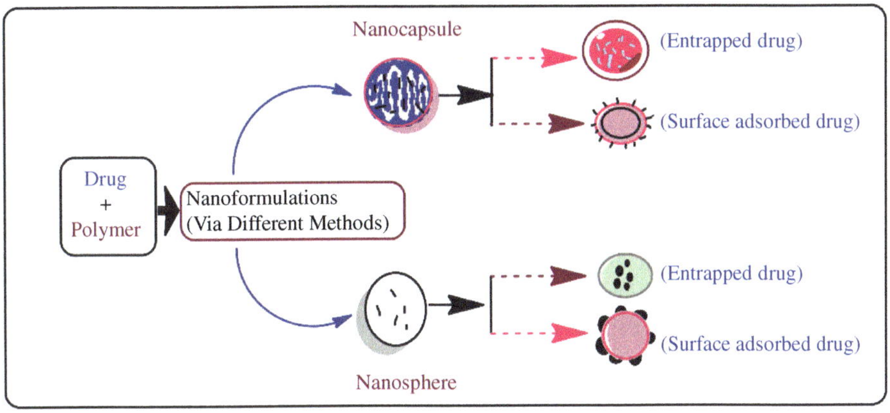

Figure 11.2 Mechanisms of drug entrapment and surface adsorption in polymeric nanoformulations.

11.3.2 Nanoparticles Based Drug Delivery System

Macromolecular substances that range from 10 to 1,000 nm are called NPs; these are solid and colloidal [27]. The two different types of NPs include nanocapsules and nanospheres. The confinement of a drug to the cavity that is enclosed by a membrane that is polymeric gives rise to the formation of nanocapsules. A matrix system that comprises of physically and uniformly dispersed drugs is called a nanosphere [27]. Figure 11.2 presents the mechanisms of drug entrapment.

There exists evidence that biodegradable NPs have improved the therapeutic effect of a number of water-insoluble/soluble drugs and bioactive moieties in terms of solubility, bioavailability, and time of residence at action site [28]. The NP–drug formulation is relatively cheap and also minimizes the possibility of toxicity. The grinding of the medicinal drugs at the nanoscale improves the therapeutic efficacy, ratio, patient friendliness, and selectivity of the related drugs [29]. The major applications of NPs as drug delivery systems (DDSs) depend on the following characteristics: the small size of the particle and biodegradable polymers used. NPs can invade through the endothelium in tumor, epithelial tissue, damaged tissue, or inflammatory tissue or diffuse through micro-capillaries due to their small size. At the target sites, the commercially available NPs having a size of nanometers enable the selective deposition of the drug besides enhancing the uptake of the drug by many a variety of cells [30]. Among both the delivery systems, NPs took favor over micro-particles because of their size, which is smaller in comparison to the micro-particles. The smallest cross sectional diameter of the capillaries is 5–6 microns. The particle size should be less than 5 μm in order to be distributed in the bloodstream without aggregation and subsequent embolism. The preparation of NPs through biodegradable carriers allows the system to maintain the drug release at a particular size for an extended period of time, such as several days or even weeks. The NPs developed from biodegradable polymers are found to be very effective for a range of deliveries applications with intracellular action [31].

11.3.3 Overview of Nanocarrier Types in Therapeutic Delivery

The advent and development of nanotherapies have brought out disease's diagnostic, therapeutic, and preventative aspects, which were not much of a concern earlier. These are mentioned as nanomedicines and are expected to deliver services of genomics and proteomics to patients. Each kind of NPs is able to simulate or to regulate any biological state, for instance, infection, tissue formation, synthesis of new tissues and organs, etc. These include functionalized carbon nanotubes; DNA-based nanodevices constructed from assembled nucleic acid components and frameworks; nanofibrous structures; self-assembling polymeric nanosystems; as well as nanomembranes and silicon-based nanoscale plates designed for the delivery of drugs or nucleic acids. In addition, lab-on-a-chip platforms are utilized for diagnostic applications [27].

11.3.3.1 Nanocrystals

Nanocrystals are large particles that aggregate together in one mass, and their estimated size ranges from 10 to 400 nm. The differences in their physical and chemical characteristics are between bulk solids and particles. The loading efficiency has increased, meaning it is soluble in water and has better adsorption in the body. They can be formulated using the contexts that come from the bottom-up method and top-down style used in the knowledge hierarchy. As the name suggests, they are produced in crystalline state although they may be undefined depending on the mode of production. Nano-sized crystals have an importance in the range of 10–400 nm for wound healing as it has better dissolution behavior and enhanced bioavailability [32].

11.3.3.2 Liposomes

Liposomes are preferred nanocarriers in the drug delivery. Because liposomes are circular and spherical particles with particle size of between 0.05–0.5 micrometer's because they are formed by emulsion of natural or artificial lipids. Liposomes were first noticed in the 1960s by Bengham and In subsequent years, it has developed into one of the most high-cost drug delivery systems [33]. Nano emulsions have applied for nanomedicine due to aspects, which include high biocompatibility, stability of nano emulsions, methods of synthesis, high drugs encapsulation efficiency and high bioavailability with the aid of safe components in formulation. The advantageous properties of liposomes include their small size and the amphiphilic nature of phospholipids, which possess both hydrophilic and hydrophobic characteristics. This allows drugs to be incorporated either within the aqueous core of the liposome or embedded in the lipid bilayer [34]. These features make liposomes highly effective and versatile carriers for drug delivery applications. To date, several liposome-based drug delivery systems have received approval from the U.S. Food and Drug Administration for the treatment of various diseases. Also, liposomes are good for diagnostic applications applied by the following methods: ocular, oral, pulmonary, transdermal, and parenteral. Liposomes, therefore, are mainly composed of natural lipids like soybean phosphatidylcholine [35].

11.3.3.3 Polymeric NPs

Polymeric NPs have emerged in the past years because of their characteristics that arise from their relatively small size. The benefits related to polymeric NPs have been identified as follows: their use in controlled release, protection from the environment of drugs as well as molecules with biological activity, and enhancement of bioavailability and the therapeutic index [36, 60]. The term "nanoparticle" has been used in this study to refer to both the nanocapsules as well as the nanospheres in spite of the fact that the two differ in terms of structure [37]. Nanocapsules, containing an oily phase in which it is generally dissolved, are located in the polymeric shell that determines drug rate from the core. Among the polymeric systems, nanospheres have the property of a continuous solid phase with pores in which the drug can be dispersed and entrapped within [38].

11.3.3.4 Dendrimers

Dendrimers are well-defined and highly symmetric macromolecules that are polydisperse and tree-like in nature, with their branches radiating from a central point [39]. These hyperbranched molecules were first synthesized in 1978, later by Donald Tomalia and his coworkers in the first decade of the 1980s, and independently by Newkome at the same time. These polymers might also be called "cascade molecules," yet this name does not seem to be as set as "dendrimers" [40]. Dendrimers are highly uniform (monodisperse) macromolecules composed of symmetrically branched units radiating from a central core, along with linear polymer segments [41]. Contrary to the architectural use that is evident in this article, "dendrimer" is simply not a compound. Polyionic dendrimers are nonspherical, and their size and structure can change with the increase in the generations [42]. The end groups, defined as the arms, which are at the periphery of dendrimer structure, can be operated on, meaning that particular characteristic physicochemical or biological characteristics of dendrimers can be altered [43]. Dendrimers have found the multiplicity of uses in many fields and disciplines of supramolecular chemistry especially in host-guest interactions as well as self-assembly. In view of these properties, dendrimers have some features that make them suitable for many applications. Dendrimers are synthetic and well-defined polymer structures, and they are known to possess several functional groups along with a symmetry of structure. These dendritic structures are increasingly important in oncology treatments as well as imaging-based diagnostics [44]. It has been observed that with an increase in the dendrimer generation, dendritic macromolecules are characterized by a progressive increase in the diameter and a change in the overall shape from linear to more spherical. Thus, dendrimers made of cation biopolymers have been chosen to apply an exclusive platform for the study of the effects of the corrected parameters of polymer architecture on important physiological properties in cells, including interaction with lipid bilayers, cytotoxicity, uptake of dendrimers, their serum stability, biodistribution, and filtration [45].

11.3.3.5 Metallic NPs

Metals and metalloids have attracted scientific interest for over a century and are now widely utilized in both biomedical and engineering fields. They are of particular concern because of their application in the great nanotechnology system. Today, such material can be prepared and coupled with bioactive chemical groups that will allow it to interact selectively with specific antibodies, ligands, and drugs, given the number of applications that the material can find its use in fields including biotechnology and drug delivery. Other imaging techniques have been discovered after some time, for example, magnetic resonance imaging (MRI), positron emission tomography (PET), surface-enhanced Raman scattering (SERS), computed tomography (CT), and optical imaging in diagnosing several diseases. These imaging agents work with different techniques, and as for the tools they require, they more prominently need the contrast agent along with specific physiochemical properties that are most suitable for the chosen imaging technique. This has proved to the formation of different nanoparticulated contrast agents of Fe_3O_4 [46], gold [47], and silver [48] for the said imaging techniques. In addition, for getting different types of images simultaneously, newer multifunctional nanoshells and nanocages have been developed [49].

11.3.3.6 Nanogels

According to IUPAC, a nanogel (NG) is an amorphous form of gel particle of this form or any other form and is 1–100 nm in diameter. This seems to be less profound when observed across the various literature, and thus a more comprehensive definition can be given: an NG can include a number of interpenetrating, crosslinked hydrogel particles, whose size is submicron [50]. As mentioned earlier, it is understood that NGs are a subcategory of NPs, and very often the word "nanoparticle" is related to solid and non-susceptible structures and formations, while NGs may also contain the deformable structure. They are soft nanomaterials based on swellable polymeric groups that preferably shape themselves as fluid absorbents, including water. The characteristics and their applicability stem from the specifics, such as increase in size in different solvents, ability to swell and to be deformed, and the structure that can be regarded as dynamic, porous, and containing a net [51]. As the 3D structures of NGs, the hydrophilic polymer chains in the NGs can be physically crosslinked by physical bonds or chemically crosslinked, namely covalently. This type of organization enables NGs to be leached with liquid media without dissolving within the liquid media [52].

They exhibit some favorable characteristics suitable for biomedical uses, including swelling/shrinking characteristics, structural adjustability, improved colloidal stability [53], and biocompatibility. Such characteristics elicited the use of NGs in sensing, diagnostics, vaccines, and drug delivery [54]. The effect on the latter continues to be most pronounced to the present time, as characterizing NGs as having the "right" characteristics for therapeutics. These systems can transport small molecules, which can be either small chemical molecules or particles, proteins, nucleic acids, as well as minor inorganic NPs [55, 56]. The modularity of the NG platform is enormous within the chemical context and can help to visualize new "intelligent" NGs with the ability to provide cargo in response to external or internal signals.

11.3.3.7 Exosomes/Nanovesicles

Exosomes are small membrane vesicles originating from endosomes, having received much interest in the last decade, as described below. Actually, the extracellular appearance of exosomes was first reported approximately a decade earlier during the period of the late 80s. Nevertheless, the productions of exosomes were once believed to be a phenomenon that resulted in cell death or injured cells as a consequence of disease or physiological procedures such as autophagy that were harmless to the neighboring cells. It is only of recent years that such extracellular are functional carriers that carry proteins, lipids, and nucleic acids to be able to deliver such loads to other cells they come across and alter the targeted cells far from where they were shed. Consequently, the exosomes can be regarded as a new trend in intercellular signaling, and they perform various functions in the cell, such as immune response, signaling, and antigen presentation. Since exosomes are secreted by any eukaryotic cell, it is rational that the contents of the exosomes can vary depending on the functions of the cells that produced them and their condition (normal, transformed). Thus, circulating exosomes' active content may offer diagnostic value in other human's diseases, including chronic inflammation and renal diseases, as well as tumors [57–59].

11.4 Functionalization and Surface Modifications of NPs for Targeted Delivery

When drugs are associated with ordinary carriers, then the biodistribution routes favor the delivery to the MPS, which includes the liver, bone marrow, and lungs. The administered NPs are detected by the host's systemic immunogenic responses and can be identified when NPs are injected intravenously and subsequently removed from circulation by phagocytes. In addition to the size of NPs, the extent of the NP hydrophobicity will define the amount of blood components that will bind on this surface, for instance, opsonin particles. Therefore, the effect of surface hydrophobicity is that it plays a role in the in vivo behavior of NPs. However, when in a stream of blood, the non-modified NPs' surface undergoes opsonization and is almost removed by MPS [60]. To enhance the effectiveness of the drug aiming, one needs to decrease the rate of opsonization and extend NPs' circulation time in blood plasma. This can be made possible by using surfactants/polymers with hydrophilic natures or using nonbiodegradable copolymers possessing hydrophilic characteristics. Different investigations indicate that functionalization of NPs with PEG results in the absence of opsonization through other serum proteins [107]. Zeta is a typical parameter of the charge of the surface for the description of NPs, as a rule [61]. It stands for particles' electrical potential and depends on type of particle and type of the medium it is dispersed in. The particles with zeta potential greater than $\pm$30 mV was found to be well-dispersed as the above statement shows that the surface charge keeps the particles apart.

In the recent past, design/functionalization and application of NPDDS have been the greatest shift in technology; there have been conjugations between polymers and surfaces. NPDDS are altered in a number of ways to act as depots or delivery systems for differential diseases [62]. Some of the functionalization processes applied on NPs are targeting ligands and agents for diagnostics and imaging in order to enhance the functionality of NPs or to introduce new functionality for targeting with precision. In today's functionalized NPs, all physicochemical properties, specific targeting skills, bioavailability, biocompatibility factors, anti-corrosiveness, non-agglomeration, and noninvasive nature. Thus, the trend of functionalizing NPs is on the rise in the studies in an attempt to improve the gross efficiency as well as the mode of functionalization of these particles [63].

11.5 Drug Repurposing: Concept and Role of Nanotechnology

11.5.1 Definition and Significance of Drug Repurposing in Reducing Time and Cost for Drug Development

The process of developing a new drug for market approval is highly expensive and can take approximately 10–15 years. This lengthy and resource-intensive timeline has driven interest in drug repurposing, where existing drugs are utilized to shorten the overall development period. The utilization of a drug, which has already been approved by various authorities like the FDA, the EMA, the MHRA, among others, in a different disease or condition for treatment. Due to the potential of reducing the time of drug development, many pharmaceutical companies are recently using drug repurposing to diversely redevelop the existing FDA approved but failed pipeline molecules. This section provides an overview of methods used at the present stage of utilizing and developing approaches for drug repurposing and shows the effectiveness and a perspective of the usage of drug repurposing based on the reduction of the time needed to develop the drug due to the presence of all data on its clinical and toxicological effects [64].

The following steps show that for a new drug to get into the market it has to meet some conditions that are set Alison (2010). To discover a drug and build it requires time and capital, primarily because of variations in drug inputs' physio-chemical properties as well as in the challenges of increasing productivity [65]. This limitation also provides pharmaceutical companies or an academic center with the ability to put in effort and effectively adapt any drug that has already been approved by authority for some other diseases to be used for an indication that isn't yet available for patients who are suffering from that disease. Those investigational molecules, which have been of no avail for a specific indication, are usually good candidates as far as repurposing is concerned. They can be again identified for a new purpose or indication and thus they can be developed as potential treatment options, especially for use in the treatment of genetic disorders, which are often difficult to diagnose and manage [66]. For instance, certain autoimmune diseases, bacterial diseases, as well as certain forms of cancer are not genetically inherited and therefore are of idiopathic nature; therefore, they are difficult to manage. Drug repurposing is clearly defined to be efficient in terms of money and time as compared to other tiresome long-time-taking procedures involved with new drug discovery and development. Also, it proves to be a sigh of relief for the patient, "as" he has to spend almost directly on the out-of-pocket expenses, which in return makes the actual cost of therapy drastically low. Ideally, for an investigational molecule, it is sometimes difficult to present adequate data on safety and efficacy; hence, there is a high dropout rate during drug development, contributing to most of the failure either in safety or efficacy. On the other hand, all aspects of safety, preclinical, and efficacy are known in the case of a repurposed molecule, and an investigator in this case has all the details for decision-making in each step of the drug development process [67]. Information regarding the safety and efficacy of the identified drug also cut down the amount of expenses and efforts that are needed for the repositioning of a drug and its development into the market and the time required as well [68].

For instance, sildenafil, which is a phosphodiesterase type 5 (PDE5) inhibitor, is one of the examples of success in the repurposing process. Though originally introduced for the treatment of hypertension, it was found also to provide great assistance in cases of erectile dysfunction; hence, the approval of the FDA for the same. It was subsequently referred to as a treatment for a disorder, a rare illness, pulmonary hypertension [69]. In the current world, accompanied by several technologies and computational facilities, the drugs are less costly, sometimes beginning from a more efficient molecule of a trait drug. This has contributed to providing approximately 30% of the new FDA-approved drugs in several years prior to now. Out-licensing is another factor that is positively affected by drug repositioning, as these drugs are endowed with some desirable attributes for the purchaser [70].

However, there are two things also associated with the above advantages, which makes it also important that identification of a new disease target should not adversely affect the marketing of the drug for the originally targeted disease. Knowledge regarding DCS and disorders touches upon the theme of uncommon diseases, which do not have well-known available treatments, and many cases have a prognosis [71].

11.5.2 Limitations of Conventional Drug Repurposing and Nanotechnology as a Solution

Drug repurposing strategies aim to discover new medical indications for previously approved medications. Those drugs that were initially indicated only for one disease but later exemplified and used in the treatment of some other diseases are on the rise. Drug repurposing is also highlighted by Kirtonia et al. as an example of innovative approaches; its potential to become transformative, especially in the context of oncology, is described in detail [72]. Drug repurposing also has several advantages, which include less time and cost as compared to other methods in the development of new drugs since there is prior information on the toxicity, doses, etc. of the existing drugs. Hence, there has been an increasing interest in drug repurposing in the recent past. Chlorambucil and bufulfone, for instance, were initially designed as alkylating agents but derived from the lethal toxic chemical warfare agent mustard gas, although they were later found useful in the treatment of leukemias [73]. Drug targeting or repurposing for cancer has mainly been occasioned either by an insight into the disease mechanisms or accidentally. Therefore, if greater focus is set on developing new approaches on how the discovered drug can be related to a new disease, then drug repurposing would yield better results. There are two approaches that can be used in identifying a drug that could be repurposed to treat a certain condition: computational and experimental. As for the experimental one they include models like induced pluripotent stem cells and first-function phenotypic screenings, also referred as reverse chemical biology [74]. While on the side of computational methods, include target-centric, knowledge-directed, signature-aligned, pathway-focused, and mechanism-specific strategies [75]. Computational approaches include target-based, knowledge-driven, signature-based, pathway-oriented, and mechanism-specific strategies [73], Current in silico methods for drug repurposing in cancer integrate drug effects with clinical profiles and disease models to identify disease-modifying candidates. These tools can predict ligands, characterize drug–target binding patterns, and efficiently filter promising compounds from large datasets [76]. Current computational techniques applied to drug repurposing in cancer involve integrating drug response data with clinical disease profiles and biological disease models to identify potential disease-modifying effects. These approaches are capable of identifying potential ligands, elucidating drug–target interaction patterns, and narrowing down candidates from large compound libraries for further investigation [77].

11.5.3 Examples of Repurposed Drugs Delivered via Nanocarriers

Curcumin is a yellow polyphenolic compound derived from *Curcuma longa* Linn, known for its diverse bioactive properties, particularly in the treatment of skin disorders [78]. It has demonstrated notable antitumor activity in both in vitro and in vivo studies, acting through multiple cellular pathways. Curcumin is also among the few natural compounds that have progressed to phase III clinical trials [79]. Recent findings suggest that its anticancer effects may involve modulation of cell membrane properties. Furthermore, Jalilian et al. reported that curcumin downregulates the expression of α-SMA and COX-2, along with reducing PGE2 levels in cancer-associated fibroblasts (CAFs). This study also supported the proof that curcumin directly changed the pro-tumor state of CAFs by downregulating PGE2 gene followed by the upregulation of T-bet gene and interferon-gamma thereby overcoming inflammation in TME. Thus, all these changes contribute to improving the antitumor functions of immune cells [80]. Also, curcumin activity in the inhibition of metastasis of breast cancer can also be seen from its effect of down-regulation of the CXCL12/CXCR4 arrest of ADMSCs' differentiation towards CAFs. Furthermore, Mao et al. showed that curcumin inhibited the number and tumorigenic ability of LGR5(+) colorectal cancer stem cells (CSCs) through increasing the autophagy level and acting on the TFAP2A-regulated ECM pathway. While it is already ascertained that curcumin possesses varied antitumor activities experimentally and clinically, this golden spice has not made its way into mainstream anticancer therapy mainly due to its poor pharmacokinetic profile. Moreover, most clinical trials related to curcumin contain a small number of patients, which results in clinical fluctuations. Therefore, for future reference, broader and better-designed clinical trials are needed to ascertain that curcumin possesses potential as an anticancer agent.

Chloroquine, primarily known as an antimalarial agent, has also shown therapeutic potential in rheumatoid arthritis (RA), systemic lupus erythematosus (SLE), and various cancers and viral infections. It is widely used in research to investigate autophagy, nanoparticle uptake, endocytosis, and multiple signaling pathways in vivo across different disease models, particularly in cancer [81].

It is important to understand how Chloroquine inhibits cancer cell growth. Its primary mechanism involves disruption of autophagy, where it blocks autophagosome degradation by accumulating in lysosomes due to its pH-sensitive nature. This inhibition can promote apoptosis, partly through ROS-mediated endoplasmic stress and depolarization of lysosomal and mitochondrial membranes. Although chloroquine alone may not significantly alter mitochondrial membrane potential, it is often used alongside chemotherapy or radiotherapy, acting through both autophagy-dependent and independent pathways. However, its use has been associated with adverse effects, including renal and organ toxicity. Recent studies also indicate that chloroquine interferes with key metabolic pathways, such as those involving cholesterol, glucose, amino acids, and mitochondrial function. Additionally, it may help overcome multidrug resistance by inhibiting drug efflux mechanisms through interactions with ATP-binding cassette (ABC) transporters and related membrane proteins [82, 83].

The disaster caused by viral diseases to the human race has been present right from the beginning of mankind. Every virus differs in its virulent factors, and the pathological effects vary in most of the viral diseases, especially in the case of the new pathogenic SARS-CoV-2 virus. Single-stranded RNA-engulfed viruses belonging to the family Coronaviridae cause an ongoing pandemic, COVID-19 [84]. This disease is associated with fever, cough, fatigue, and exhaustion and has many complications in patients and has a 5.7% mortality rate for this disease. Out of the COVID-19 positive patients, 50% do not show any symptoms. The early symptoms of this disease include tingling in the ears, loss of taste and smell, nasal congestion, headache, and red eyes. Cough, fever, and dyspnea typical to the period of fruiting are described as characteristic of the multipurpose disease by many commentators; however, they are noticeable only after 8–10, in the last days of the infected period [85]. Chloroquine is an effective anti-viral drug and from the mechanisms learnt from several diseases, here are the probable mechanism of the chloroquine against SARS-CoV-2 Now that there are trials ongoing to see if chloroquine will inhibit SARS-CoV-2. The following outlines the key mechanisms by which Chloroquine acts against COVID-19

11.5.3.1 Mechanisms Preventing Viral Entry into Cells

Chloroquine may inhibit viral entry by interfering with the interaction between the viral spike glycoprotein and the ACE2 receptor on host cells. It has also been reported to suppress enzymes such as glycosyltransferases and quinone reductase involved in sialic acid synthesis on ACE2, which may further limit viral attachment [86]. Molecular docking studies by Wu et al. (2020) suggested that chloroquine may preferentially bind to viral proteins such as Nsp3b or the E-channel, although these findings (mfScores −130.355 and −107.889) require further validation. SARS-CoV-2 particles, typically 60–140 nm in size, share similarities with nanoparticles and are primarily internalized via endocytosis [87]. Chloroquine can inhibit cellular uptake by disrupting clathrin-mediated endocytosis through interference with phosphatidylinositol-binding clathrin assembly protein (PICALM) [85]. A similar mechanism may limit the internalization of SARS-CoV-2 particles into host cells. Chloroquine also contributes to reducing viral entry by inhibiting endocytic pathways. It is well established that viral entry requires both ACE2 and TMPRSS2 [88]. Notably, combining chloroquine with a serine protease inhibitor further impairs viral entry, as TMPRSS2 activity can be blocked by agents such as camostat mesylate [89].

11.5.3.2 Inhibition of Viral Replication and Post-translational Processing

It disrupts the release of the viral RNA genome into the host cell cytoplasm and inhibits endosomal and lysosomal acidification, thereby blocking viral progression. Additionally, it interferes with endosome–lysosome fusion and intracellular trafficking processes essential for viral replication [87].

11.5.3.3 Suppression of Autophagy in Viral Infection

Many human viruses exploit autophagy to support their replication within host cells. In the case of SARS-CoV-2, evidence suggests that autophagy plays a role in viral replication [87]. For instance, Prentice et al. [90] reported that the endogenous LC3 protein, essential for autophagosome formation, is involved in SARS-CoV replication. Since Chloroquine is known to inhibit autophagy, it may serve as a potential agent for interfering with coronavirus replication.

11.5.3.4 Immune Regulation and Cytokine Storm Inhibition

Chloroquine is currently used in the treatment of rheumatoid arthritis (RA) and systemic lupus erythematosus (SLE) due to its immunomodulatory properties. It disrupts pH-dependent Toll-like receptor signaling within endosomes, thereby reducing excessive cytokine release. By inhibiting this pathway, antigen recognition by dendritic cells is impaired [91]. Additionally, chloroquine can enhance CD8+ T-cell responses to viral antigens and promote antibody-related activity within dendritic cells. It also interferes with antigen processing in lysosomes, limiting major histocompatibility complex class II (MHC II)–antigen interactions and reducing inflammatory responses [92]. Furthermore, it suppresses the production of tumor necrosis factor-alpha (TNF-α) and its signaling pathways, contributing to the prevention of cytokine storm [93].

11.5.3.5 Interference with the Metabolic Pathways

It has been observed that when chloroquine is used in handling glioblastoma, it can in a certain way regulate the metabolic processes especially the lipid metabolism in a cell. From all the above points, it could be observed that lipid metabolism plays a role in virus getting into host cells as well as their replication. Infection with COVID-19 alters lipid metabolism by elevating free fatty acids and glycerol levels while reducing total cholesterol, HDL, and LDL concentrations in the serum [87]. As in GBM and SLE, this report shows that chloroquine can modulate metabolic processes during SARS-CoV-2 infection because that is how it functions.

11.6 Combination Therapy: Challenges and Nanotechnology Solutions

11.6.1 Definition and Benefits of Combination Therapy

The use of one or more active agents to develop new therapeutic strategies for disease prevention and treatment has emerged as an effective therapeutic approach. Previous studies have demonstrated that synergistic drug interactions can significantly enhance therapeutic outcomes in a wide range of diseases. Consequently, combination therapies are increasingly being explored for conditions such as cancer, atherosclerosis, pulmonary hypertension, myocarditis, rheumatoid arthritis, inflammatory bowel disease, metabolic disorders, and neurodegenerative diseases because they can target multiple pathological pathways while reducing dosage requirements, toxicity, and drug resistance. In these therapeutic systems, two or more agents are delivered simultaneously to treat a specific disease. Among the various delivery platforms, liposomal nanoparticles, polymeric nanoparticles, and nanocrystals are considered some of the most important carriers in combinatorial therapy due to their targeting capability, sustained drug release, and enhanced drug stability, which help prevent rapid clearance from the diseased site [94].

Also, the combined therapy is a novel view on utilizing the principle of "drug repositioning," which presupposes the utilization of the drugs that are already permitted for use, which in turn reduces business risk and expenses to design new drugs [95]. But it can also be attributed to the negative treatment outcomes like antagonism and increased toxicity of the drug due to restrictions such as pharmacokinetic differences, asynchronous tissue biodistribution, poor penetration across the barrier, and attainment of intracellular delivery [96]. For example, small molecular drugs and active protein blends techniques have a good effect on presenting normal functioning of cells in the form of Petri dishes. However, the combination of all those elements, if dosed, shows low therapeutic efficacy since such molecules are quickly metabolized by livers and poorly internalized by cells.

Due to these advantages, multifunctional nanoparticle-based drug delivery systems are effective in combination therapy, as they allow co-encapsulation of active agents, enhance solubility, protect drugs from degradation, control biodistribution, improve tissue penetration, reduce clearance, prolong half-life, and lower toxicity. Moreover, the DDSs developed above mentioned could administer two or more drugs concurrently or separately, which enables the pharmacokinetics optimization for the drugs and achieve the most efficient drug-drug interaction [97, 98]. For example, stimuli-responsive drug delivery systems, including enzyme-responsive and pH-responsive nanoparticles, are capable of transporting therapeutic agents to specific target sites, including particular organelles or lesion regions, and enabling their controlled release in a regulated manner [99]. Moreover, even higher levels of synergy may be attained due to the fact that NAD and L549 are released into the cytosol from different DDSs after endocytosis because spatial velocity is incompatible with each other at the intracellular target site. For instance, as compared to the active compounds, biological drugs need more time to describe the activity after taking the drugs. The structure formed by combining drug crystals with microRNAs enables sustained drug release over a defined period. This design also facilitates rapid cellular uptake of the biologic components, thereby enhancing overall therapeutic efficacy in eliminating cancer cells or reducing inflammation [100]. Altogether, these NP preparations can be given orally, parenterally or by the use of skin patch and inhalation which would increase its possibilities of being used clinically. To date, only one liposomal formulation, Vyxeos, co-encapsulating daunorubicin (DNR) and cytarabine (ara-C), has received approval in 2017.

11.6.2 Challenges in Combination Therapy

Combining multiple drugs can lead to overlapping or synergistic toxicities, increasing the risk of adverse events and potentially requiring dose adjustments or treatment interruptions [102]. This may occur either through changing the pharmacokinetics and pharmacodynamics of these drugs, resulting in undesired and variable drug exposure and accountability of the dosage form [103]. Evaluating the safety and efficacy of multiple drugs, including their interactions, requires complex and often resource-intensive clinical trials [104]. Cancer cells may acquire resistance to targeted treatments, making it essential to understand these resistance mechanisms when developing effective combination therapies [105]. Access to combination therapies, particularly those involving multiple manufacturers and indications, can be challenging due to pricing complexities, regulatory hurdles, and the difficulty of attributing value to the combination [106]. Tumor heterogeneity can make it difficult to identify the optimal combination of therapies for a given patient, as different tumor cells may respond differently to treatment [107]. Finding robust biomarkers to predict treatment response and guide the selection of appropriate combination therapies is also a challenge [103]. Combination therapies can be costly due to the need for multiple drugs, extensive clinical trials, and potential complications [104]. The complexity of combination therapy regimens, including multiple medications and potential side effects, can lead to poor patient compliance [108].

11.6.3 Nanotechnology's Contributions: Examples of Nano-enabled Combination Therapies

Over the past several decades, the clinical use of combination chemotherapy has been guided by a set of well-established principles. Typically, each drug included in a combination regimen should demonstrate proven antitumor efficacy as a single agent. In addition, the selected drugs should exhibit non-overlapping toxicity profiles, possess distinct resistance mechanisms, act through different biological pathways, and ideally target separate phases of the cell cycle. In clinical practice, sequential administration of anticancer agents is commonly recommended in treatment guidelines, whereas simultaneous (concomitant) administration is less frequently adopted [109]. This preference is largely based on observations that, although concurrent combination therapies may produce higher initial response rates, they often fail to significantly improve overall survival and are associated with increased adverse effects [110]. More recently, phase II clinical studies using low-dose metronomic chemotherapy schedules have demonstrated moderate improvements in tumor response with manageable toxicity profiles [111]. The variability in clinical outcomes associated with conventional drug combinations has prompted the exploration of nanocarrier-based delivery systems, which can help overcome limitations encountered with free drug administration. Nanotechnology offers several strategies for co-delivery of multiple therapeutic agents. These include: (i) administration of one drug in its free form alongside another delivered via a nanocarrier (free + nano), (ii) delivery of both drugs using separate nanocarriers (nano + nano), and (iii) co-encapsulation of multiple drugs within a single nanocarrier system.

In the first two approaches (free + nano and nano + nano), drugs are typically administered sequentially, whereas co-encapsulation enables simultaneous delivery, enhancing local drug concentration and improving coordination of therapeutic action against cancer cells. For the purposes of current clinical applications, the free + nano strategy remains particularly relevant. While the nano + nano approach may offer advantages in terms of formulation simplicity and dosing flexibility, there is no guarantee that different nanocarriers will reach the same target site to achieve synergistic effects. In contrast, co-encapsulation may provide superior efficiency by ensuring that multiple agents are delivered to the same cellular location. Clinically approved nanomedicine formulations such as albumin-bound paclitaxel (Abraxane) and PEGylated liposomal doxorubicin (Doxil) can be used in combination with other therapeutic agents, either in free form or within separate delivery systems [112, 113]. An important consideration in such strategies is that multiple agents are often introduced sequentially rather than simultaneously. For example, cisplatin-based nanoparticles have been applied in breast cancer treatment, followed by administration of PI828, a phosphoinositide 3-kinase (PI3K) inhibitor, resulting in enhanced therapeutic efficacy [114, 115].

The combination of these drugs in the same carrier system and the fact that the drugs used in the present study have been reported to exhibit synergistic effects by many researchers both in vitro and in vivo have been supported [115]. An early preclinical proof-of-concept study by Liu et al. demonstrated that even direct intratumoral administration of the anticancer drug doxorubicin (DOX) together with the P-glycoprotein (P-gp) inhibitor verapamil, when delivered in separate nanoparticle formulations, resulted in only a modest delay in tumor growth and was accompanied by notable toxicity in a multidrug-resistant (MDR) breast tumor-bearing murine model [116]. In a similar context, studies involving the co-encapsulation of DOX with either a P-gp inhibitor or another chemotherapeutic agent within a single nanocarrier showed improved therapeutic outcomes compared to both the "nano + nano" strategy and free drug combinations [117]. In both wild-type and drug-resistant MDA-MB-435/LCC6 human breast cancer cell lines, hybrid polymer–lipid nanoparticles co-loaded with DOX and mitomycin C (MMC) (DMPLN) exhibited significantly enhanced efficacy against resistant cells compared to the co-administration of individually loaded nanoparticles (DOX-PLN + MMC-PLN). This finding suggests that co-encapsulation can enhance drug–drug interactions and promote synergistic effects [118]. Numerous drugs have been incorporated into nanocarriers of different types to provide codelivery, and the purpose of the formation of nanocarriers is to have a synergistic effect. It also aims at addressing the following limitations: poor solubility, poor compliance, and high toxicity when given systematically and to the MDR strains [119, 120].

Recently, a large part of such co-encapsulated nanocarriers was developed with strains that are known to be resistant to antibiotics: *S. aureus*, *Mycobacterium tuberculosis*, *Escherichia coli*, *Pseudomonas aeruginosa*, *Klebsiella pneumoniae*, and *Chlamydia trachomatis*. This was done for the encapsulation of up to four anti-tubercular drugs: isoniazid, rifampicin, pyrazinamide, and streptomycin in *M. tuberculosis* [212]. These are the encapsulated oily solutions of isoniazid, as mentioned above, using the co-encapsulation method for rifampicin and pyrazinamide. Additionally, ethambutol, which is one of the first-line drugs in treating tuberculosis, was not co-encapsulated because of its effect on the disintegration of the nanocarrier [121]; this explains how chemical bonding played a paramount role in co-encapsulation.

11.7 Drug Repurposing and Combination Therapy

11.7.1 Introduction

Combination therapy, defined as the use of two or more active agents within a single treatment regimen, has emerged as an increasingly adopted strategy for managing complex human diseases [122]. Clinical mainstream diseases, such as cancer, cardiovascular disorders, IBD, lung diseases, RA, and metabolic disorders, possess complex microenvironments and interrelated pathological pathways, and thus most traditional monotherapies always have moderate efficacy. With the advantages, such as the targeting of multiple signaling pathways, high treatment efficacy, reduced administration dose and side effects, and low drug resistance [123]. Accordingly, combination therapies have emerged as a promising approach for addressing these multifaceted diseases (Figure 11.3). Moreover, such strategies provide a valuable framework for drug repurposing, facilitating the application of existing approved drugs to new therapeutic indications while potentially lowering development costs and associated risks [95]. Despite their potential, cocktail-based drug combinations can also lead to undesirable outcomes, including antagonistic interactions and heightened toxicity. These limitations often arise from disparities in pharmacokinetics, non-uniform tissue distribution, inadequate penetration of biological barriers, and inefficient

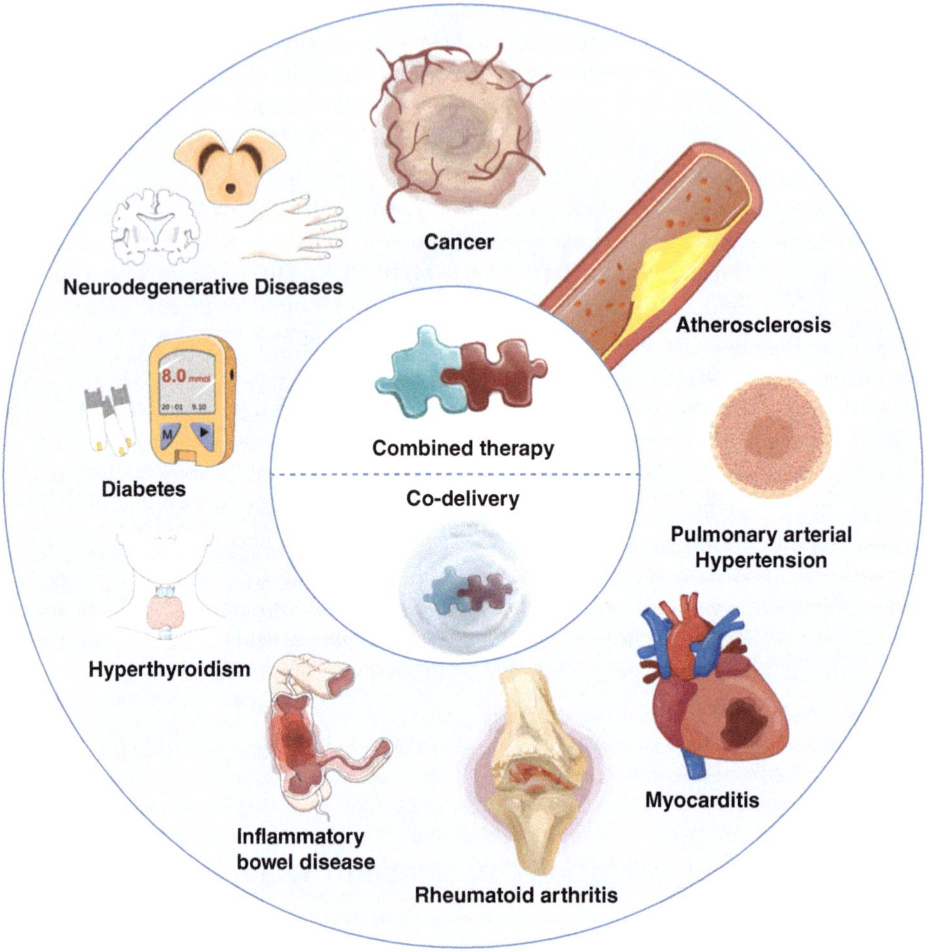

Figure 11.3 Combination therapies and nanoparticle-based co-delivery approaches for treating human diseases. *Source*: [94]/Springer Nature/CC BY 4.0.

intracellular delivery. For instance, while the co-administration of small-molecule agents and therapeutic proteins may demonstrate favorable effects on normal cellular functions in vitro, their combined use *in vivo* frequently results in suboptimal efficacy. This is largely attributable to rapid protein degradation in the liver and insufficient cellular uptake, which together compromise the overall therapeutic performance [124].

Multifunctional nanoparticle (NP)-based DDSs have gained considerable attention as an effective platform for combination therapies, owing to their capacity to co-encapsulate multiple therapeutic agents within a single carrier. These systems can enhance drug solubility, protect labile compounds from degradation, regulate biodistribution, improve tissue penetration, evade rapid systemic clearance, extend circulation half-life, and minimize off-target effects. Notably, NP-based DDSs enable the co-delivery of multiple agents either concurrently or in a spatially controlled manner, thereby promoting synchronized pharmacokinetic profiles and maximizing therapeutic synergy [125]. For example, stimuli-responsive DDSs—such as enzyme- and pH-sensitive nanoparticles—can achieve controlled and sequential release of their payloads, facilitating targeted delivery to specific pathological sites or subcellular compartments [126]. Moreover, the non-synchronized intracellular release of co-delivered agents from DDSs after endocytosis can further enhance therapeutic synergy, as these agents often act on distinct targets with different temporal requirements. A codelivery system formed from drug crystals and micro-RNAs facilitated long-term sustained release of the drug and, meanwhile, fast release of the biologics, improving the synergy to kill cancer cells or alleviate inflammation [100]. Additionally, nanoparticle-based drug formulations are compatible with diverse administration routes—such as oral, parenteral, transdermal, and inhalational delivery—thereby enhancing their translational and clinical potential [101]. A representative example is the liposomal product Vyxeos, which co-delivers DNR and ara-C and was approved in 2017 for the treatment of therapy-related acute myeloid leukemia (t-AML) and AML with myelodysplasia-related changes [127]. This milestone underscores the clinical advancement and translational potential of multifunctional NP-based combination therapies.

11.7.2 Nanotechnology in Drug Repurposing and Combination Therapies

Nanotechnology has emerged as a transformative strategy in drug discovery and development, particularly in the areas of drug repurposing and combination therapy. Preclinical studies have clarified efficacy, underlying mechanisms, and possible clinical applications of nanotechnology in these areas, providing valuable information for future therapeutic interventions. The use of NPs not only improves the pharmacokinetic and pharmacodynamic properties of existing drugs but also provides new paths for synergistic effects through combined therapies. Recent studies have underlined the promise of the reuse of drugs in cancer therapy, with emphasis on using nanotechnology to increase the therapeutic potential of previously approved medications. For example, Xia et al. [128] described the advantages of drug reuse, particularly in oncology, where the urgent need for new therapeutic strategies is exacerbated by the increase in resistance to existing treatments. The incorporation of nanotechnology in this area allows for improving supply mechanisms, targeting specific tumor sites, and minimizing systemic toxicity, thus improving general effectiveness.

A notable illustration of this approach is the use of nanotechnology in the topical management of skin cancer. The authors demonstrated that NPs could facilitate the localized supply of reused antitumor agents, which leads to a significant improvement in treatment results. By directly attacking the tumor microenvironment (TME), the NPs exceeds the barriers for penetration and deliver greater concentrations of therapeutics at the site of action [129]. This strategy highlights the capacity of nanotechnology to reposition established therapeutics within combination regimens, thereby enhancing their antitumor efficacy. In recent years, several topical and transdermal nanoformulations incorporating repurposed agents with recognized anticancer potential have been developed for the treatment of various forms of skin cancer. These formulations span multiple pharmacological classes, including antibacterial agents such as Doxycycline, antifungal drugs like Itraconazole, anthelmintics such as Niclosamide, lipid-lowering agents like Simvastatin, antirheumatic drugs such as Leflunomide, antidiabetic agents like Metformin, and nonsteroidal anti-inflammatory drugs (NSAIDs) such as Celecoxib. These compounds are incorporated into

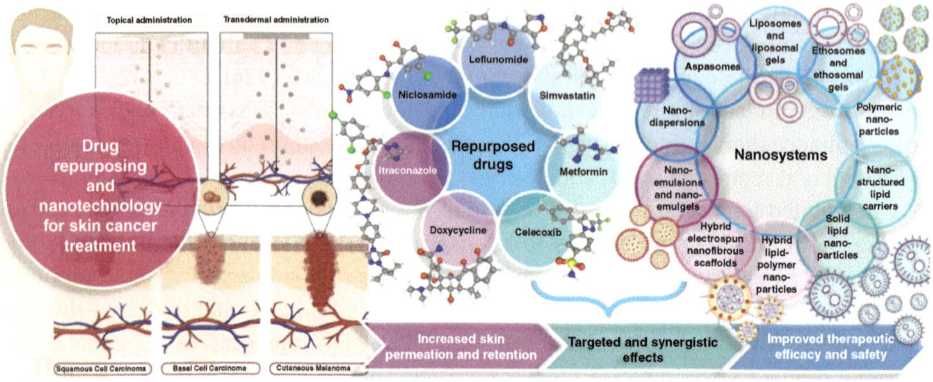

Figure 11.4 Drug repurposing and nanotechnology for skin cancer: Repurposed medications and cutting-edge nanosystems work together to increase skin penetration, more precisely target cancer, and better treatment results. *Source*: Reproduced from [129] with permission of American Chemical Society.

nanoscale delivery systems to enhance their localization, penetration, and overall therapeutic efficacy within skin cancer tissues [130]. Figure 11.4 presents the framework of drug repurposing and nanotechnology for skin cancer.

By taking advantage of the ability of NPs to encapsulate multiple agents, studies showed potential to improve therapeutic results through combination strategies, thus racing the way for advanced clinical applications in the management of this aggressive malignancy [131]. WNT/β-catenin signaling pathways have been involved in various types of cancer, including liver carcinoma. Mohapatra and Chandrasekaran [132] explored how the reuse of drugs based on nanotechnology has attacked these pathways, revealing that the encapsulation of reused drugs within the transporters not only improves the efficiency of the supply but also facilitates the selective orientation of malignant cells. Such mechanisms highlight the sophisticated interaction between nanotechnology and the pharmacological modalities established in reuse efforts. Another approach area is pancreatic cancer, which presents significant therapeutic challenges. Lellis et al. [131] analyzed strategies to reuse existing medications for the treatment of pancreatic cancer and emphasized the role of nanotechnology in improving the effectiveness of these therapies.

The concept of combined drug therapies is particularly relevant in the context of breast cancer, as Correia et al. [133] demonstrate. His research highlighted how the combination of reused drugs with novel nanocarriers can amplify the therapeutic responses while minimizing side effects. This strategy is based on the synergistic effects of medicines, where the lowest doses are enough to achieve the desired results, thus reducing the risk of adverse events. The findings support the need for preclinical studies to evaluate these combined therapies together with the integration of nanotechnology. Another study described the development of multifunctional NPs designed to combine therapeutic agents, emphasizing that such constructions could significantly improve treatment paradigms for various human diseases. When enabling the concurrent delivery of multiple medications, these NPs can provide not only renewed agents, but also new therapies that work synergistically. This approach highlights the versatility of nanoporters in a large number of therapeutic contexts, establishing a critical link between drug reuse and combination strategies [94]. The reuse of antimicrobial agents in cancer therapy represents another innovative approach where nanotechnology has made significant advances. The repositioning of antibiotics and their derivatives within oncological environments: a review [134]. The unique mechanisms of action associated with these agents, often overlooked in conventional cancer therapies, are used through nanotechnology, producing new anti-cancer effects. As their findings demonstrate, the potential of reused agents, when combined with nanotechnology, establishes a convincing argument for broader explorations in preclinical and clinical environments. Simvastatin, mainly known for its cholesterol-reducing effects, exemplifies a drug with remarkable versatility. ARA et al. [135] provided an updated review on the pharmacological and nanotechnological facets of the reuse of simvastatin in the treatment of cancer. Their findings illustrated the synergistic roles of simvastatin when combined with nanoformulations that improve bioavailability and target specificity. Therapeutic implications are deep, suggesting that such reused agents can revolutionize treatment paradigms when administered through nanocarriers. Preclinical studies that focus on the intersection of nanotechnology, reuse of drugs, and combined therapies illuminate the potential for greater efficiency and innovative mechanisms of action in the treatment of cancer. The varied applications of nanocarriers, from facilitating localized delivery to qualifying combination strategies, are under the promise of nanotechnology as a cornerstone in modern therapeutic approaches. As preclinical findings continue to mature, it will be essential to transition from these ideas to clinical applications, paving the way for more effective and specific cancer therapies. The integration of nanotechnology not only revitalizes interest in established agents but also transforms them into vital components of new treatment regimes, satisfying the medical needs not satisfied in a spectrum of diseases.

Although combination chemotherapy has been shown to some degree effective in improving the outcome of cancer therapy, the overall effectiveness of it is still significantly hampered by a range of various pharmacological and physiological constraints. The problems at hand are inadequate drug concentrations in tumor tissues, which can drastically limit the therapeutic effectiveness, and the need for high systemic doses that are not always safe or practical. In addition, rapid elimination of drugs from the system, poor aqueous solubility, and variable bioavailability contribute to the complexities involved. These multifactorial problems generally result in suboptimal therapeutic outcomes and contribute to increased systemic toxicity, which can lead to serious complications for treated patients. In light of the present limitations outlined, there is a growing focus placed on the development of sophisticated DDSs that are optimized to provide maximum therapeutic effectiveness while, at the same time, minimizing any unwanted effects that may occur. In this regard, nanotechnology has proven to be a revolutionary tool, providing a wealth of innovative solutions particularly well-suited for cancer therapy. A diverse array of nanocarrier platforms including liposomes, dendrimers, micelles, carbon nanotubes (CNTs), polymer–drug conjugates, and nanoparticles (NPs) has been extensively investigated for their transformative potential in drug delivery. These nanoscale systems can shield therapeutic agents from premature degradation and reduce rapid elimination by the reticuloendothelial system (RES), thereby extending their circulation time within the bloodstream. Such improvements in pharmacokinetic behavior facilitate more effective transport across biological barriers, promote enhanced accumulation at target intracellular sites, and ultimately contribute to lowering systemic toxicity and the adverse effects typically associated with conventional chemotherapy. Over the past few years, there has been a very thorough investigation of different nanocarrier systems that have been meticulously designed to co-deliver more than one chemotherapeutic agent at a time, and these systems have demonstrated an extraordinary enhancement in their effectiveness as anticancer agents. The targeted and deliberate application of nanoparticle-based co-delivery systems for the exclusive purpose of combination chemotherapy is a visionary and cutting-edge area of research [125, 136].

11.7.2.1 Liposomes

Liposomes consist of spherical lipid vesicles made of amphiphilic phospholipids with bilayer membrane structure with central aqueous space. Closed bilayers are produced by shielding the hydrophobic groups from the aqueous phase while there is simultaneous contact of the aqueous phase with the hydrophilic head group [137]. A number of liposomal formulations have been successfully translated into clinical oncology, underscoring the importance of nanocarrier-based drug delivery systems. Prominent examples include liposomal doxorubicin (Doxil), liposomal daunorubicin (DaunoXome), liposomal cytarabine (DepoCyt), and liposomal vincristine (Onco TCS) [138]. In recent years, increasing attention has been directed toward exploiting liposomes as carriers for combination therapies. For example, liposomal co-encapsulation of curcumin and resveratrol has been shown to exert chemopreventive effects after systemic administration in prostate-specific PTEN knockout mouse models, likely linked to PTEN loss and activation of the p-Akt signaling pathway [22]. In addition, fine-tuning the molar ratios of co-delivered agents has been recognized as a key parameter for optimizing therapeutic outcomes in combination strategies [24]. Consistent with this, a Phase I clinical study investigated a liposomal formulation containing irinotecan and floxuridine at a fixed 1:1 ratio. The results demonstrated that this dual-drug system achieved an acceptable safety profile, established pharmacokinetic parameters, and showed enhanced anticancer activity in patients with advanced solid tumors [139]. Another study was conducted by Wu et al., where a transferrin-conjugated liposome was prepared by co-entrapment of DOX and verapamil (P-gp inhibitor). They investigated the efficacy in DOX-resistant K562 cells and demonstrated enhanced cytotoxicity by overcoming P-glycoprotein-mediated MDR [140]. The radiometric mechanism involving the codelivery of two dissimilar drugs with different molar ratios after intrasystemic administration is another essential consideration for enhancing the therapeutic efficacy of the dual drug in a combination therapy of cancer [141]. On this line, a Phase I study was conducted using a liposomal formulation entrapped with two dissimilar drugs [i.e. irinotecan and floxuridine in the ratio of (1:1)], and the maximum tolerated dose and pharmacokinetic parameters of the liposomal formulation were established in advanced solid tumor patients. The results showed the above dual-drug-loaded liposomal formulation was well tolerated with enhanced anticancer activity in patients [142].

Co-delivery nanoparticle (NP) systems have the potential to enhance the cytotoxic impact of anticancer agents by delivering them simultaneously as a unified formulation. In this context, the concept of "guard" drugs has been explored, whereby one agent modulates the response of normal cells while a second cytotoxic drug preferentially targets malignant cells. Careful dose optimization is critical in such strategies, as it allows exploitation of subtle biological differences between healthy and cancerous tissues to achieve improved therapeutic selectivity. For example, Doxorubicin (DOX), a well-established inducer of p53, exhibits pronounced cytotoxicity at high doses. However, at lower concentrations, it can induce cell cycle arrest at the G1/G2 phases in normal cells [143]. This temporary suppression of proliferation in healthy cells provides a protective effect, enabling the concurrent administration of a second cytotoxic agent that more selectively targets tumor cells. As a result, the combined regimen can reduce damage to normal tissues and mitigate treatment-associated adverse effects.

Beyond the co-administration of multiple chemotherapeutic agents, liposomal delivery systems also offer the capability to simultaneously transport nucleic acids (e.g., genes or RNA therapeutics) alongside small-molecule drugs, further expanding the scope and effectiveness of combination treatment strategies. Li et al. engineered liposomes to co-deliver VEGF siRNA and etoposide (ETO) [144]. This action inhibits tumor cell proliferation by silencing VEGF gene expression and synergistically kills tumor cells by the pro-apoptotic activity exhibited by ETO. In particular, the simultaneous delivery. The product had a definite layer made up of PEGylated polymer, which fulfilled a specific role. Histidine-conjugated chitosan-lipoic acid on the surface of cationic liposomes allows negatively charged and augments the general stability and functionality of blood circulation. In a noticeable contrast to this, the coating was effectively activated. This charge transition enhances cellular internalization, facilitates deeper tumor penetration, and promotes more efficient lysosomal escape. As a result, the dual-delivery platform provides effective drug protection, targeted tumor accumulation, and significantly greater inhibition of tumor growth and angiogenesis compared to control formulations. Such an approach may also help address limitations of conventional gene co-delivery systems, which often exhibit adverse effects due to persistent surface electropositivity. Reactive oxygen species (ROS) within the TME further play a critical role in modulating therapeutic responses. These highly reactive molecules act as secondary messengers in cellular signaling and are involved in regulating growth-related pathways and biosynthetic processes [145]. Accumulating evidence indicates that ROS exert dual functions during cancer progression, acting as both oncogenic drivers and antiproliferative factors depending on their concentration and context [146]. In early stages, oxidative stress can initiate molecular alterations that promote oncogenic mutations and pathological remodeling of cellular pathways. During tumor progression, ROS contribute to enhanced proliferation by amplifying mitotic signaling cascades. However, at advanced stages, excessive ROS levels can exceed a critical threshold, resulting in DNA damage—such as double-strand breaks and 8-oxodG formation—and ultimately triggering apoptotic cell death [147, 148].

Tang et al. developed a reduction-sensitive liposomal system modified with cleavable octa-arginine (R8) and polyethylene glycol (PEG), referred to as CL-R8-LP, for the co-delivery of Doxorubicin (DOX) and the P-glycoprotein (P-gp) inhibitor Verapamil (VER). The PEGylated surface contributed to enhanced nanoparticle stability and prolonged systemic circulation. Following internalization into tumor cells, the elevated reductive conditions—often associated with intracellular reactive oxygen species—trigger cleavage of disulfide bonds, resulting in the exposure of the positively charged R8 moiety. This charge transition enhances cellular internalization, facilitates endosomal/lysosomal escape, and promotes efficient intracellular drug release. Following release, intracellular

Verapamil (VER) inhibits P-glycoprotein (P-gp)-mediated drug efflux, leading to increased nuclear accumulation of Doxorubicin (DOX). This coordinated action markedly enhances cytotoxic effects through the induction of both apoptosis and necrosis. Importantly, the CL-R8-LP system exhibited greater DOX uptake, improved synergistic antitumor efficacy, and lower systemic toxicity in both MCF-7/ADR and MCF-7 cancer cell line [149]. Wang et al. found that co-loading a ROS-activated PTX prodrug and a low-molecular-weight compound with PD-1/PD-L1 pathway inhibitory activity, i.e., BMS-202, into the liposomal cores produced effective tumor targeting; ROS triggered the release of PTX and long-term release of BMS-202 upon the introduction of the cells. The liposomal formulation yielded positive chemo-immunotherapy owing to time-differentiated treatment of these two drugs [150]. In parallel, reactive oxygen species (ROS)-based pro-oxidative therapy has emerged as an alternative anticancer strategy, wherein intracellular ROS levels are elevated beyond a critical threshold to activate ROS-mediated cell death pathways [33, 151]. In this context, Yuan et al. demonstrated a ROS-responsive delivery system based on a thioacetal (TA)-containing polymer integrated with the ROS-generating agent cinnamaldehyde (CA). This system exhibited a self-amplifying degradation mechanism, in which endogenous ROS initiated cleavage of the TA linkages, leading to the release of CA. The liberated CA further disrupted mitochondrial function, generating additional ROS and accelerating polymer degradation in a positive feedback manner. Furthermore, functionalization of the polymer with Doxorubicin (DOX) enhanced therapeutic efficacy by integrating chemotherapeutic and oxidative stress mechanisms. This combined approach not only intensified cytotoxic effects in tumor cells but also promoted chemoimmunotherapeutic outcomes, as elevated oxidative stress is known to facilitate immunogenic cell death (ICD) [152].

Surface conjugation of prodrugs onto liposomes has gained attention as a strategy to improve tumor-specific delivery and support combination treatment approaches. In this design, agents such as hyaluronic acid (HA)–Paclitaxel (PTX), HA–oridonin, and cholesterol–mitoxantrone conjugates are incorporated onto drug-loaded liposomes, thereby enhancing delivery efficiency and target selectivity [153]. For instance, HA–PTX prodrugs have been conjugated onto thermosensitive liposomes encapsulating the matrix metalloproteinase (MMP) inhibitor marimastat, resulting in dual-targeted nanoparticles capable of addressing both tumor cells and the tumor microenvironment [154]. This surface modification enhances drug-loading capacity and promotes selective accumulation, particularly when combined with localized thermal therapy. Similarly, HA–oridonin-based systems have been incorporated into checkpoint blockade strategies using anti-CTLA-loaded thermosensitive liposomes. This co-delivery approach has demonstrated improved immunotherapeutic outcomes by reducing tumor-induced immune suppression and enhancing lymphocyte activation. Mechanistic studies suggest that these agents act synergistically by inhibiting thrombospondin-1 (THBS1) secretion and disrupting its interaction with CD47, thereby promoting antitumor immune responses [155]. In addition, liposomes combining imaging or photothermal agents with drugs were thoroughly detailed and widely documented in relation to both the diagnosis and the treatment of cancer (Figure 11.5).

11.7.2.2 Polymeric Micelle Systems

Polymeric micelles are nanoscale assemblies formed from amphiphilic block copolymers composed of both hydrophobic and hydrophilic segments. When dispersed in aqueous media, these copolymers spontaneously organize into core–shell structures,

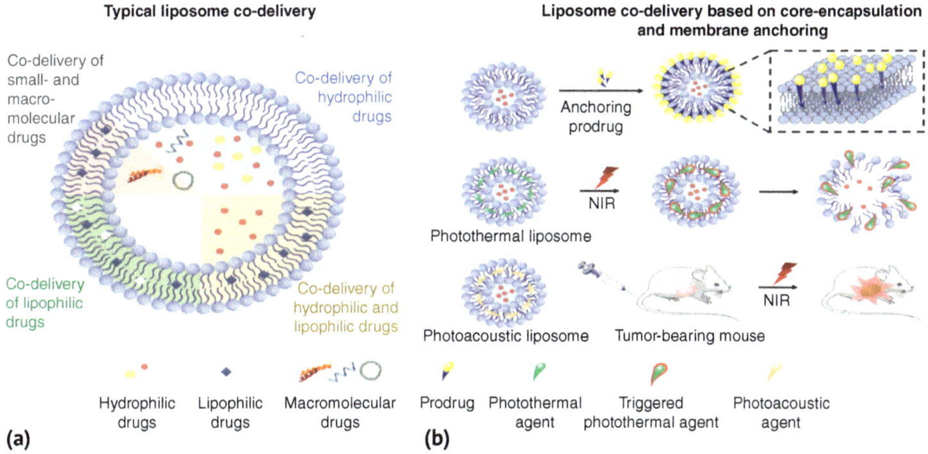

Figure 11.5 Liposome-based co-delivery systems. (a) Conventional loading of drugs within the aqueous core or lipid bilayer. (b) Combined core encapsulation and membrane anchoring, where one drug is encapsulated and others (e.g., prodrugs or photothermal agents) associate with the membrane via non-covalent interactions. *Source*: [94]/Springer Nature/CC BY 4.0.

with a hydrophobic inner core and a hydrophilic outer corona that stabilizes the system [137]. Owing to their small size and physicochemical properties, these carriers can preferentially accumulate in tumor tissues via enhanced permeation and retention (EPR), offering advantages over many conventional nanocarriers. Consequently, numerous single- and multi-drug-loaded micellar systems are being explored in preclinical studies to improve anticancer efficacy. For example, Wang et al. investigated a micellar platform for co-delivery of Paclitaxel with or without Verapamil, demonstrating enhanced cytotoxicity in multidrug-resistant tumor cells through reversal of drug resistance [156]. Similarly, Katragadda et al. developed PEG-DSPE/TPGS mixed micelles co-loaded with paclitaxel and 17-AAG, which significantly inhibited proliferation in SKOV-3 ovarian cancer cells, supporting the utility of micelles in combination chemotherapy. In another study, Shin et al. engineered PEG-b-PLA micelles capable of co-encapsulating Paclitaxel (PTX), Rapamycin (RAPA), and 17-AAG into a single "three-in-one" nanocarrier, demonstrating efficient multi-drug solubilization [157, 158]. The resulting formulation exhibited strong synergistic anticancer effects in MCF-7 and 4T1 breast cancer cell lines, highlighting its potential as a simple and effective nanomedicine platform. Using a solvent evaporation method, the three hydrophobic drugs were physically co-loaded into PEG-b-PLA micelles in different combinations. The tri-drug-loaded formulation, referred to as Triolimus markedly improved the aqueous solubility of PTX, RAPA, and 17-AAG, reaching concentrations of approximately 7.2, 3.3, and 7.3 mg/mL, respectively. In vitro evaluations indicated a more sustained release profile compared to micelles carrying a single agent. *In vivo*, intravenous administration (q4d × 3) of these micelles (particle size ~30–40 nm) at doses of 60, 30, and 60 mg/kg for PTX, RAPA, and 17-AAG, respectively, resulted in prolonged antitumor efficacy in A549 non-small cell lung cancer and MDA-MB-231 breast cancer xenograft models, while maintaining acceptable acute toxicity. Notably, this dosing exceeds the maximum tolerated dose of Taxol® (20 mg/kg, q4d × 3 in mice). Mechanistically, RAPA and 17-AAG act synergistically by inhibiting PI3K/Akt/mTOR and Ras/Raf/MEK/ERK signaling pathways, thereby overcoming feedback resistance and enhancing PTX-mediated cytotoxicity. Pharmacokinetic evaluation further showed that co-delivery increased the area under the curve (AUC) of PTX and RAPA by 1.7- and 1.6-fold, respectively, compared to single-drug micelles, while the AUC of 17-AAG remained unchanged [159, 160].

11.7.2.3 Polymeric NPs

Nanoparticles (NPs) are colloidal systems within the submicron size range, in which therapeutic agents may be dissolved, dispersed, adsorbed, or encapsulated within a structured matrix [137]. The potential applications in the future of using these vehicles as drug-delivery systems must be considered, taking into account their different properties and capabilities. These polymeric NPs can be targeting specific spots through the process of employing surface alterations through different processes or methods [161]. Misra Sahoo and coauthors carried out a study that proved the existence of synergy in the co-formulation process. The combination of DOX and curcumin is coated in a polymeric matrix of poly (D, L-lactide-co-glycolide). The use of poly(lactic-co-glycolic acid) (PLGA) NPs greatly increases the cytotoxic effects of both drugs in this process. leukemic K562 cells in vitro by circumventing the MDR phenotype [159].

Platinum and taxane combination drugs have been thoroughly investigated and researched in terms of effectively treating ovarian cancer. Various dosing schedules can be utilized strategically to optimize the overall efficacy of treatment regimens. Translating, however, the results of preclinical research, which entails sequential administration of drugs for hours to days, to clinical studies, which typically span durations of days to weeks, is a huge challenge owing to the huge disparity in timescales. Apart from this challenge, there are numerous constraints associated with the use of free drug formulations. For instance, the use of free drug formulation is greatly constrained by the high levels of systemic toxicity they induce as well as their poorly water-soluble nature. Thus, achieving a drug dose that is both effective and safe is a huge challenge that needs to be addressed in the field [162, 163]. The development of these specific combinations of drugs in the form of NPs for the treatment of ovarian cancer has the potential to surpass many of the current challenges in this field of research and treatment [163]. The encapsulation of these drugs into NPs has the potential to improve their solubility while, at the same time, limiting their extent of toxicity, a key component of enhancing the effects of treatment [164]. In addition, the controlled release of drugs from these NPs could offer a more sequential manner of drug delivery, enabling more effective control over the pharmacokinetics of the administered therapy [165, 166].

11.8 Artificial Intelligence in Nanoparticle Design for Drug Repurposing

The integration of artificial intelligence (AI) into nanoparticle design represents a significant advancement in pharmaceutical science, particularly for the optimization of drug delivery systems. Traditional characterization and optimization of NPs are usually time-consuming and resource-intensive and heavily dependent on time- and resource-consuming experimental procedures. In contrast, AI methods like ML algorithms render the process simple by enabling sophisticated molecular modeling and predictive analysis of NP behavior and properties. Such an approach is of extreme importance in designing NPs for particular therapeutic applications, as it enables rapid identification of optimum NP formulations and their respective physicochemical properties. Recent studies focus on the efficiency of AI in carrying out progress in drug reuse programs. Drug reuse is a strategy of identifying new therapeutic uses for approved drugs, a strategy that has found particular significance in the context of urgent global health crises such as the COVID-19 pandemic. For example, research focuses on the utilization of AI approaches in being capable of forecasting the likely utilization of approved drugs and thus accelerating their availability against new health risks. This risk-free strategy utilizes current security data and has the capability of dramatically shortening the calendar to bring effective treatments to market [167]. Drug discovery is a high-risk, long and costly process [9, 10]. As per a report by the Eastern Research Group (ERG), whereas it takes 10–15 years to find a new molecular entity, the success rate is a mere 2.01% [6]. The mechanism of drug repurposing reuses old drugs for the therapeutic indication of a never-considered disease. It is an experimental method of discovering the preapproved, shelved, investigational, and discontinued drugs for approved restatement for the treatment of other diseases [167].

Furthermore, the intersection of AI with bioinformatics makes it possible to make stunning advancements in optimizing drug administration through nano therapy, specifically cancer treatment. Other advancements in AI technologies unlock new ways of designing NPs. Self-learning algorithms can comb through vast amounts of data, determine the models and relationships that human scientists could not possibly find [168]. The capability of AI to learn and continuously improve based on new data enhances its capacity to revolutionize the design of NP processes. Despite the promising progress being made by AI in this field, there remain certain issues. The intricate nature of biological systems and the heterogeneity of NP interactions may complicate predictive model building processes. In addition, the integration of AI-based methodologies into standard pharmaceutical practices requires rigorous validation techniques to establish the efficacy and safety of the predicted outcomes associated with AI. The regulatory framework for the application of AI in pharmaceutical products is a new field that requires close attention, particularly in training managers who can ensure the efficacy and safety of drug delivery methods supplemented by AI. As the convergence of AI, NP technology, and drug reuse intensifies, future research directions in this field lean toward increasing focus on multidisciplinary research. Integrative strategies that incorporate computational biology, nanotechnology, and clinical research are set to provide new treatments while simplifying drug formulations. The function of ongoing innovation and interdisciplinary collaboration will be key to unlocking the full potential of AI in NP design, with significant influence on the revolutionary of pharmaceutical applications [169].

11.9 Preclinical Studies Demonstrating the Effectiveness of Nanotechnology-based Drug Repurposing and Combination Therapy

11.9.1 Introduction

Nanotechnology emerged as a transformative approach within the pharmaceutical sciences, in particular, the kingdoms of the reuse of drugs and combination therapy. By manipulating the materials at nanoscale, researchers can improve the solubility, stability, and bioavailability of existing pharmacological agents, thus facing the limits associated with traditional drug development paradigms. The need for innovative strategies is underlined by the growing occurrence of drug-resistant diseases, which place significant challenges on effective treatment regimes. The development of conventional drugs is often a long and expensive process, creating a critical question for re-proposing existing drugs that could provide therapeutic benefits against various disorders without the need for large clinical studies. The mechanisms with which nanotechnology improves the reuse of drugs and combination therapy are multifaceted. NPs can facilitate the modulation of the extracellular matrix, cellular absorption, and the intracellular release of drugs, promoting an efficient interaction with cell targets. In addition, the incorporation of elements sensitive to stimuli such as the pH or the sensitive material to temperature allows space-time control over the release of drugs, further refining the therapeutic results. Research has shown that these nanocarriers can exploit the enhanced permeability and retention (EPR) effect to preferably accumulate in tumor tissues, thus maximizing the localized delivery of drugs and mitigating systemic toxicity [170].

As the field advances, there is a growing imperative in moving from preclinical success to clinical application. While preliminary studies have promising results, it is necessary to face challenges such as the downsizing of production, the maintenance of regulatory compliance, and the guarantee of patient safety. The profiles of effectiveness and safety of drugs and combination therapies enhanced by nanotechnologies must undergo a rigorous evaluation to validate their clinical use. In addition, the interaction between nanotechnology and personalized medicine (PM) has a compelling path for future research, potentially enabling tailor-made therapeutic regimes based on the individual profiles of patients and the characteristics of the disease. Nanotechnology emerged as a transformative approach to improve drug release systems, demonstrating significant potential in the context of the reuse of drugs and combination therapy. The mechanisms below through which NPs improve the delivery of drugs can be attributed to different distinctive characteristics, such as their size, surface properties, and the ability to encapsulate therapeutic agents. The small dimensions of the NPs, typically in the interval from 1 to 100 nanometers, facilitate greater bioavailability, allowing drugs to bypass biological barriers, subsequently increasing their absorption and distribution within the target tissues [171]. This advantageous feature is particularly relevant in the reuse of drugs, in which existing drugs can have limited effectiveness due to nonoptimal bioavailability in their original formulations.

Above all, ideas taken from these preclinical models will be crucial to translate combined therapies focused on nanotechnology at the clinic. By elucidating the specific action mechanisms and elucidating their pharmacodynamic profiles thanks to a robust experimental design, researchers can better navigate the regulatory pathways necessary for the approval and implementation of these innovative therapeutic strategies. Overall, taking advantage of nanotechnology in combination with the reuse of the drug contains immense potential to solve the pressing problem of antibiotic resistance, ultimately improving the results of patients in the treatment of bacterial infections of MDR. The application of nanotechnology in drug redirect offers promising advances for the treatment of neurological disorders, particularly neurodegenerative diseases such as Alzheimer's (DA) disease and Parkinson's disease (PD). The blood brain barrier (BBB) has a major challenge in the effective delivery of therapies for these diseases. However, specialized nanoparticle platforms have emerged as innovative solutions to improve drug bioavailability and local nervous system (CNS).

Chakraborty et al. [172] conducted a study demonstrating the effectiveness of biodegradable NPs loaded with the drug-reused donepezil, traditionally used for AD. The findings indicated that the nanoscale delivery system not only increased the solubility and stability of the drug but also facilitated the delivery directed to neuronal cells, significantly improving cognitive functions in AD transgenic mouse models. The underlying mechanism involved receiver-mediated endocytosis, where NPs were designed to interact specifically with low density lipoprotein receptor 1 (LRP-1), allowing efficient BBB intersection. This targeted approach has the potential to reduce systemic side effects and improve the therapeutic indices associated with the AD.

Similarly, another study explored the use of lipid-based NPs to provide a combined curcumin and α-synuclein inhibitor therapy regime in PD models. The study stressed that curcumin, a compound known for its neuroprotective properties, usually faces challenges in bioavailability and metabolism. By encapsulating it in lipid NPs, the researchers observed substantial improvements in the pharmacological activity and bioavailability of curcumin. The results showed that this formulation in nanoscale allowed the effective crossing of the BBB while simultaneously providing sustained release of medicine. Mechanically, the approach operates by modulation of autophagic pathways, which are critical for protein homeostasis and neuronal survival [173]. A remarkable study by Pawar et al. [174] has shown that the functionalization of NPs with immuno-educational agents can improve the activation of dendritic cells, leading to a stronger anticancer immune response. In their research, the NPs loaded with a repurposed drug, metformin, were combined with a toll receptor agonist (TLR). This double-action approach has not only improved the pharmacokinetics and pharmacodynamics of metformin but has also increased the recruitment and maturation of dendritic cells in the tumor micro-room. The study reported a greater presentation of the tumor antigen and the subsequent activation of T cells, showing a synergistic effect between the reuse of the drug through nanotechnology and immunotherapy.

The implications of these results extend beyond the preclinical settings; the integration of nanotechnology with immunotherapy hosts the potential to revolutionize treatment protocols. While researchers continue to clarify the mechanisms behind this synergy, the path to clinical application becomes more and more promising. The ability to reuse existing drugs in combination with immunotherapy agents through nanotechnology offers an approach layered to oncology, potentially leading to personalized treatment regimens that maximize effectiveness by minimizing the risks associated with conventional chemotherapy and immunotherapy. This integration not only increases the therapeutic paraphernalia against different types of cancer but also places significant implications for the results of patients in future clinical studies and therapeutic paintings. Although the integration of nanotechnology in the reuse of drugs and combined therapies has shown promising results in preclinical studies, there are significant challenges in the translation of these findings to clinical practice. Factors such as biocompatibility, regulatory obstacles, and the need for extensive clinical validation have substantial obstacles that researchers must address before these innovative therapies can be applied widely in clinical environments.

11.9.2 Clinical Trials on Drug Repurposing and Combination Therapy

The convergence of oncology and nanotechnology has revolutionized the techniques used in the delivery, repurposing, and optimization of chemotherapeutic agents for therapeutic use. Several nanoformulated drugs have shown enhanced efficacy, reduced toxicity, and improved clinical utility due to innovative designs such as liposomal encapsulation, particle stabilization, and thermosensitive release. The current discussion presents a detailed overview of five significant nanodrug formulations that illustrate these revolutionary improvements.

11.9.2.1 Vyxeos (CPX-351)

Vyxeos represents a liposomal co-formulation of two well-established chemotherapeutic drugs: cytarabine and daunorubicin. The combination is held at a synergistic fixed molar ratio of 5:1 (cytarabine:daunorubicin), which was optimized to maximize anti-leukemic potency. Encapsulation in a liposomal carrier provides synchronized delivery of both drugs directly to the leukemia cell, maximizing therapeutic effect and reducing systemic toxicity. Vyxeos® was designed for the management of high-risk acute myeloid leukemia (AML), including therapy-related AML and AML with myelodysplasia-related changes. The formulation was approved by the FDA in 2017 based on a pivotal Phase III clinical trial that demonstrated superior overall survival compared with conventional chemotherapy. This system highlights the potential of nanotechnology to precisely regulate drug ratios, optimize pharmacokinetic behavior, and improve the therapeutic index of established anticancer drugs [175].

11.9.2.2 Doxil (Liposomal DOX)

Doxil, the first FDA-approved nanomedicine, is a PEGylated liposomal formulation of doxorubicin. Originally approved for the treatment of ovarian cancer and AIDS-related Kaposi's sarcoma, it has since gained attention as a promising candidate for drug repurposing owing to its distinctive pharmacokinetic and pharmacological properties. Liposomal entrapment and surface PEGylation enable Doxil to remain in the bloodstream for a longer time and selectively accumulate in tumors through the EPR effect. These characteristics minimize the infamous cardiotoxicity of free DOX, as well as its cytotoxicity to normal tissues. Beyond its initial applications, Doxil has been explored in combination regimens, such as with bortezomib against multiple myeloma and with cyclophosphamide in breast cancer, among others. Its repurposing potential is a testament to the versatility of nanocarriers for reformulating established drugs for novel oncologic applications. Abraxane (albumin-bound PTX NPs) Abraxane is a NP preparation of PTX conjugated to human albumin, intended to circumvent PTX's solubility problem and eliminate the use of toxic solvents such as Cremophor EL. Approved for breast cancer, NSCLC, and pancreatic cancer, Abraxane provides enhanced bioavailability and decreased hypersensitivity reactions. Its albumin-bound structure promotes receptor-mediated transcytosis through the gp60 receptor and SPARC (secreted protein acidic and rich in cysteine) binding in tumor tissue, resulting in increased drug concentration in tumors. Abraxane is also under investigation in clinical trials for triple-negative breast cancer and melanoma, highlighting its potential as a platform for the extension of PTX's clinical utility to a range of tumor types [176].

11.9.2.3 Onivyde (Liposomal Irinotecan)

Onivyde is a liposomal preparation of irinotecan developed initially to treat metastatic pancreatic cancer in patients with failure of gemcitabine-based treatment. Liposomal entrapment of irinotecan preserves against premature metabolism, with extended release and enhanced tumor targeting. Onivyde was approved by the FDA in 2015 based on Phase III clinical proof of enhanced survival when given together with 5-fluorouracil and leucovorin. Outside of pancreatic cancer, Onivyde is explored for activity in colorectal and gastric malignancies. Its success highlights the potential of liposomal delivery to increase the safety and efficacy of topoisomerase inhibitors in difficult TMEs [177].

11.9.2.4 ThermoDox (Thermosensitive Liposomal DOX)

ThermoDox is a preclinical prototype for a new paradigm in targeted drug delivery with thermally induced release. It is a thermosensitive liposome formulation of the chemotherapeutic DOX, formulated to release its payload at moderately elevated temperatures of approximately 42 °C, which are commonly achieved by external radiofrequency ablation (RFA). This strategy allows for highly localized drug release at the tumor site, with maximal tumoricidal effects and minimal systemic exposure. ThermoDox is being tested in Phase III trials for HCC with RFA, with variable but encouraging results. Its discovery represents an innovative advance in precision nanomedicine, combining physical stimuli to control drug release with spatial and temporal accuracy [178].

11.10 Regulatory Considerations for Nanoformulations in Repurposed Drugs and Combination Therapies

Regulatory obstacles also represent a significant barrier for the clinical translation of NP-based therapies. The current regulatory frameworks, such as those established by the United States Food and Medicines Administration (FDA) and the European Medicines Agency (EMA), are mainly designed for conventional pharmaceutical products and may not adequately address the unique characteristics and complexities associated with nanomedicine [128]. As a result, there is an urgent need for the development of specific guidelines and regulatory routes adapted to the evaluation of nanotechnology applications in medicine. The lack of standardized test methods further complicates the evaluation of safety and efficiency, which leads to delays when bringing these innovative therapies to the clinic. The clinical validation of nanotechnology in the reuse of drugs and combined therapy implies extensive evidence to ensure that the promising results observed in preclinical models can replicate in human subjects. The discrepancies that are often observed between in vitro results and clinical results require rigorous clinical trial designs that may precisely reflect the complexities of the real world [128]. This includes the need to understand several pharmacokinetic and pharmacodynamic parameters associated with nanocarriers. In addition, the optimization of parameters such as particle size, surface load, and functionalization is crucial to improve the effectiveness of drug administration while minimizing possible adverse effects.

When addressing these limitations, researchers and doctors must prioritize a multidisciplinary understanding of the broader implications of nanotechnology in therapeutic environments. Future studies should aim to elucidate the mechanisms of action of the NPs in the reuse of drugs and combined therapy, since this knowledge will close the gap between the promising preclinical results and the realities of the clinical application.

Only through addressing these challenges can nanotechnology's potential be fully realized in the field of medicine. The deployment of nanotechnology in the fields of the reuse of drugs and combined therapy has a potential transformer which deserves more in-depth exploration in clinical environments. As evidenced by recent preclinical studies, the use of nanocarriers has facilitated improved bioavailability, targeted delivery, and improved pharmacokinetics of existing therapeutic agents, thus stressing the importance of this technology to approach complex diseases such as cancer, neurodegenerative and disorders, and infectious diseases [179].

Given the promising results obtained so far, it is crucial to describe several key areas that require future research efforts. A major involvement is the need for comprehensive clinical trials that rigorously evaluate the safety and efficiency of nanotechnology—compatible drug formulations. Current preclinical research has mainly focused on mechanistic ideas and therapeutic results, but the translation of these results towards human systems remains a great challenge. To fill this gap, multicenter clinical trials designed to assess pharmacodynamics, pharmacokinetics, and long-term safety profiles of reused drugs issued via nanocarriers are essential. These studies must also adapt to the populations of various patients, taking into account variations in genetic history, pathological states, and environmental factors [135]. In addition, the development of regulatory frameworks adapted to the unique challenges posed by nanotechnology in pharmaceutical products is imperative. As nanomédicins are progressing towards clinical application, the definition of standardized protocols for their evaluation will be essential to guarantee patient safety and efficiency. This includes the establishment of guidelines for quality control, manufacturing processes, and regulatory approval processes specific to nanocarrier systems [179].

Overall, the intersection of nanotechnology with therapies of reuse and combination of drugs announces a new border in clinical pharmacology; however, significant research and development efforts are necessary to translate preclinical success into clinical realities. The enlargement of the emphasis placed on mechanistic investigations, the optimization of administration strategies, the evaluation of synergistic processing approaches, and the formulation of robust regulatory frameworks will be essential stages to exploit the full potential of nanotechnology in future therapeutic applications.

11.11 Challenges and Limitations

Nanotechnology enables precise manipulation of materials at the nanoscale (1–100 nm), revolutionizing targeted therapies, imaging, and biomaterials. However, its medical applications raise significant safety concerns due to the unique physicochemical properties of nanomaterials, which may lead to unforeseen toxicity, bioaccumulation, and complex interactions with biological systems. NPs exhibit enhanced therapeutic efficacy but pose risks related to biocompatibility, immune activation, and long-term effects. Their small size, high surface area-to-volume ratio, and reactivity contribute to these challenges. Diverse nanomaterials—such as liposomes, polymeric NPs, carbon-based structures, and metal-based NPs—vary in cytotoxicity and biodistribution, necessitating rigorous safety assessments.

11.11.1 Safety Concerns–Nanotoxicology

Nanotoxicology evaluates the adverse effects of engineered nanomaterials on health and the environment, requiring a multidisciplinary approach (toxicology, chemistry, medicine, etc.). Despite their benefits, NPs raise concerns akin to past carcinogens like asbestos. Some NPs share fibrous structures with asbestos, prompting questions about their long-term safety. Toxicity depends on factors like exposure time, dose, particle size, shape, surface charge, and aggregation state.

11.11.1.1 Key Toxicity Factors

The toxicity of NPs is influenced by multiple factors, including their size, shape, surface charge, composition, and functionalization. These factors determine how NPs interact with biological systems at the cellular and molecular levels, affecting biodistribution, cellular uptake, oxidative stress generation, and inflammatory responses (Figure 11.1). Additionally, exposure routes and dosage play a crucial role in their toxicity profile. Table 11.1 shows some of the key factors:

11.11.1.2 Toxicity Mechanisms

NPs interact with biological systems in complex ways, often leading to unintended toxic effects. Their small size allows them to penetrate cells, cross biological barriers, and accumulate in tissues, which can disrupt normal cellular functions. The toxicity mechanisms of NPs are largely dependent on their physicochemical properties, including size, shape, surface charge, and composition. Several key mechanisms contribute to NP-induced toxicity, including oxidative stress, inflammation, genotoxicity, protein corona formation, and membrane disruption. These mechanisms can lead to cellular damage, impaired organ

Table 11.1 Key factors driving NPs' toxicity profiles.

Factor	Description
Size	Smaller NPs have larger surface areas, increasing reactivity and toxicity. Size also affects lung deposition and clearance rates
Surface chemistry and charge	Surface properties (e.g., coatings, charge) influence NP interactions with biological systems. Positive or negative charges can alter toxicity, as shown by electron microscopy studies
Dosage	Dose-dependent toxicity is well-documented, but mass concentration alone may not fully explain NP-tissue interactions
Chemical composition	The core material determines dissolution, redox activity, and toxicity. For example, **nano-ZnO** is highly toxic due to Zn^{2+} release, and **nano-TiO_2** and **nano-SiO_2** are less toxic but vary according to tissue response (e.g., embryotoxicity) [180]. **Gold NPs (Au NPs)** are biocompatible, unlike soluble metal NPs (e.g., CuO, Ag)
Shape	Shape impacts biodistribution and cellular uptake. **Spherical NPs** are internalized faster than rods or disks [181]. **Sharp-edged NPs** (e.g., prismatic Au) cause more cell damage [182]

Table 11.2 Core mechanism regulating the toxicity of NPs.

Mechanism	Description
Aggregation	NPs aggregate in biological environments, altering their apparent size and toxicity. Aggregates may reduce cellular uptake but can persist in organs like the liver, causing long-term damage [183]
Biodegradation	NPs degrade in harsh environments (e.g., lysosomes, GIT): **Toxic Degradation**: Ag NPs release Ag^+, causing lysosomal damage and apoptosis [184]. **Benign Degradation**: MoS_2 breaks into molybdate, used in metabolic pathways. **Biopersistence**: Nondegradable NPs (e.g., CNTs) accumulate, causing chronic inflammation [185]

function, and systemic toxicity. Additionally, NPs may interact with biomolecules, altering signaling pathways and immune responses (Table 11.2).

11.11.2 Exposure Routes and ADME

Inhalation is the most characteristic and perhaps the most famous among all the ways of nanomaterial exposure. It can also penetrate the human body through skin, ingestion, and using needles and syringes. NPs are believed to be involved in the development of some diseases through affecting the lungs and the other systems through diverse pathogenic processes. With respect to aerosol penetration to the lower respiratory tract and its deposition, particles with an aerodynamic diameter less than 0.1 μm may reach the distal airways with respiratory units. NPs inhaled reach the respiratory epithelium, and the NPs then pass through the pores of the alveoli-capillary membrane and reach the interstitium and then the systemic circulation through blood and lymphatic circulation. According to some studies on rats, it has been proved experimentally that, for instance, when NPs are instilled in the trachea of mice, they enter the systemic circulation in this manner. In experiments aimed at exploring the toxic effects of NPs on human health, NPs were used in various ways—through inhalation, intratracheal, intravenous and intraperitoneal ones, etc., and in different doses, while parameters such as the capability of NPs to enter the systemic circulation of living organisms, accumulation in the tissues, the inflammation that occurs in the tissues, other immunological reactions, and excretion of NPs from the body are also detected. The specific ultra-fine carbon particles with size less than 100 nm have been observed in the study to freely enter the systemic circulation within a period of 10 minutes and sustain their concentration in systemic circulation for an hour in five healthy people. In a study of mice, tissue distribution of magnetoelectric NPs of various sizes administered intravenously was observed through electron microscopy, and in about one week from the time of administration, NPs achieved maximum deposition in the lung, but the elimination rate of large particles of 600 nm in the lung is slower than that of small particles.

NPs that are short in size and of spiral shape and that are introduced into the body are engulfed in tissues by macrophages. Nevertheless, nanotubes often get to the pleural area like asbestos fibers and are accumulated at the pores of that area. These fibers are not phagocytable, and proinflammatory, genotoxic, and mitogenic agents are produced by mesothelial cells. Thus, there is the development of an inflammation, and damage process starts. This, on one side, causes the endothelial dysfunction in the lungs and stimulates the pulmonary reflexes; on the other hand, it activates the platelets and raises the thrombotic activity. Furthermore, inflammation in the vascular region leads to the dysfunction of vascular endothelial cells, and cardiovascular diseases include abnormal heartbeats and rhythms, the formation of the atherosclerotic plaque, and rupture. NPs therefore elicit both innate and adaptive immunity, as well as inflammation. All the natural immunity cells involved in macrophage/monocyte, neutrophil, dendritic, and natural killer cells are released along with the proinflammatory cytokines, lipid mediators, and free radicals that cause neutrophilic or eosinophilic lung inflammation for the acquired immunity involving dendritic cells and lymphocytes. It is also noteworthy that immunomodulatory impacts of NPs might also depend on the size, structure, electric charge, and degree of aggregation.

11.11.3 Entry Routes of NPs into the Human Body

Mortimer states that it is not possible to avoid contact with nanomaterials since the human being is a social creature. There is much literature about the nanomaterials that can cause harm to the various organs of the human body. Nanomaterials primarily penetrate the body through the respiratory system and are in close contact with the lungs. This route too is frequently used by the nanomaterials by direct contact and through the ingestion of food substances. Moreover, they said that implants and injections are the ways by which nanomaterials can easily penetrate the body system.

11.11.3.1 Inhalation Exposure

It is the size of the NPs; its ability to fight gravity and its distribution will ultimately decide the zone it is found in the respiratory tract. Particles inhaled through the respiratory system were cleared from the respiratory system by the mucociliary layer or engulfed by macrophages, or they settled in the lungs and transported to other parts of the body through the bloodstream. It was reported that about 33% of the inhaled NPs can be dealt by the defense mechanism of the respiratory tract [186]. There are studies on animals that showed that CNTs cause lung fibrosis, inflammation, and granuloma, and these toxic effects affect the cardiovascular system. The impact of inhaled NPs deposited in different organs of the body, including the brain, and the risk assessment of association with prostate cancer have also been assessed in many studies.

11.11.3.2 Dermal Exposure

The three approaches to improving penetration of NPs into skin are the properties of NPs, properties of equipment dispersing the penetrating molecule, and the zone of infiltration and skin condition. Cosmetic cream, lotion, and toothpaste are some of the NP-based instruments that are frequently employed for exposure on skin. NPs tend to penetrate the horny layer and the dermis region. It is also said that some of these NPs entering the skin can pass through to the blood veins.

11.11.3.3 Ingestion

NPs are directly or indirectly ingested from the gastrointestinal tract (GIT) or through secondary ingestion from inhaled particles. It should also be noted that NPs with high possibility for inadvertent ingestion include metal compounds and pesticides. This can more often be left unmentioned, as it is believed to happen only deliberately or through gross carelessness. This is also contributed to by low uptake of NPs from the intestines and metabolism in the liver.

11.11.4 Toxic Effects of NPs on Systems

Because NPs can penetrate the body in three principal routes, it is established by experimental investigations that they produce toxic impacts within various systems. This section elaborates on the harms of nanomaterial in the systems primarily through animal testing.

11.11.4.1 Circulatory System

Studies have determined cardiac oxidative stress and DNA RAPs damage after using intravenous iron oxide NPs in mice [187]. When Ni NPs were administered intravenously, cardiac toxicity-arrhythmia was found and toxic effects on the organs, including liver, spleen, and lung.

11.11.4.2 Digestive System

Numerous studies have stated that the 100 ppm zirconia oxide NP exerts toxicity to the liver tissue in rats. Moreover, it has been revealed that naturally generated iron oxide NPs are toxic to the liver in mice.

11.11.4.3 Endocrine System

Oral form of iron oxide NPs affects the hormonal imbalances of the thyroid glands in rats.

11.11.4.4 Immune System

According to numerous studies, the number of white blood cells boosted critically due to the presence of TiO_2 NPs in mice. Moreover, it has been reported that the iron oxide NPs lead to an increase in the number of WBCs, and out of all immunologically affected organs, the greatest impact is on the liver and spleen.

11.11.4.5 Respiratory System

The five different metal oxide ENPs, cobalt oxide, nickel oxide, and titanium oxide, exhibit pulmonary complications determined by oropharyngeal aspiration in rats. Likewise, it has also been concluded that iron oxide NPs have lung toxicity in rats.

11.11.4.6 Urinary System

The nanotoxicity of zinc oxide, iron oxide, and copper in different monkeys, pigs, and bovine kidney cells has been determined in many studies. Furthermore, it has been studied that the intraperitoneal administration of TiO_2 NPs precipitates the degeneration in kidneys in rats.

11.11.4.7 Nervous System

Skin irritation and irritation effects were also conducted in animal ears and eyes regarding vision and hearing toxicity and usually proved very low toxicity or no toxicity at all.

11.12 Scale Up and Manufacturing Challenges

Even today, nanotechnology is in a relatively early stage of commercialization, but the frequency of technology enabling continues to rise, for which there is considerable formal government funding for nanotechnology. Commercialization as the next step to technology development is the process through which new technologies are taken through a wheels and gears process with the help of businessmen, marketers, and a host of other people with commercial and economic backgrounds to transform the technological products into marketable goods or services. This entails a process of following a sequence from idea generation and technological advancement to its commercial disposition, nurturing and transforming ideas to technology, construction of a model to determine the feasibility of the idea, bringing out a new procedure or enhancing the general procedures, and finally, the supply of proposed commodities to the market, marketing, and development of structures for the easy provision of supply and demand. As a result of this great concern and disapproval, the nanotech-based pharmaceutical firms have many financial characteristics and options to turn to, while the start-up biotech company founding members are using their own personal effort and capital in it. Today's world economy is not friendly in terms of nano start-ups, and nano-firms are experiencing much difficulty in acquiring seed capital. Concerning the risks connected with funding companies, the investors do not invest large sums of money in such companies until they meet sure-shot technologies, first movers ready with their products, proprietary protection, new markets in view, high profit opportunities, and strong leadership teams.

11.12.1 Challenges to Product Development and Commercialization

The current generation is generally a "no or low controversy generation," which tends to avoid or hardly express a negative attitude towards the use of new technology. As a result, such a general attitude is quite positive and receptive to the newer technologies, for instance, nanotechnology. It has been seen that the commercially viable nanotech-based products appear to have a number of hurdles that they have to cross in order to achieve this. A good product together with its precise formulation that does not contain toxic chemicals must be among the basic requirements for any product to be marketed. Many of the proposed nanotech-based products, however, do not meet any one or both of the requirements, and as such, the product does not make it to the market.

11.12.2 Safety

According to the guidance document of the US FDA, if a test product gives plasma concentrations much higher than the reference product, which is a frequent feature with nanobased systems, then the issue of concern is not therapeutic failure with the test product but the safety. Due to the nature of these products being of nano-size, they will always face problems of bioaccumulation, super-optimal bioactivity, biocompatibility, and metabolism and elimination from the body. In addition to this, there are some researches that indicate that nanomaterials are not safe for human consumption. This has been recently noticed by some Chinese researchers when they noticed that during animal testing, the effect of bioaccumulation of nano-silver disrupted replication of DNA molecules and rerouted the molecular networks that are capable of causing mutations.

As far as the cosmetic industry is concerned, the presence of a large number of nanocosmetic products in the markets worldwide can be attributed to the fact that FDA oversight for cosmetics neither includes premarket evaluation nor requires reporting on post-market surveillance, such that most adverse events may go unnoticed unless directly reported to the FDA. While a similar oversight cannot be granted to other nanoproducts, especially those catering to life-threatening diseases like cancers, as these products would by default require more stringent guidelines. The intensity of these safety threats increases manifold when NPs are applied to disease therapeutics in oral and parenteral dosage forms rather than topical application usually followed for cosmetic products. Liposomes, solid lipid nanoparticles (SLNs), as well as polymeric NPs are reported to be the most popular DDSs in therapeutics. However, there is a good number of works within the last 5–10 years that aim at proving the applicability and reliability of such systems. Challenges relating to the usage of organic solvents as well as nonbiodegradable polymers for the synthesis of polymeric NPs have received much concern with the development of more appropriate techniques for preparation using safe solvents and environmentally friendly polymers. The lipids having attributes for being employed in the formulation of SLNs are also to be biologically safe. Compared with Abelcet (Amphotericin B lipid complex), the adverse effects and nephrotoxicity of Ambisome (liposomal Amphotericin B) are even lesser in side effects. Many studies assessed the cytotoxicity of risperidone incorporating SLNs prepared with glyceryl monostearate at 40–55% using Caco-2 cells through an MTT assay. Apart from the cell death marker tested above, the MTT assay measures the mitochondrial function of viable cells in a cell culture because it can identify damaged cells prior to the point when they lose their structural integrity. Therefore, it can be concluded that most of the evaluated for mutations had little or no toxicology impact on Caco-2 cells, thus being safe in the GIT. Further research on nanotherapeutics, however, needs to be undertaken, and this should focus on the challenges of safety and toxicity of the products. The experts in this industry do agree that the question of toxicity is set to be answered soon enough, and, at that time, the investors will be far more inclined. However, the therapeutic NPs should undergo practical safety assessment to determine the safety of utilizing the formulation for both narrow and long-term usage. This consideration cannot be overlooked, as it calls for a bearing action on the part of the regulatory authorities.

11.12.3 Potential Barriers to Successful Commercialization of Nanotechnology

Some of the possible challenges that can be encountered during the commercialization of nanotechnology include the following:

- Failure to consider or failing to incorporate the features that will help in creating a model product
- Insert—Capability of performing an adequate safety test or an inadequate safety test.
- Difficulty for large-scale manufacturing or fabrication
- Lack of funding

- Adverse public perception of nanotechnology
- Due to the concept known as "niche failure or mismatch," faced competition from technologies having a lower content for the application of interest.

11.12.3.1 Inability to Adapt or Design the Technology to Produce a Practical Protoype

This is not at all surprising when in the developmental stage of a project there could be failures when a new technology is proposed in order to solve any real-life problem. This is quite expected, as knowledge from shortcomings and shortcomings from others is part of the discovery and presents interest and opportunity. Sometimes though, the new technology application never fully gets to the practical prototype, or it stays so long in the practical prototype stage that a failure to move to product development occurs, and the project is left in determining the feasibility of discovery for so long that interest fades, funding is used up, or others can address the critical problems more efficiently and/or effectively; thus, an opportunity is missed.

In commercialization decisions, this can be mitigated by allowing adequate and suitable planning and management of the project during the initial development phase. Leadership of the project management team for discovery should be someone respected by those doing the innovation, such as a scientist who understands the fluctuations in the discovery and prototype development processes. Having worked earlier within the project, they are capable of leading the innovative process with a dynamic, highly effective team. At this stage of the project, marketing-led leadership often brings problems such as misunderstanding and exclusion of participants who put a stop to the innovative process and significantly reduce productivity.

An example in the process of developing a medical nanoproduct where it may be used is, for instance, when the nanodevices are used to perform a particular task, such as the controlled release of a drug substance such as a hormone, for instance. The advancement of the tactic could move along several tracks. One of the tracks includes a biodegradable delivery technique, implantation of a nanodevice or another modified technique. Being able to control the release of a hormone such as testosterone or estrogen, as examples, to better simulate natural release profiles and cycles is desirable, but if the use of nanotechnology to provide a solution is incapable of producing a product that is substantially more effective, cost-efficient, or safe than lower-tech controlled release technologies, a failure to provide a prototype suitable for transition to development will result until either the opportunity is finally realized or it must be abandoned to other projects. The ability to deliver human insulin is one such project that will either be accomplished in a few years' time or will remain as a project on the shelf until a commercial prototype of the product is developed.

The best approach when working on such a project would be to refrain from inhibiting the creativity and productivity of the discovery team and do anything that is possible and reasonable to allow the team to reach a solution with regard to the commercialization aspects and their timelines.

11.12.3.2 Ability to Perform Adequate Safety-testing or Failure of Safety-testing

Yet another challenge is the health effects of NPs and ENMs in terms of toxicity. Whether these existing regulatory frameworks regarding medical devices and pharmaceuticals and those used for making environmental impact assessments will suffice or not to cater to any special characteristics that nanomaterials may possess or whether there will be a need for new strategies and solutions is yet to be answered. At present, it is up to the various organizations such as the US Food and Drug Administration (FDA) and the Environmental Protection Agency to consider and approve new products based on nanotechnology or those containing nanomaterials on a product-by-product basis and without general guidelines on how to assess the safety of both nanotechnology products and products that contain nanomaterials. It is only logical to wonder whether NPs can be dangerous when taken intravenously or orally, as other properties of materials transform with their dispersion down to very minuscule sizes. In fact, some nanoscale particulate matter (asbestos fibers and some other emissions from automobile exhaust) can cause negative effects related to their size. However, many nanoscaled materials have been around and in use for many years appended to the human body (e.g., nanoparticulate albumin in nuclear medicine and dispersions of nanoparticulates of natural origin, for instance, milk). This has been proven: any attempt to generalize the toxicity or the reverse, the nontoxic nature of any class of nanomaterials, is unwarranted because of the high possibility of exposure through so many routes and types of nanomaterials available in the market or being used in various industries. Toxicology studies just simply have to be well designed and conducted on any new nanoscaled material so that more concrete information is available for deregulation processes to permit marketing of the items long before their production goes into volume. There is, hence, a clear need for improved toxicology testing on nanotechnology products that are in the market for medical and health use.

11.12.3.3 Difficulty for Large-scale Manufacturing or Fabrication

There are financial and strategic barriers, where some of them are expressed directly. Some of these are lack of critical mass injection into research and development, absence of interdisciplinary facilities and skills, and the possibility of a shortage of skilled employees. The second is relatively less direct; it implies that there is massive acceptance required in nanotechnology, but opposition would be detrimental and fatal. This is, however, slowly changing as the manufacturing technology advances and comes with the movement of manufacturing to an extent where large-scale manufacturing is more economical and efficient.

11.12.3.4 Lack of Funding

The lack of project or bad spending planning that is caused by inadequate initial capital or inability to control spending rate is also one of the key hurdles in the large-scale manufacturing of nanomaterials.

11.12.3.5 Adverse Public Perception of Nanotechnology

With all emerging technologies that promise that the public becomes aware of, there are perceived issues regarding the potential of the technology to offer a solution in a proximate time that may act against a project when the perception turns into a no-go factor or when the technology contains fatal issues no matter whether their inception is either safety or efficacy. This leads to some kind of public relations activity on the part of the innovator or loss of investment due to general condoning of the application for some reasons that may not be anchored in scientific facts but in assumption.

As in the case of the commercialization of any new technology, including nanotechnology for its application in the medical field, the groundbreaking development of a useful product is normally preceded by some amount of media hype that contains a statement to that effect from a known figure in the medical field. Some of the companies that are good at this process normally exercise a very cautious approach to the way they disseminate information to the public on this emerging technology to ensure it and do what is within their power to keep the public expectancy realistic and to refrain from releasing information to the media that will probably shock the public. It is required to develop public enlightenment campaigns targeting the business community, universities, and the media, among others, about the implications and opportunities expected from nanotechnology. Thus, if a country has not developed public awareness on nanotechnology, the country loses out on the economic possibilities of nanotechnology.

11.12.3.6 Competition with Lower-content Technologies Due to "Niche Failure or Mismatch" for the Application of Interest

It is crucial to make sure that all the lower technologies have their solutions looked for adequately to prove the value and backing of the high-content and possibly higher-price-tag imaginative solutions. If not done, then this gives rise to "niche failure and mismatch," which ultimately leads to commercialization failure of the new, high-content technology. Two important and closely related matters that arise when carrying out analytical characterization and quality control are the following: Insufficient data on methods for in-process control as well as quality control (QC) and issues of validating such methods are major challenges to both the developers and regulators. Classification methods used for nanomedicines are size, size distribution, morphology, surface charge and functionality, particle interactions and behavior, porosity, drug loading, solubility, pH, viscosity, toxicity, and stability. Therefore, the characterization of nanomedicines differs and involves dynamic light scattering, electron microscopy, fluorescence spectroscopy, etc., while most of the industries possess normal characterization equipment. It is expensive to analyze nanomedicines; besides, the analysis and data interpretation require engaging the services of experts. This goes a long way in increasing the cost of producing and testing the nanomedicines in the market. One significant issue is to select an ideal method for sterilization that does not affect the physicochemical properties of these nanomedicines, which are very important for their action. There is no robust method that can be used to evaluate the nanomedicine stability in vivo and particularly under good manufacturing practice conditions, which remains one of the major challenges toward the translation of most nanomedicines.

11.12.3.7 Pharmacological and Safety Challenges

As it stands, nanomedicine clinical translation lacks suitable and/or credible screening platforms that would be useful in determining the efficacy and potential toxicity of the compounds involved with a good correlation to the in vivo models. This is particularly more relevant in case of the preclinical screening techniques and models of a realistic value mirroring the human clinical data. Minor variation in the physicochemical characteristics of nanomedicine caused by variation in the manufacturing process may lead to differences in pharmacokinetics, biodistribution, drug release profile, and toxicity/safety profile. Due to the size of the nanocarriers, the dealing of nanomedicines is a matter of absorption, distribution metabolism, and elimination. In vivo and in vitro are the two approaches that tend to assess the behavior of the product in biological systems; however, the data obtainable from tests carried out on the current preclinical models are sometimes not translatable to humans. Most in vitro tests such as drug release, cell uptake, cytotoxicity, and immunological activities are typically conducted using the cells grown on a two-dimensional culture system [188]. However, the transport of the nanoformulations with the cells and tissues depends on interactions within the physicochemical properties of the preparations as well. Hence, utilization of 3D cell systems may lead to better results [189].

Currently, the utility of the preclinical in vivo small-molecule drugs as a measure to evaluate the efficacy, toxicity, and absorption, distribution, metabolism, and excretion (ADME) of the nanomedicines is sufficient when performed on at least two animal models, including multiple doses over long-term treatment. However, immunotoxicity and other preclinical results could not exhibit the safety profile of nanomedicines in the human body [190]. The other concern used in the preparation of nanomedicines is environmental concern during the preparation process of the nanomedicines. The nanosized powder, if not well handled, poses a danger in that it can penetrate into the skin or dermal exposure and may also have pulmonary toxicities.

11.12.3.8 Regulatory Challenges

Due to the research and development barriers, there are no explicit rules in the standard regulatory guidelines for nanomedicine products' preclinical development and characterization at the biophysiological level, limiting the nanomedication's potential usage in clinical practice. Challenges in the regulatory pathways for translating nanomedicines are mainly because of the product characteristics and the processes of small-molecule drugs, which are often borrowed [191]. Therefore, there is a significant deficit fraction that needs to be addressed concerning checklists that encompass characterization, pharmacological, and toxicological testing of nanomedicines. Another question that should be posed regarding nanomedicine translation is whether or not the product would pass the scientifically socially acceptable definition of what can be termed nanomedicine. Currently, when considering whether an FDA-regulated product involves the application of nanotechnology, the FDA parameters are the following: Whether a material or end product is engineered to have at least one external dimension or an internal or surface structure in the nanoscale range (~1–100 nm); and whether a material or end product is engineered to exhibit properties or phenomena, including physical or chemical properties or biological effects that are attributable to its dimension(s), even if these dimensions fall outside the nanoscale range, up to 1,000 nm [14]. Due to the ever-growing and complex nature of nanomedicine, both the FDA and EMA, along with other such regulatory authorities, recommend that the regulation and approval of nanomedicine products be taken on a case-by-case basis. As a result, the strategy to be employed shall be versatile, cyclic, positivistic, cooperative, and product-centered in order to foster new development [192].

11.12.3.9 Regulatory Hurdles and Standardization Issues

There is a growing need to encourage research and development of nanomedicines; however, there are important challenges that have been posed to the industry and regulatory agencies. Indeed, there has been an overall dearth of standard procedures to describe these nanomedicines at, for instance, molecular, bioactive, and physiologic levels, which in several instances might have been accountable for the nanomedicines' failure in the last stages of clinical studies. Therefore, the traditional regulatory environment for those revolutionary therapeutic nanomedicines has been raised by main matters, which is an opportunity to outline the further direction of nanomedicines' creation [193].

11.12.3.10 Regulatory Perspective on the Development of Nanomedicines

Despite a growing number of approved nanomedicines for biomedical uses in the past few decades, general guidelines for preclinical studies and characterization of these products remain grossly underdeveloped, which limits the progress of nanomedicines further in clinical practice. It is still to be noted that there have been still no global regulatory trends established, although there have been several attempts conducted already. An increase in the cooperation between the regulatory agencies is still required, but a lot of progress was made within the last five years. Otherwise, the strategies that have been applied for the development of the "conventional" medicinal products have been often employed to assess the safety/toxicity and compatibility of nanomedicines. It is the active principal ingredient (API) of nanomedicines that brings specifications that must be assessed within the framework of regulation. New chemical entities such as proteins, peptides, or antibodies have to meet the requirements of the regulations for biological medicinal products and new chemical entities (NCEs).

In general, one of the hurdles underlying the regulation of nanomedicines is clearly related to their particular characteristics. Following the large scope of evidence assembled for liposomes, polymer therapeutics, and other polymeric systems (including polymeric micelles) and considering the regulatory issues related to their design and development, the clinical use of these sophisticated and complex nanoproducts is strongly dependent on extensive assessment, characterization, and understanding of crucial properties. In fact, their properties are not hard to change; they can be changed by smooth variation of raw material and slight changes of the production process also. However, these changes may lead to minimal changes in the structure aspect, while the biological properties and biodistribution patterns may be slightly changed. Moreover, as molecules and targets, trackers and imaging agents, drugs and prodrugs, and entities are coupled to nanomedicine on a regular basis, new adequate quality control tests and stable methods have to be defined in order to evaluate and characterize not only physical parameters like size and its variation, shape, and charge but also their performance, including release of the drug, metabolic profile and protein binding, and specificity of the cellular uptake [194]. These methods would further allow relating the modifications with overall physicochemical alterations and consequent effects on biological properties, biocompatibility, and therapeutic effect.

It has been realized that another limitation for the further development and clinical application of these nanomedicines is the manufactural engineering. Several associated issues concerning their manufacturing are currently impacting pharmaceutical innovation at production facilities, likely compromising its scale-up, for the most part because the new materials present a very broad range of properties. There is a need to point out and control the significant points of each manufactured process. This can assist in achieving a further synchronized growth of more complicated medications and in lessening the effects of the hurdles when creating them. EMA has formed a working group to address more focused issues, including quality, safety, and efficacy aspects of nanomedicines. This group has produced documents that are the "orientation documents," referring to the aspects intended for consideration by the applicants while developing nanoproducts. Although the lack of specific protocols for nanomedicines can be observed, since 2009–2010, EMA, the FDA, and PDMA/MHLW have already started to cooperate to establish comparable approaches to nanomedicine development. Meanwhile, major pharmaceutical companies have also increased their interest in "proof of concept" and clinical development of these complex systems. Overall, all these aspects will encourage more transparent processes of establishing the safety and efficacy of the new nanotechnology-based drugs. Also, before these nanomedicines are commercialized, pharmaco-economic studies should be conducted with a view of establishing social and economic returns on these new and complex nanotherapeutic products when compared to existing conventional treatments. It shall be necessary to review some variables such as increased in QALYs, that is, quality-adjusted life expectancy years, or costs of consecutive hospitalizations.

11.12.3.11 Nanomedicines in Market

The pharmaceutical regulatory environment, the healthcare strategies, the demographics, and the wide economic environment affect the competitive nanomedicine-related drug market approval.

As for nanomedical products existing in the market today, about 40% of them are protein–polymer conjugates and liposomal formulations [193]. One of them is in the treatment of tumor diseases as one of the major principal therapeutic objectives of these nanomedicines. But there has been much research toward the development of new-generation nanomedicines with specific targeting with the help of surface ligands and available receptors on the outer membrane of targeted cells. Furthermore, new nanomedicines face challenges in entering the market; the cost of clinical development and regulatory approval is expensive, but returns may be marginal through niche indications [195].

11.12.3.12 Regulatory Development for "Next-generation" of Nanomedicines

Since ICH began in the early 90s, the regulatory environment around the development of nanomedicines has been under increased challenge. The anticipated evolution calls for more and more proof of greater clinical efficacy from new innovative approaches as well as an increased pressure for pharmacoeconomic evaluation. The notion of non-inferiority has been used in the past to justify nephrological treatments in many clinical settings, but due to the evolution of more clinical standards, most such applications are continually reviewed for updates. In addition, the new trends in healthcare management and the introduction of health technologies assessment (HTA) involve the integration of other related treatments apart from the traditional chemotherapy treatment of cancer.

In several clinical areas, there have been great strides of biopharmaceuticals that also show their inherent capability of surpassing some of the nanotechnology-based approaches. We need to be aware that most of the currently developed systems refer to molecules that were new in the late 70s/early 80s but not for today, such as DOX, alike many others. Another important problem of the current regulatory discourse is the problem of nanosimilars, which are made of new generic drugs and nanocarriers as innovative excipients. Similarly, controversy surrounding pharmaceutical options within NBCDs led to putting forward a number of problems in interaction with special formulations (iron oxide NPs, liposomes, polymeric micelles). In the year 2009, the ad hoc expert group in nanomedicines initiated a number of activities, including the first nanomedicines conference that involved regulators and other stakeholders of health from the United States and Japan, among others. At the same time, different issues have been triggered from the European Union debate on nanomaterials' classification related to the "similarity" discussion, namely from a Top Institute Pharma (TI Pharma)-based international expert group in the Netherlands (NBCDs expert group) composed of a number of experts from academia and industry. Recent advancements in the area of nanomedicine research shifted the focus to the fact that although it has become an already well-defined area of clinical care, it now poses some questions for which NCEs and biologicals were answerable. In fact, the progress made in clinical practice in relation to nanomedicines in the last 30 years will indeed permit further improvement in the next phase experiencing a continuous growth of both the fundamental research and the enormous amount of clinical data being gathered across different technologies and therapies. New advancements are enabling the emergence of both targeted medicine and combined therapy as the sources of innovations in practice.

11.13 Stability and Storage of Nanocarriers

Nanocarriers, such as liposomes, polymeric NPs, SLNs, dendrimers, micelles, and inorganic NPs, are widely used in drug delivery, diagnostics, and therapeutics. However, their stability and storage conditions significantly impact their efficacy, safety, and shelf life. Maintaining the integrity of nanocarriers during storage is crucial to prevent aggregation, degradation, or premature drug release. This makes their behavior influenced by physical stability and chemical and biological stabilization, all of which need certain methods to enhance equipment usability.

11.13.1 Physical Stability of Nanocarriers

Stability of physical state is one of the significant problems of nanocarrier storage, which results in aggregation, sedimentation, or fusion of particles. NPs have a tendency to aggregate by van der Waals forces, hydrophobic interactions, or through variation in surface charge. This aggregation can change the size distribution of the particles, thus affecting its further efficiency in drug delivery. However, regarding lipid-based nanocarriers such as liposomes and SLNs, it is possible that particles can fuse, leading to the disruption of the carriers and release of the drug. Another problem, sedimentation or creaming, can be linked to the case when the density of the used NPs is drastically different from the density of the dispersion medium. To address these concerns, agents that help to stabilize colloidal structures are added, for instance, surfactants such as polysorbates or poloxamers or cryoprotectants such as trehalose or sucrose.

11.13.2 Chemical Stability of Nanocarriers

There are several disadvantages of chemical degradation; for instance, lipid- and polymer-based nanocarriers are probably the most delicate when it comes to this aspect. Lipid peroxidation is a major challenge that affects most lipid-based systems, in that unsaturated lipids become vulnerable to peroxidation and cause rancidity and structural changes. Some of the additives, such as α-tocopherol (vitamin E) or butylated hydroxytoluene (BHT), are added to help avoid the process of oxidation. Hydrolysis influences the stability of polymeric NPs, for instance PLGA, where ester linkages split in contact with water to trigger the degradation of the polymer and shift in the release profile of the drug. Further, the desired drug may leach out from the microcapsule due to instability of the capsule because different conditions affecting the capsule's stability in terms of liposomes or micellar structures are unstable by changes in temperature or pH. Conditions like minimal humidity, prevention from contact with air through nitrogen blanketing, and relatively low temperature also assist in conserving chemical stability.

11.13.3 Biological Stability and Sterility Considerations

Biomedical nanocarriers, as another aspect, must also be biostable, that is, not be thereby degradable by enzymatic action and prone to microbial infection. In particular, proteins in biological fluids can adhere to the surface of the NPs, a phenomenon known as the protein corona that can affect further biodistribution and targeting efficiency. In addition, it is important to store the samples in a sterile manner to reduce the chance of bacterial or fungal growth, which may be manifested in aqueous dispersions of samples. The method widely used to improve long-term stability while preserving sterility is lyophilization, or freeze-drying. However, the process of freeze-drying per se must be carefully controlled in order to avoid NP degradation, sometimes with the use of agents such as trehalose or mannitol.

Temperature and Humidity Range for Different Nanocarriers: The latter also applies to storage temperature and humidity; the conditions will vary with the type of nanocarrier:

- **Liposomes and Lipid-based Carriers**: Store at **4 °C** under inert gas (e.g., argon or nitrogen) to prevent oxidation. Lyophilization with cryoprotectants can extend shelf life.
- **Polymeric NPs (PLGA, Chitosan)**: Best stored as a **freeze-dried powder** at **−20 °C** to prevent hydrolysis.
- **SLNs and Nanostructured Lipid Carriers (NLCs)**: Store at **4–8 °C**; some may remain stable at **room temperature** if properly stabilized.
- **Inorganic NPs (Gold, Silica, Iron Oxide)**: Generally more stable; can be stored as **aqueous dispersions at 4 °C** or in **lyophilized form**.
- The stability and storage of nanocarriers are critical for their successful application in medicine. One has to prevent physical inactivation (aggregation, sedimentation), chemical inactivation (oxidation, hydrolysis), and also biological inactivation like protein adsorption and microbial growth. Some of the techniques for enhancing long-term storage include techniques like lyophilization, use of cryoprotectants, antioxidants, and temperature control during storing. Some of the factors that can lead to alteration in properties of nanocarriers include improper storage conditions. Appropriate storage conditions for the particular type of nanocarriers have been earlier found to improve shelf life and therapeutic efficacy.

11.14 Future Directions and Opportunities

Recent progress in nanomedical applications guarantees a promising perspective for the further evolution of the science, such as the application of individualized medicine and the use of nanotechnologies for the concentration of medicine. Nanoformulation design is experiencing a world of changes with the assistance of AI and ML to enhance both speed and accuracy rates. Nanotechnology can also be applied in reviving drugs and disease diagnoses to target a different market of diseases. Such innovation fields, such as nanotheranostics, are good examples of interrelated diagnostic and therapeutic capacities of the new generation, which shifts to precision medicine. As with most of the emerging technologies therein, there are several factors of ethical consideration in nanomedicine, some of which are discussed below.

11.15 Integration of Nanotechnology to the Medical Field, Specifically to PM

Personalization of medicine can be explained as the ability to create a correct and effective treatment plan in terms of the right chemical composition, right concentration, and right individual. It was dictated by several factors, which are prejudice adverse consequences of the drugs that many patients have experienced and inconsistency of the efficacy of the treatment using several sorts of the drugs, ranging from 25 to 80% according to the classes.

Proteomics, genomics, epigenetics, special health status, patient conditions, and the environment are the components of PM [196]. Nanotechnology, on the other hand, is a broader definition because it is systems that go in the range of 10–100 nm. It also suggests the condition in which one is capable of directing what emerges within this nanorange to the intended outcomes. This was possible because of nanoline molecules, as it would not be possible when the molecules are larger than 1 μm in size, beyond the capability of the particles to interact with the cells and subcellular and molecular components. Nanotechnology is being used in the prevention, detection, diagnosis, and even treatment of diseases, and most of these products are currently being used in current practice [197]. There is a blurring of the horizon between the two, which are intertwined in many a way; it may appear somewhat difficult to identify exactly where one stops and the other starts. First, the diagnostic area, where nanotechnology can be of much use in the determination of the status of particular drug targets, pharmacogenetic testing, and the ability to conduct tissue culture tests and animal tests. Second, the field of application, as nanomedicine can deliver the drug at the desired position for a certain duration for a specific disease in the concerned patient [198]. In addition, nanomedicine is a targeted system and allows accurate dosage and delivery of the drug in the body, which is less than the tolerable maximum dose of the non-formulated drug; for this reason, it is possible to use dose titration relative to the patient's status. In addition, nanomedicine can eliminate two aspects in the pharmacogenomics of drug response regarding the diversity of drug transporters and cytochrome P enzymes (CYP) in various ethnicities. By altering enzymes and the endocytic process, which is unrelated to transporters, nanomedicine drug formulation could make

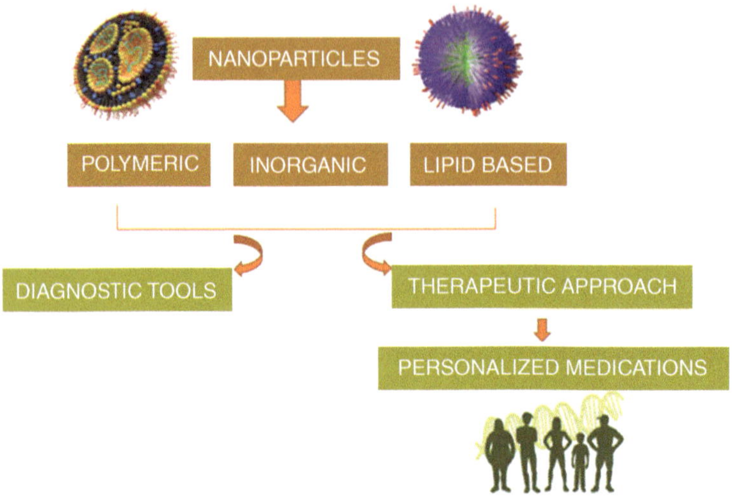

Figure 11.6 Representation of the use of nanoparticles for personalized medication.

the medications in question indigestible. Figure 11.6 gives an overview of potential pathways for NP adoption for personalized medication.

Nanodrugs were originally developed as an enhancement of a current drug or diagnostic agent's characteristics. Today, NPs are manufactured to have fewer side effects on the target region and the rest of the body, to increase the extent of drug absorption after oral administration, and to increase the biological activity, as well as the kinetic and dynamic properties. Furthermore, NPs can minimize the dosing frequency and hence improve patients' compliance as well as disease outcomes [199].

In comparing a nanodrug with the traditional DDSs, the following pharmacokinetic properties can be expected: major solubility and absorption, capability to control the release, enhanced stability and metabolism, low toxicity or side effects, longer circulation time, and better targeted delivery. The ability of nanomedicine to tailor a drug's ability to reach the site of action in light of its pharmacokinetics and pharmacodynamics to overcome obstructions of bodily barriers is a feasible personalized therapeutic opportunity for an individual or a group of people with specific genomes, even though the major pharmaceutical industries and the commercialization of NPs continue to focus on improving the one-size-fits-all approach [200]. Consequently, the use of NPs in customized medicine involves using the uniqueness of the genome data of the patient, coexisting diseases, and sensitiveness to the environment to develop a new treatment with high drug selectivity and appropriate dosage of the drug administered at a particular site. According to the pharmacokinetic processes, a drug may go through four stages, which are ADME. The traits of NPs and their impact on pharmacokinetic and pharmacodynamics will appear as follows.

11.15.1 Absorption

The absorption process is initiated at the level of getting nanodrugs into the bloodstream by various physiological routes. The usual routes of drug administration include intravenous, oral, topical, and other routes of drug administration. Among them, the intravenous route of administration has been considered the most common and effective one. Influence of the surface charge, hydrophobicity, and size of NPs has a significant effect on mucosal absorption. However, it could be postulated that smaller NPs are likely to have a better transcellular uptake as compared to the larger ones [201]. Additionally, circulatory and targeting capabilities are better in NPs with a diameter not exceeding 500 nm, and the danger of capillary thrombosis and embolism is lower. Additionally, the opsonization mechanism and action of the RES macrophages cause the bigger nanodrugs to be rapidly removed from the circulation [199]. Nanodrugs must first pass through biological barriers in non-IV vehicles in order to enter the bloodstream. Because typical medications have unfavorable characteristics including poorer log P, solubility, and stability, this isn't always feasible. This is because they are better able to penetrate biological barriers and be absorbed; they use alternative routes than regular medications.

For instance, many conventional medications taken orally may degrade due to low stomach pH or enzyme activity, or they may be insoluble in bodily fluids, which would prevent absorption and reduce the medication's therapeutic efficacy. All these problems are solved as soon as the drug is incorporated into NPs and a NP's absorption occurs solely based on its physical and chemical characteristics. When NPs are administered orally, the NPs for transport in the GIT use the paracellular transport system, transcytosis with a carrier, passive diffusion across cell membranes, and M cell pathways. Low absorption of NPs can be a benefit, and it involves the anticancer drugs like olaparib (Ola). Its mechanism of action is majorly based on the PARP inhibition. It also exhibits good activity in tumor tissue with BRCA gene mutational anomalies. However, Ola is comparatively less absorbed through the GIT. Liposomal entrapment into a liposphere enhances its oral bioavailability to a great extent and also minimizes its hematological toxicity [202]. The most critical barrier to penetrate in the case of the nanodrugs intended for transdermal application is the skin. The skin has lipophilic as well as hydrophilic layers, and this depiction has an impact on challenging conventional drug adsorption. Wang et al.'s investigation showed that imidazole-based ionic liquid microemulsions can decrease the complexities of the skin barrier and have an impact on the SC surface's characteristics and corneocyte alignment [203]. Nanodrugs administered via nasal administration can either be breathed out of the lungs or be taken up and broken down by macrophages. This is accomplished by the mucociliary escalator. The rest can pass through lung epithelium and mucus and be absorbed. Additionally, it has been demonstrated that positively charged nanodrugs interact with mucin's sugar and negatively charged with sulfate sialic acid groups, improving their absorption through the lung mucosa [204].

11.15.2 Distribution

When the nanodrugs move out of circulation and reach tissues and cells, the process of distribution starts. Usually, sinusoidal endothelial capillaries cause NPs to spread quickly after absorption and deposit in the liver, spleen, and bone marrow. Additionally, the BBB makes it difficult to transfer nanodrugs into brain tissue. Since the majority of endothelia in the human body are roughly 5 nm in fenestration, NPs can typically be larger than 5 nm, which broadens their dissemination. Consequently, two distinct distribution patterns of nanodrugs, known as passive and active targeting, can be distinguished.

11.15.3 Passive Targeting and EPR Effect

The size of the NPs and the intact fenestration of endothelia have the biggest effects on passive targeting. EPR is predicated on the theory put forth by Maeda H et al., according to which a medication has greater permeability and retention within the tumor mass and the infection site [205].

Tumor microvasculature and inflammatory tissues, on the other hand, have an ineffective lymphatic drainage system and a vasculature system with pores that range in size from a few hundred nanometers to 800 nm, depending on the type of tumor cells or the condition of the inflammatory tissues. This could be because of the fact that the tissue has a different vascular pattern. Further, normal tissues have a vasculature structure which has pore sizes that fail to permit penetration by NPs; for this reason, nanodrugs that conform to the size reported above can, without intervention, reach the site of the impaired area and release the drug, have a much more targeted and efficient action, as well as decrease the general dosage and therefore the onset of side effects.

Greish et al. used styrene-co-maleic acid (SMA) as a micellar nanocarrier for a number of anticancer drugs in their investigation. In SMA-DOX, the team demonstrated that SMA-DOX accumulated in tumor masses 13-fold more than the free DOX at a similar dose. Likewise, the same micellar system was used for loading of nano-micelles of Dasatinib combined with targeting ligands of PDGF, KIT, and ABl. Therefore, it makes sense to say that the EPR effect might have certain potential disadvantages. Specifically, leukemia and other nonsolid malignancies are unable to benefit from the EPR effect. Further, the pores in tumors may be of different sizes, and the nanodrug may reach only some parts of the tumor, which leads to an ineffective therapeutic effect [206].

11.15.4 Active Targeting

Even though the EPR effect enhances the targeting of NPs to the tumor interstitium, it does not contribute to the cellular uptake of these particles. The conjugation of ligands that have an affinity not only reduces the negative impacts of passive targeting for particular cancer cells or subcellular compartments, but it also improves the internalization of the NPs into the cells [207]. This is due to the fact that particular receptors exist/are overexpressed across the cell surface, which can be engaged by implication through a particular binding entity. Also, the active targeting may be an excellent approach to apply in PM to perform the treatment using the best ligand, which leads to the right localization of the nanodrug dose. Various ligands have been attached to NPs, including aptamers, proteins and peptides, antibodies or antibody fragments, and other receptor-specific ligands.

11.15.4.1 Targeting Ligands

There are many ligands available for the decoration of the nanodrug, including adenosine, folate ligands, glucose, and many others. Adenosine is a naturally occurring nucleoside, which replaces the function of the cell and activates the adenosine receptors. Some of the tumors expressing the adenosine A1 receptor include colorectal, prostatic, lymphoma, and breast cancer. Swami et al. studied the ability of adenosine in guiding SLN laden with docetaxel towards human breast and prostate carcinoma [208]. The results proved that the developed adenosine-conjugated SLN charged with docetaxel (ADN-SLN-DTX) possessed better properties regarding cell inhibition and its pharmacokinetic properties in comparison to SLN-DTX. Folate receptors refers to a high-affinity folate-binding protein class that participates in the process of transport. For the folate receptors, normal tissues have no folate receptors only on the surface of tissues, but some types of malignant tumors have a larger number of folate receptors, such as testicular, brain, endometrium, breast, and colon; the worse the clinical stage, the higher the expression of folate receptors. According to Patil et al., when liposomes were used for the grafting of mitomycin C prodrug pegylated and folic acid conjugate, there was a higher nanodrug uptake of the tumor cells, hence higher cytotoxicity as observed [209]. Glucose can be also used for the active targeting approach. The cells which are abnormal are known to be metabolically active than other normal cells, thus they have higher glucose requirement. For this reason, glucose-coated nanodrugs, due to the glucose transporter channel 1, have a higher permeation into cancer cells and this could be helpful for the theragnostic approach. Specifically, Gromnicova et al referred that glucose-coated gold NPs can enhance the transport of loperamide across the BBB selectively and effectively [210].

11.15.4.2 Protein-based Active Targeting

Some of the proteins and glycoproteins are known to interact with the cellular receptors, for example, transferrin. Transferrin has been described as a LDL glycoprotein that binds to transferrin receptors and participates in the transportation of iron. Namely, in cancer and metastatic cells, the transferrin receptors are 100-fold higher while being devoid of any interaction with normal cells; for this reason, transferrin can be considered the most suitable ligand for the active targeting strategy. According to Cui et al., reports on the transferrin-decorated nanodrugs carrying DOX and curcumin, the latter diminished the side toxic effects and enhanced the effectiveness in attacking the cancer cells [211].

Antibodies, due to their fragment antigen binding (Fab), have high selectivity and binding capability to surface receptors present in the cells. Of these targets, it is obligatory to mention the epidermal growth factor receptor 2 (HER2) targeted by trastuzumab. According to Arya et al., it was also found that utilizing gemcitabine with chitosan NPs and trastuzumab conjugate could be more antiproliferative and cytotoxic than unconjugated ones for treating pancreatic cancer [212].

Other forms of antibodies can be used instead of the whole antibody, which are fragment antigen-binding (Fab) or single-chain variable fragments (scFV). The readiness of fragments might increase the specificity of the binding, and also, because the particles cut from the larger molecules are significantly smaller, it is possible to attach several fragments of antibodies at the same place, enhancing the therapeutic effect. For instance, it was observed by Sapra et al. that DOX and vincristine-loaded stealth immunoliposome formulations, which were attached with anti-CD19 Fab fragments, had longer circulation times and better therapeutic efficacy compared to formulations attached with entire antibodies [213].

11.15.4.3 Aptamers, Gapmers, and siRNA Active Targeting

Aptamers are single-stranded RNA or DNA molecules that have the ability to bind to proteins or biological targets, and they are produced using the in vitro technique. Owing to their sensitivity and specificity for their target, they can be said to be a viable replacement for antibodies for the active targeting strategy with some advantages. For instance, aptamers are less immunogenic and much more convenient as well as cheaper in production as compared to traditional antibodies. They also permit the use of routine synthetic chemistry reactions for bioconjugation and have higher tissue penetrance, which results in improved therapeutic outcomes.

Lopez-Nunes et al. described the use of gold NPs modified with the AS1411 G-quadruplex DNA aptamer ligand for the delivery of imiquimod, a proapoptotic molecule for cervical cancer. The selected aptamer has a high specificity toward nucleolin; therefore, it gets engulfed into the tumor cells and discharges the payload at the site of malignancy. These findings have established that this format of aptamer development was useful in offering a reduced viability of HeLa cells while displaying a lower cytotoxic response toward normal cells since nucleolin is less expressed in normal cells and has a lesser affinity, therefore, to the AS1411 DNA aptamer [214].

The possible application of siRNA in the treatment of recurrent GBM was demonstrated in a clinical phase 0 trial study (NCT03020017) [215]. The specific forms of pharmaceutical treatment administered are lomustine, carmustine, temozolomide, and bevacizumab in combination with surgery and radiotherapy. This type, though, is relatively rare, but it is known to be very dangerous, and very little hope of survival has been seen. Thus, the development of new anticancer therapies employing the process of unveiling new small molecule or nanoscale medications is already required as a potentially advantageous precision medication approach to recurrent GBM treatment.

11.15.5 Protection Against Degradation Enzymes and Metabolism

The pre-absorption metabolism and the first-pass metabolism contribute to variation in the pharmacokinetics of the drug when administered orally. Orally administrated drugs have several barriers within the host's GIT, including pH variability, motility, a dense mucus layer, mucosa, microbiota, as well as trypsin, lipase, CYP450 enzymes, and P-gp efflux pumps. For example, one of the CYP450 family, particularly the CYP3A4, is involved in the major inter-population variability and has major effects on drug response in special group members such as children, pregnant women, elderly people, and different ethnic groups, leading toward unpredictable therapeutic effects as well as multiple side effects. For these reasons, the pre-uptake metabolism of the drugs that are taken orally should not be overlooked. It was found that CYP3A4 is significantly distributed from the proximal and luminal aspect of the duodenum to the distal and serosal side of the ileum; the amounts of the enzymes could significantly vary from person to person. However, it is important relating to this perspective that CYP3A4 content could be affected by diseases including obesity, cancer, infection, and inflammation, because CYP3A4 expression is highly regulated by the pregnane X receptor.

There are several approaches for stabilizing the NP formulation that have been tried to obliterate the pre-uptake metabolism as well as the variation arising from it in an attempt to have the pharmacokinetic characteristics of the once orally administered drugs enhanced. Mucoadhesive polymers could permit a better residence time and resist the peristaltic movements, so enhancing the absorption of the nanodrug through the GIT. For instance,

Han et al. promoted the adsorption at the mucus layer by the chitosan coating of the alendronate-containing liposomes as the chitosan interacts with the mucin that is invariably present in the proximal tract due to its negative charge. Another interesting approach is the application of highly lipophilic lipid NPs thereof with triglycerides to transport the drug across the enterocytes and the vessels of the lymphatic system. In fact, the HLLN is reduced in the intestinal lumen, taken up by enterocytes, and then reassembled into chylomicrons. The event shall also result in chylomicron bearing the incorporated drug exiting through lymphatic vessels and avoiding both pre-tooltip uptake metabolism and first-pass metabolism.

The other alternative is to develop NPs conjugated with CYP3A4 inhibitors that can block this isoenzyme during transcytosis [60]; however, the detrimental effect of this inhibitory process might be the toxic side effects that come along with this shutting down the practical function of the enzyme. Conversely, it has been established that the common surfactants, cosolvents, and oils, namely Tween, PEG, Poloxamer Cremophor, Polysorbate 80, and oleic acid, were shown to significantly decrease the activity of CYP3A4. However, it is also to be noted that another delivery strategy is employing M-cell-targeting NPs to transport the nanodrug into the lymphatic capillaries. Peyer's patches contain M cells special to transport and to promote endocytosis of antigens into these ones. Some approaches are based on the mimicry of the invasion of the host tissues by the pathogenic bacteria like Salmonella and Yersinia or the interaction of these M cells with the special receptors like the integrins that are present on the surface of these cells. NPs conjugated with vitamin B12 receptor–ligand can be utilized in such a manner with a view of eluding the CYP3A4 metabolism. Liu et al. synthesized polymeric NPs that are curcumin incorporated with vitamin B12-conjugated chitosan known as H/VC-LPNs.

Nanomedicine might be useful in avoiding such challenges, shielding the drug from wiping enzymes, enhancing the half-life, and delivering the appropriate amount of the drug to the affected area. Nanomedicines for the management of HIV/AIDS are, for instance, promising in increasing the bioavailability of an antiretroviral drug that is beyond the CYP system, which is restricted by the therapeutic plan that requires the use of a pharmacobooster and potentially provoking drug–drug interactions. For example, Atazanavir, a potent HIV protease inhibitor, faces great first-pass hepatic metabolism by CYP3A4, resulting in low bioavailability.

11.15.6 NPs Interaction with the Microenvironment

As the goals of PM are to use exact genetic and environmental information as well as comorbidities in patient databases for the correct patient segmentation and disease condition treatment. It could therefore be useful in precision medicine because it enables delivery of the encapsulated drug in a safer manner and in concert with specific cells. However, several drawbacks have established the NP's inefficiency, majorly due to the heterogeneity of the biological barriers and environments of the human body tissues, much as regarding variability attributable to morbidity and comorbidity. However, clinical trials for NPs are currently still conducted in a general patient population, and the latter will possibly shift in the future due to the rationale for targeted treatment. In oncology patient stratification was deemed important in order to achieve good results in treatment that included nanomedicine. As a result, the patient stratification and the changes in the NPs dependent on the patient's information have to be closely connected with successfully conducting the PM approach. However, while active targeting brings so many advantages, diseased cell markers can be different with different patients, and thus the target selection factors are quite limiting. Furthermore, it appears that the microenvironment has a strong effect on the successful accomplishment of drug delivery processes. NPs must penetrate and cross the local microenvironment to deliver the drug in the target cells, where the potential barriers can be physical or a change in chemical environment. Consequently, it appears that microenvironment seems to be very important in order to target the required tissues with nanomedicines. Thus, it can be presupposed that the features of microenvironment are different from the features characteristic of circulation, which can lead to changes in the stability and other properties of the NPs.

Typically, the free drug available to reach the site of action is a mere 0.1%, while only 15% of the NPs administered can reach the said site. This increase in tumor accumulation is normally associated with the EPR effect, as we have discussed above. In recent years, there are evidences that have significantly downplayed and disputed the role of EPR effect in the accumulation, clearly indicating that only a small part of the nanodrugs that penetrate can be attributed to the EPR effect. Indeed, an apparent significance appears to be associated with immune cells and the protein corona mechanism. In addition, irregular distribution of new vessels in and around the tumor is sometimes affected by personal factors that comprise diet, heredity, age, chemotherapeutic medicines, and disorders. To carry out targeted treatments, NPs have to be chosen from the disease's vascularization of the certain patient. Another challenge that affects the absorbable positively charged NPs is the repulsive forces due to the negative charges of the ECM components that would hinder them from penetrating the target site. Indeed, the restricted NP penetration in the brain can also be associated with restricted space available for extracellular movement in the brain and nonspecific interaction with ECM. Moreover, the action of biofilms and mucus layers can alter the distribution of NPs and tend to entrap them within the mesh pore size or by other nonspecific interactions, and hence, the NPs are cleared from the epithelial surfaces. Mucus is a complex secretion that is composed mainly of a gel structure with other substances; its physicochemical properties, such as elasticity and concentration, rely on the physiological status and position of the body. For example, in cystic fibrosis, increasing the MUC5B polymers leads to a decrease in mucus clearance and pore sizes. Therefore, as it appears that the microenvironment is potentially extremely important for the fate of a NP, it is required to develop new kinds of NPs, or to perform changes to existing ones, in order to make use of this stochastic characteristic. Trying to achieve similar effects based on the presence of certain enzymes such as matrix metalloproteinases (MMPs) or proteases or a hypoxic, or acidic environment within the tumor could enhance the NPs' degradation and drug release. Another advantage that pertains to the regulation of release of NPs is the application of near-infrared (NIR) light, radiofrequencies, or magnetic fields from outside the body.

Isolation of macrophage or leukocyte cell membranes from the patient and using them within the NPs also enhances their efficiency on the targeted cancer cells.

Another employed strategy is to address the challenges within the context of the microenvironment attributed to NPs concerned with the modulation of the microenvironment.

It may be advantageous for enhancing NP distribution and increasing the tumor's response to a particular therapy. In addition, the high variability of NPs with regard to shape, size, and charge and with regard to active targeting ligands attached to the surface of the carrier may be beneficial for precision medicine, as it can create more specific and better-suited DDSs to the TME.

11.15.7 Intracellular Internalization and Subcellular Organelles Targeting

This is because the efficacy of the applied nanotherapy depends only on the designed formulation because, by altering the properties of the NPs, it is achievable to differentiate between healthy and unhealthy cells as well as the uptake and targeting of cells. Nevertheless, the transition from in vitro to in vivo and from ideas to real-world application is often not straightforward due to the increased difficulty in dealing with the biological systems NPs are engaging with. These factors have to be taken into consideration even with much more concern in fields of PM; indeed, drug targeting can be highly influenced depending on age, sex, target tissues or organs, and metabolic differences as well as other diseases, which can change different biological parameters. For example, after systemic administration, change occurs in the NPs' outer surface, and serum proteins are adsorbed, which can mask the ligands grafted on the NPs for targeting a particular cell. The captured proteins can also alter the properties of the nanoformulation, which could result in reduced cellular uptake, inactivation, and removal. The formation of a protein corona, of course, has its influence and can be slightly countered if NPs are designed to be zwitterionic at the surface and hydrophobic overall or by the introduction of low molecular weight PEG chains [216].

11.15.8 Overcoming MDR Mechanisms

MDR is a critical factor that contributes to reduced effectiveness in the recovery of patient health. The mechanisms, most of which are linked to cancer treatment failure, include several mechanisms used by cancer cells to prevent cell death and promote tumor growth, like decreased drug accumulation, alteration of the targets, and increased drug degradation or alteration of the targets. The awareness on pharmacoproteomics, pharmacogenomics, pharmacogenetics, and pharmacometabolomics might be more beneficial to advance new significant refinements for the characteristic therapeutic regimens. Indeed, the concepts of nanomedicine and PM are extremely related in this context. NPs are one of the most effective multifunctional systems, which can incorporate more than one drug, diagnostic agents, and/or biotechnological drugs with the intention of targeting those elements that are involved in MDR development even in a personalized manner.

Among the most relevant mechanisms concerning MDR, one of the fundamental aspects is linked to the overexpression of molecule transporters commonly known as ABC ATP-binding cassettes. These transporters eject the drugs from the cell and thus reduce the rate at which this concentration of the drug is effective in killing the cancer cells. Thus, cells that developed chemotherapeutic agent resistance inevitably become resistant to drugs from different chemical classes or drugs with different mechanisms of action, which universally aggravates the MDR phenomena. There is pharmacogenomic research in the context of MDR-associated proteins in order to forecast the efficacy of a particular treatment regimen in a subset of patients, in which interpatient variability can directly contribute to the effectivity of pharmacologic intervention.

Among the transporters associated with MDR, P-gp is the most widely known. This protein is also involved in the expulsion of DOX, Paclitxel, ETO, and Vinblatine from the cancer cells. Nanomedicine strategies targeting the P-gp predominantly involve the modern-day administration with a single nano formulation of a cytotoxic agent and a P-gp blocker like verapamil, cyclosporine, or curcumin. Another strategy is to use siRNA to target some of the genes that are customary in the synthesis of ABC transporters. The result would be a decreased ability of the drug to diffuse out of the cell and therefore an accumulation of the drug within the cell. It is also known that there are poor chemotherapy responders, which are defined by genetic variants. Some of the additional alterations are alterations in proteins that are involved in apoptosis, over expression of pro-apoptotic proteins and down regulation of anti-apoptotic proteins as the other grounds that lead to the development of MDR. It also revealed that the TME and cancer stem cells are equally significant for the development of MDR. Cancer stem cells are quiescent, and they also over express drug efflux proteins due to which, they are inherently resistant to cytotoxic drugs because they can fix any possible damage in their DNA at any given time. Together with that, growth factors secreted by the TME promote the proliferation of cancer cells. CSCs are highly plastic post radiotherapy and chemotherapy, hence they can change into another phenotype making them be able to create a new subpopulation of tumor cells. Therefore, the identification of specific TME and cancer stem cells biomarkers, which may be useful for defining the new, more effective anticancer therapies, is of crucial importance.

11.15.9 Solubility

The identification of a new drug and the possible administration of this drug for the treatment of a specific disease is usually prevented by some physicochemical characteristics of the drug. Usually drugs which are lipophilic and have a high molecular weight are usually known to be poorly soluble, this may just lead to the exclusion of the drug with consequent formulation challenges. High lipophilicity drugs have an imperfection in their absorption since they may be lodged within the phospholipidic layer of cells. On the other hand, hydrophilic drugs such as proteins and nucleic acids cannot be uptaken easily through simple passive diffusion because these cannot cross the cell membrane and at the same time they can be deteriorated in aqueous media. Such highly charged drugs like DNA, miRNA and siRNA could be delivered to the target site after they were encapsulated in polymeric NPs prepared from cationic monomers. These issues can be solved by drug encapsulation into NPs, which subsequently increases the drug's bioavailability. By delivering the drugs in amphiphilic or hydrophobic NPs, the variables that affect the solubility and rate of dissolution of lipophilic substances can be addressed. One of the drugs with promising effects as anticancer treatment is curcumin. This substance is a lipid soluble, photoreactive, labile in acidic and alkaline media, and efficiently metabolized and excreted in the organism. There are numerous reports on curcumin NPs and the highest values of curcumin load efficiency, improved stability and solubility were achieved in many studies through encapsulation in solid–lipid nanocarriers [217]. Another drug that could be regarded as an example of drugs with low aqueous solubility is salinomycin, which is an anticancer antibiotic. Ni et al. synthesized poly (ethylene glycol) functionalized salinomycin-containing poly (lactic-co-glycolic acid) NPs coupled with a CD133 aptamer. When applied to osteosarcoma cancer stem cells that expressed the CD133 protein, the developed nanoformulations shown great specificity and cytotoxicity, which can eliminate the MDR linked to the stem-like activity of these cancer stem cells [218].

These instances collectively demonstrated that poorly soluble medications could be reconstituted following conjugation with specific designed nanoconstructs. aiming to develop new potentials in nanomedicine. Furthermore, the functionalization of these NPs with ligands able to selectively bind a receptor overexpressed on a particular dwellers' population shall be a good approach to be considered for further experimentation in the framework of PMs treatment regimens.

11.16 Role of AI and Machine Learning in Optimizing Nanoformulation Design

AI has helped the systematic drug discovery process for all the classes of drugs, right from small molecule ones to biologics. Only in the recent decade has this technology been employed to design nanomedicine formulations. This approach has been considered the conventional way of discovering new nanomedicine formulations since the advent of the technique. An attempt has been made in the past decade to use AI in nanomedicine design, especially in identifying new formulations that yield the desired particle size distribution, drug payload, release profile, PK, targeting, biocompatibility, and efficacy.

So far, the emphasis in this context has been made on evaluating big molecule drugs and other excipients like lipids and polymers with an intention to predict molecular interactions for the identification of new nanomedicines exhibiting suitable characteristics using deep learning algorithms. In particular, recurrent neural networks (RNNs), convolutional neural networks (CNNs), generative adversarial networks (GANs), and Transformers have been utilized to develop the ML platforms that have the ability to predict and obtain the favorable nanomedicine characteristics. For instance, attaining a high drug loading in nanoformulations has always been a problem, particularly for hydrophobic drugs. Previous work demonstrated that the designed high-throughput experimentation with the help of ML can predict the self-assembly of hydrophobic drugs with small molecular dyes or polymers. These efforts have resulted in the development of fast and huge identification of nanomedicines with high drug loading capability. The application of AI and ML in long-acting nanomedicine formulation and advancement is also in progress. From AI-based imaging and ML algorithms, the accurate and rapid prediction of the parameters, including long-term drug release, stability, and in vivo behavior, has been reported, thus improving the formulation development exercise. Likewise, in the context of adeno-associated viruses (AAVs) for gene delivery, some AI approaches that can accelerate the discovery of AAVs loops and select variants that are likely to improve the overall success rate of AAVs have also been envisioned. One area that is reported as significant progress on the use of AI/ML is in the development of formulations of messenger RNA (mRNA) lipid nanoparticles (LNPs).

Some researchers have also employed AI/ML-based approaches based on theories of deep learning and combinatorial chemistry to screen for suitable lipid molecules to formulate

LNPs that would target cells or organs rather than liver and lymph nodes. It is also important to note that some recently developed centralized web-based AI-based systems, including Formulation AI and AGILE, are based on existing data on drugs and excipients. AGILE is the first formulation-based platform, which was developed based on the GNN algorithm in terms of encoding the highly flexible but simple structures of the ionizable lipid pharmacophore. It has been applied to predict other ionizable lipids, which may be used in formulation of LNPs for delivering mRNA to muscle and immune cells. In the case of the formulation AI, the models that were developed by the team for selection were more than 10 of the algorithms that were available. In a similar manner, the use of random forest, support vector machines (SVM), and light gradient-boosting machine (LightGBM) as well as the eXtreme gradient boosting (or XGBoost) models was seen to be most effective. It enables the formulation design as soon as the basic information of drugs and excipients is inputted. Most of the work in this field, particularly for the application of AI in formulation design, relies on either preformulation data obtained from an experimental setup within an organization or data from molecular computations (e.g., molecular descriptors, intermolecular distances, and energies).

11.17 Potential for Nanotechnology in Tackling Rare Diseases with Repurposed Drugs

The term rare diseases is used for diseases that affect a small number of people; these affects about 300–400 million people worldwide. This comes at a time when the field of genomics and molecular medicine has greatly improved, and yet 95% of rare diseases remain untreatable by the drugs that have been approved by the Food and Drug Administration. This can be attributed to several factors, such as extremely expensive costs of drug development, which stand at an average of $2.6b per drug; small and genetically diverse populations of patients that are complex to enroll for trials; and sometimes only slightly understood disease etiology. Drug repurposing, which involves finding new uses for existing drugs, has become the ideal approach due to the mentioned barriers. Secondly, the safety profiles of compounds used for repurposing can be well evidenced, thus leading to shorter time and money required for development. Many of the candidates show significant issues such as limited bioavailability (e.g., poor solubility, short half-life), toxicity to off-target tissues/organs, or an inability to penetrate critical barriers like the BBB. Advanced engineering at the nanoscale offers a promising solution to these problems in perfecting drug solubility, increasing and prolonging drug stability, targeting diseased tissues, adjusting release rates and timing of the drug delivery, and enabling multiple loadings of different drugs at sites of action and occasionally the diagnostic and therapeutic roles of the drug. This section aims at discussing how nanotechnology can be employed for repositioning of drugs for OFDs with reference to nanocarrier systems, mechanistic benefits, preclinical and clinical examples, and future prospects.

11.17.1 Nanocarrier Systems for Drug Repurposing

Lipid-based NPs, polymeric NPs, inorganic NPs, and extracellular vesicles are some of the most popular nanocarriers applied for drug repurposing of rare diseases. Lipid NPs, particularly liposomes, SLNs, and nanoemulsions, are more favorable due to biocompatibility, higher drug entrapment efficiency, and the capability for surface modification. For example, PEGylation will help to prolong their circulation time, while ligander conjugation will help target tissues. One of which is a case of Patisiran (Onpattro®), which is siRNA therapeutic conjugated with lipid NPs used to treat hereditary transthyretin amyloidosis and designed to silence a mutant transthyretin mRNA. Likewise, liposomal amphotericin B has been used in the treatment of mucolipidosis, hoping to minimize the nephrotoxicity that is exhibited by the free drug. Therefore, using polymeric NPs like PLGA or chitosan, one can control the degradation rate that affects drug release rate. These carriers can be additionally conjugated with targeting ligands like transferrin, which helps in overcoming the barriers of BBB. For instance, the PLGA NPs containing nitisinone showed better targeting toward the liver without side effects in treating tyrosinemia type 1. Dendrimer-encapsulated dexamethasone has also been used in Duchenne muscular dystrophy due to its ability to increase delivery of the drug to affected regions of muscles in the body.

Gold NPs, mesoporous silica NPs, and iron oxide NPs have these merits of high surface area for drug conjugation and multifunctionality. Some modified nanogold with drug applied to the treatment of Sjögren's syndrome avoided systemic immunosuppression while retaining treatment efficacy by using cyclosporine. In the same way, deferoxamine-loaded silica NPs have better iron chelation in neurodegenerative rare diseases. Extracellular vesicles (EVs), especially exosomes, are producing significant attention in recent years as they are derived from the natural source, recognized by the immune system minimally, and possess inherent tissue-targeting ability. Concerning the administration of the antisense oligonucleotides, exosome delivery has proved to be outstanding remarkably in SMA treatment due to the capability of crossing the BBB. Other EVs containing rapamycin are also under consideration for lysosomal storage disorders on the basis of their targeted delivery to affected cells.

11.17.2 Mechanistic Advantages of Nanocarriers in Rare Diseases

Nanocarriers overcome some serious challenges associated with repurposed drugs such as poor ADMET parameters, nonselective biodistribution, and low solubility. For instance, hydroxypropyl-β-cyclodextrin (HPβCD), which has been reinvented for Niemann-Pick Type C disease, has the problem of low renal clearance time. Liposomal encapsulation has been known to have the advantage of increasing its half-life up to 10 times, which boosts its therapeutic value. Targeted delivery is another advantage, which involves ligand-mediated approaches or stimuli-responsive systems for releasing the drug dosage. For instance, transferrin-coated NPs have been helpful in crossing the BBB in Batten disease, and pH-sensitive release of NPs helps in releasing drug in the acidic environment of lysosomes in lysosomal storage disorder. Nanocarriers also have the advantage of multicomponent synergy as a major factor for use in complicated and rare diseases. To begin on the molecular level, the combination therapy of co-encapsulation of ivacaftor and lumacaftor in NLCs has shown increased treating power for several pathological pathways in case of the cystic fibrosis.

11.17.3 Preclinical and Clinical Case Studies

There is good evidence at preclinical and clinical levels of the shift that is witnessed with nanotechnology in the repurposing of drugs that target rare diseases. In Friedreich's ataxia, liposomal resveratrol resulted in threefold higher brain permeability compared to the free drug, as established in the recent study of 2021 (PMID: 33493622). The use of ambroxol for Gaucher disease, due to its encapsulation on solid lipid NPs, demonstrated increased lysosomal uptake in vitro and offers a way forward for clinical application. For amyotrophic lateral sclerosis (ALS), riluzole-loaded polymeric micelles took half of the dosing interval of the conventional approach but were equally effective. The present treatment of TSC involves PEGylated liposomes with rapamycin, a therapy that is already in a Phase I trial (NCT04595513).

11.17.4 Challenges and Future Perspectives

However, some critical challenges have emerged in the business translation context. This is because there are no standard regulatory guidelines that have been set by the FDA or EMA on nano-repurposed drugs. Another challenge relates to manufacturing scalability, given nanocarrier synthesis may not have the same outcome for each batch, which is a concern for clinicians. Some of them are design tools for the specific drug-nanocarrier combinations relying on AI and organ-on-a-chip models replicating the pathology of rare diseases in preclinical studies. Further down the line, for example, CRISPR-NP complexes offer the hopes of correct gene editing in monogenic rare diseases; 3D-printed individualized nanopharmaceuticals may, on the other hand, target the different mutations of the same disease in a given patient.

11.18 Nanotheranostics

It is a relatively new field of research that combines nanotechnology imaging and DDSs as a single entity. Nanotheranostics is used in order to optimize the effectiveness of the treatment, as well as to control its results in real time. It utilizes imaging and drug delivery and therapy in one system that enhances the precision medicine strategies and minimizes side effects, which are evident in conventional treatments.

- **Key Concepts in Nanotheranostics:** Theranostic agents are multifunctional NPs that are capable of diagnosing a disease or condition and treating the same at the same instance. There are several potential benefits of nanotheranostic systems, as listed below:
- **Better Circulating Lifetime:** The observations of tumor-selective uptake, which takes place in consequence of high vascular permeability, cause a higher circulating lifetime of rather large particles.
- **Trastuzumab:** Surface functionalization and targeting ligands include PEG coatings and receptor-specific ligands that enhance circulation time and provide organ targeting.
- **Imaging Capability:** The same nanotheranostic platform may possess optical fluorescence, MRI, CT, and ultrasound imaging ability to provide an indication of drug biodistribution and therapeutic effect in real-time.
- **pH-sensitive Drug Delivery:** Nanocarriers change the release of the entrapped agents in response to the internal body environment, ensuring a better therapeutic impact together with the lowest toxicity.

The major types of nanotheranostic materials are enlisted below.

11.18.1 Gold-based Nanomaterials

Gold NPs, nanorods, nanoshells, and nanocages have been quite popular because of their LSPR feature. This allows their use in PTT, where laser light produces heat, thus eliminating cancer cells belonging to the affected area. Gold NPs have the function of being conjugated with chemotherapeutic drugs such as DOX, which gives control so that the drug is slowly released. Besides, gold NPs have been used as a contrast agent in both the CT imaging and photoacoustic imaging to improve the accuracy of tumor detection. Gold nanostructures are also used for radiotherapy, where they work as radiosensitizers, thereby increasing the efficiency of radiation cancer therapy [219].

11.18.2 Magnetic Nanomaterials

In the course of this study, superparamagnetic iron oxide NPs SPIONs are well accepted in MRI-guided therapy. They help to improve image contrast of MRI so that tumors can be visualized accurately before the start of the treatment. SPIONs are also applied in magnetic hyperthermia treatment, by which localized heat is created in response to an AC magnetic field. It is a method that is used to ensure that the targeted cancer cells are irreversibly and specifically killed without affecting normal cells. Polymer-coated SPIONs have better biocompatibility and stability in the body, which enables dual imaging and drug delivery. SPIONs can be functionalized on the surface and can be encapsulated with chemotherapeutic agents, as well as targeting ligands can be attached to them for reaching specific sites [220].

11.18.3 Polymeric Nanomaterials

Among several nanocarriers, including polymeric micelles, dendrimers, and PEGylated NPs, are well known for use in nanotheranostics due to their biocompatibility and controlled release of drugs. Stimuli-sensitive polymeric nanocarriers display drug release in the form of pH-sensitive and enzyme-sensitive stimuli and those that are responsive to stimuli such as temperature and light. These systems allow activation of the drug only at the tumor location, and hence, the side effects are reduced. DOX-, PTX-, or rapamycin-containing polymeric micelles have been confirmed to exhibit longer circulation time, better tumor accumulation, and greater therapeutic effect [221].

11.18.4 Silica-based Nanomaterials

Mesoporous silica nanoparticles (MSNs) are reported to have high solubility, a tremendous surface area, easily adjustable pore size, and the ability to encapsulate large amounts of the drug molecules. The use of MSNs is based on easy functionalization of the particles with fluorescent dyes, MRI contrast agents, and targeting ligands, making MSNs suitable for various theranostic applications. Functionalized MSNs have been used for the codelivery of chemotherapy and siRNA, which can be considered as a novel strategy in the modification of chemotherapy for drug-resistant cancer. These NPs also allow the prodrug to be released in a controlled and/or sustained way, which proves to be optimal for therapy [222].

11.18.5 Carbon-based Nanomaterials

Graphene, CNTs, and nanodiamonds have high NIR absorbance, heat conductivity, and mechanical strength, which enables them to be applied in photothermal therapy and delivery of drugs. Pendant functionalizing with PEG keeps GO with high tumor accumulation, and thermal therapy is achieved when the tissue is exposed to NIR laser irradiation. Carbon nanomaterials are also explored for the concurrent imaging; nevertheless, questions about their permanent biocompatibility and possible toxicity have to be elucidated to some extent [223].

11.18.6 Composite Nanomaterials

Each of the nanovehicles in hybrid nanotheranostic platforms can have several functional modules and can be used in a multipurpose therapy and imaging. For example, gold-magnetic nanostructures offer the enhancement of MRI, in addition to providing the therapy including photothermal and chemotherapy. In parallel, silica-coated gold nanorods having magnetic properties are also useful for releasing drugs in a controlled manner and imaging in real time [224].

11.19 Ethical Dimensions of Nanomedicine

Nanomedicine is considered one of the most groundbreaking developments in the contemporary healthcare system, which is based on the use of particles sized 1–100 nm for disease diagnosis, therapy, and prevention. Unlike most of the conventional courses of treatment, nanoscale manipulations occur at the cellular and molecular scale faculties, and this is the reason that incredibly sophisticated treatments like controlled drug delivery, biosensing, and tissue engineering, among others, have been made possible. But this transformation brings with it quite some ethical issues that put to question the current medical, legal, and even philosophical understanding. The ethical concern for nanomedicine can be regarded as safety, social justice, human enhancement, informational privacy, and environmental sustainability, all of which should be taken into account as the tools of nanotechnology are moved from the laboratory setting to areas of application. Another important ethical issue that arose in connection with nanomedicine is the issue of risk-benefit ratio regarding a therapy. NPs, unlike their bulk counterparts, exhibit high surface area-volume ratios and surface quantum effects, which may cause unpredictable biological responses to occur. For instance, liposomal doxorubicin (Doxil) has been defined as a therapy with less cardiotoxic effects compared to the other chemotherapy; on the other hand, some carbon NPs, including CNTs and graphene, have been found to exert fibrogenic effects when inhaled, similar to asbestos. Recent chronic toxicity studies prove that some metallic NPs like gold and silver deposit in the liver and spleen, thus posing potential danger. This principle of primum non nocere, or first, do no harm, is quite hard to follow in a scenario where the full risks of the various levels are yet to be fully understood; thus, there is a need for post-market criteria and adaptable legislation.

A second significant ethical issue stems from such obscurity in the difference between therapy and enhancement. Nanomedicine not only restores the faulty functions in our body but also enhances the normal body capacities beyond even the inherent potential. They raise the possibility of advancing memory and cognitive functions through the use of graphene nanoelectrodes in the brain in rats and mice, and improving muscular strength with myosin-activating nanomotors a millionfold better than humans. The existence of these two types of applications raises essential questions regarding medical ethics: Should nanotechnologies applied to healthy subjects be regulated in a different manner than the ones used for the treatment of sicknesses? In light of this, how can we ensure that exclusion of people by sticking implantable chips as well as those who opt not to take the nanotechnological enhancements does not occur? The 2023 Geneva Convention on Human Augmentation marks an effort to regulate these technologies by banning the deployment of cognitive nanotechnologies in military use; however, other general guidelines have not been thoroughly expanded.

Privacy and data security involve one of the most important ethical issues in the age of constant, highly detailed, nanoscale biomonitoring. Devices like next-generation glucose monitors that will be implantable nanosensors produce huge amounts of health information, many times what electronic health record processes and stores at the moment. The risk is wonderful; everybody wants to gain knowledge about others, and, for instance, patients with such medical nanodevices could be discriminated by their insurance firms or may be sponsored by other mysterious people. Furthermore, the practical application of DNA nanobots that map the epigenome elevates biosensing to a higher level of biological peculiarities not seen with nanobodies populated with genetic identifiers for tracking or stem cell delivery to target tissues and organs. While traditional medical data is captured only during and after a disease episode, nanodiagnostics monitors physiological conditions continuously and for a lifetime, eradicating the distance between physician and patient and requiring enhancement of security motives.

A second and equally compelling ethical issue of nanomedicine distribution pertains to equity from a global perspective. A total of 92% of the clinical trials on nanotherapies are being conducted in high-income countries, and the population in low- and middle-income countries lack access to them. Despite the enormous amount of potential, the cost of nanomedicine—such as Abraxane, a NP-bound PTX at $125,000 per cycle—poses to widen the prevailing healthcare disparities. Further, several NPs, such as lipid NPs used previously for mRNA delivery, should be stored at low temperatures (−70 °C), which is practically impossible in developing countries. This separation of nanomedicine creates a range of ethical questions, such as distributive justice, and thus the policymakers should think about the proportions of nanomedical pricing, technology appropriations, and infrastructure investments to turn nanomedicine from the luminaries' prerogative. Environmental aspects of ethics also come in handy when it comes to the development of nanomedicine. The life cycle of medical NPs—from production to patient excretion—introduces new ecological hazards. The researchers found that long-lived micro- and NPs from drug carriers have been identified in the wastewater stream, and antimicrobial silver NPs have been associated with reduced microbial populations in water. Unlike conventional pharmaceuticals, NPs have reduced biodegradability and, therefore, pose more extended ecological risks, as they persist in ecosystems. The risks are under consideration of the European Union's Sustainable Nanomedicine 2030 initiative, which has required the use of biodegradable formulations in the relevant industries but does not dictate global norms.

To ensure nanomedical advancement is sustainable and is not detrimental to the planet, a precautionary approach that will require integrations of life cycle assessment (LCA) and green chemistry principles will be useful.

There are issues with the lack of appropriate legislation that cannot regulate nanomedicine because it is still rapidly evolving and new jurisdictions allow unsafe nanomaterials to be introduced in the market. In the United States, the FDA has its Nanotechnology Task Force formed in 2006, yet it has no regulatory powers over new technologies such as living nanorobots, and no two global bodies agree on how the enhancement technologies should be categorized. Proponents of adaptable regulation suggest that instead of full approval, a therapy should be approved for a certain period on condition that its effects will be monitored; another suggested regulation model is proposed to be a treaty under WHO supervision. This issue is that there is no clear governance and that may lead either to very risky decisions because of overcaution or to the introduction of dangerous technologies into the market.

In the future, prospects such as nanoscale gene drive, or rewritable, programmable, in vivo drug synthesis, and consciousness augmentation by neural lace are only around the corner. These advancements challenge fundamental concepts of human identity, autonomy, and equality, demanding proactive ethical scrutiny. In that connection, UNESCO has called for establishing Nanomedicine Ethics Committees in all research institutions to study NP effects on future generations and track them; currently, blockchain-based systems are being experimented with for this purpose. If you will, public deliberation panels, similar to the climate change citizens' assemblies, could assume the role of giving democracy a look into enhancement technologies. Thus, the ethical aspects of nanomedicine can only be addressed by combining the principles of molecular medicine with moral values, legal enlightenment, social justice, and reason. The decisions made in the present concerning the rules of the game and funding for research and international cooperation will determine whether nanomedicine will turn into a system of radical inequity or will become an opportunity for the masses to receive better healthcare. It is in such a context that ethical concerns should be mainstreamed and integrated in all the phases of development, ranging from the bench to patients; stakeholders should use this transformational technology in a responsible manner to the benefit of everyone and not simply for the benefit of the elites. The way forward poses the challenge of a new moral compass for medicine and even for medical technologies—the paradigm for 21st-century medical practice.

11.20 Conclusion

The combination of nanotechnology with drug repurposing and combination therapy is a milestone for present-day pharmacological and biomedical sciences. The work presented in this chapter has encapsulated the removal of the fundamental challenges associated with drug solubility, bioavailability, resistance, toxicity, and targeting by incorporating nanotechnology in these areas. The ability of nanotechnology to engineer highly specialized delivery platforms, liposomes, dendrimers, polymeric NPs, NGs, and exosomes, among others, allows for their use to deliver not only repurposed and coformulated drugs but also site-specific and time-commanded release. In effect, these inventions have altered the paradigm from systemic administration of generalized treatment to targeted therapies of precision. In addition, the concept of theranostics also increases the capability of the former through simultaneous treatment and real-time monitoring of disease progression.

In case of drug repurposing, computational modeling and ML-supported strategies represent a time-efficient and comparatively cost-effective drug development approach. Nanocarriers alleviate these barriers by enhancing the solubility, maintaining the drug levels for a long period of time, and co-delivering synergistic agents. In addition, demonstrating the synergy of combination therapies achieved through the use of nanocarriers, particularly for diseases with complex pathophysiology, is crucial.

Nanotechnology has brought beyond drug delivery an invaluable benefit in enhancing diagnostic sensitivity and specificity. NP-based imaging agents and contrast-enhancing probes have contributed to the evolution of the field of molecular imaging, including the setting of earlier disease detection, targeted imaging, and tailored intervention strategies. Nanotechnology has also contributed to the advancement of both drug repurposing and combination therapy in addressing critical pharmacological problems of poor solubility, rapid clearance, and off-target toxicity. The technological innovations for nanomedicines are also unmatched by regulatory frameworks and policy, suggesting the need for standardized protocols of safety, efficacy, and quality control of nanoformulations. Additionally, to enable commercial viability and accessible worldwide application, scalability and cost-effectiveness of nanomanufacturing processes must be accomplished. In terms of future research directions, much effort should be directed to translate preclinical success into clinical application. Both repurposed drug combinations and nanotechnology-based delivery system therapeutic potentials in various patient populations need validation through large-scale, multicenter clinical trials. Nanomedicine is expected to play a significant role in the future of precision medicine, particularly through AI-driven platforms that enable the design of patient-specific nanocarriers based on patient's genetic and disease profiles. Such theranostic systems are exciting since they bring diagnostics and treatment into a single platform that allows real-time monitoring and treatment adjustment. In summary, nanomedicine enjoys the promise of transforming healthcare, which could never have been reached without the sophisticated science, enabling mechanisms, and achievable protocols through the realization of nanoscale capabilities.

References

1 Pourmadadi M, Eshaghi MM, Ostovar S, et al. Innovative nanomaterials for cancer diagnosis, imaging, and therapy: Drug delivery applications. *Journal of Drug Delivery Science and Technology* 2023;82:104357.

2 Ahmad A, Ahmad I, Khan A, Abdullah M, Chee CY, and Verpoort F. Introduction to nanomedicine an overview. *Nanomedicine Manufacturing and Applications* 2021: 1–20.

3 Silva GA. Introduction to nanotechnology and its applications to medicine. *Surgical Neurology* 2004;61:216–220.

4 Caruthers SD, Wickline SA, and Lanza, G.M. Nanotechnological applications in medicine. *Current Opinion in Biotechnology* 2007;18:26–30.

5 Logothetidis S. Nanomedicine: The medicine of tomorrow. In: *Nanomedicine and Nanobiotechnology*, (pp. 1–26). Springer; 2011.

6 Jones EF, He J, Vanbrocklin HF, Franc BL, and Seo Y. Nanoprobes for medical diagnosis: Current status of nanotechnology in molecular imaging. *Current Nanoscience* 2008;4:17–29.
7 Nahrendorf M, Zhang H, Hembrador S, et al. Nanoparticle PET-CT imaging of macrophages in inflammatory atherosclerosis. *Circulation* 2008;117:379–387.
8 Debbage P and Jaschke W. Molecular imaging with nanoparticles: Giant roles for dwarf actors. *Histochemistry and Cell Biology* 2008;130:845–875.
9 Dinić J, Efferth T, García-Sosa AT, et al. Repurposing old drugs to fight multidrug resistant cancers. *Drug Resistance Updates* 2020;52:100713.
10 Rakholiya KD, Kaneria MJ, Shaikh AM, Wandhekar SS, and Chanda SV. Medicinal plants as alternative sources of therapeutics against multidrug resistant pathogenic microorganisms based on their antimicrobial potential and synergistic properties. *Fighting Multidrug Resistance with Herbal Extracts, Essential Oils and Their Components* 2025:271–289.
11 Keith CT, Borisy AA, and Stockwell BR. Multicomponent therapeutics for networked systems. *Nature Reviews Drug Discovery* 2005;4:71–78.
12 Radhakrishnan ML and Tidor B. Optimal drug cocktail design: methods for targeting molecular ensembles and insights from theoretical model systems. *Journal of Chemical Information and Modeling* 2008;48:1055–1073.
13 Rausch M, Rutz A, Allard P-M, et al. Drug repurposing to identify a synergistic high-order drug combination to treat sunitinib-resistant renal cell carcinoma. *Cancers* 2021;13:3978.
14 Zheng W, Sun W, and Simeonov A. Drug repurposing screens and synergistic drug-combinations for infectious diseases. *British Journal of Pharmacology* 2018;175:181–191.
15 Low ZY, Farouk IA, and Lal SK. Drug repositioning: New approaches and future prospects for life-debilitating diseases and the COVID-19 pandemic outbreak. *Viruses* 2020;12:1058.
16 Kast RE, Karpel-Massler G, and Halatsch M-E. CUSP9* treatment protocol for recurrent glioblastoma: Aprepitant, artesunate, auranofin, captopril, celecoxib, disulfiram, itraconazole, ritonavir, sertraline augmenting continuous low dose temozolomide. *Oncotarget* 2014;5:8052.
17 Hu Z, Yu Q, Gao YX, et al. Drug synergistic combinations predictions via large-scale pre-training and graph structure learning. In: *Proceedings of the 27th Annual International Conference of Research in Computational Molecular Biology (RECOMB)*, (p. 265); 2023.
18 Begines B, Ortiz T, Pérez-Aranda M, et al. Polymeric nanoparticles for drug delivery: Recent developments and future prospects. *Nanomaterials* 2020;10:1403.
19 Medina OP, Zhu Y, and Kairemo K. Targeted liposomal drug delivery in cancer. *Current Pharmaceutical Design* 2004;10:2981–2989.
20 Gao L, Meng F, Yang Z, et al. Nano-drug delivery system for the treatment of multidrug-resistant breast cancer: Current status and future perspectives. *Biomedicine & Pharmacotherapy* 2024;179:117327.
21 Hussain Z, Khan S, Imran M, Sohail M, Shah SWA, and de Matas M. PEGylation: A promising strategy to overcome challenges to cancer-targeted nanomedicines: A review of challenges to clinical transition and promising resolution. *Drug Delivery and Translational Research* 2019;9:721–734.

22 Liu Z, Tabakman S, Welsher K, and Dai H. Carbon nanotubes in biology and medicine: in vitro and in vivo detection, imaging and drug delivery. *Nano Research* 2009;2:85–120.

23 Orive G, Gascon AR, Hernández RM, Domínguez-Gil A, and Pedraz JL. Techniques: New approaches to the delivery of biopharmaceuticals. *Trends in Pharmacological Sciences* 2004;25:382–387.

24 Patra JK and Baek K-H. Green nanobiotechnology: Factors affecting synthesis and characterization techniques. *Journal of Nanomaterials* 2014;2014:219.

25 Mirza AZ and Siddiqui FA. Nanomedicine and drug delivery: A mini review. *International Nano Letters* 2014;4:94.

26 Lam P-L, Wong W-Y, Bian Z, Chui C-H, and Gambari R. Recent advances in green nanoparticulate systems for drug delivery: Efficient delivery and safety concern. *Nanomedicine* 2017;12:357–385.

27 Singh R and Lillard Jr JW. Nanoparticle-based targeted drug delivery. *Experimental and Molecular Pathology* 2009;86:215–223.

28 Shenoy DB and Amiji MM. Poly (ethylene oxide)-modified poly (ε-caprolactone) nanoparticles for targeted delivery of tamoxifen in breast cancer. *International Journal of Pharmaceutics* 2005;293:261–270.

29 Kumari A, Yadav SK, and Yadav SC. Biodegradable polymeric nanoparticles based drug delivery systems. *Colloids and Surfaces B: Biointerfaces* 2010;75:1–18.

30 Panyam J and Labhasetwar V. Biodegradable nanoparticles for drug and gene delivery to cells and tissue. *Advanced Drug Delivery Reviews* 2003;55:329–347.

31 Davda J and Labhasetwar V. Characterization of nanoparticle uptake by endothelial cells. *International Journal of Pharmaceutics* 2002;233:51–59.

32 Litany RJ and Praseetha P. Tiny tots for a big-league in wound repair: Tools for tissue regeneration by nanotechniques of today. *Journal of Controlled Release* 2022;349:443–459.

33 Sercombe L, Veerati T, Moheimani F, Wu SY, Sood AK, and Hua S. Advances and challenges of liposome assisted drug delivery. *Frontiers in Pharmacology* 2015;6:286.

34 Fathi S and Oyelere AK. Liposomal drug delivery systems for targeted cancer therapy: Is active targeting the best choice? *Future Medicinal Chemistry* 2016;8:2091–2112.

35 Nsairat H, Khater D, Sayed U, Odeh F, Al Bawab A, and Alshaer W. Liposomes: Structure, composition, types, and clinical applications. *Heliyon* 2022;8:e09394.

36 Owens III DE and Peppas NA. Opsonization, biodistribution, and pharmacokinetics of polymeric nanoparticles. *International Journal of Pharmaceutics* 2006;307:93–102.

37 Schaffazick SR, Pohlmann AR, Dalla-Costa T, and Guterres SlS. Freeze-drying polymeric colloidal suspensions: Nanocapsules, nanospheres and nanodispersion. A comparative study. *European Journal of Pharmaceutics and Biopharmaceutics* 2003;56:501–505.

38 Crucho CI and Barros MT. Polymeric nanoparticles: A study on the preparation variables and characterization methods. *Materials Science and Engineering: C* 2017;80:771–784.

39 Srinivasa-Gopalan S and Yarema K. Nanotechnologies for the life sciences: Dendrimers in cancer treatment and diagnosis. In: *Treatment Diagnosis*; 2007. Wiley: Hoboken, NJ, USA.

40 Tomalia DA, Christensen JB, and Boas U. *Dendrimers, Dendrons, and Dendritic Polymers: Discovery, Applications, and the Future*; 2012. Cambridge University Press.

41 Hawker CJ and Frechet JM. Preparation of polymers with controlled molecular architecture. A new convergent approach to dendritic macromolecules. *Journal of the American Chemical Society* 1990;112:7638–7647.

42 Bhalgat MK and Roberts JC. Molecular modeling of polyamidoamine (PAMAM) Starburst™ dendrimers. *European Polymer Journal* 2000;36:647–651.
43 Gillies ER and Frechet JM. Dendrimers and dendritic polymers in drug delivery. *Drug Discovery Today* 2005;10:35–43.
44 Wolinsky JB and Grinstaff MW. Therapeutic and diagnostic applications of dendrimers for cancer treatment. *Advanced Drug Delivery Reviews* 2008;60:1037–1055.
45 Svenson S and Tomalia DA. Dendrimers in biomedical applications—Reflections on the field. *Advanced Drug Delivery Reviews* 2012;64:102–115.
46 Morales M, Bomati-Miguel O, De Alejo RP, Ruiz-Cabello J, Veintemillas-Verdaguer S, and O'Grady K. Contrast agents for MRI based on iron oxide nanoparticles prepared by laser pyrolysis. *Journal of Magnetism and Magnetic Materials* 2003;266:102–109.
47 Edwards PP and Thomas JM. Fein verteiltes Gold–Faradays Beitrag zu den heutigen Nanowissenschaften. *Angewandte Chemie* 2007;119:5576–5582.
48 Atiyeh BS, Costagliola M, Hayek SN, and Dibo SA. Effect of silver on burn wound infection control and healing: Review of the literature. *Burns* 2007;33:139–148.
49 Mody VV, Siwale R, Singh A, and Mody HR. Introduction to metallic nanoparticles. *Journal of Pharmacy and Bioallied Sciences* 2010;2:282–289.
50 Oh JK, Lee DI, and Park JM. Biopolymer-based microgels/nanogels for drug delivery applications. *Progress in Polymer Science* 2009;34:1261–1282.
51 Karg M, Pich A, Hellweg T, et al. Nanogels and microgels: From model colloids to applications, recent developments, and future trends. *Langmuir* 2019;35:6231–6255.
52 Kabanov AV and Vinogradov SV. Nanogels as pharmaceutical carriers: Finite networks of infinite capabilities. *Angewandte Chemie International Edition* 2009;48:5418–5429.
53 Atif R, Shyha I, and Inam F. Mechanical, thermal, and electrical properties of graphene-epoxy nanocomposites—A review. *Polymers* 2016;8:281.
54 Nie Y, Li L, and Wei Z. Achievements in Pt nanoalloy oxygen reduction reaction catalysts: Strain engineering, stability and atom utilization efficiency. *Chemical Communications* 2021;57:12898–12913.
55 Basak S, Khare HA, Roursgaard M, et al. Simultaneous cross-linking and cross-polymerization of enzyme responsive polyethylene glycol nanogels in confined aqueous droplets for reduction of low-density lipoprotein oxidation. *Biomacromolecules* 2020;22:386–398.
56 Wu W, Mitra N, Yan EC, and Zhou S. Multifunctional hybrid nanogel for integration of optical glucose sensing and self-regulated insulin release at physiological pH. *ACS Nano* 2010;4:4831–4839.
57 Simpson RJ, Lim JW, Moritz RL, and Mathivanan S. Exosomes: Proteomic insights and diagnostic potential. *Expert Review of Proteomics* 2009;6:267–283.
58 Gangoda L, Boukouris S, Liem M, Kalra H, and Mathivanan S. Extracellular vesicles including exosomes are mediators of signal transduction: Are they protective or pathogenic? *Proteomics* 2015;15:260–271.
59 Greening DW, Gopal SK, Xu R, Simpson RJ, and Chen W. Exosomes and their roles in immune regulation and cancer. In: *Proceedings of the Seminars in Cell & Developmental Biology*, (pp. 72–81); 2015.
60 Brigger I, Dubernet C, and Couvreur P. Nanoparticles in cancer therapy and diagnosis. *Advanced Drug Delivery Reviews* 2012;64:24–36.

61 Bhadra D, Bhadra S, Jain P, and Jain N. Pegnology: A review of PEG-ylated systems. *Die Pharmazie* 2002;57:5–29.

62 Thiruppathi R, Mishra S, Ganapathy M, Padmanabhan P, and Gulyás B. Nanoparticle functionalization and its potentials for molecular imaging. *Advanced Science* 2017;4:1600279.

63 Montes-Fonseca SL, Orrantia-Borunda E, Aguilar-Elguezabal A, Horta CG, Talamás-Rohana P, and Sánchez-Ramírez B. Cytotoxicity of functionalized carbon nanotubes in J774A macrophages. *Nanomedicine: Nanotechnology, Biology and Medicine* 2012;8:853–859.

64 Aubé J. Drug repurposing and the medicinal chemist. *ACS Medicinal Chemistry Letters* 2012;3:442–444.

65 Kulkarni NS, Guererro Y, Gupta N, Muth A, and Gupta V. Exploring potential of quantum dots as dual modality for cancer therapy and diagnosis. *Journal of Drug Delivery Science and Technology* 2019;49:352–364.

66 Parvathaneni V, Kulkarni NS, Muth A, and Gupta V. Drug repurposing: A promising tool to accelerate the drug discovery process. *Drug Discovery Today* 2019;24:2076–2085.

67 Arrowsmith J. Phase II failures: 2008–2010. *Nature Reviews Drug Discovery* 2011;10:328–329.

68 Padhy B and Gupta Y. Drug repositioning: Re-investigating existing drugs for new therapeutic indications. *Journal of Postgraduate Medicine* 2011;57:153–160.

69 Hatzimouratidis K. Sildenafil in the treatment of erectile dysfunction: An overview of the clinical evidence. *Clinical Interventions in Aging* 2006;1:403–414.

70 Deotarse P, Jain A, Baile M, Kolhe N, and Kulkarni A. Drug repositioning: A review. *International Journal of Pharma Research & Review* 2015;4:51–58.

71 Sleigh SH and Barton CL. Repurposing strategies for therapeutics. *Pharmaceutical Medicine* 2010;24:151–159.

72 Kirtonia A, Gala K, Fernandes SG, et al. Repurposing of drugs: An attractive pharmacological strategy for cancer therapeutics. In: *Proceedings of the Seminars in Cancer Biology*, (pp. 258–278); 2021.

73 Schein CH. Repurposing approved drugs for cancer therapy. *British Medical Bulletin* 2021;137:13–27.

74 Rabben H-L, Andersen GT, Ianevski A, et al. Computational drug repositioning and experimental validation of ivermectin in treatment of gastric cancer. *Frontiers in Pharmacology* 2021;12:625991.

75 Tanoli Z, Seemab U, Scherer A, Wennerberg K, Tang J, and Vähä-Koskela M. Exploration of databases and methods supporting drug repurposing: A comprehensive survey. *Briefings in Bioinformatics* 2021;22:1656–1678.

76 Huang H, Zhang P, Qu XA, Sanseau P, and Yang L. Systematic prediction of drug combinations based on clinical side-effects. *Scientific Reports* 2014;4:7160.

77 Celebi R, Bear Don't Walk IV O, Movva R, Alpsoy S, and Dumontier M. In-silico prediction of synergistic anti-cancer drug combinations using multi-omics data. *Scientific Reports* 2019;9:8949.

78 Fu Y-S, Chen T-H, Weng L, Huang L, Lai D, and Weng C-F. Pharmacological properties and underlying mechanisms of curcumin and prospects in medicinal potential. *Biomedicine & Pharmacotherapy* 2021;141:111888.

79 Wang M, Jiang S, Zhou L, et al. Potential mechanisms of action of curcumin for cancer prevention: Focus on cellular signaling pathways and miRNAs. *International Journal of Biological Sciences* 2019;15:1200.
80 Jalilian E, Abolhasani-Zadeh F, Afgar A, Samoudi A, Zeinalynezhad H, and Langroudi L. Neutralizing tumor-related inflammation and reprogramming of cancer-associated fibroblasts by Curcumin in breast cancer therapy. *Scientific Reports* 2023;13:20770.
81 Varisli L, Cen O, and Vlahopoulos S. Dissecting pharmacological effects of chloroquine in cancer treatment: Interference with inflammatory signaling pathways. *Immunology* 2020;159:257–278.
82 Zhou W, Wang H, Yang Y, Chen Z-S, Zou C, and Zhang J. Chloroquine against malaria, cancers and viral diseases. *Drug Discovery Today* 2020;25:2012–2022.
83 Hanscheid T, del Portal Luyten CR, Hermans SM, and Grobusch MP. Repurposing of anti-malarial drugs for the treatment of tuberculosis: Realistic strategy or fanciful dead end? *Malaria Journal* 2024;23:132.
84 Zheng J. SARS-CoV-2: An emerging coronavirus that causes a global threat. *International Journal of Biological Sciences* 2020;16:1678.
85 Lechien JR, Chiesa-Estomba CM, Place S, et al. Clinical and epidemiological characteristics of 1420 European patients with mild-to-moderate coronavirus disease 2019. *Journal of Internal Medicine* 2020;288:335–344.
86 Devaux CA, Rolain J-M, Colson P, and Raoult D. New insights on the antiviral effects of chloroquine against coronavirus: what to expect for COVID-19? *International Journal of Antimicrobial Agents* 2020;55:105938.
87 Hu TY, Frieman M, and Wolfram J. Insights from nanomedicine into chloroquine efficacy against COVID-19. *Nature Nanotechnology* 2020;15:247–249.
88 Matsuyama S, Nao N, Shirato K, et al. Enhanced isolation of SARS-CoV-2 by TMPRSS2-expressing cells. *Proceedings of the National Academy of Sciences* 2020;117:7001–7003.
89 Hoffmann M, Kleine-Weber H, Schroeder S, et al. SARS-CoV-2 cell entry depends on ACE2 and TMPRSS2 and is blocked by a clinically proven protease inhibitor. *Cell* 2020;181:271–280.e8.
90 Prentice E, McAuliffe J, Lu X, Subbarao K, and Denison MR. Identification and characterization of severe acute respiratory syndrome coronavirus replicase proteins. *Journal of Virology* 2004;78:9977–9986.
91 Baby T, Liu Y, Yang G, Chen D, and Zhao C-X. Microfluidic synthesis of curcumin loaded polymer nanoparticles with tunable drug loading and pH-triggered release. *Journal of Colloid and Interface Science* 2021;594:474–484.
92 Kearney J. Chloroquine as a potential treatment and prevention measure for the 2019 novel coronavirus: a review. 2020. https://doi.org/10.20944/preprints202003.0275.v1
93 Touret F and de Lamballerie X. Of chloroquine and COVID-19. *Antiviral Research* 2020;177:104762.
94 Li X, Peng X, Zoulikha M, et al. Multifunctional nanoparticle-mediated combining therapy for human diseases. *Signal Transduction and Targeted Therapy* 2024;9:1.
95 Zhang Z, Zhou L, Xie N, et al. Overcoming cancer therapeutic bottleneck by drug repurposing. *Signal Transduction and Targeted Therapy* 2020;5:113.
96 Shrestha B, Wang L, Zhang H, Hung CY, and Tang L. Gold nanoparticles mediated drug-gene combinational therapy for breast cancer treatment. *International Journal of Nanomedicine* 2020:8109–8119.

97 Guo M, Sun X, Chen J, and Cai T. Pharmaceutical cocrystals: A review of preparations, physicochemical properties and applications. *Acta Pharmaceutica Sinica B* 2021;11:2537–2564.

98 Li Y, Teng C, Azevedo HS, Yin L, and He W. Cocrystallization-like strategy for the codelivery of hydrophobic and hydrophilic drugs in a single carrier material formulation. *Chinese Chemical Letters* 2021;32:3071–3075.

99 Liu R, Hu C, Yang Y, Zhang J, and Gao H. Theranostic nanoparticles with tumor-specific enzyme-triggered size reduction and drug release to perform photothermal therapy for breast cancer treatment. *Acta Pharmaceutica Sinica B* 2019;9:410–420.

100 Du X, Hou Y, Huang J, et al. Cytosolic delivery of the immunological adjuvant Poly I: C and cytotoxic drug crystals via a carrier-free strategy significantly amplifies immune response. *Acta Pharmaceutica Sinica B* 2021;11:3272–3285.

101 Zhang S, Langer R, and Traverso G. Nanoparticulate drug delivery systems targeting inflammation for treatment of inflammatory bowel disease. *Nano Today* 2017;16:82–96.

102 Wei Q, Li P, Yang T, et al. The promise and challenges of combination therapies with antibody-drug conjugates in solid tumors. *Journal of Hematology & Oncology* 2024;17:1.

103 Lopez JS and Banerji U. Combine and conquer: Challenges for targeted therapy combinations in early phase trials. *Nature Reviews Clinical Oncology* 2017;14:57–66.

104 Humphrey RW, Brockway-Lunardi LM, Bonk DT. et al. Opportunities and challenges in the development of experimental drug combinations for cancer. *Journal of the National Cancer Institute* 2011;103:1222–1226.

105 Jin H, Wang L, and Bernards R. Rational combinations of targeted cancer therapies: Background, advances and challenges. *Nature Reviews Drug Discovery* 2023;22:213–234.

106 Webster RM. Combination therapies in oncology. *Nature Reviews. Drug Discovery* 2016;15:81.

107 Rodon J, Perez J, and Kurzrock R. Combining targeted therapies: Practical issues to consider at the bench and bedside. *The Oncologist* 2010;15:37–50.

108 Singh SK, Mohammed A, Alghamdi OA, and Husain SM. New approaches for targeting drug resistance through drug combination. In: *Combination Therapy Against Multidrug Resistance* (pp. 221–246). Elsevier; 2020.

109 Shuhendler AJ, O'Brien PJ, Rauth AM, and Wu XY. On the synergistic effect of doxorubicin and mitomycin C against breast cancer cells. *Drug Metabolism and Drug Interactions* 2007;22:201–234.

110 Alba E, Martín M, Ramos M, et al. Multicenter randomized trial comparing sequential with concomitant administration of doxorubicin and docetaxel as first-line treatment of metastatic breast cancer: A Spanish Breast Cancer Research Group (GEICAM-9903) phase III study. *Journal of Clinical Oncology* 2004;22:2587–2593.

111 Cancello G, Bagnardi V, Sangalli C, et al. Phase II study with epirubicin, cisplatin, and infusional fluorouracil followed by weekly paclitaxel with metronomic cyclophosphamide as a preoperative treatment of triple-negative breast cancer. *Clinical Breast Cancer* 2015;15:259–265.

112 Pignata S, Lauraine EP, du Bois A, and Pisano C. Pegylated liposomal doxorubicin combined with carboplatin: A rational treatment choice for advanced ovarian cancer. *Critical Reviews in Oncology/Hematology* 2010;73:23–30.

113 Megerdichian C, Olimpiadi Y, and Hurvitz SA. Nab-Paclitaxel in combination with biologically targeted agents for early and metastatic breast cancer. *Cancer Treatment Reviews* 2014;40:614–625.

114 Pandey A, Kulkarni A, Roy B, et al. Sequential application of a cytotoxic nanoparticle and a PI3K inhibitor enhances antitumor efficacy. *Cancer Research* 2014;74:675–685.

115 Hasenstein JR, Shin H-C, Kasmerchak K, Buehler D, Kwon GS, and Kozak KR. Antitumor activity of Triolimus: A novel multidrug-loaded micelle containing Paclitaxel, Rapamycin, and 17-AAG. *Molecular Cancer Therapeutics* 2012;11:2233–2242.

116 Liu Z, Ballinger JR, Rauth AM, Bendayan R, and Wu XY. Delivery of an anticancer drug and a chemosensitizer to murine breast sarcoma by intratumoral injection of sulfopropyl dextran microspheres. *Journal of Pharmacy and Pharmacology* 2003;55:1063–1073.

117 Zhang RX, Cai P, Zhang T, et al. Polymer–lipid hybrid nanoparticles synchronize pharmacokinetics of co-encapsulated doxorubicin–mitomycin C and enable their spatiotemporal co-delivery and local bioavailability in breast tumor. *Nanomedicine: Nanotechnology, Biology and Medicine* 2016;12:1279–1290.

118 Shuhendler AJ, Cheung RY, Manias J, Connor A, Rauth AM, and Wu XY. A novel doxorubicin-mitomycin C co-encapsulated nanoparticle formulation exhibits anti-cancer synergy in multidrug resistant human breast cancer cells. *Breast Cancer Research and Treatment* 2010;119:255–269.

119 Nørgaard SM, Jensen CS, Aalestrup J, Vandenbroucke-Grauls CM, De Boer MG, and Pedersen AB. Choice of therapeutic interventions and outcomes for the treatment of infections caused by multidrug-resistant gram-negative pathogens: A systematic review. *Antimicrobial Resistance & Infection Control* 2019;8:1–13.

120 Serra-Burriel M, Keys M, Campillo-Artero C, et al. Impact of multi-drug resistant bacteria on economic and clinical outcomes of healthcare-associated infections in adults: Systematic review and meta-analysis. *PloS One* 2020;15:e0227139.

121 Niu N-K, Yin J-J, Yang Y-X, et al. Novel targeting of PEGylated liposomes for codelivery of TGF-β1 siRNA and four antitubercular drugs to human macrophages for the treatment of mycobacterial infection: A quantitative proteomic study. *Drug Design, Development and Therapy* 2015:4441–4470.

122 He C, Tang Z, Tian H, and Chen X. Co-delivery of chemotherapeutics and proteins for synergistic therapy. *Advanced Drug Delivery Reviews* 2016;98:64–76.

123 Da Silva C, Rueda F, Löwik C, Ossendorp F, and Cruz LJ. Combinatorial prospects of nano-targeted chemoimmunotherapy. *Biomaterials* 2016;83:308–320.

124 Shrestha B, Tang L, and Romero G. Nanoparticles-mediated combination therapies for cancer treatment. *Advanced Therapeutics* 2019;2:1900076.

125 Greco F and Vicent MJ. Combination therapy: Opportunities and challenges for polymer–drug conjugates as anticancer nanomedicines. *Advanced Drug Delivery Reviews* 2009;61:1203–1213.

126 Xie X, Sun T, Xue J, et al. Ag nanoparticles cluster with pH-triggered reassembly in targeting antimicrobial applications. *Advanced Functional Materials* 2020;30:2000511.

127 Krauss AC, Gao X, Li L, et al. FDA approval summary:(daunorubicin and cytarabine) liposome for injection for the treatment of adults with high-risk acute myeloid leukemia. *Clinical Cancer Research* 2019;25:2685–2690.

128 Xia Y, Sun M, Huang H, and Jin W-L. Drug repurposing for cancer therapy. *Signal Transduction and Targeted Therapy* 2024;9:92.

129 Martins M, Veiga F, Paiva-Santos AC, and Pires PC. Drug repurposing and nanotechnology for topical skin cancer treatment: Redirecting toward targeted and synergistic antitumor effects. *ACS Pharmacology & Translational Science* 2025;8:308–338.

130 Turabi KS, Deshmukh A, Paul S, et al. Drug repurposing—An emerging strategy in cancer therapeutics. *Naunyn-Schmiedeberg's Archives of Pharmacology* 2022;395:1139–1158.

131 De Lellis L, Veschi S, Tinari N, et al. Drug repurposing, an attractive strategy in pancreatic cancer treatment: Preclinical and clinical updates. *Cancers* 2021;13:3946.

132 Mohapatra P and Chandrasekaran N. Wnt/β-catenin targeting in liver carcinoma through nanotechnology-based drug repurposing: A review. *Biomedicine & Pharmacotherapy* 2022;155:113713.

133 Correia AS, Gärtner F, and Vale N. Drug combination and repurposing for cancer therapy: The example of breast cancer. *Heliyon* 2021;7:e05948.

134 Pfab C, Schnobrich L, Eldnasoury S, Gessner A, and El-Najjar N. Repurposing of antimicrobial agents for cancer therapy: What do we know? *Cancers* 2021;13:3193.

135 Ara N, Hafeez A, and Kushwaha SP. Repurposing simvastatin in cancer treatment: An updated review on pharmacological and nanotechnological aspects. *Naunyn-Schmiedeberg's Archives of Pharmacology* 2024;397:7377–7393.

136 Parhi P, Mohanty C, and Sahoo SK. Nanotechnology-based combinational drug delivery: An emerging approach for cancer therapy. *Drug Discovery Today* 2012;17:1044–1052.

137 Sahoo SK and Labhasetwar V. Nanotech approaches to drug delivery and imaging. *Drug Discovery Today* 2003;8:1112–1120.

138 Hu C-MJ, Aryal S, and Zhang L. Nanoparticle-assisted combination therapies for effective cancer treatment. *Therapeutic Delivery* 2010;1:323–334.

139 Narayanan NK, Nargi D, Randolph C, and Narayanan BA. Liposome encapsulation of curcumin and resveratrol in combination reduces prostate cancer incidence in PTEN knockout mice. *International Journal of Cancer* 2009;125:1–8.

140 Wu J, Lu Y, Lee A, et al. Reversal of multidrug resistance by transferrin-conjugated liposomes co-encapsulating doxorubicin and verapamil. *Journal of Pharmacy & Pharmaceutical Sciences* 2007;10:350–357.

141 Dicko A, Mayer LD, and Tardi PG. Use of nanoscale delivery systems to maintain synergistic drug ratios in vivo. *Expert Opinion on Drug Delivery* 2010;7:1329–1341.

142 Batist G, Gelmon KA, Chi KN, et al. Safety, pharmacokinetics, and efficacy of CPX-1 liposome injection in patients with advanced solid tumors. *Clinical Cancer Research* 2009;15:692–700.

143 Blagosklonny MV. "Targeting the absence" and therapeutic engineering for cancer therapy. *Cell Cycle* 2008;7:1307–1312.

144 Li F, Wang Y, Chen W-L, et al. Co-delivery of VEGF siRNA and etoposide for enhanced anti-angiogenesis and anti-proliferation effect via multi-functional nanoparticles for orthotopic non-small cell lung cancer treatment. *Theranostics* 2019;9:5886.

145 Nakamura H and Takada K. Reactive oxygen species in cancer: Current findings and future directions. *Cancer Science* 2021;112:3945–3952.

146 Kohan R, Collin A, Guizzardi S, Tolosa de Talamoni N, and Picotto G. Reactive oxygen species in cancer: A paradox between pro-and anti-tumour activities. *Cancer Chemotherapy and Pharmacology* 2020;86:1–13.

147 AruIJothi K, Kumaran K, Senthil S, et al. Implications of reactive oxygen species in lung cancer and exploiting it for therapeutic interventions. *Medical Oncology* 2022;40:43.

148 Sarmiento-Salinas FL, Perez-Gonzalez A, Acosta-Casique A, et al. Reactive oxygen species: Role in carcinogenesis, cancer cell signaling and tumor progression. *Life Sciences* 2021;284:119942.

149 Mirzaei S, Gholami MH, Hashemi F, et al. Advances in understanding the role of P-gp in doxorubicin resistance: Molecular pathways, therapeutic strategies, and prospects. *Drug Discovery Today* 2022;27:436–455.
150 Wang Y, Yu J, Li D, et al. Paclitaxel derivative-based liposomal nanoplatform for potentiated chemo-immunotherapy. *Journal of Controlled Release* 2022;341:812–827.
151 Zhang J, Duan D, Song ZL, Liu T, Hou Y, and Fang J. Small molecules regulating reactive oxygen species homeostasis for cancer therapy. *Medicinal Research Reviews* 2021;41:342–394.
152 Zong Q, Li J, Xiao X, Du X, and Yuan Y. Self-amplified chain-shattering cinnamaldehyde-based poly (thioacetal) boosts cancer chemo-immunotherapy. *Acta Biomaterialia* 2022;154:97–107.
153 Xiao Q, Li X, Liu C, et al. Liposome-based anchoring and core-encapsulation for combinatorial cancer therapy. *Chinese Chemical Letters* 2022;33:4191–4196.
154 Lv Y, Xu C, Zhao X, et al. Nanoplatform assembled from a CD44-targeted prodrug and smart liposomes for dual targeting of tumor microenvironment and cancer cells. *ACS Nano* 2018;12:1519–1536.
155 Xiao Q, Li X, Liu C, et al. Improving cancer immunotherapy via co-delivering checkpoint blockade and thrombospondin-1 downregulator. *Acta Pharmaceutica Sinica B* 2023;13:3503–3517.
156 Wang F, Zhang D, Zhang Q, et al. Synergistic effect of folate-mediated targeting and verapamil-mediated P-gp inhibition with paclitaxel-polymer micelles to overcome multi-drug resistance. *Biomaterials* 2011;32:9444–9456.
157 Katragadda U, Teng Q, Rayaprolu BM, Chandran T, and Tan C. Multi-drug delivery to tumor cells via micellar nanocarriers. *International Journal of Pharmaceutics* 2011;419:281–286.
158 Shin H-C, Alani AW, Cho H, Bae Y, Kolesar JM, and Kwon GS. A 3-in-1 polymeric micelle nanocontainer for poorly water-soluble drugs. *Molecular Pharmaceutics* 2011;8:1257–1265.
159 Misra R and Sahoo SK. Coformulation of doxorubicin and curcumin in poly (D, L-lactide-co-glycolide) nanoparticles suppresses the development of multidrug resistance in K562 cells. *Molecular Pharmaceutics* 2011;8:852–866.
160 Krakovičová H, Etrych T, and Ulbrich K. HPMA-based polymer conjugates with drug combination. *European Journal of Pharmaceutical Sciences* 2009;37:405–412.
161 DeAngelis LM, Seiferheld W, Schold SC, Fisher B, and Schultz CJ. Combination chemotherapy and radiotherapy for primary central nervous system lymphoma: Radiation Therapy Oncology Group Study 93-10. *Journal of Clinical Oncology* 2002;20:4643–4648.
162 Rajora AK, Ravishankar D, Zhang H, and Rosenholm JM. Recent advances and impact of chemotherapeutic and antiangiogenic nanoformulations for combination cancer therapy. *Pharmaceutics* 2020;12:592.
163 Li Y, Gao Y, Zhang X, Guo H, and Gao H. Nanoparticles in precision medicine for ovarian cancer: From chemotherapy to immunotherapy. *International Journal of Pharmaceutics* 2020;591:119986.
164 Orlandi L, Zaffaroni N, Bearzatto A, Villa R, De Marco C, and Silvestrini R. Lonidamine as a modulator of taxol activity in human ovarian cancer cells: Effects on cell cycle and induction of apoptosis. *International Journal of Cancer* 1998;78:377–384.

165 Taghipour-Sabzevar V, Sharifi T, and Moghaddam MM. Polymeric nanoparticles as carrier for targeted and controlled delivery of anticancer agents. *Therapeutic Delivery* 2019;10:527–550.

166 Shim G, Kim M-G, Kim D, Park JY, and Oh Y-K. Nanoformulation-based sequential combination cancer therapy. *Advanced Drug Delivery Reviews* 2017;115:57–81.

167 Mohanty S, Rashid MHA, Mridul M, Mohanty C, and Swayamsiddha S. Application of Artificial Intelligence in COVID-19 drug repurposing. *Diabetes & Metabolic Syndrome: Clinical Research & Reviews* 2020;14:1027–1031.

168 Tanoli Z, Vähä-Koskela M, and Aittokallio T. Artificial intelligence, machine learning, and drug repurposing in cancer. *Expert Opinion on Drug Discovery* 2021;16:977–989.

169 Aspatwar A, Nath R, Bhowmik R, and Jesudasan R. Exploring the applications of formulation-based drug development strategies in neurological disorders using artificial intelligence and machine learning approaches. *ChemRxiv* 2024.

170 Chehelgerdi M, Chehelgerdi M, Allela OQB, et al. Progressing nanotechnology to improve targeted cancer treatment: Overcoming hurdles in its clinical implementation. *Molecular Cancer* 2023;22:169.

171 Yousefnezhad M, Davaran S, Babazadeh M, Akbarzadeh A, and Pazoki-Toroudi H. PCL-based nanoparticles for doxorubicin-ezetimibe co-delivery: A combination therapy for prostate cancer using a drug repurposing strategy. *BioImpacts: BI* 2023;13:241.

172 Chakraborty S, Vishwas S, Harish V, et al. Exploring nanoparticular platform in delivery of repurposed drug for Alzheimer's disease: Current approaches and future perspectives. *Expert Opinion on Drug Delivery* 2024;21:1771–1792.

173 Hernández-Parra H, Cortés H, Avalos-Fuentes JA, et al. Repositioning of drugs for Parkinson's disease and pharmaceutical nanotechnology tools for their optimization. *Journal of Nanobiotechnology* 2022;20:413.

174 Pawar VA, Tyagi A, Verma C, et al. Unlocking therapeutic potential: Integration of drug repurposing and immunotherapy for various disease targeting. *American Journal of Translational Research* 2023;15:4984.

175 Chambial P, Thakur N, Preethi L, and Mehta P. Anti-cancer nanotherapeutics in clinical trials: The work behind clinical translation of nanomedicine. In: *Nanoparticles in Cancer Therapy* (pp. 203–222); Taylor & Francis Group; 2024.

176 Gabizon AA, Gabizon-Peretz S, Modaresahmadi S, and La-Beck NM. Thirty years from FDA approval of pegylated liposomal doxorubicin (Doxil/Caelyx): An updated analysis and future perspective. *BMJ Oncology* 2025;4:e000573.

177 Milano G, Innocenti F, and Minami H. Liposomal irinotecan (Onivyde): Exemplifying the benefits of nanotherapeutic drugs. *Cancer Science* 2022;113:2224–2231.

178 Lyon PC, Griffiths LF, Lee J, et al. Clinical trial protocol for TARDOX: A phase I study to investigate the feasibility of targeted release of lyso-thermosensitive liposomal doxorubicin (ThermoDox®) using focused ultrasound in patients with liver tumours. *Journal of Therapeutic Ultrasound* 2017;5:1–8.

179 Malla R, Viswanathan S, Makena S, et al. Revitalizing cancer treatment: Exploring the role of drug repurposing. *Cancers* 2024;16:1463.

180 Farcal L, Torres Andón F, Di Cristo L, et al. Comprehensive in vitro toxicity testing of a panel of representative oxide nanomaterials: First steps towards an intelligent testing strategy. *PloS One* 2015;10:e0127174.

181 Li Y, Kröger M, and Liu WK. Shape effect in cellular uptake of PEGylated nanoparticles: Comparison between sphere, rod, cube and disk. *Nanoscale* 2015;7:16631–16646.
182 Carnovale C, Bryant G, Shukla R, and Bansal V. Size, shape and surface chemistry of nano-gold dictate its cellular interactions, uptake and toxicity. *Progress in Materials Science* 2016;83:152–190.
183 Liu Y, Zhu S, Gu Z, Chen C, and Zhao Y. Toxicity of manufactured nanomaterials. *Particuology* 2022;69:31–48.
184 Zhou X, Jin W, Sun H, Li C, and Jia J. Perturbation of autophagy: An intrinsic toxicity mechanism of nanoparticles. *Science of the Total Environment* 2022;823:153629.
185 Mocan T, Clichici S, Agoşton-Coldea L, et al. Implications of oxidative stress mechanisms in toxicity of nanoparticles. *Acta Physiologica Hungarica* 2010;97:247–255.
186 Sajid M, Ilyas M, Basheer C, et al. Impact of nanoparticles on human and environment: Review of toxicity factors, exposures, control strategies, and future prospects. *Environmental Science and Pollution Research* 2015;22:4122–4143.
187 Nemmar A, Al-Salam S, Beegam S, Yuvaraju P, and Ali BH. Aortic oxidative stress, inflammation and DNA damage following pulmonary exposure to cerium oxide nanoparticles in a rat model of vascular injury. *Biomolecules* 2019;9:376.
188 Ragelle H, Danhier F, Préat V, Langer R, and Anderson DG. Nanoparticle-based drug delivery systems: A commercial and regulatory outlook as the field matures. *Expert Opinion on Drug Delivery* 2017;14:851–864.
189 Björnmalm M, Yan Y, and Caruso F. Engineering and evaluating drug delivery particles in microfluidic devices. *Journal of Controlled Release* 2014;190:139–149.
190 Desai N. Challenges in development of nanoparticle-based therapeutics. *The AAPS Journal* 2012;14:282–295.
191 Sainz V, Conniot J, Matos AI, et al. Regulatory aspects on nanomedicines. *Biochemical and Biophysical Research Communications* 2015;468:504–510.
192 Hamburg MA. FDA's approach to regulation of products of nanotechnology. *Science* 2012;336:299–300.
193 Gaspar RS, Florindo HF, Silva LC, et al. Regulatory aspects of oncologicals: nanosystems main challenges. In: *Nano-oncologicals: New Targeting and Delivery Approaches* (pp. 425–452). Springer; 2014.
194 Tinkle S, McNeil SE, Mühlebach S, et al. Nanomedicines: Addressing the scientific and regulatory gap. *Annals of the New York Academy of Sciences* 2014;1313:35–56.
195 Hafner A, Lovrić J, Lakoš GP, and Pepić I. Nanotherapeutics in the EU: An overview on current state and future directions. *International Journal of Nanomedicine* 2014;9:1005–1023.
196 Alghamdi MA, Fallica AN, Virzì N, Kesharwani P, Pittalà V, and Greish K. The promise of nanotechnology in personalized medicine. *Journal of Personalized Medicine* 2022;12:673.
197 Auffan M, Rose J, Bottero J-Y, Lowry GV, Jolivet J-P, and Wiesner MR. Towards a definition of inorganic nanoparticles from an environmental, health and safety perspective. *Nature Nanotechnology* 2009;4:634–641.
198 Ventola CL. The nanomedicine revolution: Part 1: Emerging concepts. *Pharmacy and Therapeutics* 2012;37:512.
199 Onoue S, Yamada S, and Chan H-K. Nanodrugs: Pharmacokinetics and safety. *International Journal of Nanomedicine* 2014;9:1025–1037.

200 Pereira MC, Reshetnyak YK, and Andreev OA. Advanced targeted nanomedicine. *Journal of Biotechnology* 2015;202:88–97.

201 Roger E, Lagarce F, Garcion E, and Benoit J-P. Biopharmaceutical parameters to consider in order to alter the fate of nanocarriers after oral delivery. *Nanomedicine* 2010;5:287–306.

202 Pathade AD, Kommineni N, Bulbake U, Thummar MM, Samanthula G, and Khan W. Preparation and comparison of oral bioavailability for different nano-formulations of olaparib. *AAPS PharmSciTech* 2019;20:1–13.

203 Wang C, Zhu J, Zhang D, et al. Ionic liquid–microemulsions assisting in the transdermal delivery of Dencichine: Preparation, in-vitro and in-vivo evaluations, and investigation of the permeation mechanism. *International Journal of Pharmaceutics* 2018;535:120–131.

204 Tashima T. Delivery of orally administered digestible antibodies using nanoparticles. *International Journal of Molecular Sciences* 2021;22:3349.

205 Haider M, Elsherbeny A, Pittalà V, Fallica, AN, Alghamdi MA, and Greish K. The potential role of sildenafil in cancer management through EPR augmentation. *Journal of Personalized Medicine* 2021;11:585.

206 Markman JL, Rekechenetskiy A, Holler E, and Ljubimova JY. Nanomedicine therapeutic approaches to overcome cancer drug resistance. *Advanced Drug Delivery Reviews* 2013;65:1866–1879.

207 Ambade AV, Savariar EN, and Thayumanavan S. Dendrimeric micelles for controlled drug release and targeted delivery. *Molecular Pharmaceutics* 2005;2:264–272.

208 Swami R, Singh I, Jeengar MK, Naidu V, Khan W, and Sistla R. Adenosine conjugated lipidic nanoparticles for enhanced tumor targeting. *International Journal of Pharmaceutics* 2015;486:287–296.

209 Patil Y, Amitay Y, Ohana, P, Shmeeda, H, and Gabizon, A. Targeting of pegylated liposomal mitomycin-C prodrug to the folate receptor of cancer cells: Intracellular activation and enhanced cytotoxicity. *Journal of Controlled Release* 2016;225:87–95.

210 Gromnicova R, Davies HA, Sreekanthreddy P, et al. Glucose-coated gold nanoparticles transfer across human brain endothelium and enter astrocytes in vitro. *PloS One* 2013;8:e81043.

211 Cui T, Zhang S, and Sun H. Co-delivery of doxorubicin and pH-sensitive curcumin prodrug by transferrin-targeted nanoparticles for breast cancer treatment. *Oncology Reports* 2017;37:1253–1260.

212 Arya G, Vandana M, Acharya S, and Sahoo SK. Enhanced antiproliferative activity of Herceptin (HER2)-conjugated gemcitabine-loaded chitosan nanoparticle in pancreatic cancer therapy. *Nanomedicine: Nanotechnology, Biology and Medicine* 2011;7:859–870.

213 Sapra P, Moase EH, Ma J, and Allen TM. Improved therapeutic responses in a xenograft model of human B lymphoma (Namalwa) for liposomal vincristine versus liposomal doxorubicin targeted via anti-CD19 IgG2a or Fab' fragments. *Clinical Cancer Research* 2004;10:1100–1111.

214 Lopes-Nunes J, Agonia AS, Rosado T, et al. Aptamer-functionalized gold nanoparticles for drug delivery to gynecological carcinoma cells. *Cancers* 2021;13:4038.

215 Kumthekar P, Ko CH, Paunesku T, et al. A first-in-human phase 0 clinical study of RNA interference–based spherical nucleic acids in patients with recurrent glioblastoma. *Science Translational Medicine* 2021;13:eabb3945.

216 Moyano DF, Saha K, Prakash G, et al. Fabrication of corona-free nanoparticles with tunable hydrophobicity. *ACS Nano* 2014;8:6748–6755.

217 Gupta T, Singh J, Kaur S, Sandhu S, Singh G, and Kaur IP. Enhancing bioavailability and stability of curcumin using solid lipid nanoparticles (CLEN): A covenant for its effectiveness. *Frontiers in Bioengineering and Biotechnology* 2020;8:879.

218 Ni M, Xiong M, Zhang X, et al. Poly (lactic-co-glycolic acid) nanoparticles conjugated with CD133 aptamers for targeted salinomycin delivery to CD133+ osteosarcoma cancer stem cells. *International Journal of Nanomedicine* 2015;10:2537–2554.

219 Gao Q, Zhang J, Gao J, Zhang Z, Zhu H, and Wang D. Gold nanoparticles in cancer theranostics. *Frontiers in Bioengineering and Biotechnology* 2021;9:647905.

220 Gul S, Khan SB, Rehman IU, Khan MA, and Khan M. A comprehensive review of magnetic nanomaterials modern day theranostics. *Frontiers in Materials* 2019;6:179.

221 Peng H, Liu X, Wang G, et al. Polymeric multifunctional nanomaterials for theranostics. *Journal of Materials Chemistry B* 2015;3:6856–6870.

222 Knežević NŽ and Kaluđerović GN. Silicon-based nanotheranostics. *Nanoscale* 2017;9:12821–12829.

223 Patel KD, Singh RK, and Kim H-W. Carbon-based nanomaterials as an emerging platform for theranostics. *Materials Horizons* 2019;6:434–469.

224 Dykman LA and Khlebtsov NG. Multifunctional gold-based nanocomposites for theranostics. *Biomaterials* 2016;108:13–34.

12

Nanotechnology-based Repurposing and Combination Therapy with Case Studies and Mechanisms

Amol D. Gholap[1,2], Pankaj R. Khuspe[3], Sadikali F. Sayyad[2*], Navnath Hatvate[4], and Machindra Chavan[5]*

[1]*Department of Pharmaceutics, St. John Institute of Pharmacy and Research, Palghar, India*
[2]*Department of Pharmaceutics, Amrutvahini College of Pharmacy, Sangamner, India*
[3]*Department of Pharmaceutics, College of Pharmacy, Paniv, Solapur, India*
[4]*Department of Pharmaceutical Sciences and Technology, Institute of Chemical Technology Mumbai, Marathwada Campus, Jalna, India*
[5]*Department of Pharmacognosy, Amrutvahini College of Pharmacy, Sangamner, India*

*Corresponding authors: amolgholap16@gmail.com; sadiksayyad@gmail.com

CONTENTS

12.1 Introduction

In contemporary pharmacology, combination therapy and drug repurposing have become viable approaches to boost treatment effectiveness and speed up drug development. Drug repurposing, sometimes referred to as drug repositioning, is the process of finding new therapeutic uses for already-approved medications in order to get around the exorbitant expenses, drawn-out processes, and legal restrictions that come with de novo drug discovery [1]. When it comes to treating unmet medical requirements like rare diseases, antibiotic resistance, and complicated pathologies like cancer and neurological disorders, this strategy is especially beneficial. However, inherent difficulties such as low aqueous solubility, quick systemic clearance, low bioavailability, and inferior pharmacokinetics frequently impede the practical translation of repurposed medications. Similarly, combination therapy provides synergistic effects that improve therapeutic outcomes and reduce drug resistance by using various pharmacological agents to target different disease pathways. Despite its potential, pharmacokinetic mismatches, off-target toxicities, and challenges in attaining synchronized and regulated drug release sometimes limit combination therapy [2].

By utilizing nanoscale carriers including liposomes, polymeric NPs, dendrimers, micelles, and inorganic nanomaterials for precision drug administration, nanotechnology offers a creative and adaptable way to overcome these constraints. These nanocarriers optimize the therapeutic index and reduce side effects by providing improved drug solubility, extended circulation time, targeted distribution to particular tissues or cellular compartments, and regulated release kinetics [3]. By encapsulating current medications in biocompatible nanocarriers that improve permeability across physiological barriers, such as the tumor microenvironment and the blood–brain barrier (BBB), nanotechnology enables the repurposing of existing medications. Furthermore, in combination therapy, sophisticated nanocarrier systems allow for the regulated spatial and temporal loading and delivery of numerous medications, increasing synergy and decreasing systemic toxicity. For example, lipid-based nanostructures and polymeric NPs have been effectively used to deliver anticancer medications with several modes of action, resulting in increased effectiveness against tumors that are resistant to multiple therapies. Furthermore, by guaranteeing site-specific release and reducing systemic exposure, stimuli-responsive nanocarriers, which release therapeutic agents in response to endogenous (such as pH, enzymes, or redox gradients) or exogenous (such as light, magnetic fields, or ultrasound) triggers, further enhance drug action [4].

The practical application of medication repurposing and combination therapy based on nanotechnology still faces a number of obstacles in spite of these developments. Thorough preclinical and clinical research is required because of the scalability of nanocarrier manufacturing, batch-to-batch repeatability, stability during storage, and possible immunogenicity [5]. To expedite the process of bringing nanopharmaceuticals from the bench to the bedside, regulatory regimes that oversee their clearance also need to be harmonized. In order to optimize nanoparticle (NP) design, improve predictive modeling of drug–nanocarrier interactions, and enable personalized nanomedicine techniques catered to patient-specific pathophysiology, future research should concentrate on combining AI and machine learning (ML) as presented in Figure 12.1 [6]. Furthermore, it is anticipated that nanotechnology's significance in medication repurposing and combination therapy

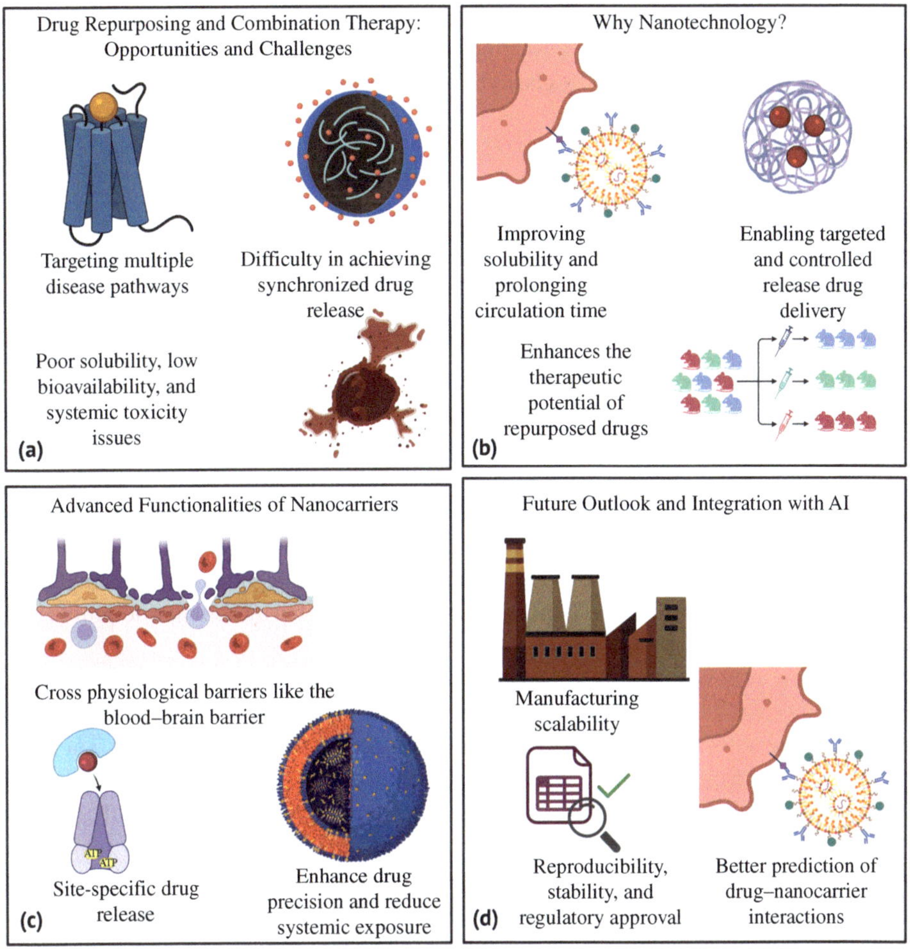

Figure 12.1 This figure elucidates the contributions of nanotechnology and artificial intelligence (AI) in the domains of drug repurposing and combination therapy. (a) Traditional challenges encompass issues such as poor solubility, low bioavailability, systemic toxicity, and the difficulty of achieving synchronized drug release. (b) Nanotechnology addresses these challenges by enhancing solubility, prolonging circulation, and enabling targeted, controlled release, thereby improving therapeutic efficacy. (c) Advanced nanocarriers are capable of traversing physiological barriers, such as the blood–brain barrier, facilitating site-specific delivery and minimizing systemic exposure through precise drug release. (d) Future research directions emphasize the importance of manufacturing scalability, regulatory compliance, and the integration of AI to enhance the prediction of drug–nanocarrier interactions, thereby expediting approval processes and improving therapeutic outcomes in repurposed and combination drug therapies. Created by using Biorender.com.

will grow as it converges with cutting-edge disciplines including gene therapy, immunotherapy, and precision medicine [7].

By exploring recent developments, technological advancements, current difficulties, and prospective future directions, this chapter seeks to present a thorough examination of the relationship between nanotechnology, medication repurposing, and combination therapy.

This talk will provide important insights into how nanoscale engineering is changing the field of contemporary therapies and opening the door for more efficient, individualized, and focused treatment modalities by clarifying the basic ideas and translational possibilities of nanomedicine [8].

12.2 Nanotechnology in Drug Repurposing

12.2.1 Advantages

12.2.1.1 Improved Solubility

The low water solubility of many pharmacological substances, which can significantly limit their therapeutic efficiency by lowering bioavailability and absorption, is one of the biggest obstacles to medication repurposing [9]. Conventional medication formulations frequently have trouble dissolving in physiological fluids, which leads to unpredictable pharmacokinetics, inadequate dosage, and the requirement for high drug concentrations that could cause systemic toxicity. By providing nanoscale delivery systems that improve the solubility and dispersion of hydrophobic medications, nanotechnology has completely changed this field and improved the pharmacokinetic and pharmacodynamic profiles of these medications [10].

By encasing poorly water-soluble medications in hydrophilic polymeric matrices, lipid bilayers, or inorganic nanostructures, NPs act as effective delivery systems. Composed of biocompatible substances like chitosan, poly(lactic-co-glycolic acid) (PLGA), and polyethylene glycol (PEG), polymeric NPs create hydrophilic shells around lipophilic drug cores, enhancing aqueous solubility and prolonging drug release [11]. Similarly, hydrophobic compounds are solubilized by lipid-based nanocarriers like liposomes, solid lipid nanoparticles (SLNs), and nanoemulsions, which produce an amphiphilic environment that speeds up systemic circulation and facilitates the passage of these molecules across biological membranes. Long-term drug retention in the bloodstream is ensured by surface-modifying NPs with hydrophilic polymers or surfactants, which also lessen aggregation and early clearance [12].

The capacity of nanocarriers to solubilize hydrophobic molecules is very helpful in therapeutic repurposing since it allows for the rediscovery and reformulation of existing medications that previously failed clinical trials because of their poor solubility. For example, repurposed anticancer drugs with strong bioactivity but low water solubility, like resveratrol and curcumin, have been successfully nanoformulated to enhance their therapeutic efficacy. Furthermore, an alternate method to increase dissolution rates and boost oral bioavailability is to use nanosuspensions, which stabilize poorly soluble medications in sub-micron colloidal dispersions. By encasing hydrophobic medications in amphiphilic copolymers, recent developments in self-assembling nanomicelles have further aided in their solubilization, providing for improved stability and targeted delivery [13].

In addition to improving solubility, drug repurposing techniques based on nanotechnology guarantee better medication absorption at the site of action, especially for oral and intravenous delivery. Drug-loaded NPs' nanoscopic size increases their ability to interact with biological membranes, promoting transcellular transport through receptor-mediated uptake, passive diffusion, or endocytosis. Additionally, site-specific drug release is made possible by stimuli-responsive nanocarriers, which change their physicochemical characteristics in response to pH, temperature, or enzyme activity. This maximizes therapeutic efficacy while reducing off-target effects. In conditions like cancer, neurological illnesses, and infectious diseases, where medication solubility and controlled distribution are essential to the effectiveness of treatment, this is especially important [14].

Despite these developments, there are still issues with assuring batch-to-batch consistency, sustaining long-term stability, and scaling up NP-based formulations for commercial manufacturing. Future studies should concentrate on improving the design of nanocarriers using sophisticated production methods such as continuous flow synthesis and microfluidics, ML-guided drug-excipient selection, and computational modeling. In order to guarantee the safety, effectiveness, and translational viability of nanopharmaceuticals, regulatory frameworks must also change. All things considered, using nanotechnology in drug repurposing provides a revolutionary way to get over solubility restrictions, unleash the therapeutic potential of hitherto underutilized medications, and hasten their clinical use in a variety of illness scenarios [15].

12.2.1.2 Targeted Delivery

One of the most revolutionary benefits of nanotechnology in drug repurposing is targeted drug delivery, which makes it possible to precisely and effectively localize therapeutic agents to sick tissues while reducing systemic toxicity. Conventional drug delivery techniques can lead to broad distribution, which increases unpleasant reactions, causes off-target effects, and results in less than ideal drug accumulation at the illness site. By using functionalized nanocarriers such as liposomes, polymeric NPs, dendrimers, mesoporous silica NPs, and metallic nanocarriers that can be created with surface alterations to provide active or passive targeting, nanotechnology gets beyond these restrictions [16]. The enhanced permeability and retention (EPR) effect is used by passive targeting, especially in tumors and inflammatory tissues where the accumulation of NPs is facilitated by leaky vasculature. Nevertheless, active targeting entails functionalizing NPs with ligands like aptamers, peptides, antibodies, or small molecules that attach to overexpressed receptors on sick cells, guaranteeing selective drug internalization and improving specificity.

When it comes to drug repurposing, targeted nanocarriers are essential for re-engineering current medications to get around delivery issues related to certain illnesses. For example, the therapeutic efficiency of many repurposed medications is hampered by their poor solubility, short half-lives, or insufficient bioavailability. When these drugs are encapsulated in NPs, their circulation period can be extended, they can be protected from premature degradation, and their regulated release can be facilitated once they reach the target site [17]. Additionally, by avoiding enzymatic degradation and efflux pumps, functionalized NPs improve the bioactivity of repurposed medications by facilitating their intracellular transport. This is especially helpful for treating infectious diseases, malignancies, and neurological conditions when selective tissue penetration and medication resistance are significant issues. Repurposed chemotherapeutics, for instance, can be incorporated into ligand-functionalized NPs in oncology to target tumor cells specifically, increasing cytotoxicity while preserving healthy tissues. Similar to this, nanocarriers coupled with transferrin or low-density lipoprotein (LDL) receptor ligands help to penetrate the BBB in neurodegenerative illnesses, guaranteeing efficient medication delivery to the central nervous system (CNS) [18].

Multifunctional nanocarriers that can release drugs dynamically and in response to stimuli have been developed as a result of recent developments in nanotechnology. The safety and effectiveness of repurposed medications are further enhanced by these systems, which include pH-sensitive, enzyme-responsive, or redox-triggered processes that enable precise control over drug activation at the target site [19]. Furthermore, theranostic NPs that incorporate imaging modalities like fluorescence, magnetic resonance imaging (MRI), and positron emission tomography (PET) allow for real-time drug distribution tracking, which offers vital information about therapy efficacy and biodistribution [20].

Targeted nanomedicine has enormous potential, but clinical translation is fraught with difficulties, including immune system recognition, large-scale repeatability, and regulatory complexity. Innovations in nanomaterial fabrication, personalized nanomedicine strategies, and thorough preclinical testing to guarantee safety and effectiveness are needed to address these problems [21]. In order to create more efficient targeted delivery systems, future research should concentrate on maximizing surface changes for NPs, investigating hybrid nanocarriers, and utilizing AI-driven predictive modeling. Nanotechnology-driven targeted drug repurposing has enormous potential to transform therapeutic approaches for complicated and refractory diseases by improving precision medicine tactics [22].

12.2.1.3 Prolonged Circulation

The capacity of nanotechnology to prolong the systemic circulation duration of therapeutic compounds by shielding them from enzymatic degradation and premature clearance is one of its main benefits in drug repurposing [23–28]. The bioavailability and therapeutic efficacy of conventional small-molecule medicines are greatly diminished by their fast metabolism, renal filtration, and hepatic clearance. Liposomes, polymeric NPs, dendrimers, and lipid-based systems are examples of nanocarriers that serve as protective reservoirs, encasing repurposed medications to prevent oxidative metabolism and enzymatic hydrolysis [29–33]. By guaranteeing a prolonged release profile, this encapsulation enhances the drug's pharmacokinetic profile and therapeutic index by enabling it to stay in circulation for a longer amount of time [34].

The main way that nanocarriers extend circulation time is by altering their surfaces to increase stability and prevent quick immune clearance. For example, grafting PEG chains onto the nanocarrier surface to produce a hydrophilic steric barrier is a commonly used technique known as PEGylation. The mononuclear phagocyte system (MPS) reduces macrophage-mediated absorption by blocking opsonization, which is the mechanism by which plasma proteins tag foreign particles for immune detection. Consequently, PEGylated NPs have a longer plasma half-life, which makes it easier for repurposed medications to get to their intended locations. Additionally, biomimetic strategies that encapsulate NPs in coatings derived from cell membranes, such as leukocyte or erythrocyte membranes, have shown great promise in extending circulation and improving immune evasion [35].

The size, shape, and surface charge of nanocarriers are important factors that affect circulation dynamics in addition to their stealth characteristics [36–38]. Through the EPR effect, NPs in the ideal size range of 10–200 nm show decreased renal filtration and increased accumulation in sick tissues, especially in tumors and inflammatory tissues where vascular permeability is impaired [39]. Furthermore, circulation duration and biodistribution can be further altered by surface functionalization using ligands that target particular transport proteins or receptors. Transferrin-coated NPs, for instance, have been demonstrated to take advantage of receptor-mediated transcytosis, allowing for extended circulation and effective entry into the CNS for repurposed medications intended to treat neurodegenerative illnesses [40].

Long-term circulation has several advantages; however, there are still difficulties in maximizing nanocarrier stability and reducing possible immunogenicity. Sustained drug delivery is limited by the accelerated blood clearance (ABC) phenomenon, in which repeated administration of PEGylated NPs causes an immunological response that speeds up removal. Advanced techniques are being investigated to lessen this problem and extend the durability of drug repurposing methods based on nanocarriers, including zwitterionic coatings, biomimetic surface engineering, and dynamic surface alterations [41].

In the field of drug repurposing, the potential of nanocarriers to prolong drug circulation time is a revolutionary development that enables current therapies to obtain improved tissue penetration, increased systemic persistence, and improved therapeutic effects. These systems will continue to be improved by the incorporation of novel surface modifications and biomimetic techniques, opening the door for future medication repurposing efforts based on nanomedicine that are more successful [42].

12.2.2 Case Studies

12.2.2.1 Cancer Therapy: Lipid Nanoparticles Co-loaded with Paclitaxel and Doxorubicin for Metastatic Breast Cancer

Because of its aggressiveness, high recurrence rate, and the development of MDR, metastatic breast cancer continues to pose a significant problem in oncology. Despite their effectiveness, conventional chemotherapy regimens are frequently linked to systemic toxicity, fast drug clearance, and inadequate tumor penetration. Lipid nanoparticles (LNPs) have become a very effective nanoscale delivery method that can improve the therapeutic index of combination chemotherapy in order to overcome these drawbacks. A promising method for achieving synergistic anticancer effects while minimizing off-target toxicities is the co-loading of doxorubicin (DOX), an anthracycline that inhibits topoisomerase II and intercalates DNA, and paclitaxel (PTX), a microtubule-stabilizing agent, into lipid-based nanocarriers.

When PTX and DOX are delivered together, LNPs offer a number of benefits. Due to the EPR effect, these NPs have improved tumor accumulation, regulated release kinetics, and good biocompatibility. The systemic stability of PTX and DOX is greatly enhanced by enclosing them in a lipid bilayer or core-shell nanostructure, which delays early breakdown and increases circulation time. Additionally, PEG can be used to surface-functionalize lipid-based nanocarriers to give them stealth qualities. This will help them avoid being recognized and cleared by the MPS, which will boost drug accumulation at the tumor site. Furthermore, receptor-mediated endocytosis is made possible by targeted LNPs coupled with ligands like folic acid or monoclonal antibodies, which improves cancer cells' selective absorption and reduces exposure to healthy tissues.

When compared to free drug formulations, preclinical research has shown that LNPs co-loaded with PTX and DOX have greater cytotoxicity against metastatic breast cancer cells. More efficiently than monotherapy, the combined action of PTX and DOX induces apoptosis and inhibits tumor proliferation by interfering with several hallmarks of cancer, including mitotic progression and DNA replication. Additionally, by facilitating both temporal and spatial drug release, these nanosystems guarantee that both medications reach their intracellular targets at the ideal quantities. The safety profile of combination therapy is enhanced by the controlled drug release profile, which also lessens the cardiotoxicity linked to DOX and the neurotoxicity linked to PTX.

The large-scale production, reproducibility, and regulatory approval of LNP-based combination medicines continue to present difficulties in spite of these developments. To preserve stability and effectiveness, LNPs' physicochemical characteristics such as size, charge, and drug loading capacity must be carefully adjusted. Further development of NP engineering is also necessary to overcome tumor heterogeneity and MDR mechanisms. This includes adding stimuli-responsive components that allow drug release in response to tumor-specific triggers like pH, enzymes, or redox gradients. To further improve clinical outcomes in the treatment of metastatic breast cancer, future research should concentrate on combining nanotechnology with patient-specific nanomedicine, formulation design driven by artificial intelligence, and combination therapies that use immunomodulators or gene therapy techniques [43]. The role of nanotechnology in cancer is represented in Figure 12.2.

12.2.2.2 Thalidomide: Repurposed for Multiple Myeloma, Liposomal Formulations Improve Its Therapeutic Index

Originally created as a sedative and then discontinued because of its teratogenic effects, thalidomide has been effectively repurposed as an immunomodulatory and anti-angiogenic drug for the treatment of multiple myeloma (MM). Conventional thalidomide therapy has serious pharmacokinetic drawbacks, such as poor water solubility, fast systemic clearance, and dose-dependent neurotoxicity, despite its therapeutic effectiveness. In order to improve its therapeutic index and reduce side effects, these restrictions call for creative delivery methods. One intriguing strategy to maximize thalidomide's bioavailability, targeted distribution, and controlled release kinetics in combination therapy for MM is the use of nanotechnology, especially liposomal formulations.

For the administration of thalidomide, liposomal nanocarriers, which are made of phospholipid bilayers encasing hydrophobic or hydrophilic drug payloads, offer a flexible and biocompatible platform. By increasing the drug's solubility and stability, liposomal encapsulation improves systemic circulation and permits continuous drug release. More significantly, liposomes can be functionalized with targeted ligands, like peptides or monoclonal antibodies, to produce EPR effects that allow for selective accumulation at the tumor microenvironment. When used with immunomodulatory drugs like lenalidomide and proteasome inhibitors like bortezomib, liposomal thalidomide formulations have shown greater therapeutic efficacy in the treatment of MM. By utilizing thalidomide's capacity to suppress angiogenesis and increase the cytotoxic effects of concurrently delivered drugs, this combinatorial strategy improves patient outcomes and lowers systemic toxicity.

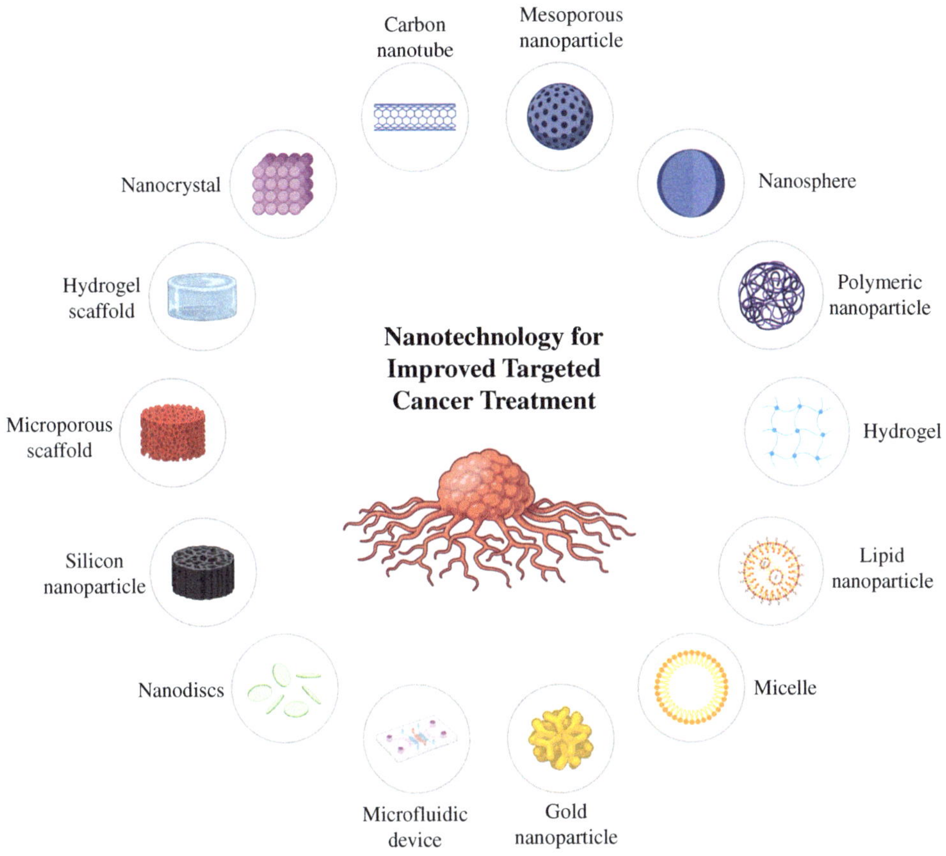

Figure 12.2 This illustration presents a range of nanotechnology platforms engineered to enhance targeted cancer treatment. Various nanostructures, such as carbon nanotubes, mesoporous nanoparticles, nanospheres, polymeric nanoparticles, hydrogels, lipid nanoparticles, micelles, gold nanoparticles, and nanodiscs, exhibit distinct properties conducive to drug delivery. Further innovations, including microfluidic devices, silicon nanoparticles, microporous and hydrogel scaffolds, and nanocrystals, contribute to improved therapeutic efficacy. These nanosystems facilitate enhanced solubility, controlled release, site-specific targeting, and diminished systemic toxicity. Collectively, these advanced materials and technologies signify a transformative approach to cancer therapy by enhancing precision, reducing side effects, and optimizing treatment outcomes through tailored, nanotechnology-driven delivery systems. Created by using Biorender.com.

Liposomal thalidomide formulations have been further improved by advances in nanotechnology through the incorporation of stimuli-responsive mechanisms, such as pH-sensitive or enzyme-triggered drug release systems. These cutting-edge liposomes maximize treatment efficacy while reducing off-target effects by enabling targeted drug unloading in the lysosomal compartments of malignant plasma cells or in the acidic tumor microenvironment. PEG-coated stealth liposomes have also been used to increase medication accumulation at tumor locations, decrease immune recognition, and extend systemic circulation. Clinical and preclinical research indicates that liposomal thalidomide has improved synergistic benefits in MM treatment regimens in addition to reducing neurotoxic side effects.

Notwithstanding these developments, there are still issues with the clinical translation, regulatory approval, and large-scale production of liposomal thalidomide formulations. Important topics that need more research include resolving possible immunogenicity related to nanocarrier systems, improving drug-loading efficiency, and guaranteeing batch-to-batch reproducibility. To further improve the accuracy and efficacy of thalidomide-based combination therapy, future research should concentrate on hybrid nanocarrier systems that combine liposomal platforms with polymeric NPs or exosome-based delivery vehicles. Furthermore, the creation of next-generation liposomal formulations suited for personalized medicine techniques in the treatment of MM may be made easier by the incorporation of AI into the design and optimization of NPs.

Thalidomide's function in MM therapy has been greatly expanded by utilizing nanotechnology, namely liposomal drug delivery, which opens up new possibilities for combination regimens that minimize side effects while improving patient prognosis. This is a prime example of the revolutionary potential of nanomedicine in repurposing current medications for more focused and efficient cancer treatment approaches [44].

12.3 Nanotechnology in Combination Therapy

12.3.1 Mechanisms

12.3.1.1 Co-delivery Systems

Combination therapy has been transformed by nanotechnology, which makes it possible to precisely and effectively co-deliver several therapeutic drugs with complimentary modes of action [45–50]. A sophisticated method for resolving pharmacokinetic differences, guaranteeing synchronized biodistribution, and boosting synergistic interactions between co-encapsulated medications is provided by co-delivery systems that make use of nanocarriers. The therapeutic efficacy of conventional medication combinations is frequently compromised by problems such as differential metabolism, varying half-lives, and off-target effects. By encapsulating several medications in a single nanostructure, nanocarrier-based co-delivery devices overcome these drawbacks by offering spatial co-localization at the targeted site of action and controlled release kinetics. By reducing off-target accumulation, this selective co-encapsulation improves therapeutic index, decreases systemic toxicity, and increases drug synergy [51].

Liposomes, polymeric NPs, dendrimers, micelles, and inorganic nanostructures like mesoporous silica NPs and gold-based nanocarriers are among the classes of nanocarriers that have been thoroughly investigated for co-delivery applications [52–57]. It is possible to customize these nanoplatforms to concurrently encapsulate hydrophilic and hydrophobic medications, guaranteeing the best possible solubility and bioavailability. To increase cytotoxicity against drug-resistant tumors, polymeric NPs with dual-drug loading capabilities, for example, have been developed to encapsulate chemotherapeutic drugs in addition to chemosensitizers. Similarly, to enhance the immunogenic response in cancer treatment, liposomal co-delivery methods have been created to encapsulate immunological checkpoint modulators and small-molecule inhibitors. To further improve therapeutic accuracy, these nanocarriers can be functionalized with targeting ligands as aptamers, peptides, or monoclonal antibodies to enable site-specific accumulation [58].

Co-delivery nanocarriers can be designed for active targeting, where surface alterations allow particular interactions with disease-specific receptors, in addition to passive targeting through the increased EPR effect [59]. This is especially important in oncological applications, because tumor-specific ligands allow malignant cells to preferentially absorb substances while protecting healthy organs. Additionally, stimuli-responsive nanocarriers have been created to accomplish on-demand drug release, in which site-specific drug unloading is triggered by external triggers (such as light, ultrasound, or magnetic fields) or physicochemical changes like pH shifts, redox gradients, or enzyme activity. By ensuring that both medications reach their intracellular targets in the ideal ratio and sequence, this regulated and synchronized release maximizes therapeutic efficacy while reducing side effects [60].

Co-delivery nanocarriers have benefits, but a number of obstacles prevent their practical application. To guarantee batch-to-batch reproducibility, considerable optimization is required due to the intricacy of nanocarrier manufacturing, possible instability during storage, and variability in drug release profiles. Commercialization is also severely hampered by regulatory obstacles pertaining to the approval of multidrug nanopharmaceuticals. In order to forecast drug–NP interactions, optimize nanocarrier composition, and expedite the design of next-generation co-delivery systems, future research should concentrate on combining AI and high-throughput screening (HTS) techniques. Clinical outcomes should be further improved by the fusion of nanotechnology with personalized medicine, where patient-specific biomarkers direct the formulation and selection of customized nanocarrier-based combination therapies [61].

Nanocarrier-based co-delivery systems have the potential to alter combination therapy by tackling these issues and providing a revolutionary method of treating multifactorial illnesses like cancer, infectious diseases, and neurological disorders [62–67]. This development demonstrates how important nanotechnology is to contemporary therapies and opens the door to safer, more efficient, and more precisely tailored treatment approaches [68].

12.3.1.2 Controlled Release

One essential component of combination therapy made possible by nanotechnology is controlled release, which enables the accurate manipulation of drug release kinetics to maximize pharmacokinetics, reduce toxicity, and improve therapeutic synergy [69]. Nanocarrier-based systems provide customized release patterns, guaranteeing that each therapeutic agent reaches its target at the most efficient concentration and time, in contrast to traditional drug formulations where the simultaneous administration of multiple agents frequently results in non-synchronized pharmacokinetic profiles [70–76]. This is especially important in combination therapy, when the simultaneous or sequential administration of medications needs to be carefully planned to minimize systemic toxicity or antagonistic interactions and maximize synergistic effects [77].

Diffusion-mediated, erosion-controlled, stimuli-responsive, and hybrid release techniques are some of the several processes used by NPs designed for controlled release [78–83]. For example, biodegradable polymers like PLGA, chitosan, or PEG are used in polymeric NPs to allow for prolonged and adjustable drug release. The progressive release of encapsulated medications is determined by the breakdown of these polymers under physiological conditions, guaranteeing a longer duration of therapeutic activity and a lower frequency of dosage [6, 9]. The structural dynamics of lipid bilayers or surfactant assemblies provide controlled release in liposomal and micellar systems, where changes in lipid composition or the addition of phase-transition lipids enable prolonged and environment-responsive drug discharge [84].

A significant advancement in the achievement of spatially and temporally regulated release is represented by stimuli-responsive nanocarriers. To initiate drug release precisely at the target site, these nanosystems take advantage of endogenous cues found in diseased tissues, such as pH gradients, enzymatic activity, redox potential, and hypoxic conditions. For example, in acidic environments, pH-sensitive NPs made for tumor microenvironments experience structural destabilization, which enables targeted drug unloading at cancerous locations while preserving healthy tissues [85–91]. Similarly, redox-responsive NPs trigger targeted intracellular medication release by taking advantage of the increased glutathione levels in cancer cells. The accuracy of NP-mediated drug release is further improved by exogenous stimuli such as temperature changes, magnetic fields, light irradiation, and ultrasonic waves. This enables externally controlled activation of therapeutic drugs with less systemic exposure [92].

In combination therapy, hybrid release mechanisms that incorporate several control techniques are also becoming more popular. For example, hydrophilic and hydrophobic medications can be encapsulated in separate compartments by core-shell NPs, enabling differential release rates that are customized to the pharmacodynamic profiles of the co-delivered medicines [93–99]. By selectively reacting to biomolecular triggers, layered nanocarriers such as mesoporous silica nanoparticles (MSNs) functionalized with molecular gates offer further control and guarantee that every drug component is administered in a coordinated manner to maximize therapeutic efficacy [100].

Even while controlled-release nanocarriers have a lot of promise for combination therapy, there are still a number of obstacles in the way of their clinical application. Further improvement is needed in the areas of potential immunogenicity, batch-to-batch consistency in large-scale manufacturing, reproducibility of release kinetics across various patient groups, and regulatory approval obstacles. Future developments should concentrate on using biomimetic and multifunctional nanoplatforms to improve drug release precision and using AI-driven predictive modeling to optimize NP formulations. A new era of highly effective, patient-specific combination medicines is expected to be ushered in by the convergence of nanotechnology with gene therapy, immunotherapy, and precision medicine. This will further revolutionize controlled drug release techniques [101].

12.3.1.3 Reduced Toxicity

By encapsulating therapeutic chemicals in nanocarriers, nanotechnology can reduce systemic toxicity, which is one of its key benefits in combination therapy [102, 103]. Because pharmaceuticals biodistribute indiscriminately throughout the body, conventional drug delivery, especially in combination regimens, frequently produces severe off-target effects. In addition to decreasing therapeutic efficacy, this systemic exposure causes dose-limiting toxicities that jeopardize treatment results and patient compliance. Nanotechnology tackles these obstacles by integrating bioactive molecules within designed nanocarriers, permitting controlled and site-specific medication release while reducing unwanted interactions with healthy tissues [104].

Liposomes, polymeric NPs, dendrimers, and lipid-polymer hybrid NPs are examples of nanocarrier-based drug delivery systems that act as protective reservoirs, preventing encapsulated medications from degrading too quickly and entering the bloodstream until they arrive at their destination. The incidence of side effects like nephrotoxicity, hepatotoxicity, and cardiotoxicity, which are frequently linked to systemic chemotherapy and combination medication regimens, is decreased by this tailored delivery, which dramatically reduces the exposure of healthy organs to harmful chemicals. Additionally, stimuli-responsive nanocarriers improve drug release kinetics to guarantee less systemic leakage and more precise therapeutic action by releasing their payload in response to particular physiological triggers like pH gradients, enzyme activity, or redox potential [105].

The use of PEGylated liposomal doxorubicin (PLD), which has shown better safety profiles than free DOX in the treatment of cancer, is a perfect illustration of this strategy. Doxorubicin's reduced cardiotoxicity and preservation of its anticancer activity due to its encapsulation in a liposomal matrix demonstrate how nanotechnology can be used to optimize combination therapy. Similar to this, polymeric micelles have been used to co-deliver molecularly targeted medications and chemotherapeutic agents, guaranteeing synchronized release and avoiding the immediate toxicity that comes with free drug formulations. Furthermore, because of their high drug-loading capacity and adjustable porosity, metal–organic framework (MOF)-based nanocarriers present a viable platform for controlled and prolonged drug release, which further reduces systemic toxicity [106].

Beyond oncology, antibacterial and antiviral combination medicines have also investigated nanotechnology-mediated toxicity reduction, where nanocarriers reduce cytotoxic effects on host cells and off-target immune responses. Long-term patient adherence has been improved by nanoformulated antiretroviral medication combinations for HIV treatment, which have shown improved pharmacokinetic profiles with decreased hepatotoxicity and gastrointestinal adverse effects. Lipid-based nanocarriers have also made it easier to distribute antibiotic combinations under controlled conditions, lowering nephrotoxicity while preserving strong antibacterial activity [107].

Even with these developments, there are still several obstacles to overcome in order to guarantee the long-term safety and full biocompatibility of nanocarrier systems. Thorough preclinical and clinical assessments are required due to possible worries about NP buildup, unforeseen immunological reactions, and long-term biodistribution. To further improve safety while preserving the effectiveness of combination therapy, future studies should concentrate on creating biodegradable nanocarriers with adjustable clearance mechanisms. Furthermore, including AI into the design of NPs may speed up the practical translation of safer nanomedicine-based treatment approaches by making it easier to forecast and optimize toxicity profiles [108].

Because it provides targeted and regulated drug administration, nanotechnology is essential in lowering the systemic toxicity linked to combination medication therapy. Systemic exposure to harmful substances is reduced by encapsulating therapeutic drugs in carefully designed nanocarriers, which raises the therapeutic index and improves patient safety. The incorporation of innovative biomaterials and clever drug delivery techniques will further enhance nanotechnology's ability to produce safer and more efficient combination therapies as nanomedicine continues to progress [109].

12.3.2 Examples

Disulfiram (DSF), a medication commonly used to treat alcoholism, may be able to prevent pyroptosis, a type of programmed cell death linked to inflammatory disorders, according to research by Xueming Xu et al. According to the study, Cu^{2+} greatly increased DSF's anti-pyroptotic action by forming a copper ethylthiocarbamate (CuET) complex, which had more potent inhibitory effects than DSF by itself. Nevertheless, CuET showed low permeability and poor solubility, which made therapeutic use difficult. A six-fold increase in bioavailability was achieved by synthesizing CuET nanocrystals (CuET NCs) using a metal coordination-driven self-assembly technique in order to overcome these constraints. CuET NCs have a high biodistribution in the intestine, according to pharmacokinetic studies, which suggests that they may be useful in the treatment of inflammatory bowel diseases (IBDs). CuET NCs successfully decreased inflammation and eased colitis symptoms in a Dextran Sulfate Sodium (DSS)-induced colitis mouse model, demonstrating their promise as a treatment. Mechanistically, CuET differed from DSF, which primarily targets inflammasome-induced pyroptosis, by specifically inhibiting NLRP3 inflammasome activation by interfering with the NLRP3-NEK7 connection. These results demonstrated CuET's potential as an anti-pyroptotic agent, and the creation of CuET NCs effectively increased its druggability, opening the door for use in inflammatory diseases [110].

Shi-He Liu et al. used a tumor-specific super-promoter reporter assay for HTS of Food and Drug Administration (FDA)-approved medications and found that baculoviral inhibitor of apoptosis repeat-containing 5 (BIRC5) is a major therapeutic target for pancreatic ductal adenocarcinoma (PDAC). Metformin (MET), simvastatin, and digoxin (C3 treatment) were found to be a synergistic medication combination that successfully targets PDX-1 and BIRC5 in PDAC cells as a result of the screening. Studies conducted both in vitro and in vivo showed that C3 treatment dramatically reduced tumor growth in PDAC models without showing any signs of toxicity. Mechanistic studies showed that C3 treatment increased apoptotic signaling in cancer cells by interfering with survival pathways linked to BIRC5 expression. The study demonstrated how repurposing FDA-approved medications can hasten the development of tailored PDAC medicines, providing a viable approach to enhance patient outcomes. The effectiveness and translational potential of C3 therapy in the treatment of PDAC require additional clinical validation [111].

In order to prevent antimicrobial-resistant (AMR) wound infections and encourage wound healing, Nashath Kalangadan et al. created a dual-functional electrospun composite nanofiber (NF) that contains repurposed ivermectin (IVM) and ciprofloxacin (CIP). The successful creation of NFs with a diameter range of 225–295 nm and better mechanical qualities than Polyvinyl alcohol (PVA) was validated by characterization using FTIR, SEM, and tensile strength analysis. With CIP (1.6–2.5 μg/mL) and IVM (0.3–0.5 μg/mL) present, the optimized CIP-IVM NF formulation showed strong biofilm breakdown against multidrug-resistant (MDR) bacteria such as *Pseudomonas aeruginosa*, *Staphylococcus aureus*, and *Enterococcus faecalis*. The pro-healing qualities of the NFs were supported by in vitro wound healing assays that showed increased fibroblast migration and in ovo angiogenesis investigations that showed a 60% increase in vascular junctions and a 50% increase in vessel length. The results demonstrated the potential of CIP-IVM NFs as a successful wound care approach, providing both rapid tissue regeneration and antibacterial activity against resistant microorganisms. To confirm their clinical use for infected and chronic wounds, more in vivo research is required [112].

Using both in vitro and in silico methods, Kamal A. Qureshi et al. examined the antibacterial and anti-biofilm properties of eugenol (EU) and cinnamon aldehyde (CN) against MDR microorganisms. While CN demonstrated strong effects against *Staphylococcus epidermidis* (29.6 ± 0.4 mm) and *Candida albicans* (36.6 ± 0.4 mm), EU demonstrated notable antimicrobial activity, especially against *Aspergillus niger* (20.5 ± 0.5 mm inhibitory zone). Strong binding affinities of EU and CN with important microbial targets, such as peptide deformylase and β-carbonic anhydrase, were found by molecular docking and MM-GBSA studies. Their binding free energies (ΔG) ranged from −12.75 to −0.60 kcal/mol. While CN showed modest inhibition in the micromolar range, EU had better inhibitory capacity, as evidenced by the dissociation constant (K_d) values, which correlated with low IC_{50} values. The stability of the protein-ligand complexes through essential polar and hydrophobic interactions was validated by molecular dynamics (MD) simulations. Furthermore, ADMET profiling indicated that both drugs had good pharmacokinetic and safety profiles. These findings highlighted EU and CN's therapeutic promise as promising antimicrobial drugs, especially against illnesses linked to biofilms and MDR bacteria. The development of innovative antimicrobial techniques was made possible by the complementary antibacterial characteristics of EU and CN, which indicated their potential for structural optimization and combinatorial medicinal applications [113].

For improved breast cancer treatment, Pallavi Kumari et al. created and refined a nanostructured lipid carrier (NLC) co-loaded with thymoquinone (THY) and MET. Effective drug encapsulation was ensured by the MET-THY NLC formulation, which had a nanoscale size of 78.99 nm and high entrapment efficiencies of 90.14 ± 1.25% for MET and 91.63 ± 2.05% for THY. Due to increased cellular absorption and prolonged release, in vitro cytotoxicity experiments showed a markedly improved anticancer effect when compared to free medicines. A greater rate of apoptosis and cell cycle arrest in breast cancer cells treated with MET-THY NLCs was confirmed by flow cytometry analysis, suggesting enhanced therapeutic efficacy. The effective uptake of the nanocarrier was further confirmed by cellular internalization tests, confirming its potential for targeted medication delivery. These results demonstrated the benefits of combination therapy and medication repurposing in the treatment of breast cancer, and they showed that MET-THY NLCs would be a viable strategy for enhancing therapeutic results while lowering systemic toxicity. Additional in vivo research was recommended to confirm its clinical suitability [114].

12.4 Challenges and Limitations

Notwithstanding their benefits, nanotechnologies have challenges related to manufacturing, regulation, and scalability:

12.4.1 Toxicity Concerns

Even though nanotechnology holds great potential for combination therapy and medication repurposing, toxicity is still a major obstacle to clinical adoption. Size, shape, surface charge, composition, and functionalization are some of the physicochemical characteristics of nanomaterials that have a major impact on their biodistribution, cellular uptake, and possible toxicological effects. Significant safety concerns are raised by unintentional accumulation in off-target tissues, longer retention, and unknown metabolic pathways, even while nanocarriers improve targeted drug delivery and lessen the systemic toxicity of conventional therapies. Nonbiodegradable inorganic NPs, like those based on gold and silica, for example, can linger in key organs like the kidneys, liver, and spleen and cause oxidative stress, chronic inflammatory reactions, and even organ malfunction. Similar to this, cationic NPs are frequently linked to cytotoxicity, mitochondrial damage, and the induction of apoptosis, despite their high cellular absorption resulting from electrostatic interactions with negatively charged cellular membranes [115].

The intricacy of NP clearance pathways, which are impacted by elements including protein corona formation, immunological identification, and renal or hepatic excretion, is a major obstacle in nanomedicine. A key factor in NP sequestration is the MPS, which can restrict therapeutic efficacy by causing quick clearance and possible immunogenicity. Furthermore, nothing is known about the long-term effects of recurrent NP delivery, especially when it comes to managing chronic diseases when a continued dose is necessary. Degradation products of NPs may cause immunomodulatory effects, disturb homeostasis, or show unexpected toxicities; hence, long-term studies are necessary to evaluate cumulative exposure hazards [116].

Recent developments in surface modification techniques, including PEGylation, zwitterionic coatings, and cell membrane-based biomimetic cloaking, have shown promise in lowering immunogenicity generated by NPs and lengthening circulation half-life. The necessity for creative approaches in nanocarrier design is underscored by worries that the "stealth effect" will eventually wear off and cause ABC syndrome after repeated administration. Furthermore, although lipid-based and biodegradable polymeric NPs have demonstrated good biocompatibility and clearance characteristics, their clinical translation is made more difficult by batch-to-batch variability in synthesis, the possibility of NP aggregation, and interactions with biological fluids.

To evaluate the safety of nanomaterials, regulatory bodies such as the FDA and European Medicines Agency (EMA) place a strong emphasis on standardized procedures and thorough preclinical toxicological examination. However, expedited approval processes are hampered by the absence of generally recognized standards for evaluating nanomedicine. In order to forecast nanotoxicity and improve formulation design, future studies should concentrate on combining HTS platforms, in silico modeling, and organ-on-a-chip technologies. Additionally, to clarify NP–host interactions and ensure the development of safe, effective, and commercially feasible nanotherapeutics for drug repurposing and combination therapy, interdisciplinary approaches combining nanotechnology with immunology, pharmacokinetics, and systems biology are crucial [117].

12.4.2 Cost

Widespread clinical use and commercialization of combination therapy and medication repurposing based on nanotechnology are still significantly hampered by their high cost. The complex costs result from many phases of development, such as obtaining raw materials, fabricating nanocarriers, surface functionalization, scaling, and regulatory compliance [118]. The synthesis of nanopharmaceuticals, in contrast to traditional drug formulations, frequently calls for specialized machinery, high-purity reagents, and sophisticated manufacturing processes like high-pressure homogenization, microfluidics, or supercritical fluid technology, all of which raise production costs. Furthermore, to ensure reproducibility and efficacy, exact control over NP size, shape, surface charge, and functionalization is essential, which makes large-scale production much more difficult and raises operating costs [2].

Beyond production, there are significant expenses associated with preclinical and clinical testing of combination and repurposing medicines based on nanomedicine. Assessing the pharmacokinetics, biodistribution, toxicity, and immunogenicity of nanoformulations necessitates extensive in vitro and in vivo research, which adds to the expensive burden. Additionally, reformulation to satisfy good manufacturing practice (GMP) requirements is frequently required when moving from laboratory-scale synthesis to industrial-scale manufacturing, requiring further investment in process optimization, quality control, and regulatory compliance. Longer approval times and higher expenses are the results of the difficulty in negotiating regulatory procedures for nanomedicines, which includes the requirement for substantial safety and efficacy data. These financial limitations hinder pharmaceutical companies' capacity to develop novel nanoscale drug delivery technologies into clinically viable treatments, especially for small and medium-sized businesses (SMEs) and academic startups [119].

Developing scalable and environmentally friendly synthesis techniques, such as green nanotechnology approaches that use biopolymers and plant-derived nanomaterials, is one way to address these issues at a reasonable cost. Furthermore, developments in HTS and formulation design aided by AI can expedite the drug development process and lessen the cost of iterative experimental optimization. In order to create standardized procedures that reduce manufacturing costs while preserving product integrity, cooperation between academic institutions, business, and government organizations is essential. Government financing programs and public-private partnerships can help reduce costs by promoting translational research and providing financial support for early-stage clinical trials. The accessibility and affordability of drug repurposing and combination therapy driven by nanotechnology will be improved by sustained innovation in cost-efficient nanomanufacturing and targeted financial investments, notwithstanding the current economic obstacles as presented in Figure 12.3 [120].

12.4.3 Regulatory Barriers

Due in large part to the intricacy of nanoscale formulations, their distinct physicochemical characteristics, and their interactions with biological systems, integrating nanotechnology into medication repurposing and combination therapy presents significant regulatory obstacles. The evaluation of nanomedicines is more complex because traditional pharmaceutical regulatory frameworks were mostly created for traditional small-molecule medications and biologics. The absence of standardized procedures for describing nanocarriers' size, shape, surface chemistry, charge, aggregation behavior, and drug release kinetics—all of which have a substantial impact on therapeutic efficacy and biocompatibility—is one of the main obstacles. Regulatory organizations like the United States, The EMA,

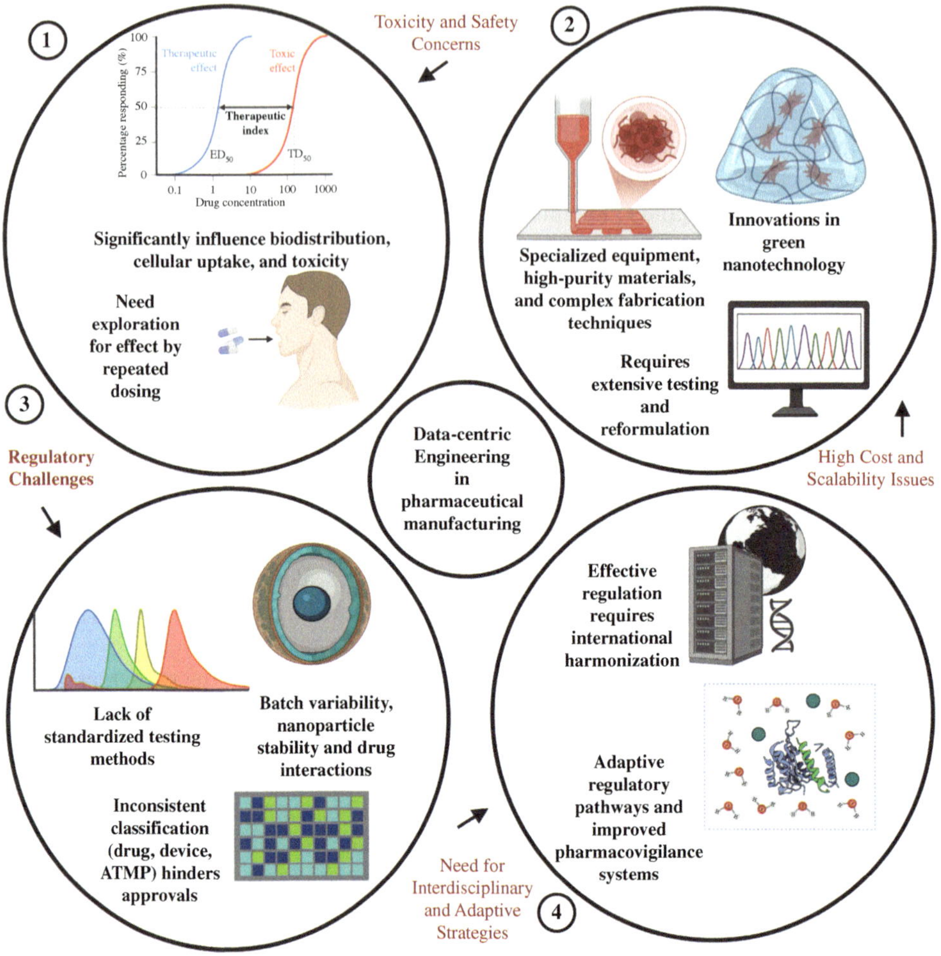

Figure 12.3 This figure delineates the principal challenges associated with the implementation of data-centric engineering in pharmaceutical manufacturing. (1) Toxicity and safety concerns: The impact of repeated dosing on biodistribution, cellular uptake, and toxicity necessitates further investigation. (2) High cost and scalability issues: The employment of specialized equipment, high-purity materials, and green nanotechnology demands extensive testing and reformulation. (3) Regulatory challenges: Discrepancies in standardization, inconsistent classifications, and batch variability complicate approval processes and drug interactions. (4) Need for interdisciplinary and adaptive strategies: Global regulatory harmonization and advanced pharmacovigilance systems are crucial for effective governance. These barriers underscore the necessity for innovative, collaborative, and adaptive approaches to advance pharmaceutical manufacturing. Created by using Biorender.com.

the FDA, and other international counterparts demand thorough preclinical and clinical assessments of nanomedicines; nevertheless, there are currently few standardized criteria designed especially for nanoformulations [121].

The development of reliable analytical methods and proven approaches to evaluate the stability, safety, and effectiveness of nanopharmaceuticals is a significant regulatory problem. In contrast to traditional medications, NPs interact dynamically with biomolecules, resulting in biodistribution and pharmacokinetics that are unpredictable. To make sure there are no unforeseen bioaccumulations or off-target effects, this calls for thorough toxicological evaluations, such as long-term biodistribution studies, immunogenicity profiling, and NP breakdown kinetics. The regulatory environment is further complicated by combination therapies that use nanocarriers since co-encapsulation and synchronized release of numerous medications require careful consideration for possible drug-drug interactions, changed metabolic pathways, and cumulative toxicity. There is also regulatory ambiguity around the classification of nanomedicines; different countries have different criteria for determining whether they should be classified as advanced therapy medical products (ATMPs), new molecular entities, or drug-device combinations, which might cause differences in the licensing procedures [122].

Two important regulatory issues are batch-to-batch repeatability and manufacturing scalability. Maintaining therapeutic consistency requires exact control over the manufacturing, functionalization, and drug loading of NPs. But even little changes to the formulation characteristics can have a big impact on the biological reactions and drug release profiles. The special production needs of nanomedicines necessitate the adaptation of current good manufacturing practice (cGMP) rules, which calls for the creation of sophisticated in-line monitoring methods and quality-by-design (QbD) strategies to improve reproducibility. Since long-term clinical results and actual safety profiles of nanotherapeutics are still not well documented, a major obstacle is the absence of comprehensive post-marketing surveillance data. The broad adoption of combination medicines and medication repurposing based on nanotechnology is slowed by this pharmacovigilance gap, which further undermines regulatory confidence [123].

A multifaceted strategy is needed to overcome these regulatory obstacles, including international cooperation to create standardized rules, funding for sophisticated characterization methods, and the application of AI-driven predictive modeling for NP behavior. In order to maintain a balance between innovation and patient safety, the regulatory framework must change to keep up with the quick advances in nanomedicine [124]. In order to guarantee long-term efficacy and safety, future regulatory frameworks should combine enhanced post-market surveillance with adaptive approval paths, such as accelerated review procedures for repurposed nanomedicines with proven safety profiles. Overcoming these regulatory obstacles will allow nanotechnology to be used more successfully to revolutionize combination therapy and medication repurposing, which will eventually improve patient outcomes in complicated and multifactorial diseases [125].

12.5 Future Perspectives

The incorporation of bioresponsive nanocarriers, which can react dynamically to particular biological stimuli, including pH changes, enzyme activity, redox potential, and external triggers like ultrasound or electromagnetic fields, is one of the most exciting advancements. By enabling precise, on-demand drug release, these intelligent delivery devices improve therapeutic efficacy while lowering systemic toxicity. Such nanocarriers can be designed to encapsulate and transport current medications to illness locations that were previously inaccessible, extending their clinical usefulness beyond their initial indications in the context of drug repurposing. Similar to this, bioresponsive nanocarriers can synchronize the delivery of several therapeutic drugs in combination therapy, maximizing synergistic effects and optimizing pharmacokinetics [126].

The logical design and optimization of nanocarriers for medication delivery is about to undergo a radical change due to the convergence of AI and nanotechnology. The translation of nanomedicine from bench to bedside can be accelerated by using ML algorithms to evaluate large datasets in order to predict drug-nanocarrier interactions, optimize the physicochemical properties of NPs, and improve biocompatibility. Through the identification of synergistic interactions between repurposed pharmaceuticals and nanocarrier-mediated delivery systems, AI-driven computational modeling can also aid in the development of novel medication combinations. Since AI-guided NP engineering makes it possible to create patient-specific nanotherapies that are suited to unique genetic, metabolic, and illness profiles, personalized medicine stands to gain a great deal from these developments. Furthermore, combining nanotechnology with precision medicine techniques like immune checkpoint inhibitors and CRISPR-based gene editing could result in highly focused, next-generation therapeutic treatments [127].

The clinical translation of medication repurposing and combination medicines based on nanotechnology still faces obstacles, despite these encouraging developments. The broad use of nanopharmaceuticals requires addressing regulatory barriers, the complexity of large-scale manufacturing, and long-term safety evaluations. The manufacture of nanocarriers should be made more scalable and reproducible in the future; regulatory frameworks should be standardized, and long-term clinical trials should be carried out to evaluate biocompatibility and efficacy. To overcome these obstacles and realize the full promise of treatments driven by nanotechnology, interdisciplinary cooperation involving materials scientists, pharmacologists, bioengineers, and clinicians will be essential [128].

Treatment paradigms for a wide range of illnesses, including cancer, neurological diseases, and infectious diseases, are anticipated to be redefined as the area develops through the integration of nanotechnology with medication repurposing and combination therapy. More efficient, secure, and individualized treatment approaches will be made possible by the development of multifunctional, responsive, and precisely targeted nanocarriers. In the end, as nanotechnology develops further, a new era of medicine will emerge, marked by better patient outcomes, decreased medication resistance, and increased treatment precision [127].

12.6 Conclusion

Drug repurposing and combination therapy have seen a paradigm shift thanks to nanotechnology, which provides focused, accurate, and effective drug delivery systems that overcome significant drawbacks of traditional pharmaceutical methods. Nanotechnology makes it possible to effectively reposition current medications with improved pharmacokinetics, bioavailability, and therapeutic index by utilizing the special physicochemical characteristics of nanoscale carriers, such as improved surface-area-to-volume ratios, tunable drug release kinetics, and superior biocompatibility. Additionally, in combination therapy, nanocarriers help transport several therapeutic drugs simultaneously and under control, maximizing synergistic effects and reducing off-target toxicity and drug resistance. In order to expand the therapeutic landscape for complex and multifactorial diseases like cancer, neurodegenerative disorders, and infectious diseases, advanced nanoplatforms such as liposomes, polymeric NPs, dendrimers, and inorganic nanocarriers have shown remarkable promise in overcoming biological barriers like the BBB, tumor microenvironment, and intracellular compartments. Notwithstanding these encouraging developments, a number of obstacles stand in the way of the broad clinical application of combination therapy and drug repurposing made possible by nanotechnology. Large-scale manufacturing limitations, batch-to-batch variability, possible immunogenicity, and the regulatory difficulties surrounding the approval processes for nanopharmaceuticals are some of the main challenges. To overcome these obstacles, pharmacologists, physicians, regulatory bodies, and nanotechnologists must work closely across disciplinary boundaries to provide standardized procedures for the production, characterization, and safety assessment of NPs. Furthermore, it is anticipated that the combination of HTS methods, ML, and AI can hasten the logical development of nanocarriers with improved targeting effectiveness and patient-specific customization. As the field develops, the combination of nanotechnology and cutting-edge biomedical breakthroughs like gene editing, immunotherapy, and personalized medicine will further transform therapeutic approaches and open the door for next-generation treatment modalities that are unmatched in accuracy and effectiveness. Nanotechnology will continue to be at the forefront of contemporary drug development by supporting ongoing research, regulatory harmonization, and translational activities, ultimately changing the course of medicine and enhancing patient outcomes.

References

1 Najlah M. Drug repurposing supported by nanotechnology: A promising strategy to fight cancer. *Therapeutic Delivery [Internet]* 2021;12(4):267–269. Available from: https://www.tandfonline.com/doi/full/10.4155/tde-2021-0009

2 Martins M, Veiga F, Paiva-Santos AC, and Pires PC. Drug repurposing and nanotechnology for topical skin cancer treatment: Redirecting toward targeted and synergistic antitumor effects. *ACS Pharmacology & Translational Science Journal [Internet]* 2025;8(2):308–338. Available from: https://pubs.acs.org/doi/10.1021/acsptsci.4c00679

3 Liu Y, Tong Z, Shi J, Li R, Upton M, and Wang Z. Drug repurposing for next-generation combination therapies against multidrug-resistant bacteria. *Theranostics [Internet]* 2021;11(10):4910–4928. Available from: https://www.thno.org/v11p4910.htm

4 Correia AS, Gärtner F, and Vale N. Drug combination and repurposing for cancer therapy: The example of breast cancer. *Heliyon [Internet]* 2021;7(1):e05948. Available from: https://linkinghub.elsevier.com/retrieve/pii/S2405844021000530

5 Joseph T, Kar Mahapatra D, Esmaeili A, et al. Nanoparticles: Taking a unique position in medicine. *Nanomaterials [Internet]* 2023;13(3):574. Available from: https://www.mdpi.com/2079-4991/13/3/574

6 Gholap AD and Omri A. Advances in artificial intelligence-envisioned technologies for protein and nucleic acid research. *Drug Discovery Today [Internet]* 2025; 104362. Available from: https://linkinghub.elsevier.com/retrieve/pii/S1359644625000753

7 Malik S, Muhammad K, and Waheed Y. Emerging applications of nanotechnology in healthcare and medicine. *Molecules [Internet]* 2023;28(18):6624. Available from: https://www.mdpi.com/1420-3049/28/18/6624

8 Liu L-C, Chen Y-H, and Lu D-W. Overview of recent advances in nano-based ocular drug delivery. *International Journal of Molecular Sciences [Internet]* 2023;24(20):15352. Available from: https://www.mdpi.com/1422-0067/24/20/15352

9 Gholap AD, Khuspe PR, Bharati DK, et al. Computational neuropharmacology in psychiatry. In: *Computational Neuropharmacology [Internet]*, (pp. 207–244). Wiley; 2025. Available from: https://onlinelibrary.wiley.com/doi/10.1002/9781394242474.ch7

10 Zhuo Y, Zhao Y-G, and Zhang Y. Enhancing drug solubility, bioavailability, and targeted therapeutic applications through magnetic nanoparticles. *Molecules [Internet]* 2024;29(20):4854. Available from: https://www.mdpi.com/1420-3049/29/20/4854

11 Sultana A, Zare M, Thomas V, Kumar TSS, and Ramakrishna S. Nano-based drug delivery systems: Conventional drug delivery routes, recent developments and future prospects. *Medicine Drug Discovery [Internet]* 2022;15:100134. Available from: https://linkinghub.elsevier.com/retrieve/pii/S259009862200015X

12 Sivadasan D, Sultan MH, Madkhali O, Almoshari Y, and Thangavel N. Polymeric lipid hybrid nanoparticles (PLNs) as emerging drug delivery platform—A comprehensive review of their properties, preparation methods, and therapeutic applications. *Pharmaceutics [Internet]* 2021;13(8):1291. Available from: https://www.mdpi.com/1999-4923/13/8/1291

13 Narvekar M, Xue HY, Eoh JY, and Wong HL. Nanocarrier for poorly water-soluble anticancer drugs—Barriers of translation and solutions. *AAPS PharmSciTech [Internet]* 2014;15(4):822–833. Available from: http://link.springer.com/10.1208/s12249-014-0107-x

14 Séguy L, Groo A-C, and Malzert-Fréon A. How nano-engineered delivery systems can help marketed and repurposed drugs in Alzheimer's disease treatment? *Drug Discovery Today [Internet]* 2022;27(6):1575–1589. Available from: https://linkinghub.elsevier.com/retrieve/pii/S1359644622000800

15 Đorđević S, Gonzalez MM, Conejos-Sánchez I, et al. Current hurdles to the translation of nanomedicines from bench to the clinic. *Drug Delivery and Translational Research [Internet]* 2022;12(3):500–525. Available from: https://link.springer.com/10.1007/s13346-021-01024-2

16 Patra JK, Das G, Fraceto LF, et al. Nano based drug delivery systems: Recent developments and future prospects. *Journal Nanobiotechnology [Internet]* 2018;16(1):71. Available from: https://jnanobiotechnology.biomedcentral.com/articles/10.1186/s12951-018-0392-8

17 Sim S and Wong N. Nanotechnology and its use in imaging and drug delivery (Review). *Biomedical Reports [Internet]* 2021;14(5):42. Available from: http://www.spandidos-publications.com/10.3892/br.2021.1418

18 Rashidi N, Davidson M, Apostolopoulos V, and Nurgali K. Nanoparticles in cancer diagnosis and treatment: Progress, challenges, and opportunities. *Journal Drug Delivery*

Science Technology [Internet] 2024;95:105599. Available from: https://linkinghub.elsevier.com/retrieve/pii/S1773224724002685

19 Mitchell MJ, Billingsley MM, Haley RM, Wechsler ME, Peppas NA, and Langer R. Engineering precision nanoparticles for drug delivery. *Nature Reviews Drug Discovery [Internet]* 2021;20(2):101–124. Available from: https://www.nature.com/articles/s41573-020-0090-8

20 Majumder J and Minko T. Multifunctional and stimuli-responsive nanocarriers for targeted therapeutic delivery. *Expert Opinion Drug Delivery [Internet]* 2021;18(2):205–227. Available from: https://www.tandfonline.com/doi/full/10.1080/17425247.2021.1828339

21 Sun Y and Davis E. Nanoplatforms for targeted stimuli-responsive drug delivery: A review of platform materials and stimuli-responsive release and targeting mechanisms. *Nanomaterials [Internet]* 2021;11(3):746. Available from: https://www.mdpi.com/2079-4991/11/3/746

22 Das SS, Bharadwaj P, Bilal M, et al. Stimuli-responsive polymeric nanocarriers for drug delivery, imaging, and theragnosis. *Polymers (Basel) [Internet]* 2020;12(6):1397. Available from: https://www.mdpi.com/2073-4360/12/6/1397

23 Rajurkar VG. Recent technologies in nanomedicine for treatment of cancer. *Inventi Rapid Pharm Technology [Internet]* 2012;4(1):1–3. Available from: https://www.inventi.in/journal/article/126/9966/Inventi Rapid: Pharm Tech/Pharmaceutical

24 Gholap AD and Musmade DS. Advances in the field of nanooncology: An overview. *Inventi Impact NDDS [ISSN 0976-3791] [Internet]* 2011;4(1):1–3. Available from: https://www.inventi.in/journal/article/2/1078/Inventi Impact: NDDS/Pharmaceutical

25 Gholap AD and Rajurkar VG. Lipid nanocapsules: A novel promising tool as drug delivery system. *Inventi Rapid NDDS [ISSN 0976-3791] [Internet]* 2012;1(1):1–4. Available from: https://www.inventi.in/journal/article/127/9969/Inventi Rapid: NDDS/Pharmaceutical

26 Gholap AD. Translational nanomedicine: Status assessment and opportunities. *International Journal Research Pharmaceutical Biomedicine Science* 2012;4(3):1454–1463.

27 Rajurkar VG. New threat of clinical toxicities evoked from nanocarrier systems. *Inventi Rapid NDDS [ISSN 0976-3791] [Internet]* 2012;1(4):1–4. Available from: https://www.inventi.in/journal/article/127/9982/Inventi Rapid: NDDS/Pharmaceutical

28 Anand M. Smart crystals technology: A review. *Pharmacologyonline [Internet]* 2011; 1(1):238–243. Available from: https://pharmacologyonline.silae.it/files/newsletter/2011/vol3/028.amol.pdf

29 Borude S. Aquasomes: A potential drug delivery carrier. *Pharmacologyonline [Internet]* 2011;1(3):230–237. Available from: https://pharmacologyonline.silae.it/files/newsletter/2011/vol3/027.amol.pdf

30 Rajurkar VG. Potential applications of capillary electromigration methods for the improvement of separation efficiency. *Inventi Rapid Pharm Analysis Quality Assurance [ISSN 0976-3813] [Internet]* 2012;1(4):1–3. Available from: https://www.inventi.in/journal/article/136/9968/Inventi Rapid: Pharm Analysis /Pharmaceutical

31 Dattatray T. Potential applications of botanical extract. *Inventi Rapid Ethnopharmacol [ISSN 0976-3805] [Internet]* 2012;1(4):1–4. Available from: https://www.inventi.in/journal/article/135/9972/Inventi Rapid: Ethnopharmacolo/Pharmaceutical

32 Dighe SB. Dandruff: Pathogenesis and treatment. *Inventi Rapid Cosmeceuticals [ISSN 0976-3864] [Internet]* 2012;1(4):1–4. Available from: https://www.inventi.in/journal/article/141/9975/Inventi Rapid: Cosmeceuticals/Pharmaceutical

33 Gholap AD and Tupe D. The threat of photoaging: Clinical conditions, evaluation and treatment. *Inventi Rapid Cosmeceuticals [ISSN 0976-3864] [Internet]* 2012;1(4):1–3. Available from: https://www.inventi.in/journal/article/141/9971/Inventi Rapid: Cosmeceuticals/Pharmaceutical

34 Liu Y, Liang Y, Yuhong J, et al. Advances in nanotechnology for enhancing the solubility and bioavailability of poorly soluble drugs. *Drug Design, Development and Therapy [Internet]* 2024;18:1469–1495. Available from: https://www.dovepress.com/advances-in-nanotechnology-for-enhancing-the-solubility-and-bioavailab-peer-reviewed-fulltext-article-DDDT

35 Santhanakrishnan KR, Koilpillai J, and Narayanasamy D. PEGylation in pharmaceutical development: Current status and emerging trends in macromolecular and immunotherapeutic drugs. *Cureus [Internet]* 2024. Available from: https://www.cureus.com/articles/276399-pegylation-in-pharmaceutical-development-current-status-and-emerging-trends-in-macromolecular-and-immunotherapeutic-drugs

36 Dighe SB. Role of sebum in human health & acne. *Inventi Rapid Cosmeceuticals [ISSN 0976-3864] [Internet]* 2012;1(4):1–3. Available from: https://www.inventi.in/journal/article/141/9980/Inventi Rapid: Cosmeceuticals/Pharmaceutical

37 Dattatray T. Artificial viral envelopes: Emerging vehicle for gene delivery. *Pharmatutor-Art-1439* 2012;1(4):1–3.

38 Rajurkar VG. Potential applications of superparmagnetic nanoparticles technology. *Inventi Rapid NDDS [ISSN 0976-3791] [Internet]* 2012;1(4):1–3. Available from: https://www.inventi.in/journal/article/127/9983/Inventi Rapid: NDDS/Pharmaceutical

39 Cong X, Zhang Z, Li H, Yang Y-G, Zhang Y, and Sun T. Nanocarriers for targeted drug delivery in the vascular system: Focus on endothelium. *Journal Nanobiotechnology [Internet]* 2024;22(1):620. Available from: https://jnanobiotechnology.biomedcentral.com/articles/10.1186/s12951-024-02892-9

40 Wu J. The enhanced permeability and retention (EPR) effect: The significance of the concept and methods to enhance its application. *Journal of Personalized Medicine [Internet]* 2021;11(8):771. Available from: https://www.mdpi.com/2075-4426/11/8/771

41 van den Berg AIS, Yun C-O, Schiffelers RM, and Hennink WE. Polymeric delivery systems for nucleic acid therapeutics: Approaching the clinic. *Journal Control Release [Internet]* 2021;331:121–141. Available from: https://linkinghub.elsevier.com/retrieve/pii/S0168365921000237

42 Pan J, Wang Y, Chen Y, et al. Emerging strategies against accelerated blood clearance phenomenon of nanocarrier drug delivery systems. *Journal Nanobiotechnology [Internet]* 2025;23(1):138. Available from: https://jnanobiotechnology.biomedcentral.com/articles/10.1186/s12951-025-03209-0

43 Zheng G, Zheng M, Yang B, Fu H, and Li Y. Improving breast cancer therapy using doxorubicin loaded solid lipid nanoparticles: Synthesis of a novel arginine-glycine-aspartic tripeptide conjugated, pH sensitive lipid and evaluation of the nanomedicine in vitro and in vivo. *Biomedicine Pharmacother [Internet]* 2019;116:109006. Available from: https://linkinghub.elsevier.com/retrieve/pii/S0753332219311989

44 Palumbo A, Facon T, Sonneveld P, et al. Thalidomide for treatment of multiple myeloma: 10 years later. *Blood [Internet]* 2008;111(8):3968–3977. Available from: https://ashpublications.org/blood/article/111/8/3968/24563/Thalidomide-for-treatment-of-multiple-myeloma-10

45 Puri AV. Comparative evaluation of papaya and banana starch as disintegrant in diclofenac sodium tablet formulation. *International Journal of Pharmacy and Biological Sciences* 2018;8(3):512–522.

46 Singh T, Ravi Kumar V, Gholap AD, Sanap G, Anjaiah B, and Roja K. In vitro antipyrial activity of psidium guajava leaf extract. *European Journal Pharmaceutical and Medical Research* 2016;3(1):312–315.

47 Bothiraja C, Gholap KS, and Pawar AP. Investigation of ethyl cellulose microsponge gel for topical delivery of eberconazole nitrate for fungal therapy. *Therapeutic Delivery [Internet]* 2014;5(7):781–794. Available from: https://www.future-science.com/doi/10.4155/tde.14.43

48 Borude SS and Mahajan AM. Recent updates on Q fever: Diagnosis and treatment. *Pharmacologyonline [Internet]* 2011;1(3):254–264. Available from: https://pharmacologyonline.silae.it/files/newsletter/2011/vol3/030.gholap.pdf

49 Bhagat V. Gold nanorods: A promising drug delivery vehicle. *Inventi Rapid NDDS [ISSN 0976-3864] [Internet]* 2012;1(4):1–4. Available from: https://www.inventi.in/journal/article/127/9964/Inventi Rapid: NDDS/Pharmaceutical

50 Gholap AD, Sayyad SF, Hatvate NT, et al. Drug delivery strategies for avobenzone: A case study of photostabilization. *Pharmaceutics [Internet]* 2023;15(3):1008. Available from: https://www.mdpi.com/1999-4923/15/3/1008

51 Yao Y, Zhou Y, Liu L, et al. Nanoparticle-based drug delivery in cancer therapy and its role in overcoming drug resistance. *Frontiers in Molecular Biosciences [Internet]* 2020; 7. Available from: https://www.frontiersin.org/article/10.3389/fmolb.2020.00193/full

52 Hatvate NT, Khuspe PR, Mandhare TA, Kashid P, and Gaikwad VD. Proteomics in oncology: Retrospect and prospects. In: Kulkarni S, Haghi AK, and Manwatkar S, (Eds.). *Novel Technologies in Biosystems, Biomedical & Drug Delivery [Internet]*, (pp. 243–69). Singapore: Springer Nature Singapore; 2023. Available from: https://link.springer.com/10.1007/978-981-99-5281-6_10

53 Vora LK, Jetha K, Thakur RRS, Solanki HK, and Chavda VP. Artificial intelligence in pharmaceutical technology and drug delivery design. *Pharmaceutics [Internet]* 2023;15(7):1916. Available from: https://www.mdpi.com/1999-4923/15/7/1916

54 Magar TV, Gholap AD, and Musmade DS. Elastic vesicles as drugs carriers through the skin. *Inventi Rapid NDDS [ISSN 0976-3791] [Internet]* 2011;1(4):1–2. Available from: https://www.inventi.in/journal/article/127/9611/Inventi Rapid: NDDS/Pharmaceutical#

55 Gholap AD, Rajurkar VG, and Dighe SB. Phonophoresis: Ultrasound enhanced transdermal drug delivery. *Inventi Rapid NDDS [ISSN 0976-3791] [Internet]* 2012;1(4):1–3. Available from: https://www.inventi.in/journal/article/127/9981/Inventi Rapid: NDDS/Pharmaceutical

56 Khedkar Ajay TD. Cubosomes: Emerging drug delivery system. *Inventi Rapid NDDS [ISSN 0976-3791] [Internet]* 2012;1(4):1–3. Available from: https://www.inventi.in/journal/article/127/9987/Inventi Rapid: NDDS/Pharmaceutical

57 Khedkar Ajay TD. Pharmacosomes: An emerging vesicular system. *Inventi Rapid NDDS [ISSN 0976-3791] [Internet]* 2012;1(4):1–3. Available from: https://www.inventi.in/journal/article/127/9986/Inventi Rapid: NDDS/Pharmaceutical

58 Ma L, Kohli M, and Smith A. Nanoparticles for combination drug therapy. *ACS Nano [Internet]* 2013;7(11):9518–9525. Available from: https://pubs.acs.org/doi/10.1021/nn405674m

59 Salahpour Anarjan F. Active targeting drug delivery nanocarriers: Ligands. *Nano-Structures & Nano-Objects [Internet]* 2019;19:100370. Available from: https://linkinghub.elsevier.com/retrieve/pii/S2352507X19302926

60 Deivayanai VC, Thamarai P, Karishma S, et al. A comprehensive review on advances in nanoparticle-mediated cancer therapeutics: Current research and future perspectives. *Cancer Pathogenesis and Therapy[Internet]* 2024. Available from: https://linkinghub.elsevier.com/retrieve/pii/S2949713224000880

61 Buya AB, Mahlangu P, and Witika BA. From lab to industrial development of lipid nanocarriers using quality by design approach. *International Journal of Pharmaceutics: X [Internet]* 2024;8:100266. Available from: https://linkinghub.elsevier.com/retrieve/pii/S2590156724000380

62 Borude Santosh S and Mahajan AM. POEMS syndrome: A review. *Pharmacologyonline [Internet]* 2011;1(3):265–274. Available from: https://pharmacologyonline.silae.it/files/newsletter/2011/vol3/031.gholap.pdf

63 Rojekar S, Kapare HS, Vishwakarma N, et al. Chitosan scaffolds: Expanding horizons in biomedical applications. *Carbohydrate Polymers [Internet]* 2024;323:121394. Available from: https://linkinghub.elsevier.com/retrieve/pii/S0144861723008597

64 Kapare HS, Pagar S, Kamandar P, et al. Exploring modified chitosan-based gene delivery technologies for therapeutic advancements. *International Journal of Biological Macromolecules [Internet]* 2024;260:129581. Available from: https://linkinghub.elsevier.com/retrieve/pii/S0141813024003842

65 Pardeshi SR, Hatvate NT, Dhorkule N, Sayyad SF, Faiyazuddin M, and Khalid M. Environmental implications and nanotechnological advances in octocrylene-enriched sunscreen formulations: A comprehensive review. *Chemosphere [Internet]* 2024;358:142235. Available from: https://linkinghub.elsevier.com/retrieve/pii/S0045653524011287

66 Gupta J, Kamandar P, Bhowmik DD, et al. Harnessing nanovaccines for effective immunization—A special concern on COVID-19: Facts, fidelity, and future prospective. *ACS Biomaterials Science Engineering [Internet]*. 2024;10(1):271–297. Available from: https://pubs.acs.org/doi/10.1021/acsbiomaterials.3c01247

67 Parit S, Manchare A, Mundhe P, Hatvate N, Rojekar S, and Patravale V. Antibody-drug conjugates: A promising breakthrough in cancer therapy. *International Journal of Pharmaceutics [Internet]* 2024;659:124211. Available from: https://linkinghub.elsevier.com/retrieve/pii/S0378517324004459

68 Alshawwa SZ, Kassem AA, Farid RM, Mostafa SK, and Labib GS. Nanocarrier drug delivery systems: Characterization, limitations, future perspectives and implementation of artificial intelligence. *Pharmaceutics [Internet]* 2022;14(4):883. Available from: https://www.mdpi.com/1999-4923/14/4/883

69 Teklehaimanot Y. Nanotechnology-enhanced controlled release systems in topical therapeutics. *Precision Nanomedicine [Internet]* 2024;7(4). Available from: https://precisionnanomedicine.com/article/127336-nanotechnology-enhanced-controlled-release-systems-in-topical-therapeutics

70 Sawant SD, Sayyad SF, Hatvate NT, Chavan M, Rojekar S, and Faiyazuddin M. Food biopackaging for human benefits: Status and perspectives. In: Ahmed S. (Ed.). *Biobased Packaging Materials [Internet]*, (pp. 1–35). Singapore: Springer Nature Singapore; 2023. Available from: https://link.springer.com/10.1007/978-981-99-6050-7_1

71 Hatvate NT, Sawant SD, Khemka PN, and Vora LK. Unlocking the potential of nanobiohybrids to combat environmental pollution. In: Lens P. (Ed.). *Nanobiohybrids for Advanced Wastewater Treatment and Energy Recovery [Internet]*, (pp. 169–190). UK: IWA Publishing; 2023. Available from: https://iwaponline.com/ebooks/book/886/chapter/3516052/Unlocking-the-potential-of-nanobiohybrids-to

72 Pardeshi S, More M, Togre N, Rebello N, and Giram P. Dendrimers based antibacterial and antiviral materials. In: *Antibacterial and Antiviral Functional Materials [Internet]*, (pp. 139–169). USA: ACS Publications; 2023. Available from: https://pubs.acs.org/doi/abs/10.1021/bk-2023-1458.ch005

73 Pardeshi S and Giram P. Preparation, antibacterial and antiviral activity measurements and detection methods. In: Deshmukh K. (Ed.). *Antibacterial and Antiviral Functional Materials [Internet]*, (pp. 33–64). USA: ACS Publications; 2023. Available from: https://pubs.acs.org/doi/abs/10.1021/bk-2023-1458.ch002

74 Hatvate NT, Khuspe PR, Mandhare TA, Kashid P, and Vora LK. Decoding of nanoneutraceuticals and their delivery. In: Mohanta YK. (Ed.). *Sustainable Green Nanotechnology [Internet]*, (pp. 284–309). Florida: New York: CRC Press; 2024. Available from: https://www.taylorfrancis.com/books/9781003389408/chapters/10.1201/9781003389408-16

75 Uddin MJ, Faiyazuddin M, Omri A, Gowri S, and Khalid M. Advances in artificial intelligence for drug delivery and development: A comprehensive review. *Computers in Biology and Medicine [Internet]* 2024;178:108702. Available from: https://linkinghub.elsevier.com/retrieve/pii/S001048252400787X

76 S J, Kamandar PA, Banchhod G V, and Hatvate NT. Antibody-drug conjugates for cancer therapy: An up-to-date review on the chemistry and pharmacology. In: Verma SK and Das AK. (Eds.). *Bioconjugated Materials Part 2 Applications in Drug Delivery, Vaccine Formulations and Important Conjugates for Cancer Therapy [Internet]*, (Comprehensive Analytical Chemistry; vol. 103 pp. 105–190). Amsterdam, Netherlands: Elsevier; 2023. Available from: https://www.sciencedirect.com/science/article/pii/S0166526X23000259

77 Bai X, Smith Z, Wang Y, Butterworth S, and Tirella A. Sustained drug release from smart nanoparticles in cancer therapy: A comprehensive review. *Micromachines [Internet]* 2022;13(10):1623. Available from: https://www.mdpi.com/2072-666X/13/10/1623

78 Bhowmik DD, Deshmukh AY, and Hatvate NT. Quintessential impact of dendrimer bioconjugates in targeted drug delivery. In: Verma SK and Das AK. (Eds.). *Bioconjugated Materials Part 2 Applications in Drug Delivery, Vaccine Formulations and Important Conjugates for Cancer Therapy [Internet]*, (Comprehensive Analytical Chemistry; vol. 103 pp. 257–302). Amsterdam, Netherlands: Elsevier; 2023. Available from: https://www.sciencedirect.com/science/article/pii/S0166526X23000272

79 Gholap AD, Said RP, Pawar RD, Ambore GS, and Hatvate NT. Importance of carbohydrate-drug conjugates in vaccine development: A detailed review. In: Verma SK and Das AK. (Eds.). *Bioconjugated Materials Part 2 Applications in Drug Delivery, Vaccine Formulations and Important Conjugates for Cancer Therapy [Internet]*, (Comprehensive

Analytical Chemistry; vol. 103 pp. 191–256). Amsterdam, Netherlands: Elsevier; 2023. Available from: https://www.sciencedirect.com/science/article/pii/S0166526X23000296
80 Khuspe PR, Pardeshi SR, Uddin MJ, et al. Achieving optimal health with host-directed therapies (HDTs) in infectious diseases—A new horizon. *Advances in Therapy [Internet]* 2024; 2400169. Available from: https://onlinelibrary.wiley.com/doi/10.1002/adtp.202400169
81 Rojekar S, Togre N, Bhoj P, et al. Current status of mannose receptor-targeted drug delivery for improved anti-HIV therapy. *Journal Control Release [Internet]* 2024;372:494–521. Available from: https://linkinghub.elsevier.com/retrieve/pii/S0168365924003444
82 Shaikh MS, Faiyazuddin M, Khan MS, et al. Chikungunya virus vaccine: A decade of progress solving epidemiological dilemma, emerging concepts, and immunological interventions. *Frontiers in Microbiology [Internet]* 2024; 15. Available from: https://www.frontiersin.org/articles/10.3389/fmicb.2024.1413250/full
83 Rojekar S, Parit S, Manchare A, et al. Revolutionizing eye care: Exploring the potential of microneedle drug delivery. *Pharmaceutics [Internet]* 2024;16(11):1398. Available from: https://www.mdpi.com/1999-4923/16/11/1398
84 Yang J, Zeng H, Luo Y, et al. Recent applications of PLGA in drug delivery systems. *Polymers (Basel) [Internet]* 2024;16(18):2606. Available from: https://www.mdpi.com/2073-4360/16/18/2606
85 Vardikar A, Das U, Mandal S, et al. Surface engineered multimodal magnetic nanoparticles for neurodegenerative diseases. In: *Targeted Therapy for the Central Nervous System [Internet]*, (pp. 121–153). USA: Elsevier; 2025. Available from: https://linkinghub.elsevier.com/retrieve/pii/B9780443238413000078
86 Srivastava N, Ibrahim SA, Nasr MHA, editor. Manipulating the rhizosphere microbiome for plant health. In: *Microbiome Engineering*, 1st ed. (p. 19). Florida: CRC Press; 2024.
87 Mesut B, Al-Mohaya M, Yeşilkaya E, et al. Demystifying the potential of lipid-based nanocarriers in targeting brain malignancies. *Naunyn Schmiedebergs Archieves Pharmacology [Internet]* 2024;397(12):9243–9279. Available from: https://link.springer.com/10.1007/s00210-024-03212-6
88 Pardeshi S, Pillai D, Giram P, Pardeshi C, and Naik J. Nanocarriers for ocular drug delivery. In: *Nanocarrier Drug Delivery Systems [Internet]*, (pp. 469–500). Germany: De Gruyter; 2024. Available from: https://www.degruyter.com/document/doi/10.1515/9783111320847-015/html
89 Patil M, Dhumal D, Rojekar S, Jain H, Pingale P, and Kumar Prajapati M. Nanocarriers for antimicrobial therapy. In: *Nanocarrier Drug Delivery Systems [Internet]*, (pp. 561–602). Germany: De Gruyter; 2024. Available from: https://www.degruyter.com/document/doi/10.1515/9783111320847-018/html
90 Sawant SD, Pardeshi SR, Khuspe PR, et al. The role of inulin in altering the animal microbiome: A comprehensive review. *Microbiome Engineering* 2024;1:79–102.
91 Sawant SD, Pardeshi SR, Khuspe PR, et al. Omics tools for determining microbial composition: A systematic review. *Microbiome Engineering* 2024;1:35–50.
92 Hajebi S, Chamanara M, Nasiri SS, et al. Advances in stimuli-responsive gold nanorods for drug-delivery and targeted therapy systems. *Biomedicine Pharmacother [Internet]* 2024;180:117493. Available from: https://linkinghub.elsevier.com/retrieve/pii/S0753332224013799

93 Kumar D, Kumar S, Kumari R, et al. A comprehensive review on the biomass-mediated synthesis of silver nanoparticles: Opportunities and challenges. *Journal of Environmental Chemical Engineering [Internet]* 2024; 115133. Available from: https://linkinghub.elsevier.com/retrieve/pii/S2213343724032652

94 Faiyazuddin M, Maqsood S, Abbas S, Chaudhari YS, Sharmila S, and Webster TJ. Prospects and ethics of ChatGPT in biomedical research and medical practice: An emphasis on nanotechnology. *Journal of Biomedical Nanotechnology [Internet]* 2024;20(11):1661–1678. Available from: https://www.ingentaconnect.com/content/10.1166/jbn.2024.3925

95 Pardeshi S, Kapare H, More A, Rebello N, and Giram P. Coating approaches of functionalized magnetic nanoparticles for theranostic applications. In: *Functionalized Magnetic Nanoparticles for Theranostic Applications [Internet]*, (pp. 131–171). Wiley; 2024. Available from: https://onlinelibrary.wiley.com/doi/10.1002/9781394172917.ch5

96 Pardeshi SR, Hatvate NT, Gharat KD, Naik JB, and Omri A. Advances in dorzolamide hydrochloride delivery: Harnessing nanotechnology for enhanced ocular drug delivery in glaucoma management. *Discover Nano [Internet]* 2024;19(1):199. Available from: https://link.springer.com/10.1186/s11671-024-04154-x

97 Hatvate NT, Bhowmik D, Sawant SD, Londhe KM, Rojekar S, and Faiyazuddin M. Engineered artificial MicroRNA in plant: An overview. In: *MicroRNA Advances and Application in Plant Biology [Internet]*, (pp. 311–341). Elsevier; 2025. Available from: https://linkinghub.elsevier.com/retrieve/pii/B9780443216824000130

98 Uddin MJ, Faiyazuddin M, Saeed SM, Sadiq SB, and Koly IAC. Lipid-based drug delivery for systemic effect through pulmonary route of administration. In: *Lipids in Pulmonary Drug Delivery [Internet]*, (pp. 255–299). Elsevier; 2025. Available from: https://linkinghub.elsevier.com/retrieve/pii/B9780443223747000128

99 Khan AR, Grewal NS, Jun Z, Khalid M, and Zhang H-J. Advances in smart hybrid scaffolds: A strategic approach for regenerative clinical applications. *Engineered Regeneration [Internet]* 2025;6:85–110. Available from: https://linkinghub.elsevier.com/retrieve/pii/S2666138125000039

100 Hawthorne D, Pannala A, Sandeman S, and Lloyd A. Sustained and targeted delivery of hydrophilic drug compounds: A review of existing and novel technologies from bench to bedside. *Journal of Drug Delivery Science and Technology [Internet]* 2022;78:103936. Available from: https://linkinghub.elsevier.com/retrieve/pii/S1773224722008474

101 Ghezzi M, Pescina S, Padula C, et al. Polymeric micelles in drug delivery: An insight of the techniques for their characterization and assessment in biorelevant conditions. *Journal Control Release [Internet]* 2021;332:312–336. Available from: https://linkinghub.elsevier.com/retrieve/pii/S0168365921001061

102 Khuspe PR, Faiyazuddin M, Uddin MJ, et al. Current scenario and challenges of lipid-based nanomedicines for pulmonary drug delivery. In: *Lipids in Pulmonary Drug Delivery [Internet]*, (pp. 21–47). Elsevier; 2025. Available from: https://linkinghub.elsevier.com/retrieve/pii/B9780443223747000025

103 Khuspe PR, Faiyazuddin M, Uddin MJ, et al. Lipid-based vaccines against viruses including COVID-19. In: *Lipids in Pulmonary Drug Delivery [Internet]*, (pp. 409–433). Elsevier; 2025. Available from: https://linkinghub.elsevier.com/retrieve/pii/B9780443223747000190

104 Din F, Aman W, Ullah I, et al. Effective use of nanocarriers as drug delivery systems for the treatment of selected tumors. *International Journal of Nanomedicine [Internet]* 2017;12:7291–7309. Available from: https://www.dovepress.com/effective-use-of-nanocarriers-as-drug-delivery-systems-for-the-treatme-peer-reviewed-article-IJN

105 Lu H, Zhang S, Wang J, and Chen Q. A review on polymer and lipid-based nanocarriers and its application to nano-pharmaceutical and food-based systems. *Frontiers in Nutrition [Internet]* 2021: 8. Available from: https://www.frontiersin.org/articles/10.3389/fnut.2021.783831/full

106 Aloss K and Hamar P. Recent preclinical and clinical progress in liposomal doxorubicin. *Pharmaceutics [Internet]* 2023;15(3):893. Available from: https://www.mdpi.com/1999-4923/15/3/893

107 Ahmad MZ, Mustafa G, Abdel-Wahab BA, et al. From bench to bedside: Advancing liposomal doxorubicin for targeted cancer therapy. *Results in Surfaces and Interfaces [Internet]* 2025;19:100473. Available from: https://linkinghub.elsevier.com/retrieve/pii/S2666845925000601

108 Kyriakides TR, Raj A, Tseng TH, et al. Biocompatibility of nanomaterials and their immunological properties. *Biomedical Materials [Internet]* 2021;16(4):042005. Available from: https://iopscience.iop.org/article/10.1088/1748-605X/abe5fa

109 Al-Thani AN, Jan AG, Abbas M, Geetha M, and Sadasivuni KK. Nanoparticles in cancer theragnostic and drug delivery: A comprehensive review. *Life Sciences [Internet]* 2024;352:122899. Available from: https://linkinghub.elsevier.com/retrieve/pii/S0024320524004892

110 Xu X, Han Y, Deng J, et al. Repurposing disulfiram with CuET nanocrystals: Enhancing anti-pyroptotic effect through NLRP3 inflammasome inhibition for treating inflammatory bowel diseases. *Acta Pharmaceutica Sinica B [Internet]* 2024;14(6):2698–2715. Available from: https://linkinghub.elsevier.com/retrieve/pii/S2211383524000832

111 Liu S-H, Yu J, Creeden JF, et al. Repurposing metformin, simvastatin and digoxin as a combination for targeted therapy for pancreatic ductal adenocarcinoma. *Cancer Letters [Internet]* 2020;491:97–107. Available from: https://linkinghub.elsevier.com/retrieve/pii/S0304383520304134

112 Kalangadan N, Mary AS, Mani K, et al. Repurposing ivermectin and ciprofloxacin in nanofibers for enhanced wound healing and infection control against MDR wound pathogens. *Journal of Drug Delivery Science and Technology [Internet]* 2023;90:105166. Available from: https://linkinghub.elsevier.com/retrieve/pii/S1773224723010183

113 Qureshi KA, Parvez A, and Jaremko M. Repurposing eugenol and cinnamaldehyde as potent antimicrobial agents: A comprehensive in-vitro and in-silico study. *Bioorganic Chemistry [Internet]* 2025;156:108199. Available from: https://linkinghub.elsevier.com/retrieve/pii/S0045206825000793

114 Kumari P and Dang S. Dual drug loaded nanostructured lipid carrier for cytotoxic effect against breast cancer – A drug repurposing approach. *Surfaces and Interfaces [Internet]* 2023;40:103138. Available from: https://linkinghub.elsevier.com/retrieve/pii/S2468023023005084

115 Havelikar U, Ghorpade KB, Kumar A, et al. Comprehensive insights into mechanism of nanotoxicity, assessment methods and regulatory challenges of nanomedicines. *Discover*

Nano [Internet] 2024;19(1):165. Available from: https://link.springer.com/10.1186/s11671-024-04118-1

116 He Y, Wang Y, Wang L, Jiang W, and Wilhelm S. Understanding nanoparticle-liver interactions in nanomedicine. *Expert Opinion on Drug Delivery [Internet]* 2024;21(6): 829–843. Available from: https://www.tandfonline.com/doi/full/10.1080/17425247.2024.2375400

117 Pondman K, Le Gac S, and Kishore U. Nanoparticle-induced immune response: Health risk versus treatment opportunity? *Immunobiology [Internet]* 2023;228(2):152317. Available from: https://linkinghub.elsevier.com/retrieve/pii/S0171298522001437

118 Torres-Herrero B, Armenia I, Ortiz C, de la Fuente JM, Betancor L, and Grazú V. Opportunities for nanomaterials in enzyme therapy. *Journal Control Release [Internet]* 2024;372:619–647. Available from: https://linkinghub.elsevier.com/retrieve/pii/S0168365924003845

119 Farjadian F, Ghasemi A, Gohari O, Roointan A, Karimi M, and Hamblin MR. Nanopharmaceuticals and nanomedicines currently on the market: Challenges and opportunities. *Nanomedicine [Internet]* 2019;14(1):93–126. Available from: https://www.tandfonline.com/doi/full/10.2217/nnm-2018-0120

120 Mikušová V and Mikuš P. Advances in chitosan-based nanoparticles for drug delivery. *International Journal of Molecular Sciences [Internet]* 2021;22(17):9652. Available from: https://www.mdpi.com/1422-0067/22/17/9652

121 Liu Q, Zou J, Chen Z, He W, and Wu W. Current research trends of nanomedicines. *Acta Pharmaceutica Sinica B [Internet]* 2023;13(11):4391–4416. Available from: https://linkinghub.elsevier.com/retrieve/pii/S2211383523001636

122 Petrovic S, Bita B, and Barbinta-Patrascu M-E. Nanoformulations in pharmaceutical and biomedical applications: Green perspectives. *International Journal of Molecular Sciences [Internet]* 2024;25(11):5842. Available from: https://www.mdpi.com/1422-0067/25/11/5842

123 Herdiana Y, Wathoni N, Shamsuddin S, and Muchtaridi M. Scale-up polymeric-based nanoparticles drug delivery systems: Development and challenges. *OpenNano [Internet]* 2022;7:100048. Available from: https://linkinghub.elsevier.com/retrieve/pii/S2352952022000111

124 Huanbutta K, Burapapadh K, Kraisit P, et al. Artificial intelligence-driven pharmaceutical industry: A paradigm shift in drug discovery, formulation development, manufacturing, quality control, and post-market surveillance. *European Journal of Pharmaceutical Sciences [Internet]* 2024;203:106938. Available from: https://linkinghub.elsevier.com/retrieve/pii/S0928098724002513

125 Khatoon UT and Velidandi A. An overview on the role of government initiatives in nanotechnology innovation for sustainable economic development and research progress. *Sustainability [Internet]* 2025;17(3):1250. Available from: https://www.mdpi.com/2071-1050/17/3/1250

126 Pokrajac L, Abbas A, Chrzanowski W, et al. Nanotechnology for a sustainable future: Addressing global challenges with the international network4sustainable nanotechnology. *ACS Nano [Internet]* 2021;15(12):18608–18623. Available from: https://pubs.acs.org/doi/10.1021/acsnano.1c10919

127 Bhange M and Telange D. Convergence of nanotechnology and artificial intelligence in the fight against liver cancer: A comprehensive review. *Discover Oncology [Internet]* 2025;16(1):77. Available from: https://link.springer.com/10.1007/s12672-025-01821-y

128 Das KP and J C. Nanoparticles and convergence of artificial intelligence for targeted drug delivery for cancer therapy: Current progress and challenges. *Frontiers in Medical Technology [Internet]* 2023; 4. Available from: https://www.frontiersin.org/articles/10.3389/fmedt.2022.1067144/full

13

Nano-enabled Drug Repurposing and Chemotherapy

*Sareer Ahmad, Jawairia Ali, Salah Uddin**, *Safia Ali, and Kefayat Ullah*

Department of Applied Physical and Material Sciences, University of Swat, Swat, Pakistan

*Corresponding author: salahphysics@uswat.edu.pk

CONTENTS

13.1 Introduction

Nanotechnology is a diverse domain that focuses on the synthesis of substances at the nanoscale, typically between 1 and 100 nm. Materials in this range have unique quantum properties that differ significantly from their mass equivalents. Nanotechnological improvements in every field are growing drastically. Over the years, advancements in nanotechnology have revolutionized multiple scientific fields, particularly medicine, through the exploration of nanoscale manufacturing technologies such as nanomachines, nanobots, and nanomedicine to develop novel therapeutic and diagnostic solutions [1]. One of the most favorable applications of nanotechnology in medicine is targeted drug delivery, where nanomaterials help transport therapeutic agents across biological barriers, improve molecular targeting, and enhance drug stability. In addition to its role in diagnostics, this technique is useful in several medical fields, including surgery, oncology, odontology, dermatology, neuroscience, and other fields, to facilitate the controlled release of drugs, improve bioavailability, and reduce toxicity [2]. Historically, the aim of pharmaceuticals was focused on developing small-molecule medications owing to their effortless synthesis and large-scale production. However, despite their fruitful applications, many drugs sometimes face poor specificity for the required target, resulting in the dropping of active substances to off-target areas, leading to loss of efficiency, high dosage, toxicity, and unwanted side effects. These limitations not only impact treatment outcomes but also increase the risk of drug resistance, patient noncompliance, and increased cost of treatment [3]. Thus, there remains a critical need to implement innovative strategies to optimize drug delivery, enhance treatment effectiveness, and provide safer treatment alternatives. In addition to these worldwide efforts, one promising approach that has gained widespread attention in recent years is drug repurposing (DR), which involves identifying new therapeutic applications for existing clinically approved drugs, unlike traditional de novo drug development, which takes up to 15 years and costs approximately 2 billion dollars, requiring clinical and preclinical testing. Once the drug's therapeutic efficacy has been observed in an in vitro model, the drug has moved into three alternative phases of human clinical trials. Even with these efforts, less than 1% of formulations are likely to progress to clinical trials, and it is nearly impossible for them to reach the market [4]. DR significantly reduces both time and expenses, as the safety, toxicity, and pharmacokinetics of these drugs are already well established [5]. Despite its advantages, DR still faces challenges, such as poor solubility, poor stability, poor bioavailability, and limited targeting efficiency. This is where nanotechnology offers an effective solution. Nanotechnology for DR involves the use of nanoscale materials and technologies to modify existing drugs so that they can be used for new therapeutic applications [6]. This approach combines the principles of DR with the advantages of nanotechnology to address these limitations effectively. By utilizing nanocarriers, repurposed drugs can overcome strong obstacles such as the blood–brain barrier (BBB), which restricts all types of bulky molecule drugs from approaching the brain [7]. Compared with conventional methods, nanomaterial-based encapsulated drugs offer improved pharmacokinetic and pharmacodynamic properties, including drug stability, targeted drug delivery, and precise distribution, due to their unique dimensional profile and composition [8]. On a broader scale, nanoscale medical technologies improve efficiency, provide a fast response rate, and enhance functionality across many biological and chemical processes, especially in the case of repurposing conventional drugs.

In recent years, the treatment of complex diseases has increasingly shifted toward the use of combination therapies, which involve multiple therapeutic agents that act on different aspects of a disease's pathways to enhance treatment outcomes. In addition to standard therapy, combination therapy integrates other advanced therapeutic modalities, such as immunotherapy, phototherapy, and magnetotherapy. Accumulating evidence indicates that relying on a single therapeutic approach tends to promote drug resistance and cell tolerance, eventually leading to relapse and spread [9]. Therefore, combined therapeutic approaches might help overcome these difficulties, while some of these combinations have been successfully tested in animal models; only a few of them have reached clinical trials because of issues such as drug resistance, poor pharmacokinetics, and limited cell penetration [10].

One of the so-called challenges of combination therapy is drug delivery. Many therapeutic agents have short half-lives, limited solubility, and difficulty reaching target tissues, leading to insufficient drug accumulation and reduced therapeutic effectiveness. Additionally, inconsistent drug release profiles make it difficult to ensure that multiple drugs reach their intended targets simultaneously [11]. Nanotechnology has the potential to control these problems and improve therapeutic outcomes. Nanocarrier-based combination therapy offers several advantages over conventional approaches. First, nanoformulations improve the solubility, stability, and bioavailability of therapeutic agents, ensuring effective drug delivery. Second, nanomaterials can be engineered for precise targeting, allowing multiple drugs to be codelivered with spatial and temporal control. This ensures that different therapeutic agents act synergistically, enhancing treatment efficacy while reducing the required dosage [12]. Furthermore, nanocarriers might be functionalized by targeting ligands, including antibodies, peptides, or aptamers, to achieve highly selective drug delivery. These delivery systems are categorized as lipid-based nanoparticles (NPs), polymeric NPs, lipopolymer NPs, inorganic NPs, carbon-based NPs, and hybrid NPs. These nanocarriers can enable the loading of numerous therapeutic agents simultaneously and codeliver them to the desired specific targeting agent [13]. By leveraging nanotechnology-based combination therapy, researchers aim to address drug resistance, improve patient compliance, and minimize systemic toxicity. The synergistic effects of combining repurposed drugs with nanocarriers provide an efficient strategy for managing complex diseases such as cancer, infectious diseases, and neurological disorders.

This chapter embarks on nanotechnology-based strategies for DR and combination therapy, focusing on targeted drug delivery systems using nanocarriers with antibiotics, antivirals, and antifungals and biological control agents across various fields of medicine. Figure 13.1 illustrates the nanotechnology repurpose for different medical treatments.

13.1.1 Nanotechnology in DR

DR is a tactic for finding novel applications for authorized or experimental drugs that exceed their initial indications in medicine. With many benefits, including established pharmacology, a proven safety record, less financial contribution, and a quicker time frame for bringing the medicine to market, DR is a useful adjunct to traditional pharmacology [14]. For a particular indication, this approach has several benefits over creating a new medication. First, and perhaps most importantly, there is a low chance of failure; early-stage trials have revealed that repurposed medication is suitably harmless in preclinical models [15]. Second, since much of the preclinical testing, safety protocols, and formulation development will be finalized, the duration of drug development can be reduced. Finally, it has low expenses; however, it significantly depends upon the repurposing candidate's level of development [16]. The idea behind similar adverse effects may involve targeting the same protein or pathway. Moreover, a drug's adverse effect profile might

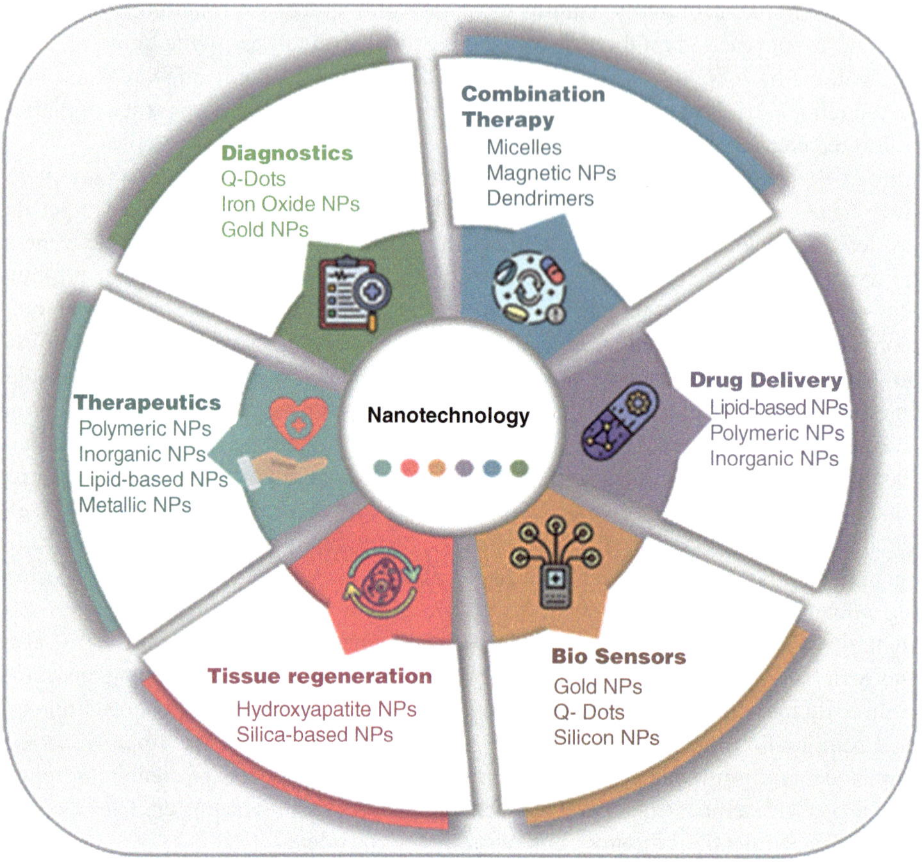

Figure 13.1 Illustration of nanotechnological repurposing for different medical treatments.

resemble that of a disease, suggesting shared physiological pathways. Hence, a drug's unique adverse effects can act as a proxy for its phenotype [17]. Nanotechnology can play a crucial role in this field by enabling innovative drug delivery systems, enhancing bioavailability, and creating new formulations, which strengthen patentability and increase therapeutic efficacy, offering new pathways to overcome challenges in DR [18]. It is now more feasible to achieve this goal because many nanotherapeutics and nanodiagnostics are on the market or have advanced to the clinical stage in recent years [19]. Additionally, by enabling regulated drug release and circumventing cellular resistance mechanisms, nanocarriers can make repurposed medications more successful in treating complicated tumors [20]. The use of nanocarriers, such as liposomes, polymeric NPs, and nanostructured lipid carriers (NLCs), improves the solubility, bioavailability, and targeted delivery of repurposed drugs, overcoming issues such as poor pharmacokinetics and drug resistance by enabling site-specific delivery, reducing off-target effects, and enhancing therapeutic efficacy [21]. Nanotechnology has made significant contributions to DR, including the following sections.

13.1.2 Targeted Drug Delivery

Among the most important uses of nanotechnology in DR is targeted drug delivery, especially for conditions such as cancer and neurological illnesses. By carefully targeting impacted tissues or cells, this technique reduces systemic exposure and improves treatment efficacy [22]. The formulations meant for intravenous administration entail effective drug loading into a delivery vehicle of some kind, adequate circulation to reach targeted body sites, retention by particular properties, and drug release at the right time frame that permits the drug to function effectively. Owing to medication delivery, current drug delivery systems are unable to direct themselves to a target [13]. Nanocarriers can prevent a drug from degrading by avoiding the environment. Using nanotechnology for drug transportation appears to be a wise strategy for site-specific drug molecule delivery. The blood flow profile increases drug accessibility at the targeted intracellular compartments and facilitates transport via biological barriers [23]. Nanodelivery technologies have been shown to enhance the biodistribution and administration routes of medications with minimal adverse effects and immunogenicity. Nanocarriers can actively or passively transport medicines to the location of illness. In the former scenario, the receptor, lipids, or antigens at the targeted region are attached with peptides and antibodies connected to the delivery system through direct chemical conjugation. The self-assembled nanostructured substance transports the medicine via passive delivery, releasing the encapsulated drug at the intended location. The building blocks for self-assembled nanostructures that can encapsulate medications in their hydrophobic environment include metals, polymers, proteins, lipids, nucleic acids, and ceramics [24].

The capacity of NPs to target diseased tissue without harming healthy tissue is the primary reason why creating NP drug delivery systems increases the effectiveness of medications [25]. Environmentally sensitive polymeric NPs have been actively designed to enhance drug release with specific spatial or temporal control, which respond to internal stimuli (e.g., pH, glucose, and enzymes) or external stimuli (e.g., temperature, magnetic fields, ultrasound, and light) by dissolving, swelling, or collapsing [26]. Depending on the type of stimulus, reaction rate, and precise location of drug release triggering, these multistimuli-responsive polymeric NPs may provide unparalleled regulation over drug delivery and release, resulting in increased anticancer potency both in vitro and in vivo. On the other hand, they may also make it easier to prepare and load drugs under mild conditions. For example, redox-sensitive drug release polymersomes are created on the basis of temperature and reduction dual-responsive PEG-PAA-PNIPAAm triblock copolymers by simply increasing the solution temperature above the lower critical solution temperature and then crosslinking with cystamine via carbodiimide chemistry [27]. Stimulus-responsive nanocarriers typically have a hydrophilic or amphiphilic shell and a hydrophobic core. Amphiphilic segments, such as lipids or surfactants, respond to pH, redox, temperature, light, enzymes, or magnetic fields. Hydrophobic cores can be modified with cationic groups to conjugate anionic agents such as nucleic acids [28]. Redox-sensitive linkers, such as disulfide bonds, accomplish drug release from nanocarriers through redox reactions. In tumor and intracellular environments, high glutathione (GSH) levels cleave these linkers and induce drug release [29]. The ability of nanocarriers to alter their characteristics in response to endogenous or external stimuli has led to a variety of biomedical uses, such as the regulated delivery of medications to the intended location, which minimizes adverse effects on nearby healthy tissue [30].

13.1.3 Improved Solubility and Bioavailability

Older medications with established safety profiles are frequently reassessed for novel therapeutic applications in pharmacological repurposing. Clinical effectiveness is limited due to problems associated with many of these medications, such as low absorption and poor solubility. For novel applications to maximize their therapeutic potential, these problems must be resolved. Nanotechnology has revolutionized the repurposing of drugs by significantly improving their solubility and bioavailability by reducing drug size to the nanoscale, which increases their surface area and improves their dissolution rate. For example, nanocrystals have been developed to increase the bioavailability of poorly soluble drugs [31]. These nanocrystals are pure drug elements stabilized with minimal surface-active agents, confirming better absorption and stability. Another approach is the use of lipid-based NPs, such as solid lipid nanoparticles (SLNs) and NLCs, which increase their solubility in aqueous environments. These systems have been successfully applied in the delivery of drugs such as curcumin and paclitaxel, which are otherwise poorly soluble [32]. Additionally, polymeric NPs, such as those made from PLGA (poly(lactic-co-glycolic acid)), have been used to increase the bioavailability of drugs such as docetaxel by protecting them from degradation and ensuring controlled release [33]. Niclosamide has been thoroughly investigated for repositioning to produce formulations with improved oral bioavailability and aqueous solubility [34]. Several drug delivery technologies, such as cocrystallization systems, cyclodextrins, emulsion-based drug delivery systems, and lipid-based formulations, have been developed to expand the aqueous solubility and oral utilization of niclosamide for therapeutic repositioning [35]. An increase in the solubility and bioavailability of repurposed niclosamide drugs can be achieved through the use of nanotechnology-based techniques, such as lipid formulations or nanocarriers, which are crucial for optimizing the efficacy of repurposed drugs in managing diseases such as severe acute respiratory syndrome coronavirus 2 (SARS-CoV-2) [36].

13.1.4 Controlled and Sustained Release

A drug delivery system is an assembly that can deliver therapeutic materials to a target in a way that improves its safety and effectiveness. A longer duration of action, increased bioavailability, or better drug localization may all contribute to improved efficacy [37]. The absence of a standard technique for measuring drug release in vitro is a noteworthy problem in assessing drug release from NPs. This made it more difficult to compare various categories, including liposomes, micelles, and polymeric NPs, and to interpret in vitro release data. Separating the released drugs from the encapsulated drugs is difficult, which poses a challenge, as the primary goal of nanocarrier research and development is to take advantage of their improved cellular penetration. The justification for employing NPs is to extend the lifespan of the blood circulation, which makes passive tissue targeting easier. The solubility of extremely hydrophobic medications, such as paclitaxel, can be increased, and the injectability of extremely hydrophobic medications can be maintained [38]. In contrast to nanocarriers, microparticles nearly never penetrate cells (delivering plasmid DNA or siRNA) and do not remain in the bloodstream long enough for passively targeted chemotherapeutic particles when administered intravenously [39]. Accordingly, obtaining sustained release from NPs is more challenging than from microparticles for two reasons: (a) drug loading is typically ineffective and limited for nanocarriers; and (b) release is quicker, with a noticeable burst release typically occurring for nanocarriers.

The absence of authorized nanomedicine products in that category may be due to the challenge of obtaining sustained release from nanocarriers, but it is also probable that there are no strong arguments in favor of using nanocarriers in some of the applications mentioned earlier. For example, subcutaneous injections of nanoencapsulated leuprolide acetate are unlikely to provide any advantages over microspheres in terms of bioavailability or pharmacokinetics, as the peptide's likely route to systemic circulation involves the transfer of molecules from the particles to the dermal blood vessels through molecular diffusion. If everything else is equal, nanospheres are likely to release leuprolide acetate more quickly than microspheres, which could even work against them in this application [40]. These dosage forms are attractive for a number of reasons, including the fact that many diseases have effective treatments, but their use is often limited by side effects or the need for clinical administration. Sustained or controlled drug delivery systems help overcome these challenges by reducing the dosing frequency, enhancing the efficacy through targeted delivery, lowering the required doses, or ensuring consistent drug release [41]. Drugs that are unstable at the pH of the stomach can therefore be made into slow-release dosage forms, with the release of the drug being postponed until the dosage reaches the gut. Orally regulated drug delivery systems cannot be used with medications that are unstable in the small intestine and undergo gut wall metabolism. It is challenging to convert medications that are unstable in the intestinal environment into prolonged-release systems because the majority of orally regulated systems are designed to discharge their components throughout the GI tract. When medications with an elimination half-life of less than two hours are provided in large amounts, a controlled-release dosage configuration might require a significant amount of the drug [42].

13.2 Nanotechnology in Combination Therapy

In recent years, the treatment of complex diseases has increasingly shifted toward the use of combination therapies, which combine multiple therapeutic agents to target different aspects of a disease's pathways. Many times, complex diseases with progressive pathological changes cannot be addressed completely through monotherapy, as monotherapy is designed to address a single target or symptom, which often leads to poor prognosis [43]. It enables a real-time combination of therapeutic and diagnostic functions in a programmed sequence. Compared with conventional predicted treatment, it can significantly minimize the amount of required therapeutic dose, reduce unhealthy side effects, enhance drug efficacy, and slow biological resistance through synergistic and complementary molecular mechanisms [44]. This strategy is beneficial in treating dreadful diseases such as cancer, AIDS, and neurological, autoimmune, and cardiovascular diseases, where single-agent therapies may be insufficient owing to resistance mechanisms. However, due to their variation, free drugs do not maintain their desired properties, such as solubility, permeation, stability, half-life, distribution, and metabolism, in conventional drug combinations, which compromises their synergy [45]. Therefore, combination therapy needs serious improvement to achieve the desired outcomes. One of the efficient strategies for combination therapy is the use of nanotechnology, which integrates multiple therapeutic agents or modulates a single nanocarrier system to enhance treatment outcomes. This is due to the active and passive targeted delivery of NPs, which are designed to release drugs in response to pH, temperature, enzymes, glucose concentration, light, or external stimuli [46]. Nanocarriers such as liposomes, polymeric NPs, dendrimers, micelles, and inorganic NPs such as quantum dots (QDs), silica, gold, and metal oxides are designed for encapsulation within a single system, enabling precise control of drug release, distribution, and the drug delivery ratio [47]. Compared with free drugs, nanomedicines extend the drug half-life, reduce undesirable side effects, and prevent drugs from escaping into the bloodstream. Additionally, nanomedicines play a key role in ensuring that their pharmacokinetics are similar to those of coencapsulated drugs [48]. Some major combinations improved by nanotechnology are discussed in the following sections.

13.2.1 Combined Chemotherapies

Combined chemotherapy is the most common and traditional way to treat almost all types of diseases, but currently mutating pathogens often show strong resistance to conventional chemotherapy, leading to major patient loss. Nanotechnological improvements have led to significant improvements in reducing these negative factors. In the case of cancer, chemotherapy is the most popular therapy; however, single-agent chemotherapy often fails to stop tumor growth. Combining many drugs at a specific ratio can significantly improve their efficacy [49]. However, several clinical trials on combination therapies have been unable to achieve the desired results because of the biological barriers in tumors. [50]. Consequently, many nanovehicles are designed to cross resistive barriers to obtain better therapeutics. Guo and colleagues developed a liposome (LP) composed of cholesterol, phosphatidylcholine, and a lipid polymer conjugate, which was used to encapsulate individual doxorubicin and tumor necrosis factor-related apoptosis-inducing ligand (TRAIL) [51]. The designed system (DOX-LP+TRAIL-LP) increased the median survival time to 48 days, whereas the median survival time was 32 days for TRAIL-LP and 39 days for DOX-LP. In addition, better outcomes were achieved in a mouse model by significantly suppressing tumor growth more efficiently than either treatment alone. To increase drug accumulation in brain tumors and reduce systemic toxicity, advanced nanosystems with active targeting have been established. Locatelli and colleagues developed polymeric NPs (Ag/Ali@PNP–Cltx-99mTc) coloaded with silver NPs and alisertib via a PLGA-block-polyethylene glycol copolymer [52]. This system is coupled with a chlorotoxin peptide that specifically binds to the overexpressed receptor matrix metalloproteinase-2, which increases drug levels in tumor cells.

According to the WHO, the use of specific antituberculosis drug (ATD) combinations is always necessary in all regimens of treatment to avoid the development of drug-resistant TB. Monotherapy should never be used in any stage of treatment. The only effective treatment for TB consists of an appropriate combination of ATDs. The WHO recommended a combined treatment of three drugs, rifampicin (RIF)+INH+pyrazinamide (PZA), for the treatment of TB and recommended treatment with four ATDs, RIF+INH+PZA+ethambutol (ETB). However, the adverse effects caused by first-line tuberculostatics may result in the interruption of treatment and have serious consequences for the control of the disease. The adverse effects related to RIF, PZA, and ETB treatment are also well reported. In general, the main adverse effects include irritative, allergic, and toxic reactions [53]. Many studies have suggested that the use of specially designed nanocarriers can improve the bioavailability and targeted delivery of these drugs at the proper dose ratio. For example, moxifloxacin (MX) is a fluoroquinolone antibiotic that is a dynamic counter to *Mycobacterium tuberculosis* (Mtb) and is as intense as RIF [54]. However, it has low intracellular activity against Mtb in macrophages. Kisich and colleagues developed MX-poly(butyl cyanoacrylate) (PBCA) NPs to improve outcomes [55]. Moxifloxacin-loaded PBCA nanoparticles (MX-NPs) were obtained via anionic polymerization of n-butyl-2-cyanoacrylate in the presence of a drug. The encapsulated MX accumulated in the cells nearly three times more capably than the free drug did, which was maintained for at least six times longer than that of free MX at an equivalent concentration. In another study, Pandey and coauthors also assessed the probability of directing ATDs through the respiratory route, with the goal of increasing the bioavailability of ATDs [56]. This study reports the formulation of three frontline ATDs (RIF, INH, and PZA) encapsulated in PLG-NPs appropriate for nebulization, which is favorable for bronchio-alveolar lung delivery. The pharmacokinetics of each drug and its chemotherapeutic efficacy in Mtb-infected pigs were assessed. A single nebulization maintained therapeutic drug levels in the plasma for 6–8 days and in the lungs for up to 11 days. Figure 13.2 illustrates the nanotechnological empowerment of combined chemotherapy with other modalities.

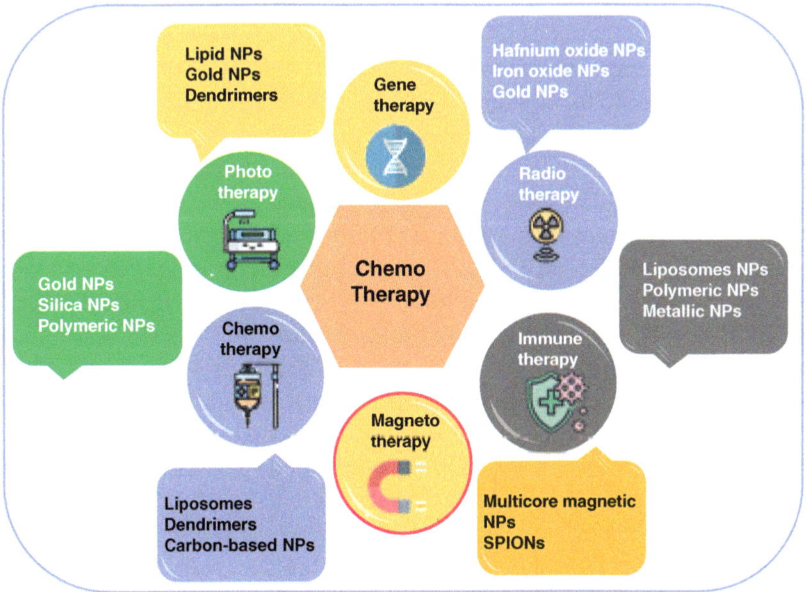

Figure 13.2 Illustration of the nanotechnological empowerment of combined chemotherapy with other modalities.

13.2.2 Chemotherapy vs. Immune Therapy

Chemoimmunotherapy is an advanced method of disease treatment. This combination works by unifying the strengths of both chemotherapy, which actively kills infected tissues, and immunotherapy, which enhances the immune response. When used together, these treatments can have a synergistic effect. For example, chemotherapeutics (e.g., PTX) displayed surprising effects under high-dose chemotherapy, impairing immune cell function, whereas lower doses supported and enhanced dendritic cell activation, demonstrating dual regulation of suppression and activation. One study investigated the polymeric nanoassembly arrangement for codelivery of the pTRP-2 vaccine to dendritic cells and low-dose PTX to improve cancer immunotherapy [57]. This therapy has also led to notable advancements in rheumatoid arthritis (RA) treatment. One of the most common and harmful symptoms of RA is chronic inflammation, which leads to pain, swelling, and cartilage and bone erosion, ultimately resulting in joint deformities and loss of function. To reduce inflammation and prevent immune system attacks, Pu and colleagues explored a novel approach for treating RA [58]. They developed a molecular separation gel composed of phospholipids (PPSGs) that codelivers methotrexate (MTX) and dexamethasone sodium phosphate (DSP) directly into the joint cavity. This method aims to enhance therapeutic efficacy while reducing the rate of drug administration and minimizing related side effects. The treatment led to significant improvements, including reduced joint swelling and decreased levels of inflammatory markers such as tumor necrosis factor (TNF)-α and interleukin (IL)-1β.

13.2.3 Chemotherapy vs. Phototherapy

Phototherapy is a promising treatment technique for many disorders, especially cancer. It consists of photodynamic therapy and photothermal therapy, which are nonsurgical methods. Nanotechnology-enhanced phototherapy addresses these barriers since old thermoresponsive and photoreactive substances are hydrophobic and often have poor tumor affinity [59]. Moreover, phototherapy alone fails to eliminate tumor cells because of irregular light penetration and the hypoxic environment within tumor tissues. Hence, the combination of chemotherapy and phototherapy might be a possible method for cancer treatment. For example, Yu and his group developed an organoplatinum(II) metallica (M) coordinated with a disodium terephthalate, a pyridyl-substituted porphyrin, and the therapeutic drug cis-(PEt3)2Pt(OTf)2 [60]. This metallica is autoassembled with a PEGylated bisphosphonate polymer and a tumor-targeting PEG-bisphosphonate conjugate to create metallica NPs designed for accurate targeting and prolonged blood circulation. This nanosystem has improved anticancer effects without recurrence after monodose delivery. In another study, Liu's team used human serum albumin (HSA) to deliver a chlorin-based photosensitizer (Ce6) and paclitaxel (PTX). They used a cRGDyK peptide to target $\alpha\nu\beta3$ integrin, which is highly expressed on tumor angiogenic endothelial cells, creating the HSA-Ce6-PTX-RGD system [61]. This targeted approach results in high NP accumulation, which is nearly two and a half times greater than that of other NPs.

MTX, the first-line disease-modifying antirheumatic drug (DMARD), can be recommended individually or in combination with additional DMARDs or biological agents for better therapeutic outcomes for RA patients. Along with high therapeutic efficacy, over time, the use of MTX can cause systemic difficulties, including infections, hepatitis, and myelosuppression. To address this concern, Lee and colleagues established MTX-loaded poly(DL-lactic-co-glycolic acid) gold half-shell nanocarriers (MTX-PLGA-Au) and coupled arginine-glycine-aspartic acid (RGD) peptides to their exterior [62]. When injected into mice, these RGD peptides target inflammation in collagen-induced arthritis (CIA), where they promote the accumulation of NPs in inflamed joints. After exposure to near-infrared (NIR) light, the gold half-shells generated heat, which triggered rapid drug release from the PLGA NPs, resulting in the required drug delivery through photothermal activation. Compared with traditional treatments, these multifunctional NPs showed greater therapeutic efficacy with a significantly reduced quantity of MTX.

13.2.4 Chemotherapy vs. Magnetotherapy

Hyperthermia has been established as a harmless and effective anticancer treatment because of its ability to promote tumor oxygenation. However, magnetothermal therapy has certain challenges, which are associated with heating tumor sites to the ideal temperature and keeping healthy tissues from being damaged. Thus, nanotechnology can improve therapy by targeting and persuading extra thermal sensitizers to infected cells, promoting antitumor properties [63]. Numerous magnetic NPs have been designed for hyperthermia, involving iron oxide, iron platinum, cobalt ferrite, etc. [64]. Many studies have shown that magnetothermal therapy combined with chemotherapy is more efficient in resisting tumors. It can improve the responsiveness of infected cells to therapeutic drugs. For example, Marino's team created lipid magnetic nanovectors (LMNVs) via PEGylated distearoyl phosphoethanolamine. These nanovectors were functionalized with antibodies against the transferrin receptor (TfR) to achieve targeting of both BBB endothelial and Glioblastoma Multiforme (GBM) cells [65]. The anti-TfR antibodies significantly enhanced the targeting of GBM in in vitro models, achieving 40.5% targeting compared with 8.1% with nontargeted nanovectors by inducing significant tumor cell death. In addition to lipid nanovectors, various synthetic polymeric nanosystems have been developed as delivery systems for combined outcomes. Yin and colleagues utilized highly magnetic ZnFe2O4 NPs combined with polyethyleneimine to deliver tumor-suppressor microRNAs, which inhibit tumor growth by targeting effectors of heat shock proteins [66]. The designed nanosystem (MNP-PEI/miRNA/PEI) demonstrated significant apoptosis in brain tumors when combined with magnetic hyperthermia.

13.2.5 Immunotherapy vs. Gene Therapy

Gene therapy has become an impressive approach for treating numerous complicated disorders. It has the ability to specifically control cancer-promoting genes in antitumor treatments [67]. However, despite the therapeutic potential of gene therapy, its inherent features limit efficient delivery to tumor sites, especially those crossing the BBB, which hinders clinical progress [68]. Combining immunotherapy with gene therapy can significantly improve tumor treatment with the help of nanotechnology [69]. For example, Erel-Akbaba and colleagues designed a peptide called iRGD-adorn solid lipid

NPs to transfer small interfering RNAs (siRNAs) that target two proteins, epidermal growth factor receptor (EGFR) and programmed death-ligand 1 (PD-L1) [70]. These NPs (f(SLN)-iRGD:siRNA) resulted in sequential reductions of 54.7% and 58.6% in EGFR and PD-L1 expression, respectively, and increased the median lifespan of the treated mice to 38 days. This combination therapy significantly enhanced the survival of the murine glioma model mice.

Combined immunotherapy and gene therapy with the help of nanocarriers can successfully address major RA-related issues, such as dysregulated inflammation and the failure of inflammation resolution, both of which contribute to RA incidence and progression. For example, Deng and colleagues developed inflammation-instructed nanocomplexes (NCs) to codeliver microRNA (miR-21) and the antiarthritic cytokine IL-4 into swollen synovial membranes and macrophages via a skillful method [71]. These NCs consist of an internal core of membrane-piercing polypeptide (PG) and miR-21, an external coating of acid-reactive charge reversal polymer (PLL-CA), and an IL-4 adsorbed exterior. Later, the negatively charged NCs target the irritated synovium via the EPR effect. In the slightly acidic environment of the swollen synovium, the PLL-CA layer undergoes transmission into positively charged poly-l-lysine (PLL), releasing IL-4 and uncovering the PG/miR-21 complex to effectively transfect macrophages. This combination of miR-21 and IL-4 reduces inflammation, promotes macrophage polarization, and aids in tissue restoration in zymosan A-induced arthritis (ZIA) mice, a trial model for studying RA resolution.

13.2.6 Photodynamic Therapy vs. Immunotherapy

Cytosine–phosphate–guanine oligodeoxynucleotides (CpG ODNs) are synthetic single-stranded DNA molecules containing unmethylated CpG motifs, with a hexamer sequence as their core [72]. These molecules can function both as immune adjuvants and as immunotherapeutic agents. Nevertheless, they can be simply recognized by the immune system, leading to a strong immune response [73]. This unique property ensures that the use of CpG ODNs is very attractive for immunotherapy. The problem is associated with crossing cell membranes because of their negative charge and susceptibility to disruption by nucleases [74]. Effective internalization of CpG ODNs into cells is important to achieve an immune response. Nanotechnology can significantly improve cellular uptake. To address this concern, Tian and his team developed a nanocomposite designed for a synergistic approach that combines photodynamic therapy with immunotherapy [75]. They used a porphyrin-based zirconium metal–organic framework (Zr-MOF), PCN-224 (PCN), both as a photosensitizer (PS) and a carrier for CpG ODNs. To further optimize delivery, the CpG ODN-loaded PCN was encapsulated with phosphatidylserine (PS) to obtain the required nanomaterial PCN-CpG@PS. This distribution not only enhances the cellular uptake of CpG ODNs but also leverages PCN as a photosensitizer for PDT. Additionally, the phosphatidylserine coating enabled targeted delivery to macrophages, the primary host cells for Mtb. Combining PDT with immunotherapy results in antituberculosis activity while simultaneously reducing toxic drug reactions and resistance.

13.3 Applications

There are significant drawbacks to the current methods of diagnosing and treating several illnesses, including low sensitivity or specificity and drug toxicity. Improved and newer NP-based diagnostic techniques are being developed [76]. The use of nanotechnology can provide a superior safety profile for medications with high potential for toxicity, such as cancer chemotherapy treatments. Through both active and passive methods, they can behave precisely at the target tissue [76]. Owing to their nanoscale dimensions and high surface area-to-volume ratio, NPs can absorb large quantities of drugs and circulate rapidly through body fluids. They can be utilized in more pharmaceutical applications because of their unique mechanical, magnetic, optical, and catalytic properties, which are enhanced by their larger surface area [77]. To date, nanotechnology has played a key role in treating many chronic disorders through DR and combination therapy, which has been shown to be impossible to cure completely. Nevertheless, research on the weak aspects of pharmaceuticals is needed to find better and healthier ways to treat multiresistant and mostly death-causing diseases. Some of the major diseases for which researchers have obtained fruitful results are discussed below. Figure 13.3 gives an overview of diverse applications of nanotechnology in complicated disorders.

13.3.1 Cancer and Tumors

According to the World Health Organization (WHO), malignant diseases have a high death ratio, resulting in 10 million patient deaths worldwide each year [78]. Malignant diseases can harm neighboring healthy cells and spread throughout organs via the fluid circulating in the body. They are caused by uncontrolled cell growth and failure of apoptosis. Chemotherapy, radiation therapy, surgery, or a combination of these are the current treatment options for different stages of different malignancies. Although these various treatment approaches can successfully lower cancer rates, patients may also experience adverse effects, such as cardiovascular difficulties, neurological imbalance, and subinfertility [79]. Nanotechnology plays a crucial role in this approach by improving drug solubility, targeting tumors more precisely, and reducing systemic side effects, which is revolutionizing DR and combination therapy in cancer treatment. A growing number of medications that were first created for nononcological illnesses are now being repurposed to target signaling pathways for cancer treatment [80]. For that purpose, nanotechnology-designed nanocarriers are used to encapsulate less soluble pharmaceuticals, protect therapeutic molecules, and modify drugs for blood circulation and tissue distribution [81]. These characteristics attract cancer patients because they permit the encapsulation of cytotoxic agents with multiple toxicities and physicochemical qualities. This study utilized the AS1411 aptamer to functionalize and direct NPs loaded with therapeutic nucleotides or chemotherapeutic chemicals toward cancer cells with Nucleolin (NCL) overexpression [82]. These nanosystems show how aptamer functionalization can minimize off-target effects in conventional cancer therapies by concentrating payload delivery exclusively to tumor cells [83]. In malignancies treated with targeted therapies, the flexibility of signaling networks and cross-talk between pathways

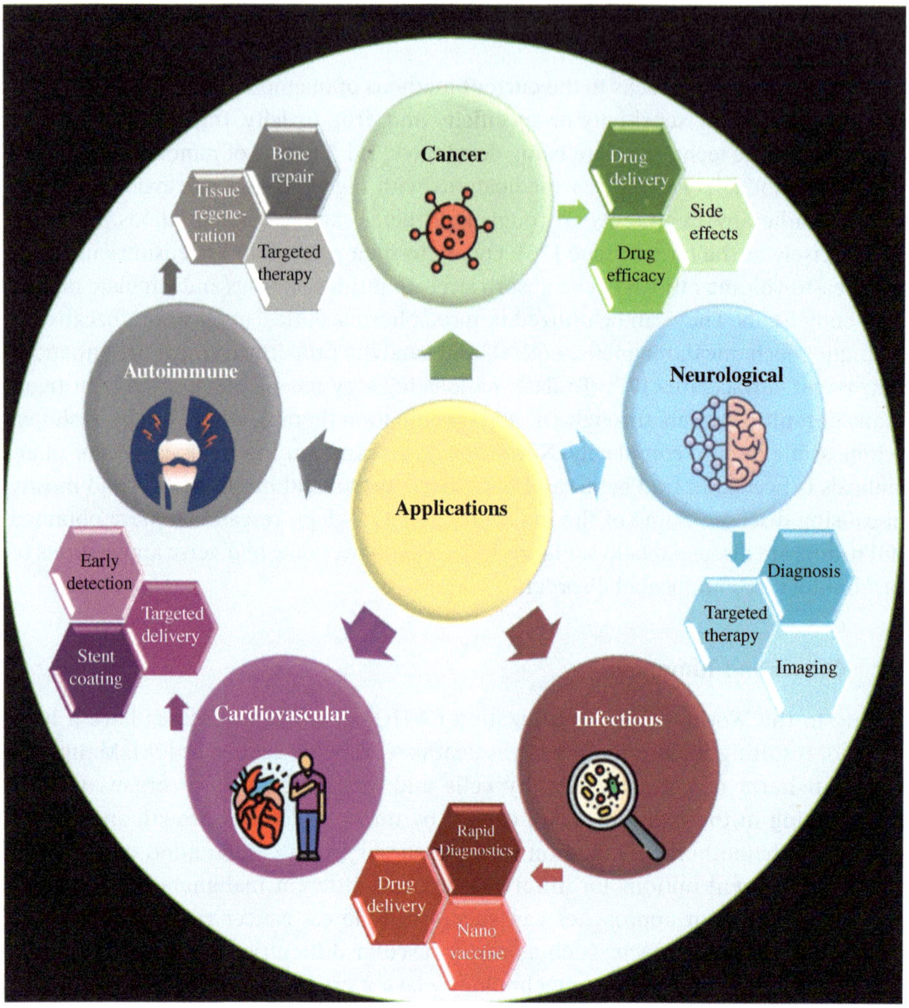

Figure 13.3 Illustration of diverse applications of nanotechnology in complicated disorders.

regularly increase resistance [84]. Maintaining sufficient plasma levels is important for the on-target activity of several kinase inhibitors, as pathways can reactivate when plasma levels decrease. Using NPs to accurately regulate protein kinase inhibitor levels can avoid this issue, resulting in better efficacy [85]. NPs aid in improving drug stability from being damaged by catalytic degradation, phagocytic uptake, and extended retention of chemotherapeutics. They can be easily transformed into aerosols and have the ability to withstand high forces during nebulization. Coating with a liver surfactant such as Dipalmitoylphosphatidylcholine (DPPC) can also diminish phagocytosis by stopping opsonic protein adsorption. Moreover, the nanoscale size of NPs is critical for targeting specific sites of cells. For example, NPSs with a range of 100–150 nm are 8–9 times more likely to be internalized into liver cancer cells than microparticles with a size of 3–5 μm [86].

Compared with monotherapy, combination therapy accompanied by nanotechnology has been shown to enhance treatment efficiency in cancer patients. Nanotechnology provides a significant tool for coencapsulation that confirms identical spatial and sequential drug distributions in target tumor cells, which effectively resolves these problems [87]. To overcome chemoresistance, a smart codelivery system has been designed that delivers small drugs with other agents, such as proteins, peptides, and DNA, and facilitates the delivery of loaded agents to tumor cells, with approximately identical pharmacokinetics and fixed-dose ratios of the combined therapeutics [88]. For example, the combined use of irinotecan and cisplatin results in conflicting interactions when the molar ratio ranges from 1:2 to 4:1, whereas outside this range, the two drugs exhibit a cooperative effect [89]. When both drugs are encapsulated in liposomes, the fixed ratio of the drugs can be preserved for a whole day. In addition to these improvements, many NPs might be used as P-gp inhibitors, which reestablish the antitumor actions of the loaded drugs in resistant cells. This may reverse the MDR effect in multiple ways. For example, silver NPs are resistant to drug efflux due to their size-exclusion effect [90]. NPs can also recover tumor receptivity by increasing the cellular absorption of the loaded agents. For example, Liu and colleagues delivered doxorubicin (DOX) and PTX via crosslinked multilamellar vesicle (cMLV) into drug-resistant cells [91]. These results suggest that while single drug-loaded cMLV does not suppress P-gp expression, dual drug-loaded cMLV effectively downregulates P-gp expression and inhibits cell proliferation by increasing cellular uptake, resulting in high intracellular drug concentrations.

13.3.2 Neurological Disorders

Central nervous system (CNS) disorders, which include Parkinson's disease (PD), Alzheimer's disease, strokes, and brain malignancies, are the primary causes of disability worldwide. The prevalence of PD [92], which has increased recently, is the second most common neurological disease worldwide after Alzheimer's disease. PD is an advanced neurodegenerative disease in which both motor and nonmotor symptoms result from dopamine loss, which has a major negative impact on patients' quality of life. Current therapies, such as dopamine agonists and levodopa, reduce symptoms but do not stop disease progression. Long-term pharmacological therapies also frequently lead to serious side effects, such as medication tolerance, impulse control issues, and dyskinesias [93]. Numerous repositioned medications that target protein aggregation, oxidative stress, and mitochondrial dysfunction have demonstrated neuroprotective effects in PD. Enhancing the effectiveness of repurposed medications is a critical function of nanotechnology. By lowering systemic adverse effects, increasing bioavailability, and overcoming the BBB, NPs can enhance drug transport to the brain. For the delivery of neuroprotective drugs, nano carriers such as liposomes, polymeric NPs, and lipid-based NPs have been investigated. Repurposed medications such as antioxidants, dopamine agonists, and MAO inhibitors can be encapsulated in these nanosystems, guaranteeing targeted transport and prolonged release to relevant brain areas [94]. By increasing therapeutic results, decreasing toxicity, and improving medication stability, combining nanotechnology with DR has the potential to completely transform the treatment of PD [95]. Nanotechnology provides novel strategies for repurposing neuroprotective drugs by focusing on mitochondrial dysfunction in neuroinflammatory and neurodegenerative disorders (NI&NDDs). In terms of cellular functions, energy production, and the control of oxidative stress, mitochondria are essential [96]. However, when they malfunction, bioenergetic deficiencies, elevated ROS levels, and protein aggregation occur, hastening dementia. It has been demonstrated that restoring mitochondrial function can reverse cell damage. To maximize treatment efficacy and reduce systemic side effects, advanced nanomaterials, nanozymes, and drug-loaded nanosystems offer tailored delivery to the CNS. NI&NDDs may be treated with these methods to enhance cellular health and mitochondrial balance [97].

The BBB, which is a membrane barrier created by the brain microvascular system, separates the extracellular fluid in the CNS from the blood [98]. One of the major barriers for drugs is crossing the BBB due to their complex penetration mechanism, which leads to the failure of many specialized medicines, and the amount of drugs transported varies depending on the NP and mechanism. For example, the carrier-mediated transport system's small and stereospecific pores limit the transport of large-molecule drugs [99]. Polymeric NPs such as PBCA, PLGA, and poly(lactic acid) (PLA) NPs, LPs, and inorganic composites such as gold, silver, and zinc oxide NPs are the major types of NPs that have been created and used to cross the BBB. However, because of their possible inherent toxicity, drug delivery methods that use QDs to cross the BBB have gradually lost favor over the years. Many models suggest that NPs may induce neurodegenerative amyloid diseases by absorbing amyloidogenic proteins on the surface of NPs, and depending on their kinetics, proteins may associate with and dissociate from the NP surface, causing fibril formation [100]. Carbon dots (CDs) are a novel family of carbon nanomaterials that show promise for drug administration across the BBB owing to their strong photoluminescence, low toxicity, and great biocompatibility. Their surface functional groups, polymeric structure, and tiny size allow for effective drug conjugation and tracking [101]. CDs made from L-aspartic acid and D-glucose demonstrated effective tumor penetration and robust PL when injected into mice with gliomas. Research has verified that these proteins are taken up by C6 glioma cells through the transporters ACST2 and GLUT1, which are widely distributed on the BBB and in brain tumors. Using TfR overexpression on the BBB and cancer cells, Li and colleagues coupled CDs with transferrin for DOX delivery to improve the treatment of juvenile brain tumors. The CD/Trans/DOX combination demonstrated increased lethality at a 10 nM concentration and better absorption by pediatric tumor cells (SJGBM2) [102]. Subsequent research using zebrafish models showed that CDs by themselves were unable to penetrate the BBB. Nevertheless, fluorescence spectroscopy verified that the transferrin-conjugated CDs effectively entered the CNS via TfR-mediated endocytosis. This study confirmed that CDs, which use specific transport mechanisms to efficiently cross the BBB, are a viable drug delivery method for the treatment of brain cancer [103].

13.3.3 Autoimmune Diseases

Autoimmune diseases are a multipart collection of syndromes in which the immune system mistakenly attacks the body's own healthy cells, eventually resulting in chronic inflammation and long-term impairment of organs. RA, lupus, multiple sclerosis, and inflammatory bowel disease (IBD) are major autoimmune diseases that affect millions of patients annually. The development of autoimmune diseases is multidimensional and includes genetic predispositions, environmental influences, and the dysregulation of immune reactions [104]. Although these diseases are common features of immune system dysfunction, they differ greatly in their clinical presentations and have varying mechanisms, which makes their treatment challenging [105]. Management becomes more problematic, as the traditional therapeutic options currently available are limited and depend on a wide range of immunosuppressive agents, including corticosteroids and DMARDs, which are designed to suppress immune reactions and minimize inflammation [106]. However, many times, these agents may be unsatisfactory for addressing multifaceted and advanced pathological variations, as they are designed to target a single drug target or indication, which might result in poor diagnosis. For example, MTX, the frontline DMARD, can reduce and interrupt RA development but does not relieve joint aches or swelling [107]. Similarly, Glucocorticoids (GCs) and nonsteroidal anti-inflammatory drugs (NSAIDs) help reduce inflammation and pain but do not prevent disease progression [108]. Nanotechnology has extraordinary potential to overcome these challenges in conventional formulations, empowering combination therapy and enhancing the codelivery of combined drugs to sore sites [109]. It can significantly improve the therapeutic properties of agents loaded in nanocarriers, such as the solubility of drugs, through encapsulation in a hydrophobic cavity and efficient delivery. For example, liposomes with twin layers were established to coencapsulate hydrophobic prednisolone and MTX in the center, and an external layer was added to stop early escape [110]. After administration, the solubility was increased, and the absorption of both drugs was greater in the inflamed joints than in the other joints, resulting in synergistic anti-inflammatory properties. Additionally, the nanosystem can increase the lifespan of codelivered drugs and decrease the drug dosage. For example, both nimesulide and sulfasalazine were loaded into the halloysite lumen and incorporated with a polymer hydrogel [111]—the halloysite lumen and hydrogel work to slow the release of both drugs. The operations were quite different when the drugs were released together compared with when they were released alone. Nimesulide did not show an initial rapid release when released on its own. However, during corelease with sulfasalazine, nimesulide was quickly released. Many specialized nanocarriers can codeliver multiple drugs to sore joints via different targeting modes, and the development of endothelial gaps benefits passive targeting [112]. Active targeting takes place because ligands are overexpressed in the affected joints or associated tissues. For example, owing to the presence of many 1,3-β-glucan groups on the surface of macrophages, MTX and ortho-vanillin were targeted and delivered via β-glucan microspheres as carriers. This carrier enhances receptor-mediated phagocytosis by interacting with dectin-1 or β-glucan receptors, which are overexpressed on macrophages and dendritic cells [113]. Tailored ligands and variations in the microenvironment of tissues provide targets for designing responsive nanosystems. Several indicators can trigger appropriate release of the drug to minimize premature diffusion before it reaches the joints. In addition to their sustained release profile, NPs with varying deprivation patterns can avoid unwanted interactions and regulate the control mechanisms of drugs. To enhance the efficacy of both dexamethasone and leflunomide, they were encapsulated in lipid carriers and PLGA NPs and subsequently integrated into a hydrogel to form a nanocomposite hydrogel [114]. Nanocomposite hydrogel codelivery significantly prolonged the period for joint residence, ultimately promoting joint healing and reducing RA development.

13.4 Conclusion

Nanotechnology has shown a great revolutionary approach by avoiding conventional complications in DR and combination therapy, which has led to significant improvements in the precise delivery of therapeutic materials. Specialized NPs are designed to incorporate old drug features with new modifications to achieve stability and extend lifespan. Through the development of sophisticated nanomaterials, scientists have achieved improved bioavailability, lessened toxicity, and bypassed drug resistance through the BBB and the tumor microenvironment. Recent research has revealed how liposomes, dendrimers, and gold NPs aid in the delivery of several therapeutic modalities on one platform, with synergistic efficacy for the treatment of complex ailments such as cancer, neurological disorders, and infectious diseases. However, in addition to incredible nanotechnological improvements, several challenges related to the scalability, biocompatibility, and toxicity of NPs need to be addressed to fully realize their therapeutic potential, only a few of which have been evaluated in in vitro models rather than in vivo models. Hence, further research on human models to meet the commercialization of these novel formulations based on nanotechnology is essential.

13.5 Future Recommendations

Future efforts must develop an interdisciplinary approach that integrates the design of NPs, their therapeutic use, and their actual implementation in the clinical setting. There needs to be comprehensive collaboration among researchers, clinicians, and pharmaceutical companies to ensure that these technologies are successfully integrated into clinical practice. The next research steps include the development of novel nanomaterials with unique features in terms of target selectivity and multitasking capabilities, such as carbon nanotubes and QDs. Combining artificial intelligence and genomics with nanotechnology will be able to assist in personalizing treatments for specific patients. There is great potential for treating chronic disorders multifocally through the simultaneous use of multiple therapeutically active NPs. More detailed toxicological studies and research on the biocompatibility of NPs should be carried out to guarantee that these materials are safe for human use. The speed at which therapies based on nanotechnology are produced and their affordability should be given special attention to allow ease of access, especially in poorer regions, which suffer from a high rate of drug resistance.

References

1 Modi S, Prajapati R, Inwati GK, et al. Recent trends in fascinating applications of nanotechnology in allied health sciences. *Crystals* 2021;12(1):39.

2 Malakar A, Kanel SR, Ray C, Snow DD, and Nadagouda MN. Nanomaterials in the environment, human exposure pathway, and health effects: A review. *Science of the Total Environment* 2021;759:143470.

3 Al-Nemrawi NK, AbuAlSamen MM, and Alzoubi KH. Awareness about nanotechnology and its applications in drug industry among pharmacy students. *Currents in Pharmacy Teaching and Learning* 2020;12(3):274–280.

4 Pillai J, Ray A, Maan M, and Dutta M. Repurposing drugs targeting metabolic diseases for cancer therapeutics. *Drug Discovery Today* 2023;28(9):103684.

5 Ashburn TT and Thor KB. Drug repositioning: Identifying and developing new uses for existing drugs. *Nature Reviews Drug Discovery* 2004;3(8):673–683.

6 Najlah M. Drug repurposing supported by nanotechnology: A promising strategy to fight cancer. *Therapeutic Delivery* 2021;12(4):267–269, Taylor & Francis.

7 Ding S, Khan AI, Cai X, et al. Overcoming blood–brain barrier transport: Advances in nanoparticle-based drug delivery strategies. *Materials Today* 2020;37:112–125.

8 Malik S, Ghosh A, Kerry RG, and Rout JR. Nanotechnology in preclinical pharmacokinetics. In: *Advances in Pharmaceutical Biotechnology: Recent Progress and Future Applications*, (pp. 461–478). Springer; 2020.

9 Tilija Pun N and Jeong C-H. Statin as a potential chemotherapeutic agent: Current updates as a monotherapy, combination therapy, and treatment for anticancer drug resistance. *Pharmaceuticals* 2021;14(5):470.

10 Mokhtari RB, Homayouni TS, Baluch N, et al. Combination therapy in combating cancer. *Oncotarget* 2017;8(23):38022.

11 Zhou Z, Edil BH, and Li M. Combination therapies for cancer: Challenges and opportunities. *BMC Medicine* 2023;21(1):171.

12 Fumoto S and Nishida K. Codelivery systems of multiple drugs using nanotechnology for future cancer therapy. *Chemical and Pharmaceutical Bulletin* 2020;68(7):603–612.

13 Mukherjee A, Waters AK, Kalyan P, Achrol AS, Kesari S, and Yenugonda VM. Lipid–polymer hybrid nanoparticles as a next-generation drug delivery platform: State of the art, emerging technologies, and perspectives. *International Journal of Nanomedicine* 2019;14:1937–1952.

14 Vaidya B, Kulkarni NS, Shukla SK, et al. Development of inhalable quinacrine loaded bovine serum albumin modified cationic nanoparticles: Repurposing quinacrine for lung cancer therapeutics. *International Journal of Pharmaceutics* 2020;577:118995.

15 Pushpakom S, Iorio F, Eyers PA, et al. Drug repurposing: Progress, challenges and recommendations. *Nature Reviews Drug Discovery* 2019;18(1):41–58.

16 Breckenridge A and Jacob R. Overcoming the legal and regulatory barriers to drug repurposing. *Nature Reviews Drug Discovery* 2019;18(1):1–2.

17 Dudley JT, Deshpande T, and Butte AJ. Exploiting drug–disease relationships for computational drug repositioning. *Briefings in Bioinformatics* 2011;12(4):303–311.

18 Cousin MA, Ebbert JO, Wiinamaki AR, et al. Larval zebrafish model for FDA-approved drug repositioning for tobacco dependence treatment. *PLoS One* 2014;9(3):e90467.

19 Liu Y, Miyoshi H, and Nakamura M. Nanomedicine for drug delivery and imaging: A promising avenue for cancer therapy and diagnosis using targeted functional nanoparticles. *International Journal of Cancer* 2007;120(12):2527–2537.

20 Al-Dimassi S, Abou-Antoun T, and El-Sibai M. Cancer cell resistance mechanisms: A mini review. *Clinical and Translational Oncology* 2014;16:511–516.

21 Sertkaya A, Wong H-H, Jessup A, and Beleche T. Key cost drivers of pharmaceutical clinical trials in the United States. *Clinical Trials* 2016;13(2):117–126.

22 Florence AT. Pharmaceutical nanotechnology: More than size. Ten topics for research. *International Journal of Pharmaceutics* 2007;339(1-2):1–2.

23 Patil A, Mishra V, Thakur S, et al. Nanotechnology derived nanotools in biomedical perspectives: An update. *Current Nanoscience* 2019;15(2):137–146.

24 Lu H, Wang J, Wang T, Zhong J, Bao Y, and Hao H. Recent progress on nanostructures for drug delivery applications. *Journal of Nanomaterials* 2016;2016(1):5762431.

25 Bamrungsap S, Zhao Z, Chen T, et al. Nanotechnology in therapeutics: A focus on nanoparticles as a drug delivery system. *Nanomedicine* 2012;7(8):1253–1271.

26 Ganta S, Devalapally H, Shahiwala A, and Amiji M. A review of stimuli-responsive nanocarriers for drug and gene delivery. *Journal of Controlled Release* 2008;126(3):187–204.

27 Cheng R, Meng F, Ma S, et al. Reduction and temperature dual-responsive crosslinked polymersomes for targeted intracellular protein delivery. *Journal of Materials Chemistry* 2011;21(47):19013–19020.

28 Castillo B, Bromberg L, Lopez X, et al. Intracellular delivery of siRNA by polycationic superparamagnetic nanoparticles. *Journal of Drug Delivery* 2012;2012(1):218940.

29 Liu M, Du H, Zhang W, and Zhai G. Internal stimuli-responsive nanocarriers for drug delivery: Design strategies and applications. *Materials Science and Engineering: C* 2017;71:1267–1280.

30 An X, Zhu A, Luo H, Ke H, Chen H, and Zhao Y. Rational design of multistimuli-responsive nanoparticles for precise cancer therapy. *ACS Nano* 2016;10(6):5947–5958.

31 Gigliobianco MR, Casadidio C, Censi R, and Di Martino P. Nanocrystals of poorly soluble drugs: Drug bioavailability and physicochemical stability. *Pharmaceutics* 2018;10(3):134.

32 Rahman M, Afzal O, Altamimi ASA, et al. Paclitaxel and curcumin as dual-drug-loaded lipid nanocapsules in the management of brain tumor. *Journal of Cluster Science* 2023;34(4):1927–1938.

33 Rafiei P and Haddadi A. Pharmacokinetic consequences of PLGA nanoparticles in docetaxel drug delivery. *Pharmaceutical Nanotechnology* 2017;5(1):3–23.

34 Kirtonia A, Gala K, Fernandes SG, et al. editors. Repurposing of drugs: An attractive pharmacological strategy for cancer therapeutics. *Seminars in Cancer Biology* 2021;68:258–278, Elsevier.

35 Shaikh MAJ, Alharbi KS, Almalki WH, et al. Sodium alginate based drug delivery in management of breast cancer. *Carbohydrate Polymers* 2022;292:119689.

36 Andrews P, Thyssen J, and Lorke D. The biology and toxicology of molluskicides, Bayluscide. *Pharmacology & Therapeutics* 1982;19(2):245–295.

37 Henriksen I, Sande SA, Smistad G, Ågren T, and Karlsen J. In vitro evaluation of drug release kinetics from liposomes by fractional dialysis. *International Journal of Pharmaceutics* 1995;119(2):231–238.

38 Vasir JK and Labhasetwar V. Biodegradable nanoparticles for cytosolic delivery of therapeutics. *Advanced Drug Delivery Reviews* 2007;59(8):718–728.

39 Natarajan JV, Nugraha C, Ng XW, and Venkatraman S. Sustained-release from nanocarriers: A review. *Journal of Controlled Release* 2014;193:122–138.

40 Jantzen GM and Robinson JR. Sustained-and controlled-release drug delivery systems. *Drugs and The Pharmaceutical Sciences* 2002;121:501–528.

41 Tao R, You C, Qu Q, et al. Recent advances in the design of controlled-and sustained-release micro/nanocarriers of pesticide. *Environmental Science: Nano* 2023;10(2):351–371.

42 Gupta M, Agrawal U, and Vyas SP. Nanocarrier-based topical drug delivery for the treatment of skin diseases. *Expert Opinion on Drug Delivery* 2012;9(7):783–804.

43 Nayak RR, Alexander M, Deshpande I, et al. Methotrexate impacts conserved pathways in diverse human gut bacteria leading to decreased host immune activation. *Cell Host & Microbe* 2021;29(3):362–377.e11.

44 Chou T-C. Theoretical basis, experimental design, and computerized simulation of synergism and antagonism in drug combination studies. *Pharmacological Reviews* 2006;58(3):621–681.

45 Wang H and Huang Y. Combination therapy based on nano codelivery for overcoming cancer drug resistance. *Medicine in Drug Discovery* 2020;6:100024.

46 Wu W, Luo L, Wang Y, et al. Endogenous pH-responsive nanoparticles with programmable size changes for targeted tumor therapy and imaging applications. *Theranostics* 2018;8(11):3038.

47 Sher EK, Alebić M, Boras MM, et al. Nanotechnology in medicine revolutionizing drug delivery for cancer and viral infection treatments. *International Journal of Pharmaceutics* 2024: 124345.

48 Scioli Montoto S, Muraca G, and Ruiz ME. Solid lipid nanoparticles for drug delivery: pharmacological and biopharmaceutical aspects. *Frontiers in Molecular Biosciences* 2020;7:587997.

49 Lun X, Wells JC, Grinshtein N, et al. Disulfiram when combined with copper enhances the therapeutic effects of temozolomide for the treatment of glioblastoma. *Clinical Cancer Research* 2016;22(15):3860–3875.

50 Galanis E, Anderson SK, Lafky JM, et al. Phase II study of bevacizumab in combination with sorafenib in recurrent glioblastoma (N0776): A north central cancer treatment group trial. *Clinical Cancer Research* 2013;19(17):4816–4823.

51 Guo L, Fan L, Pang Z, et al. TRAIL and doxorubicin combination enhances anti-glioblastoma effect based on passive tumor targeting of liposomes. *Journal of Controlled Release* 2011;154(1):93–102.

52 Locatelli E, Naddaka M, Uboldi C, et al. Targeted delivery of silver nanoparticles and alisertib: In vitro and in vivo synergistic effect against glioblastoma. *Nanomedicine* 2014;9(6):839–849.

53 Maciel ELN, Guidoni LM, Favero JL, et al. Adverse effects of the new tuberculosis treatment regimen recommended by the Brazilian Ministry of Health. *Jornal Brasileiro de Pneumologia* 2010;36:232–238.

54 Gosling RD, Uiso LO, Sam NE, et al. The bactericidal activity of moxifloxacin in patients with pulmonary tuberculosis. *American Journal of Respiratory and Critical Care Medicine* 2003;168(11):1342–1345.

55 Kisich K, Gelperina S, Higgins M, et al. Encapsulation of moxifloxacin within poly (butyl cyanoacrylate) nanoparticles enhances efficacy against intracellular Mycobacterium tuberculosis. *International Journal of Pharmaceutics* 2007;345(1-2):154–162.

56 Pandey R, Zahoor A, Sharma S, and Khuller G. Nanoparticle encapsulated antitubercular drugs as a potential oral drug delivery system against murine tuberculosis. *Tuberculosis* 2003;83(6):373–378.

57 Xu J, Xu B, Tao J, Yang Y, Hu Y, and Huang Y. Microneedle-assisted, DC-targeted Codelivery of pTRP-2 and adjuvant of paclitaxel for transcutaneous immunotherapy. *Small* 2017;13(28):1700666.

58 Pu Q, Wang K, Peng B, et al. In situ preparation of a phospholipid gel Co-loaded with methotrexate and dexamethasone for synergistic rheumatoid arthritis treatment. *International Journal of Nanomedicine* 2022;17:5153.

59 Ghosh S, Carter KA, and Lovell JF. Liposomal formulations of photosensitizers. *Biomaterials* 2019;218:119341.

60 Yu G, Yu S, Saha ML, et al. A discrete organoplatinum (II) metallacage as a multimodality theranostic platform for cancer photochemotherapy. *Nature Communications* 2018;9(1):4335.

61 Chen Q, Wang X, Wang C, Feng L, Li Y, and Liu Z. Drug-induced self-assembly of modified albumins as nanotheranostics for tumor-targeted combination therapy. *ACS Nano* 2015;9(5):5223–5233.

62 Lee S-M, Kim HJ, Ha Y-J, et al. Targeted chemo-photothermal treatments of rheumatoid arthritis using gold half-shell multifunctional nanoparticles. *ACS Nano* 2013;7(1):50–57.

63 Beik J, Abed Z, Ghoreishi FS, et al. Nanotechnology in hyperthermia cancer therapy: From fundamental principles to advanced applications. *Journal of Controlled Release* 2016;235:205–221.

64 Thiesen B and Jordan A. Clinical applications of magnetic nanoparticles for hyperthermia. *International Journal of Hyperthermia* 2008;24(6):467–474.

65 Marino A, Camponovo A, Degl'Innocenti A, et al. Multifunctional temozolomide-loaded lipid superparamagnetic nanovectors: Dual targeting and disintegration of glioblastoma spheroids by synergic chemotherapy and hyperthermia treatment. *Nanoscale* 2019;11(44):21227–21248.

66 Yin PT, Shah BP, and Lee K-B. Combined magnetic nanoparticle-based microRNA and hyperthermia therapy to enhance apoptosis in brain cancer cells. *Small* 2014;10(20):4106.

67 Kozielski KL, Ruiz-Valls A, Tzeng SY, et al. Cancer-selective nanoparticles for combinatorial siRNA delivery to primary human GBM in vitro and in vivo. *Biomaterials* 2019;209:79–87.

68 Bhaskaran V, Nowicki MO, Idriss M, et al. The functional synergism of microRNA clustering provides therapeutically relevant epigenetic interference in glioblastoma. *Nature Communications* 2019;10(1):442.

69 Kim SS, Harford JB, Moghe M, Slaughter T, Doherty C, and Chang EH. A tumor-targeting nanomedicine carrying the p53 gene crosses the blood–brain barrier and enhances anti-PD-1 immunotherapy in mouse models of glioblastoma. *International Journal of Cancer* 2019;145(9):2535–2546.

70 Erel-Akbaba G, Carvalho LA, Tian T, et al. Radiation-induced targeted nanoparticle-based gene delivery for brain tumor therapy. *ACS Nano* 2019;13(4):4028–4040.

71 Deng Y, Zhou Y, Liang Q, et al. Inflammation-instructed hierarchical delivery of IL-4/miR-21 orchestrates osteoimmune microenvironment toward the treatment of rheumatoid arthritis. *Advanced Functional Materials* 2021;31(33):2101033.

72 Fehér K. Single stranded DNA immune modulators with unmethylated CpG motifs: Structure and molecular recognition by toll-like receptor 9. *Current Protein and Peptide Science* 2019;20(11):1060–1068.

73 Kayraklioglu N, Horuluoglu B, and Klinman DM. CpG oligonucleotides as vaccine adjuvants. *DNA Vaccines: Methods and Protocols* 2021;2197:51–85.

74 Zhang H, Cheng T, Lai L, et al. BN nanospheres functionalized with mesoporous silica for enhancing CpG oligodeoxynucleotide-mediated cancer immunotherapy. *Nanoscale* 2018;10(30):14516–14524.

75 Tian N, Duan H, Cao T, et al. Macrophage-targeted nanoparticles mediate synergistic photodynamic therapy and immunotherapy of tuberculosis. *RSC Advances* 2023;13(3):1727–1737.

76 Surendiran A, Sandhiya S, Pradhan S, and Adithan C. Novel applications of nanotechnology in medicine. *Indian Journal of Medical Research* 2009;130(6):689–701.

77 Huang B, Chen F, Shen Y, et al. Advances in targeted pesticides with environmentally responsive controlled release by nanotechnology. *Nanomaterials* 2018;8(2):102.

78 Sun L, Liu H, Ye Y, et al. Smart nanoparticles for cancer therapy. *Signal Transduction and Targeted Therapy* 2023;8(1):418.

79 Jiang Y, Jiang Y, Li M, and Yu Q. Will nanomedicine become a good solution for the cardiotoxicity of chemotherapy drugs? *Frontiers in Pharmacology* 2023;14:1143361.

80 Wu Z, Cheng F, Li J, Li W, Liu G, and Tang Y. SDTNBI: An integrated network and chemoinformatics tool for systematic prediction of drug–target interactions and drug repositioning. *Briefings in Bioinformatics* 2017;18(2):333–347.

81 Chen L, Zhao L, Hu G, et al. Tumor-specific nanomedicine via sequential catalytic reactions for accurate tumor therapy. *Journal of Materials Chemistry B* 2020;8(31):6857–6865.

82 Kozani PS, Kozani PS, and Malik MT. AS1411-functionalized delivery nanosystems for targeted cancer therapy. *Exploration of Medicine* 2021;2:146.

83 Wang X, Wang Y, Chen ZG, and Shin DM. Advances of cancer therapy by nanotechnology. *Cancer Research and Treatment: Official Journal of Korean Cancer Association* 2009;41(1):1–11.

84 Fausto N and Campbell JS. The role of hepatocytes and oval cells in liver regeneration and repopulation. *Mechanisms of Development* 2003;120(1):117–130.

85 Gonciar D, Mocan T, and Agoston-Coldea L. Nanoparticles targeting the molecular pathways of heart remodeling and regeneration. *Pharmaceutics* 2022;14(4):711.

86 Nygårdas P and Hinkkanen A. Upregulation of MMP-8 and MMP-9 activity in the BALB/c mouse spinal cord correlates with the severity of experimental autoimmune encephalomyelitis. *Clinical & Experimental Immunology* 2002;128(2):245–254.

87 Livney YD and Assaraf YG. Rationally designed nanovehicles to overcome cancer chemoresistance. *Advanced Drug Delivery Reviews* 2013;65(13-14):1716–1730.

88 Ma L, Kohli M, and Smith A. Nanoparticles for combination drug therapy. *ACS Nano* 2013;7(11):9518–9525.

89 Tardi PG, Dos Santos N, Harasym TO, et al. Drug ratio–dependent antitumor activity of irinotecan and cisplatin combinations in vitro and in vivo. *Molecular Cancer Therapeutics* 2009;8(8):2266–2275.

90 Liu J, Zhao Y, Guo Q, et al. TAT-modified nanosilver for combating multidrug-resistant cancer. *Biomaterials* 2012;33(26):6155–6161.

91 Liu Y, Fang J, Joo K-I, Wong MK, and Wang P. Codelivery of chemotherapeutics via crosslinked multilamellar liposomal vesicles to overcome multidrug resistance in tumor. *PLoS One* 2014;9(10):e110611.

92 Saavedra Moreno JS, Millán PA, and Buriticá Henao OF. Introducción, epidemiología y diagnóstico de la enfermedad de Parkinson. *Acta Neurológica Colombiana* 2019;35:2–10.

93 National Institute of Neurological Disorders and Stroke. *Parkinson's Disease: Hope Through Research*. National Institutes of Neurological Disorders and Stroke, National Institutes of Health 1994.

94 Martínez-Fernández R, Gasca-Salas C, Sánchez-Ferro Á, and Obeso JÁ. Actualización en la enfermedad de Parkinson. *Revista Médica Clínica Las Condes* 2016;27(3):363–379.

95 Hernández-Parra H, Cortés H, Avalos-Fuentes JA, et al. Repositioning of drugs for Parkinson's disease and pharmaceutical nanotechnology tools for their optimization. *Journal of Nanobiotechnology* 2022;20(1):413.

96 Archer SL. Mitochondrial dynamics—mitochondrial fission and fusion in human diseases. *New England Journal of Medicine* 2013;369(23):2236–2251.

97 Chan DC. Mitochondrial dynamics and its involvement in disease. *Annual Review of Pathology: Mechanisms of Disease* 2020;15(1):235–259.

98 Mayer F, Mayer N, Chinn L, Pinsonneault RL, Kroetz D, and Bainton RJ. Evolutionary conservation of vertebrate blood–brain barrier chemoprotective mechanisms in Drosophila. *Journal of Neuroscience* 2009;29(11):3538–3550.

99 Bhaskar S, Tian F, Stoeger T, et al. Multifunctional Nanocarriers for diagnostics, drug delivery and targeted treatment across blood–brain barrier: Perspectives on tracking and neuroimaging. *Particle and Fiber Toxicology* 2010;7:1–25.

100 Linse S, Cabaleiro-Lago C, Xue W-F, et al. Nucleation of protein fibrillation by nanoparticles. *Proceedings of the National Academy of Sciences* 2007;104(21):8691–8696.

101 Zhou Y, Sharma SK, Peng Z, and Leblanc RM. Polymers in carbon dots: A review. *Polymers* 2017;9(2):67.

102 Li S, Skromne I, Peng Z, Det al. "Dark" carbon dots specifically "light-up" calcified zebrafish bones. *Journal of Materials Chemistry B* 2016;4(46):7398–7405.

103 Zhou Y, Peng Z, Seven ES, and Leblanc RM. Crossing the blood–brain barrier with nanoparticles. *Journal of Controlled Release* 2018;270:290–303.

104 Nautiyal G, Sharma I, Pandey P, and Sharma SK. Autoimmune diseases: Recent insights on epidemiology, pathogenesis, and prevalence rate. In: *Artificial Intelligence and Autoimmune Diseases: Applications in the Diagnosis, Prognosis, and Therapeutics*, (pp. 33–58). Springer; 2024.

105 Black RJ, Cross M, Haile LM, et al. Global, regional, and national burden of rheumatoid arthritis, 1990–2020, and projections to 2050: A systematic analysis of the Global Burden of Disease Study 2021. *The Lancet Rheumatology* 2023;5(10):e594–e610.

106 Yang X, Xia H, Liu C, et al. The novel delivery-exosome application for diagnosis and treatment of rheumatoid arthritis. *Pathology-Research and Practice* 2023;242:154332.

107 Liu L, Yan Y-D, Shi F-H, Lin H-W, Gu Z-C, and Li J. Comparative efficacy and safety of JAK inhibitors as monotherapy and in combination with methotrexate in patients with active rheumatoid arthritis: A systematic review and meta-analysis. *Frontiers in Immunology* 2022;13:977265.

108 Frazzei G, Musters A, de Vries N, Tas SW, and van Vollenhoven RF. Prevention of rheumatoid arthritis: A systematic literature review of preventive strategies in at-risk individuals. *Autoimmunity Reviews* 2023;22(1):103217.

109 Zhao J, Chen X, Ho K-H, et al. Nanotechnology for diagnosis and therapy of rheumatoid arthritis: Evolution toward theranostic approaches. *Chinese Chemical Letters* 2021;32(1):66–86.

110 Verma A, Jain A, Tiwari A, et al. Folate conjugated double liposomes bearing prednisolone and methotrexate for targeting rheumatoid arthritis. *Pharmaceutical Research* 2019;36:1–13.

111 Taktak F and Yiğen APT. Sulfasalazine/Nimesulide dual-loaded halloysite based poly (sulfobetaine methacrylate) hydrogels: Development and drug release characteristics. *Journal of Molecular Structure* 2022; 1252:132133.

112 Shen Q and Du Y. A comprehensive review of advanced drug delivery systems for the treatment of rheumatoid arthritis. *International Journal of Pharmaceutics* 2023;635:122698.

113 Soto ER, Caras AC, Kut LC, Castle MK, and Ostroff GR. Glucan particles for macrophage targeted delivery of nanoparticles. *Journal of Drug Delivery* 2012;2012(1):143524.

114 Zewail M, Nafee N, and Boraie N. Intra-articular dual drug delivery for synergistic rheumatoid arthritis treatment. *Journal of Pharmaceutical Sciences* 2021;110(7):2808–2822.

14

Data-driven Performance Evaluation of Drug Development and Pharmaceutical Manufacturing

Amol D. Gholap[1,2], Pankaj R. Khuspe[3], Navnath Hatvate[4], and Md. Faiyazuddin[5]*

[1]*Department of Pharmaceutics, St. John Institute of Pharmacy and Research, Palghar, India*
[2]*Department of Pharmaceutics, Amrutvahini College of Pharmacy, Sangamner, India*
[3]*Department of Pharmaceutics, College of Pharmacy, Paniv, Solapur, India*
[4]*Department of Pharmaceutical Sciences and Technology, Institute of Chemical Technology Mumbai, Marathwada Campus, Jalna, India*
[5]*Department of Pharmacy, ARAM Institute, Ranchi, India*

*Corresponding author: amolgholap16@gmail.com

CONTENTS

14.1 Introduction

Pharmaceutical production and drug development are intrinsically complicated, highly regulated, and capital-intensive processes with many scientific, technological, and financial obstacles. Predetermined process control strategies, empirical modeling, and retrospective analysis are the mainstays of the conventional approaches for assessing performance in these areas [1]. Although these methods have historically helped pharmaceuticals progress, they frequently have drawbacks in terms of flexibility, accuracy of predictions, and instantaneous decision-making. Data-driven performance evaluation that improves efficiency, scalability, and regulatory compliance has been made possible by the emergence of big data analytics, AI, and ML, which has completely changed the pharmaceutical industry [2].

A data-driven strategy analyzes multidimensional datasets from clinical trials, preclinical research, and real-time industrial processes by utilizing predictive modeling and high-throughput computer algorithms. In order to get actionable insights that maximize medication formulation, production output, and quality assurance, these approaches combine statistical process control (SPC), multivariate analysis, and deep learning algorithms. By making it easier to identify critical quality attributes (CQAs) and critical process parameters (CPPs), advanced analytics help to minimize batch variability and ensure process design resilience. Additionally, by facilitating real-time monitoring, proactive risk assessment, and adaptive process control, AI-powered process automation and digital twins are revolutionizing the pharmaceutical manufacturing industry [3].

Notwithstanding these developments, issues with data integration, standardization, and regulatory approval still exist. Concerns like data security, interoperability, and openness may be addressed with the help of cloud computing infrastructures and blockchain-based traceability systems as presented in Figure 14.1 [4]. Future drug development paradigms could be redefined by the combination of Industry 4.0 technology and pharmaceutical production, leading to faster innovation, shorter time-to-market, and better patient outcomes. The revolutionary influence of data-driven performance evaluation is examined in this chapter, along with its implications for pharmaceutical research and production going forward [5].

14.2 Data-driven Approaches in Drug Development

14.2.1 Preclinical Research

Preclinical research bridges the gap between theoretical drug design and clinical evaluation, making it a crucial stage in the drug development pipeline [6]. Drug discovery's predictive accuracy has increased dramatically as a result of the incorporation of data-driven approaches throughout this phase, which has decreased dependency on conventional in vitro and in vivo models. Researchers can identify promising drug candidates with enhanced specificity and efficacy by using bioinformatics, cheminformatics, and computational modeling to analyze intricate biological interactions at the molecular level. When combined with multi-omics technologies like proteomics, metabolomics,

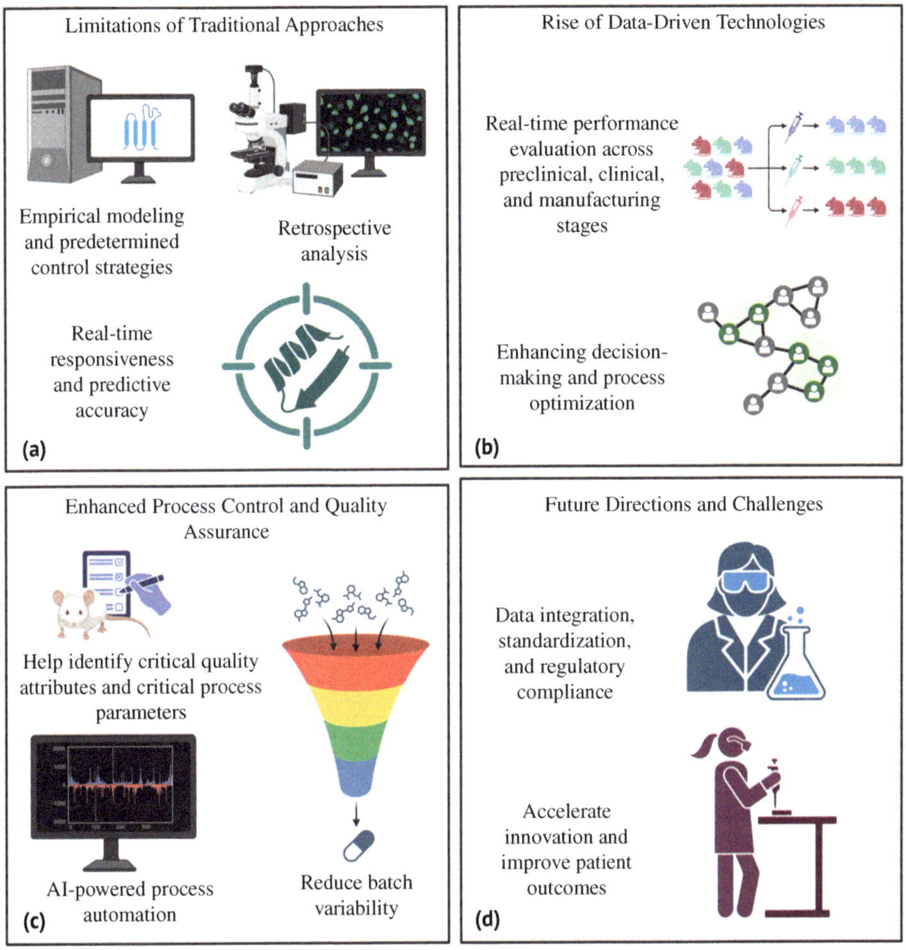

Figure 14.1 This figure represents the distinctions between traditional and data-driven methodologies in pharmaceutical manufacturing. (a) Traditional approaches are characterized by reliance on empirical modeling and retrospective analysis, which constrains real-time responsiveness and predictive precision. (b) In contrast, data-driven technologies facilitate real-time performance evaluation throughout the stages of drug development, thereby enhancing decision-making and process optimization. (c) Advanced analytics contribute to improved process control and quality assurance by identifying critical parameters, automating processes through artificial intelligence, and minimizing batch variability. (d) Future directions emphasize addressing challenges related to data integration, standardization, and regulatory compliance to expedite innovation and enhance patient outcomes. Collectively, these insights underscore a transformative shift toward intelligent and adaptive pharmaceutical manufacturing systems. Created in BioRender by Md. Faiyazuddin.

and genomes, high-throughput screening (HTS) platforms produce enormous datasets that are methodically examined using ML and AI techniques. Lead optimization is greatly accelerated by these computational tools, which make it easier to identify pharmacokinetic characteristics, structure-activity relationships (SAR), and drug-target interactions [7].

Molecular docking, quantitative structure-activity relationship (QSAR) modeling, and deep learning-driven virtual screening are examples of advanced predictive modeling techniques that improve candidate selection accuracy while lowering experimental costs. De novo drug creation is made possible by the use of AI-powered generative models, which forecast new chemical structures with the best possible therapeutic potential. Additionally, ADMET (absorption, distribution, metabolism, excretion, and toxicity) profiling and in silico toxicity prediction improve the early-stage evaluation of drug safety by reducing the risks related to late-stage failures. Data heterogeneity, the scarcity of high-quality training datasets, and the need for regulatory validation of AI-driven predictions continue to be obstacles in spite of these developments. Future directions in preclinical research include integrating blockchain-secured data repositories, federated learning models, and cloud-based computing platforms to improve data integrity, reproducibility, and cooperation. The pharmaceutical industry may accelerate the journey from bench to bedside by adopting data-driven tactics and bringing about a paradigm shift in preclinical drug development [8].

14.2.2 Clinical Trials

A crucial phase of medication research, clinical trials verify the pharmacokinetic characteristics, safety, and effectiveness of experimental treatments prior to regulatory approval. Longer drug development timelines are caused by the numerous problems that traditional clinical trial frameworks frequently confront, such as sluggish patient recruiting, high expenses, protocol deviations, and ineffective monitoring procedures [9]. Clinical trial execution has been profoundly changed by the incorporation of data-driven methodologies, which make use of AI, ML, big data analytics, and real-world evidence (RWE). These methodologies optimize a number of aspects, including participant selection, real-time monitoring, and predictive modeling for trial outcomes [10].

14.2.2.1 AI-driven Patient Recruitment and Cohort Selection

Patient recruitment is one of the biggest bottlenecks in clinical trials; traditional approaches rely on manual screening and out-of-date medical information, which results in low enrolment rates and skewed study groups. This procedure is being revolutionized by AI-powered analytics, which use wearable medical technology, genomic databases, and electronic health records (EHRs) to precisely identify qualified applicants [11]. Cohort selection is further improved using natural language processing (NLP) and predictive modeling, which stratify individuals according to risk profiles and therapy response by examining patient demographics, comorbidities, and genetic predispositions. Furthermore, real-world data sources like disease registries and social media health forums are being used more and more to enhance outreach and engagement, creating trial populations that are more inclusive and representative. While upholding strict ethical and legal requirements, this data-driven recruitment approach decreases screening failures, improves study diversity, and expedites patient enrolment [12].

14.2.2.2 Real-time Trial Monitoring and Safety Assessment

Manual data collection and sporadic site visits are the mainstays of traditional clinical trial monitoring, which frequently leads to delayed adverse event detection and protocol nonadherence. Using telemedicine platforms, digital biomarkers, and Internet of Things (IoT) enabled biosensors to power real-time data monitoring systems allows for continuous data collection on patient vitals, medication adherence, and treatment outcomes. By quickly identifying anomalies using automated anomaly detection algorithms, these systems guarantee early action and reduce the risk of drug toxicity or inadequate dose [13]. Furthermore, blockchain-based frameworks are becoming a transparent and safe way to protect trial data integrity, enhancing traceability and thwarting fraud. Multiple research institutes can work together without jeopardizing patient confidentiality due to the incorporation of federated learning models, which further improves data privacy and interoperability. By eliminating the need for manual audits and in-person site inspections, this move toward remote, data-driven monitoring lowers operating expenses while simultaneously improving patient safety [14].

14.2.2.3 Predictive Modeling for Trial Optimization and Adaptive Designs

ML-based prediction modeling, which uses biomarker profiles, historical trial data, and multi-omics integration to forecast patient responses, dropout rates, and trial success probability, is another example of how data-driven approaches are revolutionizing trial design [15]. These models make it possible to create adaptive clinical trial designs, in which endpoint evaluations, patient stratification criteria, and dosage schedule adjustments are informed by real-time analytics without sacrificing statistical validity. Researchers can optimize resource allocation and success rates by simulating different trial scenarios using Bayesian modeling, deep learning algorithms, and reinforcement learning techniques. Furthermore, AI-powered simulations of past patient data are creating synthetic control arms, which are eliminating the need for extensive placebo groups and speeding up morally and financially sound medication evaluations [14].

Notwithstanding these developments, problems still exist, including legal restrictions, data standardization problems, and moral dilemmas with AI-driven decision-making [16]. But ongoing cooperation between pharmaceutical firms, regulatory bodies, and AI researchers is opening the door to a clinical trial ecosystem that is more effective, scalable, and data-driven. Big data analytics, cloud computing, and AI-powered automation will all be seamlessly integrated in the future of clinical trial optimization, guaranteeing quicker medication approvals, better therapeutic precision, and improved patient-centric trial designs [17].

14.2.3 Post-market Surveillance

Following regulatory clearance and market introduction, post-market surveillance (PMS) is essential for evaluating a drug's safety, effectiveness, and long-term therapeutic impact. PMS assesses drug effectiveness in real-world settings, capturing diversity across demographics, comorbidities, and treatment regimens, in contrast to premarket clinical trials, which are conducted under controlled conditions with relatively small and homogeneous patient groups [18]. Conventional pharmacovigilance systems, such as the US surveillance database and the Food and Drug Administration's (FDA) Adverse Event Reporting System (FAERS), mostly depend on patients and healthcare providers reporting adverse events on their own initiative. However, underreporting, recall bias, and delayed signal detection frequently impede these traditional approaches, making it difficult for them to offer a thorough evaluation of post-market medication safety. With automated, predictive, and scalable methods for identifying adverse drug reactions (ADRs) and improving risk management tactics, the rise of big data analytics, ML, and RWE has completely transformed PMS [19].

PMS can now analyze large datasets in real time, recognizing safety signs with greater accuracy and reduced latency, thanks to the integration of EHRs, insurance claims data, wearables, patient registries, and social media analytics. Beyond conventional structured datasets, NLP techniques improve ADR detection by extracting valuable information from unstructured medical notes, physician reports, and patient comments. A popular statistical method in pharmacovigilance, proportionality analysis has been enhanced with deep learning models and ML classifiers to increase the sensitivity and specificity of drug safety evaluations. In order to detect uncommon, chronic, or multidrug interaction-induced side events that could otherwise go undetected in premarket trials, these algorithms are able to discover intricate patterns and correlations in ADR reporting [20].

Benefit-risk analyses have been improved by the use of Bayesian inference and predictive modeling approaches, which have allowed pharmaceutical companies and regulators to proactively reduce possible dangers. The reliability of safety signal identification is increased by researchers' ability to distinguish between real ADRs and confounding variables thanks to sophisticated causal inference techniques. By continuously updating medication safety profiles based on changing data inputs, real-time pharmacovigilance dashboards driven by AI and cloud computing enable dynamic risk appraisal. Additionally, digital twin technology is becoming a technique for predicting adverse effects before they materialize on a wide scale, where virtual patient cohorts mimic drug responses [21].

The potential of blockchain technology to improve PMS data integrity, traceability, and regulatory compliance is being investigated more and more. Blockchain promotes increased trust in pharmacovigilance data by offering tamper-proof, decentralized ledgers that guarantee adverse event reports, drug batch information, and safety evaluations stay transparent and unchangeable. However, issues with data privacy, interoperability, and regulatory harmonization arise when these cutting-edge technologies are widely used. Standardized frameworks for data curation, validation, and safe sharing are required since real-world datasets frequently display format heterogeneity, missing values, and reporting biases. Moreover, the dependence on AI-driven pharmacovigilance models requires continual validation and management to minimize false positives or algorithmic biases that could lead to unwarranted regulatory actions or misinterpretations of drug safety profiles [22].

Post-market medication monitoring could be greatly enhanced by the convergence of global PMS networks, AI-driven pharmacovigilance platforms, and federated learning models. These developments have the potential to improve the reliability and generalizability of drug safety evaluations by permitting cooperative, decentralized analysis of global datasets while protecting data privacy. By offering real-time physiological and behavioral health data, the combination of wearable biosensors and Internet of Medical Things (IoMT) devices would enhance PMS capabilities and provide a more individualized approach to drug safety assessment. In order to guarantee that medications continue to be safe, effective, and available to a wide range of patient populations around the globe, the future of PMS ultimately rests in coordinating regulatory regimes, encouraging cross-industry collaboration, and utilizing cutting-edge technologies [23].

14.3 Data-driven Pharmaceutical Manufacturing

14.3.1 Process Optimization

Utilizing predictive modeling and process analytical technology (PAT) to improve efficiency, dependability, and compliance, the incorporation of data-driven approaches in pharmaceutical manufacturing has completely reinterpreted process optimization as presented in Figure 14.2. Conventional production methods frequently depend on static, predetermined control mechanisms that might not take into consideration changes in raw

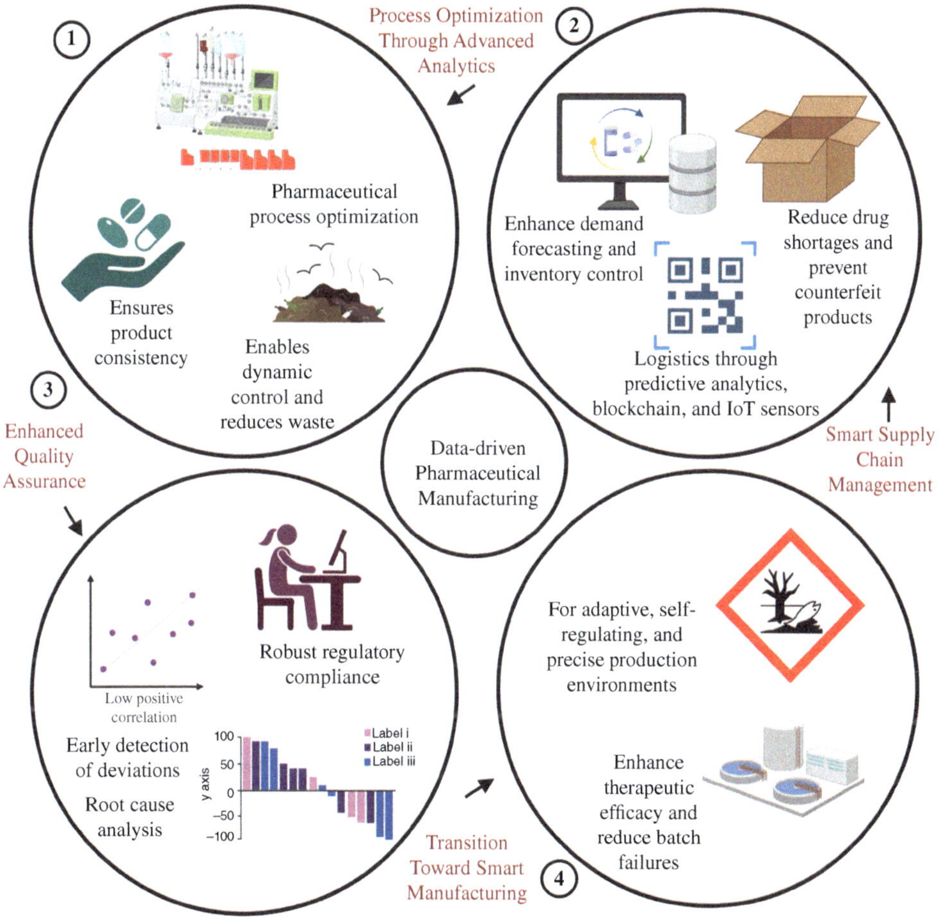

Figure 14.2 The figure presents a data-driven framework for pharmaceutical manufacturing, encompassing four principal domains: (1) process optimization through advanced analytics, which enhances product consistency and reduces waste via dynamic control; (2) smart supply chain management, which utilizes predictive analytics, blockchain, and IoT to improve inventory control and prevent counterfeiting; (3) enhanced quality assurance, which facilitates early deviation detection and root cause analysis to ensure robust regulatory compliance; and (4) transition toward smart manufacturing, which promotes adaptive, self-regulating production environments to enhance therapeutic efficacy and minimize batch failures. Collectively, these innovations contribute to a smart, efficient, and compliant pharmaceutical production ecosystem. Created in BioRender by Md. Faiyazuddin.

material characteristics, process parameters, and environmental influences in real time. Data-driven process optimization, on the other hand, uses multivariate statistical process control (MSPC), ML, and advanced analytics to dynamically modify process parameters, guaranteeing stable and flexible manufacturing workflows [24].

The real-time monitoring and control of CQAs and CPPs, which are crucial for preserving product integrity, batch-to-batch consistency, and regulatory compliance, is one of the primary components of data-driven optimization. To enable continuous evaluation of the physicochemical properties, dissolution profiles, and stability characteristics of pharmaceutical formulations, PAT frameworks are increasingly utilizing Raman spectroscopy, near-infrared spectroscopy (NIR), and real-time sensor-based feedback loops. Manufacturers may prevent deviations, minimize hazards, and cut down on waste related to out-of-specification (OOS) batches by combining these analytical tools with digital twins and AI-driven control systems [25].

Proactively identifying possible process inefficiencies and failure areas is made possible by predictive modeling approaches like Bayesian networks, deep learning algorithms, and chemometric modeling. By reducing process downtime, allocating resources optimally, and improving scale-up tactics for both continuous and batch production systems, these models increase throughput. But for these data-driven tactics to be implemented successfully, strong data infrastructure, safe cloud-based architectures, and adherence to legal frameworks like the FDA's Quality by Design (QbD) and ICH Q8/Q9 criteria are required. Pharmaceutical manufacturing's future rests in the broad implementation of Industry 4.0 principles, which will allow for self-regulating, highly adaptive, and precision-driven production environments made possible by AI, IoT-enabled smart sensors, and autonomous process control systems. This will ultimately transform drug manufacturing for improved therapeutic efficacy and patient safety [26].

14.3.2 Supply Chain Management

A complex and interconnected system, the pharmaceutical supply chain necessitates exact coordination between production, distribution, procurement, and regulatory compliance. Conventional supply chain models frequently depend on deterministic planning techniques and historical data, which can result in inefficiencies like stockouts, overproduction, or delays in the distribution of medications. By improving demand forecasting, inventory optimization, and real-time logistics tracking, the incorporation of data-driven analytics into supply chain management has revolutionized operational efficiency. Pharmaceutical firms can lower the risk of medicine shortages and surplus inventory by using sophisticated ML algorithms and predictive analytics to precisely predict changes in market demand [27].

The use of blockchain technology, which offers a decentralized, unchangeable ledger for tracking pharmaceutical products across the whole supply chain, is a significant development in data-driven supply chain management. By preserving verifiable records of transactions, storage conditions, and product identification, this system improves transparency, reduces the risk of counterfeit drugs, and guarantees regulatory compliance. By providing real-time monitoring of temperature-sensitive biologics and vaccines, the combination of radio-frequency identification and IoT sensors further enhances traceability and guarantees product integrity during transit [28].

Cloud-based data platforms reduce lead times and increase supply chain agility by enabling smooth information sharing between suppliers, manufacturers, and regulatory bodies. But issues like cybersecurity risks, data standardization, and integration difficulties continue to be major worries. It is anticipated that future developments in blockchain-based smart contracts and AI-driven supply chain automation will further optimize pharmaceutical logistics, improving security, efficiency, and responsiveness to demands in the global healthcare market [29].

14.3.3 Quality Assurance and Compliance

Pharmaceutical production relies heavily on regulatory compliance and QA to guarantee that medicinal products fulfil strict safety, effectiveness, and consistency requirements. By enabling proactive rather than reactive approaches, the combination of ML, real-time process monitoring, and powerful data analytics has revolutionized conventional quality control paradigms [30]. Big data-driven quality assurance systems create a thorough framework for quality evaluation by utilizing high-throughput data collection from real-time PAT, laboratory information management system (LIMS), and manufacturing execution system (MES). Anomaly detection techniques, chemometric modeling, and MSPC enable manufacturers to detect deviations from CPPs and CQAs with previously unheard-of accuracy [31].

The capacity of data-driven QA to manage deviations and conduct root cause analysis (RCA) using automated pattern recognition and predictive analytics is one of its main benefits. By comparing process variables across several production batches, AI-driven fault detection models evaluate departures from regulatory specifications and help manufacturers find underlying systemic inefficiencies. Additionally, by offering unchangeable recordings of batch histories, blockchain-based traceability systems improve compliance by enabling end-to-end visibility and audit readiness. QbD and real-time release testing (RTRT), both of which rely on sophisticated data-driven procedures to ensure consistent product quality and reduce batch failures, are being promoted more and more by regulatory agencies like the FDA and European Medicines Agency (EMA) [32].

Notwithstanding these developments, issues with data harmonization, digital platform interoperability, and compliance with changing legal frameworks still exist. The deployment of cloud-based compliance systems, artificial neural networks (ANNs) for risk-based quality assessment, and machine vision-enabled defect detection are promising future prospects for increasing QA in pharmaceutical manufacturing. Utilizing data-driven strategies will be crucial as the sector moves toward sector 4.0 and smart manufacturing in order to preserve regulatory compliance, maximize product quality, and promote ongoing process improvement [33].

14.4 Challenges and Solutions

Nanotechnology has fundamentally transformed the field of human therapeutics by providing highly precise, targeted, and efficient solutions for the diagnosis, treatment, and prevention of various diseases [34–38]. By manipulating materials at the nanoscale, typically below 100 nm, researchers and clinicians can exploit unique physicochemical properties absent in bulk materials, leading to significant advancements in drug delivery, cancer therapy, tissue engineering, and regenerative medicine [39–43]. One of the most transformative applications of nanotechnology is in drug delivery systems. Traditional drug administration methods often encounter limitations such as poor solubility, rapid degradation, systemic toxicity, and nonspecific distribution. Nanocarriers, including liposomes, dendrimers, solid lipid nanoparticles, and polymeric nanoparticles, offer controlled and sustained drug release, enhancing therapeutic efficacy while minimizing side effects [44–48]. These nanocarriers can be engineered to target specific tissues or cells, such as cancerous tumors, through surface modifications or ligand conjugation, thereby enhancing selectivity and reducing damage to healthy tissues. In oncology, nanotechnology plays a crucial role in both diagnosis and treatment [49–54]. Nanoparticles can be designed to preferentially accumulate in tumor tissues through the enhanced permeability and retention (EPR) effect. Moreover, multifunctional nanoparticles are being developed that combine therapeutic agents with diagnostic imaging modalities, giving rise to theranostic platforms that allow simultaneous disease diagnosis, monitoring, and treatment. For instance, gold nanoparticles and magnetic nanoparticles are being investigated for their ability to deliver drugs, enhance imaging contrast, and enable photothermal or magnetic hyperthermia therapies [3, 55–59]. Nanotechnology is also advancing gene therapy, where nanoscale carriers protect genetic material such as DNA or RNA from degradation and facilitate their transport into target cells. Lipid-based nanoparticles have been instrumental in delivering mRNA vaccines, as demonstrated during the COVID-19 pandemic [60–65]. These platforms provide a foundation for future therapies aimed at correcting genetic disorders or modulating gene expression in diseases such as cancer, cardiovascular conditions, and neurodegenerative disorders [66–71].

Beyond drug and gene delivery, nanomaterials are extensively utilized in tissue engineering and regenerative medicine. Nanoscale scaffolds and hydrogels create biomimetic environments conducive to cell growth, differentiation, and tissue regeneration. For example, nanofiber-based scaffolds replicate the natural extracellular matrix of tissues, thereby enhancing cellular interactions and promoting healing in applications such as skin grafts, bone repair, and neural regeneration [72–78]. Biosensors and nanodiagnostics are advancing disease detection and monitoring. Nanosensors are capable of detecting biomarkers at ultra-low concentrations, facilitating early diagnosis of diseases such as cancer, diabetes, and infectious diseases. These tools hold significant promise for point-of-care testing and personalized medicine, where treatments can be tailored to an individual's unique biological profile. Despite its vast potential, the clinical translation of nanotechnology continues to face challenges, including issues of toxicity, biocompatibility, scalability, and regulatory hurdles [79–86]. Ongoing research endeavors to address these challenges to ensure that nanotechnology-based therapies are safe, effective, and accessible. AI, ML, and nanotechnology are transforming human therapeutics by enhancing diagnostics, drug delivery, and personalized medicine. AI and ML facilitate the analysis of large datasets to uncover patterns in disease progression, predict patient responses, and identify potential drug candidates [87–93].

14.4.1 Data Integration

The heterogeneity of data sources and the absence of system-wide standardization are one of the biggest obstacles to data-driven performance evaluation in pharmaceutical research and manufacture. Preclinical research, clinical trials, MES, LIMS, PAT, and regulatory submissions all contribute to the massive volumes of data generated by pharmaceutical businesses. However, these statistics often exist in various formats, private databases, and isolated infrastructures, limiting their accessibility and interoperability. Lack of standardized frameworks for data interchange causes problems with real-time monitoring and predictive analytics, delays in decision-making, and inefficiencies in cross-functional collaboration. Furthermore, smooth integration and harmonization are made more difficult by the variety of data formats, including as sensor-based process monitoring outputs, unstructured clinical notes, structured numerical data, and HTS results [94].

Adopting standardized data exchange protocols and interoperable data platforms has become essential to addressing these issues. Real-time access to unified, high-fidelity data is made possible by the use of cloud-based data lakes that combine and standardize diverse information. Additionally, structured data exchange and improved compatibility across various pharmaceutical workflows are made possible by the adoption of common data standards like the clinical data interchange standards consortium (CDISC), health level seven (HL7), fast healthcare interoperability resources (FHIR), and the allotrope data format (ADF). Interoperability is further strengthened by blockchain-based data sharing frameworks and application programming interfaces, which guarantee safe, unchangeable, and verifiable data transactions across stakeholders. Furthermore, data transformation can be automated through the combination of NLP algorithms and AI, allowing for a more effective use of unstructured data sources [95].

Notwithstanding these developments, the process of fully implementing standardized data integration is still intricate and dynamic, necessitating cooperation from technology suppliers, pharmaceutical companies, and regulatory bodies. Future advancements in federated learning architectures, ontology-based knowledge graphs, and semantic data modeling could lead to smooth data harmonization while preserving security and compliance. Building strong data integration frameworks will be essential for improving operational effectiveness, regulatory compliance, and data-driven innovation in pharmaceutical sciences as the sector continues its shift to smart production and digitalized drug development [96].

14.4.2 Data Security and Privacy

Data security, patient confidentiality, and regulatory compliance are now major challenges as pharmaceutical research and manufacturing depend more and more on big data analytics, cloud computing, and AI. The pharmaceutical business is a top target for cyberattacks, data breaches, and unauthorized access because it manages enormous volumes of sensitive data, such as patient health records, clinical trial data, and patented drug formulas [97]. Maintaining the integrity and dependability of data-driven pharmaceutical operations while protecting patient privacy requires strong data governance frameworks. Advanced encryption approaches, multifactor authentication (MFA), and zero-trust security architectures must be implemented since traditional security models frequently fail to handle the complexity of decentralized data environments [96].

Tokenization, end-to-end encryption, and differential privacy techniques are essential answers to these problems because they mask personally identifiable information (PII) while enabling safe data processing and analysis. Regulatory frameworks such as the General Data Protection Regulation (GDPR) and the Health Insurance Portability and Accountability Act (HIPAA) mandate strict guidelines for data anonymization, access control, and secure data transmission, ensuring that pharmaceutical companies uphold ethical and legal obligations. Additionally, blockchain-based data management systems improve security and transparency by offering a decentralized, impenetrable ledger for tracking the drug supply chain, clinical trial documentation, and real-time regulatory auditing [98].

Despite these developments, there are still issues with managing cross-border data transfers, maintaining compliance with changing legal environments, and striking a balance between data accessibility and security. Adopting federated learning models protects patient privacy while promoting innovation by enabling cooperative AI-driven pharmaceutical research without disclosing raw data. A proactive approach to cybersecurity that incorporates behavioral threat analytics, AI-powered intrusion detection systems (IDS), and quantum-resistant encryption will be crucial in reducing risks and guaranteeing robust, secure, and compliant pharmaceutical manufacturing ecosystems as the industry embraces Industry 4.0 and digital transformation [99].

14.4.3 Regulatory Acceptance

The absence of clear regulatory frameworks makes it difficult to integrate big data analytics, ML, and AI into pharmaceutical development and manufacturing [100]. The dynamic, self-learning nature of AI/ML-driven models is frequently incompatible with traditional regulatory paradigms, which are built for deterministic process validation and preset control measures. Uncertainty in regulatory filings and compliance strategies arises from the lack of standardized standards for assessing AI-based prediction models, real-time process monitoring tools, and algorithm-driven decision-making systems. Furthermore, the opaque nature of deep learning algorithms complicates regulatory approvals for data-driven approaches by raising questions about explainability, reproducibility, and reliability in crucial pharmaceutical applications [101].

In order to overcome these obstacles, industry participants, regulatory bodies, and academic institutions must work together to create clear, risk-based rules specifically designed for AI/ML applications in pharmaceutical manufacturing. Regulatory agencies like the US ICH of technical requirements for pharmaceuticals for human use, the EMA, and the FDA are all actively attempting to incorporate AI-specific frameworks into the current good manufacturing practice and QbD guidelines. The acceptability of data-driven approaches can be improved by implementing explainable AI (XAI) techniques, digital twin validation procedures, and RWE-driven regulatory filings. Additionally, the safe application of AI-powered analytics in pharmaceutical quality assurance can be facilitated by the establishment of standardized validation procedures, regulatory sandboxes for AI research, and ongoing PMS. Encouraging interdisciplinary cooperation and regulatory harmonization will be essential as regulatory frameworks change in order to fully realize the potential of data-driven pharmaceutical innovation while maintaining adherence to international safety and efficacy standards [102].

14.5 Future Directions

14.5.1 Real-time Data Integration

The smooth integration of digital twin technology and IoT devices is set to transform data-driven performance evaluation in pharmaceutical manufacturing and medication development in the future. These advancements enable real-time data collecting, enhanced analytics, and automated decision-making, ensuring dynamic process control and continual performance optimization. Cloud-based infrastructures, edge computing, and IoT-enabled smart sensors make it easier to gather high-frequency, multidimensional data from clinical trial monitoring systems, preclinical research labs, and massive pharmaceutical manufacturing lines. Predictive analytics, process automation, and anomaly detection are all supported by this real-time data stream, which improves operational effectiveness and lowers process variances [103].

By facilitating proactive risk assessment, process optimization, and predictive simulations, digital twins' high-fidelity virtual copies of actual drug development and manufacturing processes further enhance real-time data integration. Digital twins offer a dynamic depiction of pharmaceutical processes by utilizing ML algorithms, multivariate statistical modeling, and cyber-physical systems. This enables researchers and manufacturers to pinpoint bottlenecks, optimize process parameters, and anticipate system failures before they happen. By promoting a QbD methodology, this integration guarantees regulatory compliance and reduces variability in the formulation and manufacturing of pharmaceuticals [104].

Data standardization, cybersecurity, and compatibility between historical systems and contemporary IoT frameworks continue to be issues despite the revolutionary potential of real-time data integration. To overcome these obstacles, federated learning models for decentralized data processing, blockchain-based safe data interchange, and AI-driven decision support systems will be essential. Pharmaceutical production will be redefined in the future by the confluence of IoT, digital twins, and AI-driven analytics, which will lead to increased precision, a shorter time to market, and better therapeutic results [105].

14.5.2 Predictive and Prescriptive Analytics

The pharmaceutical business is undergoing a revolutionary change as a result of the integration of predictive and prescriptive analytics, which makes data-driven decision-making possible and improves productivity, quality, and regulatory compliance. Predictive analytics uses past and current data to foresee possible problems in drug research, clinical trials, and pharmaceutical production [106]. It is driven by ML, deep learning algorithms, and AI. To find patterns suggestive of batch failures, supply chain interruptions, or variations in CQAs and CPPs, these approaches employ multivariate regression models, Bayesian networks, and time-series forecasting. By giving early warnings for potential inefficiencies, predictive analytics mitigates risks, boosts process robustness, and optimizes resource allocation across the drug development life cycle [107].

Prescriptive analytics complements predictive skills by providing data-driven recommendations for the best possible decision-making, going beyond forecasting. Prescriptive analytics assesses several intervention options and determines the best course of action by utilizing optimization algorithms, decision-support systems, and reinforcement learning. Pharmaceutical businesses can use this method to improve clinical trial design, optimize formulation parameters, and dynamically modify manufacturing settings based on real-time data insights. Additionally, prescriptive models incorporate regulatory intelligence and compliance frameworks to guarantee that suggested activities conform to the constantly changing criteria for PAT, QbD, and GMP [108].

Predictive and prescriptive analytics have great potential, but their widespread use in pharmaceutical operations is hampered by issues with data standardization, model interpretability, and legacy system interaction. Future developments will concentrate on improving AI models' explainability, creating cloud-based infrastructure for the deployment of real-time analytics, and utilizing edge computing to facilitate decentralized decision-making. Predictive and prescriptive analytics will be essential innovation enablers as the pharmaceutical sector adopts Industry 4.0 and smart manufacturing paradigms. These will promote ongoing process optimization, better regulatory compliance, and improved patient-centric drug development strategies [109].

14.5.3 Personalized Medicine

The emergence of data-driven approaches in pharmaceutical research is transforming personalized medicine by making it possible to create patient-specific treatments that maximize benefits and reduce side effects. The one-size-fits-all approach used in traditional drug development frequently ignores interindividual differences in medication response [110]. However, precision-driven treatment approaches have been made possible by the combination of RWE, multi-omics data (genomics, transcriptomics, proteomics, and metabolomics), and sophisticated bioinformatics tools. Pharmaceutical researchers can customize drug formulations and dosage schedules at the individual level by utilizing ML algorithms, AI, and predictive analytics to examine large datasets that include genetic predispositions, phenotypic expressions, and environmental influences [111].

Pharmacogenomics, which uses high-throughput sequencing and computational modeling to predict medication metabolism, efficacy, and toxicity based on a patient's genetic profile, is one of the main enabling technologies of data-driven personalized medicine. This strategy is further improved by AI-driven biomarker discovery, which finds new molecular targets that may be used to create drugs tailored to individual patients. Additionally, wearable biosensors and EHRs offer constant streams of real-time patient data, enabling dynamic treatment modifications via clinical interventions driven by AI and decision-support systems. Researchers, physicians, and regulatory bodies can work together seamlessly thanks to the safe, scalable, and interoperable data management provided by the combination of blockchain technology and cloud computing [112].

Notwithstanding these developments, issues with data standardization, moral dilemmas, and the intricate regulatory framework pertaining to customized treatments still exist.

A possible way to use decentralized patient data while strictly adhering to HIPAA, GDPR, and other data protection laws is to use federated learning frameworks, which enable collaborative AI training without jeopardizing patient privacy. Data-driven personalized medicine has the potential to revolutionize the drug development landscape as the discipline develops further. This will allow for the development of next-generation therapies that will improve patient outcomes, increase treatment precision, and shape the future of individualized healthcare [113].

14.5.4 Regulatory Innovation

The swift growth of data-driven approaches in pharmaceutical production and medication development calls for flexible regulatory frameworks that maintain product integrity and patient safety while keeping up with technology breakthroughs. The dynamic nature of AI, ML, real-time analytics, and digital twins in pharmaceutical operations is posing a growing threat to traditional regulatory paradigms, which are frequently defined by strict norms and drawn-out approval processes. Regulatory agencies like the US ICH of technical requirements for pharmaceuticals for human use, the EMA, and the FDA are all pushing for frameworks that incorporate RWE, predictive analytics, and continuous monitoring as part of their gradual transition to risk-based, data-centric regulatory models [114].

QbD, RTRT, and adaptive manufacturing controls are examples of a fundamental shift in regulatory science that uses big data analytics to improve product quality assurance and speed up market access. Furthermore, efforts are underway to improve data security, traceability, and transparency throughout international pharmaceutical supply chains by implementing blockchain-enabled regulatory compliance solutions. The time needed for regulatory clearances is decreased by the use of digital regulatory sandboxes, AI-driven dossier evaluation, and cloud-based regulatory submissions, which enable iterative testing and validation of innovative manufacturing techniques [115].

There are still issues with addressing ethical issues with AI-driven decision-making, standardizing regulatory requirements across jurisdictions, and protecting data privacy in accordance with the GDPR and HIPAA regulations. To promote more flexible, adaptable, and innovation-driven pharmaceutical ecosystem, future regulatory environments must include machine-readable regulatory frameworks, AI-assisted compliance monitoring, and real-time audit capabilities. The shift to flexible, data-driven governance models will be necessary as regulatory bodies adopt Industry 4.0 and Pharma 4.0 in order to support the upcoming generation of pharmaceutical manufacturing, research, and development [116].

14.6 Conclusion

By increasing process efficiency, cutting costs, and encouraging innovation, the incorporation of data-driven performance evaluation is radically changing pharmaceutical manufacturing and medication development. Unprecedented levels of precision in drug research, clinical trials, and production workflows have been made possible by the convergence of big data analytics, AI, ML, and real-time monitoring systems. Pharmaceutical companies can reduce variability and guarantee robust product quality by proactively identifying CQAs and

CPPs through the use of predictive modeling, multivariate analysis, and PAT. Additionally, decision-making throughout the pharmaceutical value chain has been greatly enhanced by the use of digital twins for real-time process simulation, cloud computing for scalable infrastructure, and blockchain for data security. Notwithstanding these developments, a number of obstacles still exist, such as the intricacies of data integration, interoperability problems, cybersecurity threats, and regulatory conformity among various countries. For wider adoption to be facilitated, data framework standardization, regulatory sandbox development, and AI-assisted compliance methods must be developed. To maintain confidence and responsibility in data-driven approaches, ethical issues pertaining to algorithmic bias, patient data privacy, and transparency in AI-driven judgements must also be addressed. In order to expedite medication research and approvals, real-time analytics, personalized medicine, and flexible regulatory frameworks are anticipated to be the main areas of future innovation. A paradigm change toward patient-centric drug discovery and intelligent pharmaceutical production is anticipated as a result of the integration of wearable biosensors, IoT devices, and decentralized clinical trials. Data-driven strategies will remain essential in creating a pharmaceutical ecosystem that is more effective, scalable, and responsive as the sector transitions to Industry 4.0 and Pharma 4.0, ultimately leading to better global healthcare outcomes.

References

1 Kumar SA, Ananda Kumar TD, Beeraka NM, et al. Machine learning and deep learning in data-driven decision making of drug discovery and challenges in high-quality data acquisition in the pharmaceutical industry. *Future Medicinal Chemistry [Internet]* 2022;14(4):245–270. Available from: https://www.tandfonline.com/doi/full/10.4155/fmc-2021-0243

2 Dong Y, Yang T, Xing Y, Du J, and Meng Q. Data-driven modeling methods and techniques for pharmaceutical processes. *Processes [Internet]* 2023;11(7):2096. Available from: https://www.mdpi.com/2227-9717/11/7/2096

3 Vora LK, Jetha K, Thakur RRS, Solanki HK, and Chavda VP. Artificial intelligence in pharmaceutical technology and drug delivery design. *Pharmaceutics [Internet]* 2023;15(7):1916. Available from: https://www.mdpi.com/1999-4923/15/7/1916

4 Kolluri S, Lin J, Liu R, Zhang Y, and Zhang W. Machine learning and artificial intelligence in pharmaceutical research and development: A review. *AAPS Journal [Internet]* 2022;24(1):19. Available from: https://link.springer.com/10.1208/s12248-021-00644-3

5 Sadybekov AV and Katritch V. Computational approaches streamlining drug discovery. *Nature [Internet]* 2023;616(7958):673–685. Available from: https://www.nature.com/articles/s41586-023-05905-z

6 Gholap AD, Khuspe PR, Bharati DK, et al. Computational Neuropharmacology in Psychiatry. In: *Computational Neuropharmacology* [Internet], (pp. 207–244). Wiley; 2025. Available from: https://onlinelibrary.wiley.com/doi/10.1002/9781394242474.ch7

7 Llorente-Saguer I and Oxtoby NP. A data-driven framework for biomarker discovery applied to optimizing modern clinical and preclinical trials on Alzheimer's disease. *Brain Communications [Internet]* 2024;6(6). Available from: https://academic.oup.com/braincomms/article/doi/10.1093/braincomms/fcae438/7919619

8 Derk J, Dzirasa K, and Locklear T. A novel cross-institutional college internship program to train future diverse leaders in clinical research with data-driven approaches to assess impact. *Frontiers in Pharmacology [Internet]* 2023; 14. Available from: https://www.frontiersin.org/articles/10.3389/fphar.2023.1294535/full

9 Sharma A, Virmani T, Pathak V, et al. Artificial intelligence-based data-driven strategy to accelerate research, development, and clinical trials of COVID vaccine. Ashfaq UA, editor. *BioMed Research International [Internet]* 2022;2022(1). Available from: https://onlinelibrary.wiley.com/doi/10.1155/2022/7205241

10 Bourgeois FT. Data-driven approaches to maximize the impact of pediatric clinical trials. *Pediatrics [Internet]* 2022;149(4). Available from: https://publications.aap.org/pediatrics/article/149/4/e2021055815/185585/Data-Driven-Approaches-to-Maximize-the-Impact-of

11 Lu X, Yang C, Liang L, Hu G, Zhong Z, and Jiang Z. Artificial intelligence for optimizing recruitment and retention in clinical trials: a scoping review. *Journal of the American Medical Informatics Association [Internet]* 2024;31(11):2749–2759. Available from: https://academic.oup.com/jamia/article/31/11/2749/7755392

12 Chorev M, Haderlein J, Chandra S, et al. A multi-modal AI-driven cohort selection tool to predict suboptimal non-responders to aflibercept loading-phase for neovascular age-related macular degeneration: PRECISE study report 1. *Journal of Clinical Medicine [Internet]* 2023;12(8):3013. Available from: https://www.mdpi.com/2077-0383/12/8/3013

13 Chia WMD, Keoh SL, Goh C, and Johnson C. Risk assessment methodologies for autonomous driving: A survey. *IEEE Transactions on Intelligent Transportation Systems [Internet]* 2022;23(10):16923–16939. Available from: https://ieeexplore.ieee.org/document/9756641/

14 Fountzilas E, Tsimberidou AM, Vo HH, and Kurzrock R. Clinical trial design in the era of precision medicine. *British Medical Journal [Internet]* 2022;14(1):101. Available from: https://genomemedicine.biomedcentral.com/articles/10.1186/s13073-022-01102-1

15 Maguire R, McCann L, Kotronoulas G, et al. Real time remote symptom monitoring during chemotherapy for cancer: European multicentre randomised controlled trial (eSMART). *British Medical Journal [Internet]* 2021;n1647. Available from: https://www.bmj.com/lookup/doi/10.1136/bmj.n1647

16 Gholap AD and Omri A. Advances in artificial intelligence-envisioned technologies for protein and nucleic acid research. *Drug Discovery Today [Internet]* 2025; 104362. Available from: https://linkinghub.elsevier.com/retrieve/pii/S1359644625000753

17 Lei B, Kirk TQ, Bhattacharya A, et al. Bayesian optimization with adaptive surrogate models for automated experimental design. *npj Computational Materials [Internet]* 2021;7(1):194. Available from: https://www.nature.com/articles/s41524-021-00662-x

18 Huanbutta K, Burapapadh K, Kraisit P, et al. Artificial intelligence-driven pharmaceutical industry: A paradigm shift in drug discovery, formulation development, manufacturing, quality control, and post-market surveillance. *European Journal of Pharmaceutical Sciences [Internet]* 2024;203:106938. Available from: https://linkinghub.elsevier.com/retrieve/pii/S0928098724002513

19 Badnjević A, Pokvić LG, Deumić A, and Bećirović LS. Post-market surveillance of medical devices: A review. *Technology and Health Care [Internet]* 2022;30(6):1315–1329. Available from: https://journals.sagepub.com/doi/full/10.3233/THC-220284

20 Thomas L, Hyde C, Mullarkey D, Greenhalgh J, Kalsi D, and Ko J. Real-world post-deployment performance of a novel machine learning-based digital health technology for skin lesion assessment and suggestions for post-market surveillance. *Frontiers in Medicine [Internet]* 2023; 10. Available from: https://www.frontiersin.org/articles/10.3389/fmed.2023.1264846/full

21 Reniewicz J, Suryaprakash V, Kowalczyk J, et al. Artificial intelligence / machine-learning tool for post-market surveillance of in vitro diagnostic assays. *New Biotechnology [Internet]* 2024;79:82–90. Available from: https://linkinghub.elsevier.com/retrieve/pii/S1871678423000687

22 Pokvic LG, Deumic A, Lutovac B, and Badnjevic A. Possibility of managing medical device post-market surveillance using artificial intelligence and standardized methodology. In: *2021 10th Mediterranean Conference on Embedded Computing (MECO) [Internet]*, (pp. 1–4). IEEE; 2021. Available from: https://ieeexplore.ieee.org/document/9460138/

23 Merdović N, Spahić L, Hundur M, Pokvić LG, and Badnjević A. Advancement of post-market surveillance of medical devices leveraging artificial intelligence: Infusion pumps case study. *Technology and Health Care [Internet]* 2025;33(2):915–921. Available from: https://journals.sagepub.com/doi/10.1177/09287329241291415

24 Castaldello C, Facco P, Bezzo F, Georgakis C, and Barolo M. Data-driven tools for the optimization of a pharmaceutical process through its knowledge-driven model. *AIChE Journal [Internet]* 2023;69(4). Available from: https://aiche.onlinelibrary.wiley.com/doi/10.1002/aic.17925

25 van de Berg D, Savage T, Petsagkourakis P, Zhang D, Shah N, and del Rio-Chanona EA. Data-driven optimization for process systems engineering applications. *Chemical Engineering Science [Internet]* 2022;248:117135. Available from: https://linkinghub.elsevier.com/retrieve/pii/S0009250921007004

26 Tsopanoglou A and Jiménez del Val I. Moving towards an era of hybrid modelling: advantages and challenges of coupling mechanistic and data-driven models for upstream pharmaceutical bioprocesses. *Current Opinion in Chemical Engineering [Internet]* 2021;32:100691. Available from: https://linkinghub.elsevier.com/retrieve/pii/S221133982100023X

27 Rolf B, Jackson I, Müller M, Lang S, Reggelin T, and Ivanov D. A review on reinforcement learning algorithms and applications in supply chain management. *International Journal of Production Research [Internet]* 2023;61(20):7151–7179. Available from: https://www.tandfonline.com/doi/full/10.1080/00207543.2022.2140221

28 Tirkolaee EB, Sadeghi S, Mooseloo FM, Vandchali HR, and Aeini S. Application of machine learning in supply chain management: A comprehensive overview of the main areas. Ahmadian A, editor. *Mathematical Problems in Engineering [Internet]* 2021;2021: 1–14. Available from: https://www.hindawi.com/journals/mpe/2021/1476043/

29 Pournader M, Ghaderi H, Hassanzadegan A, and Fahimnia B. Artificial intelligence applications in supply chain management. *International Journal of Production Economics [Internet]* 2021;241:108250. Available from: https://linkinghub.elsevier.com/retrieve/pii/S0925527321002267

30 Javaid M, Haleem A, Pratap Singh R, and Suman R. Significance of quality 4.0 towards comprehensive enhancement in manufacturing sector. *Sensors International [Internet]* 2021;2:100109. Available from: https://linkinghub.elsevier.com/retrieve/pii/S2666351121000309

31 Alzahrani M, Broadbent DA, Teh I, Al-Qaisieh B, Johnstone E, and Speight R. A novel multimodality anthropomorphic phantom enhances compliance with quality assurance guidelines for magnetic resonance imaging in radiotherapy. *Physics and Imaging in Radiation Oncology [Internet]* 2025;33:100707. Available from: https://linkinghub.elsevier.com/retrieve/pii/S2405631625000120

32 Molęda M, Małysiak-Mrozek B, Ding W, Sunderam V, and Mrozek D. From corrective to predictive maintenance—A review of maintenance approaches for the power industry. *Sensors [Internet]* 2023;23(13):5970. Available from: https://www.mdpi.com/1424-8220/23/13/5970

33 Li J, Maiti A, and Fei J. Features and scope of regulatory technologies: Challenges and opportunities with industrial Internet of Things. *Future Internet [Internet]* 2023;15(8):256. Available from: https://www.mdpi.com/1999-5903/15/8/256

34 Gholap AD and Musmade DS. Advances in the field of nanooncology: An overview. *Inventi Impact: NDDS [ISSN 0976-3791] [Internet]* 2011;4(1):1–3. Available from: https://www.inventi.in/journal/article/2/1078/Inventi Impact: NDDS/Pharmaceutical

35 Anand M. Smart crystals technology: A review. *Pharmacologyonline [Internet]* 2011;1(1):238–243. Available from: https://pharmacologyonline.silae.it/files/newsletter/2011/vol3/028.amol.pdf

36 Borude S. Aquasomes: A potential drug delivery carrier. *Pharmacologyonline [Internet]* 2011;1(3):230–237. Available from: https://pharmacologyonline.silae.it/files/newsletter/2011/vol3/027.amol.pdf

37 Borude SS and Mahajan AM. Recent updates on Q fever: diagnosis and treatment. *Pharmacologyonline [Internet]* 2011;1(3):254–264. Available from: https://pharmacologyonline.silae.it/files/newsletter/2011/vol3/030.gholap.pdf

38 Magar TV, Gholap AD, and Musmade DS. Elastic vesicles as drugs carriers through the skin. *Inventi Rapid: NDDS [ISSN 0976-3791] [Internet]* 2011;1(4):1–2. Available from: https://www.inventi.in/journal/article/127/9611/Inventi Rapid: NDDS/Pharmaceutical#

39 Borude SS and Mahajan AM. POEMS syndrome: A review. *Pharmacologyonline [Internet]* 2011;1(3):265–274. Available from: https://pharmacologyonline.silae.it/files/newsletter/2011/vol3/031.gholap.pdf

40 Rajurkar VG. Recent technologies in nanomedicine for treatment of cancer. *Inventi Rapid: Pharmaceutical Technology [Internet]* 2012;4(1):1–3. Available from: https://www.inventi.in/journal/article/126/9966/Inventi Rapid: Pharm Tech/Pharmaceutical

41 Rajurkar VG. Lipid nanocapsules: A novel promising tool as drug delivery system. *Inventi Rapid: NDDS [ISSN 0976-3791] [Internet]* 2012;1(1):1–4. Available from: https://www.inventi.in/journal/article/127/9969/Inventi Rapid: NDDS/Pharmaceutical

42 Gholap AD. Translational nanomedicine: Status assessment and opportunities. *International Journal of Research in Pharmaceutical and Biomedical Sciences* 2012;4(3):1454–1463.

43 Rajurkar VG. New threat of clinical toxicities evoked from nanocarrier systems. *Inventi Rapid: NDDS [ISSN 0976-3791] [Internet]* 2012;1(4):1–4. Available from: https://www.inventi.in/journal/article/127/9982/Inventi Rapid: NDDS/Pharmaceutical

44 Rajurkar VG. Potential applications of capillary electromigration methods for the improvement of separation efficiency. *Inventi Rapid: Pharmaceutical Analysis & Quality Assurance [ISSN 0976-3813] [Internet]* 2012;1(4):1–3. Available from: https://www.inventi.in/journal/article/136/9968/Inventi Rapid: Pharm Analysis /Pharmaceutical

45 Dattatray T. Potential applications of botanical extract. *Inventi Rapid: Ethnopharmacology [ISSN 0976-3805] [Internet]* 2012;1(4):1–4. Available from: https://www.inventi.in/journal/article/135/9972/Inventi Rapid: Ethnopharmacolo/Pharmaceutical
46 Dighe SB. Dandruff: Pathogenesis and treatment. *Inventi Rapid: Cosmeceuticals [ISSN 0976-3864] [Internet]* 2012;1(4):1–4. Available from: https://www.inventi.in/journal/article/141/9975/Inventi Rapid: Cosmeceuticals/Pharmaceutical
47 Tupe D. The threat of photoaging: Clinical conditions, evaluation and treatment. *Inventi Rapid: Cosmeceuticals [ISSN 0976-3864] [Internet]* 2012;1(4):1–3. Available from: https://www.inventi.in/journal/article/141/9971/Inventi Rapid: Cosmeceuticals/Pharmaceutical
48 Dighe SB. Role of sebum in human health & acne. *Inventi Rapid: Cosmeceuticals [ISSN 0976-3864] [Internet]* 2012;1(4):1–3. Available from: https://www.inventi.in/journal/article/141/9980/Inventi Rapid: Cosmeceuticals/Pharmaceutical
49 Dattatray T. Artificial viral envelopes: Emerging vehicle for gene delivery. *PHARMATUTOR-ART-1439* 2012;1(4):1–3.
50 Rajurkar VG. Potential applications of superparmagnetic nanoparticles technology. *Inventi Rapid: NDDS [ISSN 0976-3791] [Internet]* 2012;1(4):1–3. Available from: https://www.inventi.in/journal/article/127/9983/Inventi Rapid: NDDS/Pharmaceutical
51 Bhagat V. Gold nanorods: A promising drug delivery vehicle. *Inventi Rapid: NDDS [ISSN 0976-3864] [Internet]* 2012;1(4):1–4. Available from: https://www.inventi.in/journal/article/127/9964/Inventi Rapid: NDDS/Pharmaceutical
52 Gholap AD, Rajurkar VG, and Dighe SB. Phonophoresis: Ultrasound enhanced transdermal drug delivery. *Inventi Rapid: NDDS [ISSN 0976-3791] [Internet]* 2012;1(4):1–3. Available from: https://www.inventi.in/journal/article/127/9981/Inventi Rapid: NDDS/Pharmaceutical
53 Khedkar A, Tupe D, and Gholap A. Cubosomes: Emerging drug delivery system. *Inventi Rapid: NDDS [ISSN 0976-3791] [Internet]* 2012;1(4):1–3. Available from: https://www.inventi.in/journal/article/127/9987/Inventi Rapid: NDDS/Pharmaceutical
54 Khedkar A, Tupe D, and Gholap A. Pharmacosomes: An emerging vesicular system. *Inventi Rapid: NDDS [ISSN 0976-3791] [Internet]* 2012;1(4):1–3. Available from: https://www.inventi.in/journal/article/127/9986/Inventi Rapid: NDDS/Pharmaceutical
55 Bothiraja C, Gholap KS, and Pawar AP. Investigation of ethyl cellulose microsponge gel for topical delivery of eberconazole nitrate for fungal therapy. *Therapeutic Delivery [Internet]* 2014;5(7):781–794. Available from: https://www.future-science.com/doi/10.4155/tde.14.43
56 Singh T, Ravi Kumar V, Gholap AD, Sanap G, Anjaiah B, and Roja K. In vitro antipyrial activity of psidium guajava leaf extract. *European Journal Pharmaceutical and Medical Research* 2016;3(1):312–315.
57 Puri AV and Gholap AD. Comparative evaluation of papaya and banana starch as disintegrant in diclofenac sodium tablet formulation. *International Journal of Pharmaceutical and Biological Sciences* 2018;8(3):512–522.
58 Gholap AD, Sayyad SF, Hatvate NT, et al. Drug delivery strategies for avobenzone: A case study of photostabilization. *Pharmaceutics [Internet]* 2023;15(3):1008. Available from: https://www.mdpi.com/1999-4923/15/3/1008
59 Hatvate NT, Khuspe PR, Mandhare TA, Kashid P, and Gaikwad VD. Proteomics in oncology: Retrospect and prospects. In: Kulkarni S, Haghi AK, and Manwatkar S. editors.

Novel Technologies in Biosystems, Biomedical & Drug Delivery [Internet], (pp. 243–269). Singapore: Springer Nature Singapore; 2023. Available from: https://link.springer.com/10.1007/978-981-99-5281-6_10

60 Sawant SD, Sayyad SF, Hatvate NT, Chavan M, Rojekar S, and Faiyazuddin M. Food biopackaging for human benefits: Status and perspectives. In: Ahmed S, editor. *Biobased Packaging Materials [Internet]*, (pp. 1–35). Singapore: Springer Nature Singapore; 2023. Available from: https://link.springer.com/10.1007/978-981-99-6050-7_1

61 Hatvate NT, Sawant SD, Khemka PN, and Vora LK. Unlocking the potential of nanobiohybrids to combat environmental pollution. In: Lens P, editor. *Nanobiohybrids for Advanced Wastewater Treatment and Energy Recovery [Internet]*, (pp. 169–90). UK: IWA Publishing; 2023. Available from: https://iwaponline.com/ebooks/book/886/chapter/3516052/Unlocking-the-potential-of-nanobiohybrids-to

62 Pardeshi S, More M, Togre N, Rebello N, and Giram P. Dendrimers based antibacterial and antiviral materials. In: *Antibacterial and Antiviral Functional Materials [Internet]*, (pp. 139–69). USA: ACS Publications; 2023. Available from: https://pubs.acs.org/doi/abs/10.1021/bk-2023-1458.ch005

63 Pardeshi S and Giram P. Preparation, antibacterial and antiviral activity measurements and detection methods. In: Deshmukh K, editor. *Antibacterial and Antiviral Functional Materials [Internet]*, (pp. 33–64). USA: ACS Publications; 2023. Available from: https://pubs.acs.org/doi/abs/10.1021/bk-2023-1458.ch002

64 Gholap AD, Gupta JS, Kamandar PA, Banchhod G V, and Hatvate NT. Antibody-drug conjugates for cancer therapy: An up-to-date review on the chemistry and pharmacology. In: Verma SK and Das AK, editors. *Bioconjugated Materials Part 2 Applications in Drug Delivery, Vaccine Formulations and Important Conjugates for Cancer Therapy [Internet]*, (pp. 105–190) (Comprehensive Analytical Chemistry; vol. 103). Amsterdam, Netherlands: Elsevier; 2023. Available from: https://www.sciencedirect.com/science/article/pii/S0166526X23000259

65 Bhowmik DD, Deshmukh AY, and Hatvate NT. Quintessential impact of dendrimer bioconjugates in targeted drug delivery. In: Verma SK and Das AK, editors. *Bioconjugated Materials Part 2 Applications in Drug Delivery, Vaccine Formulations and Important Conjugates for Cancer Therapy [Internet]*, (pp. 257–302) (Comprehensive Analytical Chemistry; vol. 103). Amsterdam, Netherlands: Elsevier; 2023. Available from: https://www.sciencedirect.com/science/article/pii/S0166526X23000272

66 Gholap AD, Said RP, Pawar RD, Ambore GS, and Hatvate NT. Importance of carbohydrate-drug conjugates in vaccine development: A detailed review. In: Verma SK, Das AK, editors. *Bioconjugated Materials Part 2 Applications in Drug Delivery, Vaccine Formulations and Important Conjugates for Cancer Therapy [Internet]*, (pp. 191–256) (Comprehensive Analytical Chemistry; vol. 103). Amsterdam, Netherlands: Elsevier; 2023. Available from: https://www.sciencedirect.com/science/article/pii/S0166526X23000296

67 Rojekar S, Kapare HS, Vishwakarma N, et al. Chitosan scaffolds: Expanding horizons in biomedical applications. *Carbohydrate Polymers [Internet]* 2024;323:121394. Available from: https://linkinghub.elsevier.com/retrieve/pii/S0144861723008597

68 Kapare HS, Pagar S, Kamandar P, et al. Exploring modified chitosan-based gene delivery technologies for therapeutic advancements. *International Journal of Biological Macromolecules [Internet]* 2024;260:129581. Available from: https://linkinghub.elsevier.com/retrieve/pii/S0141813024003842

69 Pardeshi SR, Hatvate NT, Dhorkule N, Sayyad SF, Faiyazuddin M, and Khalid M. Environmental implications and nanotechnological advances in octocrylene-enriched sunscreen formulations: A comprehensive review. *Chemosphere [Internet]* 2024;358:142235. Available from: https://linkinghub.elsevier.com/retrieve/pii/S0045653524011287

70 Gupta J, Kamandar P, Bhowmik DD, et al. Harnessing nanovaccines for effective immunization—A special concern on COVID-19: Facts, fidelity, and future prospective. *ACS Biomaterials Science Engineering [Internet]* 2024;10(1):271–297. Available from: https://pubs.acs.org/doi/10.1021/acsbiomaterials.3c01247

71 Parit S, Manchare A, Mundhe P, Hatvate N, Rojekar S, and Patravale V. Antibody-drug conjugates: A promising breakthrough in cancer therapy. *International Journal of Pharmaceutics [Internet]* 2024;659:124211. Available from: https://linkinghub.elsevier.com/retrieve/pii/S0378517324004459

72 Hatvate NT, Khuspe PR, Mandhare TA, Kashid P, and Vora LK. Decoding of nanoneutraceuticals and their delivery. In: Mohanta YK, editor. *Sustainable Green Nanotechnology [Internet]*, (pp. 284–309). Florida, New York: CRC Press; 2024. Available from: https://www.taylorfrancis.com/books/9781003389408/chapters/10.1201/9781003389408-16

73 Uddin MJ, Faiyazuddin M, Omri A, Gowri S, and Khalid M. Advances in artificial intelligence for drug delivery and development: A comprehensive review. *Computers in Biology and Medicine [Internet]* 2024;178:108702. Available from: https://linkinghub.elsevier.com/retrieve/pii/S001048252400787X

74 Khuspe PR, Pardeshi SR, Uddin MJ, et al. Achieving optimal health with host-directed therapies (HDTs) in infectious diseases—A new horizon. *Advances Therapy [Internet]* 2024; 2400169. Available from: https://onlinelibrary.wiley.com/doi/10.1002/adtp.202400169

75 Rojekar S, Togre N, Bhoj P, et al. Current status of mannose receptor-targeted drug delivery for improved anti-HIV therapy. *Journal of Controlled Release [Internet]* 2024;372:494–521. Available from: https://linkinghub.elsevier.com/retrieve/pii/S0168365924003444

76 Shaikh MS, Faiyazuddin M, Khan MS, et al. Chikungunya virus vaccine: A decade of progress solving epidemiological dilemma, emerging concepts, and immunological interventions. *Frontiers Microbiology [Internet]* 2024; 15. Available from: https://www.frontiersin.org/articles/10.3389/fmicb.2024.1413250/full

77 Rojekar S, Parit S, Manchare A, et al. Revolutionizing eye care: Exploring the potential of microneedle drug delivery. *Pharmaceutics [Internet]* 2024;16(11):1398. Available from: https://www.mdpi.com/1999-4923/16/11/1398

78 Gholap AD, Pardeshi SR, Khuspe PR, et al. Manipulating the rhizosphere microbiome for plant health. In: Nimmy S, Salam A Ibrahim, Nasr MHA (Eds.). *Microbiome Engineering*, 1st ed. (p. 19). Florida: CRC Press; 2024.

79 Mesut B, Al-Mohaya M, Yeşilkaya E, et al. Demystifying the potential of lipid-based nanocarriers in targeting brain malignancies. *Naunyn-Schmiedeberg's Archives of Pharmacology [Internet]* 2024;397(12):9243–9279. Available from: https://link.springer.com/10.1007/s00210-024-03212-6

80 Pardeshi S, Pillai D, Giram P, Pardeshi C, and Naik J. Nanocarriers for ocular drug delivery. In: *Nanocarrier Drug Delivery Systems [Internet]*, (pp. 469–500). Germany: De Gruyter; 2024. Available from: https://www.degruyter.com/document/doi/10.1515/9783111320847-015/html

81 Patil M, Dhumal D, Rojekar S, Jain H, Pingale P, and Kumar Prajapati M. Nanocarriers for antimicrobial therapy. In: *Nanocarrier Drug Delivery Systems [Internet]*, (pp. 561–602). Germany: De Gruyter; 2024. Available from: https://www.degruyter.com/document/doi/10.1515/9783111320847-018/html

82 Sawant SD, Pardeshi SR, Khuspe PR, et al. The role of inulin in altering the animal microbiome: A comprehensive review. *Microbiome Engineering* 2024;1:79–102.

83 Sawant SD, Pardeshi SR, Khuspe PR, et al. Omics tools for determining microbial composition: A systematic review. *Microbiome Engineering* 2024;1:35–50.

84 Kumar D, Kumar S, Kumari R, et al. A Comprehensive review on the biomass-mediated synthesis of silver nanoparticles: Opportunities and challenges. *Journal of Environmental Chemical Engineering [Internet]* 2024; 115133. Available from: https://linkinghub.elsevier.com/retrieve/pii/S2213343724032652

85 Faiyazuddin M, Maqsood S, Abbas S, Chaudhari YS, Sharmila S, and Webster TJ. Prospects and ethics of ChatGPT in biomedical research and medical practice: An emphasis on nanotechnology. *Journal Biomedical Nanotechnology [Internet]* 2024;20(11):1661–1678. Available from: https://www.ingentaconnect.com/content/10.1166/jbn.2024.3925

86 Pardeshi S, Kapare H, More A, Rebello N, and Giram P. Coating approaches of functionalized magnetic nanoparticles for theranostic applications. In: *Functionalized Magnetic Nanoparticles for Theranostic Applications [Internet]*, (pp. 131–171). Wiley; 2024. Available from: https://onlinelibrary.wiley.com/doi/10.1002/9781394172917.ch5

87 Pardeshi SR, Hatvate NT, Gharat KD, Naik JB, and Omri A. Advances in dorzolamide hydrochloride delivery: Harnessing nanotechnology for enhanced ocular drug delivery in glaucoma management. *Discover Nano [Internet]* 2024;19(1):199. Available from: https://link.springer.com/10.1186/s11671-024-04154-x

88 Vardikar A, Das U, Mandal S, et al. Surface engineered multimodal magnetic nanoparticles for neurodegenerative diseases. In: *Targeted Therapy for the Central Nervous System [Internet]*, (pp. 121–53). USA; Elsevier; 2025. Available from: https://linkinghub.elsevier.com/retrieve/pii/B9780443238413000078

89 Hatvate NT, Bhowmik D, Sawant SD, Londhe KM, Rojekar S, and Faiyazuddin M. Engineered artificial MicroRNA in plant: An overview. In: *MicroRNA Advances and Application in Plant Biology [Internet]*, (pp. 311–341). Elsevier; 2025. Available from: https://linkinghub.elsevier.com/retrieve/pii/B9780443216824000130

90 Uddin MJ, Faiyazuddin M, Saeed SM, Sadiq SB, and Koly IAC. Lipid-based drug delivery for systemic effect through pulmonary route of administration. In: *Lipids in Pulmonary Drug Delivery [Internet]*, (pp. 255–299). Elsevier; 2025. Available from: https://linkinghub.elsevier.com/retrieve/pii/B9780443223747000128

91 Khan AR, Grewal NS, Jun Z, Khalid M, and Zhang H-J. Advances in smart hybrid scaffolds: A strategic approach for regenerative clinical applications. *Engineered Regeneration [Internet]* 2025;6:85–110. Available from: https://linkinghub.elsevier.com/retrieve/pii/S2666138125000039

92 Khuspe PR, Faiyazuddin M, Uddin MJ, et al. Current scenario and challenges of lipid-based nanomedicines for pulmonary drug delivery. In: *Lipids in Pulmonary Drug Delivery [Internet]*, (pp. 21–47). Elsevier; 2025. Available from: https://linkinghub.elsevier.com/retrieve/pii/B9780443223747000025

93 Khuspe PR, Faiyazuddin M, Uddin MJ, et al. Lipid-based vaccines against viruses including COVID-19. In: *Lipids in Pulmonary Drug Delivery [Internet]*, (pp. 409–433). Elsevier; 2025. Available from: https://linkinghub.elsevier.com/retrieve/pii/B9780443223747000190

94 Khuat TT, Bassett R, Otte E, Grevis-James A, and Gabrys B. Applications of machine learning in antibody discovery, process development, manufacturing and formulation: Current trends, challenges, and opportunities. *Computers and Chemical Engineering [Internet]* 2024;182:108585. Available from: https://linkinghub.elsevier.com/retrieve/pii/S0098135424000036

95 Ahluwalia K, Abernathy MJ, Beierle J, et al. The future of CMC regulatory submissions: Streamlining activities using structured content and data management. *Journal of Pharmaceutical Sciences [Internet]* 2022;111(5):1232–1244. Available from: https://linkinghub.elsevier.com/retrieve/pii/S0022354921005323

96 Le Sueur H, Bruce IN, and Geifman N. The challenges in data integration – heterogeneity and complexity in clinical trials and patient registries of Systemic Lupus Erythematosus. *BMC Medical Research Methodology [Internet]* 2020;20(1):164. Available from: https://bmcmedresmethodol.biomedcentral.com/articles/10.1186/s12874-020-01057-0

97 Bhattamisra SK, Banerjee P, Gupta P, Mayuren J, Patra S, and Candasamy M. Artificial intelligence in pharmaceutical and healthcare research. *Big Data and Cognitive Computing [Internet]* 2023;7(1):10. Available from: https://www.mdpi.com/2504-2289/7/1/10

98 McGeveran W and Schmitz C. General-purpose privacy regulation and translational genomics. *Journal of Law, Medicine & Ethics [Internet]* 2020;48(1):142–150. Available from: https://www.cambridge.org/core/product/identifier/S1073110500023810/type/journal_article

99 Williamson SM and Prybutok V. Balancing privacy and progress: A review of privacy challenges, systemic oversight, and patient perceptions in AI-driven healthcare. *Applied Sciences [Internet]* 2024;14(2):675. Available from: https://www.mdpi.com/2076-3417/14/2/675

100 Mirakhori F and Niazi SK. Harnessing the AI/ML in drug and biological products discovery and development: The regulatory perspective. *Preprints [Internet]* 2024. Available from: https://www.preprints.org/manuscript/202410.2510/v1

101 Fu C and Chen Q. The future of pharmaceuticals: Artificial intelligence in drug discovery and development. *Journal of Pharmaceutical Analysis [Internet]* 2025; 101248. Available from: https://linkinghub.elsevier.com/retrieve/pii/S2095177925000656

102 Sinha S and Lee YM. Challenges with developing and deploying AI models and applications in industrial systems. *Discover Artificial Intelligence [Internet]* 2024;4(1):55. Available from: https://link.springer.com/10.1007/s44163-024-00151-2

103 Papachristou K, Katsakiori PF, Papadimitroulas P, Strigari L, and Kagadis GC. Digital twins' advancements and applications in healthcare, towards precision medicine. *Journal of Personalized Medicine [Internet]* 2024;14(11):1101. Available from: https://www.mdpi.com/2075-4426/14/11/1101

104 Haleem A, Javaid M, Pratap Singh R, and Suman R. Exploring the revolution in healthcare systems through the applications of digital twin technology. *Biomedical Technology [Internet]* 2023;4:28–38. Available from: https://linkinghub.elsevier.com/retrieve/pii/S2949723X23000065

105 Javed AR, Hassan MA, Shahzad F, et al. Integration of blockchain technology and federated learning in vehicular (IoT) networks: A comprehensive survey. *Sensors [Internet]* 2022;22(12):4394. Available from: https://www.mdpi.com/1424-8220/22/12/4394

106 Nguyen A, Lamouri S, Pellerin R, Tamayo S, and Lekens B. Data analytics in pharmaceutical supply chains: State of the art, opportunities, and challenges. *International Journal of Production Research [Internet]* 2022;60(22):6888–6907. Available from: https://www.tandfonline.com/doi/full/10.1080/00207543.2021.1950937

107 Alharthi H. Healthcare predictive analytics: An overview with a focus on Saudi Arabia. *Journal of Infection and Public Health [Internet]* 2018;11(6):749–756. Available from: https://linkinghub.elsevier.com/retrieve/pii/S1876034118300303

108 Brandt T, Wagner S, and Neumann D. Prescriptive analytics in public-sector decision-making: A framework and insights from charging infrastructure planning. *European Journal of Operational Research [Internet]* 2021;291(1):379–393. Available from: https://linkinghub.elsevier.com/retrieve/pii/S0377221720308389

109 Wissuchek C and Zschech P. Prescriptive analytics systems revised: A systematic literature review from an information systems perspective. *Information Systems and E-Business Management [Internet]* 2024; 23: 279–353. Available from: https://link.springer.com/10.1007/s10257-024-00688-w

110 Goetz LH and Schork NJ. Personalized medicine: motivation, challenges, and progress. *Fertility and Sterility [Internet]* 2018;109(6):952–963. Available from: https://linkinghub.elsevier.com/retrieve/pii/S0015028218304072

111 Su J, Yang L, Sun Z, and Zhan X. Personalized drug therapy: Innovative concept guided with proteoformics. *Molecular & Cellular Proteomics [Internet]* 2024;23(3):100737. Available from: https://linkinghub.elsevier.com/retrieve/pii/S1535947624000276

112 Sadee W, Wang D, Hartmann K, and Toland AE. Pharmacogenomics: Driving personalized medicine. *Pharmacological Reviews [Internet]* 2023;75(4):789–814. Available from: https://linkinghub.elsevier.com/retrieve/pii/S0031699724007798

113 Taherdoost H and Ghofrani A. AI's role in revolutionizing personalized medicine by reshaping pharmacogenomics and drug therapy. *Intelligent Pharmacy [Internet]* 2024;2(5):643–650. Available from: https://linkinghub.elsevier.com/retrieve/pii/S2949866X2400087X

114 Reddy S. Global harmonization of artificial intelligence-enabled software as a medical device regulation: Addressing challenges and unifying standards. *Mayo Clinic Proceedings: Digital Health [Internet]* 2025;3(1):100191. Available from: https://linkinghub.elsevier.com/retrieve/pii/S294976122400124X

115 Isoko K, Cordiner JL, Kis Z, and Moghadam PZ. Bioprocessing 4.0: A pragmatic review and future perspectives. *Digital Discovery [Internet]* 2024;3(9):1662–1681. Available from: https://xlink.rsc.org/?DOI=D4DD00127C

116 Rajora A and Chhabra G. Quality by design approach: Regulatory need, current, and future perspective. *Asian Journal of Pharmaceutical and Clinical Research [Internet]* 2021;14(6):29–35. Available from: https://innovareacademics.in/journals/index.php/ajpcr/article/view/33733

Part III

AI for Performance Evaluation and Regulatory Compliance

15

A Comprehensive Survey of Advanced Control Chart Applications in Data-driven Pharmaceutical Manufacturing

Raiha Mumtaz[1], Tahir Mahmood[2], and Naeem Ramzan[2]*

[1] *National Center for Bioinformatics, Quaid-i-Azam University, Islamabad, Pakistan*
[2] *School of Computing, Engineering and Physical Sciences, University of the West of Scotland, Paisley, UK*

*Corresponding author: tahir.mahmood@uws.ac.uk

CONTENTS

15.1 Introduction

Ensuring high-quality outcomes and operational efficiency in pharmaceutical manufacturing and healthcare delivery has become increasingly dependent on the integration of data-driven, evidence-based methodologies [1]. In recent years, the emphasis on value-based care, regulatory compliance, and patient safety has fueled the adoption of statistical tools for process monitoring and quality assurance. Among these, statistical process control (SPC) techniques have emerged as a foundational framework for improving the consistency and reliability of processes in both manufacturing and clinical settings [2]. Control charts, a core component of SPC, have been widely adopted due to their capacity to visualize process behavior, detect variation in real time, and support early intervention [3]. As one of the original seven quality tools, control charts offer healthcare professionals and pharmaceutical engineers a practical method to continuously monitor key performance metrics, enabling timely corrective actions and promoting stability in outcomes [4]. Their implementation supports not only compliance with stringent regulatory requirements but also strengthens systems for improving patient safety and clinical effectiveness.

The broader discipline of statistical process monitoring (SPM) plays a vital role in transforming raw data into actionable insights [5]. It enhances process understanding, supports decision-making, and optimizes operations across pharmaceutical manufacturing, hospital workflows, laboratory diagnostics, and even emerging fields like precision oncology. Existing literature demonstrates the utility of SPM in a wide array of applications from reducing medication errors and improving discharge procedures to monitoring vaccine uptake and laboratory turnaround times (TATs). Despite this widespread use, comprehensive reviews that capture the full scope of control chart applications across traditional and nontraditional healthcare environments, including complementary medicine and veterinary care, remain limited.

This chapter addresses the gap by presenting an integrated review of 195 peer-reviewed studies that apply control charts across diverse pharmaceutical and healthcare contexts. The analysis is organized around four key dimensions: the type of data (discrete or continuous), the statistical parameter being monitored (location or variability), the structural design of the chart (memory-less or memory-type), and the dimensionality of the monitored quality characteristics (univariate or multivariate).

In synthesizing this literature, the chapter demonstrates the following key contributions:

- It presents the most comprehensive and structured overview of control chart applications within healthcare and pharmaceutical systems.
- It offers a classification framework that helps differentiate chart usage based on data type, statistical focus, structural design, and dimensional complexity.
- It identifies a strong reliance on univariate, memory-less control charts (such as P, $X - MR$, and $\overline{X}$ charts), and highlights the underutilization of memory-type (e.g., CUSUM, EWMA) and multivariate charts.
- It uncovers a critical gap in the application of regression-based control charts and machine learning-enhanced monitoring tools, despite their potential for adaptive, real-time quality control.
- It showcases real-world applications of control charts across a broad spectrum of healthcare functions, including traditional Chinese medicine (TCM), diagnostics, emergency care, pharmacy operations, and veterinary health underscoring their versatility.
- It lays the groundwork for future research directions by advocating for the integration of advanced analytics into next-generation pharmaceutical quality systems.
- It aligns with industry goals for real-time process control and compliance, supporting the transition to smarter, more agile quality management practices.

By connecting these findings to practical use cases, this chapter underscores the transformative potential of control charts in advancing quality assurance within the framework of data-driven healthcare. The insights presented here support the broader goals of efficiency, innovation, and safety in drug development and delivery, reinforcing the relevance of SPC in today's rapidly evolving pharmaceutical landscape.

15.2 Study Selection and Eligibility Criteria

The PRISMA 2020 statement, which is structured around 27-item checklist, integrates methodological advancements in study identification, selection, appraisal, and synthesis [6]. In this study, the emphasis lies on the exploring the application of control chart in pharmaceutical manufacturing and clinical pharmacy. A comprehensive literature search was performed using two reliable search engines, including Scopus and Web of Science. Searches were limited to articles published in English and spanned an undefined publication period. The search was conducted on 8 January 2025. The initial step in the search strategy was incorporating keywords related to SPM and pharmaceutical manufacturing, which are presented in Figure 15.1.

The query presented in Figure 15.1 is comprised of "Control Charts," "Pharmaceutical Manufacturing," "Medicine," and "Pharmacy." The keywords are utilized to conduct comprehensive search across title, abstract, keywords section, and the full text of the sources. A total of 488 articles were extracted in the search and year wise count is provided in Figure 15.2. It is found that the number of publications on control chart and pharmaceuticals notably increased in the last five years and suggest sustained research activity in this field. The possible reasons can be advancements in technology and the emerging importance of pharmaceutical and healthcare studies.

To ensure transparency, the entire period from 1984 to 2024 is considered in this study. Consequently, a screening process is conducted to select the most relevant studies from the preliminary repository. Hence, the initial screening process was performed by including articles, conference papers, and journals focusing on the application of SPM techniques in healthcare settings and pharmaceutical manufacturing. The studies were excluded based on criteria such as (i) duplicated versions, (ii) absence of control charts application, (iii) unrelated studies based on this study's theme, (iv) editorials, book chapters and reviews, (v) non-English articles, and (vi) studies with inaccessible full texts. The PRISMA flow diagram used for the study is given in Figure 15.3. From 488 articles, 134 duplicated

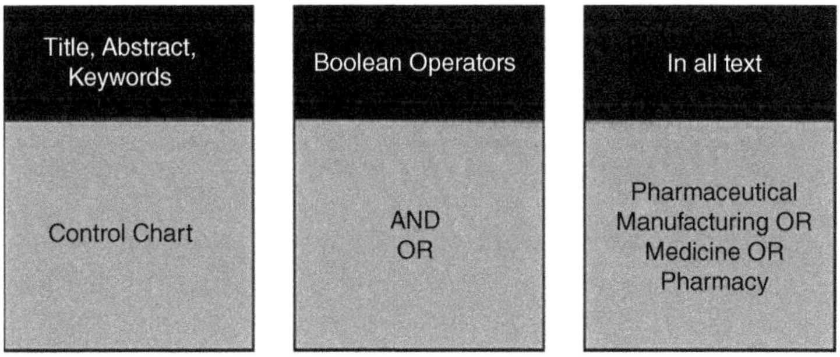

Figure 15.1 Search string used for literature search.

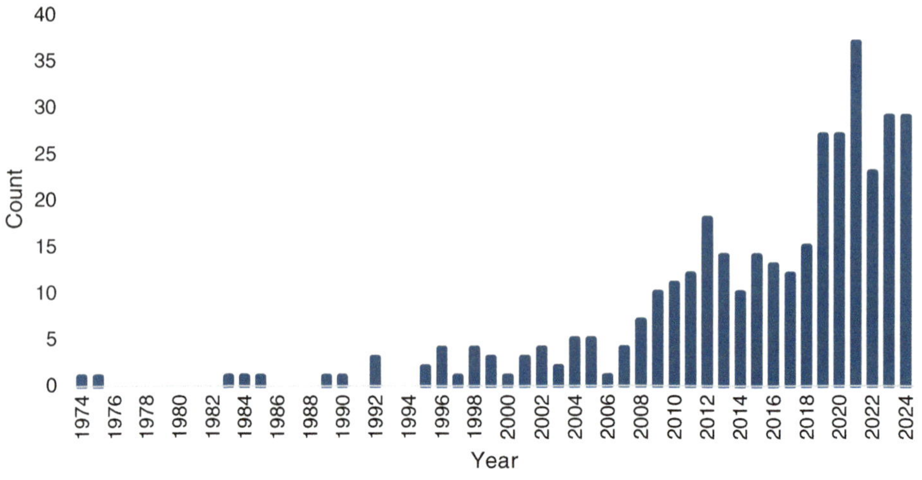

Figure 15.2 Published papers in the field of pharmaceutical manufacturing and healthcare.

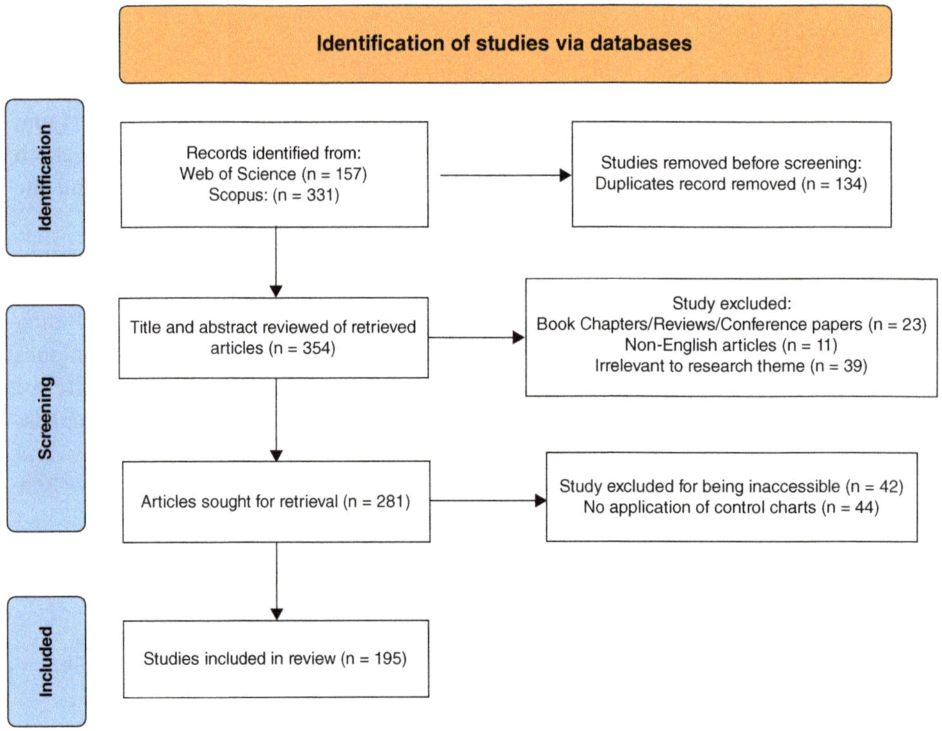

Figure 15.3 PRISMA flow diagram.

articles were removed, and the remaining were subjected to first screening. In the next step, 73 studies were rejected on the basis of our initial exclusion criteria. Furthermore, articles were screened on the basis of their accessibility and application of control chart. As a result, 159 articles were rejected apart from the duplicates and only 195 articles fully filled our inclusion criteria and subjected to content analysis.

15.3 Statistical Process Control

The concept of SPC has become a cornerstone of quality assurance across various sectors, including healthcare and pharmaceutical production. Control charts are essential tools within SPC, a methodology introduced by Shewhart [7] in the early 20th century to improve quality and efficiency in industrial processes. These charts provide a visual representation of process variation over time, enabling organizations to monitor performance, maintain stability, and detect deviations that may impact quality. Widely used across sectors such as healthcare, manufacturing, and pharmaceuticals, control charts support timely decision-making by identifying trends and unusual patterns [8–10]. In tightly regulated industries, they play a crucial role in ensuring product consistency, compliance with standards, and protection of end-user outcomes. The choice of control chart depends on the type and distribution of data, making their strategic application increasingly important as data-driven practices become central to operational and clinical decision-making. The literature on control charts is broadly categorized across several dimensions, each reflecting a different aspect of process monitoring and statistical application, particularly relevant in tightly regulated sectors such as pharmaceuticals. The four major divisions are discussed in the following sections.

15.3.1 Classification of Control Charts by Data Type

First, control charts are often classified based on the type of data they handle: attribute control charts are used for monitoring discrete data such as defect counts or proportions; common types include the *P* chart, *np* chart, *c* chart, and *u* chart, which track proportions of defectives, number of defectives, number of defects, and defects per unit, respectively [11, 12]. Variable control charts handle continuous measurement data [13] and include the $\overline{X} - R$ chart, $\overline{X} - S$ chart, individuals-moving range chart, and moving average charts commonly used to monitor parameters like weight, concentration, or pH in pharmaceutical processes.

Based on the inclusion and exclusion criteria outlined in Section 16.2, a total of 195 research articles were selected for detailed analysis. To better understand trends in control chart usage, these articles were categorized according to the type of data each chart was designed to handle, and the resulting frequency distribution is illustrated in Figure 15.4. The analysis reveals a relatively balanced application between variable and attribute control charts, with only a marginal difference of 6% observed between the two, suggesting that both data types are nearly equally represented in current literature. Notably, a subset of studies (14%) employed both chart types within the same investigation, highlighting the growing recognition of hybrid approaches in process monitoring, especially in complex or multifaceted environments such as pharmaceutical

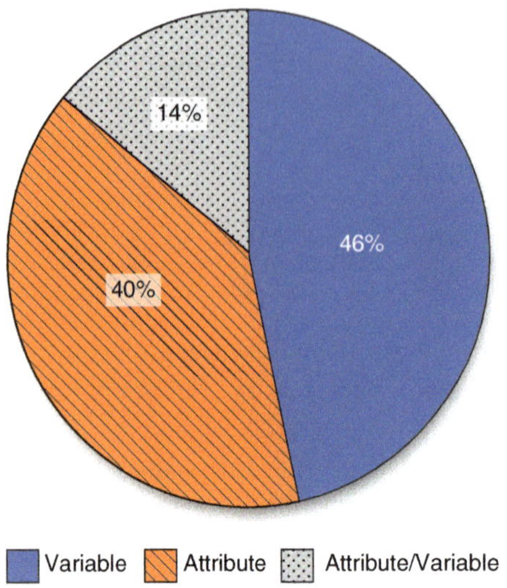

Figure 15.4 Percentage of papers regarding classification of control charts.

manufacturing, where both discrete quality indicators (e.g., defect counts) and continuous variables (e.g., ingredient concentrations) are critical for ensuring product quality and compliance.

15.3.2 Classification of Control Charts by Monitored Statistical Parameters

Second, control charts can be categorized based on the statistical parameter they are designed to monitor, either the central tendency or the variability of a process. Location charts [14], such as the $\overline{X}$ chart, focus on tracking the process mean over time, making them ideal for monitoring average outcomes like the dissolution rate of tablets. In contrast, dispersion charts, including R and s charts [15], are used to assess the spread or variability in data, such as fluctuations in tablet weight or solution viscosity. In pharmaceutical and healthcare environments, ensuring both the stability of the process average and the control of variability is fundamental for maintaining high standards of product quality and patient care. A consistent process mean supports uniform dosing, predictable therapeutic effects, and timely laboratory diagnostics. However, focusing solely on the average is not sufficient; uncontrolled variation in attributes such as tablet weight [9], solution viscosity, or assay values can introduce inconsistencies that affect drug performance, lead to regulatory challenges, and potentially endanger patient safety. Therefore, effective quality control demands simultaneous attention to both central tendency and dispersion to achieve reliability, compliance, and optimal clinical outcomes. Figure 15.5 illustrates the distribution of studies based on the parameter monitored. A majority of the reviewed articles emphasized the detection of shifts in central tendency, while only 2% addressed process variability using dispersion-focused charts. This highlights a prevailing preference in the literature for mean-centric monitoring approaches, reflecting their importance in maintaining process control and regulatory compliance in pharmaceutical production and clinical applications.

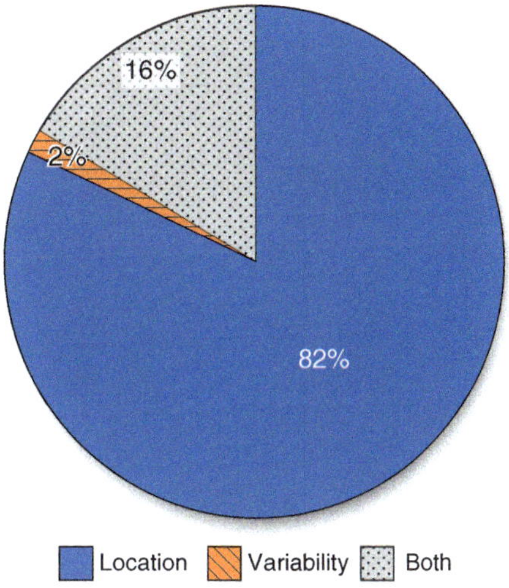

Figure 15.5 Percentage of papers related to parameter monitoring.

15.3.3 Classification of Control Charts by Structural Design

A third way to classify control charts is based on their structural design, specifically how they incorporate data over time. Memory-less charts, such as the traditional Shewhart charts (e.g., $\overline{X}$, $X - MR$, and P charts), rely solely on current data points without considering previous observations. This makes them well-suited for detecting substantial and abrupt process shifts, such as sudden changes in filling volumes during pharmaceutical production and average TAT for blood test results [16]. In contrast, memory-type charts, including Cumulative Sum (CUSUM) [17] and exponentially weighted moving average (EWMA) charts [18], accumulate historical data to enhance sensitivity to small and gradual changes, particularly valuable for monitoring parameters like pH levels, microbial contamination, or other subtle variations in sterility and quality [19]. Figure 15.6 illustrates the distribution of the reviewed literature based on the statistical structure of the control charts employed. A significant 86% of the studies utilized memory-less charts, indicating their widespread acceptance and practical simplicity, especially in routine monitoring scenarios within pharmaceutical and healthcare settings. Only 8% of the articles implemented memory-type control charts, reflecting their more specialized use where early detection of minor process deviations is critical. This trend suggests that while memory-less charts remain the standard in medical sciences due to ease of implementation and interpretation, the role of memory-type charts is growing in areas requiring enhanced sensitivity and proactive quality control.

15.3.4 Control Charts Based on Dimensionality of Quality Characteristics

A fourth classification of control charts is based on the dimensionality of the quality characteristics being monitored specifically, whether the focus is on a single variable or multiple interrelated variables. Univariate control charts are designed to track one quality attribute at a time [20], making them especially suitable for straightforward processes where only a single parameter requires monitoring. In pharmaceutical and

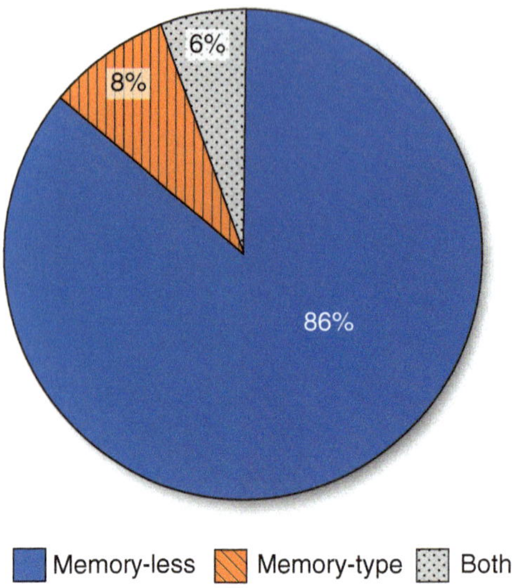

Figure 15.6 Percentage of papers based on type of statistics.

healthcare settings, univariate charts such as $\overline{X}$, $X - MR$, and P charts are extensively applied to monitor individual quality metrics like tablet hardness, dosage volume, or defect rates. These charts are particularly valuable in pharmaceutical production lines, hospital performance improvement programs, and clinical laboratories where maintaining consistency in a specific parameter is vital for product efficacy, patient safety, and regulatory compliance. On the other hand, multivariate control charts are employed in more complex environments where multiple process variables interact and jointly influence quality outcomes. These charts simultaneously evaluate the behavior of several correlated variables, providing a comprehensive overview of the process. Common examples include Hotelling's T^2 chart [21], multivariate exponentially weighted moving average (MEWMA) [22], and multivariate cumulative sum (MCUSUM) charts [23]. These tools are particularly useful in contexts such as cleanroom monitoring, where factors like temperature, humidity, and particle count must be jointly assessed to maintain strict environmental control. Multivariate charts are more sensitive to subtle shifts that may go undetected when monitoring variables individually, thereby enabling earlier intervention and improved process robustness.

Figure 15.7 presents the distribution of the reviewed literature based on the type of quality characteristics assessed. The analysis reveals a strong dominance of univariate control charts, which appeared in 93% of the selected studies, reflecting their widespread adoption and ease of implementation. In contrast, multivariate control charts were featured in only 6% of the papers, suggesting that while their potential for enhanced process insight is recognized, their complexity and data requirements may limit their routine use. Nonetheless, as healthcare and pharmaceutical systems grow increasingly data-driven and integrated, the role of multivariate control charts is expected to expand, offering more nuanced and proactive quality control solutions.

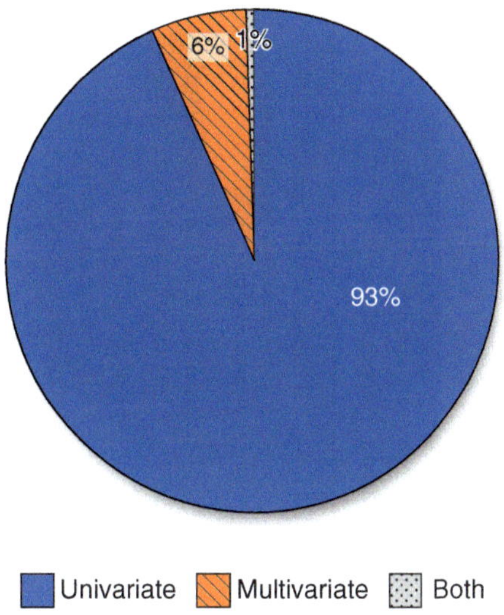

Figure 15.7 Percentage of papers corresponding to quality characteristics.

15.4 Application Domains of Control Charts in Healthcare and Pharmaceutical Sciences

To provide a broader perspective on the practical use of control charts within healthcare and pharmaceutical sciences, the final set of 195 selected articles was further organized based on their specific application domains. This domain-level classification builds upon the earlier technical categorization of control charts, which included distinctions by data type, monitored statistical parameters, structural design, and the dimensionality of quality characteristics. While those previous sections focused on the methodological and statistical features of control charts such as whether they monitor discrete or continuous data, track mean or variability, use memory-based structures, or assess single vs. multiple variables. This section shifts focus to where and how these charts are applied in real-world settings. The reviewed studies span a wide range of sectors, including pharmaceutical manufacturing, pharmacy and inventory management, clinical care and quality improvement, laboratory and diagnostic testing, medical equipment and imaging quality assurance, veterinary applications, and public health monitoring. Exploring these domains helps illustrate how control charts are tailored to meet the unique quality and process control demands of each field, supporting regulatory compliance, operational efficiency, and improved patient and product outcomes.

15.4.1 Pharmaceutical Manufacturing Process Monitoring

Control charts are vital for maintaining quality and compliance in pharmaceutical manufacturing. Table 15.1 shows that most studies use variable control charts, with a strong focus on Hotelling's T^2 and DModX for multivariate process control, particularly in TCM production. Univariate charts like $X - MR$, $\overline{X} - S$, and Shewhart are also common for

Table 15.1 Control chart application in pharmaceutical manufacturing.

Ref	Chart name	Control chart by data type	Monitored statistical parameter		Structural design		Dimensionality of quality characteristics		Application
			Location	Variability	Memory-less	Memory-type	Univariate	Multivariate	
[8]	Hotelling's T^2 and DModX	Variable	*		*			*	Traditional Chinese medicine (TCM) manufacturing process monitoring
[9]	$X - MR$	Variable	*	*	*		*		Evaluation of the manufacturing process of antiretroviral tablets
[19]	CUSUM	Variable	*			*	*		Energy performance of compressed air systems of pharmaceutical production plant
[55]	Hotelling's T^2 and DModX	Variable	*		*			*	TCM manufacturing process
[56]	Hotelling's T^2 and DModX	Variable	*	*	*			*	TCM manufacturing process monitoring (liquid–liquid extraction)
[57]	Hotelling's T^2 and SPE	Variable	*	*	*			*	Quality control of TCM concentrating process
[58]	Chi-Squared, Hotelling's T^2	Variable	*	*	*			*	Metabolite profiling in herbal medicine injections
[59]	Hotelling's T^2 and DModX	Variable	*		*			*	Quality consistency of botanical drug products (TCM)
[60]	Hotelling's T^2 and DModX	Variable	*	*	*			*	Quality evaluation of Chinese herbal injections (TCM)

[61]	$X - MR$	Variable	*	*	*		*		Quality control of Chinese medicine
[62]	EWMA	Variable		*		*	*		Monitoring the quality of bacterial colonies in medicine product
[63]	$X - MR$	Variable	*	*	*		*		Monitoring of pharmaceutical product
[64]	$X - MR$	Variable	*	*	*		*		Monitoring of water treatment plants in pharmaceutical manufacturing
[65]	$X - R$	Variable	*	*	*		*		Monitoring the TCM extraction process
[66]	Shewhart	Variable	*		*		*		Inline monitoring and process control of twin-screw granulator
[67]	$\overline{X} - S$, Hotelling's T^2	Variable	*	*	*		*	*	Improving the performance of tableting process
[68]	Hotelling's T^2 and DModX	Variable	*	*	*			*	Batch monitoring in herbal medicine (TCM)

Note: An asterisk (*) represents the respective concept was explicitly identified in the corresponding article.

monitoring single parameters such as tablet weight. Some studies combine both chart types to enhance monitoring effectiveness. Overall, there is a clear preference for memory-less charts, while memory-type charts like EWMA and CUSUM are used more selectively for detecting small process shifts.

15.4.2 Pharmacy and Inventory Process Monitoring

Control charts play a key role in pharmacy operations by supporting quality assurance, improving workflow, and managing inventory. As shown in Table 15.2, attribute charts like P charts are widely used to monitor categorical outcomes such as dispensing errors and system performance, while variable charts like EWMA and $X - MR$ are used for tracking continuous data, including inventory levels and pharmaceutical care. Most studies rely on memory-less charts for their simplicity, though memory-type charts such as CUSUM and run sum are occasionally used for detecting gradual trends in more complex scenarios.

15.4.3 Public Health and Other Monitoring Applications

Control charts are increasingly applied beyond traditional clinical settings, supporting public health surveillance and other non-conventional healthcare domains such as aerospace medicine and sports injury monitoring. As shown in Table 15.3, studies have used these tools to track epidemiological trends, system performance, and medication-related events. Memory-type charts like EWMA, EWMA(G), and CUSUM(G) are especially useful in detecting gradual shifts over time, such as changes in blood pressure, glucose levels, or patterns of drug misuse [24]. While most studies rely on univariate, memory-less charts, the growing complexity of public health data highlights the need for more multivariate and variability-focused methods to better capture interrelated trends and support early intervention strategies.

15.4.4 Medical Equipment, Radiotherapy, and Diagnostic Imaging Quality Assurance

Control charts are widely used to ensure the accuracy and reliability of medical devices, radiotherapy systems, and diagnostic imaging. Table 15.4 shows that $X - MR$ charts are commonly applied for monitoring patient-specific parameters in treatments like intensity-modulated radiotherapy (IMRT) and volumetric modulated arc therapy (VMAT) [25, 26], while X, $\overline{X}$, and $\overline{X} - S$ charts support routine equipment checks [27]. Memory-type charts such as CUSUM and EWMA offer greater sensitivity to subtle shifts, especially in advanced applications like precision oncology [10]. Some studies integrate both chart types to track trends and variability together [28, 29]. Though still limited, multivariate charts like Hotelling's T^2 are being explored for complex treatment evaluations [30], reflecting the growing role of control charts in improving clinical quality and safety.

Table 15.2 Studies related to statistical process monitoring in pharmacy.

Ref	Chart Name	Control chart by data type	Monitored statistical parameter		Structural design		Dimensionality of quality characteristics		Application
			Location	Variability	Memory-less	Memory-type	Univariate	Multivariate	
[69]	EWMA	Variable	*			*	*		Evaluation of medicine inventory management approaches
[70]	*P*	Attribute	*		*		*		Monitoring pharmacy automated expert system performance
[71]	*P*, *run*, CUSUM	Attribute/ Variable	*		*	*	*		Quality assurance in pharmacy
[72]	*P*	Attribute	*		*		*		Monitoring prescribing and dispensing practices
[73]	*P*	Attribute	*		*		*		Reducing inpatient pharmacy dispensing error
[74]	*P*	Attribute	*		*		*		Monitoring the quality of pharmacotherapeutic process
[75]	*P*	Attribute	*		*		*		Reducing inappropriate payer formulary alerts
[76]	$X - MR$	Variable	*		*		*		Monitoring pharmaceutical care

Note: An asterisk (*) represents the respective concept was explicitly identified in the corresponding article.

Table 15.3 Control chart application in public health.

Ref	Chart name	Control chart by data type	Monitored statistical parameter		Structural design		Dimensionality of quality characteristics		Application
			Location	Variability	Memory-less	Memory-type	Univariate	Multivariate	
[24]	P, EWMA(G), CUSUM(G)	Attribute	*		*	*	*		Monitoring prescription drug abuse
[77]	Shewhart, MA, EWMA	Variable	*		*	*	*		Monitoring system for public health surveillance data
[78]	X	Variable	*		*		*		Seasonal injury counts in netball
[79]	c	Attribute	*		*		*		Hospital-based violence intervention programs (HVIPs)
[80]	X	Variable	*		*		*		Medication approval in aerospace environment
[81]	$\overline{X}$	Variable	*		*		*		Evaluating healthcare simulation research
[82]	EWMA	Variable	*			*	*		Monitoring epidemiological time-series and vital signs data (BP, glucose, temperature)
[83]	P, $\overline{X}$, CUSUM	Attribute/ Variable	*		*	*	*		Electronic physiological surveillance system on hospital mortality
[84]	u	Attribute	*		*		*		Injury incidence rates in cricket

Note: An asterisk (*) represents the respective concept was explicitly identified in the corresponding article.

Table 15.4 Control chart application in radiation therapy and nuclear medicine.

Ref	Chart name	Control chart by data type	Monitored statistical parameter		Structural design		Dimensionality of quality characteristics		Application
			Location	Variability	Memory-less	Memory-type	Univariate	Multivariate	
[10]	$X - MR$	Variable	*	*	*		*		Monitoring of ionization chamber
[25]	$X - MR$	Variable	*	*	*		*		Multileaf collimator performance monitoring system for radiotherapy
[26]	$X - MR$	Variable	*		*		*		Patient-specific quality assurance in cancer treatment (IMRT/VMAT)
[25]	$X - MR$	Variable	*	*	*		*		Multileaf collimator performance monitoring system for radiotherapy
[27]	$\overline{X}$	Variable	*		*		*		Evaluation of gamma spectrometers in nuclear medicine
[28]	$\overline{X}$ – S, EWMA	Variable	*	*	*	*	*		Quality assurance in radiation therapy
[29]	$\overline{X}$ – MR, EWMA	Variable	*		*	*	*		Quality assurance in radiation therapy method
[30]	Hotelling'sT^2	Variable	*		*			*	Evaluate and monitor radiotherapy plan quality
[85]	X, EWMA	Variable	*		*	*	*		Quality assurance in radiology
[86]	X	Variable	*		*		*		Patient specific quality assurance in proton therapy
[87]	CUSUM	Variable	*			*	*		Application in targeted radiotherapy (prostate)

(Continued)

Table 15.4 (continue)

Ref	Chart name	Control chart by data type	Monitored statistical parameter		Structural design		Dimensionality of quality characteristics		Application
			Location	Variability	Memory-less	Memory-type	Univariate	Multivariate	
[88]	$X - MR$, EWMA	Variable	*	*	*	*	*		Monitoring dose delivery process in radiotherapy
[89]	$X - MR$	Variable	*	*	*		*		Monitoring radiation therapy
[90]	$\overline{X} - R$	Variable	*	*	*		*		Quality assurance in radiation therapy
[91]	$X - MR$	Variable	*		*		*		Quality improvement in radiopharmaceutical maladministration
[92]	CUSUM, EWMA	Variable	*			*	*		Monitoring in precision oncology
[93]	$\overline{X} - \mathrm{S}$	Variable	*		*		*		Improve the timeliness of chemotherapy administration

Note: An asterisk (*) represents the respective concept was explicitly identified in the corresponding article.

15.5 Clinical and Healthcare Process Monitoring and Quality Improvement

Apart from manufacturing, SPC is widely utilized in enhancing clinical care and hospital settings to reduce errors and improve patient outcomes. These papers collectively address operational efficiency, patient safety, effective communication, clinical outcomes, and care delivery across hospitals, emergency departments, clinics, and other patient-care environments.

15.5.1 Hospital and Emergency Operations

In hospital and emergency care settings, control charts are increasingly employed to support quality improvement initiatives, enhance clinical performance, and streamline operational efficiency. As shown in Table 15.5, studies have utilized a diverse set of SPC tools such as X, u, P, $X - MR$, and $\overline{X} - S$ charts to monitor various performance indicators across departments. For instance, X and u charts have been applied to evaluate general hospital quality improvement programs [31, 32], while P and $X - MR$ charts were used to assess key emergency department metrics, such as door-to-physician time and patient flow [33]. Additionally, $\overline{X} - S$ charts have been used to evaluate surgeon performance in operating rooms [34], reflecting the growing role of SPC in tracking clinical outcomes and procedural efficiency. These applications demonstrate how control charts can facilitate data-driven decision-making in high-pressure environments, ultimately supporting safer and more effective hospital care delivery.

15.5.2 Clinical Workflow, Communication, and Discharge

Table 15.6 presents studies using control charts to enhance hospital communication, streamline discharge processes, and improve clinical workflow. A variety of charts such as P, $\overline{X}$, u, and $\overline{X} - S$ are applied to monitor key performance indicators across care settings. Both attribute and variable charts support performance improvement by tracking factors like discharge efficiency, handoff communication, and patient flow. Notably, P charts and $\overline{X} - S$ charts are frequently used to improve discharge protocols and communication quality in pediatric and adult care [35–38].

15.5.3 Medication, Vaccination, and Laboratory Test Utilization

Table 15.7 presents studies that apply control charts to improve medication safety, vaccination uptake, and laboratory testing efficiency. Most initiatives use attribute control charts, particularly P charts, to monitor categorical outcomes such as medication errors, vaccination rates, and unnecessary lab tests [39–41]. Variable charts like $X - MR$ are also employed for continuous data in pharmaceutical care, primarily for monitoring mean values. Notably, only a few studies, such as Morton et al. [42], assess variability using both P and $\overline{X} - S$ charts, indicating limited focus on dispersion parameters. The absence of multivariate

Table 15.5 Control chart application in hospital and emergency operations.

			Monitored statistical parameter		Structural design		Dimensionality of quality characteristics		
Ref	Chart name	Control chart by data type	Location	Variability	Memory-less	Memory-type	Univariate	Multivariate	Application
[31]	X	Variable	*		*		*		Quality improvement program in hospital
[32]	u	Attribute	*		*		*		Quality improvement program in hospital
[33]	$P, X - MR$	Attribute/ Variable	*		*		*		Emergency department operation (door-to-physician time, left-without-being-seen)
[34]	$X, \overline{X} - S$	Variable	*	*	*		*		Monitoring operating room surgeon's performance
[94]	run	Variable	*			*	*		Quality improvement program in hospital
[95]	$X - MR$	Variable	*		*		*		Monitoring operating room setup time
[96]	Shewhart	Variable	*		*		*		Evaluation of videolaryngoscopy (VL) in emergency practice medicine
[97]	$X - MR$	Variable	*	*	*		*		Resident-centered quality improvement in general medicine
[98]	$X - MR$	Variable	*		*		*		Competency-based medical education (CBME) simulation program (pediatrics)
[99]	CUSUM	Variable		*		*	*		Real-time cardiac anomaly detection
[100]	P	Attribute	*		*		*		Proportion of patients with febrile neutropenia receiving an antibiotic in an emergency department
[101]	P	Attribute	*		*		*		Increasing isotonic maintenance intravenous fluids use
[102]	run	Variable	*		*		*		Incident of ventilator-associated pneumonia

Note: An asterisk (*) represents the respective concept was explicitly identified in the corresponding article.

Table 15.6 Application of control charts in hospital quality improvement areas.

Ref	Chart name	Control chart by cata type	Monitored statistical parameter		Structural design		Dimensionality of quality characteristics		Application
			Location	Variability	Memory-less	Memory-type	Univariate	Multivariate	
[16]	$\overline{X}$	Variable	*		*		*		Improving preanalytical phlebotomy services (blood sampling)
[35]	P	Attribute	*		*		*		Improving 2-way discharge communication in hospitals
[36]	$P, \overline{X} - S$	Attribute/ Variable	*	*	*		*		Improving translated discharge instructions (children)
[37]	$P, \overline{X} - S$	Attribute/ Variable	*		*		*		Quality improvement of discharge criteria
[38]	P	Attribute	*		*		*		Improving readable discharge instructions
[103]	u, P	Attribute	*		*		*		Quality improvement in hospital communication
[104]	$P, \overline{X} - S$	Attribute/ Variable	*	*	*		*		Quality improvement methodology to improve discharge efficiency
[105]	Shewhart	Variable	*		*		*		Effect of triage nurse-initiated radiographs
[106]	$\overline{X}$	Variable	*		*		*		Decreasing outpatient infusion clinic wait time
[107]	Mean Absolute Deviation	Variable		*	*		*		Partner notification of chlamydia in genitourinary medicine clinics
[108]	Shewhart chart	Variable	*		*		*		Genitourinary medicine clinics performance
[109]	$\overline{X}$	Variable	*		*		*		Quality improvement protocol in asthmaticus patients

(Continued)

Table 15.6 (continue)

Ref	Chart name	Control chart by cata type	Monitored statistical parameter		Structural design		Dimensionality of quality characteristics		Application
			Location	Variability	Memory-less	Memory-type	Univariate	Multivariate	
[110]	P	Attribute	*		*		*		Improving communication efficiency of primary care providers
[111]	$\overline{X}$	Variable	*		*		*		Effectiveness of telecommunication in improving patient outcomes (children)
[112]	$X - MR$	Variable	*		*		*		Quality Improvement approach to decreasing patient dwell times for cardiac nuclear medicine exam
[113]	P, u	Attribute	*		*		*		Evaluating discharge criteria (pediatrics)
[114]	P	Attribute	*		*		*		Increasing hospital medicine team workflow efficiency (pediatrics)
[115]	P	Attribute	*		*		*		Improving communication during inter-hospital transfer
[116]	P	Attribute	*		*		*		Emergency department handover tool
[117]	$X - MR$	Variable	*	*	*		*		Monitoring hospital length of stay
[118]	P	Attribute	*		*		*		Improving resident verbal handoffs in children hospital
[119]	P	Attribute	*		*		*		Optimizing pediatric discharge efficiency
[120]	P, X	Attribute/ Variable	*		*		*		Quality improvement in clinical genetics services
[121]	P	Attribute	*		*		*		Quality improvement during family centered rounds (pediatrics)

Note: An asterisk (*) represents the respective concept was explicitly identified in the corresponding article.

Table 15.7 Control chart applications in medication safety, vaccination, and laboratory test.

Ref	Chart name	Control chart by data type	Monitored statistical parameter		Structural design		Dimensionality of quality characteristics		Application
			Location	Variability	Memory-less	Memory-type	Univariate	Multivariate	
[39]	P	Attribute	*		*		*		Reducing repetitive routine laboratory testing (COVID-19)
[40]	P, np	Attribute	*		*		*		Quality improvement to reduce medication errors
[41]	P,T,C	Attribute	*		*		*		Reducing antibiotic usage in neonates
[42]	$P, \overline{X} - S$	Attribute/Variable	*	*	*		*		Quality improvement project in reducing benzodiazepine exposure during sedation (NICU)
[122]	P	Attribute	*		*		*		Quality improvement to reduce medication errors (urinary tract infections)
[123]	P	Attribute	*		*		*		Increase influenza and pneumococcal vaccination in chemotherapy patients
[124]	P	Attribute	*		*		*		Increase prescription of narrow spectrum antibiotics (pediatrics)
[125]	P	Attribute	*		*		*		Improving Influenza vaccination in asthma patients (children)

(Continued)

Table 15.7 (continue)

Ref	Chart name	Control chart by data type	Monitored statistical parameter		Structural design		Dimensionality of quality characteristics		Application
			Location	Variability	Memory-less	Memory-type	Univariate	Multivariate	
[126]	$P, X - MR$	Attribute/Variable	*		*		*		Improving medication-in hands process after discharge (pediatrics)
[127]	P	Attribute	*		*		*		Reducing inappropriate prescription of proton pump inhibitors (PPIs)
[128]	$X - MR$	Variable	*		*		*		Reducing inappropriate medicine (Acetaminophen) usage in academic medical center
[129]	P	Attribute	*		*		*		Quality improvement to reduce medication errors
[130]	P	Attribute	*		*		*		Quality improvement to increase vaccination rates
[131]	P, X	Attribute/Variable	*		*		*		Application of SPC charting in stroke medicine
[132]	P	Attribute	*		*		*		Quality improvement of Hepatitis B Vaccine administration
[133]	u, X	Attribute/Variable	*		*		*		Quality improvement to reduce unnecessary ICU phlebotomy

[134]	P	Attribute	*	*		*	Practice-based intervention for thiazides-based treatment of hypertension
[135]	EWMA	Variable	*		*	*	Predictive biomarker for pancreatic cancer therapy
[136]	P	Attribute	*	*		*	Increasing meds-in-hands with discharged patients
[137]	X	Variable	*	*		*	Improve timely administration of antibiotics
[138]	$P, G, X\overline{X} - S$	Attribute/Variable	*	*		*	Increasing enterally administered antibiotics (pediatrics)
[139]	P	Attribute	*	*		*	Increasing meds-in-hands with discharged patients (asthma)
[140]	P	Attribute	*	*		*	Increasing meds-in-hands with discharged patients (pediatrics)
[141]	P	Attribute	*	*		*	Monitoring medication error

Note: An asterisk (*) represents the respective concept was explicitly identified in the corresponding article.

charts suggests a predominant reliance on univariate analysis, with little exploration of interrelated process variables, despite the complexity of medication and testing workflows.

15.5.4 Surgical, Perioperative, and Specialty Care

Table 15.8 highlights the use of control charts in surgical, perioperative, and specialty care to monitor clinical processes and improve patient outcomes. While advanced methods like CUSUM, EWMA, and Hotelling's $T^2 - SPE$ are applied in some studies for complex surgical monitoring [43, 44], most research relies on univariate, memory-less, and attribute-based charts such as P, u, and $X - MR$. These are effective for routine tracking but may miss subtle shifts or interactions between variables, indicating a need for broader adoption of multivariate and memory-type approaches in surgical quality improvement.

15.5.5 Pediatrics, Geriatrics, and Specific Clinical Programs

Table 15.9 shows that control charts are widely used in pediatric and geriatric care to support quality improvement. P charts are common in pediatrics, while $\overline{X}$, $X - MR$, and $\overline{X} - S$ charts monitor process trends in neonatal and geriatric settings [45, 46]. Most studies rely on univariate, memory-less charts, with limited use of multivariate or memory-type methods, which may restrict deeper analysis of complex quality trends.

15.5.6 Additional Examples (Among Many Others)

Table 15.10 showcases diverse applications of control charts across various medical specialties. Most studies utilize P and $\overline{X}$ charts to monitor mean-centered, univariate quality indicators. Notably, Minne et al. [47, 48] incorporated advanced models like EWMA and zone charts for mortality monitoring, reflecting a shift toward more data-driven decision-making in clinical and hospital settings. While traditional charts remain prevalent, these examples highlight emerging interest in more sophisticated SPC methods for healthcare improvement.

15.6 Laboratory and Diagnostic Process Monitoring

Table 15.11 highlights the use of control charts to improve the accuracy and efficiency of laboratory and diagnostic processes, which are critical for informed medical decision-making. Studies primarily employ P, u, run, Shewhart, and $X/\overline{X}$ charts to monitor sample labeling, defects per million opportunities, and test TATs [49–52]. These applications rely heavily on univariate, memory-less, and central tendency-based approaches, with no consideration of variability, memory-type, or multivariate analyses. As a result, while effective for basic error tracking, these methods may overlook the more complex dynamics involved in laboratory quality control.

Table 15.8 Application of control charts in surgical, perioperative, and specialty care.

Ref	Chart name	Control chart by data type	Monitored statistical parameter		Structural design		Dimensionality of quality characteristics		Application
			Location	Variability	Memory-less	Memory-type	Univariate	Multivariate	
[43]	CUSUM, MA, EWMA	Variable	*			*	*		Monitoring the inclination of acetabular after hip replacement surgery
[44]	Hotelling's T^2 and SPE	Variable	*	*	*			*	Monitoring hemodynamic trajectory maps after cardiac surgery (pediatrics)
[142]	$G, X-MR$	Attribute/Variable	*		*		*		Decreasing unnecessary hospitalization (febrile infants—number of low risks and length of stay)
[143]	$\overline{X}-S$	Variable	*	*	*		*		Quality improvement in Pediatric emergency department length of stay (LOS) for sedated fracture reductions
[144]	Shewhart	Variable	*		*		*		Quality improvement to ventriculostomy placement
[145]	$X-MR$	Variable	*		*		*		Quality improvement to reduce door to Incision time for pediatric testicular torsion
[146]	$G, X-MR$	Attribute/Variable	*		*		*		Cataract surgery process improvement
[147]	P, np, c, u	Attribute	*		*		*		Evaluation of surgical operations

(Continued)

Table 15.8 (continue)

Ref	Chart name	Control chart by data type	Monitored statistical parameter		Structural design		Dimensionality of quality characteristics		Application
			Location	Variability	Memory-less	Memory-type	Univariate	Multivariate	
[148]	*P*	Attribute	*		*		*		Evidence-based surgical site infection reduction
[149]	CUSUM	Variable	*			*	*		Quality assessment in thoracic surgery
[150]	*G*	Attribute	*		*		*		Monitoring ventilator-associated pneumonia (VAP) events
[151]	EWMA	Variable	*			*	*		Quality assurance activities in hospital (wound infection, neonatal arterial switch surgery death rate)
[152]	Shewhart	Variable	*		*		*		Quality improvement to reduce in-hospital time for inflammatory bowel disease patients
[153]	Shewhart, EWMA	Variable	*		*	*	*		Process improvement in total joint replacement
[154]	*P*	Attribute	*		*		*		Improve proportion of postoperative extended venous thromboembolism prophylaxis (eVTEp)
[155]	$P, \overline{X} - S$	Attribute/Variable	*		*		*		Reducing pre-procedural fasting time

[156]	P	Attribute	*		*	*	Quality improvement of osteoarticular infections in children
[157]	P,run	Attribute/Variable	*		*	*	Decreasing mechanical ventilation (MV) duration
[158]	u,P	Attribute	*		*	*	Exploring stability of incidents, events, and complications documentation of perioperative anesthesia
[159]	P	Attribute	*		*	*	Implementation of a shared baseline protocol to reduce inhaled nitric oxide overuse (neonates)
[160]	P	Attribute	*		*	*	Quality improvement initiative to reduce surgical site infection
[161]	$P,X-MR$	Attribute/Variable	*		*	*	Evaluation of geriatric care models
[162]	$P,\overline{X}-\mathrm{S}$	Attribute/Variable	*	*	*	*	Quality Improvement effort to standardizing opioid prescribing
[163]	P	Attribute	*		*	*	Improve evidence-based inflammatory bowel disease care (pediatrics)
[164]	P	Attribute	*		*	*	Enhancing airway clearance therapy
[165]	P,T	Attribute/Variable	*		*	*	Monitoring of continuous renal replacement therapy filter life

Note: An asterisk (*) represents the respective concept was explicitly identified in the corresponding article.

Table 15.9 Control charts for quality improvement in pediatrics, geriatrics, and medical care.

Ref	Chart name	Control chart by data type	Monitored statistical parameter		Structural design		Dimensionality of quality characteristics		Application
			Location	Variability	Memory-less	Memory-type	Univariate	Multivariate	
[45]	$\overline{X} - S$	Variable	*	*	*		*		Quality improvement in NICU feeding guideline (pediatrics/neonates)
[46]	$u, X - MR$	Attribute/ Variable	*	*	*		*		Quality improvement in geriatric medicine service in rehabilitation
[166]	$\overline{X}$	Variable	*		*		*		Mobility assessment in cardiology clinic
[167]	$P, \overline{X}$	Attribute/ Variable	*		*		*		Decreasing ICU and hospital LOS of status asthmaticus patients (pediatrics)
[168]	$P, \overline{X} - S$	Attribute/ Variable	*	*	*		*		Improving pediatric resident admission workflow in emergency department
[169]	G, P	Attribute	*		*		*		Performance evaluation rapid sequence intubation (pediatrics)
[170]	P	Attribute	*		*		*		To improve counseling of overweight and obese children
[171]	P	Attribute	*		*		*		Point-of-care ultrasound impact on radiology ultrasound volumes in pediatric emergency department
[172]	$X - MR$	Variable	*	*	*		*		Efficiency of care in pediatric emergency department
[173]	P	Attribute	*		*		*		Quality improvement to reduce overnight vital sign monitoring (pediatrics)
[174]	P	Attribute	*		*		*		Improving penicillin allergy documentation (pediatrics)

[175]	$P, \overline{X}, T$	Attribute/ Variable	*		*	*	Patient-centered Quality improvement (pediatrics)
[176]	P	Attribute	*		*	*	Optimizing resource utilization (pediatrics cardiac)
[177]	P, u	Attribute	*		*	*	Quality improvement to reduce respiratory viral testing (pediatrics)
[178]	P	Attribute	*		*	*	Quality improvement methodology for implementation of brace treatment (pediatrics)
[179]	P	Attribute	*		*	*	Identifying patients with inappropriate penicillin allergies (pediatrics)
[180]	P	Attribute	*		*	*	Decreasing inaccurate penicillin allergy patients (pediatrics)
[181]	u	Attribute	*		*	*	Increase physician reporting of suspected diagnostic errors (pediatric)
[182]	u	Attribute	*		*	*	Reducing unplanned extubating events (UPEs) in neonates
[183]	P	Attribute	*		*	*	Increasing identification and treatment of influenza (children)
[184]	P	Attribute	*		*	*	Febrile infant management
[185]	$P, G, \overline{X}$	Attribute/ Variable	*		*	*	Pediatrics clinical practice guideline
[186]	$\overline{X} - S$	Variable	*	*	*	*	High-flow nasal cannula application in bronchiolitis

Note: An asterisk (*) represents the respective concept was explicitly identified in the corresponding article.

Table 15.10 Control chart application in other hospital settings.

Ref	Chart name	Control chart by data type	Monitored statistical parameter		Structural design		Dimensionality of quality characteristics		Application
			Location	Variability	Memory-less	Memory-type	Univariate	Multivariate	
[47]	EWMA	Variable	*			*	*		Monitoring of mortality data
[48]	Zone	Variable	*		*		*		Exploring a mortality model's prospective performance
[187]	X	Variable	*		*		*		Baseline performance measurement of clinical practice benchmarks
[188]	P	Attribute	*		*		*		Quality improvement of handoff tool in hospital
[189]	P	Attribute	*		*		*		Quality improvement of antimicrobial adherence
[190]	P	Attribute	*		*		*		Quality improvement to improve BP control in clinic
[191]	P	Attribute	*		*		*		Quality improvement to increase breast cancer screening rates
[192]	P	Attribute	*		*		*		Quality improvement in learning incorporation in clinical work
[193]	$P,\overline{X}$	Attribute/ Variable	*		*		*		Evaluation of patient satisfaction measures
[194]	P,T	Attribute/ Variable	*		*		*		Quality improvement initiative to reduce physical restraints in hospital (children)
[195]	X	Variable	*		*		*		Improving physician compliance with breast cancer screening

[196]	P	Attribute	*	*		*	Increasing situation awareness for high-risk therapies (children)
[197]	P	Attribute	*	*		*	Reducing inappropriate usage of oxygen supplementation in hospital
[198]	EWMA	Variable	*		*	*	Management of in hospital cardiac arrest
[199]	P	Attribute	*	*		*	Quality initiative project to increase delivery of reproduction health care (adolescent boys)
[200]	P	Attribute	*	*		*	Increasing the one-unit red cell transfusion orders in oncology ward
[201]	P	Attribute	*	*		*	Quality improvement to increase evidence-based care guideline (pediatrics)
[202]	$P, \overline{X}$	Attribute/ Variable	*	*		*	Monitoring of tele-clinics for kidney transplant center
[203]	X	Variable	*	*		*	Improving combat casualty care
[204]	P	Attribute	*	*		*	Improving diabetes management

Note: An asterisk (*) represents the respective concept was explicitly identified in the corresponding article.

Table 15.11 Control chart applications in laboratory settings.

			Monitored statistical parameter		Structural design		Dimensionality of quality characteristics		
Ref	Chart name	Control chart by data type	Location	Variability	Memory-less	Memory-type	Univariate	Multivariate	Application
[49]	P	Attribute	*		*		*		Monitoring the blood sampling and labeling process
[50]	u	Attribute	*		*		*		Monitoring of defects per million opportunities in laboratory
[51]	X	Variable	*		*		*		Quality monitoring of laboratory
[52]	$\overline{X}$	Variable	*		*		*		Reducing inappropriate laboratory testing
[205]	P	Attribute	*		*		*		Quality improvement of TAT in laboratory
[206]	*run*	Variable	*		*		*		Quality improvement to reduce mislabeled laboratory specimens
[207]	Shewhart	Variable	*		*		*		Improving laboratory test utilization
[208]	P	Attribute	*		*		*		Monitoring of drug quantification in human plasma
[209]	Shewhart	Variable	*		*		*		Quality control at a medical laboratory (Bilirubin analysis data)

Note: An asterisk (*) represents the respective concept was explicitly identified in the corresponding article.

15.7 Animal and Veterinary Applications

Table 15.12 presents the application of control charts in animal health monitoring, disease surveillance, and food safety within veterinary settings. Most studies focus on variable data, employing tools like Shewhart, EWMA, and $\overline{X} - R$ charts to monitor animal health and food quality. Notably, Faverjon et al. [53] demonstrate the use of advanced multivariate techniques MEWMA and MCUSUM for syndromic surveillance, enabling early detection of disease outbreaks. In contrast, Morris et al. [54] apply a c chart for monitoring drug residue in aquaculture, the only study using attribute data. While these approaches reflect growing sophistication in veterinary epidemiology, they remain limited in addressing process variability and broader system interactions.

15.8 Summary and Future Recommendations

This chapter explores the evolving role of control charts in pharmaceutical manufacturing and healthcare, emphasizing their application in quality assurance, process monitoring, and clinical decision support. Through a systematic literature review of 195 studies, the chapter categorizes control chart usage based on data type, statistical parameters monitored, structural design, and dimensionality of quality characteristics. Applications span pharmaceutical production, hospital operations, diagnostics, public health, veterinary care, and more. The findings reveal a dominant reliance on univariate and memory-less charts, such as P, $X - MR$, and $\overline{X}$ charts, with limited use of multivariate or memory-type approaches like CUSUM and EWMA. While traditional SPC methods remain prevalent, there is growing interest in hybrid models that combine univariate and multivariate features. However, the adoption of regression-based and machine learning-enhanced control charts is notably absent in the reviewed literature, highlighting an important gap and setting the stage for future innovation in data-driven pharmaceutical quality monitoring.

Given the limitations identified in the current literature, future research in control chart methodologies for pharmaceutical manufacturing and healthcare systems should prioritize the following scientifically grounded directions:

- Advancement of regression-based control charts: Future studies should explore the integration of both linear and nonlinear regression techniques into control chart frameworks to model covariate-dependent quality metrics [214]. These methods could enhance the detection of deviations by accounting for underlying relationships among explanatory variables, thus improving sensitivity and specificity in quality monitoring.
- Application of machine learning and deep learning models: Incorporating supervised and unsupervised learning algorithms into control chart systems can enable the identification of complex, nonobvious patterns and multivariate anomalies [215]. Techniques such as recurrent neural networks (RNNs) or convolutional neural networks (CNNs) could be tailored to monitor temporal dynamics and spatial quality indicators in real time.
- Real-time data analytics integration: Combining advanced control chart methodologies with real-time analytics platforms [216] (e.g., cloud-based or edge-computing

Table 15.12 Application of control charts in quality monitoring of animal health.

Ref	Chart name	Control chart by data type	Monitored statistical parameter		Structural design		Dimensionality of quality characteristics		Application
			Location	Variability	Memory-less	Memory-type	Univariate	Multivariate	
[53]	MEWMA, MCUSUM	Variable	*			*		*	Multivariate syndromic surveillance (SyS) in animal health
[54]	*c*	Attribute	*		*		*		Drug residue in aqua animal farming
[210]	Shewhart, EWMA	Variable	*		*	*	*		Disease monitoring in animals (pig)
[211]	$\overline{X} - R$	Variable	*	*	*		*		Animal food monitoring
[212]	Shewhart, EWMA	Variable	*		*	*	*		Animal health surveillance
[213]	Shewhart-CUSUM	Variable	*			*	*		Monitoring of ELISA performance in veterinary laboratory

Note: An asterisk (*) represents the respective concept was explicitly identified in the corresponding article.

environments) will facilitate prompt anomaly detection and immediate corrective action, thereby enhancing the responsiveness of pharmaceutical quality assurance systems.

- Expansion of multivariate monitoring approaches: There is a pressing need to adopt multivariate control charts such as Hotelling's T^2, MEWMA, and MCUSUM to concurrently monitor multiple interrelated quality variables [217]. This will support a more comprehensive understanding of process dynamics and improve decision-making in complex manufacturing and clinical workflows.
- Development of adaptive and self-learning control systems: Future control charts should incorporate adaptive algorithms capable of dynamically adjusting control limits based on evolving process behavior and incoming data [218]. This would allow continuous learning from operational feedback and more robust handling of nonstationary environments.
- Strengthening of infrastructure and regulatory alignment: Research should focus on designing scalable computational infrastructures, robust data governance models, and interoperable architectures that align with regulatory standards (e.g., FDA's CFR Part 11 [219], GAMP 5 [220]). This will ensure smooth integration of advanced monitoring tools within regulated healthcare and pharmaceutical environments.
- Integration with IoT and sensor networks: The deployment of Internet of Things (IoT) devices and sensor technologies provides continuous, high-frequency data streams [221]. Future research should explore how to effectively leverage these data sources for predictive maintenance, process optimization, and real-time quality control using intelligent control chart systems.

References

1 Saini JPS, Thakur A, and Yadav D. AI driven innovations in pharmaceuticals: Optimizing drug discovery and industry operations. *Royal Society of Chemistry Pharmaceutics* 2025. https://doi.org/10.1039/D4PM00323C

2 Ahmed F, Mahmood T, Riaz M, and Abbas N. Comprehensive review of high-dimensional monitoring methods: trends, insights, and interconnections. *Quality Technology & Quantitative Management* 2024. https://doi.org/10.1080/16843703.2024.2395745

3 Rakhmawan SA, Mahmood T, and Abbas N. Deep learning-based mortality surveillance: Implications for healthcare policy and practice. *Journal of Population Research* 2025;42(1):7.

4 Mahmood T. Generalized linear modelling based monitoring methods for air quality surveillance. *Journal of King Saud University-Science* 2024;36(4):103145.

5 Mahmood T and Erem A. A bivariate exponentially weighted moving average control chart based on exceedance statistics. *Computers & Industrial Engineering* 2023;175:108910.

6 Page MJ, McKenzie JE, Bossuyt PM, et al. The PRISMA 2020 statement: An updated guideline for reporting systematic reviews. *British Medical Journal* 2021;372:n71.

7 Shewhart WA. Quality control charts. *The Bell System Technical Journal* 1926;5(4):593–603.

8 Zeng S, Chen T, Wang L, and Qu H. Monitoring batch-to-batch reproducibility using direct analysis in real time mass spectrometry and multivariate analysis: A case study on precipitation. *Journal of Pharmaceutical and Biomedical Analysis* 2013;76:87–95.

9 Rocha NPd, Silva OCd, Barbosa EJ, et al. Statistical process control of manufacturing tablets for antiretroviral therapy. *Brazilian Journal of Pharmaceutical Sciences* 2023;59:e22099.

10 Trindade Filho OL, Conceição DAd, da Silva CJ, et al. A study to assess the long-term stability of the ionization chamber reference system in the LNMRI. *Journal of Physics: Conference Series: IOP Publishing* 2018;975:012059.

11 Mahmood T and Xie M. Models and monitoring of zero-inflated processes: The past and current trends. *Quality and Reliability Engineering International* 2019;35(8):2540–2557.

12 Xie M, Goh TN, and Mahmood T. *Statistical Models for Monitoring the High-Quality Processes.* Springer handbook of engineering statistics, (pp. 261–274). Springer; 2023.

13 Riaz M, Alshammari H, Abbas N, and Mahmood T. Navigating process drift: The power of CUSUM in monitoring air quality processes and maintenance operations. *Arabian Journal for Science and Engineering* 2024: 1–22. https://doi.org/10.1007/s13369-024-09453-0

14 Hyder M, Raza SMM, Mahmood T, Moeen M, and Riaz M. On moving average based location charts under modified successive sampling. *Hacettepe Journal of Mathematics and Statistics* 2024;53(2):506–523.

15 Hyder M, Raza SMM, Mahmood T, and Abbas N. Enhanced dispersion monitoring structures based on modified successive sampling: Application to fertilizer production process. *Symmetry* 2023;15(5):1108.

16 Wagar EA, Phipps R, Del Guidice R, et al. Inpatient preanalytic process improvements. *Archives of Pathology and Laboratory Medicine* 2013;137(12):1753–1760.

17 Iqbal A and Mahmood T. CUSUM charts utilizing reparametrized Birnbaum-Saunders model for fault detection and process control. *Measurement and Control* 2024;58(9):00202940241282533.

18 Ahmad S, Riaz M, Mahmood T, and Abbas N. Change point detection to analyze air pollution and its economic effects: an exponentially weighted moving average perspective. *Journal of Applied Statistics* 2025. https://doi.org/10.1080/02664763.2025.2455636

19 Benedetti M, Bonfà F, Introna V, Santolamazza A, and Ubertini S. Real time energy performance control for industrial compressed air systems: Methodology and applications. *Energies* 2019;12(20):3935.

20 Hussain S, Mahmood T, Riaz M, and Nazir HZ. A new approach to design median control charts for location monitoring. *Communications in Statistics-Simulation and Computation* 2023;51(7):3553–3577. https://doi.org/10.1080/03610918.2020.1716245

21 Mahmood T, Wittenberg P, Zwetsloot IM, Wang H, and Tsui KL. Monitoring data quality for telehealth systems in the presence of missing data. *International Journal of Medical Informatics* 2019;126:156–163.

22 Zwetsloot IM, Mahmood T, and Woodall WH. Multivariate time-between-events monitoring: An overview and some overlooked underlying complexities. *Quality Engineering* 2020;33(1):13–25. https://doi.org/10.1080/08982112.2020.1788717

23 Riaz M, Ajad JO, Mahmood T, and Abbasi SA. Multivariate mixed EWMA-CUSUM control chart for monitoring the process variance-covariance matrix. *IEEE Access* 2019;7:100174–100186. https://doi.org/10.1109/access.2019.2928637

24 Benneyan JC, Villapiano A, Katz N, Duffy M, Budman SH, and Butler SF. Illustration of a statistical process control approach to regional prescription opioid abuse surveillance. *Journal of Addiction Medicine* 2011;5(2):99–109.

25 Létourneau D, Wang A, Amin MN, et al. Multileaf collimator performance monitoring and improvement using semiautomated quality control testing and statistical process control. *Medical Physics* 2014;41(12):121713.

26 Xiao Q, Bai S, Li G, et al. Statistical process control and process capability analysis for non-normal volumetric modulated arc therapy patient-specific quality assurance processes. *Medical Physics* 2020;47(10):4694–4702.

27 Febriana S and Mirda I. Gamma spectrometer calibration of liquid standard sources of Mix ^{133}Ba, ^{137}Cs and ^{60}Co toward the measurement of ^{99}Mo breakthrough of ^{99}Mo/^{99}mTc generator. *Journal of Physics: Conference Series: IOP Publishing* 2023;2596:012032.

28 Palaniswaamy G, Scott Brame R, Yaddanapudi S, Rangaraj D, and Mutic S. A statistical approach to IMRT patient-specific QA. *Medical Physics* 2012;39(12):7560–7570.

29 Komori S, Takeuchi A, Kato R, et al. Long-term beam output stability of an accelerator-based boron neutron capture therapy system. *Medical Physics* 2024;51(12):9250–9263.

30 Roy A, Widjaja R, Wang M, Cutright D, Gopalakrishnan M, and Mittal BB. Treatment plan quality control using multivariate control charts. *Medical Physics* 2021;48(5):2118–2126.

31 James K, Growdon ME, Orkaby AR, and Schwartz AW. One step at a time: improving gait speed measurement in a geriatric medicine clinic. *Geriatrics* 2023;8(4):81.

32 Mendel A, Lott A, Lo L, and Wu R. A matter of urgency: Reducing clinical text message interruptions during educational sessions. *Journal of Hospital Medicine* 2018;13(9):616–622.

33 Pimentel L and Barrueto Jr F. Statistical process control: Separating signal from noise in emergency department operations. *The Journal of Emergency Medicine* 2015;48(5):628–638.

34 Chen TT, Chang YJ, Ku SL, and Chung KP. Statistical process control as a tool for controlling operating room performance: Retrospective analysis and benchmarking. *Journal of Evaluation in Clinical Practice* 2010;16(5):905–910.

35 Clark NA, Simmons J, Etzenhouser A, and Pallotto EK. Improving outpatient provider communication for high-risk discharges from the hospitalist service. *Hospital Pediatrics* 2021;11(10):1033–1048.

36 Choe AY, Schondelmeyer AC, Thomson J, et al. Improving discharge instructions for hospitalized children with limited English proficiency. *Hospital Pediatrics* 2021;11(11):1213–1222.

37 Christianson K, Kalinowski A, Bauer S, et al. Using quality improvement methodology to increase communication of discharge criteria on rounds. *Hospital Pediatrics* 2022;12(2):156–164.

38 Unaka NI, Statile A, Jerardi K, et al. Improving the readability of pediatric hospital medicine discharge instructions. *Journal of Hospital Medicine* 2017;12(7):551–557.

39 Abdul-Moheeth M, Valencia V, Schaefer S, Brode WM, Nieto K, and Moriates C. Improving healthcare value: Effectiveness of a program to reduce laboratory testing for non-critically-ill patients with COVID-19. *Journal of Hospital Medicine* 2021;16(8):495–498.

40 Carey RG and Teeters JL. CQI case study: Reducing medication errors. *The Joint Commission Journal on Quality Improvement* 1995;21(5):232–237.

41 Goel N, Cannell S, Davies G, et al. Implementation of an adapted Sepsis Risk Calculator algorithm to reduce antibiotic usage in the management of early onset neonatal sepsis: A multicentre initiative in Wales, UK. *Archives of Disease in Childhood-Fetal and Neonatal Edition* 2022;107(3):303–310.

42 Morton SU, Labrecque M, Moline M, Hansen A, and Leeman K. Reducing benzodiazepine exposure by instituting a guideline for dexmedetomidine usage in the NICU. *Pediatrics* 2021;148(5):e2020041566.

43 Tomak L, Bek Y, and Tomak Y. To compare time-weighted graphs to evaluate the inclination of the acetabular component of patients who had total hip replacement surgery. *BioMed Research International* 2015;2015(1):129610.

44 Howsmon DP, Mikulski MF, Kabra N, et al. Statistical process monitoring creates a hemodynamic trajectory map after pediatric cardiac surgery: A case study of the arterial switch operation. *Bioengineering & Translational Medicine* 2024;9(6):e10679.

45 Olowoyeye A, Basile E, Kim S, and Thompson-Branch A. A quality improvement project to reduce time to full enteral feeds for very low birth weight neonates. *Hospital Pediatrics* 2022;12(5):515–521.

46 Fitzgerald SR, Norman R, Sinha SK, and Romanovsky L. Quality improvement outcomes from the introduction of a geriatrician into a rehabilitation setting. *Journal of the American Geriatrics Society* 2021;69(9):2648–2658.

47 Minne L, Eslami S, de Keizer N, de Jonge E, de Rooij SE, and Abu-Hanna A. Statistical process control for validating a classification tree model for predicting mortality–A novel approach towards temporal validation. *Journal of Biomedical Informatics* 2012;45(1):37–44.

48 Minne L, Eslami S, De Keizer N, De Jonge E, De Rooij SE, and Abu-Hanna A. Effect of changes over time in the performance of a customized SAPS-II model on the quality of care assessment. *Intensive Care Medicine* 2012;38:40–46.

49 Dzik WS, Beckman N, Selleng K, et al. Errors in patient specimen collection: Application of statistical process control. *Transfusion* 2008;48(10):2143–2151.

50 Salinas M, Lopez-Garrigos M, Flores E, Gutiérrez M, Lugo J, and Uris J. Three years of preanalytical errors: quality specifications and improvement through implementation of statistical process control. *Scandinavian Journal of Clinical and Laboratory Investigation* 2009;69(8):822–826.

51 Zarbo RJ, Jones BA, Friedberg RC, et al. Q-Tracks: A College of American Pathologists program of continuous laboratory monitoring and longitudinal performance tracking. *Archives of Pathology & Laboratory Medicine* 2002;126(9):1036–1044.

52 Basuita M, Ethier C, and Soong C. Reducing inappropriate laboratory testing in the hospital setting: How low can we go. *Journal of Clinical Oncology Medicine* 2020;27(6): 261–264. https://doi.org/10.12788/jcom.0027

53 Faverjon C, Carmo LP, and Berezowski J. Multivariate syndromic surveillance for cattle diseases: Epidemic simulation and algorithm performance evaluation. *Preventive Veterinary Medicine* 2019;172:104778.

54 Morris DJ, Gray AJ, Kay J, and Gettinby G. EU sampling strategies for the detection of veterinary drug residues in aquaculture species: Are they working? *Drug Testing and Analysis* 2012; 4(S1):1–9.
55 Li W, Han H, Cheng Z, Zhang Y, Liu S, and Qu H. A feasibility research on the monitoring of traditional Chinese medicine production process using NIR-based multivariate process trajectories. *Sensors and Actuators B: Chemical* 2016;231:313–323.
56 Xiong H, Gong X, and Qu H. Monitoring batch-to-batch reproducibility of liquid–liquid extraction process using in-line near-infrared spectroscopy combined with multivariate analysis. *Journal of Pharmaceutical and Biomedical Analysis* 2012;70:178–187.
57 Jin Y, Du W, Liu X, and Wu Y. Application of near infrared spectroscopy and real time release testing combined with statistical process control charts for on-line quality control of industrial concentrating process of traditional Chinese medicine "Jinyinhua." *Infrared Physics & Technology* 2022;123:104135.
58 Jiang M, Jiao Y, Wang Y, et al. Quantitative profiling of polar metabolites in herbal medicine injections for multivariate statistical evaluation based on independence principal component analysis. *PloS one* 2014;9(8):e105412.
59 Xiong H, Yu LX, and Qu H. Batch-to-batch quality consistency evaluation of botanical drug products using multivariate statistical analysis of the chromatographic fingerprint. *American Association of Pharmaceutical Scientists PharmsciTech* 2013;14:802–810.
60 Huang H and Qu H. A comparative fingerprint study using high-performance liquid chromatography, ultraviolet, and near-infrared spectroscopy to evaluate the quality consistency of Danshen injections produced by different manufacturers. *Analytical Methods* 2013;5(2):474–482.
61 Zhang Z-Y, Zhao Y-J, Guo F-J, and Wang H-Y. Identification of Radix bupleuri from different geographic origins using matrix-assisted laser desorption/ionization time-of-flight mass spectrometry and support vector machine algorithm. *Journal of AOAC International* 2023;106(6):1682–1688.
62 Prabawani N, Mashuri M, and Irhamah. Performance of robust EWMA control chart for variability process using non-normal data. *Journal of Physics: Conference Series: IOP Publishing* 2020;1511(1):012054.
63 Eissa ME and Abid AM. Application of statistical process control for spotting compliance to good pharmaceutical practice. *Brazilian Journal of Pharmaceutical Sciences* 2018;54:e17499.
64 Eissa M, Seif M, and Fares M. Assessment of purified water quality in pharmaceutical facility using six sigma tools. *International Journal of Quality Assurance* 2015;6(2):54–72.
65 Lu P, Jiao M, and Hou BG. Quality control oriented data analysis and application of traditional Chinese medicine extraction process. *Applied Mechanics and Materials* 2014;548:1089–1092.
66 Sayin R, Martinez-Marcos L, Osorio JG, et al. Investigation of an 11 mm diameter twin screw granulator: Screw element performance and in-line monitoring via image analysis. *International Journal of Pharmaceutics* 2015;496(1):24–32.
67 Al-Refaie A, Diabat A, and Li M-H. Optimizing tablets' quality with multiple responses using fuzzy goal programming. *Proceedings of the Institution of Mechanical Engineers, Part E: Journal of Process Mechanical Engineering* 2014;228(2):115–126.

68 Li W, Zhao F, Pan J, and Qu H. Process monitoring and characterization for extraction of herbal medicines based on proton (1H) nuclear magnetic resonance spectroscopy and multivariate batch modeling: A case study. *Journal of Pharmaceutical Innovation* 2023;18(1):102–117.

69 Kalaya P, Termsuksawad P, and Wasusri T. Forecasting and inventory planning for irregular demand patterns: The case of community hospitals in Thailand. *International Journal of Knowledge and Systems Science (IJKSS)* 2023;14(1):1–21.

70 Doherty JA, Reichley RM, Noirot LA, et al. Monitoring pharmacy expert system performance using statistical process control methodology. *AMIA Annual Symposium Proceedings* 2003; 2003:205-209.

71 Hassall TH and Daniels CE. Evaluation of three types of control chart methods in unit dose error monitoring. *American Journal of Hospital Pharmacy* 1983;40(6):970–975.

72 Al-Hussein FA. Guideline implementation in clinical practice: Use of statistical process control charts as visual feedback devices. *Journal of Family and Community Medicine* 2009;16(1):11–17.

73 Trakulsunti Y, Antony J, Edgeman R, Cudney B, Dempsey M, and Brennan A. Reducing pharmacy medication errors using Lean Six Sigma: A Thai hospital case study. *Total Quality Management & Business Excellence* 2022;33(5-6):664–682.

74 Fernández-Megía MJ, Noguera IF, Sanjuán MM, and Andrés JLP. Monitoring the quality of the hospital pharmacoterapeutic process by sentinel patient program. *Farmacia Hospitalaria* 2018;42(2):45–52.

75 Dewan M, Wolfe H, Young C, and Desai B. Payer formulary alerts as a cause of patient harm and the journey to change them. *Hospital Pediatrics* 2016;6(9):529–535.

76 Chien T, Chan AL, and Leung HW. Using statistical process control chart techniques to ensure quality of care in pharmacy department of a hospital. In: *2011 IEEE International Conference on Quality and Reliability: IEEE*, pp. (273–277). 2011.

77 Williamson GD and Weatherby Hudson G. A monitoring system for detecting aberrations in public health surveillance reports. *Statistics in Medicine* 1999;18(23):3283–3298.

78 Belcher S, Whatman C, Brughelli M, and Borotkanics R. Ten-year nationwide review of netball ankle and knee injuries in New Zealand. *Journal of Science and Medicine in Sport* 2020;23(10):937–942.

79 Watkins J, Scoggins Ni, Cheaton BM, et al. Assessing improvements in emergency department referrals to a hospital-based violence intervention program. *Injury Epidemiology* 2021;8:1–5.

80 Prudhomme MB, Ropp LG, Sauer SW, and LaVan JT. Aeromedical risk assessment of pharmaceuticals using evidence-based medicine. *Aerospace Medicine and Human Performance* 2015;86(9):824–829.

81 Pirie J, Cardenas S, Seleem W, et al. The use of statistical process control charts to evaluate interprofessional education sessions embedded into a pediatric emergency in situ resuscitation program. *Simulation in Healthcare* 2019;14(2):121–128.

82 Grillenzoni C and Fornaciari M. On-line peak detection in medical time series with adaptive regression methods. *Econometrics and Statistics* 2019;10:134–150.

83 Schmidt PE, Meredith P, Prytherch DR, et al. Impact of introducing an electronic physiological surveillance system on hospital mortality. *BMJ Quality and Safety* 2015;24(1):10–20.

84 Goggins L, Peirce N, Ranson C, et al. Injuries in England and Wales elite men's domestic cricket: A nine season review from 2010 to 2018. *Journal of Science and Medicine in Sport* 2020;23(9):836–840.
85 Stevens NT, Steiner SH, Smith IR, and MacKay RJ. Monitoring radiation use in cardiac fluoroscopy imaging procedures. *Medical Physics* 2011;38(1):317–326.
86 Rah JE, Shin D, Manger RP, et al. Tolerance design of patient-specific range QA using the DMAIC framework in proton therapy. *Medical Physics* 2018;45(2):520–528.
87 Ung N and Wee L. Cumulative sum method in nonzero fixed action level setup correction strategy: An application of statistical process control for targeted prostate radiotherapy. *Medical Physics* 2012;39(5):2746–2753.
88 Gérard K, Grandhaye JP, Marchesi V, Kafrouni H, Husson F, and Aletti P. A comprehensive analysis of the IMRT dose delivery process using statistical process control (SPC). *Medical Physics* 2009;36(4):1275–1285.
89 Sanghangthum T, Suriyapee S, Kim G-Y, and Pawlicki T. A method of setting limits for the purpose of quality assurance. *Physics in Medicine & Biology* 2013;58(19):7025.
90 Pawlicki T, Whitaker M, and Boyer AL. Statistical process control for radiotherapy quality assurance. *Medical Physics* 2005;32(9):2777–2786.
91 Larcos G, Collins L, Georgiou A, and Westbrook J. Nuclear medicine incident reporting in Australia: control charts and notification rates inform quality improvement. *Internal Medicine Journal* 2015;45(6):609–617.
92 Jain SR, Sim W, Ng CH, et al. Statistical process control charts for monitoring next-generation sequencing and bioinformatics turnaround in precision medicine initiatives. *Frontiers in Oncology* 2021;11:736265.
93 Bhandari R, Orgel E, Rushing T, et al. Improving the timeliness of chemotherapy administration in the bone marrow transplant unit. *Biology of Blood and Marrow Transplantation* 2020;26(1):150–156.
94 McKnight ER, Ebersole A, Gallup J, and Haamid FW. Developing a sustainable quality improvement program in an academic center: The experience of an adolescent unit. *Pediatric Quality & Safety* 2024;9(3):e742.
95 Deutsch ES, Malekzadeh S, and Schmalbach CE. Patient safety/quality improvement primer, Part III: The role of simulation. *Otolaryngology–Head and Neck Surgery* 2022;166(1):23–34.
96 Steel A, Haldane C, and Cody D. Impact of videolaryngoscopy introduction into prehospital emergency medicine practice: A quality improvement project. *Emergency Medicine Journal* 2021;38(7):549–555.
97 Walker M, Gay L, Raynaldo G, et al. Impact of a resident-centered interprofessional quality improvement intervention on acute care length of stay. *The Journal for Healthcare Quality (JHQ)* 2019;41(4):212–219.
98 Pirie J, Fayyaz J, Prinicipi T, et al. Impact and effectiveness of a mandatory competency-based simulation program for pediatric emergency medicine faculty. *AEM Education and Training* 2023;7(2):e10856.
99 Lang M. A low-complexity model-free approach for real-time cardiac anomaly detection based on singular spectrum analysis and nonparametric control charts. *Technologies* 2018;6(1):26.

100 Kline JN and Payne AS. Improvement science is a partner in basic and clinical research. *Journal of Investigative Medicine* 2020;68(3):724–727.

101 Akinsola B, Cheng J, Iyer SB, and Jain S. Improving isotonic maintenance intravenous fluid use in the emergency department. *Pediatrics* 2021;148(1):e2020022947.

102 Morris AC, Hay AW, Swann DG, et al. Reducing ventilator-associated pneumonia in intensive care: Impact of implementing a care bundle. *Critical Care Medicine* 2011;39(10):2218–2224.

103 Carson J, Gottheil S, Dyck B, and Rice T. Paging the right resident the first time on general internal medicine: A quality improvement project. *The Joint Commission Journal on Quality and Patient Safety* 2019;45(10):711–716.

104 Yang CJ, Bottalico D, Philips K, et al. Improving the pediatric floor discharge process following tonsillectomy. *The Laryngoscope* 2022;132(1):225–233.

105 Harries B, Filiatrault L, and Abu-Laban RB. Application of quality improvement analytic methodology in emergency medicine research: A comparative evaluation. *Canadian Journal of Emergency Medicine* 2019;21(2):253–260.

106 Plourde CL, Varnado WT, Gleaton BJ, and Das DG. Reducing infusion clinic wait times using quality improvement. *Journal of Clinical Oncology: Oncology Practice* 2020;16(8): e807–e813.

107 Herzog SA, McClean H, Carne CA, and Low N. Variation in partner notification outcomes for chlamydia in UK genitourinary medicine clinics: Multilevel study. *Sexually Transmitted Infections* 2011;87(5):420–425.

108 Low N, Welch J, and Radcliffe K. Developing national outcome standards for the management of gonorrhoea and genital chlamydia in genitourinary medicine clinics. *Sexually Transmitted Infections* 2004;80(3):223–229.

109 Melendez E, Dwyer D, Donelly D, et al. Standardized protocol is associated with a decrease in continuous albuterol use and length of stay in critical status asthmaticus. *Pediatric Critical Care Medicine* 2020;21(5):451–460.

110 Ostermeier A, Ferro E, Voet S, et al. Improving efficiency of primary care provider communication for uncomplicated admissions. *Hospital Pediatrics* 2021;11(10):1043–1049.

111 Antony J, Parameswaran N, Kathavarayan R, Pothapregada S, and Kumar S. Improving the outcome of sick children referred from district hospitals in and around Puducherry by establishing a communication network: A community-based quality improvement initiative. *Indian Journal of Critical Care Medicine: Peer-reviewed, Official Publication of Indian Society of Critical Care Medicine* 2024;28(12):1153.

112 Sibley C, Ayers C, King B, Browning T, and Kwon JK. Decreasing patient dwell times for outpatient cardiac nuclear medicine studies: The benefits of SMART goals, scope limitations, and society guidelines in quality improvement. *Current Problems in Diagnostic Radiology* 2020;49(5):333–336.

113 Schefft M, Lee C, and Munoz J. Discharge criteria decrease variability and improve efficiency. *Hospital Pediatrics* 2020;10(4):318–324.

114 Unaka NI, Herrmann LE, Parker MW, et al. Improving efficiency of pediatric hospital medicine team daily workflow. *Hospital Pediatrics* 2019;9(11):867–873.

115 Mueller S, Murray M, Schnipper J, and Goralnick E. An initiative to improve advanced notification of inter-hospital transfers. *Healthcare* 2020;8(2):100423.

116 Kwok ES, Clapham G, White S, Austin M, and Calder LA. Development and implementation of a standardised emergency department intershift handover tool to improve physician communication. *British Medical Journal Open Quality* 2020;9(1):e000780.

117 Azadeh-Fard N, Megahed FM, and Pakdil F. Variations of length of stay: A case study using control charts in the CRISP-DM framework. *International Journal of Six Sigma and Competitive Advantage* 2019;11(2-3):204–225.

118 Warrick D, Gonzalez-del-Rey J, Hall D, et al. Improving resident handoffs for children transitioning from the intensive care unit. *Hospital Pediatrics* 2015;5(3):127–133.

119 White CM, Statile AM, White DL, et al. Using quality improvement to optimise paediatric discharge efficiency. *British Medical Journal Quality & Safety* 2014;23(5):428–436.

120 McCann E, Baines EA, Gray JR, and Procter AM. Improving service delivery by evaluation of the referral pattern and capacity in a clinical genetics setting. *American Journal of Medical Genetics Part C: Seminars in Medical Genetics* 2009; 151C(3):200–206.

121 Glick AF, Foster LZ, Goonan M, Hart LH, Alam S, and Rosenberg RE. Using quality improvement science to promote reliable communication during family-centered rounds. *Pediatrics* 2022;149(4):e2021050197.

122 Sofeso ST, Plasencia E, Safri AA, et al. Pharmacist review of discharge antibiotics for urinary tract infections in the emergency department. *Journal of the American College of Clinical Pharmacy* 2024;7(10):996–1003.

123 McGinnis JM, Jones R, Hillis C, et al. A pneumococcal pneumonia and influenza vaccination quality improvement program for women receiving chemotherapy for gynecologic cancers at a major tertiary cancer centre. *Gynecologic Oncology* 2021;161(1):236–243.

124 Sundberg M, Sexton JR, Smith A, and Toomey SL. Improving use of antibiotics for pediatric pneumonia in community hospital emergency departments. *Pediatrics* 2023;152(1):e2022059201.

125 Foradori DM, Sampayo EM, Fanny SA, Namireddy MK, Kumar AM, and Lo H-y. Improving influenza vaccination in hospitalized children with asthma. *Pediatrics* 2020;145(3).

126 Choi JH, Barrett MC, Michel N, Bouvay K, Schwartz H, and Vukovic AA. A dedicated outpatient pharmacy improves access to discharge medications in a pediatric emergency department: A quality improvement study. *Annals of Emergency Medicine* 2024;83(6):552–561.

127 Nallapeta N, Reynolds JL, and Bakhai S. Deprescribing proton pump inhibitors in an academic, primary care clinic: Quality improvement project. *Journal of Clinical Gastroenterology* 2020;54(10):864–870.

128 Nguyen LP, Nguyen L, and Austin JP. A quality improvement initiative to decrease inappropriate intravenous acetaminophen use at an academic medical center. *Hospital Pharmacy* 2020;55(4):253–260.

129 George D, Supramaniam ND, Abd Hamid SQ, Hassali MA, Lim W-Y, and Hss A-S. Effectiveness of a pharmacist-led quality improvement program to reduce medication errors during hospital discharge. *Pharmacy Practice (Granada)* 2019;17(3).

130 Moore E, Bauer SC, Rogers A, and McFadden V. An opportunity in cancer prevention: Human papillomavirus vaccine delivery in the hospital. *Hospital Pediatrics* 2022;12(5): e157–e162.

131 Henderson G, Mead G, Van Dijke M, Ramsay S, McDowall M, and Dennis M. Use of statistical process control charts in stroke medicine to determine if clinical evidence and changes in service delivery were associated with improvements in the quality of care. *British Medical Journal Quality & Safety* 2008;17(4):301–306.

132 Hayashi M, Grover TR, Small S, Staples T, and Roosevelt G. Improving timeliness of hepatitis B vaccine administration in an urban safety net level III NICU. *British Medical Journal Quality & Safety* 2021;30(11):911–919.

133 Bodley T, Levi O, Chan M, Friedrich JO, and Hicks LK. Reducing unnecessary diagnostic phlebotomy in intensive care: a prospective quality improvement intervention. *British Medical Journal Quality & Safety* 2023;32(8):485–494.

134 Ashton CM, Khan MM, Johnson ML, et al. A quasi-experimental test of an intervention to increase the use of thiazide-based treatment regimens for people with hypertension. *Implementation Science* 2007;2:1–13.

135 Zhao A, Tu D, He Y, Liu L, Wu B, and Ren Y. Identification and validation the predictive biomarkers based on risk-adjusted control chart in gemcitabine with or without erlotinib for pancreatic cancer therapy. *Frontiers in Genetics* 2024;15:1497254.

136 Hatoun J, Bair-Merritt M, Cabral H, and Moses J. Increasing medication possession at discharge for patients with asthma: The meds-in-hand project. *Pediatrics* 2016;137(3).

137 Ma C, Levin G, Panda SK, Sambalingam D, and Singh AP. Improving timing of antibiotics in neonates with early onset sepsis–Quality improvement project. *Journal of Neonatal-Perinatal Medicine* 2020;13(2):239–246.

138 Girdwood SCT, Sellas MN, Courter JD, et al. Improving the transition of intravenous to enteral antibiotics in pediatric patients with pneumonia or skin and soft tissue infections. *Journal of Hospital Medicine* 2020;15(1):9–15.

139 Parikh K, Perry K, Pantor C, and Gardner C. Multidisciplinary engagement increases medications in-hand for patients hospitalized with asthma. *Pediatrics* 2019;144(6).

140 Mallory LA, Diminick NP, Bourque JP, et al. Pediatric patient-centered transitions from hospital to home: improving the discharge medication process. *Hospital Pediatrics* 2017;7(12):723–730.

141 Campbell GM and Facchinetti NJ. Using process control charts to monitor dispensing and checking errors. *American Journal of Health-System Pharmacy* 1998;55(9):946–952.

142 Foster LZ, Beiner J, Duh-Leong C, et al. Implementation of febrile infant management guidelines reduces hospitalization. *Pediatric Quality & Safety* 2020;5(1):e252.

143 Paydar-Darian N, Goldman MP, Michelson KA, et al. Improving efficiency and communication around sedated fracture reductions in a pediatric emergency department. *Pediatric Quality & Safety* 2019;4(1):e135.

144 Chang H, Silva MA, Giner A, et al. Effects of an external ventricular drain alert protocol on venticulostomy placement time in the emergency department. *Neurosurgical Focus* 2021;51(5):E4.

145 Heckscher D, Jalfon M, Buck MB, et al. Implementation of a health system intervention to reduce time from presentation to surgical intervention for pediatric testicular torsion. *Journal of Pediatric Urology* 2024;20(2):254 e1–254 e7.

146 Cordero-Coma M, Yilmaz T, Padula WV, Rodriguez E, and Lanier ST. Statistical process control charts for ophthalmology. *Graefe's Archive for Clinical and Experimental Ophthalmology* 2011;249:1103–5.

147 Tomak L and Bek Y. Evaluation of operation and death statistics at surgical procedures by using attribute control charts. *Journal of Experimental and Clinical Medicine* 2017;34(2):111–116.

148 Johnson MP, Kim SJ, Langstraat CL, et al. Using bundled interventions to reduce surgical site infection after major gynecologic cancer surgery. *Obstetrics & Gynecology* 2016;127(6):1135–1144.

149 Gómez-Hernández MT, Novoa NM, Varela G, and Jiménez MF. Quality control in anatomical lung resection. Major postoperative complications vs failure to rescue. *Archivos de bronconeumologia* 2021;57(4):251–255.

150 Nathanson BH and Higgins TL. The use of scan statistics and control charts in assessing ventilator-Associated pneumonia quality control programs. *Journal of Healthcare Engineering* 2010;1(4):579–594.

151 Spliid H. Monitoring medical procedures by exponential smoothing. *Statistics in Medicine* 2007;26(1):124–138.

152 Sandberg KC, Lucien JN, Stoll D, Yanney E, and Mezoff A. Decreasing door-to-door times for infliximab infusions in a children's hospital observation unit. *Pediatric Quality & Safety* 2019;4(1):e131.

153 Benneyan JC and Taşeli A. Exact and approximate probability distributions of evidence-based bundle composite compliance measures. *Health Care Management Science* 2010;13:193–209.

154 McGinnis JM, Helpman L, Gundayao M, et al. Evaluating compliance of extended venous thromboembolism prophylaxis following abdominopelvic surgery for cancer: A multidisciplinary quality improvement project. *Journal of Surgical Oncology* 2022;125(3):437–447.

155 Carroll AR, McCoy AB, Modes K, et al. Decreasing pre-procedural fasting times in hospitalized children. *Journal of Hospital Medicine* 2022;17(2):96–103.

156 Dimo J, Hahn D, Schlidt K, et al. Use of clinical practice guideline to improve management of osteoarticular infections in children. *Hospital Pediatrics* 2023;13(7):579–591.

157 Chumpia MM, Ganz DA, Chang ET, and de Peralta SS. Reducing the rare event: lessons from the implementation of a ventilator bundle. *British Medical Journal Open Quality* 2019;8(2):e000426.

158 Bothner U, Georgieff M, and Schwilk B. Validation of routine incidence reporting of one anaesthesia provider institution within a nation-wide quality of process assessment program. *Journal of Clinical Monitoring and Computing* 1998;14:305–311.

159 Hughes Driscoll CA, Davis NL, Miles M, and El-Metwally D. A quality improvement project to improve evidence-based inhaled nitric oxide use. *Respiratory Care* 2018;63(1):20–27.

160 Bollo RJ, Gross PH, Rocque BG, et al. A multicenter initiative to reduce intrathecal baclofen pump surgical site infection: A Cerebral Palsy Research Network quality improvement project. *Journal of Neurosurgery: Pediatrics* 2023;31(5):444–452.

161 Pohnert AM, Schiltz NK, Pino L, et al. Achievement of age-friendly health systems committed to care excellence designation in a convenient care health care system. *Health Services Research* 2023;58:89–99.

162 Donado C, Solodiuk JC, Mahan ST, et al. Standardizing opioid prescribing in a pediatric hospital: A quality improvement effort. *Hospital Pediatrics* 2022;12(2):164–173.

163 Breton J, Witmer CM, Zhang Y, et al. Utilization of an electronic medical record–integrated dashboard improves identification and treatment of anemia and iron deficiency in pediatric inflammatory bowel disease. *Inflammatory Bowel Diseases* 2021;27(9):1409–1417.

164 Ernst MM, Wooldridge JL, Conway E, et al. Using quality improvement science to implement a multidisciplinary behavioral intervention targeting pediatric inpatient airway clearance. *Journal of Pediatric Psychology* 2010;35(1):14–24.

165 Mottes T, Owens T, Niedner M, Juno J, Shanley TP, and Heung M. Improving delivery of continuous renal replacement therapy: Impact of a simulation-based educational intervention. *Pediatric Critical Care Medicine* 2013;14(8):747–754.

166 Orkaby AR, James K, Leuchtenburg J, Solooki E, Gaziano JM, and Driver JA. Taking prevention to the next step: implementation of a brief, sustainable frailty assessment in a cardiology clinic. *British Medical Journal Open Quality* 2021;10(1):e001140.

167 Lopez M, Wilson M, Cobbina E, et al. Decreasing ICU and hospital length of stay through a standardized respiratory therapist-driven electronic clinical care pathway for status Asthmaticus. *Pediatric Quality & Safety* 2023;8(6):e697.

168 Kouo T, Kleinman K, Fujii-Rios H, et al. A resident-led QI initiative to improve pediatric emergency department boarding times. *Pediatrics* 2020;145(6).

169 Kerrey BT, Mittiga MR, Rinderknecht AS, et al. Reducing the incidence of oxyhaemoglobin desaturation during rapid sequence intubation in a paediatric emergency department. *British Medical Journal Quality & Safety* 2015;24(11):709–717.

170 Harahsheh AS, Sabati A, Anderson J, et al. American college of cardiology body mass index counseling quality improvement initiative. *Pediatric Cardiology* 2021;42:1190–1199.

171 Kaplan SL, Chen AE, Rempell RG, Brown N, Velez-Florez MC, and Khwaja A. Impact of emergency medicine point-of-care ultrasound on radiology ultrasound volumes in a single pediatric emergency department. *Journal of the American College of Radiology* 2020;17(12):1555–1562.

172 Emerson BL, Prozora S, Jacob A, et al. An initiative to decrease time to antibiotics for patients with fever and neutropenia. *American Journal of Medical Quality* 2019;34(2):158–164.

173 Lin HM, Macias C, McGee C, et al. "Help me sleep": A quality initiative to reduce overnight vital signs. *Hospital Pediatrics* 2022;12(2):142–147.

174 Hampton LL, DeBoy JT, Gunaratne A, et al. Improving the documentation of penicillin allergy labels among pediatric inpatients. *Hospital Pediatrics* 2023;13(9):811–821.

175 Chiem JL, Hansen EE, Fernandez N, et al. Transforming into a learning health system: A quality improvement initiative. *Pediatric Quality & Safety* 2024;9(3):e724.

176 Del Grippo EL, Baskar S, Gray S, et al. A standardized cardiac protocol for pediatric drug ingestion hospital admissions. *Pediatric Quality & Safety* 2019;4(6):e223.

177 Ostrow O, Savlov D, Richardson SE, and Friedman JN. Reducing unnecessary respiratory viral testing to promote high-value care. *Pediatrics* 2022;149(2):e2020042366.

178 Little KJ, Godfrey J, Cornwall R, Carr P, Dolan K, and Samora JB. Increasing brace treatment for pediatric distal radius buckle fractures: Using quality improvement methodology to implement evidence-based medicine. *Journal of Pediatric Orthopaedics* 2019;39(8):e586–e591.

179 Bauer ME, MacBrayne C, Stein A, et al. A multidisciplinary quality improvement initiative to facilitate penicillin allergy delabeling among hospitalized pediatric patients. *Hospital Pediatrics* 2021;11(5):427–434.

180 McLaughlin DJ, Grayson M, and Toth C. Quality improvement to engage general pediatrics in reducing inaccurate penicillin allergy labels. *Academic Pediatrics* 2022;22(7):1175–1183.

181 Marshall TL, Ipsaro AJ, Le M, et al. Increasing physician reporting of diagnostic learning opportunities. *Pediatrics* 2021;147(1).

182 Fontánez-Nieves T, Frost M, Anday E, Davis D, Cooperberg D, and Carey A. Prevention of unplanned extubations in neonates through process standardization. *Journal of Perinatology* 2016;36(6):469–473.

183 Murphy A, Lindegren ML, Schaffner W, et al. Improving influenza testing and treatment in hospitalized children. *Hospital Pediatrics* 2018;8(9):570–577.

184 Biondi EA, McCulloh R, Staggs VS, et al. Reducing variability in the infant sepsis evaluation (REVISE): A national quality initiative. *Pediatrics* 2019;144(3).

185 Johnson DP, Arnold DH, Gay JC, et al. Implementation and improvement of pediatric asthma guideline improves hospital-based care. *Pediatrics* 2018;141(2).

186 Charvat C, Jain S, Orenstein EW, Miller L, Edmond M, and Sanders R. Quality initiative to reduce high-flow nasal cannula duration and length of stay in bronchiolitis. *Hospital Pediatrics* 2021;11(4):309–318.

187 Kocher KE, Arora R, Bassin BS, et al. Baseline performance of real-world clinical practice within a statewide emergency medicine quality network: The Michigan Emergency Department Improvement Collaborative (MEDIC). *Annals of Emergency Medicine* 2020;75(2):192–205.

188 Vega MCF, Aiss MA, Smith M, et al. Improving handoff with the implementation of I-PASS at a tertiary oncology hospital. *British Medical Journal Open Quality* 2023;12(4):e002481.

189 Gupta S, Winckler B, Lopez MA, et al. A quality improvement initiative to improve postdischarge antimicrobial adherence. *Pediatrics* 2021;147(1).

190 Sadeghi C, Khan HA, Gudleski G, Reynolds JL, and Bakhai SY. Multifaceted strategies to improve blood pressure control in a primary care clinic: A quality improvement project. *International Journal of Cardiology Hypertension* 2020;7:100060.

191 Bakhai S, Ansari M, Sadeghi C, and Reynolds JL. Advancing health equity in improving breast cancer screening with the use of a mobile mammography bus in marginalised population: Quality improvement project. *British Medical Journal Open Quality* 2024;13(1):e002482.

192 Ogrinc G, Ercolano E, Cohen ES, et al. Educational system factors that engage resident physicians in an integrated quality improvement curriculum at a VA hospital: A realist evaluation. *Academic Medicine* 2014;89(10):1380–1385.

193 Piccirillo JF. The use of patient satisfaction data to assess the impact of continuous quality improvement efforts. *Archives of Otolaryngology–Head & Neck Surgery* 1996;122(10):1045–1048.

194 Dalton EM, Herndon AC, Cundiff A, et al. Decreasing the use of restraints on children admitted for behavioral health conditions. *Pediatrics* 2021;148(1).

195 Shannon M and El-Nigoumi SE-N. Improving physician compliance with mammography screening by implementing a breast cancer screening guideline at a hospital in Bahrain. *British Medical Journal Open Quality* 2021;10(3):e001351.

196 Sosa T, Mayer B, Chakkalakkal B, et al. Improving shared situation awareness for high-risk therapies in hospitalized children. *Hospital Pediatrics* 2022;12(1):37–46.

197 Daniels E, Lamb GC, and Beckius A. Reducing inappropriate oxygen use in hospitalized medicine patients. *Hospital Practice* 2023;51(4):205–210.

198 Paul RA, Beaman C, West DA, and Duke GJ. COBRA: COde Blue Retrospective Audit in a metropolitan hospital. *Internal Medicine Journal* 2023;53(5):745–752.

199 McFadden V, Bauer SC, Porada K, Mehta S, and Pickett ML. Quality initiative to increase delivery of adolescent hospital-based reproductive health care. *Hospital Pediatrics* 2022;12(1):53–61.

200 Gob A, Bhalla A, Aseltine L, and Chin-Yee I. Reducing two-unit red cell transfusions on the oncology ward: A choosing wisely initiative. *British Medical Journal Open Quality* 2019;8(2):e000521.

201 Riney LC, Frey TM, Fain ET, Duma EM, Bennett BL, and Murtagh Kurowski E. Standardizing the evaluation of nonaccidental trauma in a large pediatric emergency department. *Pediatrics* 2018;141(1).

202 Udayaraj UP, Watson O, Ben-Shlomo Y, et al. Establishing a tele-clinic service for kidney transplant recipients through a patient-codesigned quality improvement project. *British Medical Journal Open Quality* 2019;8(2).

203 Eastridge BJ, Wade CE, Spott MA, et al. Utilizing a trauma systems approach to benchmark and improve combat casualty care. *Journal of Trauma and Acute Care Surgery* 2010;69(1):S5–S9.

204 Al-Hussein FA. A tale of two audits: Statistical process control for improving diabetes care in primary care settings. *Quality in Primary Care* 2008;16(1):53–60.

205 White E, Proudlove N, and Kallon D. Improving turnaround times for HLA-B* 27 and HLA-B* 57: 01 gene testing: A Barts Health NHS Trust quality improvement project. *British Medical Journal Open Quality* 2021;10(3):e001538.

206 Seferian EG, Jamal S, Clark K, et al. A multidisciplinary, multifaceted improvement initiative to eliminate mislabelled laboratory specimens at a large tertiary care hospital. *British Medical Journal Quality & Safety* 2014;23(8):690–697.

207 Harb R, Hajdasz D, Landry ML, and Sussman LS. Improving laboratory test utilisation at the multihospital Yale New Haven Health System. *British Medical Journal Open Quality* 2019;8(3):e000689.

208 Baldelli S, Marrubini G, Cattaneo D, Clementi E, and Cerea M. Application of quality by design approach to bioanalysis: Development of a method for elvitegravir quantification in human plasma. *Therapeutic Drug Monitoring* 2017;39(5):531–542.

209 Inada M and Terano T. QC chart mining: Extracting systematic error patterns from quality control charts. In: *2005 IEEE International Conference on Systems, Man and Cybernetics: IEEE*, (pp. 3781–3787); 2005.

210 Merca C, Lindell IC, Ernholm L, et al. Veterinary syndromic surveillance using swine production data for farm health management and early disease detection. *Preventive Veterinary Medicine* 2022;205:105659.

211 Barrientos-Blanco JA, Moraes L, Waddell J, and Reed KF. Farm-specific estimates of forage variability result in larger optimal control limits for forage quality monitoring. *Journal of Dairy Science* 2024;107(10):7822–7841.

212 Dórea FC, McEwen BJ, McNab WB, Sanchez J, and Revie CW. Syndromic surveillance using veterinary laboratory data: Algorithm combination and customization of alerts. *PLoS One* 2013;8(12):e82183.

213 Blacksell SD, Cameron AR, Chamnanpood C, et al. Implementation of internal laboratory quality control procedures for the monitoring of ELISA performance at a regional veterinary laboratory. *Veterinary Microbiology* 1996;51(1-2):1–9.

214 Amin M, Noor A, and Mahmood T. Beta regression residuals-based control charts with different link functions: an application to the thermal power plants data. *International Journal of Data Science and Analytics* 2024: 1–13. https://doi.org/10.1007/s41060-023-00501-w

215 Ibrahim M, Zhang C, and Mahmood T. Surveillance of high-yield processes using deep learning models. *Quality and Reliability Engineering International* 2024;40(8):4365–4393. https://doi.org/10.1002/qre.3635

216 Gadde H. AI-augmented database management systems for real-time data analytics. *Revista de Inteligencia Artificial en Medicina* 2024;15(1):616–649.

217 Zwetsloot IM, Mahmood T, Taiwo FM, and Wang Z. A real-time monitoring approach for bivariate event data. *Applied Stochastic Models in Business and Industry* 2023;39(6):789–817.

218 Kan C and Yang H. Dynamic network monitoring and control of in situ image profiles from ultraprecision machining and biomanufacturing processes. *Quality and Reliability Engineering International* 2017;33(8):2003–2022. https://doi.org/10.1002/qre.2163

219 Lee MH, Whittlesey KJ, Dang JM, et al. Overview of the FDA regulatory process. In: *Essentials of Stem Cell Biology*, (pp. 605–634). Elsevier; 2014.

220 Peklaj K and Cerovšek M. Implementing laboratory computerized systems in pharmaceutical industry: Regulatory compliance. *Ventil: revija za fluidno tehniko in avtomatizacijo* 2025;31(1):26–32.

221 Shirvanian N, Shams M, and Rahmani AM. Internet of Things data management: A systematic literature review, vision, and future trends. *International Journal of Communication Systems* 2022;35(14):e5267.

16

Artificial Intelligence in the Management of Pharmaceutical Manufacturing

Rida Sagheer[1], Samra Siddiqui[2], and Aisha Badruddin[3]**

[1]*Department of Chemistry Integral University, Lucknow, India*
[2]*Department of Computer Science and Engineering, Integral University, Lucknow, India*
[3]*Department of Business Management, Integral Business School, Integral University, Lucknow, India*

*Corresponding authors: rida.sageer@gmail.com; aishabadruddin@gmail.com

CONTENTS

16.1 Introduction

Present advancements in technology coupled with fundamental scientific research is influencing all areas of research and medicine and making it possible to produce highly designed products.

Artificial intelligence (AI) and machine learning (ML) have emerged as critical components in the field of computing, covering all scientific and technical domains, from basic science, engineering to medicine. The significant growth transforms numerous industries, including finance, where it improves risk assessment, portfolio management, and fraud detection [1]. The transportation sector benefited from AI and the advancement of self-driving cars [2], traffic management [3], and autonomous farming trucks [4]. Furthermore, AI in healthcare gives support in disease diagnosis [5] and has a major impact on automation and clinical decision-making [6], and robotic surgery 2021 [7]. It provides personalized treatment regimens, drug discovery, and accelerated drug manufacturing [8, 9]. The pharmaceutical industry is a complex and highly regulated sector that requires efficient management to ensure the production of high-quality medicines. Computer aided drug design, which easily recognize target proteins, has overtaken the traditional methods for preparing pharmaceuticals in the last few years.

AI is an amalgamation of several intelligent behavior and procedures generated by computational models, algorithms, or a set of rules that allow the machine to mimic human cognitive processes such as learning and problem solving. AI methods include computer vision [10], robots, and natural language processing (NLP) 2018) [11]. However, ML focuses on developing algorithms that learn from data and make predictions without the need for explicit programming. Deep learning (DL), a subfield of ML, is distinguished using artificial neural networks (ANNs) that mimic the human brain structure to detect deep patterns and relationships in large data (Figure 16.1). Nowadays, AI has been extensively used to improve drug design techniques and has the potential to manage pharmaceutical manufacturing within the specified time frame, allowing advancement in the industrial-driven Revolution 5.0. It offers cutting-edge technologies for data analysis and decision-making, resulting in more efficient processes, lower costs, and improved product quality and increasing the possibility of a successful therapeutic approach as well. Therefore, AI is being employed by the pharmaceutical industry at each step of drug development, such as collection of clinical data, drug discovery, accelerating drug design, and boosting optimization, along with clinical trials 2022 [12]. Furthermore, AI also regulates pharmaceutical manufacturing, controlling quality maintenance and optimization

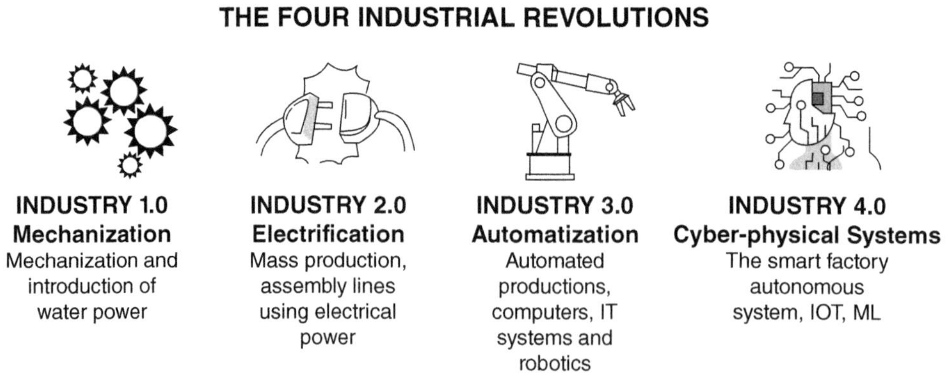

Figure 16.1 An illustration depicted an overview of the four industrial revolutions. Created in Canva by Md. Faiyazuddin.

of supply chain management as well as accelerating drug delivery to patients. Moreover, AI significantly reduces the manufacturing cost and the health concerns associated with preclinical studies. Pharmaceutical manufacturing requires careful attention to design to guarantee product safety and effectiveness; therefore, it focusses primarily on maintenance of quality control, like ensuring the safety, efficacy, and consistency of drugs through rigorous testing, inspections, and regulatory compliance with GMP standards. AI techniques can actively track manufacturing processes such as drug design, formulation, continuous production workflows, packaging, and labeling, while leveraging anomalies in addition to enhancing workflow efficiency and performance measures. AI enables pharmaceutical companies to make the most informed decisions regarding which medications to avoid and how to arrange and handle them while balancing both financial and social concerns. However, AI is often employed to study ML in an approach that simulates human cognitive processes, delivering insights and interpretations that tend to be accurate and insightful Generally, NLP [11], physical robots, robotic process automation, ML, and other AI technologies are used in the pharmaceutical sector [11]. The aim of this book chapter is to cover applications of AI (ML, DL, and ANNs) in pharmaceutical manufacturing, drug design, formulation, packaging, process optimization, and supply chain management. Additionally, it offers a quick review of the current advances, challenges and limitations of AI adoption in the pharmaceutical industry by pharmaceutical companies, and future directions.

16.2 Industrial Revolution in Pharmaceutical Manufacturing

The industrial revolutions are periods of significant technological advancements and societal shifts on the history of human civilization spanning from the agrarian societies to industrialized ones. Its impact has been profound and far-reaching, transforming the economic growth, social change, technological innovations, as well as globalization, resulting in improved quality, additional flexibility, and reduction in waste.

The initial phases of industrial growth, referred to as Industrial Revolution 1.0, marked the use of large-scale technology that could crush, mill, mix, and press a larger number of pharmaceuticals, substituting the time-consuming preparation of botanical, mineral, and animal-based materials [13]. Whereas Industrial Revolution 2.0 was driven by electricity, the first electronic gadgets, process controls, and assembly lines with basic automation. Because of these developments, manufacturers now could set basic process parameters [14]. The Industrial Revolution 3.0 began with the advancement and broad availability of computer technology and communication platforms, such as wireless communication, networked computing, and the internet. These advancements made it simpler for automating machinery and more tasks [15]. The primary characteristic of Industrial Revolution 4.0 is the combination of robotics, AI, and connectivity to enable integrated, self-sufficient, and self-regulating production systems that function without human intervention 2021 [15, 16]. Pharma 4.0 is the term used by the pharmaceutical industry to describe the revolution in manufacturing processes that occurs during this time (Figure 16.1). Moreover, Industrial Revolution 5.0 is a human-centric, AI-driven shift encompassing automation, collaboration between humans and AI powered systems, empowering computing capabilities, embracing sustainable, and green manufacturing for the smart industries of the future.

16.3 Pharmaceutical Manufacturing

Several industries employ several strategies for enhancing growth and to satisfy the customer demands and expectations. The pharmaceutical industry plays an essential role for lifesaving medications. Pharmaceutical manufacturing is a complex and multifaceted sector that involves the drug discovery, drug development, production of pharmaceutical products, and their distribution. Deep research and development in numerous areas of the pharmaceutical industry is usually based on innovative approaches including production technology, packaging considerations, and customer-oriented marketing strategies. From tiny pharmacological molecules to biological substances, novel pharmaceutical developments have been selected for their outstanding stability and efficacy to fulfill major healthcare needs 2021) [8]. Understanding the elevated doses of toxicity associated with new medicines is a matter of a great concern that will demand in-depth research and analysis soon. One of the foremost objectives is to develop therapeutic compounds with the best potential benefits and compatibility for application within the field of medicine. Therefore, the pharmaceutical manufacturing industry is at the forefront of innovation, safety, and precision, manufacturing life-saving medications and therapeutics for both humans and animals that have a direct influence on healthcare across the globe. It is essential for maintaining the safety and quality of pharmaceutical products, leading to the need for strict compliance to regulations and standards. Pharmaceutical manufacturing is based on two main stages, such as primary manufacturing and secondary manufacturing, in which primary manufacturing deals with creating chemical compounds with therapeutic value (active pharmaceutical ingredients, or APIs), while secondary pharmaceutical manufacturing creates medications by properly combining APIs and excipients to create a finished product (such as tablets and capsule). Usually complex chemical entities, APIs are generated by a series of unit operations under meticulously tracked and validated settings [17]. Chemical synthesis (paracetamol and acetylsalicylic acid), fermentation (penicillin, human insulin), extraction from natural sources (morphine, heparin), or a combination of these methods (amoxicillin, a semisynthetic penicillin, codeine, a morphine derivative) can all yield them in large quantities. Most of conventional pharmaceutical manufacturing procedures comprise either batch or semi-batch. Due to their ability to satisfy varying market demands through manufacturing several products in a single plant, these are excellent choices for the manufacturing of low-volume, high-value products.

Referring to the limitations of conventional approaches, pharmaceutical manufacturing faces several challenges when it comes to ensuring quality control and decision-making. Despite this, the pharmaceutical industry facing several challenges that need additional development, leveraging technology-driven solutions to meet global medical and healthcare needs. The health care industry has a constant need for competent employees, which implies that healthcare professionals must undergo ongoing training in order to boost their participation in routine duties. It is an essential task to find skill gaps in pharmaceutical by understanding and offering beneficial training that successfully closes the gaps, which have been discovered through suitable reformative action. The product's efficacy, regulatory compliance, and operational efficiency can overcome significant challenges related to patients' safety. A report from a few states that in June 2022, supply chain disruptions accounted for almost 41% of all incidents. Additionally, the research highlights that supply chain disruption has emerged as the second most challenging task to accomplish 2023 [18]. To overcome the challenges, new development and innovative solution in several pharmaceutical enterprises for the supply chain might strengthen the resilience of their businesses. Natural disasters, price adjustments, cyberattacks, problems with products, logistical delays, and pandemics all contribute to supply chain disruptions. The epidemic's transportation problems severely impacted the supply chain system and global manufacturing sectors. Price fluctuations are caused by decision-induced delays for price updates from suppliers because of confusion over whether to use the new price or the current price for commodities or materials.

Traditional pharmaceutical manufacturing confronts several challenges, complex manufacturing procedures in which the quality of finished pharmaceutical products is influenced by several factors in the pharmaceutical industry. The difficulty of managing these factors and maintaining uniformity throughout the production process might result in differences in the quality of the final product. Furthermore, to ensure the safety of patients, the pharmaceutical industry is subject to strict regulations, which include good manufacturing practices (GMP), which must be followed by manufacturers in order to meet regulatory requirements and quality standards. There could be severe consequences for any departure from these standards. The huge volume of data from various sources is integrated and analyzed, including supply chain data and quality control tools, as well as production equipment generated during pharmaceutical manufacturing. However, it is difficult to integrate and analyze this data successfully so that decisions are up-to-date. However, quality control requires real-time monitoring of critical development data and standard deviations. But developing technologies that can monitor these modifications and operate accordingly can be difficult. Whereas maintenance of data integrity and accountability is needed for compliance and quality assurance throughout the manufacturing process. It can be difficult to make sure that data is accurate, inaccessible, and accessible when required. Generally, the manufacturing of pharmaceuticals is produced in batches, and even minor modifications to the procedures, equipment, or raw materials might lead to variability from batch to batch. It can be difficult to control this variability while keeping the quality of the final output constant. Additionally, the pharmaceutical sector produces a wide range of products with different dosages, complex formulas, and delivery methods. In this diverse product range, it can be challenging to make decisions and maintain quality control. Innovations in technology such automation, digitalization, and AI are leading to a rapid transition significantly in the manufacturing industry. Although ensuring compliance with existing regulations and constraints, it is difficult to implement these technologies. Complex supply chains are used in the pharmaceutical production process, as they are susceptible to outages from components like shortages in raw materials or regulatory changes. To maintain a constant supply of high-quality resources, it is essential to manage the risks. Manufacturing choices about batch release, resource allocation, and process changes frequently must be made in the face of uncertainty. It can be challenging to predict how choices will impact both regulatory compliance and product quality. It can be challenging to predict how decisions will affect both product quality and regulatory compliance. Manufacturing decisions about resource allocation, batch release, and process modifications must often be made in the face of uncertainty. Manufacturers must act promptly to respond with shifting patients demands, safety regulations, and competition in the pharmaceutical industry despite compromising quality. The organizational atmosphere must evolve to reflect emerging technologies like AI. It can be difficult to train workers, develop AI skills, and encourage a data-driven approach. A comprehensive strategy that includes personnel training, process optimizing, technology adoption, and regulatory compliance is required to get over these obstacles. Pharmaceutical companies must embrace innovation while keeping quality and patient safety as their top priorities if they wish to compete in the multifaceted present market.

16.4 Artificial Intelligence

AI combines digital data and computational analysis to make decisions that are often done by humans. It helps in situation analysis by simulating the learning and analytical processes of the human brain, helping with problem-solving. AI is comprised from a few subfields that approach computer intelligence differently based on its characteristics and tasks that are needed. Computer intelligence-based tasks might include learning, reasoning, problem-solving, and decision-making, among other things. ML and DL are two of the most cutting-edge methods for prognosis and risk management among the different AI techniques, owing to advancements in technology and software development. However, DL is a branch of ML that uses ANNs. According to some researcher, ANNs are a subdiscipline of ML, which is a subdiscipline of AI [19, 20]. According to [21], ANN and ML are an exact replica of the human brain, which is an interconnected network of neurons made up of weighted communication channels among neurons [21]. These are made up of several connected nodes that transmit information to successfully execute the intended function [22]. According to McCarthy and Wright [23], ML is the process that enables computers to learn tasks without explicit external input or prompts through analysis of data and applying statistical methods to generate general knowledge through the generation of mathematical relationships. According to Chen and Decary [24], ML and ANN require implementing algorithms to handle huge and diverse datasets [24]. AI and ML currently use computer systems and small robotic systems to mimic human intelligence patterns and processes. It has the capacity to carry out tasks that are usually done by intelligent people [25] (Figure 16.2). However, ANNs are conceptual frameworks primarily designed to carry out AI algorithms. It's important to keep in mind that ML algorithms divide into three groups according to the method in which it utilizes input data: supervised learning, unsupervised learning, and reinforcement learning.

16.4.1 Supervised Learning

Supervised learning incorporates techniques like ANNs, multivariate regression, and classification analysis that learn from and connect input data to process design and control

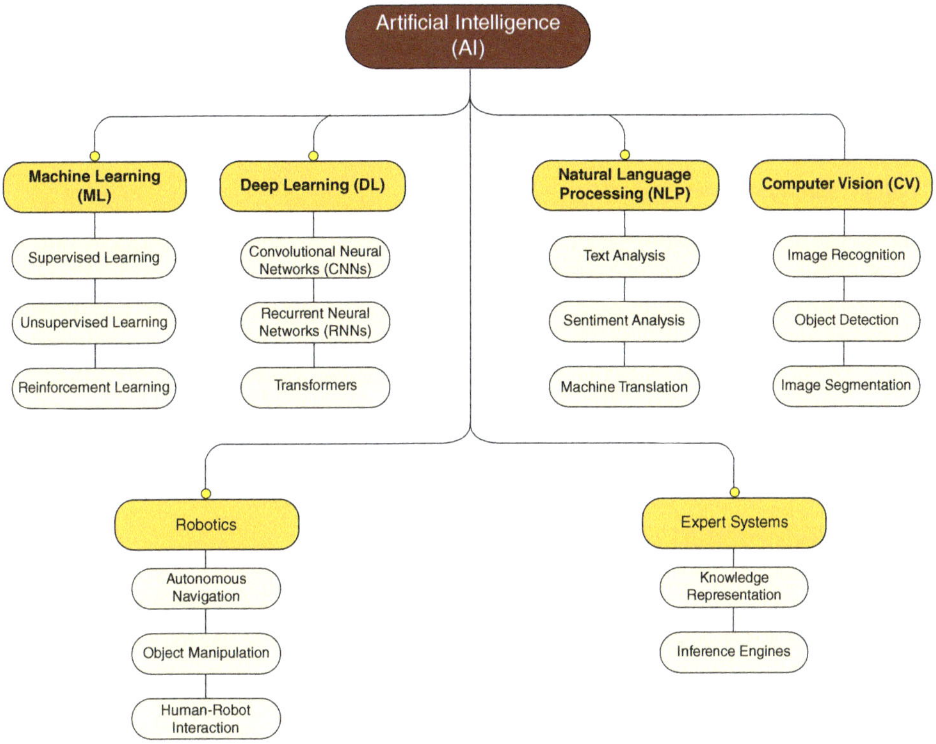

Figure 16.2 An illustration showing the artificial intelligence and its component. Created in Canva by Md. Faiyazuddin.

outcomes [26, 27]. This technique instructs algorithms for tasks like outcome assumptions and data classification using labeled information. Regression, which involves anticipating an integer, and grouping, which includes predicting a label, are the two most prevalent supervised learning challenges. Various approaches can be used to solve supervised learning tasks, depending on the type of data in a particular issue domain. Depending on the type of data in a particular problem area, supervised learning problems can be solved using a range of methods [28, 29]. Among these methods are Naïve Bayes, K-nearest neighbors, support vector machines, random forests, ensemble learning, linear regression, and support vector regression, among others.

16.4.2 Unsupervised Learning

Unsupervised learning is a subset of ML in which labeled data is not supplied to the algorithm. It gets charged with independently discovering connections and patterns within the data. This method is frequently applied in exploratory data analysis and can be helpful in identifying clusters or hidden features in a dataset [17]. Clustering, dimensionality reduction, visualization identifying association rules, and anomaly detection are a few common unsupervised tasks. Depending on the properties of the data, popular techniques like clustering algorithms (e.g., hierarchical clustering, K-means, K-medoids, single linkage, complete linkage, BOTS), association learning algorithms, and feature selection and extraction techniques (e.g., Pearson correlation, principal component analysis) can be used to address a range of unsupervised learning tasks [29].

16.4.3 Reinforcement Learning

The reinforcement learning algorithm is a category of ML model with input, output, and training. Reinforcement learning relates decisions to future goals through associating actions with delayed outcomes. It is applicable in situations involving intricate dynamics, such as logistics or plant operations. Pharmaceutical manufacturing operations could benefit from all of these ML techniques. However, supervised learning techniques are believed to bring less risk and uncertainty and have thus far garnered the most interest [16].

16.5 History of AI in Pharmaceutical Industry

The union of AI and the pharmaceutical industry has a rich and detailed history that is spanning over six decades. Interestingly in the beginning of 1960s, computer simulations were utilized to model molecular interactions, providing the framework playground for future innovations. This pioneering work of AI and pharmaceutical industry have long and fruitful collaboration that was made possible by this cutting-edge research. And as the Industry progressed in the 1960s and 1970s, toward optimization, it saw the adoption of computer-aided design (CAD) software for molecular structure design and optimization. This early fostering of AI in the pharmaceutical industry has laid tremendous foundation for future achievements, targeting researchers to explore new molecular structures and design with more effectiveness. Wherein, to facilitate decision-making in drug research and discovery, in the 1980s and 1990s rule-based expert systems were introduced, supporting decision-making in drug discovery and development phases in the manufacturing process. Predefined rule was used by these systems to analyze data and provide recommendations, in the application of AI in the pharmaceutical industry and marking a significant step forward for years. However, it was the arrival of ML and DL algorithms in the 2000s that truly incredibly revolutionized AI's role in the pharmaceutical industry. These technologies were used to enable the analysis of large datasets, identification of patterns in the development, and prediction of outcomes for the target outcome. This, in turn, has done numerous advances in drug discovery, clinical trials, and pharmacovigilance. Today, in the modern world with high-tech technologies, AI is applied in various areas of the pharmaceutical industry, namely counting, identifying potential drug targets, designing new molecular structures, predicting drug efficacy and toxicity, and streamlining clinical trials. As the pharmaceutical industry continues to grow and develop, we can expect to see even more innovative and promising applications of AI, leading to faster, more efficient, and more effective drug development.

16.6 AI in Pharmaceutical Manufacturing

The pharmaceutical industry is a vital global industry considered as one of the largest markets, annually spending billions of dollars to create novel medicines and ingredients. Pharmaceutical companies make significant investments in developing novel therapeutics, regardless of whether a guarantee that they will be profitable. However, time management presents a variety of difficulties in the field of human resource management, from pre-formulation studies to product marketing. The industry is driven by continuous innovation, regulatory frameworks, and increasing demand for effective treatments. Therefore, AI is being used by pharmaceutical companies to mitigate failure risks and expenses. In pharmaceutical research, the main benefit of AI is its aptitude to examine immense amounts of data and discover patterns and connections that researchers may overlook. Applications of AI and machine vision technology provide an example in which AI is currently being employed in pharmaceutical manufacturing monitoring visual check of packets, locks, and vials with humans; predictive equipment maintenance can reduce risks, disruptions, and production downtime [30, 31]. The pharmaceutical industry is changing because of AI, which is improving manufacturing processes, streamlining clinical trials, and transforming drug discovery [32]. It also helps in automated quality control that allow to enhance data integrity, continuous process quality assurance, along with simple scheduling of analytical testing [33]. Pharmaceutical companies are increasingly moving toward a big data paradigm for clinical trial design to overcome these shortcomings. This will mitigate the associated failures and R&D costs. As the industry continues to use AI, we can expect further developments that will transform production procedures and reduce expenditures. Also automated quality control allows the efficient management of analytical tests [33]. In addition to continuous process quality assurance [34] and improved data integrity, the adoption of AI-driven technologies enables pharmaceutical firms to boost efficiency, cut costs, and accelerate the development of ground-breaking treatments. ANN has been used for prediction and control in pharmaceutical development [28, 29] and recently to perform risk-based analysis of biomanufacturing processes [35–37], develop control schemes, perform fault detection for complex dynamic processes, and to predict outcomes for therapeutic drug pharmacokinetics and pharmacodynamics [38, 39]. In pharmaceutical manufacturing, AI helps in real-time monitoring of production lines, predicting equipment failures, and ensuring compliance with regulatory standards (Table 16.1).

One effective way to reduce batch-to-batch variability is through continuous manufacturing. To decrease variability in the final product and patient outcomes, the FDA advocates for the continuous manufacturing approach. Continuous manufacturing can be monitored using a variety of process analytical technology (PAT) techniques. These PAT tools are economical, effective, and capable of ongoing manufacturing process monitoring. AI in conjunction with PAT can help control the manufacturing process and enable process improvement through automated ML [40]. Integrating logistics into processing operations is one way data analytics might help companies that make biopharmaceutical goods. Because biopharmaceuticals are produced using modified living cells, maintaining

Table 16.1 Application of AI techniques in different areas of pharmaceutical manufacturing.

Application area	AI techniques used	Description	Benefits
Drug discovery	Machine learning (ML), deep learning (DL), Generative AI	Identifying potential drug targets, predicting drug efficacy and safety, designing novel drug molecules	Accelerated discovery process, reduced R&D costs, higher success rates
Clinical trials	ML, NLP	Optimizing trial design, patient recruitment, and monitoring, and predicting patient response to treatment	Faster and more efficient trials, reduced costs, personalized treatment approaches
Personalized medicine	ML, DL, NLP	Analyzing patient data (genomics, medical history) to tailor treatments to individual patients	Improved treatment outcomes, reduced side effects, increased patient adherence
Drug repurposing	ML, graph analysis	Identifying new uses for existing drugs	Reduced development time and cost, faster availability of treatments for new diseases
Manufacturing and quality control	ML, computer Vision	Optimizing production processes, predicting equipment failures, ensuring product quality and compliance	Increased efficiency, reduced errors, improved product safety
Supply chain management	ML, predictive analytics	Forecasting drug demand, optimizing inventory management, preventing drug shortages and counterfeiting	Reduced costs, improved reliability, increased security
Pharmacovigilance	NLP, ML	Monitoring drug safety, detecting adverse drug reactions from various data sources (e.g., social media, electronic health records)	Early detection of safety issues, improved patient safety

healthy levels of these variables is regarded as an important regulatory aspect. To keep the created product consistent and pure, several things need to be managed and observed. Therefore, big data is widely used by most the world's top chemical companies to increase vaccine production output and monitor product quality. Implementing AI in pharmaceutical manufacturing involves a strategic and well-structured approach.

These are the key steps of AI which are needed to implement in pharmaceutical manufacturing. The well-defined objectives are required for establishment of AI for integration into the pharmaceutical production process. Collect appropriate data from multiple sources during the production process. This comprises historical records, sensor readings, manufacturing data, and quality control data. While using the data to train AI models, make sure the information is properly labeled, purified, and organized.

Identify the precise area of attention, such as supply chain management, process optimization, or quality control improvement. Moreover, discover possible applications where AI can be most useful. These can include leveraging data-driven insights to enhance decision-making, refining production processes with predictive analytics, or implementing image recognition to enhance quality control. The right AI tools are needed to select technologies according to the use cases that have been found [41].

Depending on the needs, this could entail computer vision systems, NLP, DL models, or ML techniques. Moreover, AI models develop according to the specified cases used, train them to use the prepared data, and optimize them for optimum output [42]. Ensure that the model is trained to precisely detect mistakes, for instance, while using image recognition for quality control. The existing infrastructure for pharmaceutical manufacturing integrates the AI models, which might involve collaborating with IT professionals to ensure compatibility and smooth integration with manufacturing equipment and data sources. Validate the performance of the AI models through pilot tests under actual manufacturing conditions. Also, track the efficacy of the AI solutions in accomplishing the objectives set and make all necessary adjustments in considering the results. Good manufacturing practices (GMP), data privacy, validation, and regulatory compliance are critical considerations, and data security and compliance with regulatory standards must be maintained throughout the implementation process. Moreover, for smooth functioning, AI system training must be provided to the manufacturing team. Thus, the workforce can efficiently operate, manage, and maintain the AI-integrated processes if their skills are upgraded. Monitor the performance of the AI systems regularly while obtaining feedback from the manufacturing team. Update and modify the AI models regularly to deal with new problems, improve accuracy, and cope with changing conditions. It is required to scale up the AI implementation across the whole manufacturing method or other relevant areas once the pilot phase is successful. Also, it may be needed to replicate the solution across different factories or locations.

For continuous improvement, regulatory agencies, pharmaceutical professionals, and AI experts should continue to work together. Stay updated on developments in AI and upgrade the systems as needed to overlook for chances of future growth. Documentation and reporting are needed to ensure the AI implementation process, including the methodologies, models used, and outcomes achieved. Notably for regulatory compliance and future reference the documentation is vital. Regular audits and validations are carried out to make sure AI systems are performing as designed and conform to regulatory standards. Thus, the pharmaceutical industry may effectively apply AI to boost quality control, streamline operations, and make data-driven decisions that lead to enhanced performance and product quality by following the above steps.

16.6.1 Drug Design and Drug Development

The drug discovery route has been divided into multiple phases. Finding novel targets that exhibit a relationship to a specific condition is the initial step. Next, high-throughput screening (HTS) for chemical library discovers compounds that might interact with these targets and then optimized to enhance drug like properties. Before approval, it undergoes rigorous preclinical and clinal trials [43]. The pharmaceutical industry has made substantial use of computer-aided drug discovery (CADD) in the last few years, employing methods of design that are based on both structure and ligand. Researchers may now more accurately predict a compound's biological action because of the development of techniques like molecular modeling, virtual screening, and quantitative structure–activity relationship (QSAR) analysis [44, 45]. Presently, AI has become a pioneering catalyst that is profoundly altering the conventional methods used in the discovery, development, and administration of possible remedies. The shift is being driven by the ability of AI to evaluate huge amounts of data. AI-powered algorithms significantly expedite drug discovery by analyzing extensive biological datasets to identify promising drug candidates. ML models predict molecular interactions, reducing reliance on traditional trial-and-error methods [46]. Additionally, AI supports computational drug design, where DL techniques create novel molecular structures with desirable pharmacological characteristics [47]. AI was effectively used by the biotechnology company in silico medicine to find a treatment candidate for idiopathic pulmonary fibrosis in just 46 days, significantly reducing the time required for traditional drug discovery [46].

16.6.2 Identifying Drug Targets

For a drug to show its therapeutic action, it must first interact and bind to the receptor protein. The drug interaction with target protein examined using AI computational tools. Moreover, supervised learning approaches can assist in finding possible drug targets through evaluating biological data. AI can be used to investigate atomic simulations and forecast electric properties, where deep ML enables the measurement of the potential energies of tiny molecules. Accurate outputs require a combination of quantum mechanics, molecular mechanics, and deep ML. The ML algorithm can discover trends and pinpoint possible areas for more research by being trained on data that includes details on genetic, proteomic, or transcriptome characteristics and how they relate to drug response or disease advancement [47]. AI helps identify targets by establishing a causal association between the target and the disease, selecting targets with specific indicators by leveraging large, publicly accessible omics and textual datasets. Omics data, including proteomics, metabolomics, genomic transcriptomics, and epigenomics, offer significant insights on modified signaling pathways and molecular interactions and serve as additional inputs for target selection. Additionally, the approach incorporates text-based data from documents, patents, funding reports, and clinical trials [48]. The availability of multiple drug datasets from platforms like ChEMBL [49] and DrugBank [50] rendered it critical to choose the most significant datasets, and AI-based data mining tools like ANNs were later used for this purpose [51]. To accomplish specific research targets, AI attempts to streamline the list of AI-driven targets by selecting candidates based on a range of parameters, including protein family classification, stages of development, drug ability, toxicity, and novelty [48]. ML and DL algorithms are used to evaluate different kinds of data to discover and rank potential targets for drug development. A cutting-edge AI tool for thorough therapeutic target identification and biomarker development is Panda Omics, a DL platform. These strategies include statistical analysis-driven techniques that uncover putative target genes linked to disease using omics data, including genome-wide association studies and summary data-based Mendelian randomization [52]. Additionally, ML-based approaches use classifiers and regression models to predict possible therapeutic targets, whereas network-based approaches use gene co-expression and miRNA illness networks to identify complex biological relationships [53, 54]. In drug discovery, target validation is yet another crucial phase. By identifying the best model and assessing how well it predicts across various datasets, the validation procedure ensures the model's accuracy and efficacy in practical applications [55]. Currently, many target prediction models are accessible via primarily free web services [49]. However, there are wide variations in the level of rigor used to evaluate various approaches, with the majority depending on statistical studies.

16.6.3 Designing of Dosage

The human body system has been divided into several components to understand the effects of drug distribution. The biological membranes form a framework for further simplifying the components. Biological components require physicochemical barriers, which can be used in different ways depending on the mode of drug delivery within the body. Penetration rate according to the administration route is one of the most important parameters for effective drug delivery system monitoring. Once the orally administered medication entered the gastric environment, it must pass through the intestinal or gastric epithelium. This step is essential for further distribution of drug into blood stream. Through the distribution process, the medicine gets delivered to the target location, which may be tissue or any biological component. Drug outcome in the human body is greatly influenced through a drug's interactions with biological components and availability in biological environments. This process is controlled according to the molecular features of medicine. A specific drug delivery system is required for passive permeation, which is ineffective for many small molecules and biologically active compounds. The active permeation process is driven by membrane transport, which is based on complex biological interactions. Therefore, using computation and systemic modeling approach through different specific parameters, the complex procedure must be explored. The more recent computational model is employed to investigate the pharmacokinetic parameters of drug delivery system [39]. Factors such as medicinal features, patient characteristics, the route of administration, release kinetics, and manufacturing problems are all part of the design of dosage forms. Drugs must be properly formulated before being given to patients, regardless of age. Pediatric patients provide challenges for the formulator in terms of therapeutic effectiveness and compliance. Pharmaceutical companies and compounding pharmacies must produce and supply children's medications because there are not enough drug items for them. The dosage form design has a specific function in serving the various needs in modern therapies, including traditional tablets and capsules to more complex formulations like nanoparticles, liposomes, and transdermal patches. Additionally, targeted drug delivery and customized therapies are now exciting new areas in dosage form design. Remarkably, AI offers advanced technologies for multi layer data processing. Several elements, including simulation, scoring, and refinement, are used in the systematically applied model and parameter evaluation at each step of the study to figure out the best possible outcomes. The system biology type of the databases suggests that an extensive knowledge of the drug-biological interaction is needed to optimize AI training in the biological environment. Several emerging AI technologies, including ANNs, can be used to perform pharmacokinetic studies [26]. In addition, AI offers several kinds of datasets, including chemical, genomic, and phenotypic databases, to enhance comprehension of medication effects and patient outcomes [56]. To fully understand the drug's toxicity and disposition, some of the techniques are also used to investigate how the drug delivery system affects the drug's pharmacokinetics. This involves assessing the drug's rate of distribution, metabolism, excretion, and absorption. Researchers can evaluate the drug's safety and effectiveness using various delivery systems by examining these parameters. As shown in Figure 16.3, AI tools can process or digest a large quantity of data for more effective approaches to the rational design of the product.

16.6.4 Drug Formulation

The properties of formulations are determined by the processing conditions in addition to the ratios used to mix the ingredients. It is extremely difficult to quantify the interactions that exist between ingredient quantities, processing settings, and product performance; however, they are probably known anecdotally. The first formulation consists of one or more drugs combined with different ingredients (excipients). As development moves forward, the selection of these and their concentrations, as well as the manufacturing process, are altered and

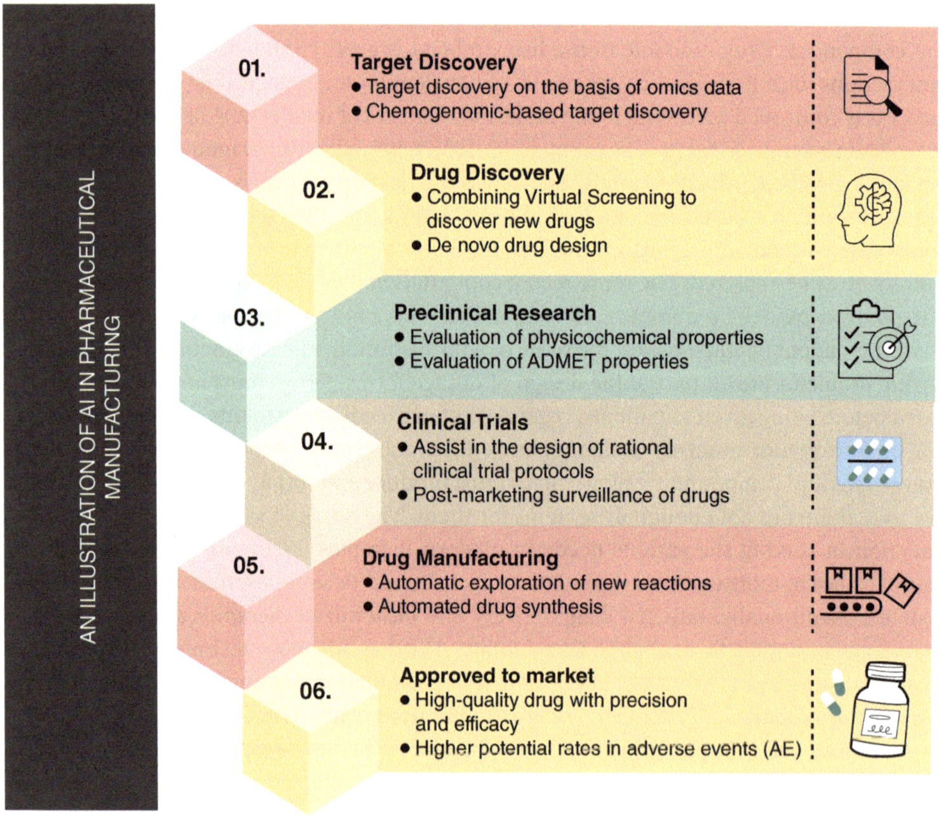

Figure 16.3 An illustration showing AI used in pharmaceutical manufacturing. Created in Canva by Md. Faiyazuddin.

optimized because of extensive, time-consuming experimentation. The repeated attempts therefore lead to the generation of huge amounts of data, which are difficult to interpret and comprehend. The formulator has to do in a multidimensional design space that is almost impossible to conceptualize [57]. In addition to standard dosage forms, pharmaceutical sciences have observed the emergence of different formulations, such as solid dispersions, extrudates, pellets, nanoparticles, and liposomes. Since AI applications in formulation techniques can effectively address a range of API issues, including low solubility, stability, bioavailability, and production power, it becomes even more worthwhile to investigate to develop next-generation drug products with the designed efficacy and health outcomes [58]. To create controlled-release formulations, researchers use pharmacokinetic simulations and ANN. Using Chem software, the ANN model learns sophisticated and specific knowledge from the input and output data units [59].

To explain the diluting agent and binder content in each formulation, processing factors (granulator type, binder addition technique), and granule and tablet attributes (disintegration time, hardness, and friability), Kesavan and Peck [60] designed a model of a caffeine tablet formulation. These two investigations showed that neural networks surpassed traditional statistical methods. So, researchers have reexamined Kesavan and Peck's findings using a range of neural networks and genetic algorithms [61]. Expert systems (ES) and ANNs are executive techniques used in the development of hard gelatin capsule. ANNs promote human cognitive functions like learning, generalization, abstraction from domain knowledge, and prediction. ANNs allow for a quick transformation of data and statistics obtained during research into knowledge, which allows the manufacturer to predict the properties of theoretical preparation networks or develop a few domain-specific strategies for future occurrences [62].

16.6.5 Quality Management and Predictive Maintenance

An approach that is rigorous and scientifically motivated is necessary to successfully navigate the complex process of developing a pharmaceutical formulation within the allotted time frame while maintaining quality. Collecting data include recording details with respect to the properties of medicinal products and excipients, their interactions with each other, equipment, and unit operations [63]. ML algorithms detect anomalies in production processes, ensuring consistent drug quality [63]. Variety of knowledge is used, such as first-principal models, decision trees, correlation, and heuristics. The manufacturing process, excipients, and equipment size are all decided based on this information and knowledge [65]. Therefore, AI, along with its networks, technologies, and tools, guarantees higher-quality products, less waste, and more profit for pharmaceutical companies. Quality management and predictive maintenance in pharmaceutical manufacturing is controlled by supervised learning model. The model can be calibrated with data from manufacturing procedures, equipment sensor data, or quality testing results to forecast equipment failure, variations in the quality of the product, or anomalies in the process, providing proactive maintenance and quality management [66]. The main goals of quality management are to stop problems with quality, identify and fix errors, and promote ongoing process and product improvement. Furthermore, the predictive power of AI has the potential to completely transform pharmaceutical production decision-making. For manufacturing companies,

AI and its networks, technologies, and tools therefore ensure improved product quality, reduced waste, and higher profits.

16.6.6 Packaging and Distribution

To preserve a product's stability, safety, purity, and quality, packaging is most important from production to consumer delivery. It is vital to ensure that the product's quality and packaging meet all requirements. Machine vision system tools, a subset of AI, leverage DL and image processing concepts to detect process defects and reject inadequate packaging and belong to several automated controls set on the machinery used in the bottle packaging process. However, machine vision (MV) is a key component of Pharma 4.0, which promotes automation and digitalization in the pharmaceutical manufacturing industry [67]. Interestingly, a medicine can receive a label with dosage-related information and directions because of automation, which speeding the packing and labeling procedure. All these are done under the supervision of sensors that ensure there is no labeling error. This minimizes human errors and enhances the safety of the product. According to a validation study published in Drug Safety, Label Comp, a cutting-edge AI tool generated to boost the precision and efficacy of post marketing drug safety surveillance, has a high accuracy rate in detecting changes in adverse events (AE) terms in updated FDA drug labeling documents [68].

16.6.7 Pharmaceutical Manufacturing and Supply Chain Optimization

Pharmaceutical manufacturing includes formulation, packaging, and quality control in addition to the finished medicine production. AI is used by pharmaceutical companies for supply chain management, quality control, and automated maintenance. AI has the potential to transform pharmaceutical supply chains. AI-based forecasting tools enhance inventory management, mitigating drug shortages and minimizing waste [69]. There are several factors that influence the use of AI and information analytics in pharmaceutical supply chain management. First, to manage stocks, control errors, and ensure timely shipment of life-saving medicines, the growing complexity of global supply chains requires the use of more advanced tools and methods. In addition, the growth of information sources, such as IoT sensors, e-health records, and real-time advertising data, offers a chance to apply AI algorithms to enhance perceivability and choice assistance. Robotics will boost warehouse efficiency as global healthcare demands continue to rise by automating sorting, packing, and distribution, ensuring that pharmaceuticals are delivered on time and in adequate quantities [70]. Furthermore, the need for supply chain management methods to advance is highlighted by the pressure to reduce costs, increase operational effectiveness, and maintain item quality while fostering competitiveness and administrative research. Whilst adopting AI and information analytics into pharmaceutical supply chain management may have advantages, it comes with drawbacks as well. These include the need for strict approval and administrative compliance, interoperability problems, information security concerns, and skill gaps. Moreover, the complexity of pharmaceutical supply chains requires specially designed arrangements that consider industry-specific quirks resulting from the temperature-sensitive products, high quality standards, and regulatory constraints. But overcoming several challenges will take ongoing work from every participant involved in the pharmaceutical ecosystem; this sector of the industry has a bright future, making healthcare affordable and convenient for more people [58].

16.6.8 Advantages and Limitations of AI in Pharmaceutical Manufacturing

One of the reputed pharmaceutical industries, Pfizer, used the technology to quickly identify millions of potential compounds in developing the COVID-19 treatment, a medication called Paxlovid [71]. For quality control improvement, AI-powered quality control systems can reliably detect defects and irregularities in pharmaceutical products, culminating with improved product quality and patient safety. Implementing AI automates manual tasks, streamlining processes and reducing the time required for tasks such as data analysis, image recognition, and documentation, which increase productivity. Interestingly, AI-powered predictive maintenance aims to prevent equipment breakdowns by anticipating maintenance needs. Consequently, strengthening overall work productivity. AI, although, analyzes intricate information to enhance production processes, resulting in less waste, higher yield, and better resource allocation. It provides statistical findings that facilitate informed decisions across all phases of pharmaceutical manufacturing, including drug development to distribution. AI helps to examine patient data to design the treatment based on individual traits, resulting in more effective targeted therapies and accelerating drug discovery through analysis of huge datasets and predicting the potential drug candidates to bind the active target sites [72]. It facilitates regulatory standards through automation, documentation, and reporting quality control process. AI helps in detecting optimization and suggesting desired modification, resulting in continuous improvement. However, different top AI models are used in pharmaceutical manufacturing which is shown in Table 16.2.

16.7 Case Studies of AI Application in Pharmaceutical Management

Some are the case studies done highlighting the application of AI in pharmaceutical manufacturing (Table 16.3). A well-known pharma industry, GlaxoSmithKline (GSK) improve its quality control process by implementing AI tools that automate the analysis of data generated during the manufacturing process. Because of this, GSK can identify variations in real time, lowering the possibility of quality problems and guaranteeing that the products' quality satisfies legal standards. Another well-known pharmaceutical company, Amgen, has revolutionized the manufacture of biologic pharmaceuticals by using AI. Leveraging ML methods, they have reduced production costs and the likelihood of product faults by detecting and fixing problems that would have otherwise gone undetected. Another global pharmaceutical corporation, Merck, uses predictive analytics driven by AI to enhance supply chain management [27, 73]. The company has developed an intelligent program that can simulate the effects of numerous scenarios on its supply chain, enabling it to make well-informed decisions about production schedules, inventory levels, and transportation logistics. AI helps Swiss multinational pharmaceutical company, Roche, run its warehouse more efficiently. To maximize inventory management, cut waste, and boost productivity, the company has created a ML-based system that tracks and evaluates real-time data from

Table 16.2 Top 10 AI models used in pharmaceutical manufacturing.

#	AI model	Description	Application in pharmaceutical manufacturing	Benefits
1	Convolutional neural networks (CNNs)	Deep learning model for processing structured array data	Quality control (defect detection in products/packaging), visual inspection, and monitoring of manufacturing processes	High accuracy in image-based quality control, automation of visual inspection, and real-time monitoring
2	Recurrent neural networks (RNNs)	Deep learning model for sequential data	Process monitoring, predicting product quality parameters, and optimizing batch processes	Effective in modeling time-dependent processes, improved process control, and enhanced product consistency
3	Long short-term memory networks (LSTMs)	A type of RNN that handles long-term dependencies	Complex time-series forecasting, optimizing drug production schedules, and ensuring consistent product quality over time	Improved handling of long sequences, better predictions in complex manufacturing scenarios, and enhanced supply chain management
4	Transformer models	Models that use attention mechanisms	Optimizing chemical synthesis, predicting reaction outcomes, and improving drug formulation processes	Handles complex relationships, accelerates drug development
5	Generative adversarial networks (GANs)	Framework for generating new data instances	Generating synthetic data for training AI models, designing new drug formulations, and optimizing manufacturing processes	Data augmentation for improved model training, creation of novel drug formulations, and process optimization
6	Autoencoders	Neural networks that learn efficient data codings	Anomaly detection, data compression, and feature extraction	Detects deviations, reduces data dimensionality, and improves data quality
7	Bayesian models	Statistical models using Bayesian inference	Quality control, risk assessment, and process optimization	Provides uncertainty quantification, robust decision-making, and adaptive control
8	Deep Q-networks (DQNs)	Reinforcement learning algorithm	Optimizing robotic control for material handling, optimizing chemical synthesis	Effective for decision-making in complex environments, automates robotic tasks, and optimizes synthetic routes

#	AI model	Description	Application in pharmaceutical manufacturing	Benefits
9	Graph neural networks (GNNs)	Neural networks that work on graph structures	Analyzing molecular structures, predicting drug interactions, and optimizing chemical reactions	Captures complex relationships between entities, accelerates drug discovery, and optimizes chemical processes
10	Predictive analytics (Regression, etc.)	Statistical techniques to predict future outcomes	Demand forecasting, predicting equipment failures, and optimizing inventory levels	Reduces downtime, optimizes stock, and improves production planning

Table 16.3 Top 5 companies using AI/ML in pharmaceutical research.

Company	Focus area	AI/ML applications	Notable achievements/ collaborations
Recursion pharmaceuticals	Drug discovery	Phenotypic drug discovery using high-content imaging and machine learning	Identifying novel uses for existing drugs, collaborations with major pharmaceutical companies
In silico medicine	Drug discovery and development	Generative AI for novel molecule design, target identification, and clinical trial prediction	Developing drugs for various therapeutic areas, partnerships for co-development of new drugs
Ex scientia	Drug discovery	AI-driven drug design platform (Centaur Chemist) to accelerate the discovery of new molecules	Developing drugs in collaboration with pharmaceutical companies, significant funding secured to advance AI-driven drug discovery
Atom wise	Drug discovery	Structure-based drug discovery using AI, particularly CNNs	Partnering with research institutions and pharmaceutical companies to accelerate drug discovery
Sanofi	Drug discovery, clinical trials	Using AI to accelerate mRNA research, improve clinical trial site selection, and aggregate internal data with AI platform "plai"	Developing AI tools in-house and collaborating with AI companies to enhance drug development

sensors placed in its warehouses. A leading pharmaceutical company, Pfizer, combined AI into supply chain management system. To optimize the movement of goods throughout its whole supply chain, the company has created an intelligent logistics platform that makes use of analytics and AI. To guarantee on-time product delivery and reduce transportation and logistics expenses, the platform can forecast demand, identify supply chain interruptions, and optimize inventory levels. With the use of AI-powered technologies, these businesses can streamline their production, logistics, and quality control procedures, which lowers expenses, improves product quality, and increases patient safety overall. In the upcoming years, it is anticipated that more pharmaceutical organizations will embrace AI as its advantages become more apparent, completely changing the pharmaceutical supply chain management industry.

16.8 Limitations of AI in Pharmaceutical Manufacturing

AI algorithms face some challenges in pharmaceutical manufacturing, they need high quality datasets to learn patterns successfully. AI algorithms need vast and high-quality datasets to learn patterns successfully. Getting to adequate and trustworthy data can be difficult, especially in specific fields. However, strong security measures are necessary when handling private and sensitive patient data to guard against breaches and guarantee compliance with data protection laws. Because regulations are constantly changing, it might be difficult to comply with them when integrating AI technology into pharmaceutical manufacturing. Moreover, the process of thoroughly testing and validating AI models for efficacy and safety can be time-consuming and resource intensive. AI skill gap where the pharmaceutical sector could not have the specialized skills need to develop, execute, maintain, and update AI systems. Furthermore, a cultural shift inside organizations may be necessary to implement AI-driven improvements, which could result in opposition or make it difficult for new technologies to be adopted. It can be difficult to integrate AI systems with current industrial processes and infrastructure; this calls for collaboration between the manufacturing and IT departments. For the successful integration of AI into pharmaceutical manufacturing processes, careful planning, skilled collaboration, and a well-defined strategy are needed to balance these advantages and limitations.

16.9 Challenges in AI Adoption

Currently, the pharmaceutical and biomedical industries have made widespread use of AI. Several AI-based technologies are extensively used by these sectors, integrating data-driven and predictive decision-making with more automated and efficient procedures. Furthermore, the second obstacle preventing the full-scale implementation of AI in the pharmaceutical and biomedical sectors is the shortage of qualified personnel to run the AI-based sites. The additional expense to a business significantly damaged the data acquisition model from several databases. Additionally, the data must be substantial and trustworthy to guarantee the pharmaceutical analysis. Generally, AI uses vast volume of patient data, which creates issues with privacy and regulatory compliance. It is essential to maintain data security and abide to regulations including the General Data Protection Regulation (GDPR) and the Health Insurance Portability and Accountability Act (HIPAA) [74]. Moreover, AI models may exhibit biases in training datasets, potentially leading to disparate healthcare outcomes. According to Obermeyer and Tóth, ensuring accessibility and reducing bias through varied data representation are essential for the ethical application of

AI algorithms. Inaccurate or inadequate data may also cause the model to make incorrect inferences, which could result in inappropriate forecasts [75]. Also, applying an AI model to clinical decision-making is one of the most challenging jobs. Pharmaceutical industries frequently struggle with implementing AI into outdated systems. Effective AI application requires a robust IT infrastructure and staff training [76]. Consequently, it requires programmers with a deep knowledge of mental capabilities of AI- and ML-based approaches, along with competent data scientists, and a specific company goal that can be updated to make full use of AI and ML platforms.

16.10 Conclusion

The pharmaceutical industry is currently undergoing a revolution because of AI and ANNs, which are being investigated in the development of various dosage pharmaceutical and healthcare sectors and can also be used with robotics. These technologies speed up drug discovery, optimize clinical trials, and improve manufacturing operations. The application of AI in pharmaceutical manufacturing will enhance output and purity while cutting costs and time. Although AI is showing offering growth for the pharmaceutical industry, it still has shortcomings. To function at their best, ML algorithms often require millions of data scores, therefore rendering them incredibly data hungry. To keep elderly patients' minds active, it is also utilized for social engagement and dialogue. When used in pharmaceutical manufacturing, AI will help improve output and purity while cutting costs and time. Considering these few challenges, AI has gained momentum in the pharmaceutical industry, and several board leaders are expressing a strong interest in integrating AI into their system. While the manufacturing of biotech products involves complex procedures, biotech industries have been especially interested in integrating AI. According to "The State of AI: AI in Business," 62% of companies are considering implementing AI, and the pharmaceutical industry shouldn't be left behind in terms of improving product quality and cutting expenses. Several larger companies have started making investments in AI and ML, notably Novartis, Pfizer, Roche, AstraZeneca, Merck, Johnson & Johnson, GlaxoSmithKline, Sanofi, AbbVie, and Bristol-Myers Squibb. To benefit from it, these companies are either developing their own AI system or collaborating with AI technology companies. Roche is implementing AI working together with Owkin, a French data science company; according to the business, AI accelerates and aids in drug study, discovery, and clinical trials. Additionally, Novartis collaborated on AI implementation with Intel, IBM Watson, Quantum Black, and the Massachusetts Institute of Technology. Successful AI adoption can be fueled by best practices including cross-functional collaboration, ongoing monitoring, and regulatory adherence, even while challenges like bias, data privacy issues, and system integration still exist. Pharmaceutical firms may promote innovation and improve patient care by strategically using AI.

Declaration of Conflict of Interest

Authors found no conflict of interest.

Data Availability and Copyright Permission

No data was used for the research described in the article. All the image used in above study are original.

References

1 Bartram, S. M., Branke, J., & Motahari, M. (n.d.). *CFA INSTITUTE RESEARCH FOUNDATION / LITERATURE REVIEW ARTIFICIAL INTELLIGENCE IN ASSET MANAGEMENT.*

2 Lugano G. Virtual assistants and self-driving cars. In: *2017 15th International Conference on ITS Telecommunications (ITST)*, (pp. 1–5). IEEE; 2017.

3 Sukhadia A, Upadhyay K, Gundeti M, Shah S, and Shah, M. Optimization of smart traffic governance system using artificial intelligence. *Augmented Human Research* 2020;5(1):13.

4 Ghobadpour A, Monsalve G, Cardenas A, and Mousazadeh H. Off-road electric vehicles and autonomous robots in agricultural sector: Trends, challenges, and opportunities. *Vehicles* 2022;4(3):843–864.

5 Szolovits P, Patil RS, and Schwartz WB. Artificial intelligence in medical diagnosis. *Annals of Internal Medicine* 1988;108(1):80–87.

6 Paul, D. (n.d.). *REVISTA DE INTELIGENCIA ARTIFICIAL EN MEDICINA Best Practices in Clinical Data Management.* Retrieved https://redcrevistas.com/index.php/Revista

7 Chang TC, Seufert C, Eminaga O, Shkolyar E, Hu JC, and Liao JC. Current trends in artificial intelligence application for endourology and robotic surgery. *Urologic Clinics* 2021;48(1):151–160.

8 Boniolo F, Dorigatti E, Ohnmacht AJ, Saur D, Schubert B, and Menden MP. Artificial intelligence in early drug discovery enabling precision medicine. *Expert Opinion on Drug Discovery* 2021;16(9):991–1007.

9 Pataranutaporn P, Danry V, Leong J, et al. AI-generated characters for supporting personalized learning and well-being. *Nature Machine Intelligence* 2021;3(12):1013–1022.

10 Kakani V, Nguyen VH, Kumar BP, Kim H, and Pasupuleti VR. A critical review on computer vision and artificial intelligence in food industry. *Journal of Agriculture and Food Research* 2020;2:100033.

11 Jain, A., Kulkarni, G., and Shah, V. Natural Language Processing. *International Journal of Computer Sciences and Engineering*, 2018; 6(1), 161–167. https://doi.org/10.26438/ijcse/v6i1.161167.

12 Selvaraj C, Chandra I, and Singh SK. Artificial intelligence and machine learning approaches for drug design: Challenges and opportunities for the pharmaceutical industries. *Molecular Diversity* 2022; 26(3):1893–1913.

13 Sharma A and Singh BJ. Evolution of industrial revolutions: A review. *International Journal of Innovative Technology and Exploring Engineering* 2020;9(11):66–73.

14 Berry, IR and Nash, RA. (2003). *Pharmaceutical process validation.* Marcel Dekker.

15 Lee J, Davari H, Singh J, and Pandhare V. Industrial Artificial intelligence for industry 4.0-based manufacturing systems. *Manufacturing Letters* 2018;18:20–23.

16 Arden NS, Fisher AC, Tyner K, Yu LX, Lee SL, and Kopcha M. Industry 4.0 for pharmaceutical manufacturing: Preparing for the smart factories of the future. *International Journal of Pharmaceutics* 2021;602:120554.

17 Menezes JC, Ferreira AP, Rodrigues LO, Brás LP, and Alves TP. Chemometrics role within the PAT context: Examples from primary pharmaceutical manufacturing. In: *Comprehensive chemometrics*, (pp. 313–355). Elsevier; 2009.

18 Vora LK, Gholap AD, Jetha K, Thakur RRS, Solanki HK, and Chavda VP. Artificial intelligence in pharmaceutical technology and drug delivery design. *Pharmaceutics* 2023;15(7):1916.

19 Russell SJ and Norvig P. *Artificial Intelligence: A Modern Approach*. Pearson; 2016.

20 Kim KG. Book review: Deep learning. *Healthcare Informatics Research* 2016;22(4):351–354.

21 Rong X and Li A. A review of research on artificial intelligence life cycle based on bibliometrics. *Frontiers in Business, Economics and Management* 2022;4(2):129–137.

22 Beneke F and Mackenrodt MO. Artificial intelligence and collusion. *IIC-International Review of Intellectual Property and Competition Law* 2019;50:109–134.

23 McCarthy J and Wright P. Technology as experience. *Interactions* 2004;11(5):42–43.

24 Chen M and Decary M. Artificial intelligence in healthcare: An essential guide for health leaders. In: *Healthcare Management Forum* (Vol. 33, No. 1, pp. 10–18). Sage CA: Los Angeles, CA: Sage Publications; 2020.

25 Davenport T and Kalakota R. The potential for artificial intelligence in healthcare. *Future Healthcare Journal* 2019;6(2):94–98.

26 Arinez JF, Chang Q, Gao RX, Xu C, and Zhang J. Artificial intelligence in advanced manufacturing: Current status and future outlook. *Journal of Manufacturing Science and Engineering* 2020;142(11): 110804.

27 Perez-Liebana D, Liu J, Khalifa A, Gaina RD, Togelius J, and Lucas SM. General video game ai: A multitrack framework for evaluating agents, games, and content generation algorithms. *IEEE Transactions on Games* 2019;11(3):195–214. https://doi.org/10.1109/TG.2019.2901021.

28 Ekins S. The next era: Deep learning in pharmaceutical research. *Pharmaceutical Research* 2016;33(11):2594–2603.

29 Korteby Y, Mahdi Y, Azizou A, Daoud K, and Regdon Jr G. Implementation of an artificial neural network as a PAT tool for the prediction of temperature distribution within a pharmaceutical fluidized bed granulator. *European Journal of Pharmaceutical Sciences* (2016);88:219–232.

30 Otto B. Interview with Reinhold Achatz on "data sovereignty and data ecosystems." *Business & Information Systems Engineering* 2019;61:635–636.

31 Rodríguez-González, A., Zanin, M., and Menasalvas-Ruiz, E. Public health and epidemiology informatics: Can artificial intelligence help future global challenges? An overview of antimicrobial resistance and impact of climate change in disease epidemiology. *Yearbook of Medical Informatics* 2019; 28(1), 224–231. https://doi.org/10.1055/s-0039-1677910.

32 Bender A and Cortés-Ciriano I. Artificial intelligence in drug discovery: What is realistic, what are illusions? Part 1: Ways to make an impact, and why we are not there yet. *Drug Discovery Today* 2021;26(2):511–524.

33 Han X, Gao T, Yao Y, Ye D, Liu Z, and Sun M. OpenNRE: An open and extensible toolkit for neural relation extraction. *arXiv preprint arXiv:1909.13078*. 2019.

34 O'Shea O, Steeg R, Chapman C, Mackintosh P, and Stacey GN. Development and implementation of large-scale quality control for the European bank for induced Pluripotent Stem Cells. *Stem Cell Research* 2020;45:101773.

35 Shirazian S, Grant CD, Aina O, Mattana J, Khorassani F, and Ricardo AC. Depression in chronic kidney disease and end-stage renal disease: Similarities and differences in diagnosis, epidemiology, and management. *Kidney International Reports* 2017; 2(1):94–107.

36 von Stosch M, Hamelink JM, and Oliveira R. Toward intensifying design of experiments in upstream bioprocess development: An industrial *Escherichia coli* feasibility study. *Biotechnology Progress* 2016;32(5):1343–1352.

37 von Stosch M, Hamelink JM, and Oliveira R. Hybrid modeling as a QbD/PAT tool in process development: An industrial *E. coli* case study. *Bioprocess and Biosystems Engineering* 2016;39:773–784.

38 Lin Z and Chou WC. Machine learning and artificial intelligence in toxicological sciences. *Toxicological Sciences* 2022;189(1):7–19.

39 Yamada S, Mochizuki M, Atobe K, and Kato Y. Pharmcokinetic and pharmacodynamic considerations for the anticholinergic burden scale of drugs. *Geriatrics & Gerontology International* (2024); 24:81–87.

40 Gams M, Horvat M, Ožek M, Luštrek M, and Gradišek A. Integrating artificial and human intelligence into tablet production process. *Aaps Pharmscitech* 2014;15:1447–1453.

41 Mishra V. Artificial intelligence: The beginning of a new era in pharmacy profession. *Asian Journal of Pharmaceutics (AJP)* 2018;*12*(02).

42 Wang WY, Li J, and He X. Deep reinforcement learning for NLP. In: Artzi Y and Eisenstein J, editors. *Proceedings of the 56th Annual Meeting of the Association for Computational Linguistics: Tutorial Abstracts*, (pp. 19–21). Association for Computational Linguistics; 2018.

43 Hughes, JP, Rees, SS, Kalindjian, SB, and Philpott, KL. Principles of early drug discovery. In *British Journal of Pharmacology* 2011; 162(6): 1239–1249. https://doi.org/10.1111/j.1476-5381.2010.01127.x.

44 Kirschner LS. The next generation of therapies for adrenocortical cancers. *Trends in Endocrinology & Metabolism* 2012;23(7):343–350.

45 Yoshida F and Topliss JG. QSAR model for drug human oral bioavailability. *Journal of Medicinal Chemistry* 2000;43(13):2575–2585.

46 Zhavoronkov A, Vanhaelen Q, and Oprea TI. Will artificial intelligence for drug discovery impact clinical pharmacology. *Clinical Pharmacology & Therapeutics* 2020;107(4):780–785.

47 Vamathevan J, Clark D, Czodrowski P, et al. Applications of machine learning in drug discovery and development. *Nature Reviews Drug Discovery* 2019;18(6):463–477.

48 Pun FW, Ozerov IV, and Zhavoronkov A. AI-powered therapeutic target discovery. *Trends in Pharmacological Sciences* 2023;44(9):561–572.

49 Gaulton A, Bellis LJ, Bento AP, et al. ChEMBL: A large-scale bioactivity database for drug discovery. *Nucleic Acids Research* 2012; 40(D1):D1100–D1107.

50 Knox C, Wilson M, Klinger CM, et al. DrugBank 6.0: The DrugBank knowledgebase for 2024. *Nucleic Acids Research* 2024; 52(D1):D1265–D1275.

51 Agatonovic-Kustrin S and Morton D. Data mining in drug discovery and design. In: *Artificial Neural Network for Drug Design, Delivery and Disposition*, (pp. 181–193). Academic Press; 2016.

52 Badia-Bringué G, Lavín JL, Casais R, and Alonso-Hearn M. Alternative splicing of pre-mRNA modulates the immune response in Holstein cattle naturally infected with Mycobacterium avium subsp. paratuberculosis. *Frontiers in Immunology* 2024;15:1354500.

53 Badwan BA, Liaropoulos G, Kyrodimos E, Skaltsas D, Tsirigos A, and Gorgoulis VG. Machine learning approaches to predict drug efficacy and toxicity in oncology. *Cell Reports Methods* 2023:3(2).

54 Jin S, Zeng X, Fang J, et al. A network-based approach to uncover microRNA-mediated disease comorbidities and potential pathobiological implications. *NPJ Systems Biology and Applications* D 2019;5(1):41.

55 Mathai, N and Kirchmair, J. Similarity-based methods and machine learning approaches for target prediction in early drug discovery: Performance and scope. *International Journal of Molecular Sciences* 2020; *21*(10). https://doi.org/10.3390/ijms21103585.

56 Chavda VP, Anand K, and Apostolopoulos V, editors. *Bioinformatics Tools for Pharmaceutical Drug Product Development*. John Wiley & Sons; 2023.

57 Rowe RC and Colbourn EA. Neural computing in product formulation. *Journal of Chemical Education* 2003;8:1–8.

58 Al Kuwaiti A, Nazer K, Al-Reedy A, et al. A review of the role of artificial intelligence in healthcare. *Journal of Personalized Medicine* 2023;13(6):951.

59 Hamet P and Tremblay J. Artificial intelligence in medicine. *Metabolism* (2017);69:S36–S40.

60 Kesavan, JG and Peck, GE. Pharmaceutical Granulation and Tablet Formulation Using Neural Networks Kesavan and Peck. In *Pharmaceutical Development and Technology* 1996; 1(4).

61 Katsila T, Spyroulias GA, Patrinos GP, and Matsoukas MT. Computational approaches in target identification and drug discovery. *Computational and Structural Biotechnology Journal* 2016;14:177–184.

62 Vyas, M., Thakur, S., Riyaz, B., Bansal, K. K., Tomar, B., & Mishra, V. (n.d.). Artificial Intelligence: The Beginning of a New Era in Pharmacy Profession. In *Asian Journal of Pharmaceutics*, 12(2).

63 Siepmann J and Siepmann F. Mathematical modeling of drug delivery. *International Journal of Pharmaceutics* 2008;364(2):328–343.

64 Choi J, Gu B, Chin S, and Lee JS. Machine learning predictive model based on national data for fatal accidents of construction workers. *Automation in Construction* 2020;110:102974.

65 Zhao H, Zhao L, Lin X, and Shen L. An update on microcrystalline cellulose in direct compression: Functionality, critical material attributes, and co-processed excipients. *Carbohydrate Polymers* 2022;278:118968.

66 Shafiq M, Thakre K, Krishna KR, Robert NJ, Kuruppath A, and Kumar D. Continuous quality control evaluation during manufacturing using supervised learning algorithm for Industry 4.0. *The International Journal of Advanced Manufacturing Technology* 2023; 1–10.

67 Reinhardt IC, Oliveira JC, and Ring DT. Current perspectives on the development of industry 4.0 in the pharmaceutical sector. *Journal of Industrial Information Integration* 2020;18:100131.

68 Soldatos TG, Kim S, Schmidt S, Lesko LJ, and Jackson DB. Advancing drug safety science by integrating molecular knowledge with post-marketing adverse event reports. *CPT: Pharmacometrics & Systems Pharmacology* 2022;11(5):540–555.

69 Joung S, Kim Y, Kim J, et al. Simultaneous quantitative analysis of ^{3}H and ^{14}C radionuclides in aqueous samples via artificial neural network with a liquid scintillation counter. *Applied Radiation and Isotopes* 2021;170:109593.

70 Stasevych M and Zvarych V. Innovative robotic technologies and artificial intelligence in pharmacy and medicine: paving the way for the future of health care—A review. *Big Data and Cognitive Computing* 2023;7(3):147.

71 Markarian RA, da Silva RLB, Burgoa S, Pinhata-Baptista OH, No-Cortes J, and Cortes ARG. Clinical relevance of digital dentistry during covid-19 outbreak: A scoped review. *Brazilian Journal of Oral Sciences* 2020;19:e200201.

72 Jain C, Rodriguez-R LM, Phillippy AM, Konstantinidis KT, and Aluru S. High throughput ANI analysis of 90K prokaryotic genomes reveals clear species boundaries. *Nature Communications* 2018;9(1):5114.

73 Swarbrick J. *Pharmaceutical Process Validation* (No. 22150). RA Nash and AH Wachter (Eds.). New York, NY: Marcel Dekker; 2003.

74 Khalid N, Qayyum A, Bilal M, Al-Fuqaha A, and Qadir, J. Privacy-preserving artificial intelligence in healthcare: Techniques and applications. *Computers in Biology and Medicine* 2023;158:106848.

75 Obermayer N and Tóth VE. The future of human knowledge and artificial intelligence. In: *European Conference on Knowledge Management*, (pp. 808-XXIV). Academic Conferences International Limited; 2019.

76 Sharma S, Singh G, Sharma CS, Kapoor S. Artificial intelligence in Indian higher education institutions: a quantitative study on adoption and perceptions. *International Journal of System Assurance Engineering and Management* 2024; 1–17.

17

Regulatory Compliance on Data-driven Pharmaceutical Research and Development

Md Akbar[1], Hasan Ali[2], Md. Faiyazuddin[3], Shrestha Sharma[4], Rakesh Kumar[4], and Ankur Sharma[5]*

[1]*School of Pharmacy, Al-Karim University, Katihar, India*
[2]*Department of Pharmacy, Meerut Institute of Technology, Meerut, India*
[3]*Department of Pharmacy, ARAM Institute, Ranchi, India*
[4]*Amity Institute of Pharmacy, Amity University, Gurugram, India*
[5]*Hyalo Technologies, Somerset, New Jersey, USA*

*Corresponding author: akbarmd86@hotmail.com

CONTENTS

17.1 Introduction

Pharmaceutical processes involve a series of interrelated steps, including drug discovery, preclinical and clinical trials, formulation development, and manufacturing. Traditionally, these processes relied on time-consuming and costly experimental approaches to optimize drug production and ensure quality. However, the pharmaceutical industry has embraced data-driven modeling to leverage the wealth of available data and enhance process understanding. This chapter essentially aims to explore the concepts and approaches that facilitate the integration of data-driven modeling in pharmaceutical processes. Data-driven modeling begins with the collection of relevant data from various sources. These sources include research and development laboratories, clinical trials, manufacturing facilities, and even social media platforms for gathering patient feedback. Additionally, data from analytical instruments, sensors, and other process monitoring tools contribute to the extensive dataset used in data-driven models [1]. The quality of data is crucial in ensuring the accuracy and reliability of data-driven models. Preprocessing and data cleaning techniques are applied to remove noise, correct errors, and handle missing data. Proper data preparation is vital for obtaining meaningful insights and establishing robust models.

Machine learning (ML) algorithms are widely used in data-driven modeling due to their ability to identify patterns, relationships, and trends within datasets. ML techniques, such as classification, regression, clustering, and reinforcement, also play an important role in visualization of data integrity and its pattern in a data set. Essentially, ML is applied to predict drug behavior, optimize formulations, and analyze drug interactions. Artificial neural network (ANN) is a subset of ML algorithms that simulate the human brain's neural connections. They are particularly useful for modeling complex and nonlinear relationships in pharmaceutical processes. Moreover, ANNs can predict drug–drug interactions, optimize drug dosages, and predict adverse drug reactions and clinical manifestations. Furthermore, deep learning, a subset of ML and ANN, uses multiple layers of artificial neurons to learn hierarchical representations from data. It has proven valuable in image analysis, drug design, and target identification [2]. Data mining techniques, such as association rule mining and anomaly detection, help uncover hidden patterns and correlations in pharmaceutical datasets. These insights can lead to a better understanding of drug interactions and side effects. Classical statistical approaches like regression analysis and hypothesis testing are used to identify significant factors influencing drug development and manufacturing processes. Data-driven models must be rigorously validated to ensure their accuracy and reliability. Validation involves comparing model predictions with experimental results and assessing model performance using metrics such as accuracy, precision, and product recall. Additionally, cross-validation techniques are applied to verify the generalization capabilities of the models. Data-driven models play a pivotal role in drug discovery by predicting drug–target interactions, identifying potential drug candidates, and optimizing lead compounds. Optimizing pharmaceutical formulations is crucial to ensure drug stability, bioavailability, and patient compliance. Data-driven models aid in predicting the ideal composition and manufacturing process for drug formulations [3]. Data-driven models facilitate the optimization of manufacturing processes, leading to increased production efficiency, reduced waste, and improved product quality. Data-driven modeling enables the development of personalized treatment plans based on a genetic makeup, lifestyle, and medical history of the patient(s). Data-driven models support real-time monitoring and quality control during the manufacturing process, ensuring adherence to regulatory standards and the delivery of safe and effective pharmaceutical products [4]. Despite the numerous advantages of data-driven modeling in pharmaceutical processes, there are several challenges and limitations to be addressed. Acquiring high-quality, diverse, and relevant data can be challenging due to privacy concerns and proprietary information. Some data-driven models, especially deep learning algorithms, may lack interpretability, making it difficult to understand the underlying mechanisms of predictions. Data-driven modeling often requires significant computational resources, especially for training complex models on large datasets [5]. Regulators may require transparency and validation of data-driven models before their integration into pharmaceutical processes. Data-driven modeling is poised to revolutionize the pharmaceutical industry, leading to the accelerated drug development, reduced costs, and improved patient outcomes. Future developments may focus on enhancing model interpretability, addressing regulatory challenges, and developing more efficient data collection and preprocessing techniques. Data-driven modeling has become an integral part of pharmaceutical processes, providing valuable insights and optimization capabilities that were previously unattainable. By leveraging vast amounts of data, pharmaceutical companies can make informed decisions, streamline drug development, and ensure the delivery of safe and effective drugs. As technology continues to evolve, data-driven modeling will play an increasingly vital role in shaping the future of pharmaceutical research and manufacturing [6].

Hence, data-driven modeling has ushered in a new era of pharmaceutical processes, revolutionizing drug development, manufacturing, and patient care. Leveraging vast amounts of data, these models offer valuable insights, optimization capabilities, and the potential for personalized medicine. However, challenges, such as data availability, model interpretability, and regulatory compliance, must be overcome to fully realize the potential of data-driven modeling in the pharmaceutical industry [7]. By addressing these challenges and continuing to advance the field, data-driven modeling will continue to be a driving force behind innovations in drug discovery, formulation development, and manufacturing, leading to better and safer pharmaceutical products for patients worldwide.

17.2 Data-driven Approaches in Drug Regulatory Compliance

Data-driven approaches in drug regulatory compliance involve data analytics and artificial intelligence (AI) to enhance efficiency, risk mitigation, and global compliance, ultimately speeding up drug development and ensuring product safety and efficacy (Figure 17.1).

17.2.1 Real-world Data

Real-world data (RWD) is information about patient health and healthcare delivery collected in routine clinical practice, outside of controlled research settings, from sources like electronic health records (EHRs), claims databases, and patient-generated data. RWD is increasingly used to inform healthcare decisions, drug development, and regulatory processes [8]. RWD can provide a broader and more diverse perspective on patient populations and outcomes compared to data from traditional clinical trials. RWD is increasingly used to inform healthcare decisions, drug development, and regulatory processes. RWD often needs to be transformed into a more usable format before it can be used in decision-making [9].

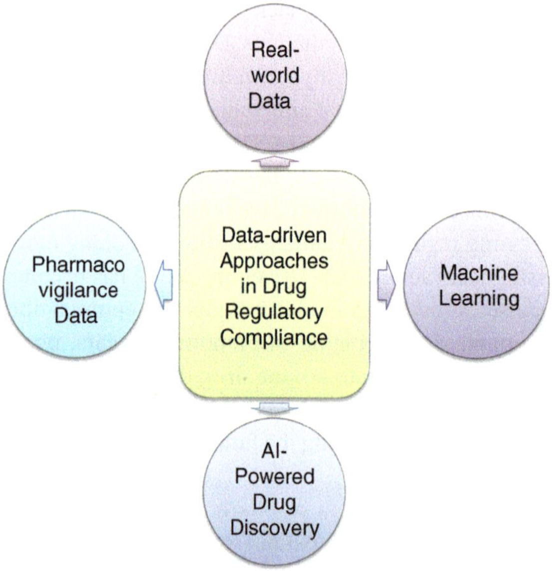

Figure 17.1 Various approaches of data-driven regulatory compliance.

This involves cleaning, validating, and organizing the data. RWD can be derived from various sources, including:

1. EHRs
2. Claims and billing activity
3. Patient registries
4. Data from mobile devices and wearables
5. Patient questionnaires
6. Social media.

Moreover, the term "real-world evidence" (RWE) is closely related to RWD. RWE is defined by the Food and Drug Administration (FDA) as "clinical evidence regarding the usage and potential benefits or risks of a medical product derived from analysis of RWD." The FDA has established a framework for evaluating the potential use of RWE to support the approval of new indications for drugs or to support post-approval study requirements [10].

17.2.2 ML Algorithms

ML algorithms are widely used in data-driven modeling due to their ability to identify patterns, relationships, and trends within datasets. ML techniques, such as classification, regression, clustering, and reinforcement learning, are applied to predict drug behavior, optimize formulations, and analyze drug interactions. ANNs are a subset of ML algorithms that simulate the human brain's neural connections. They are particularly useful for modeling complex and nonlinear relationships in pharmaceutical processes [11]. ANNs can predict drug–drug interactions, optimize drug dosages, and even predict adverse drug reactions. Deep learning, a subset of ML and ANN, uses multiple layers of artificial neurons to learn hierarchical representations from data. It has proven valuable in image analysis, drug design, and target identification. Unlike traditional ML methods, deep learning models use ANNs as their basic architecture, adding multiple layers of nonlinear processing units to learn data representations. Deep neural networks are available in a variety of different frameworks, such as recurrent neural networks and convolutional neural networks, and may benefit from a multilayer neural network architecture, which allows deep learning models to use more diverse descriptors and process larger datasets [12]. With powerful learning and data processing capabilities, deep learning methods are making a significant impact in areas such as reaction outcome prediction and inverse synthesis route planning.

17.2.3 AI-powered Drug Discovery

AI is revolutionizing drug discovery by accelerating the process, improving accuracy, and reducing costs through techniques like ML and natural language processing (NLP), enabling faster identification of drug candidates and better prediction of their efficacy and toxicity. Drug discovery, the process of identifying and developing new medications, is a complex and time-consuming endeavor that traditionally relies on labor-intensive techniques, such as trial-and-error experimentation and high-throughput screening. However, AI techniques such as ML and NLP offer the potential to accelerate and improve this process by enabling more efficient and accurate analysis of large amounts of data. AI-based methods have also been able to predict the toxicity of drug candidates. These research efforts have highlighted the capacity of AI to improve the efficiency and effectiveness of drug discovery processes. However, the use of AI in developing new bioactive compounds is not without challenges and limitations. Ethical considerations must be taken into account, and further research is needed to fully understand the advantages and limitations of AI in this area. Despite these challenges, AI is expected to significantly contribute to the development of new medications and therapies in the next few years [13].

17.2.4 Pharmacovigilance Data

Comprehensive methods for evaluating safety are needed to objectively assess the full risk profile of a medication. The confidence of the prescribing provider in the safety and effectiveness of pharmaceuticals is extremely important. Pharmacovigilance is a key component of drug safety regulatory processes and is paramount for ensuring the safety profile of medications used to treat patients. All participants in the healthcare system, including healthcare providers and consumers, should understand and meaningfully engage in the pharmacovigilance process. Healthcare providers should integrate pharmacovigilance into everyday practice, inviting feedback from patients. This narrative review aims to give an overview of the main topics underlying pharmacovigilance and drug safety in pharmaceutical research phase after the authorization of a drug in the United States. The United States Food and Drug Administration (US-FDA) guidance and post-approval regulatory actions are considered from an industry perspective [14].

17.3 Characteristics of Drug Regulatory Compliance in Data-driven Research and Development

In data-driven drug research and development, regulatory compliance hinges on ensuring data integrity, adherence to good laboratory practices and good clinical practices, and maintaining robust data governance and security measures (Figure 17.2).

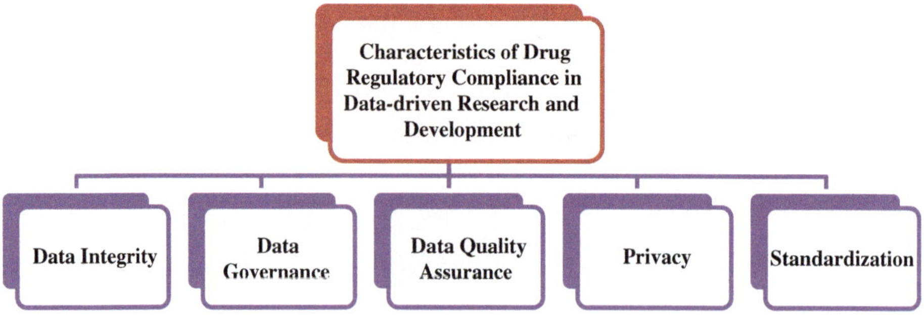

Figure 17.2 Characteristics of drug regulatory compliance in data-driven research and development.

17.3.1 Data Integrity

In the pharmaceutical industry, data is essential for preserving patient safety and product quality as well as creating better medical items for human consumption. Both manual and electronic formats may contain data. Data must comply with data integrity requirements. The accuracy, consistency, and reliability of data in a database or other information storage system are referred to as data integrity. It attests to the fact that data remains unchanged during transmission, storage, and retrieval. Companies and organizations need to safeguard data integrity while executing informed decisions based on accurate information. It can also be used to describe how accurate, consistent, thorough, and dependable the data is over the course of its existence [15]. The term "data lifecycle" refers to all of the activities that occur with data over the course of its whole existence, including creation, recording, processing (including transmission or modification), applications, retention, archiving, retrieval, and destruction. It is common practice to use "data quality" as a stand-in for "data integrity." A crucial element of drug development, manufacturing, and regulatory compliance in the ever-evolving pharmaceutical sector is data integrity assurance [16, 17]. As the pharmaceutical industry continues to evolve, ensuring the accuracy, completeness, and dependability of data has become increasingly important in maintaining the quality, safety, and efficacy of pharmaceutical products. Because regulators like the US-FDA and the European Medicines Agency (EMA) enforce strict requirements, pharmaceutical companies must create and implement thorough policies and processes to safeguard the quality of data throughout a product's life cycle. In the life sciences sector, data integrity is a crucial component of quality control [18]. Processes must follow stringent guidelines, especially in the manufacturing of pharmaceuticals, to ensure the consistency and verifiability of final product quality. Only then it would be possible to ensure the efficacy of medications and lower the possibility of adverse effects and errors in manufacturing procedures. Regulatory bodies define how and to what degree producers are required to document their processes throughout time and record the data pertaining to them [19].

The regulatory structure known as attributable, legible, contemporaneous, original, and accurate (ALCOA) is used by authorized companies to guarantee data integrity and is necessary for the execution of good documentation practices (GDP). ALCOA covers both paper and electronic data. Promoting data integrity campaigns, maintaining GDP, following current good manufacturing practices (cGMP), and maintaining an electronic and paper data handling system that satisfies standards all depend on the ALCOA principles. In the 1990s, Stan W. Woollen of the FDA's Office of Enforcement was the first to use the abbreviation ALCOA. Complete, consistent, enduring, and available (CCEA) were added to the Journal Pre-proof ALCOA in 2010 [20, 21, 22]. More often, ALCOA-C or ALCOA is used instead of ALCOA-CCEA. The ALCOA/ALCOA+ principles, their applicability in the life sciences sector, and how innovative data logging, archiving, and analysis solutions assist users satisfy data integrity standards are all covered in this article [23]. Guidelines for ensuring data integrity have been released by international regulatory bodies and associations, including the EMA, the US-FDA, and the International Society for Pharmaceutical Engineering (ISPE). Among other things, the ALCOA/ALCOOA+ documentation practice standards serve as the foundation for these rules [22]. As of November 2020, the World Health Organization and the Pharmaceutical Inspection Cooperation Scheme are developing good practice guidelines that make reference to the ALCOA/ALCOA+ principles [24]. These guidelines, in conjunction with the ALCOA/ALCOA+ principles, will serve as the foundation for a worldwide uniform strategy to guarantee data integrity [25]. Following are the ALCOA elements.

17.3.2 Attributable

Attributable provides information on who created or collected the data, when it was created or gathered, and whether the individual explicitly marked the date, time, and signature of the higher authority with a countersignature. It is necessary to store data in a manner that allows it to be uniquely linked to a person or computer system. The information generated or gathered must reveal the name of the individual who created it, ensuring accountability. This gives a record of what was done, by whom, and when. The name of the scientist who carried out the experiment should be signed or otherwise identified on every entry in a lab record. An investigation trail must be signed by the client who has the authority to change an existing limit on a process or checking model [26].

17.3.3 Legible

According to Legible, any written or computer-generated material must be readable. Errors happen frequently, and human handwriting in particular has to have a single line removed so that everyone can see what went wrong. The person who made the mistake must have an asterisk at the incorrect location and, as shown by the same asterisk at the page's vacant area, must explain why the error was committed. The individual who has gathered or produced the data must sign the table and mark any blank spaces with "NA," which stands for "Not Applicable." Once adequate print papers have been obtained, printed documents that are not readable or correctly printed must be closely monitored and, if required, documented [27]. Data must always be captured in a way that makes it simple to track down and replicate. Even in this digital age, readable data is not always readily apparent. It is true that maintaining legible and persistent records is essential to maximizing data accessibility during the course of their existence. This also holds true for human-readable metadata that can be recorded to support an electronic record. It is crucial to remember that paper records must be kept under control to prevent illegal duplication and that all records must be readable and understandable by those with the proper authorization. Buzzwords and colloquial terms should be avoided as they are likely to be changed over time and also from place to place, and they are often not understandable by a particular region [22].

17.3.4 Contemporaneous

Contemporaneous means that the event must be documented right away once it is over, not the other way around. This implies that the person must finish the task and record it right away, rather than recording and then carrying out the activity. When data is created, it needs to be tracked and recorded. Only at the moment of the observation or action must the data be documented [28]. In addition to introducing mistakes or inaccuracies, delayed documentation may cause us to undermine the data's accuracy and quality. Instead of being documented after the fact, patient data in clinical studies should be captured as the observations are made. Real-time entries or those sent as close to the event as feasible are preferred. Real-time data recording guarantees information accuracy while reducing the risk of memory errors. When data is created, it needs to be tracked and recorded [29].

17.3.5 Original

To guarantee process traceability, the original data record needs to be logged. The original format of the data must be kept and maintained. It alludes to original data. Any edits or changes must be recognized and justified, and data must be recorded in its original format [28]. This contributes to preserving the data's dependability. Original batch records should not be changed during the production process. Any changes or revisions must follow the right procedures and be properly recorded. It is important to make sure that the permitted protocol includes documentation of the validation test findings. Make sure a "true copy" is made if the original data is handwritten and is intended to be saved in an electronic format. Only after verifying that the information is accurate should the copy be moved to the computer system. Errors may occur when the results are transcribed onto a notepad for subsequent transcription [22].

17.3.6 Accurate

The information must be kept in a way that ensures its accuracy, completeness, validity, and dependability. Appropriate control mechanisms and process structures are needed for this. Data must faithfully depict the actual events that have taken place. It must accurately and truthfully represent the real data or observations [27]. This principle highlights how crucial accuracy and precision are when recording data. When a substance's weight is measured with a balance, the recorded weight should accurately reflect the measurement, free of inaccuracies or rounding mistakes. Every measurement and observation must be accurate and consistent with the real values. Data accuracy is influenced by routine instrument calibration and technique validation [30].

17.3.7 Complete

All pertinent aspects should be covered by the data. Analysis and decision-making may be hampered by missing information. All pertinent information should be included in the records to provide a thorough understanding of the procedure and experiment. Incomplete or missing records may cause the results to be interpreted incorrectly. The information must be comprehensive, such as containing data from experiments that may be repeated [31].

17.3.8 Consistent

Every component, including the chronologically ordered sequence of events, needs to be dated and time-stamped appropriately. Data uniformity across many records and systems is ensured by consistency [24]. Errors and confusion can result from inconsistencies. It is important to adhere to standardized protocols continuously across various operations or investigations. Data dependability and comparability throughout time are guaranteed by consistency [32].

17.3.9 Enduring

To guarantee that data is accessible for the duration of the retention term, it should be kept on appropriate storage media (either electronic or hard copy). Long-lasting data should maintain its integrity. Archiving, backup, and proper storage are always seen as crucial [33].

17.3.10 Available

Data should be accessible at every stage of its life cycle, for example, for inspection. Data must be accessible for validation, examination, and evaluation. Transparency and compliance depend on accessibility. Both digital and paper records must be readily available for inspections or investigations for the duration of the record. Both paper and electronic materials should be properly labeled and/or have clear indexing to facilitate retrieval [30].

17.4 Data Governance

Two fundamental ideas that guarantee data management, security, and responsible use inside an organization are ownership and responsibility. Establishing precise roles, duties, and expectations with data assets requires an understanding of these ideas. The term "data ownership" describes the power and accountability over the data. It concerns who is in charge of the data and who makes the decisions about how to use it. Making sure that those who handle, utilize, or access data are held accountable for their actions, especially with regard to data security, privacy, and compliance, is known as accountability. This guarantees that the organization can identify the accountable actors for choices, actions, and results. The chief data officer is in charge of the data strategy organization, data governance, and data quality [34]. This entails supervising the gathering, handling, and application of data throughout the company to guarantee that it is used for strategic advantage while upholding standards of quality and compliance. A regulatory compliance officer makes sure the business abides by all applicable laws, rules, and guidelines. By making sure company operations comply with industry-specific rules (such as general data protection regulation (GDPR), Health Insurance Portability and Accountability Act (HIPAA), and Sarbanes-Oxley Act), they concentrate on reducing legal risks. Information technology security teams are in charge of defending the organization's data, systems, and networks against online attacks. This can include safeguarding private information and maintaining the reliability of digital systems [35].

To guarantee that data is managed appropriately and effectively over its whole existence, data life cycle management is essential. Effective management at each stage can assist assure data quality, security, compliance, and cost-efficiency. The lifetime usually consists of multiple stages, from generation to archival or disposal. Applications, transactions, sensors, and human interaction are just a few of the sources from which data is generated or gathered. Databases, file systems, cloud storage, and data lakes are some of the places where data is kept. Most data stays here for the duration of its useful life. Access, processing, analysis, and sharing of data are done for reporting, decision-making, company operations, and other reasons. When data is no longer needed, it is safely disposed of to prevent recovery or misuse. A method for controlling user permissions according to their positions within a company or system is called role-based access control. Users are given designated roles in a role-based access control system, and each position has a set of permissions. This guarantees that specific resources or data can only be accessed or modified by authorized users. Governments enact pharmaceutical laws as a collection of legislative, administrative, and technical procedures to ensure the quality, safety, and efficacy of medications as well as the accuracy and relevance of publicly available product data [36]. The word "regulation" refers to a wide range of documents with various legal foundations and levels of authority, including guidelines, suggestions, processes, policies, etc. Additionally, these authorities are essential in guaranteeing and growing regular implementation in various regions of the planet [35, 36].

17.4.1 European Medicines Agency

In its "Reflection document on current cGMP and Marketing Authorization Holders" and "cGMP for Medicinal Products for Human and Veterinary Use" guidelines, the EMA places a strong emphasis on data integrity. Important points consist of the process of ensuring that information is unique, accurate, legible, traceable, and up-to-date, involves putting in place a quality management system that incorporates data integrity (ALCOA). Essentially, EMA is an agency of the European Union (EU) that oversees and assesses pharmaceuticals. Prior to 2004, it was called the EMA or the European Agency for the Evaluation of Medicinal Products. In order to standardize the work of the national regulatory agencies that currently oversee pharmaceuticals, the EMA was founded in 1995 with financial support from the EU, the pharmaceutical sector, and indirect subsidies from member states [37]. Prior to the exit of the United Kingdom from the EU, the EMA was a decentralized organization based in London. In March 2019, it was moved to Amsterdam. Currently, almost one-third of the new medications introduced to the global market each year come from the EU [38].

17.4.2 United States Food and Drug Administration

Guidelines for data integrity are provided by the US-FDA through various documents, including the "Data Integrity and Compliance With cGMP" guidance [39].

The key principles include:

- Implementing controls to maintain the integrity of data
- Performing periodic checks and data assessments
- Providing training on data integrity principles.

To make sure they are safe and effective for the general public to use, the US-FDA regulates food, medications, medical devices, biologics, and cosmetics. It is essential to the approval of novel medications, vaccines, and medical equipment [25]. Both the GDPR and the HIPAA are important laws intended to safeguard sensitive information, although they concentrate on distinct kinds of data and geographical areas. On May 25, 2018, the EU implemented the GDPR, a data protection law. Through the unification of EU legislation, it aims to simplify the regulatory environment for international commerce and provide EU citizens with more control over their personal data. Collaboration and interoperability refer to the smooth sharing and interchange of data across various platforms, systems, and organizations while preserving the data's integrity, security, and compliance. As businesses handle enormous volumes of data and need to facilitate communication across different departments, stakeholders, or even entire ecosystems, the concepts of interoperability and collaboration in data governance have become more and more crucial [19].

17.5 Data Quality Assurance

The process of making sure that data is accurate, consistent, dependable, and suitable for its intended use is known as data quality assurance (DQA). Fundamentally, DQA is crucial in many fields, particularly those like banking, healthcare, marketing, and research where data is used extensively to make choices. High-quality data is maintained at every stage of its lifespan, from collection and storage to analysis and reporting, with the support of a strong DQA system. Indicators are chosen, available documents and datasets are reviewed, the data collection and management system is evaluated, its implementation and operationalization are reviewed, the data is verified and validated, and a DQA report is created. A crucial component of data processing, DQA is a collection of procedures and actions meant to preserve the consistency, accuracy, and dependability of data [40]. Organizations greatly depend on high-quality data in today's increasingly data-driven world in order to make well-informed decisions, propel strategic initiatives, and obtain a competitive advantage. Fundamentally, DQA entails setting strict guidelines, carrying out methodical inspections, and putting remedial mechanisms in place to ensure the accuracy and reliability of data. Businesses can reduce errors, stop data abnormalities, and improve the overall quality of their data by following these guidelines. The objectives of the DQA are to

- establish a system for evaluating the quality of data that includes regular data monitoring, yearly discrete DQA, and recurring in-depth evaluations of priority health programs.
- determine the data management system's shortcomings and potential solutions to strengthen it.
- track the ability to generate high-quality data and the performance of data quality over time.

17.5.1 Statistical Principles for Clinical Trials

It stresses pre-specification of statistical analysis methods, including management of intercurrent events and missing data, and offers a framework for the design, conduct, analysis, and interpretation of clinical trials to guarantee that the evidence of efficacy and safety is solid and dependable. In order to more precisely quantify treatment effects, it established the idea of estimands in the ICH E9 (R1) addendum and encourages bias minimization through randomization, blinding, and adherence to a well-defined protocol [41].

17.5.2 US-FDA Quality Management System Regulation and Data Reliability Guidelines

Describes the most recent cGMP standards for quality systems and is applicable to producers of medical devices sold in the United States. Design controls: make certain that medical devices are created in a way that satisfies both user demands and legal specifications. Production and process controls: to guarantee that devices meet specifications and maintain manufacturing consistency [2]. Determine flaws, fix them, and stop them from happening again. With the implementation of the new quality management system regulation (QMSR) in February 2026, the US-FDA is bringing Part 820 into compliance with International Organization for Standardization (ISO), ISO 13485. Throughout the clinical trial process, place a strong emphasis on data reliability and integrity, regardless of whether data is gathered electronically or manually. Attributable, unique, accurate, current, and readable data are required. Robust audit trails, restricted access, and error-reducing features (such as edit checks, prompts, and validation criteria) are essential for computerized systems. Sponsors are responsible for making sure that data management procedures adhere to applicable code of federal regulation (CFR) sections (e.g., 21 CFR Part 11 for electronic records and signatures) [42].

The statistical basis provided by International Council for Harmonization (ICH) of Technical Requirements for Pharmaceuticals for Human Use E9 needs to be explicitly stated in the statistical analysis plan and clinical trial protocol. This entails adhering to the predetermined analysis, managing concurrent events appropriately, and planning for bias minimization. Adherence to 21 CFR Part 820, often known as QMSR, guarantees that the equipment utilized (or data generated in clinical trials) is manufactured in a manner that satisfies stringent quality requirements. It is essential to integrate corrective and preventive actions, procedures, risk management, and design controls. The US-FDA data dependability criteria demand that the data supporting clinical trial results be robust and verifiable, which is accomplished by following consistent quality control methods and strict computerized system requirements (as per 21 CFR Part 11). In addition to ensuring trustworthy trial results and protecting patient safety, this integrated regulatory framework offers a strong foundation for regulatory decision-making [43].

17.6 Data Privacy

Protecting personal information and giving people control over how it is gathered, processed, stored, and disseminated is known as data privacy. It includes people's right to keep their data safe and private, preventing others from accessing or using it without authorization. Typically, "data privacy" relates to how important personal data—also known as "personally identifiable information"—is handled. Social Security numbers, medical records, and financial information such as credit card and bank account numbers are examples of this type of data. Data privacy in a company setting extends beyond the personally identifiable information of clients and staff. This could include financial information, development statistics, or confidential research. Data privacy laws ensure that patient-identifiable information is protected and used solely for approved research [44].

17.6.1 Key Data Privacy Regulations

17.6.1.1 European Medicines Agency

The EU passed a comprehensive data protection law known as the GDPR. It went into effect on May 25, 2018, and its goals are to standardize data privacy rules throughout Europe and provide people greater control over their personal data [45]. Before collecting patient data, informed consent must be obtained. Grants the right to access, correct, and delete data (sometimes known as the "right to be forgotten"). Enacted in 1996, HIPAA is a federal law in the United States that establishes national standards for safeguarding private patient health information from disclosure without the patient's knowledge or agreement. It safeguards electronic medical records and clinical trials' personal health information and requires patient data to be de-identified in order to conduct secondary research [46].

17.6.1.2 United States Food and Drug Administration

It is a regulatory body from the US-FDA that establishes the criteria under which electronic records and electronic signatures are considered trustworthy, reliable, and equivalent to paper records and handwritten signatures. It governs how US-FDA-regulated industries (like pharmaceuticals, biotech, and medical devices) must manage electronic records and signatures [47]. It ensures security and traceability of electronic records in environments where compliance is critical (e.g., clinical trials, manufacturing, quality assurance/quality control processes). The California Consumer Privacy Actis a state-level privacy law in California that came into effect on January 1, 2020. It grants California residents specific rights regarding how their personal information is collected, used, and shared by businesses. Organizations use implementation measures for data privacy to safeguard personal information and guarantee adherence to national rules or data protection laws such as the California Consumer Privacy Act and GDPR. These actions may be procedural, organizational, or technical [48].

17.6.1.3 Clinical Data Interchange Standards Consortium

Clinical data interchange standards consortium (CDISC) created the standardized data format known as the study data tabulation model (SDTM). Data from clinical trials is arranged and submitted to regulatory bodies such as the US-FDA using this method that is used to format and arrange data before sending it to regulatory bodies. A fundamental component of CDISC standards is SDTM. It establishes a uniform structure for tabulating data from clinical trials [49]. To make data collection, management, analysis, and reporting procedures more efficient, SDTM offers a standard for data formatting and organization. Putting into practice SDTM allows sharing, encourages mining and reuse, supports data aggregation and warehousing, aids in due diligence and other critical data review tasks, and enhances the regulatory review and approval procedure. Additionally, SDTM is utilized in pharmacogenomics/genetics research, medical devices, and nonclinical data. A worldwide nonprofit group called the CDISC creates common standards for information gathered in clinical research [50]. In clinical research, clinical data standards are crucial frameworks that guarantee data quality, consistency, and regulatory compliance, especially when submitting to regulatory bodies such as the US-FDA or Pharmaceuticals and Medical Devices Agency (PMDA). A global nonprofit organization called CDISC creates data standards to facilitate data exchange and reuse while streamlining clinical research [51]. The reliable, consistent, and interoperable interchange of safety data pertaining to pharmaceutical goods depends on pharmacovigilance data standards. In order to monitor and report adverse drug reactions and guarantee patient safety, these standards assist regulatory bodies, marketing authorization holders, medical practitioners, and other stakeholders.

17.7 Standardization

Standardization plays a crucial role in promoting uniformity, consistency, and interoperability in global drug development by streamlining regulatory approval processes and enabling reliable and efficient data sharing across stakeholders. By ensuring uniformity, consistency, and interoperability across global drug development processes, standardization makes regulatory approvals and data transfer easier. The SDTM is a standardized data format developed by the CDISC. This strategy is used to organize and submit clinical trial data to regulatory agencies like the US-FDA and to organize and format data prior to submitting it to regulatory agencies. SDTM is an essential part of CDISC standards [25]. It creates a consistent format for clinical trial data tabulation. SDTM is one of the standards required to submit data to the PMDA in Japan and the US-FDA in the United States. SDTM provides a standard for data formatting and organizing to improve the efficiency of data collection, management, analysis, and reporting processes. Implementing SDTM facilitates data aggregation and warehousing, promotes mining and reuse, permits sharing, helps with due diligence and other crucial data review chores, and improves the regulatory review and approval process. SDTM is also used in nonclinical data, medical devices, and pharmacogenomics/genetics research. Common standards for data collected in clinical research are developed by the CDISC, a global nonprofit organization. Clinical data standards are essential standards in clinical research that ensure data uniformity, quality, and regulatory compliance, particularly when submitted to regulatory agencies like the US-FDA or PMDA [52]. To streamline clinical research and enable data exchange and reuse, a global nonprofit organization called CDISC develops data standards. Pharmacovigilance data standards are essential for the trustworthy, consistent, and interoperable exchange of safety information on pharmaceutical products. These standards support regulatory agencies, marketing authorization holders, physicians, and other stakeholders in monitoring and reporting adverse drug reactions and ensuring patient safety.

17.7.1 ICH E2B (R3)

Harmonized guidelines for the structure and content of individual case safety reports are provided by ICH E2B (R3). The objective is to enhance the quality, lucidity, and uniformity of safety data reporting on a global scale, hence promoting effective signal detection and regulatory decision-making. It allows the use of contemporary data interchange formats that facilitate enhanced data analysis and automation. Data elements are refined to gather more pertinent and detailed safety information. This gives conformity to changing global regulatory requirements to enhance the timeliness and comprehensiveness of safety data submissions [17].

17.7.2 Medical Dictionary for Regulatory Activities

A standardized medical terminology system called medical dictionary for regulatory activities (MedDRA) is used all around the world to categorize indications, medical histories, adverse events, and other safety-related data. It guarantees data comparability across various research and geographical areas, facilitating effective aggregation and analysis. A hierarchical structure makes it possible to organize data at several levels, from highly specific phrases to more general categories, which makes it easier to perform in-depth analysis and identify high-level safety signals. Extensively used by business and regulatory bodies, supporting the uniformity and caliber of safety data submissions in many countries. ICH E2B (R3) criteria can be used by sponsors to create individual case safety report that meet international requirements. The interoperability of safety data is improved when these reports are coded using MedDRA, which helps regulators better monitor, identify, and assess possible safety concerns. Robust pharmacovigilance depends on this integrated strategy, which guarantees timely identification of safety signals and well-informed decisions about patient safety [19].

17.7.3 ICH Q7

Essentially, cGMP standards for the production of active pharmaceutical ingredients (APIs) are established by the internationally standardized ICH Q7 guideline. In order to guarantee that APIs fulfill specified quality features, including identity, purity, strength, and quality, it addresses every facet of manufacturing, from facility design and raw material control to process validation and quality control. It outlines strong quality management practices (such as audits, change control, and corrective/preventive measures) as well as specifications for appropriate factory settings and calibrated, well-maintained machinery [37]. In order to reduce variability and guarantee constant product quality, it offers guidance for process design, in-process controls, and process validation. In order to verify compliance with standards, it also offers protocols for the sampling and testing of raw materials, materials used during processing, and completed APIs.

17.7.4 United States Pharmacopeia

The United States pharmacopeia (USP) is a widely accepted worldwide standard that establishes legally binding requirements for medications, excipients, and completed dosage forms in the US. In terms of identification, strength, quality, and purity, its criteria guarantee that pharmaceutical items are of the highest caliber [19]. In addition to guiding appropriate analytical processes, method validation, reference standards, and other quality tests that promote consistent, trustworthy assessments throughout the industry, it provides specific specifications, test methods, and acceptance criteria for individual substances and products. To assist manufacturers in continuously meeting regulatory quality standards, USP standards are incorporated into regulatory filings and quality assurance initiatives. USP and ICH Q7 standards complement one other. The basis for creating and sustaining high-quality API manufacturing processes is provided by ICH Q7. In order to evaluate the final quality features of pharmaceutical products, USP provides the particular compendial benchmarks (general chapters and monographs). When combined, they aid in ensuring that the product (quality specifications) and the process (cGMP) are reliable, consistent, and in compliance with legal requirements [37]. The worldwide plan to guarantee that medications are produced safely and satisfy strict quality requirements prior to being administered to patients is supported by this integrated approach.

17.8 Regulatory Compliance

For the pharmaceutical industry, regulatory compliance means following the rules, regulations, and standards that control each stage of medication development, production, and distribution. The enforcement of these regulations is supervised by international regulatory organizations, including the Medicines and Healthcare Products Regulatory Agency (MHRA), EMA, and US-FDA [52]. In pharmaceutical innovation, regulatory compliance is essential to guaranteeing that novel medications, treatments, and medical equipment are safe, efficient, and compliant with regulations prior to being released onto the market. Because of the possible hazards to the public's health, the pharmaceutical sector is heavily controlled. Throughout a drug or product's whole life cycle—from development to clinical trials, approval, manufacturing, and post-market surveillance—compliance measures are used [53].

1. Preclinical and clinical trial requirements
2. Investigational new drug and clinical trial applications
3. Regulatory approval process (new drug application, marketing authorization application)
4. cGMP
5. Post-market surveillance (pharmacovigilance)
6. Regulatory reporting and transparency
7. Data privacy and cybersecurity compliance
8. Global regulatory harmonization.

17.8.1 Code of Federal Regulations

Two sets of rules that control the use of electronic documents and signatures in the biotechnology, medical device, and pharmaceutical sectors are Annex 11 (EMA) and the US-FDA. In order for electronic records and signatures to be utilized for regulatory compliance, both regulations are designed to guarantee their dependability, security, and credibility [54, 55]. The EMA and the US-FDA are two distinct regulatory agencies, and while they both deal with related issues, their methods differ slightly [56]. To regulate the use of electronic records, electronic signatures, and audit trails in regulated businesses, the US-FDA created 21 CFR Part 11. It covers any documentation that is mandated by US-FDA rules, such as documentation produced during the production, testing, and oversight of food, medical devices, and pharmaceutical items [55].

Clinical trial audit trails are essential for guaranteeing data openness and legal compliance. Every modification made to the data is automatically recorded by these systems, which also note who made the change, when it was made, and what was changed. This automated, ongoing logging fulfills a number of vital functions. Every change is traceable thanks to audit trails, which keep an exhaustive, chronological record. Stakeholders can then confirm that the data is accurate and hasn't been changed. To make sure that any changes to clinical trial data are recorded, regulatory agencies like the US-FDA want thorough audit trails (per 21 CFR Part 11). During inspections, this offers a verifiable paper trail. Real-time modification tracking through audit trails makes it easier to spot inconsistencies or unauthorized changes so that they may be fixed and data corruption can be avoided [57]. By documenting the identification of every user making modifications, the audit log encourages responsibility and deters purposeful or inadvertent data manipulation. All parties involved, from sponsors to regulatory bodies, may rely on the integrity of the research findings thanks to these characteristics, which together strengthen the dependability and transparency of clinical trial data.

17.8.2 European Medicines Agency

To control the use of electronic documents, electronic signatures, and audit trails in the EU pharmaceutical industry, the EMA released Annex 11. It is designed to guarantee the confidentiality, integrity, and validity of electronic records and is a component of the EU cGMP requirements [36].

17.8.2.1 Electronic Records

- Electronic records must be reliable, accurate, and secure.
- They must be maintained in a way that ensures their authenticity and integrity over time.
- The regulation mandates controls to ensure that the records cannot be tampered with, and there are requirements for backup and recovery in case of system failures.

17.8.2.2 Electronic Signatures

- Electronic signatures must be as legally binding as handwritten signatures.
- They must be unique to each individual, and the system must ensure that only authorized individuals can apply their signatures.
- There must be a record linking each electronic signature to the individual who applied it, including the date and time.

17.8.2.3 Audit Trails

- The regulation requires that systems maintain an audit trail that records all changes made to electronic records, including who made the change, when, and why.
- Audit trails must be secured and maintained in a way that prevents tampering.

17.8.2.4 Electronic Data Capture

In actuality, the implementation of electronic data capture (EDC) systems has resulted in a discernible decrease in data inconsistencies, guaranteeing that clinical trial outcomes are trustworthy and accurately represent the safety and effectiveness of experimental therapies. In the end, this methodical error prevention speeds up the medication development process and enhances patient safety. As data is entered, EDC systems use built-in validation criteria to look for missing fields, inconsistent data, and values that are outside of the range. By identifying problems early on, this real-time verification lessens the need for corrections later on. By reducing the amount of human data entry, digital forms greatly reduce transcription errors. To guarantee uniformity between study sites, these forms are made with predefined answer choices and organized fields [58]. An audit trail records each modification made to the data, guaranteeing openness and allowing data managers to pinpoint the location and time of an issue. This tracking improves regulatory compliance and speeds up resolution. The ability to remotely enter data allows investigators to quickly identify and resolve any differences. Errors are prevented from spreading throughout the dataset through real-time monitoring. Cross-validating data and identifying mistakes across numerous sources is made easier by the frequent integration of EDC platforms with other clinical trial systems. Data integrity and consistency are further supported by this integration [36].

17.9 Compliance Frameworks

For a company to manage and guarantee the security, privacy, quality, and integrity of its data, data governance frameworks are crucial. In order to guarantee that data is managed properly and in accordance with applicable laws, these frameworks are usually molded by legal requirements, industry standards, and best practices [59]. It is a collection of globally accepted standards and recommendations for clinical trial design, execution, monitoring, auditing, documentation, analysis, and reporting. It guarantees the rights, safety, and welfare of trial participants as well as the scientific and ethical soundness of clinical trials. The International Council for Harmonization of Technical Requirements for Pharmaceuticals for Human Use (ICH) created this widely accepted guideline. The goal of ICH Q10 is to give the pharmaceutical industry a framework for creating and sustaining an efficient pharmaceutical quality system that guarantees the reliable manufacture of high-quality goods. It is intended to assist with pharmaceutical product development, manufacturing, and post-market life cycle management. Essentially, ISO 27001 and National Institute of Standards and Technology are both prominent frameworks that help organizations manage information security, but they have different focuses and origins.

17.10 Key Considerations for AI/ML in Drug Development

Showing that AI/ML algorithms are reliable, tested on sample datasets, and transparent enough to pass regulatory assessment is essential. This comprises comprehensive records of the model's creation, the properties of the training data, and performance indicators. The quality and representativeness of the underlying data must be guaranteed by AI/ML systems. Potential biases in data gathering and model training should be recognized and addressed. Sponsors must set up explicit change control protocols and post-market monitoring for models that change over time (adaptive algorithms) in order to continuously evaluate performance and preserve patient safety [60]. It is advised to have early and continuous communication with regulatory bodies in order to develop creative solutions, set expectations, and make sure AI/ML implementations adhere to legal requirements. In addition to speeding up the process, this integrated regulatory strategy helps guarantee that AI/ML-driven breakthroughs in medication development uphold the highest standards of safety and data dependability (Table 17.1) [61].

Table 17.1 Description and application of AI tools.

S. No	AI/ML Technology	Applications
1	Natural language processing (NLP)	• Extracts data points like adverse events, efficacy outcomes, and safety signals. • Generates summaries of lengthy reports for quicker information access. • Monitors social media for public sentiment and drug side effects.
2	Advanced analytics	• Predicts compliance issues or regulatory delays. • Forecasts market access probabilities based on clinical trial results and market conditions. • Optimizes strategies for drug development.
3	Automated document management	• Automatically classifies documents to meet compliance standards. • Generates standardized templates for submissions. • Reduces errors and accelerates the regulatory approval process.
4	Machine learning algorithms	• Detects adverse events from electronic health records and clinical trial data. • Optimizes clinical trial designs and patient populations. • Improves patient safety and development efficiency.
5	The bots and AI-driven assistants	• Answers frequently asked questions about regulatory processes. • Provides onboarding and training support for new professionals.
6	Real-time monitoring tools	• Performs signal detection for safety signals from various data sources. • Monitors compliance with regulatory requirements and reporting timelines. • Improves drug safety oversight and regulatory compliance.

17.10.1 Blockchain Technology

In the pharmaceutical and medical device industries, blockchain is becoming a potent tool for regulatory compliance by guaranteeing that vital data is both transparent and impenetrable [62]. This is how it applies in various contexts.

17.10.2 Tamper-proof Clinical Trial Records

Every entry in an immutable ledger created by blockchain technology, including clinical trial data, cannot be changed covertly. To provide transparency and traceability, every alteration is time-stamped and logged using cryptography. By ensuring the authenticity of clinical data used in regulatory filings, this technology gives regulatory bodies and inspectors a strong audit trail[63].

17.10.3 Supply Chain Transparency

Blockchain makes it possible to track medications in real time from manufacturing to delivery, guaranteeing that all supply chain transactions are recorded. This can guarantee that only authentic, superior products are delivered to patients and help stop counterfeiting. Without depending on a single authority, a number of stakeholders, such as producers, distributors, and regulators, can confirm the authenticity of records [64].

17.10.4 Adverse Event Tracking

Pharmacovigilance systems can be integrated with blockchain to guarantee that adverse event reports are irrevocably documented. This increases confidence that safety alerts are reliably recorded and accessible for prompt examination. The ability to identify, evaluate, and address possible safety concerns is enhanced by consistent, standardized data on adverse events from multiple sources (clinical trials and post-market surveillance). All things considered, blockchain's decentralized, safe, and open characteristics meet regulatory requirements [63]. Blockchain can help regulatory bodies protect public health and uphold high standards of data integrity by providing solid supply chain traceability, enabling dependable adverse event tracking, and guaranteeing tamper-proof clinical trial records [63].

17.11 Challenges of Data-driven Drug Regulatory Compliance

Data-driven modeling methods and techniques play critical roles in many aspects of pharmaceutical processes, including drug discovery, process development, and product manufacturing. However, there are still several challenges and open questions that researchers and practitioners face in implementing and advancing the data-driven drug regulatory compliance. Pharmaceutical enterprises worldwide navigate a complex ecosystem where vast amounts of sensitive datasets are central to their operations [65]. The data typically includes information about clinical trials, EHR, drug manufacturing, and internal workflows, making data management a cornerstone of patient safety and innovation. The volume and variety of this data are rising due to rapid industry growth, increasing complexity in drug development, and distributed supply chains. Data governance strategies are largely inconsistent and have failed to keep pace with technological advancements. Besides, limited standardization and underinvestment in modern analytics exacerbate inaccuracies, delaying critical patient care and drug safety decisions. As reliance on data increases, the quality of information being processed in the sector is essentially marred by four top challenges (Figure 17.3):

1. Data silos
2. Data complexity
3. Regulatory uncertainty
4. Validation of AI models.

Inefficient data management in the pharmaceutical industry does impact patient care and safety. For instance, a lack of reliable information on drug interactions or allergies might result in adverse reactions, potentially endangering lives. Similarly, errors in inventory data result in shortages of critical medications or the unintentional administration of expired drugs, jeopardizing patient health. Another threat of poor data quality is distorted research findings that cause ineffective or harmful medications to reach the market. This mismanagement has a direct impact on treatment efficacy and patient outcomes [66].

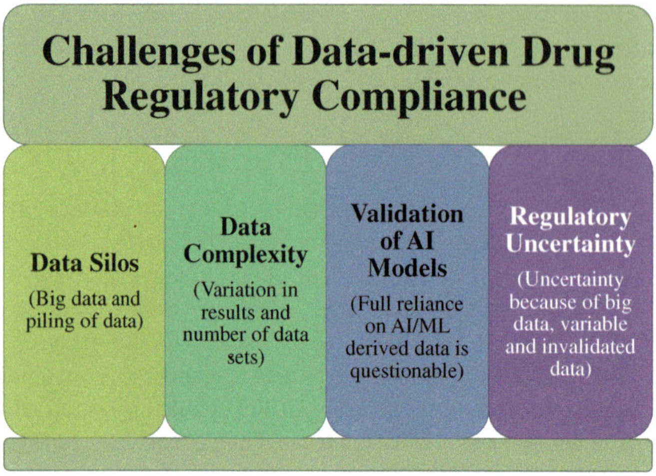

Figure 17.3 Various types of challenges of data-driven drug regulatory compliance.

17.12 Conclusion

Maintaining data integrity is a continuous process that requires the application of dependable practices, technological developments, and an organizational commitment to a culture of quality. Pharmaceutical firms have to constantly adapt their operations to meet the ever-changing regulatory constraints. Data integrity is essential for the pharmaceutical industry to ensure information accuracy, reliability, and compliance throughout the product life cycle. Drug manufacturers must adhere to stringent data management protocols in order to meet regulatory standards established by organizations such as the US-FDA, EMA, PMDA, and the MHRA. Implementing GDP, managing data with validated electronic systems, safeguarding data with audit trails and access restrictions, and fostering an organization-wide culture that values quality are all necessary to achieve this. Training programs, regular internal audits, risk assessments, and continuous improvement projects are essential for maintaining data integrity. By adopting these precautions to safeguard the security of their data, pharmaceutical companies can meet regulatory agency standards and encourage the development of safe and effective medications.

References

1 Huma T and Peng Z. Introduction to regulatory affairs and different regulatory bodies for pharmaceutical products and impact of digitalization on regulatory affairs. *Pharmacology & Pharmacy* (2023);14(11):463–477.

2 Kabir M, Rana MRH, and Debnath A. The role of quality assurance in accelerating pharmaceutical research and development: Strategies for ensuring regulatory compliance and product integrity. *Journal of Angiotherapy* (2024);8(12):1–11.

3 Ajmal CS, Yerram S, Abishek V, et al. Innovative approaches in regulatory affairs: Leveraging artificial intelligence and machine learning for efficient compliance and decision-making. *The American Association of Pharmaceutical Scientists Journal* (2025);27(1):22.

4 Freitas, AT. Data-driven approaches in healthcare: Challenges and emerging trends. In: Sousa Antunes H, Freitas PM, Oliveira AL, et al., editors. *Multidisciplinary Perspectives on Artificial Intelligence and the Law*, (pp. 65–80). Cham: Springer; 2023.
5 Liu Z, Roberts RA, Lal-Nag M, Chen X, Huang R, and Tong, W. AI-based language models powering drug discovery and development. *Drug Discovery Today* (2021);26(11):2593–2607.
6 Ajmera P, Saini K, and Jain V. The Impact of the data-driven supply chain on quality: Evidence from the medical device manufacturing industry. In: *Data-Driven Technologies and Artificial Intelligence in Supply Chain*, (pp. 174–182). CRC Press; 2023.
7 Xavier J and Picoto WN. Regulation of data-driven marketing and management theory: Bibliometric analysis, systematic literature review and research agenda. *International Journal of Law and Management* (2023);65(5);461–482.
8 Zillner S. Innovation in Times of Big Data and AI: Introducing the Data-Driven Innovation (DDI) Framework. In: Curry E, Metzger A, Zillner S, et al., editors. *The Elements of Big Data Value*, (pp. 289–310). Cham: Springer; 2021.
9 Parashar J, Jain A, and Ali S. Artificial intelligence impact on human fitness: Exploring emerging trends. *International Journal for Research in Applied Science and Engineering Technology* (2023);11(10):287–293. https://doi.org/10.22214/ijraset.
10 Guedj M, Swindle J, Hamon A, et al. Industrializing AI-powered drug discovery: Lessons learned from the patrimony computing platform. *Expert Opinion on Drug Discovery* (2022);17(8):815–824.
11 Chintala S. Accelerating drug discovery: Ai powered approaches in pharmaceutical research. In: *2024 4th International Conference on Innovative Practices in Technology and Management (ICIPTM)*, (pp. 1–6). IEEE; 2024.
12 Egon K and KARL L. *Machine Learning in Drug Safety Monitoring: Enhancing Pharmacovigilance Efforts*. OSF pre-print, 2023.
13 Pohane AL, Dighade SJ, Rithe ES, Mangwani RR, Raut AS, and Bhamburkar SS. AI-driven advancements in pharmacy: Enhancing drug discovery, optimization and clinical decision-making. *International Journal of Innovative Science and Research Technology* (2025);10(3):615–626.
14 Vignesh M and Ganesh GNK. Current status, challenges and preventive strategies to overcome data integrity issues in the pharmaceutical industry. *International Journal of Applied Pharmaceutics* (2020);12(6):19–23.
15 Xie Q, Han W, Chen Z, et al. Finben: A holistic financial benchmark for large language models. *Advances in Neural Information Processing Systems* (2024);37:95716–95743.
16 Nikam NR, Patil PR, Vakhariya RR, Mohite SK, and Magdum CS. Data integrity: An overview. *International Journal of Recent Scientific Research* (2020);11(6):38762–38767.
17 Charoo NA, Khan MA, and Rahman Z. Data integrity issues in pharmaceutical industry: Common observations, challenges and mitigations strategies. *International Journal of Pharmaceutics* (2023);63:122503.
18 Ullagaddi P. Safeguarding data integrity in pharmaceutical manufacturing. *Journal of Advances in Medical and Pharmaceutical Sciences* (2024);26(8):64–75.
19 Madabushi R, Seo P, Zhao L, Tegenge M, and Zhu H. Role of model-informed drug development approaches in the lifecycle of drug development and regulatory decision-making. *Pharmaceutical Research* (2022);39(8):1669–1680.
20 Woollen SW. Data quality and the origin of ALCOA. *Newsletter of the Southern Regional Chapter Society of Quality Assurance, Summer*. 2010.

21 Tabersky D, Woelfle M, Ruess JA, Brem S, and Brombacher S. Recent regulatory trends in pharmaceutical manufacturing and their impact on the industry. *Chimia* (2018);72(3):146–150.

22 Jaiswal H, Muddukrishna BS, and Kulyadi GP. Data integrity violations: A challenge to the pharmaceutical industry. *International Journal of Pharmaceutical Quality Assurance* (2020); 1(1):196.

23 Gokulakrishnan D and Venkataraman S. Ensuring data integrity: Best practices and strategies in pharmaceutical industry. *Intelligent Pharmacy* (2025);3(4):296–303.

24 Sembiring MH and Novagusda FN. Enhancing data security resilience in AI-driven digital transformation: Exploring industry challenges and solutions through ALCOA+ principles. *Acta Informatica Medica* (2024);32(1):65.

25 Sihite EACPB and Sumarna AD. The evaluation of the good documentation practice error using the WHO's ALCOA principles. In: *International Conference on Applied Science and Technology on Social Science 2023* (iCAST-SS 2023), (pp. 329–334). Atlantis Press; 2024.

26 Woollen SW, Quality Systems Compliance LLC, Durivage M. Data integrity for the FDA regulated industry. Quality Systems Compliance LLC. 2019. Available from: https://qscompliance.com/wp content/uploads/2019/01/ALCOA-Principles.pdf

27 Ahmad S, Kumar A, and Hafeez A. Importance of data integrity & its regulation in pharmaceutical industry. *Pharma Research* (2019);8:306–313.

28 Sabale MM, Pande VA, Tagalpallewar AA, Swami AG, Pawar AT, and Baheti AM. Maintaining data safety and accuracy through data integrity (DI): A comprehensive review. *Research Journal of Pharmacy and Technology* (2024);17(5):2431–2440.

29 Ingale MH, Tayade MC, Patil YP, and Salunkhe R. Data integrity violations in the pharmaceutical industry and regulatory measures. *International Journal of Pharmaceutical Quality Assurance* (2023);14(2):416–420.

30 Shah DB, Shah JS, and Maheshwari DG. Exploring the data integrity ordinance in pharmaceutical domain. *Journal of Young Pharmacists* (2022);14(4):377–381.

31 Agba MS and Rijami J. Data integrity in research: a fundamental prerequisite for credible research output across disciplines and professions. *International Journal of Democracy and Development Studies* (2022);5(2):31–45.

32 Miller R, Whelan H, Chrubasik M, Whittaker D, Duncan P, and Gregório J. An overview of current and new data quality dimensions under a common framework. *Data* (2024);9(12):151.

33 Leach CD. Enhancing data governance solutions to optimize ALCOA+ compliance for life sciences cloud service providers. Doctoral dissertation. Colorado Technical University. 2024.

34 McDowall RD and Burgess C. Data governance, data integrity, and data quality. *Method Validation in Pharmaceutical Analysis: A Guide to Best Practice*, pp. 13–33. 2025.

35 Kulkarni P and Kothari C. Documentation and data integrity in pharmaceutical industry. In: *Modern Aspects of Pharmaceutical Quality Assurance: Developing & Proposing Application models, SOPs, practical audit systems for Pharma Industry*, (pp. 381–403). Singapore: Springer Nature Singapore; 2024.

36 Matthews D, Lenarczyk G, and Zemła-Pacud Ż. The European Medicines Agency's path to greater access to pharmaceutical regulatory data: balancing intellectual property rights and the right to privacy. *Forthcoming in Kritika*: *Essays on Intellectual Property*, 6. 2024.

37 Ilyas A, Ali F, Sahu V, and Rady D. Regulations in the European Union. In: *Global Regulations of Medicinal, Pharmaceutical, and Food Products* (pp. 40–69). CRC Press; 2024.

38 Ehidiamen AJ and Oladapo OO. The role of electronic data capture systems in clinical trials: Streamlining data integrity and improving compliance with FDA and ICH/GCP guidelines. *World Journal of Biology Pharmacy and Health Sciences* (2024);20(01):321–334.

39 Kulyadi GP, Jaiswal H, and Muddukrishna B. Data integrity violations: A challenge to the pharmaceutical industry. *International Journal of Pharmaceutical Quality Assurance* 2020;11(01):193–195. https://doi.org/10.25258/ijpqa.11.1.30.

40 Handoko BL, Riantono IE, and Gani E. Importance and benefit of application of governance risk and compliance principle. *Systematic Reviews in Pharmacy* (2020);11(9).

41 Niv Y and Tal Y. Risk management and patient safety processes in a healthcare organization. In: *Patient Safety and Risk Management in Medicine: From Theory to Practice* (pp. 129–174). Cham: Springer Nature Switzerland; 2024.

42 Bode G. Regulatory guidance: ICH, EMA, FDA. In: Hock F, and Gralinski M, editors *Drug Discovery and Evaluation: Methods in Clinical Pharmacology*, (pp. 1085–1138). Cham: Springer; 2020.

43 Davis AL and Miller JD. The European Medicines Agency and publication of clinical study reports: a challenge for the US FDA. *Jama* (2017);317(9):905–906.

44 Mofid S, Bolislis WR, and Kühler TC. Real-world data in the postapproval setting as applied by the EMA and the US FDA. *Clinical Therapeutics* (2022);44(2):306–322.

45 Polak TB, van Rosmalen J, and Uyl–de Groot CA. Expanded access as a source of real-world data: An overview of FDA and EMA approvals. *British Journal of Clinical Pharmacology* 2020;86(9):1819–1826.

46 Mankar SD, Gholap VD, Zende TP, and Dighe RS. Drug regulatory agencies in India, USA, Europe and Japan: A review. *International Journal of Institutional Pharmacy and life sciences* (2014);4(2):288–300.

47 Huser V, Sastry C, Breymaier M, Idriss A, and Cimino JJ. Standardizing data exchange for clinical research protocols and case report forms: An assessment of the suitability of the Clinical Data Interchange Standards Consortium (CDISC) Operational Data Model (ODM). *Journal of Biomedical Informatics*, (2015);57:88–99.

48 Facile R, Muhlbradt EE, Gong M, et al. Use of clinical data interchange standards consortium (CDISC) standards for real-world data: Expert perspectives from a qualitative Delphi survey. *JMIR Medical Informatics* (2022);10(1):e30363.

49 Sushmita S, Nikhil P, Drashti S. Importance of data integrity in pharmaceutical industry. *EPRA International Journal of Economics, Business and Management Studies* 2023;10(2):100–106.

50 Simuni G and Atla A. Hadoop in enterprise data governance: Ensuring compliance and data integrity. 2024. Available from: SSRN 4982500.

51 Ronolo SC. Assuring data integrity towards regulatory compliance: A study on process improvement in data integrity compliance of computerized systems. Master thesis. Degree Programme in Business Informatics. 2023.

52 De La Vega J. Data integrity: Paper and electronic 21CFR11. In: *Good Laboratory Practice for Nonclinical Studies*, (pp. 179–186). CRC Press; 2022.

53 HV G. A review and analysis of data integrity: Pharmaceutical industry perspective. *International Journal of Pharmaceutical Research*(09752366) 2020;12(1).
54 Chavan SM, Biradar SV, Sonawane JK, et al. A review: An illustration of data integrity. *International Journal of Research Publication and Reviews* 2023;4(5):5620–5629.
55 Kumar K. Data integrity in pharmaceutical industry. *Journal of Analytical &Pharmaceutical Research* 2016;2(6). https://doi.org/10.15406/japlr.2016.02.00040.
56 Bhadrashette MS. Overview of data integrity issues in the pharmaceutical industry. *International Journal of Pharmaceutical Sciences Review and Research* 2018;50(2):95–101.
57 Mak KK and Pichika MR. Artificial intelligence in drug development: Present status and future prospects. *Drug Discov Today* 2019;24(3):773–80. https://doi.org/10.1016/j.drudis.2018.11.014.
58 Kavasidis I, Lallas E, Gerogiannis VC, and Karageorgos A. Analytics and blockchain for data integrity in the pharmaceuticals industry. In *2022 4th International Conference on Advances in Computer Technology, Information Science and Communications (CTISC)*, (pp. 1–5). IEEE; 2022.
59 Tuladhar A, Rogerson M, Engelhart J, Parry GC, and Altrichter B. Blockchain for compliance: An information processing case study of mandatory supply chain transparency in conflict minerals sourcing. *Supply Chain Management: An International Journal* (2024);29(4):755–777.
60 Abu-Elezz I, Hassan A, Nazeemudeen A, et al. The benefits and threats of blockchain technology in healthcare: A scoping review. *International Journal of Medical Informatics* 2020;142:104246.
61 Rivare A. Artificial intelligence and digitalization in pharmaceutical regulatory affairs. Doctoral dissertation. University of Helsinki. 2023.
62 Gudimitla RB, Babu MK, Bhimana S, Dasu CD, and Perla S. Artificial intelligence, machine learning, and the revival of big data in pharmaceutical sciences. Revolutionizing Drug Discovery, Process Optimization. In M. Ayyalasomayajula (Ed.), *AI and the Revival of Big Data* (pp. 275–310). IGI Global Scientific Publishing; 2025. https://doi.org/10.4018/979-8-3693-8472-5.ch01
63 US FDA, https://www.fda.gov/ Accessed on 02-05-25
64 US FDA, https://www.fda.gov/regulatory-information/search-fda-guidance-documents/e9r1-statistical-principles-clinical-trials-addendum-estimands-and-sensitivity-analysis-clinical, Accessed on 02-05-25
65 ECFR, https://www.ecfr.gov/current/title-21/chapter-I/subchapter-A/part-11, Accessed on 02-05-25
66 Ajmal CS, Yerram S, Abishek V, et al. Innovative approaches in regulatory affairs: Leveraging artificial intelligence and machine learning for efficient compliance and decision-making. *The American Association of Pharmaceutical Scientists Journal* (2025);27:22.

18

Responsible AI and Data Governance in Pharmaceutical Settings

Piyush Sewal[1], Sriti Thakur[2], and Deepak Kumar[3]*

[1]*Amity School of Engineering and Technology, Amity University Punjab, Mohali, India*
[2]*Department of Mathematics, Jaypee University of Information Technology, Waknaghat, Solan, India*
[3]*Department of Pharmaceutical Chemistry, School of Pharmaceutical Sciences, Shoolini University, Solan, India*

*Corresponding author: piyush.sewal@gmail.com

CONTENTS

18.1 Introduction

Artificial Intelligence (AI) has become a transformative force in the pharmaceutical and healthcare sectors, leading to a radical change in basic concepts of drug discovery, clinical trials, and personalized medicine. Once everywhere, AI counts a whole toolbox of machine learning (ML) algorithms and data-driven approaches that can be used to improve decision-making and efficiency, enhance the provision of healthcare, and assist in solving most of the health problems faced by the globe [1, 2]. Nevertheless, the fast development of AI in terms of the pharmaceutical industry and even healthcare is overstraining the current abilities and functions of existing systems by demanding responsible management, ethical policies, and standards. The consequences of AI-related algorithmic bias, data breaches, the lack of transparency behind the use of AI in decision-making, and IP-related concerns have been met by high-quality governance and responsible use of AI with intense concern regarding the topic in the international community [3]. The chapter will evaluate the many ways of AI is transforming innovation in today's pharmaceutical landscape, while also considering the ethical and regulatory dilemmas that are emerging as it is embraced so rapidly. This chapter will begin with discussing the governance and compliance challenges of digital pharma and will evaluate the importance of the policies, structures, and best practices around ensuring safe, fair, transparent practices that enable AI use. This chapter will also evaluate how AI has fundamentally altered data governance models, assisted regulatory compliance, facilitated data utilization, and limited the perils associated with data privacy and security [4].

The chapter proceeds to examine the bigger picture of AI in healthcare, including best practices and policy frameworks of ethical use. International provisions such as the General Data Protection Regulation (GDPR) and the Health Insurance Portability and Accountability Act (HIPAA) are reviewed in their contributions to providing an ethical AI era. The chapter also goes on to discuss the changing landscape in the pharmaceutical field with the aid of AI, including drug target discovery, digital therapeutics, and custom drug design—solid testimony to the fact that responsible AI can and will influence and maintain the future of digital pharma and hold a balance between compliance, understanding of the expectation of the state [5]. A significant portion of the chapter is dedicated to the involvement of AI in drug revolutionization and development. It addresses the process of how AI-directed solutions are streamlining pharmaceutical excipients, bringing innovations to the field of natural product science too. The chapter also talks about the dynamic nature of intellectual property rights (IPR) in the area of AI-driven-pharmaceuticals and how the research and development challenges are regarding ownership of data and how to patent AI-generated compounds. With reference to the cases that take place in real life, this chapter discussed the role of AI in the pharmaceutical ecosystem in terms of AI-based drug repurposing and optimization of clinical trials. Referring to the international case studies, including the cases in India, which exemplify the innovative practices that reflect the possibilities of the AI potential, in this chapter, both the potential advantages and the hazards and threats of the requirements of the AI adoption in this context have been revealed. The other recommendation in this chapter is the establishment of multidisciplinary teams, new governance structures, and adherence to responsible AI in developing collective trust, safeguarding information security, as well as safety, and holding accountable fair health outcomes.

Thus, the chapter gives a strategic overview of how the responsible use of AI and data governance is going to leverage the pharmaceutical ecosystem. This chapter shows how sustainable innovation and new equity in the global healthcare system may be achieved by embracing ethical, regulatory, and governance issues instead of their evasion. Throughout this chapter, the aim is to allow interested parties within pharmaceutical and healthcare environments to start taking advantage of the potential to deliver AI-powered innovation and continue to practice ethically—at least safely, responsibly, and sustainably.

18.2 Governance and Compliance in the Age of Digital Pharma

Nowadays, in a digital age, the pharmaceutical industry will continue to depend on data-based knowledge and AI to transform healthcare solutions. Nevertheless, such increasing reliance on data necessitates robust governance and compliances to guarantee ethical acts, regulatory observance, and data consequences. The section that is pointed to is Governance and Ethics in Data Governance Frameworks that state that the systems of managing data should be transparent, fair, and accountable, with emphasis on the rights of a patient and his or her privacy. It also includes transforming data governance with AI that tells us how AI-enabled data tools are making data management easier and compliance an automated experience and how it may help transform the pharmaceutical industry by bringing more effective and reliable data governance through real-time decision-making (Figure 18.1).

18.2.1 Governance and Ethics in Data Governance Frameworks

As part of the data-governance discourse, the concept of governance and ethics is also noted as one of the most important pillars as it suggests taking care of data and adhering to the existing regulatory practices. A strictly laid out governance system contributes to transparency, fairness, and responsibility, which makes the stakeholders trust the activity of pharmaceutical enterprises.

Nissenbuam's contextual integrity framework provides a foundational understanding of the data governance challenges in pharmaceutical and other healthcare contexts as the digitization and aggregation of personal health data (PHI) becomes universal [6]. Contextual

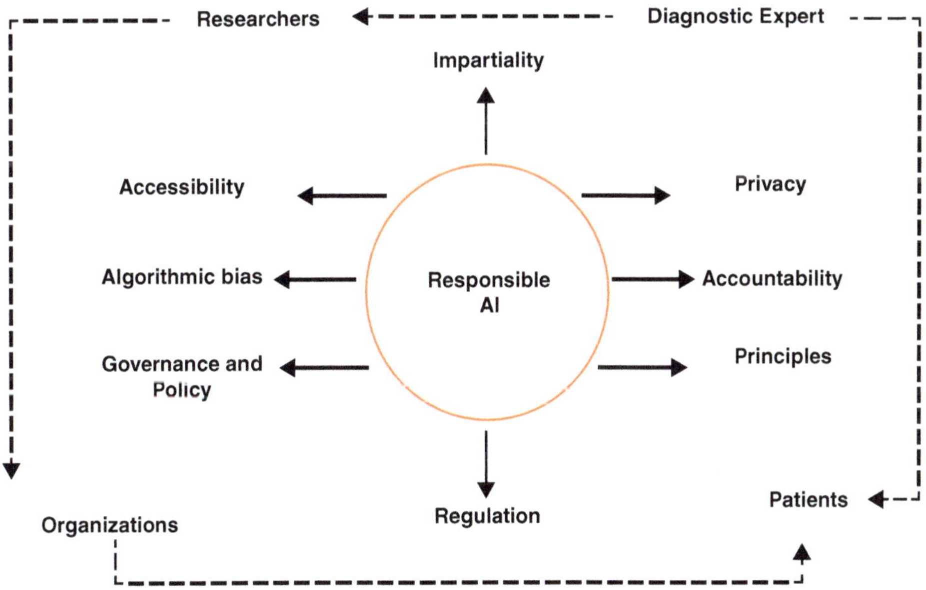

Figure 18.1 Ethical frontiers in AI pathology.

integrity becomes particularly helpful when formal governance models are challenged by breaches of contextual integrity such as PHI extracted from context and used in new contexts following very different norms of appropriateness. One example includes a public–private sector partnership between the Royal Free Trust in the UK and DeepMind Health, and new governance practices developed by the for-profit, self-identified innovation sector, compared to existing health data privacy frameworks. The analysis underscores the importance of proactive analytical tools to articulate and negotiate governance goals before conflicts emerge, striking a balance between individual privacy, the public good, and corporate interests. However, the study's reliance on publicly available sources and its focus on a single case highlight the need for future research to explore diverse governance models, enhance stakeholder representation, and improve transparency in big data analytics (BDA).

The changing nature of regulatory affairs (RA) is crucial in protecting people, ensuring quality and effectiveness for medicines and other health products, especially with recent rapid developments in AI, ML, and digital disruption [7]. The complexity of the regulatory system that is going to be established all around the world will be challenging and will drive the urgency of harmonized regulation to a national level. Findings derived out of international institutions such as the World Health Organization (WHO), Pan American Health Organization (PAHO), or International Council for Harmonization (ICH) provide a crucial basis upon which the means in which electronic transformation and innovative forms of treatment are remodeling the area of regulation can be analyzed. The analysis that follows shows that regulatory professionals must broaden their skills base and improve to nurture an aspect in order to be flexible within this ever-changing sphere. These findings are fundamental views to the study, but further research will establish a clear understanding of the long-term effect of these changes on regulation practices, pharmaceutical development, and medical practice. This therefore strengthens the centrality of RA professionals in dictating the path of pharmaceutical innovation in the coming days and in the promotion of well-being in society.

A universal implementation of big data and AI-empowered technology in the sphere of healthcare has a significant potential to increase its role in prevention, diagnostics, and therapeutics intervention; however, it is also associated with a range of complex ethical questions. [8]. Existing ethical guidance and oversight mechanisms lack sufficient, adequate consideration of issues related to privacy, equity, fairness, accountability, and the distribution of benefits. A review of health research and digital health applications reveals gaps in existing ethical guidance and a need for meaningful reform to ensure ethically informed data governance. There is a tendency in the ethical discourse in this space to focus primarily on privacy and technical factors and turn a blind eye to the larger considerations that are necessary for an all-inclusive governance dimension. In order to close the gaps, we recommend an ethical toolkit for stakeholders and a collective value-based governance framework. Collectively, this toolkit and framework can support the realization of the full potential of big data and AI in a way that protects individual and collective well-being in health.

The use of AI in the pharmaceutical industry has impacted the ways in which drugs are discovered, clinical trials are conducted, and how patients receive care, but it also brings relevant ethical issues that must also be considered. Table 18.1 identifies key ethical issues related to AI, such as algorithmic bias, data privacy, transparency, accountability, informed consent,

Table 18.1 Ethical considerations in AI-driven pharma.

Ethical principle	Description	Challenges in AI implementation	Proposed solutions
Algorithmic bias [9]	AI models may reinforce biases present in training data, leading to unfair treatment of patients	Bias in historical medical data can result in disparities in drug recommendations or treatment plans	Implement diverse and representative datasets; conduct regular bias audits and model retraining
Data privacy [10]	Ensuring patient data confidentiality and compliance with global regulations like GDPR and HIPAA	AI-driven pharma relies on large datasets, increasing the risk of data breaches	Use differential privacy techniques, secure data-sharing protocols, and strong encryption methods
Transparency [11]	AI models should be interpretable, allowing stakeholders to understand how decisions are made	Many AI algorithms, especially deep learning models, operate as "black boxes," making it difficult to explain their decisions	Develop explainable AI (XAI) models and integrate model interpretability tools
Accountability [12]	Assigning responsibility for AI-generated decisions in pharma and healthcare	Unclear legal frameworks make it difficult to determine liability in AI-related errors	Establish regulatory guidelines defining accountability in AI-driven healthcare decisions
Informed consent [13]	Patients should be aware of and consent to AI-driven medical recommendations and data usage	AI often analyzes patient data without direct patient awareness or explicit consent	Implement clear patient consent mechanisms and provide AI-generated insights in an understandable format
Fair access to AI innovations [9]	Ensuring that AI-driven pharmaceutical advancements benefit all populations, not just privileged groups	High costs and unequal access to AI-driven healthcare solutions may worsen health disparities	Develop policies promoting affordable AI-based treatments and invest in AI-driven healthcare solutions for underserved regions
IPR [9]	Addressing the ownership of AI-generated drug discoveries and medical research	Ambiguity exists over whether AI or human researchers own AI-generated pharmaceutical inventions	Update patent laws to define AI's role in intellectual property and provide guidelines for AI-assisted discoveries
Security of AI models [10]	Preventing adversarial attacks on AI models that could manipulate drug development processes	AI models can be vulnerable to cyber threats, leading to potential exploitation	Implement robust cybersecurity protocols, AI model validation, and adversarial testing frameworks

(Continued)

Table 18.1 (continue)

Ethical principle	Description	Challenges in AI implementation	Proposed solutions
Regulatory compliance [11]	Ensuring AI applications in pharma adhere to existing and evolving global laws	Regulations often lag behind technological advancements, creating compliance gaps	Engage in continuous dialogue with regulators and align AI developments with international regulatory frameworks
Environmental impact [12]	Evaluating the carbon footprint of AI-driven pharma research and data processing	High computational demands of AI models contribute to significant energy consumption	Optimize AI model efficiency, promote sustainable computing, and use green energy sources

equitable access to AI solutions, IPR, security of risk models, regulatory compliance, and environmental impacts. The spread of AI in various fields presents a set of moral issues that should be overcome so that it is possible to implement the use of AI responsibly and equally in the framework of the pharmaceutical industry. Such a framework to overcome these issues and thus develop a culture of trust among the population is needed to take advantage of the plentiful advantages of AI and keep unintended consequences in check.

The ongoing transformation in the pharmaceutical industry by the use of AI emphasizes the need to implement a broad-based data governance strategy, which will facilitate ethical behavior, safety, and transparency. Countries in the world have come up with regulatory frameworks that guide the use of AI in areas such as data stewardship and privacy of patients as well as development of therapeutics. The extent, the goals sought, and implementation challenges to these regimes are not the same, hence, the extent of adherence to legal and ethical compulsions of AI-powered pharmaceutical advancement. As presented in Table 18.2, Key Regulatory Frameworks of AI in Pharmaceuticals provides the comparative overview of major directives currently established all over the world, detailing their regional scope, target objectives, influence on the AI deployment, and challenges faced in its implementation. These frameworks become essential in the full understanding of pharmaceutical organizations that are obligated to pursue the governance of AI in a responsible manner and maintain innovation within the restriction of the law.

18.2.2 Transforming Data Governance with AI

AI is another driving effect of data governance, as it brought forward the automatization of procedural work, increased data quality, and improved the grade of decision-making. Pharmaceutical enterprises can gain higher data management, increased compliance, and scalability as the structure of the industry continues its shifting through the implementation of AI in its regular processes.

The impact of AI on data governance, in its turn, also acquires increased salience due to the necessity to utilize its benefits and to counter possible risks at the same time. The solid

Table 18.2 Comparison of key regulatory frameworks for AI in pharmaceuticals.

Regulation name	Region/ country	Scope	Impact on AI integration	Challenges in implementation
GDPR [13]	European Union (EU)	Governs data protection and privacy for individuals within the EU	Requires AI systems to ensure data privacy and obtain explicit consent for data processing	Compliance necessitates robust data management practices; noncompliance can result in significant fines
HIPAA [14]	United States (US)	Protects the privacy and security of individuals' medical information	AI systems handling protected health information (PHI) must implement stringent security measures	Ensuring AI compliance with HIPAA's complex requirements can be challenging
Digital Personal Data Protection (DPDP) Act, 2023 [11]	India	Regulates the processing of personal data to protect individuals' privacy	AI applications must adhere to data protection principles, including purpose limitation and data minimization	Organizations need to establish comprehensive data protection frameworks to ensure compliance
FDA's proposed regulatory framework for modifications to AI/ML-based software as a medical device (SaMD) [12]	United States (US)	Provides guidelines for the regulation of AI/ML-based medical devices	AI-based medical devices must demonstrate safety and effectiveness, considering their adaptive nature	Navigating the evolving regulatory landscape requires continuous monitoring and adaptation
Personal Information Protection Law (PIPL) [15]	China	Establishes comprehensive rules for personal information protection	AI systems must comply with strict data protection requirements, including data localization	Balancing data utilization with stringent compliance obligations poses challenges
Proposed Artificial Intelligence Act [16]	European Union (EU)	Aims to regulate AI systems based on their risk levels	High-risk AI applications, such as those in healthcare, face stringent requirements	Organizations must assess and mitigate risks associated with their AI systems to ensure compliance
Act on the Protection of Personal Information (APPI) [17]	Japan	Protects individuals' rights concerning their personal information	AI systems must handle personal data in accordance with APPI's principles	Implementing necessary safeguards to protect personal data can be complex

(Continued)

Table 18.2 (continue)

Regulation name	Region/ country	Scope	Impact on AI integration	Challenges in implementation
ISO/IEC 23053:2022—framework for AI systems using ML [18]	International	Provides guidelines for the application of ML in AI systems	Encourages the development of safe and reliable AI systems	Adoption is voluntary, and organizations may face challenges in implementation
Model AI governance framework [19]	Singapore	Offers guidance on the responsible use of AI	Encourages organizations to adopt ethical AI practices	As a nonmandatory framework, its adoption relies on organizational commitment
General Data Protection Law (LGPD) [20]	Brazil	Regulates the processing of personal data to protect individuals' privacy	AI systems must ensure compliance with data protection principles	Organizations may face challenges in aligning their AI practices with LGPD requirements

data governance framework acts as the skeleton around which organizations will build an ethical, safety-focused, and legislation-compliant usage of AI, as well as its overlaps with data privacy, data quality, data security, and data compliance [13]. Such structures require being tailored to be in line with company goals without ignorance of the legal requirements and technological capabilities. They also have to define the respective roles and responsibilities as well as relationships of the stakeholders such as data scientists, information technology (IT) units, lawyers, and the different lines of businesses. Continuous observation and modification are also irrefutable parts of the framework, with bias in algorithms, drift in data, and changing regulatory effects. It is the reduction of these risks in combination that facilitates trust, transparency, and reliability, and thus organizations can derive maximum value of their data. However, the areas of implementation complexity, legal sensitivity, data ownership, and maintenance of accuracy can complicate adoption efforts. Nonetheless, this research recognizes data governance as a strategic asset that not only has positive implications for risk mitigation but also assures a profit for organizations in a data-driven environment [3–5].

Implementing AI governance in organizational settings introduces ethical, legal, and technical challenges, as illustrated by AstraZeneca, a biopharmaceutical company navigating these complexities. A key focus is on translating ethical principles into actionable practices, addressing challenges such as defining governance scope, harmonizing standards across decentralized structures, and measuring governance impact [21]. AstraZeneca's strategy included leveraging existing governance frameworks, integrating risk management into development and procurement processes, and fostering employee empowerment through education and change management initiatives. The methodology involved stakeholder consultations, the drafting of ethics principles, and conducting an independent AI audit, which required significant time and financial investment. Despite the associated costs, this approach improved business metrics such as data security, brand reputation, and talent acquisition. The case of AstraZeneca presents a reproducible approach to other fields, showing that not only should the existing technologies be integrated into the organizational processes, but the entire AI life cycle should be fostered with a sense of ethical thinking to ensure that the process of seeking economic and social value is not opposed to the realization of ethical standards.

Ecaterina Pitel et al. [22] performed a systematic review of 33 landmark articles on AI in drug discovery, outlining the main ethical concerns that arise. The concerns provided under data privacy, algorithmic bias, intellectual property, transparency, and fair access to AI-driven therapies are highlighted in the work. Besides, it questions the affordability of the benefits of AI in healthcare and establishes how it could guarantee that the benefits of AI are fairly distributed all around the world.

At the same time, Virginia Sanchini et al. [23] presented a data-intensive health research (encompassing genomics) ethics and data governance framework developed at the European Institute of Oncology (IEO) in Milan. The developed framework with privacy experts, clinicians, and bioethicists corresponds to the current rules of data security and research ethics and provides a hierarchical method of harmonizing innovation with ethical conduct in health studies.

It has been realized among different data-oriented domains, such as pathology, that deep learning (DL), AI, and algorithmic decision-making (ADM) are significant forces of change [24]. The organized pathology information data repositories would be an optimal training ground of AI systems, but the discussion also highlights associated risks due to an unethical implementation process. In particular, in the absence of effective control, AI tools can contribute to strengthening existing healthcare disparities without any intent. Thus, the paper provides root ethical issues related to pathology AI, especially study design processes, danger of incorporation into the working processes, and the principles of ethical AI (transparency, accountability, and governance). All these views serve to verify that there is a need to adopt strong systems of governance that would both promote responsible adoption and guard against the bane of healthcare AI applications.

Introducing AI in healthcare is quite promising in enhancing the quality of medicine practice but also creates complex ethical–social concerns, such as bias, transparency, and iatrogenic misrepresentation. According to Theresa Willem et al. [25], embedded ethics is a feasible strategy of addressing such issues. Based on the personal experience of conducting AI-powered healthcare initiatives, the authors invite one to use a validated set of methodological tools that will enable the integration of ethical analysis into AI practice. Interdisciplinary dissemination strategies, as well as stakeholder analyses, literature reviews, ethnographic approaches, interviews, focus groups, bias analyses, and workshops will be included in the toolkit. New tools, such as responsible AI-driven maturity models, ethics impact assessments, and so-called internal ethics-reference persons, are proposed to ensure that ethical support is maintained over time in a way that is specifically responsive to the needs and contexts of the project.

The rapid deployment of the generative AI models, particularly large language models (LLMs), is introducing opportunities and threats to healthcare delivery. In [26], the ethical issues related to the adoption of LLMs in clinical settings in terms of principles, where its main sticking points beneficence, non-maleficence, autonomy, and justice are considered. In a positive note, the administrative role can be facilitated by LLMs and even enhance clinical decision-making, especially in the highly stressful, limited-resource situations.

These breakthroughs can eventually improve patient care and reduce health inequality around the world. However, Armitage notes that to be sure that these tools will not harm but will simply help, it is required to address the potential downsides associated with them, such as biases in training data, threats to patient autonomy, uncertainties on accountability, privacy issues, and insufficient transparency.

Derek Leben [27] looks at the two-faced nature of AI in healthcare, pointing out that it has both the capabilities of saving lives and the threat of causing death. In his analysis, he prioritizes inherent issues of ethics based on the principle of explainability, fairness, safety, and accountability, indicating that reliance on AI systems can emphasize the underlying inequality and compromise the patient autonomy. At the same time, the ongoing development of AI in medicine opens the perspectives of the explanation of the standards in areas that were previously ill-defined. Lebens, in his turn, claims that transparency and creation of effective policy frameworks are essential to introduce AI into clinical practice in a responsible way.

In the pharmaceutical RA environment, AI and ML have played a significant role in changing the operational practices, as shown by Seetharam Gude et al. [28]. Their study traces the way how AI-based interventions increase efficiency, streamline the complex regulatory routine, and support compliance. These technological tools would allow the professionals to navigate the regulatory frameworks with increasing agility due to automation of the repetitive operations, enabling predictive analytics, and enable real-time decisions. The ability to streamline regulatory submission approaches, compress the time to get approvals, carry out strict risk-based assessments, and ensure data integrity and provenance is a big step in pharmaceutical innovation.

In [29], the development of in-house AI research activity to commercialization of regulated medical devices is discussed. The authors single out five key steps: clinical need identification, data management, model training and validation, compliance with regulations and emphasize that the last two, validation, and post-deployment surveillance, are vital in terms of making AI resistant and adaptive to changing circumstances in the clinical environment. The article also examines the regulatory labyrinth, mentioning the EU medical device regulation (MDR) and the upcoming EU AI Act that require the adherence to the stringent certification standards and efforts to address risks proactively. The issues regarding ethics such as transparency, patient privacy, and fair access are kept as the main focal points throughout the development process.

The possible scope of the application of AI in the context of biomedicine is given in [30] by Naskar et al., revealing the way AI is transforming the industry by analyzing complex biological data using advanced analytics, increasing the accuracy of diagnoses, automation of specific procedures, and improvement of patient lives via early diagnosis and tailor-made treatment. The most advanced tools of AI: ML, DL, neural networks (NNs), and generative models (GMs) have been used in drug discovery, treatment planning, disease forecasting, and healthcare worker education. Nevertheless, the authors provide emphasis on the challenges that are entailed: ethical controversies, privacy risks, lack of transparency, and archaic regulatory oversight. One of possible solutions to these problems is interdisciplinary cooperation, creation of effective policy frameworks, and integration of new strategies like blockchain as a secure data management tool.

The integration of AI in pharmaceutical data governance will redefine how pharmaceutical data are being generated, stored, and used across the sector. As an example of such development, Table 18.3 features a survey of three major AI deployments: automated data

Table 18.3 AI-driven enhancements in pharmaceutical data governance.

AI application	Description	Benefits to data governance	Challenges and mitigation strategies
Automated data classification [31]	AI-driven systems categorize and label pharmaceutical data based on sensitivity and compliance requirements	Improves data organization, enhances security, and ensures regulatory compliance (e.g., GDPR, HIPAA)	Risk of misclassification due to training data biases; mitigate with ongoing AI audits and human oversight
AI-powered data audits [13]	AI continuously monitors and audits data usage, access logs, and compliance adherence	Enhances regulatory transparency, detects unauthorized access, and ensures policy enforcement	High computational cost; optimize with scalable cloud AI solutions
Predictive compliance monitoring [11]	ML predicts potential regulatory compliance violations before they occur	Prevents costly noncompliance penalties and maintains data integrity	Requires integration with evolving regulatory frameworks; keep models updated with the latest guidelines
Real-time anomaly detection [32]	AI detects unusual patterns in pharmaceutical data access, signaling potential security breaches	Strengthens cybersecurity by preventing data leaks and insider threats	False positives can lead to unnecessary alerts; refine models using adaptive learning techniques
Natural language processing (NLP) for policy enforcement [33]	AI reads and interprets compliance documents to ensure adherence in pharma data processes	Reduces human errors in compliance checks, speeds up regulatory reviews	NLP may misinterpret legal jargon; fine-tune models with domain-specific training datasets
Blockchain-integrated AI for data integrity [11]	AI verifies and records pharmaceutical transactions on blockchain to ensure data authenticity	Enhances traceability, prevents data tampering, and builds trust in AI-generated drug research	Blockchain scalability issues; adopt hybrid AI-blockchain solutions for efficiency
Automated data de-identification [13]	AI removes personally identifiable information (PII) from datasets while retaining research utility	Ensures compliance with privacy laws (e.g., GDPR, India's DPDP Act), facilitates ethical AI use	Risk of reidentification through AI models; enhance with differential privacy techniques
AI-driven master data management (MDM) [34]	AI consolidates and standardizes pharma data from various sources, eliminating inconsistencies	Ensures data accuracy, enhances interoperability, and improves decision-making	Integration challenges with legacy systems; implement phased AI adoption strategies

(Continued)

Table 18.3 (continue)

AI application	Description	Benefits to data governance	Challenges and mitigation strategies
Federated learning for secure data sharing [13, 35]	AI enables collaborative pharmaceutical research without sharing raw data across institutions	Protects patient confidentiality while allowing multi-center drug discovery collaborations	Computationally intensive; optimize with edge computing and decentralized AI models
AI-enabled regulatory reporting [36]	Automates documentation and submission of compliance reports to regulatory bodies.	Saves time, reduces manual errors, and ensures timely regulatory submissions	Legal interpretations of AI-generated reports may vary; include human validation in final reviews

classification, predictive compliance monitoring, and blockchain-based AI along with their definitions, benefits to data governance, and challenges with relevant mitigation strategies. Together, the innovations are improving the management of data in a more efficient and transparent manner and simultaneously addressing some deeper issues with matters of compliance, integrity, and privacy. By implementing thoughtful use of solutions driven by AI, pharmaceutical companies will be able to reinforce the governance systems, and, in turn, maintain the ethical, regulatory, and operational quality.

Technological revolution in the drug sector warrants a strong ethical background and well-defined data control plans to maintain openness, safety, and impartial end results. Use of AI in the data governance processes enhances the efficiency of operations as well as decision-making and at the same time maintains regulatory compliance. All these elements provide the preconditions of a progressive and responsible drug production.

18.3 AI in Healthcare: Best Practices and Policy Frameworks

AI is transforming the medical world by promoting diagnostic tools, treatment planning, and drug discovery. This combination, however, comes with major policy and ethical issues. Main ethical issues are those related to privacy of patients, safety of data, fairness of AI machine models, explainability of decision-making, and responsibility of AI-based medical errors. The policy structures look at ensuring compliance with regulation, ethical governance, and international cooperation to come up with an internationally commonly referred standard. The US food and drug administration (FDA), the EU, the WHO, and the organization of economic co-operation and development (OECD) are formulating rules to make the application of AI in medical care responsible. The balance of innovation, ethical protection, and legal systems is essential to avoid violating trust, equitability, and accountability in medical progress built on AI.

Ramakrishnan Neelakandan et al. [37] describe best practices for the assessment development and utilization of AI in healthcare, including practices such as ensuring data quality, the identification of bias, support interpretability of the model, conduct proper testing, allow for monitoring post-deployment into practice, and ensuring ethical and associated regulations were adhered to during testing and deployment. Their discussion covers the particular areas of data quality, validating algorithms, monitoring output following deployment, and governance. Included were strategies for minimizing bias, enhancing transparency and interpretability, protecting patient information, and upholding ethical standards of fairness, accountability, and autonomy. They also stressed the importance of rigorously testing and the use of multidisciplinary teams for testing and the systematic monitoring of algorithmic response post-deployment. The authors also spoke of stakeholder engagement, incident or error response, and public education as a way to enable trust in the AI.

The author in [38] presented a study exploring the collaboration between healthcare professionals and algorithms in public healthcare decision-making, introducing the concept of "collaborative tinkering." Building on the improvisational practices highlighted by scholars Mol and Pols, it demonstrates how both humans and algorithms adapt and interact in response to the complex and unpredictable nature of healthcare. Through an ethnographic study across hospitals worldwide, the paper reveals how agency and intelligence are distributed among humans, algorithms, and their environments, forming a hybrid decision-making entity. The study highlights the need for a critical reassessment of ethical and policy frameworks to address the realities of human–AI collaboration in healthcare, providing insights into the hybrid nature of computational practices and their implications for AI ethics and medical decision-making. Seema Yadav et al. [39] introduced how AI is revolutionizing the pharmaceutical industry and healthcare by expediting drug discovery, improving treatment outcomes, and enhancing patient care. In drug discovery, AI-powered methods such as ML and DL optimize target identification, lead compound development, drug repurposing, and clinical trials. AI integration enables faster, cost-effective medication development while improving drug design and manufacturing processes.

The AI systems will revolutionize clinical practice through enhanced diagnostics using complex imaging processing, custom-made medicine by customized drug prescription based on in-depth genomic screening and health histories. Moreover, technologies of predictive modeling and remote monitoring make it possible to provide healthcare in a distributed environment, facilitating both telemedicine platforms and management of chronic illnesses using prognostic analytics, increasing both organizational efficiency and healthcare outcomes. However, there are still a number of difficulties, such as the protection of data privacy, the necessity to provide transparency of algorithms, and ethical issues, which are essential to the just and secure implementation of such systems. In another study, Rajiv Ratan Singh et al. [35] represent a current platform with massive potential in the emergency field, which is Metaverse, a virtual reality space that is capable of interactions in real-time setting. It can be anticipated that response times can be reduced, clinical practices refined, and access to healthcare increased, especially in medically underserved regions, with such applications. However, the authors detect the significant number of law-ethical implications such as data protection, patient privacy, informed consent, liability, and adherence to regulatory frameworks, which must be covered prior to adoption in full. An orderly assimilation of Metaverse technologies with emergency medicine requires formulation of robust legal frameworks and preparation of standardized directions. Close cooperation between policymakers, medical professionals, and the law is essential to protect the rights of the patients, secure the confidentiality of sensitive information, and organize the accountability system.

18.4 Transforming Pharmaceuticals with Digital Intelligence and Responsible AI

The pharmaceutical industry has now become a leader in digital transformation (DT) driven by the development of the AI and data-driven technologies. The remaining discussion analyzes how these innovations are transforming the way drugs are developed, patients are undertaken care of, and efficiency is realized. Section 18.4.1, which is titled Navigating the Future of Digital Pharma, defines the developing field of digital pharma through the identification of current trends and opportunities within the area. A Case Study of AI in Pharmaceutical Innovations is then examined by looking at the possibilities of AI application, which include accelerated drug discovery, the simplification of clinical trial administration. Lastly, Section 18.4.3, which is titled Guiding the Future with Responsible AI discusses the need of ethical AI adoption, with a focus on transparency, fairness, and accountability to achieve sustainable growth in the sector.

18.4.1 Navigating the Future of Digital Pharma

Supporting the digital pace over data analytics, automation, and connected technologies, pharmaceutical industries are leading toward swift digitalization. The change is expected to bring about cutting-edge solutions in drug discovery, patient administration, and operational performance.

The pharmaceutical industry nowadays takes the central stage in terms of DT and new technologies (AI and ML) create a breakthrough in conventional ways of doing things and provide new opportunities to grow. The prospect of AI, ML being incorporated into pharmaceutical sciences is enormous because it is advancing drug design, discovery, reformulation, and formulation [40]. This transformation has been comprised of the unprecedented increase in data and a nonstop advancement in analytical tools and ML algorithms [41]. Such technologies provide solutions to problems that are not only efficient and cost-effectively achieved but also that could represent complex and nonlinear relationships that are common in pharmaceutical research, particularly using artificial neural networks (ANNs). However, a big reason behind the success of such applications is based on methodologies that are specific to the task, choices made in the selection of algorithms, and the nuance of curated data. The spark of AI innovating the field of pharmaceutical sciences would not be realized without proper data preparation practices.

The use of AI in industries should go beyond the technological breakthrough and merge with some acute strategic demands, such as the increased focus on sustainability and moral management. Pharmaceutical alignment in environmental, social, and governance (ESG) protocols is, therefore, essential in ensuring the implementation of the sustainable development goals (SDGs) by the United Nations [42]. Based on the dynamic capabilities (DC) theory, a framework of looking at how the BDA can be embedded into ESG management systems appears, namely in order to strengthen provenance, traceability, and data availability. Prototypical examples of pharmaceutical organizations in Germany, Portugal, and Switzerland highlight the fact that such integration can benefit the company significantly in terms of efficiency of operations, research and development (R&D), and supply chain management. However, there are still some obstacles, such as the complex character of successful BDA implementation and the poor data governance framework. Countering these challenges requires long-term innovation and interdisciplinary cooperation, as well as integrating AI in order to reinforce ESG efforts and embrace sustainable operations in the industry.

DT has become one of the most important concepts in academic inquiries into the pharmaceutical sector, going beyond the technology application to include technology-induced reorganizations of the operational as well as the strategic spheres [43]. As much as there has been a lot of focus given to the practical implementation of DT, its implications pertaining to pharmaceutical management remain entirely underrepresented in scholarly literature. A new systemic review of 404 articles identifies 35 of them concerning operations management, strategic management, organizational theory, and stakeholder engagement, which is the focus that contributes to conducting a broad research agenda. This agenda has become more topical due to the COVID-19 pandemic that revealed the role of DT in promoting operational changes, providing competitive benefits, and the driver of organizational improvement. However, the lack of comprehensive structures in outlining the impact of DT on various stakeholders forms a sustainable constraint. By filling this knowledge gap, management-based studies are required to combine empirical investigation with the theoretical explanation in order to provide an overall comprehension of DT within pharmaceutical context.

Profound shift has characterized the development of quality management systems (QMS) driven by the increasing adoption of digital programs aimed to address some of the new challenges in the industry, the most significant of which are the acceleration of the drug development process, customizable medicine, and the maximized supply-chain visibility [11]. In this regard, the digitalization of the QMS sphere is put in the focus of the data integrity and in-time management of quality, superiorization of the processes, and permanent growth. Blockchain to provide transparency to supply chain, the Internet of Things (IoT) to provide predictive maintenance, and ML to increase quality have been observed as the key drivers of this paradigm shift. Despite the apparent benefits, namely, the increase in the quality of products, the operational efficiency, and regulatory compliance, the transition is confronted by a multitude of barriers. The most important one of them is the problem linked to legacy systems integration, the risk of cultural resistance, and the increased possibility of cybersecurity threats. Such challenges require an approach to management that is both holistic and strategic, supported by scalable technologies, collaborative alliances, and sound governance systems to maintain a compliance position, as well as operational excellence and patient trust in an ever-evolving regulatory sea.

Based on the conclusion of Shashi Verma et al. [44], this systematic study observes how the pharmaceutical industry uses Facebook, Twitter, and YouTube as the fundamental social media systems. The analysis shows that the level of engagement and the choice of the platform vary across the industry, and all of them can suggest that social media facilitates narrowly targeting the marketing activities, deeper contact interaction with the customers, and comprehensive competitor analysis. The discussion also shows that analysis tools like Facebook Insights can provide great analytics to polish strategic decisions. Besides, the report reveals the level at which AI enhances sales productivity and reimbursement to help personalize campaigns, predict and foresee customer behavior as well as automate marketing assignments. Taken together, these lessons create a starting point in terms of future scholarship but also a source of strategic advice about better digital engagement in the pharmaceutical field.

The most recent technology is the DT in pharmaceutical sector, which has existed due to AI, BDA, and other emerging technologies, and poses widespread potentials of innovation and patient-centered outcomes. By understanding these possibilities, though, it is not just about using new tools; it is also about an application of strict principles in ensuring that technological developments are coupled with responsibility, and therefore, data governance, sustainability, and responsible behavior get to be at the core of every development in the future.

18.4.2 Harnessing AI for Pharmaceutical Advancements

AI is revolutionizing modern pharmaceutical science because it enhances research, optimizes clinical trial design, and makes precision medicine possible. The gains in productivity and clinical efficacy, which it produces once employed systematically, are previously unequalled.

The convergence of AI and ML in pharmaceutical R&D is now shaping a new carve-out in drug development and commercialization. The modern increase in computing performance levels has enabled automating even complex procedures, enhancing predictivity accuracy, and overall optimization [45]. As another example, during drug discovery, the algorithmic review of large amounts of data allows identifying new therapeutic candidates by trial and error. During clinical trials, AI and ML algorithms maximize trial design as well as the post-data examination. Coupled with these explosions are the innovations that help to reduce spiraling development expenses without sacrificing scientific rigor.

The rates of DT have been break-neck, and as such, the work of the pharmaceutical sector has replied. The high growth in digital solutions witnessed during the COVID-19 pandemic confirmed that AI can serve to ensure that R&D agendas get modernized. Since AI systems do not necessarily boast the inherent element of trustworthiness, observance of thoroughly tested methodologies is necessary; it is absolutely essential to put such systems beyond the reach of the hype as well as to provide their sustainable stability.

The swift AI proliferation in the context of healthcare and the pharmaceutical research area is soundly supported by the commensurate increase in the number of its applications and a mirror-like increase in its related problems. The proposed improvements in drug discovery, clinical trial design, and disease diagnosis rely on technologies, including DL, NNs, and Bayesian nonparametric models as well as natural language processing [46]. A thorough methodological approach, which included using extensive literature resources at PubMed, Science Direct, and Google Scholar with a focus on the changes over the last five years, has been taken. The main outcomes of the study imply that AI can help decrease the time and cost of drug development and better identify patients and streamline clinical trial procedures. However, utilization of this technology is limited by their data dependency, the problem with implementation, and the issues of reliability of some tools, such as quantitative structure-property relationships (QSPR) and Chem Mapper. The study hence highlights the need to balance the implementation of AI in the healthcare sector in order to maximize AI benefits as well as address ethical and practical issues [32].

The pharmaceutical industry has become a leading field that AI has turned into the catalyst of new developments and a major factor that drives the overall changes in the organizational functioning of drug enterprises of any size. Both the extent and the direction of AI-powered change are, however, highly dependent on both the size and the strategic priorities of any organization. A comparison of the involvement of AI into the activities of fifteen pharmaceutical businesses, ranging in size to include small, medium, and large firms, showed that company size examined significantly impacted the use of AI in the business [34]. In the case of smaller organizations, it lies on the optimization of homogeneous functions of research and development (R&D), management of data, and human resources. AI allows more efficiency in drug discovery, makes data processing more efficient, and allows resources to use limited resources with more reasonableness. Big businesses also focus on the implementation of AI in production, sales, marketing, and ad analytics. Within this context, AI acts as an immense tool of operational growth, decision-making, and grasping competitive advantages. The middle-sized firms are more flexible and customized, integrating AI projects with their capability of specialization and the available resources. Still, even with these improvements, the AI has yet to be integrated fully, as successful scenarios of such usage are often very narrow in the field of application. Such a condition compels the necessity of multimethod adoption policies and corporate-level cooperation to exchange best practices and use the transformational power of AI in pharmaceutical environments to full potential.

Outside of the traditional options of research and development, AI is today at the center stage of driving pharmaceutical formulation and manufacturing. ANNs, genetic algorithms, and fuzzy logic have also been used to streamline formulation processes and gain an understanding of complex correlations between parameters. Specifically, ANNs are characterized by the ability to capture the patterns and draw conclusions based on the large amount of data; as such, they have attained significant levels of performance in the context of a wide range of applications, such as disease diagnostics, classification, and parameter modeling [47]. Combining machine-learning techniques with probabilistic approaches, pharmaceutical organizations will be able to extend data exploration and to reinforce the precision of process modeling.

Despite these features, a number of practical hindrances are still in place, particularly where scalability and the day-to-day implementation of such operational systems are concerned. Companies must overcome these limitations in order to bring the potential benefits of AI demonstrated so far to the industrial context. These concerns need to be resolved in the case of reliable extension of AI-enabled innovations in formulation and processing throughout the sector.

The pharmaceutical world of manufacturing and quality assurance activities is being radically changed by convergence of automation, robotics, and AI. The accuracy of preclinical research extends to increasing the accuracy in identifying and classifying molecules through AI-powered predicting models, and automation ensures constant tracking and forecasted preventative maintenance of the production lines. RPA is altering the tradition of manufacturing workflows and leading to rich operational flexibility and adaptability to market demand [48]. All these innovations enhance the effectiveness of operations, reduce expenses, and minimize the element of human error. However, it is important to control the process of moving toward highly automated production environments. Human knowledge is still required when it comes to the maintenance of control, the correction, and validation of AI results. Although the field of custom-built solutions now prevails, the current technological developments are expected to provide a new level of transparency, reduce the number of drug shortages, and improve product quality. Collectively, these technologies emphasize the strategic need to implement the AI-driven manufacturing systems to handle the emerging needs of the pharmaceutical industry.

The use of AI in RA has grown significantly in recent years, with stakeholders interested in improving the efficiency of operations as well as automating the complex submission processes. With data processing and decision functions being automated, AI easily has the potential to significantly reduce the time it takes to provide new therapy to patients. Such speeding of the regulatory process can provide solutions to critical healthcare needs and fasten the availability of innovative options in healthcare [36]. However, the successful implementation of AI in the sphere of regulation requires cooperation among stakeholders and the development of unified regulatory systems. These structures should resolve the conflict between encouraging innovation and protecting the population. Improved risk-management procedures and standard operating procedures Transparency is vital in developing a basis of trust and being compliant with the regulations. Nevertheless, the recognition of the value of AI in regulatory processes will enable it to transform the future of RA, as well as sustainable growth due to the increasingly competitive nature of healthcare industry.

The current commentary assesses the integration of AI into the practice of pharmacy, including its benefits and the challenges that are associated with this process. AI has the potential to optimize medicine dispensing by pharmacists, thus improving patient safety and reducing human errors. Technical approach involved a critical review and analysis of governmental reports and policy papers, which was done through databases such as PubMed, Science Direct, Springer, and Google Scholar. Within the scope of key findings mentioned, it is possible to note that AI may support predictive analytics, personalized treatment, and real-time pharmacovigilance, and all of them help to improve clinical outcomes and operational performance [49]. However, data privacy issues, complicated regulations, and opposition to healthcare personnel continue to plague its implementation. The conclusion supports the need to create effective governing regulatory systems, make data governance transparent, and provide constant education to healthcare workers to allow responsible implementation of AI. The solutions to these issues will help in making pharmacy practice a more accurate, patient-oriented, and efficient healthcare system.

18.4.3 Guiding the Future with Responsible AI

Responsible AI explains the ethical, transparent, and fair manipulation of modern technologies in the pharmaceutical market. It focuses on responsibility and trust and includes such issues as bias, protection of data, and compliances.

An ethical application of AI in the medical field requires a comprehensive framework, which puts fairness, transparency, privacy, and accountability to the forefront. To realize ethical AI implementation, a broad-based set of guidelines will be needed, which brings multidisciplinary expertise to the table. The application of generative AI in ambient documentation in the clinical environment would be one of the demonstrative ones representing the benefits and the pitfalls of AI applicability to real life [12]. The most important factors related to keeping up-to-date and efficient in the presence of dynamic environments are ongoing monitoring, evaluation, and adaptation. At the same time, it is vital to promote stakeholders' participation and implement cross-institutional associations to facilitate an easier procedure of AI implementation and maintain high ethical standards. Taken together, all these measures create a systematic map that will foster just and efficient care outcomes, where AI innovations will be used with humanistic values on the front burner.

Further, effective integration of AI into the related healthcare settings requires strict consideration of the ethical principles that underpin the specific space, including fairness, transparency, inclusiveness, and human-centered design in particular. The SHIFT framework, or sustainability, human-centeredness, inclusiveness, fairness, and transparency, is more organized and offers a systematic approach in the ethical design of AI systems. The framework is developed in terms of virtue ethics theory, but it gathers its information in a systematic way based on the insight provided by a systematic analysis of 253 articles published between 2000 and 2020 [50]. Using the aspects of article selection using PRISMA and PRISMA methodology in combination with the hermeneutic approach to systematic review, the paper thus outlines the issues as well as the opportunities of ethical AI practice. It emphasizes the necessity of constant collaboration with stakeholders, creation of adaptive AI, and drafting strict ethical policies to see fair outcomes in a variety of healthcare situations. In addition, it promotes a broader worldwide research outlook by considering the specific requirements of the heterogeneous healthcare environments and populations.

In the sphere of pharmaceuticals, contextually suited use of AI brings a set of ethical issues of its own, which requires the particularized solution. The results of the latest national conclave organized in India define crucial issues, such as bias, privacy, transparency, and accountability, that need to be overcome in order to make AI comply with societal values and to protect human dignity [51]. A thematic case study in genome biology also shows how the integration of fair data governance practices and the human-centered design principles can work to enhance the AI applications in this field [52]. Creating explainable and transparent models of AI is an essential process in building trust; the proactive approaches like reskilling the workforce will be beneficial to counteract the possible impact. Being based on the Indian context, this discussion indicates that global cooperation is required in order to perfect the practices of ethical AI to their wide applicability into various parts of the world. This type of cooperation motivates sharing the best practices and reinforces the international system of AI control.

The revolution of the pharmaceutical industry using digital intelligence and responsible AI has the potential to transform the industry. Through the use of such technologies, the drug industry will be able to embrace digital-pharma future and introduce revolutionary changes. Another factor of serious importance is the support of responsible innovation activities. It is only after technological advancement comes into play with ethical governance that one will be able to see sustainable growth and better healthcare results.

18.5 Transforming Innovation and Development Through AI in Pharma and Drug Discovery

With the introduction of AI in the pharmaceutical sphere, the process of research and drug discovery is changing by helping to accelerate innovation, reduce the costs, and the efficiency of the developmental process. Increasing the effectiveness of critical stages of work, including the identification of targets, the design of molecular structures, and clinical studies, AI improves efficiency and increases success rates. AI-enhanced drug development can allow reduced timelines and cost reduction to the potential of supplying a customized treatment based on individual patient data. However, a few remaining problems based on regulatory adjustment, data security, and industry cross-collaboration are to be addressed to capitalize entirely on the ability of AI. Precisely integrated and controlled, AI can revolutionize the pharmaceutical industry, making it more agile, more accurate, more democratic [33].

The radical capabilities of AI in reconstructing drug finding, production, and conveyance throughout the pharmaceuticals sector have been given due deliberation. This trend in academic research has shown that AI solutions can overcome the inherent issues of traditional drug development, which primarily includes prohibitively high costs, extended time frame, and the inaccuracy of animal models to identify targets involved in disease pathology and optimize candidate compounds using comprehensive datasets of biological information, including genomics and proteomics [31]. Such approaches as experimental design, the prediction of pharmacokinetics, and predictive toxicity testing using AI-powered algorithms have been featured and can enable goal-specific and personalized treatment regimens. According to empirical evidence, AI enhances patient performance, reduces the costs of development, and simplifies regulatory compliance by using personalized medicine, computerized pharmaceutics, and ML-mediated modeling of pharmacokinetics/pharmacodynamics. However, there are challenges that still persisted; these are data quality, model validation, and ethical governance. The integration of AI technologies presents a promising pathway for advancing the pharmaceutical industry from era 4.0 to era 5.0, with significant implications for drug efficacy, manufacturing efficiency, and patient-centric care.

With the inclusion of AI and ML in the drug development of pharmaceuticals, technology has achieved high levels of prominence, particularly in the context of computer-aided drug design. This paradigm takes care of the ongoing challenge of producing new drug molecules in the expensive and lengthy structure of drug discovery. Incorporation of AI and ML technologies into various steps has increased the likelihood of discovering hit compounds, optimizing the results of clinical trials. These strategies utilize data-driven AI models to predict clinical trial outcomes, thus saving costs and increasing experimental success rates [53]. However, there are still some snags, the first one being the dependency on large sets of data that affect the performance of AI/ML and the difficulty to comprehend DL mechanisms. Nonetheless, techniques that rely on AI are still evolving and allow more and more automated accurate ways of drug discovery. It is an excellent direction that can be used in pharmaceutical research.

This study presented in [54] covered the challenge of costs associated with R&D and R&D inefficiency leading to a high attrition rate of new drugs to approval. AI is offered as a means of improving the efficiency of drug discovery through the analysis of large datasets and through collaboration between AI drug discovery firms and pharmaceutical companies. The focus is on training algorithms with domain knowledge to accelerate drug development and the development of AI-driven analyses and synthesis of large datasets. The authors stress the importance of closer collaboration between medicinal chemists and AI specialists. Key results elucidate how AI can potentially revolutionize drug discovery by shortening timelines and extending knowledge to integrated technologies such as gene therapy and regenerative medicine. Challenges still exist, for example, developing relationships with domain knowledge and working through real experimental validation and developing expertise at a domain level to realize the true impact of AI in drug development.

In a similar study [31], the authors introduce a perspective to the challenges faced by the pharmaceutical industry, such as traditionally high R&D costs, ineffectiveness in traditional approaches, and significant R&D inefficiency, stating that AI will maximize efficiencies during drug discovery and development. AI approaches discovery by harnessing ML algorithms to process biological data, carry out a pharmacokinetic prediction, and navigate the drug candidate space [31]. Several papers were reviewed, and key representation results reveal AI's ability to integrate drug development by lessening reliance on expensive animal models and supporting a more thorough personalized medicine approach toward analysis to real-world patient data. Although AI will impact discovery and development, the author points out limits such as needing domain-specific expertise and validation in the real-world; therefore, AI should realistically be viewed as an additional source of input rather than a substitute, and human counterparts are still invaluable despite AI advancing rapidly. Overall, the fundamental take-home is that AI provides pharmaceutical research and development potential to enable a more effective, accessible, and patient-centered approach to drug development.

18.6 Optimizing Pharmaceutical Excipients with AI Technology and AI-driven Discoveries in Natural Product Science

AI is changing the pharmaceutical industry by optimizing excipients and accelerating natural product discovery. AI improves drug formulation by predicting the interactions of excipients, enhancing drug stability, delivery, and bioavailability. In natural product research, AI can aid the development of new proteins like esmGFP, accelerate drug discovery through genome mining, and utilize AI-enhanced digital organisms to simulate biological systems. Overall, these innovations stress the fundamental importance of AI applied to drug development and biotechnology.

Ashutosh Kumar et al. present a study based on pharmaceutical excipients and its interaction with active pharmaceutical ingredients (APIs) in [55]. Traditional methods for compatibility assessment, such as Differential Scanning Calorimetry (DSC), Fourier Transform Infrared Spectroscopy (FTIR), and High-Performance Liquid Chromatography (HPLC), are time-consuming and costly. Therefore, this study analyzes the role of AI to streamline these processes by enabling predictive modeling, reducing experimental testing, and optimizing excipient selection. AI enhances drug development efficiency, minimizes costs, and ensures precise formulation choices. Additionally, AI accelerates drug discovery, automates routine tasks, and enhances workflow efficiency, ultimately improving the quality of pharmaceutical services.

David Meijer et al. [56] focus on the application of AI in natural product research, which has been limited due to challenges with multimodal, unbalanced, unstandardized, and dispersed data, which hinder pattern recognition and decision-making emulation. To address this, researchers proposed creating a knowledge graph by identifying key entities (e.g., compounds, sources, biological activities) and their relationships, enriching them with relevant attributes and metadata. This transformation of non-relational data into structured, machine-readable formats follows resource description framework (RDF) standards and links datasets using uniform resource identifiers (URIs).

18.7 Robust Framework of IPR in AI-driven Pharma

AI is transforming the pharmaceutical industry by streamlining drug discovery, enhancing formulations, and enabling personalized treatments. However, these advancements introduce significant IPR challenges, including the patentability of AI-generated inventions, ownership disputes, data privacy concerns, and the balance between proprietary research and open innovation. Current IPR regimes, such as United States Patent and Trademark Office (USPTO) and European Patent Office (EPO) patent laws, only recognize human inventors, and regulations like GDPR and HIPAA only provide regulations on data access and security. In light of these challenges, new collaborative and open innovation licensing models are developing. Moving forward, legal reform, standardization of AI regulations, and fairness of IP will all be essential to enabling equitable access to AI-derived medical innovations while protecting the ability to innovate. A strong IPR framework, as mentioned, will be key to successfully implementing this approach and to responsibly managing AI in the pharmaceutical industry.

This dialogue engages among the transformative potential of AI on the drug discovery and development processes in the life sciences sector, while identifying ethical, legal, and effectuation constraints. AI use in predictive analytics, chemical optimization, and clinical trial protocol detailing creates an opportunity for a shared platform of responsibility existing within an IPR and ethics framework [9]. Your person adopted a position of: we must study AI ethics and understand its implications on IPRs—understanding challenges caused by AI-produced inventions, data privacy, and algorithm bias. This report includes literature review and countervailing perspectives, establishing some critical conclusions from our analysis of AI and IPRs. The most notable observation is the importance of working toward synergy of educational enhancement for innovation and preservation of ethics in practice. Few identified limitations include performative consent with patients, patients having no control over sensitive health data, and health inequity caused by bias inherent in training data [57]. AI ethics conclusion included the need to work/recommend collaboration from pharmaceutical companies, regulatory bodies, healthcare providers, and patients advocacy bodies to create a unified AI ethical frame. The recommendation is to bring diverse stakeholders together to engage in a clearly established regulatory framework, to use AI in drug development responsibly and transparently while protecting patient safety and health equity—for example, by identifying the responsibility of use, concerning enforcement of the regulation of AI.

The application of AI has revolutionized the pharmaceutical industry, particularly in drug discovery and development. This research investigates the challenges pharmaceutical companies face such as, non-standardized data, insufficient labor force, and data confidentiality [58]. Through case studies and reviews of previous studies, it showcases AI's contribution to drug repurposing, substances for biological targets, optimizing clinical trials, effective distribution of drugs, etc. The principal findings outlined how companies like XtalPi, BioMap, and iCarbonX capitalized on AI to accelerate drug development processes and promote precision medicine. Nevertheless, challenges remain and require international collaboration, government regulations, and ethical considerations when leveraging AI to create drugs. Therefore, this review outlined the promising but complex integration of AI into the pharmaceutical industry with China leading the way.

18.8 Conclusion and Future Scope

The integration of Responsible AI and data governance frameworks into pharmaceutical processes has altered the way pharmaceuticals manage data, drug development, and healthcare provision. As industry begins to see an increasing reliance on AI and ML-based systems, the consideration for regulatory compliance and ethical requirements is at the forefront. Governance and compliance are critical in establishing transparency, security, and accountability when handling sensitive pharmacy data. AI has revolutionized the concept of data governance and improved efficiencies in decision-making, mitigating human error and bias but ensuring the integrity of the data has not compromised. In effect, the growing reliance on AI and ML is fundamentally changing the drug discovery and drug development process, although issues surrounding data privacy, security, and transparency of algorithms require a structured methodology for adopting AI in a manner that transforms the industry while continuing to endorse ethical values and meeting compliance requirements.

Although the application and integration of AI in pharmaceutical processes have many facets beyond just data governance and drugs. AI and related technologies have a profound effect on drug discovery, personalized medicine, and medicine development. AI-enabled technology has improved pharmaceutical excipients and has streamlined clinical trials, and AI-driven drug development has become faster, cheaper, and easier. Included in this, telehealth and the convergence of AI and digital intelligence in pharmaceuticals, has provided better diagnostics, new predictive analytics, and automated workflows, and has positively impacted the outcomes, and overall usage of limited resources of patients. As AI changes the pharmaceutical landscape, a well-established IPR framework is important for addressing issues of AI-related innovation, patent rights, and data ownership. It will be essential to ensure that any advancement in AI is consistent with ethical standards and responsible practices to build trust and long-term sustainability in our industry.

The future of responsible AI in pharmaceutical settings is largely dependent on sustained development for AI governance models, regulations, and sustainable AI guides, both separately and/or collaboratively. Other emerging technologies like quantum computing and blockchain will augment security, efficiency, and innovation in drug discovery and healthcare delivery. Federated learning will allow numerous innovative global pharmaceutical research institutions to collaborate safely and securely with each other while protecting data privacy. In support, all these developments will provide (or continue to provide) industry operational effectiveness, recognize the significance of green AI and ethically deployed AI, drive the industry toward more sustainable and social responsibility, and more broadly outline practical ethical applications from pharmaceutical research and health interventions. The pharmaceutical industry can fully realize the breadth and potential of AI through interdisciplinary collaboration and adaptive governance policies while encouraging ethical, just, and innovative emerging developments that continue to support modern global healthcare.

References

1 Singh H, Sewal P, and Verma DC. A predictive analysis of the COVID-19 pandemic for traditional and tree-based regression algorithms. In: *Impact of Digital Solutions for Improved Healthcare Delivery* (pp. 303–40). 2024 [cited 2024 Nov 21]. https://services.igi-global.com/resolvedoi/resolve.aspx?doi=104018/979-8-3693-5237-3.ch011 [Internet]. Available from: https://www.igi-global.com/chapter/a-predictive-analysis-of-the-covid-19-pandemic-for-traditional-and-tree-based-regression-algorithms/360999

2 Sewal P and Singh H. Analyzing distributed Spark MLlib regression algorithms for accuracy, execution efficiency and scalability using best subset selection approach. *Multimed Tools Appl [Internet]* 2023. Available from: https://doi.org/10.1007/s11042-023-17330-5

3 Whig P, Sharma P, Aneja N, Elngar AA, and Silva N. *Artificial Intelligence and Machine Learning for Sustainable Development.* CRC Press; 2024 [cited 2025 May 7]. Available from: https://www.routledge.com/Artificial-Intelligence-and-Machine-Learning-for-Sustainable-Development-Innovations-Challenges-and-Applications/Whig-Sharma-Aneja-Elngar-Silva/p/book/9781032778358

4 Sargiotis D. *Data Governance.* Data Gov; 2024.

5 Arunachalam P, Usharani R, Thirumal S, Swarnalatha AP, and Maheswari BU. Study on AI-powered advanced drug discovery for enhancing privacy and innovation in healthcare. In: *AI-Powered Advances in Pharmacology*, (pp. 25–60). 2024 [cited 2025 May 7]. https://services.igi-global.com/resolvedoi/resolve.aspx?doi=104018/979-8-3693-3212-2.ch003 [Internet]. Available from: https://www.igi-global.com/chapter/study-on-ai-powered-advanced-drug-discovery-for-enhancing-privacy-and-innovation-in-healthcare/356858

6 Winter JS, Davidson E. Big data governance of personal health information and challenges to contextual integrity. *Information Society* [Internet] 2019;35(1):36–51. Available from: https://doi.org/10.1080/01972243.2018.1542648

7 Huma T and Peng Z. Introduction to regulatory affairs and different regulatory bodies for pharmaceutical products and impact of digitalization on regulatory affairs. *Pharmacology & Pharmacy* 2023;14(11):463–477.

8 Ferretti A. Ethics and governance of big data in health research and digital health applications. 1992. ETH Zurich. https://doi.org/10.3929/ethz-b-000489154

9 Lal S, Singh B, and Kaunert C. Role of artificial intelligence (AI) and intellectual property rights (IPR) in transforming drug discovery and development in the life sciences: Legal and ethical concerns. Library Progress International 2024;44(3):7070–7082.

10 Prasad AN. *Introduction to Data Governance for Machine Learning Systems: Fundamental Principles, Critical Practices, and Future Trends*, (pp. 1–966). Springer; 2024.

11 Ullagaddi P. Digital transformation in the pharmaceutical industry: Enhancing quality management systems and regulatory compliance. *International Journal of Health Sciences (Qassim)* 2024;12(1):2372–5079.

12 Saenz AD, Bundela Y, Logan M, et al. Establishing responsible use of AI guidelines: A comprehensive case study for healthcare institutions. *npj Digital Medicine* 2024;7(1):8–13.

13 Thomas BG. Building a data governance framework for AI-driven organizations. *Malaysian Journal of Computer Science [Internet]* 2022;3(1):1–15. Available from: https://mzjournal.com/index.php/MZCJ/index 15

14 Ferretti A. Ethics and governance. In: *Foundations for Fintech*, (pp. 3–22). 2021. Available from: https://www.worldscientific.com/doi/abs/10.1142/9789811238819_0001

15 Personal Information Protection Law of the People's Republic of China - PIPL [Internet]. [cited 2025 May 8]. Available from: https://personalinformationprotectionlaw.com/

16 AI Act | Shaping Europe's digital future [Internet]. [cited 2025 May 8]. Available from: https://digital-strategy.ec.europa.eu/en/policies/regulatory-framework-ai

17 Act on the Protection of Personal Information - English - Japanese Law Translation [Internet]. [cited 2025 May 8]. Available from: https://www.japaneselawtranslation.go.jp/en/laws/view/4241/en

18 ISO/IEC 23053: 2022 - Framework for AI systems using machine learning [Internet]. [cited 2025 May 8]. Available from: https://www.iso.org/standard/74438.html

19 PDPC | Singapore's Approach to AI Governance [Internet]. [cited 2025 May 8]. Available from: https://www.pdpc.gov.sg/help-and-resources/2020/01/model-ai-governance-framework

20 Brazilian General Data Protection Law (LGPD, English translation) [Internet]. [cited 2025 May 8]. Available from: https://iapp.org/resources/article/brazilian-data-protection-law-lgpd-english-translation/

21 Mökander J, Sheth M, Gersbro-Sundler M, Blomgren P, and Floridi L. Challenges and best practices in corporate AI governance: Lessons from the biopharmaceutical industry. *Frontiers of Computer Science [Internet]* 2022; 4. Available from: https://www.frontiersin.org/articles/10.3389/fcomp.2022.1068361/full

22 Pitel E, Leașu F, Nicolau A, et al. Etică medicală Ethical aspects in the use of artificial intelligence in the process of drug development Aspecte etice în utilizarea inteligenței artificiale în procesul de dezvoltare a medicamentelor. *Jurnal Medical Brasovean* 2023;(2)

23 Sanchini V, Marelli L, Monturano M, et al. A comprehensive ethics and data governance framework for data-intensive health research: Lessons from an Italian cancer research institute. *Accountability in Research [Internet]* 2023;32(1):59–76. Available from: https://doi.org/10.1080/08989621.2023.2248884

24 Chauhan C and Gullapalli RR. Ethics of AI in pathology: Current paradigms and emerging issues. *American Journal of Pathology [Internet]* 2021;191(10):1673–1683. Available from: https://doi.org/10.1016/j.ajpath.2021.06.011

25 Willem T, Fritzsche MC, Zimmermann BM, et al. Embedded ethics in practice: A toolbox for integrating the analysis of ethical and social issues into healthcare AI research. *Science and Engineering Ethics [Internet]* 2025;31(1). Available from: https://doi.org/10.1007/s11948-024-00523-y

26 Armitage RC. Implications of large language models for clinical practice: Ethical analysis through the principlism framework. *Journal of Evaluation in Clinical Practice* 2025;31(1):1–6.

27 Leben D. Ethical issues for AI in medicine. In: *Digit Heal Telemed Beyond* (pp. 309–319). 2024. Elsevier. https://doi.org/10.1016/B978-0-443-23901-4.00021-0

28 Gude S and Gude YS. The synergy of artificial intelligence and machine learning in revolutionizing pharmaceutical regulatory affairs. *Translational and Regulatory Sciences* 2024;6(2):37–45.

29 Pesapane F, Hauglid MK, Fumagalli M, et al. The translation of in-house imaging AI research into a medical device ensuring ethical and regulatory integrity. *European Journal of Radiology [Internet]* 2025;182:111852. Available from: https://doi.org/10.1016/j.ejrad.2024.111852

30 Naskar S, Sharma S, Kuotsu K, et al. The biomedical applications of artificial intelligence: An overview of decades of research. *Journal of Drug Targeting [Internet]* 2025;33(5):1–32. Available from: https://doi.org/10.1080/1061186X.2024.2448711

31 Vora LK, Gholap AD, Jetha K, Thakur RRS, Solanki HK, and Chavda VP. Artificial intelligence in pharmaceutical technology and drug delivery design. *Pharmaceutics [Internet]* 2023;15(7):1916. Available from: https://www.mdpi.com/1999-4923/15/7/1916

32 Jackson FA. The technology adoption model for cloud computing, storytelling artificial intelligence, and the federal risk and authorization management program. Ph.D. in CS. Aspen University. 2024.

33 Mottaghi-Dastjerdi N, Soltany-Rezaee-Rad M. Advancements and applications of artificial intelligence in pharmaceutical sciences: A comprehensive review. *Iranian Journal of Pharmaceutical Research [Internet]* 2024 [cited 2025 May 7];23(1): e150510. Available from: https://pmc.ncbi.nlm.nih.gov/articles/PMC11787549/

34 Kulkov I. The role of artificial intelligence in business transformation: A case of pharmaceutical companies. *Technology in Society [Internet]* 2021;66:101629. Available from: https://doi.org/10.1016/j.techsoc.2021.101629

35 Singh RR, Kumar S, Pandey A, Tripathi SK, Yadav PK, and Pathak V. The role of the metaverse in emergency medicine and its legal implications : A systematic review. *Indian Journal of Forensic Medicine & Toxicology* 2025;19(1):28–37.

36 Patil RS, Kulkarni SB, and Gaikwad VL. Artificial intelligence in pharmaceutical regulatory affairs. *Drug Discovery Today* 2023;28(9).

37 Neelakandan R and Sankaran V. Ensuring safety, quality, and trust : A holistic framework for responsible. *International Journal of Artificial Intelligence Research and Development* 2024;2(1):65–75.

38 van Voorst R. Redefining intelligence: Collaborative tinkering of healthcare professionals and algorithms as hybrid entity in public healthcare decision-making. *AI & Society [Internet]* 2025 [cited 2025 May 7]; 1–12. Available from: https://link.springer.com/article/10.1007/s00146-024-02177-7

39 Yadav S, Singh A, Singhal R, and Yadav JP. Revolutionizing drug discovery: The impact of artificial intelligence on advancements in pharmacology and the pharmaceutical industry. *Intelligent Pharmacy [Internet]* 2024;2(3):367–380. Available from: https://doi.org/10.1016/j.ipha.2024.02.009

40 Damiati SA. Digital pharmaceutical sciences. *American Association of Pharmaceutical Scientists Pharmaceutical Science and Technology [Internet]* 2020;21(6):206. Available from: https://link.springer.com/10.1208/s12249-020-01747-4

41 Sewal P and Singh H. Algorithmic proficiency in spark configuration tuning: An empirical study using execution time metrics across varied workloads. *Procedia Computer Science [Internet]* 2024;235(2023):2307–2317. Available from: https://doi.org/10.1016/j.procs.2024.04.219

42 Pesqueira A and Sousa MJ. Exploring the role of big data analytics and dynamic capabilities in ESG programs within pharmaceuticals [Internet]. *Software Quality Journal* 2024;32:607–640 p. Available from: https://doi.org/10.1007/s11219-024-09666-4

43 Miozza M, Brunetta F, and Appio FP. Digital transformation of the pharmaceutical industry: A future research agenda for management studies. *Technological Forecasting and Social Change [Internet]* 2024;207:123580. Available from: https://doi.org/10.1016/j.techfore.2024.123580

44 Verma S, Tiwari RK, and Singh L. Integrating technology and trust: Trailblazing role of AI in reframing pharmaceutical digital outreach. *Intelligent Pharmacy [Internet]* 2024;2(3):435–440. Available from: https://doi.org/10.1016/j.ipha.2024.01.005

45 Kolluri S, Lin J, Liu R, Zhang Y, and Zhang W. Machine learning and artificial intelligence in pharmaceutical research and development: A review. *AAPS Journal [Internet]* 2022;24(1):19. Available from: https://journals.sagepub.com/doi/10.1177/1461444810365020

46 Bhattamisra SK, Banerjee P, Gupta P, Mayuren J, Patra S, and Candasamy M. Artificial intelligence in pharmaceutical and healthcare research. *Big Data and Cognitive Computing [Internet]* 2023;7(1):10. Available from: https://www.mdpi.com/2504-2289/7/1/10

47 Sable P and Khanvikar VV. Pharmaceutical applications of artificial intelligence. *International Journal of Pharma Research and Health Sciences [Internet]* 2018;6(2):2342–2387. Available from: www.pharmahealthsciences.net

48 Manoj Kumar T, Preethi B, Nunavath RS, Nagappan K. Future of pharmaceutical industry: Role of artificial intelligence, automation and robotics. *Journal of Pharmacology and Pharmacotherapeutics [Internet]* 2024;15(2):142–152. Available from: https://journals.sagepub.com/doi/10.1177/0976500X241252295

49 Hakimi M, Amiri GA, and Shamsi SE. Artificial intelligence and public health: Addressing pharmacy practice challenges and policy issues. *British Journal of Pharmacy and Pharmaceutical Sciences [Internet]* 2024;1(1):09–21. Available from: https://creativecommons.org/licenses/by/4.0/

50 Siala H and Wang Y. SHIFTing artificial intelligence to be responsible in healthcare: A systematic review. *Social Science & Medicine [Internet]* 2022;296:114782. Available from: https://doi.org/10.1016/j.socscimed.2022.114782

51 Balan A. AI and Legal Education: Ethical and sustainable approaches. In: *AI Leg Educ Ethical Sustain Approaches [Internet]*, (pp. 1–125). London: Routledge;

2025 [cited 2025 May 7]. Available from: https://www.taylorfrancis.com/books/mono/10.4324/9781003607397/ai-legal-education-anil-balan

52 Yadav G, Munshi A, Kumari R, Singh D, Kumari N, and Mujoo U. AI & ethics: Charting a responsible future. *Annals of Library and Information Studies [Internet]* 2024;71(04):496–505. Available from: https://or.niscpr.res.in/index.php/ALIS/article/view/14291

53 Selvaraj C, Chandra I, and Singh SK. Artificial intelligence and machine learning approaches for drug design: Challenges and opportunities for the pharmaceutical industries. *Molecular Diversity [Internet]* 2022;26(3):1893–913. Available from: https://link.springer.com/10.1007/s11030-021-10326-z

54 Mak KK and Pichika MR. Artificial intelligence in drug development: Present status and future prospects. *Drug Discovery Today [Internet]* 2019;24(3):773–780. Available from: https://doi.org/10.1016/j.drudis.2018.11.014

55 Kumar A, Gupta GD, and Raikwar S. Artificial intelligence technologies used for the assessment of pharmaceutical excipients. *Current Pharmaceutical Design* 2024;30(6):407–409.

56 Meijer D, Beniddir MA, Coley CW, et al. Empowering natural product science with AI: Leveraging multimodal data and knowledge graphs. *Natural Product Reports* 2024; 42(4):654–662. https://doi.org/10.1039/d4np00008k

57 Gupta T, Sharma A, Aryan, Rana K, and Sewal P. Utilizing twitter data and NLP to analyze and predict public sentiment trends in mental health. In: Pundir AKS, Yadav A, Das S, editors. *Recent Trends in Communication and Intelligent Systems. ICRTCIS 2024. Algorithms for Intelligent Systems*, (pp. 119–33). Singapore: Springer. 2024 [cited 2025 May 8]. Available from: https://link.springer.com/chapter/10.1007/978-981-97-7632-0_9

58 Sampene AK and Nyirenda F. Evaluating the effect of artificial intelligence on pharmaceutical product and drug discovery in China. *Future Journal of Pharmaceutical Sciences [Internet]* 2024;10(1). Available from: https://doi.org/10.1186/s43094-024-00632-2

Appendixes

Appendix A

The hyperparameters (epoch, batch size, dropout rate, etc.) of the developed deep learning model are given in Table A.1.

Appendix B

To our best knowledge, the KIND dataset [1] represents the most extensive collection of kinetic data available to the public. It consists of 3,812 small molecule structures accompanied by their association rate constants k_{on}, k_{off}, and dissociation constants K_d. BindingDB [2] is another database that gathers dissociation kinetic data for various proteins and small molecules. It collects 301 nonredundant k_{off} records belonging to 12 distinct protein types. On average, each protein type possesses approximately 25 records of its corresponding ligands. The PDBbind-koff-2020 dataset [3] comprises 680 protein-ligand k_{off} records. These records span nearly 10 orders of magnitude, ranging from −3 to 7 (log unit). The majority of the k_{off} values collected in this dataset fall between 2 and 3. The proteins included in this dataset can be classified into 155 groups based on a sequence similarity cutoff of 90%. These groups encompass a diverse range of enzymes, receptors, transporters, and other proteins [3]. Amangeldiuly et al. [4] present a kinetic dataset comprising 501 protein-ligand complexes.

Table A.1 The hyperparameters of the deep learning model.

Hyperparameter	Value
Epoch	40
Batch size	32
Dropout rate	0.4
Max learning rate	0.0005
Min learning rate	0.00005
Learning rate decay factor	0.5

These complexes are annotated to 53 distinct protein families and encompass experimental k_{off} values. The logarithmic scale of k_{off} values exhibits a variation of eight orders of magnitude, although the majority of compounds are distributed across five orders of magnitude. In Amangeldiuly's dataset [4], the pk_{off} data mainly falls within the range of −2 to 6, with a peak value biased toward the interval of 3–4. The dissociation kinetic data contributed by the above four datasets are then preprocessed according to the following criteria:

- The entries involving the case that the ligand SMILES string fails to be read by the RDKit tool [5] or the ligand Morgen fragments [6] cannot be found in the Mol2vec method [7] are deleted.
- The entries involving the protein–ligand complex whose Morgen fragment number ranging from 20 to 74 and amino acid number ranging from 40 to 500 are retained.
- The entries involving the protein–ligand complex whose pk_{off} value ranging from 0 to 5 and having no uncertain inequality sign (e.g., ">," "<") are retained.
- The entries involving the protein–ligand complex whose protein chain number is greater than one are deleted.
- The ligands that comply with the Lipinski's "Rule of Five" are retained.
- The entries involving the charged ligands are deleted.
- The pk_{off} values in duplicate entries are averaged and kept one.

Appendix C

The architecture of the generic dissociation kinetic model without the mixture-of-experts (MoE) layer (baseline model) is presented in the following. The ligand matrix $\mathbf{x}_{\mathrm{L}}$ (74 × 300) undergoes a reduction in dimensionality by first passing through a fully connected (FC) layer $\mathbf{V}$ (300 × 30). This transformation to a 30-dimension ligand feature set not only decreases model redundancy but also aids in saving on training costs. The resulting ligand matrix $\mathbf{x}'_{\mathrm{L}}$ (74 × 30) then proceeds to an attention layer, where $\mathbf{x}'_{\mathrm{L}}$ is transformed by a learnable weight matrix $\mathbf{W}$ (30 × 30) to produce the new ligand matrix $\mathbf{x}''_{\mathrm{L}}$ with notable features. The next step involves obtaining the attention coefficient e_{ij} through the equation $e_{ij} = \mathbf{x}''_{\mathrm{L},i}\mathbf{x}''^{\mathrm{T}}_{\mathrm{L},j} + \mathbf{x}''^{\mathrm{T}}_{\mathrm{L},i}\mathbf{x}''_{\mathrm{L},j}$, where e_{ij} represents the significance of the jth fragment feature to the ith fragment feature. The symmetry of e_{ij} ($e_{ij} = e_{ji}$) is enforced by summing $\mathbf{x}''_{\mathrm{L},i}\mathbf{x}''^{\mathrm{T}}_{\mathrm{L},j}$ and $\mathbf{x}''^{\mathrm{T}}_{\mathrm{L},i}\mathbf{x}''_{\mathrm{L},j}$. Subsequently, e_{ij} is input into a SoftMax activation function to obtain the normalized attention coefficient a_{ij}, expressed by the equation $a_{ij} = \exp\left(e_{ij}\right)/\sum_j \exp(e_{ij})$. Using a_{ij}, each fragment feature is updated as a linear combination of neighboring fragment features ($\mathbf{x}^{\mathrm{out}}_{\mathrm{L},i} = \sum_j a_{ij}\mathbf{x}''_{\mathrm{L},j}$). A Gelu activation function is then applied to $\mathbf{x}^{\mathrm{out}}_{\mathrm{L},i}$. Finally, the feature vectors of ligand fragments are aggregated into a single vector to represent the ligand feature ($\mathbf{x}^{\mathrm{out}}_{\mathrm{L}} = \sum_{i\in\mathrm{fragments}}\mathbf{x}^{\mathrm{out}}_{\mathrm{L},i}$).

The protein matrix $\mathbf{x}_{\mathrm{P}}$ (500 × 30) first goes through a one-dimensional convolutional (Conv1) layer (input channels = 30, output channels = 30, kernel size = 3, stride = 3) followed by a MaxPool layer (kernel size = 3, stride = 3). The kernels in these layers employ a parameter sharing mechanism [8] to scan the local features of the protein matrix and extract critical higher-order amino acids, focusing the amino acid features on the pocket area of the protein–ligand complex. This process also decreases model redundancy and saves computational costs in the subsequent attention layer. The processed protein matrix then proceeds to an attention layer similar to the one in the ligand section. The final output protein feature $\mathbf{x}_{\mathrm{P}}^{\mathrm{out}}$ is obtained through $\mathbf{x}_{\mathrm{P}}^{\mathrm{out}} = \sum_{i \in \text{higher-order amino acids}} \mathbf{x}_{\mathrm{P},i}^{\mathrm{out}}$. Next, the output ligand feature ($\mathbf{x}_{\mathrm{L}}^{\mathrm{out}}$) and the output protein feature ($\mathbf{x}_{\mathrm{P}}^{\mathrm{out}}$) are concatenated to represent the protein–ligand complex feature, which are then sent to an FC layer (60 × 30) with a Gelu activation, followed by an output layer (FC layer) (30 × 1) with another Gelu activation. Additionally, a dropout layer is incorporated before the output layer to mitigate overfitting in the training process [9].

The loss function in our dissociation kinetic model is mean square error (MSE), which is a commonly used loss function for regression tasks. To minimize the loss function, the model parameters are optimized by using an adaptive moment optimizer (Adam) [10] with the learning rates from 0.0005 to 0.00005, where a decay factor of 0.5 is used to update the learning rates when the loss function of the validation set is not reduced with epochs. The evaluation criteria of the Pearson coefficient (r) and root mean square error (RMSE) are considered as important metrics functions to evaluate the prediction results of our dissociation kinetic model in estimating pk_{off}. The closer the values of r and RMSE are to 1 and 0, respectively, the better prediction ability of the dissociation kinetic model behaves. To avoid the fortuitously exceptional training outcomes, the entire dissociation kinetic dataset undergoes ten random splits into training (2,218 samples), validation (277 samples), and test sets (278 samples) by a ratio of 8 : 1 : 1. Subsequently, for each split, the deep learning model is subjected to ten separate rounds of random training. The hyperparameters (epoch, batch size, dropout rate, etc.) are determined by the trial-and-error method with our empirical knowledge. The dissociation kinetic model is established on the PyTorch library [11] using the Python language [12].

Appendix D

The hyperparameters (epoch, batch size, dropout rate, etc.) of the developed deep learning model are determined by the trial-and-error method with our empirical knowledge and listed in Table D.1.

Table D.1 The hyperparameters of the deep learning model.

Hyperparameter	Value
Epoch	40
Batch size	32
Number of experts	16
Dropout rate	0.2
Max learning rate	0.0005
Min learning rate	0.00005
Learning rate decay factor	0.5

Appendix E

Figure E.1a, b records the values of the loss function (MSE) and the metrics function (r) varying with the epochs during the training process of the MoE-based dissociation kinetic model, respectively, based on 100 random training instances. The training and validation results indicate that the developed MoE-based dissociation kinetic model has no overfitting issue. Note that the utilization of attention mechanism is able to accelerate the learning process of our dissociation kinetic model. In the training process, the dissociation kinetic model is nearly converged at around the 40th epoch. If the attention layers are removed, the dissociation kinetic model needs around 70 epochs to obtain similar results.

Samples exhibiting an absolute prediction error larger than 1 (98 data points) are selected for analysis by mapping their protein–ligand embedding features into a chemical space, as shown in Figure E.2. Among these 98 data points, samples corresponding to proteins with 20 or fewer data points (25 samples) are also labeled. On the one hand, Figure E.2 indicates that most samples with large prediction errors are associated with proteins that have a limited number of values (i.e., 20 or fewer). This observation suggests that our MoE-based dissociation kinetic model may not have effectively learned the characteristics of these proteins. To verify this point of view, our dataset is divided into two subsets: one containing samples with 20 or fewer data points for each protein and the other one with more than 20 data points. The mean absolute errors (MAEs)/mean absolute percentage errors (MAPEs) for these subsets are 0.369/48.0% and 0.313/35.9%, respectively. It is found that both MAE and MAPE are larger for proteins with fewer data points. On the other hand, certain proteins exhibit a large number of values. For example, the sample size for neutrophil elastase (circled in red) reaches as high as 172. However, many samples for this type of protein still exhibit absolute errors greater than 1. This result can be attributed to the formation of "feature islands" in chemical space, where the protein–ligand characteristics denoted within the red circle are significantly distant from those in other regions, hindering the ability of our MoE-based dissociation kinetic model to make accurate predictions of pk_{off} values for neutrophil elastase based on features learned

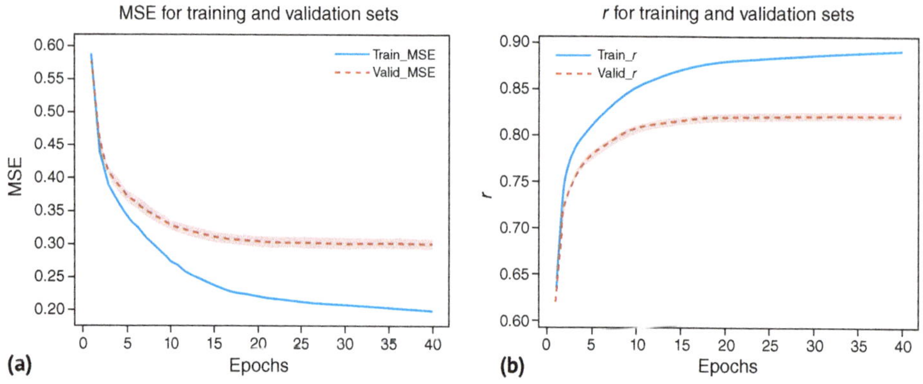

Figure E.1 The values of (a) the loss function (MSE) and (b) the metrics function (r) varying with the epochs during the training process of the MoE-based dissociation kinetic model based on 100 random training instances.

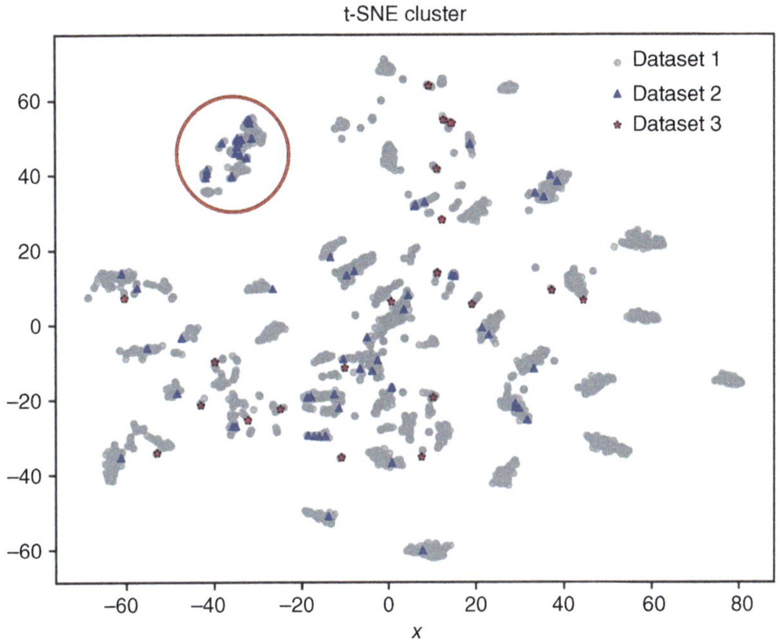

Figure E.2 The chemical space of dataset 1 (entire dataset), dataset 2 (samples exhibiting an absolute prediction error larger than 1), and dataset 3 (samples corresponding to proteins with 20 or fewer data points among dataset 2) using protein–ligand embedding features developed in this work and the t-SNE technique.

from other protein–ligand complexes. To address these two issues, future efforts should focus on enhancing the density of protein–ligand features in chemical space. This can be achieved by increasing the number of data samples, promoting a more uniform feature distribution, and eliminating feature islands. Additionally, it is essential to expand the sample size for each protein type. Implementing these strategies may further improve the prediction accuracy of our MoE-based dissociation kinetic model.

Moreover, the overall dataset, which includes pk_{off} values ranging from 0 to 5, is divided into five intervals: [0, 1), [1, 2), [2, 3), [3, 4), and [4, 5). The MAEs/MAPEs for these five intervals are 0.364/107.1%, 0.252/17.3%, 0.347/14.4%, 0.422/12.4%, and 0.474/10.6%, respectively. It is found that both the MAE and MAPE for the [0, 1) interval are larger than those for the other four intervals, indicating that the MoE model has difficulty in training and predicting small pk_{off} values. This difficulty may stem from the fact that the samples with small pk_{off} values contribute a minor portion to the loss function, leading to their negligible influence on model training. Given that our de novo drug design method aims to design drug candidates with large pk_{off} values, the prediction error in the [0, 1) interval can be reasonably omitted. Furthermore, it is found that MAEs increase as pk_{off} values rise, while MAPEs decrease with increasing pk_{off}. The MAE is larger for high pk_{off} values.

Subsequently, 100 randomly trained MoE models are used to predict the pk_{off} values for the entire dataset, calculating the average and standard deviation of the prediction results for each sample. Furthermore, a statistical analysis of the MAE of the samples is conducted

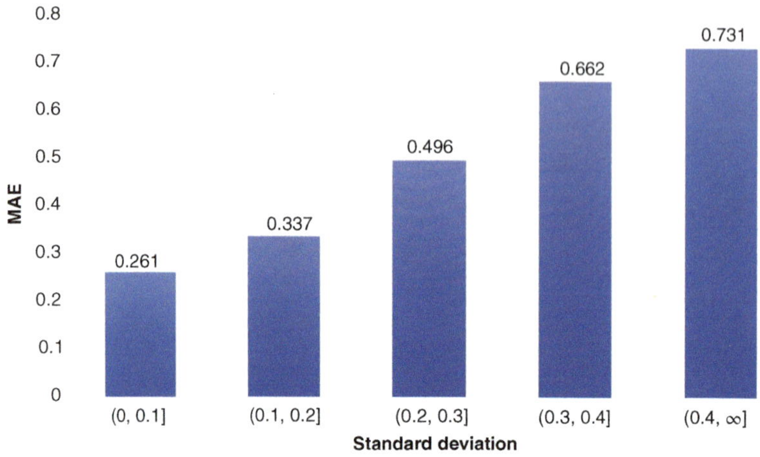

Figure E.3 The relationship between the standard deviation and MAE.

across different standard deviation ranges: (0, 0.1], (0.1, 0.2], (0.2, 0.3], (0.3, 0.4], (0.4, ∞). The results, shown in Figure E.3, indicate that as the standard deviation increases, the MAE value also increases. This finding indicates that a larger standard deviation in predictions from multiple MoE models—indicative of greater instability—corresponds to larger prediction errors. Therefore, when using MoE-based dissociation kinetic model for future pk_{off} predictions, it is suggested to consider the standard deviation as a criterion for assessing prediction reliability. If the standard deviation for a sample predicted by the 100 MoE models exceeds 0.4, it is suggested to avoid using that prediction due to the likelihood of substantial errors.

Appendix F

The remaining nine random data splits are given in Figures F.1–F.9.

Appendix G

The protein–ligand embedding features derived from our MoE model are used to plot a chemical space, enabling the analysis of sample similarity among our dataset, Liu et al. [3], and Amangeldiuly et al. [4], as shown in Figure G.1. It is found that our dataset predominantly overlapped the samples from Liu et al. [3] and Amangeldiuly et al. [4]. This overlap can be attributed to the fact that our dataset is collected from four public datasets, including those of Liu et al. [3] and Amangeldiuly et al. [4]. However, it is important to note that we have thoroughly cleaned the collected data, which may result in some samples from Liu et al. [3] and Amangeldiuly et al. [4] not being overlapped by our dataset.

100 randomly trained MoE models are used to predict pk_{off} (represented as mean ± standard deviation based on these 100 predictions) for the samples from Amangeldiuly et al. [4]

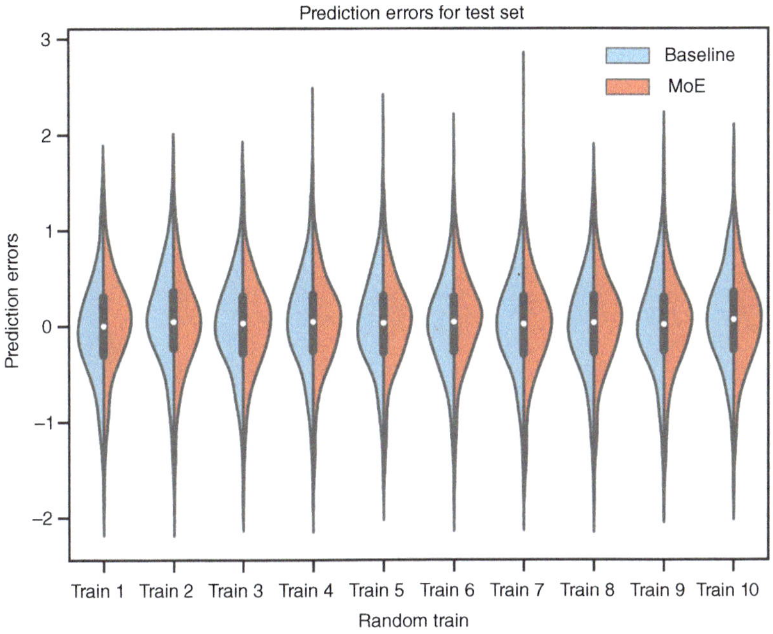

Figure F.1 The prediction errors (predicted pk_{off} minus experimental pk_{off}) of the dissociation kinetic models with MoE and without MoE (baseline model) for ten random model trainings in the second random data split.

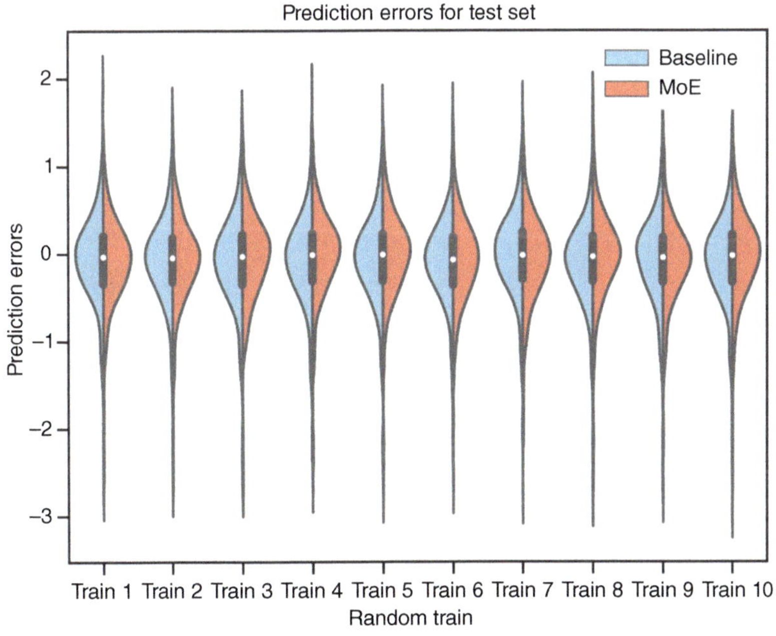

Figure F.2 The prediction errors (predicted pk_{off} minus experimental pk_{off}) of the dissociation kinetic models with MoE and without MoE (baseline model) for ten random model trainings in the third random data split.

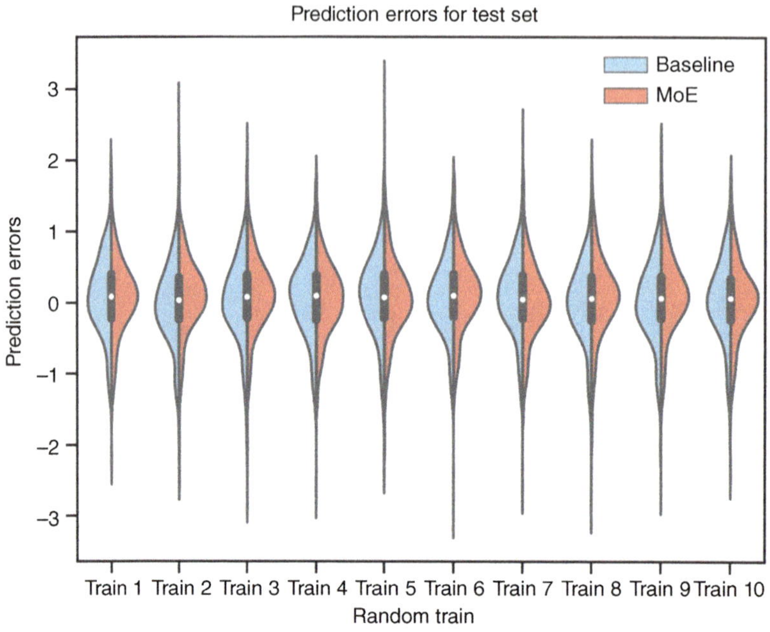

Figure F.3 The prediction errors (predicted pk_{off} minus experimental pk_{off}) of the dissociation kinetic models with MoE and without MoE (baseline model) for ten random model trainings in the fourth random data split.

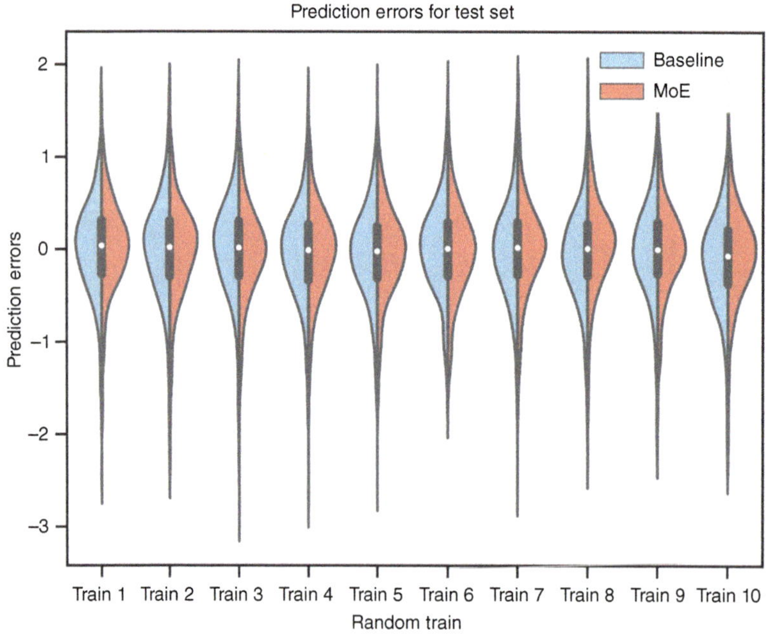

Figure F.4 The prediction errors (predicted pk_{off} minus experimental pk_{off}) of the dissociation kinetic models with MoE and without MoE (baseline model) for ten random model trainings in the fifth random data split.

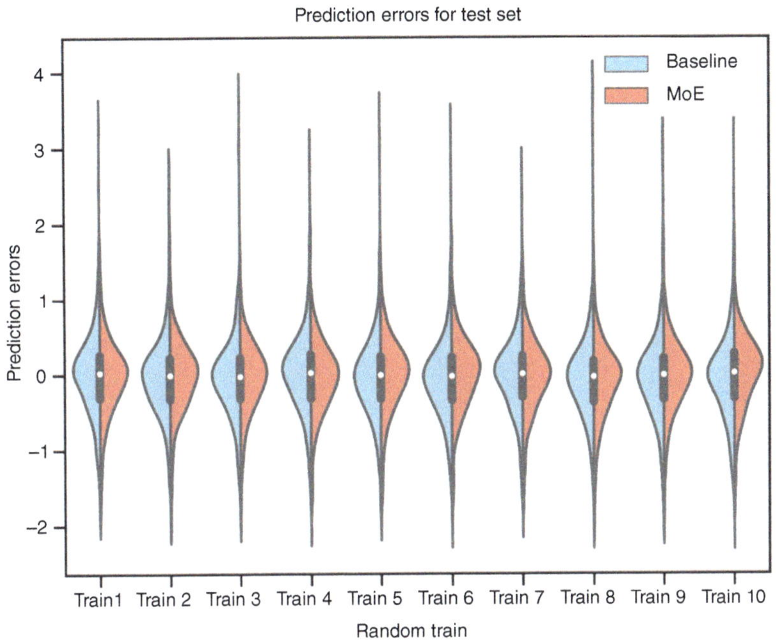

Figure F.5 The prediction errors (predicted pk_{off} minus experimental pk_{off}) of the dissociation kinetic models with MoE and without MoE (baseline model) for ten random model trainings in the sixth random data split.

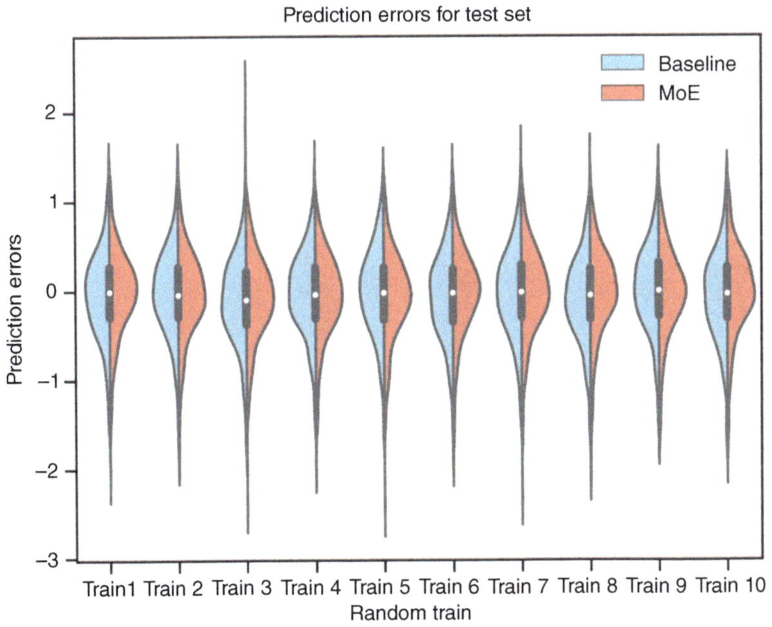

Figure F.6 The prediction errors (predicted pk_{off} minus experimental pk_{off}) of the dissociation kinetic models with MoE and without MoE (baseline model) for ten random model trainings in the seventh random data split.

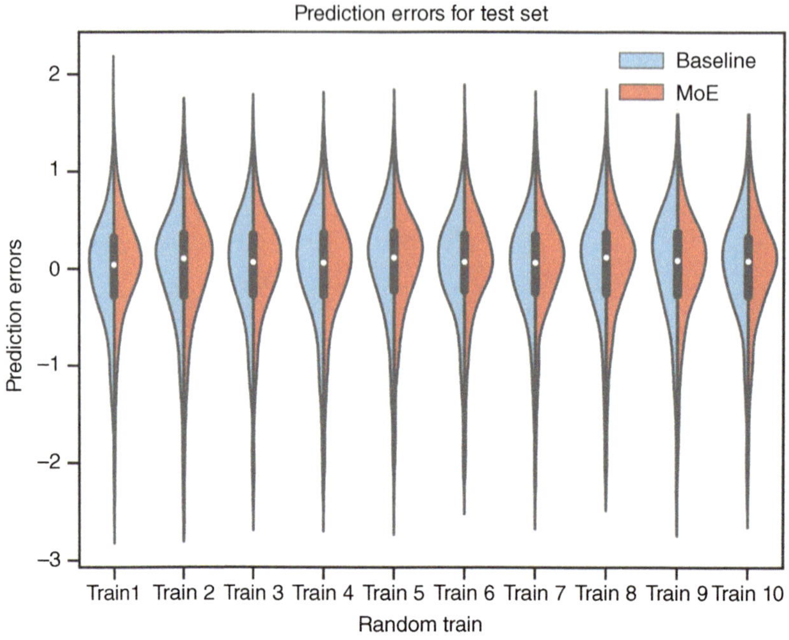

Figure F.7 The prediction errors (predicted pk_{off} minus experimental pk_{off}) of the dissociation kinetic models with MoE and without MoE (baseline model) for ten random model trainings in the eighth random data split.

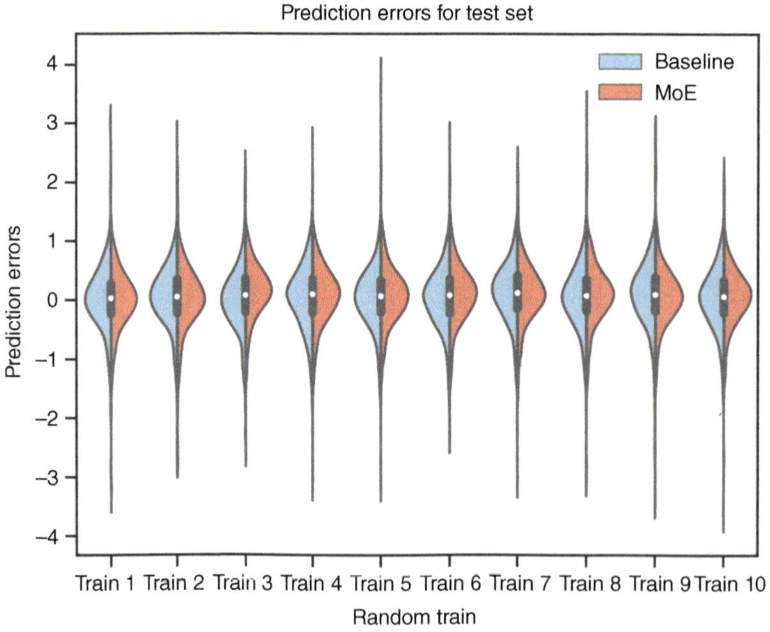

Figure F.8 The prediction errors (predicted pk_{off} minus experimental pk_{off}) of the dissociation kinetic models with MoE and without MoE (baseline model) for ten random model trainings in the ninth random data split.

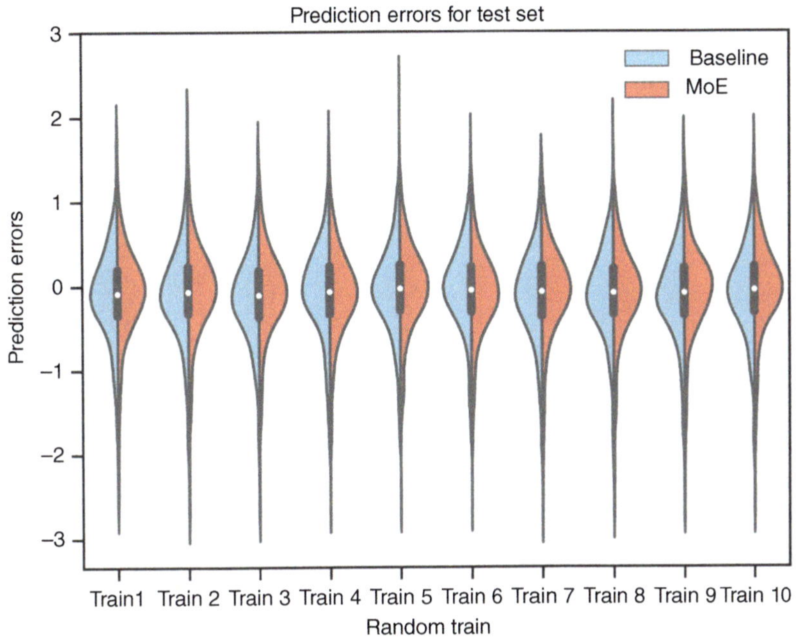

Figure F.9 The prediction errors (predicted pk_{off} minus experimental pk_{off}) of the dissociation kinetic models with MoE and without MoE (baseline model) for ten random model trainings in the tenth random data split.

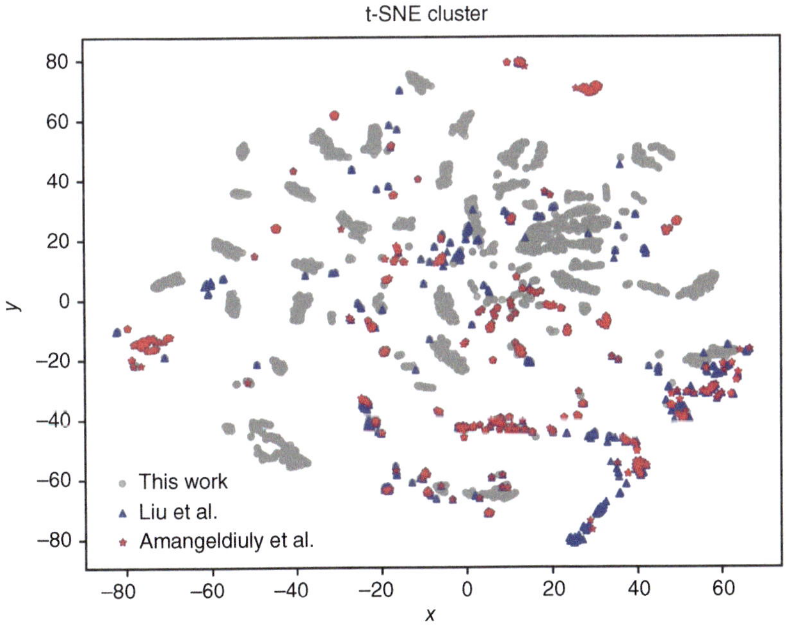

Figure G.1 The overlap of our dataset with the datasets from the literature.

and Liu et al. [3] Note that certain samples deemed unsuitable for our MoE model are excluded, including those with more than one FASTA string, those with abnormal SMILES containing "N+H," those with the Morgen fragment number exceeding the range of 20–74, those with the amino acid number exceeding the range of 40–500, and those with ligand Morgen fragments not found using the Mol2vec method. The r values for the prediction results on the samples from Amangeldiuly et al. [4] and Liu et al. [3] are 0.870 and 0.728, respectively.

Appendix H

The SMILES-based isomer generation algorithm in our previous work [13] is able to generate the SMILES strings of molecules from the functional group sets. Nevertheless, the algorithm needs the forward and reverse SMILES strings of functional groups containing "*." For example, the functional group "—CH_2—OH" needs the SMILES strings of "*CO" (forward) and "OC*" (reverse) for the SMILES-based isomer generation algorithm. Encoding the forward and reverse SMILES strings for hundreds of functional groups manually is not too much laborious, however, it is extremely difficult to manually encode the forward and reverse SMILES strings for thousands of complicated scaffolds. Therefore, a revised SMILES-based isomer generation algorithm is developed in this paper, which requires only forward or reverse SMILES strings of fragments (scaffolds and functional groups) and is more computationally efficient compared with the original one. In the following, the details of the revised SMILES-based isomer generation algorithm are described.

The diagrammatic sketch of the revised SMILES-based isomer generation algorithm is shown in Figure H.1. It is executed through the RWMol module in the RDKit library [5].

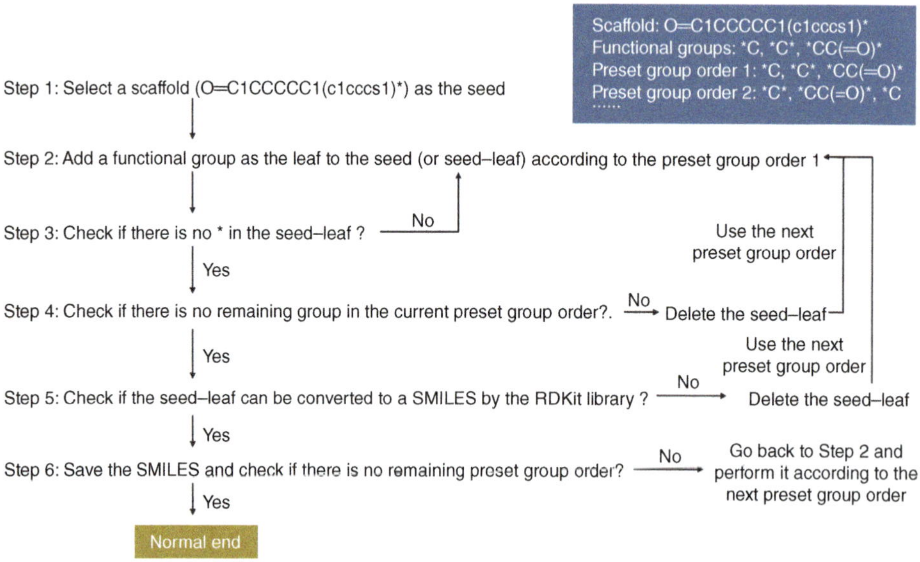

Figure H.1 The diagrammatic sketch of the revised SMILES-based isomer generation algorithm.

Here is an example to illustrate. A molecule represented by a set of fragments (Scaffold: O=C1CCCCC1(c1cccs1)*; functional groups: *C, *C*, *CC(=O)*) is the input to the algorithm. The scaffold (O=C1CCCCC1(c1cccs1)*) is first selected as the seed, which is then combined with a functional group (a leaf, *C) according to the preset group order 1 (*C, *C*, *CC(=O)*). The generated seed–leaf structure is checked if there is no * in the seed–leaf. If not, add another functional group to the seed–leaf according to the current preset group order. If so, check if there is no remaining functional group in the current preset group order. If not, delete the seed–leaf and use the next preset group order to generate a new seed–leaf. If so, convert the seed–leaf to a SMILES string by the RDKit library. If the conversion fails, delete the seed–leaf and use the next preset group order to generate a new seed–leaf. If the conversion succeeds, save the SMILES and check if there is no remaining preset group order. If so, the algorithm ends normally. Otherwise, go back to Step 2 and perform it according to the next preset group order.

Appendix I

The size of the MILP model for case study 1 is given in Table I.1.

Table I.1 The size of the MILP model for case study 1.

Name	Number
Blocks of equations	23
Single equations	107
Blocks of variables	6
Single variables	48
Nonzero elements	2,180
Discrete variables	43

Appendix J

For proteins, their three-dimensional crystal structures are collected from the research collaboratory for structural bioinformatics protein data bank (RCSB PDB) [14]. Before performing molecular docking, proteins are processed by deleting the original ligands in crystals and adding hydrogen atoms by the AutoDockTools suite [15]. For ligands, the three-dimensional molecular structures (stored in "PDB" format) are converted from their SMILES representations using the Open Babel software [16]. Afterward, the PDB files of proteins and ligands are converted to the PDBQT files using the AutoDockTools suite [15]. The PDBQT format is an extension of the PDB format with additional fields (partial charge and atom type) for ATOM and HETATM records and information about the rigid blocks of the molecule. Interaction grids are generated for each protein structure using the autogrid4 program in Auto Dock 4.2.6 [15]. The grid box of 50 × 50 × 50 with 0.375 Å is set around the binding sites. Protein–ligand docking is performed using the Lamarckian Genetic Algorithm in autodock4 program to obtain the best docking conformation.

Appendix K

The system of protein–ligand complex is solvated by applying TIP3 water model, and the necessary Na or Cl ions are added to balance the system charge. An energy minimization method is first performed for the simulated system using the steepest descent algorithm. Afterward, the system is balanced at 300 K via 100 ps NVT and NPT simulations. Finally, a 20 ns molecular dynamics (MD) simulation is carried out in 2 fs time steps to maintain the stability of the entire system. All the above simulations are performed under periodic boundary conditions with a temperature of 300 K and atmospheric pressure. The electrostatic interactions are calculated using the particle mesh Ewald (PME) method [17] during the simulation. The linear constraint solver (LINCS) algorithm [18] constrains the covalent bond involving the H atom. And the cutoff distances for calculating van der Waals interaction and Coulomb interaction are set to 1.4 nm and 1.4 nm, respectively. The isothermal compression rate is set to 4.5×10^{-5} per bar by the Parrinello-Rahman protocol while the pressure is maintained at 1.0 bar [19] and the temperature is kept constant using a Berendsen thermostat. The MD simulation results are capable of providing insights into the binding mode of protein–ligand complex.

Appendix L

The details of the MD and random acceleration molecular dynamics (RAMD) simulations are presented as follows:

1. The preparation of the solvated protein–ligand complex
   The docking structures of the complexes are utilized for the simulations. The ligand parameters are generated using the Antechamber program [20]. The BCC atomic partial charges are employed via the Amber22 software [21]. The GAFF force field [22] is implemented for the ligands, while the Amber ff14 force field [23] is applied to the protein. The crystallographic water molecules located within a 5 Å radius of the ligand are included in the system. The system is solvated in a periodic box of TIP3P water molecules [24].
2. Preparation of a minimized and equilibrated structure
   The system is first energy minimized, heated, and equilibrated using the Amber22 software [25]. The minimization process involves 500 steps of steepest descent followed by 1,000 steps of conjugate gradient minimization, where restraints are applied to all heavy atoms except water and ions in the system, with values of 500, 100, and 5 kcal/mol·Å^2 for 1,500 steps each and then the system is simulated for an additional 1,500 steps without any restraints. After that, the system is gradually heated to 300 K over a period of 50 ps, employing harmonic restraints on non-hydrogen atoms with a force constant of 50 kcal/mol·Å^2. In the NPT ensemble, restraints are gradually

decreased in three stages: 50, 10, and 2 kcal/mol·Å^2, each lasting for 20 ps. Finally, the system is simulated without any restraints. A constant temperature of 300 K is maintained using a Langevin thermostat, and a constant pressure of 1 atm is achieved with a Berendsen barostat. Throughout the simulations, the nonbonded Coulombic and Lennard-Jones interactions are truncated at a cutoff distance of 10 Å, and periodic boundary conditions are used with a PME treatment of long-range Coulombic interactions. A time step of 2 fs is used, and bonds to hydrogen atoms are constrained using the SHAKE algorithm [26].

3. Generation of starting bound-state replicas through MD simulations
   To perform simulations in GROMACS [27], the final output coordinate and topology files are transferred from Amber to GROMACS using ParmEd [28]. Then, short NVT simulations are conducted using the Berendsen thermostat for a duration of 20 ns. Subsequently, six starting bound-state replicas are generated under NPT conditions, employing the Nosé–Hoover thermostat and Parrinello–Rahman barostat, with each replica simulated for 20 ns. To ensure diversity among the replicas, various velocities are generated from the Maxwell distribution at the beginning of each replica simulation.
4. RAMD simulation of dissociation trajectories for each replica
   RAMD simulations are conducted using GROMACS for each starting bound-state replica. In these simulations, an additional randomly oriented force of 15.5 kcal/mol·Å is applied to the center of ligand mass. The force direction is changed randomly every 100 fs, unless the center of ligand mass moves further than a threshold of 0.025 Å, in which case the direction is maintained. Simulations are terminated if the center of ligand mass moves further than 40 Å from its initial position. Coordinates are saved at intervals of 10 ps.
5. Statistical analysis to provide a final τ value and insights into the dissociation mechanisms.
   Starting with each bound-state replica, a total of 20 RAMD dissociation trajectories are generated. The relative τ is defined as the time at which a dissociation event is observed in 50% of the trajectories. τ is computed for each starting bound-state replica and then averaged over all the replicas that are simulated. Sufficient sampling to compute τ is ensured by increasing the number of bound-state replicas and/or the number of dissociation trajectories [29].

Appendix M

The selected subsets of scaffolds (G_1) and functional groups (G_2) in case study 1 are given in Table M.1.

Appendix N

Lower and upper bounds for the structure and property constraints are provided in Table N.1.

The selected scaffolds and groups are provided in Table N.2.

Table M.1 The selected subsets of scaffolds and functional groups.

No.	Scaffold (G_1)	Degree	MW	HBA	HBD	ROT_{frag}
1	O=C(NC[C@H]1CN(c2ccc(N3CCOCC3=O)cc2)C(=O)O1)c1ccc(*)s1	1	400.44	5	1	5
2	O=C1C=CN(c2c(*)cc(N3C[C@H](CNc4ccon4)OC3=O)c(*)c2*)CC1	3	351.34	7	1	5
3	O=C1OC(*)CN1c1ccc(N2CCOCC2)c(*)c1	2	246.27	4	0	2
4	O=C1OC(*)CN1c1ccc(N2CCSCC2)c(*)c1	2	262.33	4	0	2
5	O=C(N/N=C/C=C/c1ccc(*)o1)c1ccc(*)s1	2	244.28	2	1	4
6	O=C1N[C@H](C2Cc3ccccc3C2)C(=O)N([C@@H](C(=O)N2CCOCC2)c2coc(*)n2)C1*	2	422.44	6	1	4
7	O=C(NC[C@@H]1CCCN1*)c1c(*)ccc(*)c1*	4	200.24	2	1	3
8	O=C1NC(=O)C2(Cc3cc4c(N5C(=O)OCC5*)noc4c(*)c3N3CC(*)OC(*)[C@@H]32)C(=O)N1	4	423.34	9	2	1
9	c1cc(*)sc1*	2	82.13	0	0	0
10	O=C1OC(*)C2(*)C3=C(C(=O)c4occ1c42)[C@@H]1CCC(=O)C1(*)CC3*	4	294.26	5	0	0
11	O=C1c2ccccc2S(=O)(=O)N1CCCCN1CCN(c2cc(*)cc3c2OCCO3)CC1	1	456.54	7	0	6
12	O=C(NC(Cc1ccc(-c2ccc3oc(=O)n(*)c3c2)cc1)*)[C@@H]1CNCCCO1	2	379.42	5	2	5
13	O=C(C(N1CC[C@H](NS(=O)(=O)/C=C/c2ccc(*)s2)C1=O)*)N1CCOCC1	2	397.48	5	1	6
14	c1sc(*)nc1*	2	83.12	1	0	0

No.	Group (G_2)	Degree	MW	HBA	HBD	ROT_{frag}
1	*C	1	15.035	0	0	0
2	*C*	2	14.027	0	0	0
3	*C(*)*	3	13.019	0	0	0
4	C=C*	1	27.046	0	0	0
5	*C=C*	2	26.038	0	0	0
6	C=C(*)*	2	26.038	0	0	0
7	*C=C(*)*	3	25.03	0	0	0
8	O*	1	17.007	1	1	0
9	OC(=O)*	1	45.017	2	1	0
10	CC(=O)*	1	43.045	1	0	0
11	*CC(=O)*	2	42.037	1	0	1
12	*C(*)C(=O)*	3	41.029	1	0	1
13	C(=O)*	1	29.018	1	0	0
14	*C(=O)O*	2	44.009	2	0	0

No.	Scaffold (G_1)	Degree	MW	HBA	HBD	ROT_{frag}
15	CO*	1	31.034	1	0	0
16	*CO*	2	30.026	1	0	1
17	*C(*)O*	3	29.018	1	0	1
18	CN*	1	30.05	1	1	0
19	*CN*	2	29.042	1	1	1
20	*C(*)N*	3	28.034	1	1	1
21	CN(*)*	2	29.042	1	0	0
22	*CN(*)*	3	28.034	1	0	1
23	*N	1	16.023	1	1	0
24	*C#N	1	26.018	1	0	0
25	*[N+](=O)[O−]	1	46.005	2	0	0
26	*I	1	126.904	0	0	0
27	*Br	1	79.904	0	0	0
28	*F	1	18.998	0	0	0
29	*Cl	1	35.453	0	0	0

Table N.1 List of structure and property constraints with upper and/or lower bounds.

Constraint	Upper and/or lower bound
Number of total scaffolds	$1 \leq \sum_{i \in G_1} n_i \leq 1$
Number of total fragments (scaffold and group)	$1 \leq \sum_{i \in G_1, G_2} n_i \leq 5$
Number of same fragments (scaffold and group)	$n_i \leq 3, i \in G_1, G_2$
Number of main groups ($-CH_3$, $-CH_2$, $-CH$)	$0 \leq \sum_{i \in G_3} n_i \leq 4$
Number of functional groups (except for $-CH_3$, $-CH_2$, $-CH$)	$0 \leq \sum_{i \in (G_2 - G_3)} n_i \leq 4$
Molecular weight	$MW \leq 500$
Number of hydrogen bond acceptors	$HBA \leq 10$
Number of hydrogen bond donors	$HBD \leq 5$
Octanol-water partition coefficient	$-2 \leq \log P \leq 5$
Number of rotatable bonds	$ROT(ROT_{frag}) \leq 10$
Synthetic accessibility score	$SA \leq 6$
Probability of high binding affinity	$Prob_{affinity} \geq 50\%$

Table N.2 The selected subsets of scaffolds and functional groups.

No.	Scaffold (G_1)	Degree	MW	HBA	HBD	ROT_{frag}
1	O=C1OC(*)C2(*)C3=C(C(=O)c4occ1c42)C1CCC(=O)C1(*)CC3*	4	294.262	5	0	0
2	O=C1Nc2ccc(*)cc2C12NC(*)Cc1c2[nH]c2cc(*)c(*)cc12	4	285.306	2	3	0
3	O=C1OCc2c(cc3n(c2=O)Cc2c–3nc3cc(*)c(*)c4c3c2C(*)CC4)C1*	4	340.338	4	0	0
4	C1=CC23CC(*)C1(*)C1Oc4c(*)ccc5c4C12CCN(*)C3C5	4	261.324	2	0	0
5	c1cc(*)c2nn(–c3ccc(C4CCCNC4)cc3)cc2c1	1	276.363	2	1	2
6	c1ccc2c(c1)CCC2Nc1ncnc2c1ccn2C1CC(*)C(*)C1	2	316.408	3	1	3
7	c1nc2c(c(NC3CCN(*)C3)n1)CN(c1cnc(*)c(*)c1)CC2	3	293.354	6	1	3
8	c1nc(*)c2c(–c3cnc4[nH]ccc4c3)nn(C3CCCC3)c2n1	1	303.349	4	1	2
9	c1nc(NC2CCC(*)N(*)C2)c2cc[nH]c2n1	2	215.26	4	2	2
10	c1ccc(–c2nc(NC3CCC(*)CC3)c3cc[nH]c3n2)cc1	1	291.378	3	2	3
11	c1c(*)cc(*)c(Nc2nc3cnc(NC4CCC(*)CC4)nc3n2C2CCOC2)c1*	4	374.448	6	2	5
12	c1cc(N2CCN(*)CC2)ccc1Nc1ncc(*)c(NC2CC2)n1	2	308.389	6	2	5
13	c1ccc2c(c1)nc(NC1CCNCC1)n2Cc1ccc(*)cc1	1	305.405	3	2	4
14	c1nc(*)c(*)cc1–n1nc(*)c2cc(*)ccc21	4	191.193	2	0	1
15	c1ccc(–c2coc3c(*)cccc23)cc1	1	193.225	1	0	1
16	O=C(c1cnc2ccccc2c1)N(Cc1cccc(*)c1)C(c1ccc(*)cc1)*	3	349.413	2	0	5
17	O=c1c(–c2ccccc2)coc2cc(*)ccc12	1	221.235	2	0	1
18	c1ccc2c(–c3ccc(*)cc3)c(*)c(C3CC3)nc2c1	2	243.309	1	0	2
19	O=C(C=Cc1ccoc1)N(C1CCC2(*)C3Cc4ccc(*)c5c4C2(CCN3CC2CC2)C1O5)*	3	427.524	4	0	5
20	c1ccc2oc(C(c3ccc(*)cc3)*)cc2c1	2	206.244	1	0	2
21	c1ccc2nc3c(*)cccc3cc2c1	1	178.214	1	0	0
22	c1nc2c(cnn2*)c(NC2CCOCC2)c1–c1nnc(Cc2sc(*)nc2*)o1	3	380.413	8	1	5
23	c1nn(C2CCNCC2)cc1Nc1ncc2sc(*)c(–c3ccc(*)cc3*)c2n1	3	373.465	5	2	4
24	O=C1C(Nc2ncnc3cc(*)nn23)CCN1C1CCC(*)CC1*	3	297.342	5	1	3
25	c1c(*)cc2c(*)c(–c3nc(*)no3)c(N3CCC(NC4CCOCC4)CC3)nc2c1*	4	375.432	7	1	4
26	c1nc(NCCCC2CCN(*)CC2)nc(*)c1–c1nc2c(*)cc(*)cc2[nH]1	4	332.411	5	2	6
27	c1cc(*)c2[nH]c(–c3nc(C4CCC(*)CC4)n4cnc(*)c34)cc2c1	3	314.372	3	1	2

No.	Scaffold (G_1)	Degree	MW	HBA	HBD	ROT_{frag}
28	O=c1c(–c2ccc(*)nc2)cc2c(*)nc(*)nc2n1C1CCC(*)CC1	4	302.337	4	0	2
29	c1nc(Nc2ccc(*)c(*)c2)c2nc(NC3CCN(*)CC3)ncc2n1	3	318.364	7	2	4
30	c1cc(*)c2nc(–c3ccc[n+](*)c3)c(C(Nc3ncnc4cccnc34)*)cc2c1	3	362.396	5	1	4
31	O=C(c1cc(*)ccc1–n1nccn1)N1CCN(c2nc3cc(*)ccc3o2)CCC1*	3	385.407	6	0	3
32	O=c1c2c(*)cccc2nc(C(Nc2ncnc3nc[nH]c23)*)n1-c1ccccc1	2	367.372	6	2	4
33	C1CC(*)CCC1*	2	82.146	0	0	0
34	C1CC1*	1	41.073	0	0	0
35	c1ccc(*)c(*)c1	2	76.098	0	0	0
36	c1ccc2occc2c1	0	118.135	1	0	0
37	c1ccc2sc(*)cc2c1	1	133.195	0	0	0
38	c1ccc2ncc(–c3cccc4c3c(*)nn4-c3ccc(*)c(NC4CCC(*)CC4)c3)cc2c1	3	415.52	3	1	4

No.	Group (G_2)	Degree	MW	HBA	HBD	ROT_{frag}
1	*C	1	15.035	0	0	0
2	*C*	2	14.027	0	0	0
3	*C(*)*	3	13.019	0	0	0
4	O*	1	17.007	1	1	0
5	OC(=O)*	1	45.017	2	1	0
6	CC(=O)*	1	43.045	1	0	0
7	*CC(=O)*	2	42.037	1	0	1
8	*C(*)C(=O)*	3	41.029	1	0	1
9	C(=O)*	1	29.018	1	0	0
10	*C(=O)O*	2	44.009	2	0	0
11	CO*	1	31.034	1	0	0
12	*CO*	2	30.026	1	0	1
13	*C(*)O*	3	29.018	1	0	1
14	CN*	1	30.05	1	1	0
15	*CN*	2	29.042	1	1	1
16	*C(*)N*	3	28.034	1	1	1
17	CN(*)*	2	29.042	1	0	0
18	*CN(*)*	3	28.034	1	0	1
19	*N	1	16.023	1	1	0
20	*C#N	1	26.018	1	0	0

(Continued)

Table N.2 (continue)

No.	Scaffold (G_1)	Degree	MW	HBA	HBD	ROT_{frag}
21	*[N+](=O)[O-]	1	46.005	2	0	0
22	*I	1	126.904	0	0	0
23	*Br	1	79.904	0	0	0
24	*F	1	18.998	0	0	0
25	*Cl	1	35.453	0	0	0

Appendix O

Details on the cleaning processes for the HSP90 datasets from Liu et al. [3] and Kokh et al. [30] are presented in the following.

The dataset from Liu et al. [3] contains a total of 89 HSP90 ligands. Note that eight ligand SMILES containing the string "N+H" cannot be read by RDKit, and one ligand SMILES cannot be predicted by our MoE model, as its Morgen fragment number exceeds the range of 20–74 or the Morgen fragments are not compatible with the Mol2vec method. The remaining 80 HSP90 ligands are tested using our MoE model, yielding r, R^2, MAE, and RMSE values of 0.895, 0.773, 0.359, and 0.458, respectively.

Some pk_{off} values in Kokh et al. [30] are excluded because the corresponding ligands contain uncertain inequality signs ("<") or cannot be predicted by our MoE model. The inability to predict these ligands arises from their Morgen fragment numbers exceeding the range of 20–74 or because their Morgen fragments are not found using the Mol2vec method. The r and R^2 for the remaining 66 ligands are 0.876 and 0.767, respectively.

Appendix P

The docking method is used to identify the binding conformation of the potential HSP90 inhibitor candidate and the compound with a novel scaffold. The PDB file of the "loop" conformation (PDB: 2BSM) is used to perform molecular docking tasks. The results show that the top-ranked conformations obtained from docking these two compounds to the "loop" conformation have yielded poor docking scores of −3.114 and −1.703, respectively. In contrast, the top-ranked conformations obtained from docking the same compounds to the "helix" conformation have achieved significantly greater docking scores of −15.121 and −10.161, respectively. Furthermore, the docking conformations reveal that both designed molecules only interact with the binding pocket surface of the "loop" conformation and do not enter the binding pocket for stable binding. Therefore, the potential HSP90 inhibitor candidate and the compound with a novel scaffold belong to the "helix" category.

The RAMD simulation results of the potential HSP90 inhibitor candidate, the reference compound, and the compound with novel scaffold are presented in Figures P.1–P.3,

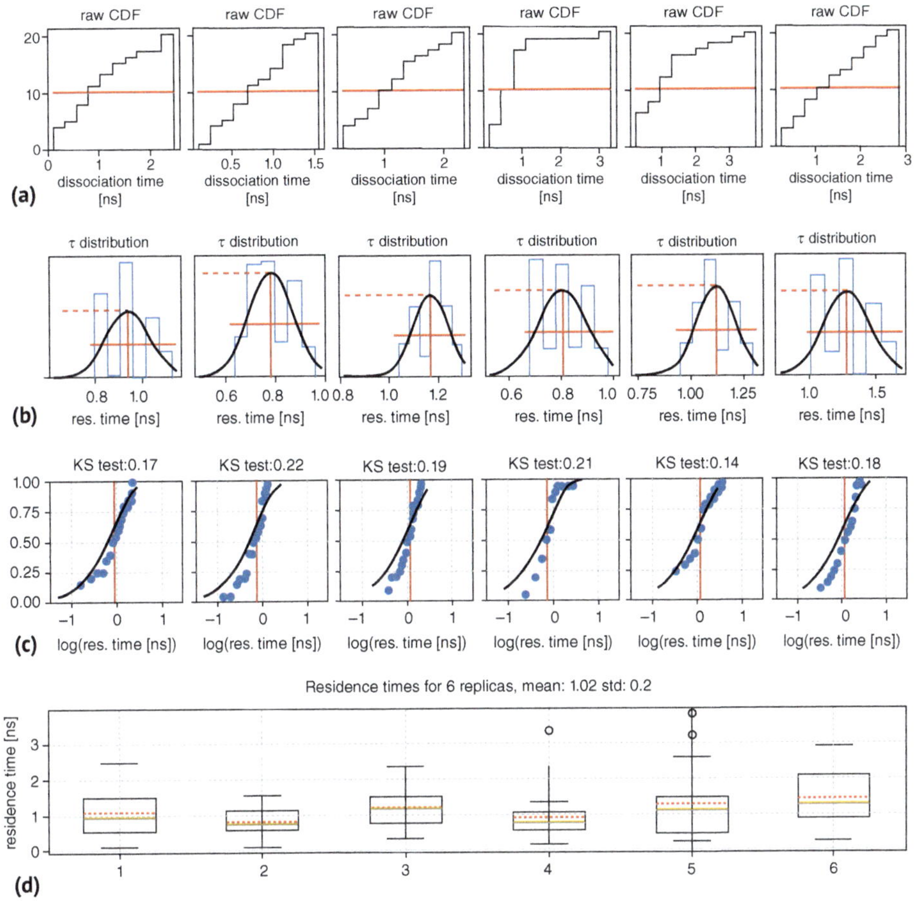

Figure P.1 The RAMD simulation results of the potential HSP90 inhibitor candidate. (a) CDF for 6 sets of 20 RAMD dissociation time, with each trajectory in a set starting from the same replica; the τ obtained from these raw data is indicated by the red solid line at the time when the ligand has dissociated in half of the trajectories. (b) Distribution function of the τ obtained after bootstrapping of the raw data along with the corresponding Gaussian distribution (black line); the mean and half-width are indicated by red lines. (c) Poisson CDF (black line); $P = 1 - \exp(-t/\tau)$ is compared with the empirical CDF (blue points) obtained from the dissociation probability distribution; the mean of $\log(\tau)$ is indicated by the red line; the results of the KS test are quantified by the supremum of the distance between the Poisson and empirical CDFs. (d) Bar plot of the relative τ obtained from each replica; the box extends from the lower to upper quartile values of the data; the whiskers show the range of the data; outliers are shown by points; the median and mean are shown by orange lines and red dashed lines, respectively; the average τ in ns computed from all replicas is shown above the plot.

respectively. Figure P.1a shows the dissociation time for each replica, which corresponds to the point where 50% of the cumulative distribution function (CDF) is reached in the set of RAMD trajectories. A bootstrapping procedure (5000 rounds with 80% of the samples selected randomly) is carried out to obtain τ for each replica. Figure P.1b presents that the Gaussian-like distribution is achieved in each replica, indicating that the sampling number

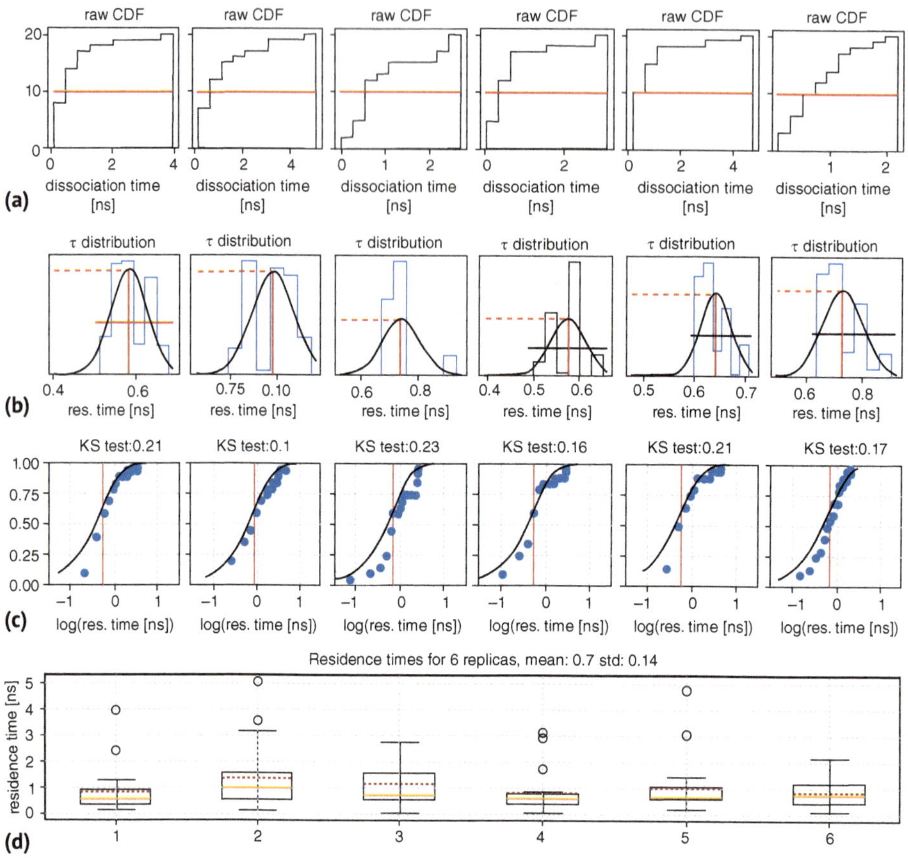

Figure P.2 The RAMD simulation results of the reference compound. (a) CDF for 6 sets of 20 RAMD dissociation time, with each trajectory in a set starting from the same replica; the τ obtained from these raw data is indicated by the red solid line at the time when the ligand has dissociated in half of the trajectories. (b) Distribution function of the τ obtained after bootstrapping of the raw data along with the corresponding Gaussian distribution (black line); the mean and half-width are indicated by red lines. (c) Poisson CDF (black line); $P = 1 - \exp(-t/\tau)$ is compared with the empirical CDF (blue points) obtained from the dissociation probability distribution; the mean of $\log(\tau)$ is indicated by the red line; the results of the KS test are quantified by the supremum of the distance between the Poisson and empirical CDFs. (d) Bar plot of the relative τ obtained from each replica; the box extends from the lower to upper quartile values of the data; the whiskers show the range of the data; outliers are shown by points; the median and mean are shown by orange lines and red dashed lines, respectively; the average τ in ns computed from all replicas is shown above the plot.

of simulated trajectories for each replica is sufficiently large. A Kolmogorov–Smirnov (KS) test is utilized to assess the sampling quality, measuring the maximum deviation of the simulated data from the Poisson distribution and typically yielding a value between 0.1 and 0.2. The values of KS test in Figure P.1c demonstrate the effectiveness of the trajectory sampling for each replica. Finally, the relative τ is computed as the mean of the τ values

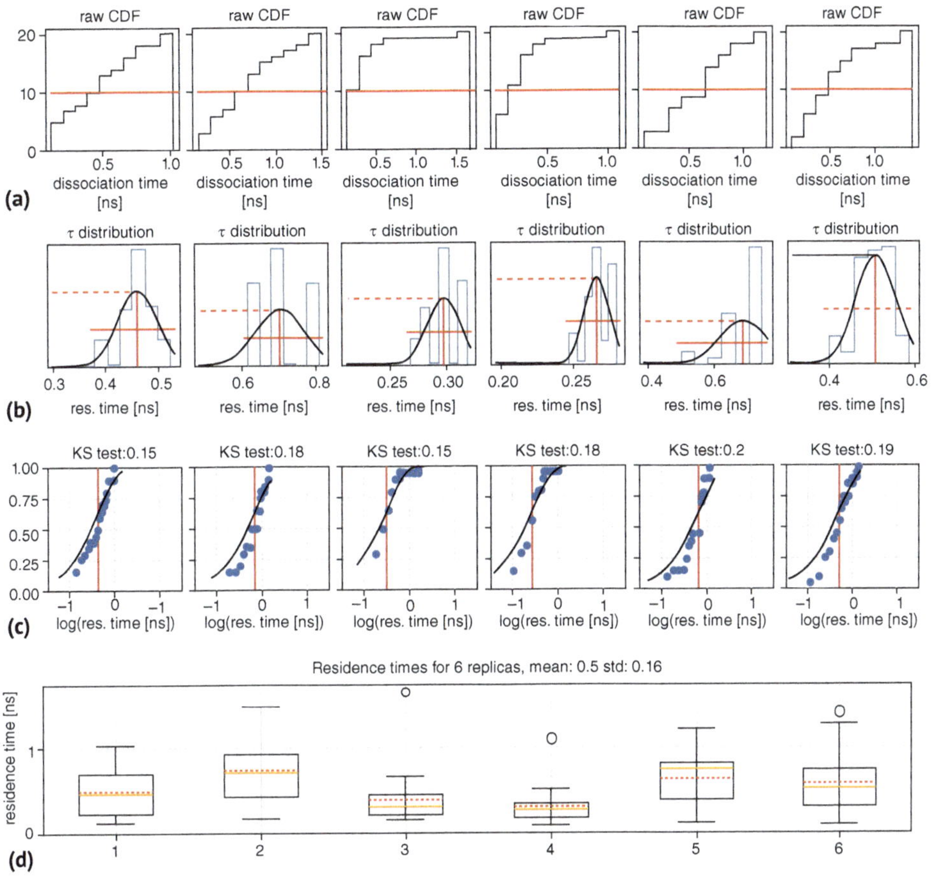

Figure P.3 The RAMD simulation results of the compound with novel scaffold. (a) CDF for 6 sets of 20 RAMD dissociation time, with each trajectory in a set starting from the same replica; the τ obtained from these raw data is indicated by the red solid line at the time when the ligand has dissociated in half of the trajectories. (b) Distribution function of the τ obtained after bootstrapping of the raw data along with the corresponding Gaussian distribution (black line); the mean and half-width are indicated by red lines. (c) Poisson CDF (black line); $P = 1 - \exp(-t/\tau)$ is compared with the empirical CDF (blue points) obtained from the dissociation probability distribution; the mean of $\log(\tau)$ is indicated by the red line; the results of the KS test are quantified by the supremum of the distance between the Poisson and empirical CDFs. (d) Bar plot of the relative τ obtained from each replica; the box extends from the lower to upper quartile values of the data; the whiskers show the range of the data; outliers are shown by points; the median and mean are shown by orange lines and red dashed lines, respectively; the average τ in ns computed from all replicas is shown above the plot.

across all replicas, as illustrated in Figure P.1d. The computed τ averaged over six replicas is 1.02 ± 0.20 ns for the potential HSP90 inhibitor candidate. The principles of Figures P.2 and P.3 are similar to Figure P.1.

The compound with a novel scaffold stably binds to the α-helix3 fragment in a "helix" conformation, and its binding mode is shown in Figure P.4.

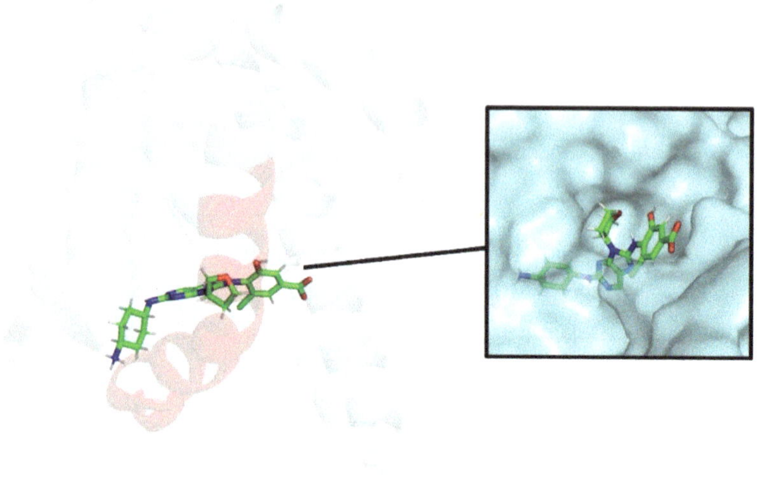

Figure P.4 The MD simulation results for the bound states of the compound with a novel scaffold.

References

1 Schuetz DA, Richter L, Martini R, and Ecker GF. A structure–kinetic relationship study using matched molecular pair analysis. *RSC Medicinal Chemistry* 2020;11(11):1285–1294.

2 Gilson MK, Liu T, Baitaluk M, Nicola G, Hwang L, and Chong J. BindingDB in 2015: A public database for medicinal chemistry, computational chemistry and systems pharmacology. *Nucleic Acids Research* 2016;44(D1):D1045–D1053.

3 Liu H, Su M, Lin HX, Wang R, and Li Y. Public data set of protein–ligand dissociation kinetic constants for quantitative structure–kinetics relationship studies. *ACS Omega* 2022;**7**(22):18985–18996.

4 Amangeldiuly N, Karlov D, and Fedorov MV. Baseline model for predicting protein–ligand unbinding kinetics through machine learning. *Journal of Chemical Information and Modeling* 2020;60(12):5946–5956.

5 Landrum G. RDKit: Open-source cheminformatics. 2006.

6 Rogers D and Hahn M. Extended-connectivity fingerprints. *Journal of Chemical Information and Modeling* 2010;50(5):742–754.

7 Jaeger S, Fulle S, and Turk S. Mol2vec: Unsupervised machine learning approach with chemical intuition. *Journal of Chemical Information and Modeling* 2018;58(1):27–35.

8 Chen H, Engkvist O, Wang Y, Olivecrona M, and Blaschke T. The rise of deep learning in drug discovery. *Drug Discovery Today* 2018;23(6):1241–1250.

9 Hinton GE, Srivastava N, Krizhevsky A, Sutskever I, and Salakhutdinov RR. Improving neural networks by preventing co-adaptation of feature detectors. *Arxiv Preprint* 2012. arxiv.1207.0580.

10 Kingma DP and Ba J. Adam: A method for stochastic optimization. *Arxiv Preprint* arxiv.1412.6980.

11 Paszke A, Gross S, Massa F, et al. Pytorch: An imperative style, high-performance deep learning library. *Advances in Neural Information Processing Systems* 2019; 32.
12 Oliphant TE. Python for scientific computing. *Computing in Science & Engineering* 2007;9(3):10–20.
13 Liu Q, Zhang L, Liu L, et al. OptCAMD: An optimization-based framework and tool for molecular and mixture product design. *Computers & Chemical Engineering* 2019;124:285–301.
14 Burley SK, Bhikadiya C, Bi C, et al. RCSB protein data bank: Powerful new tools for exploring 3D structures of biological macromolecules for basic and applied research and education in fundamental biology, biomedicine, biotechnology, bioengineering and energy sciences. *Nucleic Acids Research* 2021; 49(D1):D437–D451.
15 Morris GM, Huey R, Lindstrom W, et al. AutoDock4 and AutoDockTools4: Automated docking with selective receptor flexibility. *Journal of Computational Chemistry* 2009;30(16):2785–2791.
16 O'Boyle NM, Banck M, James CA, Morley C, Vandermeersch T, and Hutchison GR. Open Babel: An open chemical toolbox. *Journal of Cheminformatics* 2011;3(1): 33.
17 Lin JH, Perryman AL, Schames JR, and McCammon JA. Computational drug design accommodating receptor flexibility: The relaxed complex scheme. *Journal of the American Chemical Society* 2002;124(20):5632–5633.
18 Hess B, Bekker H, Berendsen HJ, and Fraaije JG. LINCS: A linear constraint solver for molecular simulations. *Journal of Computational Chemistry* 1997;18(12):1463–1472.
19 Parrinello M and Rahman A. Polymorphic transitions in single crystals: A new molecular dynamics method. *Journal of Applied Physics* 1981;52(12):7182–7190.
20 Chen J, Yin B, Wang W, and Sun H. Effects of disulfide bonds on binding of inhibitors to β-amyloid cleaving enzyme 1 decoded by multiple replica accelerated molecular dynamics simulations. *ACS Chemical Neuroscience* 2020;11(12):1811–1826.
21 Braka A, Garnier N, Bonnet P, and Aci-Seche S. Residence time prediction of Type 1 and 2 kinase inhibitors from unbinding simulations. *Journal of Chemical Information and Modeling* 2020;60(1):342–348.
22 Deb I and Frank AT. Accelerating rare dissociative processes in biomolecules using selectively scaled MD simulations. *Journal of Chemical Theory and Computation* 2019;15(11):5817–5828.
23 Maier JA, Martinez C, Kasavajhala K, Wickstrom L, Hauser KE, and Simmerling C. ff14SB: Improving the accuracy of protein side chain and backbone parameters from ff99SB. *Journal of Chemical Theory and Computation* 2015;11(8):3696–3713.
24 Azhagiya Singam ER, Tachachartvanich P, La Merrill MA, Smith MT, and Durkin KA. Structural dynamics of agonist and antagonist binding to the androgen receptor. *The Journal of Physical Chemistry B* 2019;123(36):7657–7666.
25 Gobbo D, Piretti V, Di Martino RMC, et al. Investigating drug–target residence time in kinases through enhanced sampling simulations. *Journal of Chemical Theory and Computation* 2019;15(8):4646–4659.
26 Fischer A and Smieško M. Ligand pathways in nuclear receptors. *Journal of Chemical Information and Modeling* 2019;59(7):3100–3109.
27 Pronk S, Páll S, Schulz R, et al. GROMACS 4.5: A high-throughput and highly parallel open source molecular simulation toolkit. *Bioinformatics* 2013;29(7):845–854.

28 Shirts MR, Klein C, and Swails JM. Lessons learned from comparing molecular dynamics engines on the SAMPL5 dataset. *Journal of Computer-Aided Molecular Design* 2017;31(1):147–161.

29 Kokh DB, Kaufmann T, Kister B, and Wade RC. Machine learning analysis of τRAMD trajectories to decipher molecular determinants of drug-target residence times. 2019; 6.

30 Kokh DB, Amaral M, Bomke J, et al. Estimation of drug-target residence times by τ-random acceleration molecular dynamics simulations. *Journal of Chemical Theory and Computation* 2018;14(7):3859–3869.

Index

Printed in the USA/Agawam, MA
September 16, 2026

910660.021